Vol. 28. **The Analytical Chemistry of Nitrogen and Its Compounds** *(in two parts).* Edited by C. A. Streuli and Philip R. Averell

Vol. 29. **The Analytical Chemistry of Sulfur and Its Compounds** *(in three parts).* By J. H. Karchmer

Vol. 30. **Ultramicro Elemental Analysis.** By Günther Tölg

Vol. 31. **Photometric Organic Analysis** *(in two parts).* By Eugene Sawicki

Vol. 32. **Determination of Organic Compounds: Methods and Procedures.** By Frederick T. Weiss

Vol. 33. **Masking and Demasking of Chemical Reactions.** By D. D. Perrin

Vol. 34. **Neutron Activation Analysis.** By D. De Soete, R. Gijbels, and J. Hoste

Vol. 35. **Laser Raman Spectroscopy.** By Marvin C. Tobin

Vol. 36. **Emission Spectrochemical Analysis.** By Morris Slavin

Vol. 37. **Analytical Chemistry of Phosphorus Compounds.** Edited by M. Halmann

Vol. 38. **Luminescence Spectrometry in Analytical Chemistry.** By J. D. Winefordner, S. G. Schulman, and T. C. O'Haver

Vol. 39. **Activation Analysis with Neutron Generators.** By Sam S. Nargolwalla and Edwin P. Przybylowicz

Vol. 40. **Determination of Gaseous Elements in Metals.** Edited by Lynn L. Lewis, Laben M. Melnick, and Ben D. Holt

Vol. 41. **Analysis of Silicones.** Edited by A. Lee Smith

Vol. 42. **Foundations of Ultracentrifugal Analysis.** By H. Fujita

Vol. 43. **Chemical Infrared Fourier Transform Spectroscopy.** By Peter R. Griffiths

Vol. 44. **Microscale Manipulations in Chemistry.** By T. S. Ma and V. Horak

Vol. 45. **Thermometric Titrations.** By J. Barthel

Vol. 46. **Trace Analysis: Spectroscopic Methods for Elements.** Edited by J. D. Winefordner

Vol. 47. **Contamination Control in Trace Element Analysis.** By Morris Zief and James W. Mitchell

Vol. 48. **Analytical Applications of NMR.** By D. E. Leyden and R. H. Cox

Vol. 49. **Measurement of Dissolved Oxygen.** By Michael L. Hitchman

Vol. 50. **Analytical Laser Spectroscopy.** Edited by Nicolo Omenetto

Vol. 51. **Trace Element Analysis of Geological Materials.** By Roger D. Reeves and Robert R. Brooks

Vol. 52. **Chemical Analysis by Microwave Rotational Spectroscopy.** By Ravi Varma and Lawrence W. Hrubesh

Vol. 53. **Information Theory As Applied to Chemical Analysis.** By Karel Eckschlager and Vladimir Štěpánek

Vol. 54. **Applied Infrared Spectroscopy: Fundamentals, Techniques, and Analytical Problem-solving.** By A. Lee Smith

Vol. 55. **Archaeological Chemistry.** By Zvi Goffer

Vol. 56. **Immobilized Enzymes in Analytical and Clinical Chemistry.** By P. W. Carr and L. D. Bowers

Vol. 57. **Photoacoustics and Photoacoustic Spectroscopy.** By Allan Rosencwaig

Solid Phase Biochemistry

CHEMICAL ANALYSIS

A SERIES OF MONOGRAPHS ON ANALYTICAL CHEMISTRY AND ITS APPLICATIONS

VOLUME 66

A WILEY-INTERSCIENCE PUBLICATION

JOHN WILEY & SONS

New York / Chichester / Brisbane / Toronto / Singapore

Solid Phase Biochemistry

ANALYTICAL AND SYNTHETIC ASPECTS

Edited by
WILLIAM H. SCOUTEN
Department of Chemistry
Bucknell University
Lewisburg, Pennsylvania

A WILEY-INTERSCIENCE PUBLICATION
JOHN WILEY & SONS
New York / Chichester / Brisbane / Toronto / Singapore

Library of Congress Cataloging in Publication Data:

Main entry under title:

Solid phase biochemistry.

(Chemical analysis, ISSN 0069-2883; v. 66)
"A Wiley-Interscience publication."
Includes bibliographies and index.
1. Biological chemistry—Technique. 2. Affinity chromatography. 3. Immobilized proteins. 4. Solid state chemistry. I. Scouten, William H., 1942–
II. Series.

QP519.7.S64 1983 574.19'285 82-21886
ISBN 0-471-08585-5

Printed in the United States of America

10 9 8 7 6 5 4 3 2 1

CONTRIBUTORS

Alan Bergold
Chemistry Department
Bucknell University
Lewisburg, PA

Staffan Birnbaum
Pure and Applied Biochemistry
Chemical Center
Lund Institute of Technology
University of Lund
Lund, Sweden

T. C. Bøg-Hansen
The Protein Laboratory
University of Copenhagen
Copenhagen N, Denmark

Joaquim M. S. Cabral
Laboratorio de Engenharia Bioquimica
Instituto Superior Tecnico
Universidade Tecnica de Lisboa
Lisboa, Portugal

Peter Dean
Department of Biochemistry
University of Liverpool
Liverpool, United Kingdom

Thomas O. Fox
Department of Neuroscience
Children's Hospital Medical Center
Boston, MA

George G. Guilbault
Department of Chemistry
University of New Orleans
New Orleans, Louisiana

Wolfgang Haller
National Bureau of Standards
Washington, DC

Keiichi Itakura
Department of Molecular Genetics
City of Hope Research Institute
Duarte, California

John F. Kennedy
Research Laboratory for the Chemistry of Bioactive Carbohydrates and Proteins
Department of Chemistry
University of Birmingham
Birmingham, England

J. Kohn
Department of Biophysics
Weizman Institute of Science
Rehovot, Israel

Per-Olof Larsson
Pure and Applied Biochemistry
Chemical Center
Lund Institute of Technology
University of Lund
Lund, Sweden

Catherine Lewis
Chemistry Department
Bucknell University
Lewisburg, PA

Klaus Mosbach
Pure and Applied Biochemistry
Chemical Center
Lund Institute of Technology
University of Lund
Lund, Sweden

A. Hirotoshi Nishikawa
Biopolymer Research Department
Hoffman-LaRoche, Inc.
Nutley, NJ

Firdausi Quadri
Department of Biochemistry
University of Liverpool
Liverpool, United Kingdom

Charalambos Savakis
Department of Neuroscience
Children's Hospital Medical Center
Boston, MA

William Scouten
Chemistry Department
Bucknell University
Lewisburg, PA

John M. Stewart
Department of Biochemistry
University of Colorado Medical School
Denver, Colorado

J. Tramper
Department of Process Engineering
Agricultural University
Wageningen, The Netherlands

R. Bruce Wallace
Department of Molecular Genetics
City of Hope Research Institute
Duarte, California

M. Wilchek
Fogarty Scholar-in-Residence
Fogarty International Center
National Institutes of Health
Bethesda, Maryland

PREFACE

This book is intended to bring together the major aspects of solid phase biochemistry, providing a review of the significant achievements of the field while at the same time providing the experimentalist with a source for the most useful techniques in each area. This volume supplements two previous volumes in this series that deal with solid phase biochemical techniques, namely Volume 56, *Immobilized Enzymes in Analytical and Clinical Chemistry* by P. W. Carr and L. D. Bowers and Volume 59, *Affinity Chromatography* by W. H. Scouten.

The first chapter introduces the vast area of biochemistry on solid phase supports and compares and contrasts many aspects of solid phase biochemistry. In so doing, I hope to help persons active in one aspect of solid phase biochemistry become more aware of how other areas have developed and introduce them to techniques in other aspects of solid phase biochemistry that they might want to apply to their own research. Later chapters are monographs by recognized leaders in various fields of solid phase biochemistry. Chapter 2 introduces affinity chromatography in a way that very much complements Volume 59 of this series. Chapters 3, 4, 5, and 6 deal with newer, specialized, and very promising aspects of affinity chromatography and related fields which have not been extensively reviewed in other literature. Chapter 7 introduces immobilized enzymes and proteins; it is followed by a discussion of the application of immobilized enzyme systems in Chapters 8 and 9. Chapter 10 provides an extensive and very practical introduction to solid phase protein synthesis. The final chapters in this volume deal with special topics in biochemistry on solid supports, namely, problems associated with matrices (Chapters 11 and 12) and with newer methods of applying solid supports to DNA synthesis and genetic engineering (Chapter 13), to the study of membrane and protein topology (Chapter 14), and to immobilized cells (Chapter 15).

Although it is difficult to employ a multiauthored book as a textbook in any course, it is equally difficult to find a single author capable of dealing well with the entire area of solid phase biochemistry. I became aware of the lack of a suitable text in this field while teaching a graduate seminar titled "Solid Phase Biochemistry," and it was from this awareness that the book developed. The general concept was to provide a book that would review basic concepts, give the flavor of the latest developments, and provide an introduction to the very most significant methods. The contributors, while retaining their individual styles and emphasis, have far exceeded my expectations in performing their tasks. Thanks are due to all of them for their part in making this volume what it is, namely, a practical review that can be reasonably employed as an advanced text in an area of immense importance to students of modern biochemistry.

Special acknowledgment goes to Janet Zimmerman, who bravely kept manuscripts and correspondence flowing in spite of her heavy departmental responsibilities. I salute my wife Nancy and our children, and the families of every contributor, for their patience and perseverance while we worked feverishly to meet our deadlines. Finally, I thank my colleagues at Bucknell and at Wageningen, The Netherlands, where these volumes began, for their answers to my many inquiries and their review of occasional manuscript pieces during the process of editing.

WILLIAM H. SCOUTEN

Lewisburg, Pennsylvania
March 1983

CONTENTS

Solid Phase Biochemistry

CHAPTER

1

INTRODUCTION

WILLIAM H. SCOUTEN

Chemistry Department
Bucknell University
Lewisburg, PA 17837

Solid phase biochemistry consists of a wide variety of biochemical processes in which solid phase support materials are used to perform many tasks that would be much more difficult, perhaps impossible, using solution chemistry. For example, solid phase DNA synthesis enables the genetic engineer to chemically synthesize defined sequences of deoxynucleotides cheaply and quickly, sequences that may be essential for the subsequent production of many biopolymers, such as interferon, the anti cancer drug that may provide mankind with many benefits. "Solution" (liquid phase) synthesis of DNA or protein is far more tedious and time consuming and cannot be automated as easily as solid phase systems. Thus the solid phase methods are enabling researchers to do far more than they otherwise could.

Solid phase biochemistry is at the same time an ancient aspect of biochemistry and one of its newest and most promising areas. The cell per se is highly organized, with many "matrix bound" solid phase components which early biochemists tried to separate into soluble homogeneous materials. In contrast, the "new" solid phase biochemistry begins with these soluble, often homogeneous, biological components and couples them to an insoluble, usually inert, matrix. The purposes for which the resulting insoluble biochemicals are employed are so diverse and so numerous that it is almost impossible to define solid phase biochemistry except in the very broadest of terms: Solid phase biochemistry is that area of biochemistry which concerns all biochemically related reactions or functions in which one or more components is in an insoluble solid phase.

We hope this book will demonstrate the utility of solid phase biochemical methods, show how powerful they can be, and give the reader a practical insight into new ways in which solid phase methods can be introduced into his or her own laboratory. In this introduction we quickly review some of the newer, more novel, and more important solid phase methods, many of which will be more fully discussed in later chapters by recognized experts in each

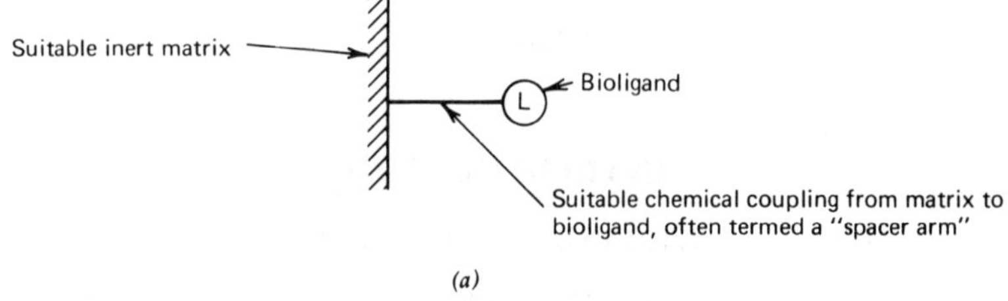

(a)

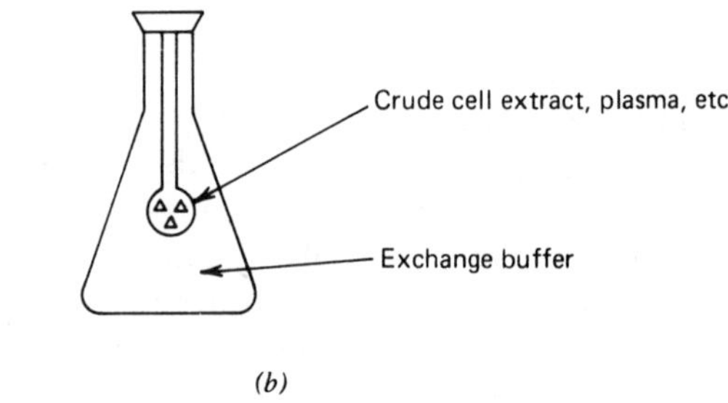

(b)

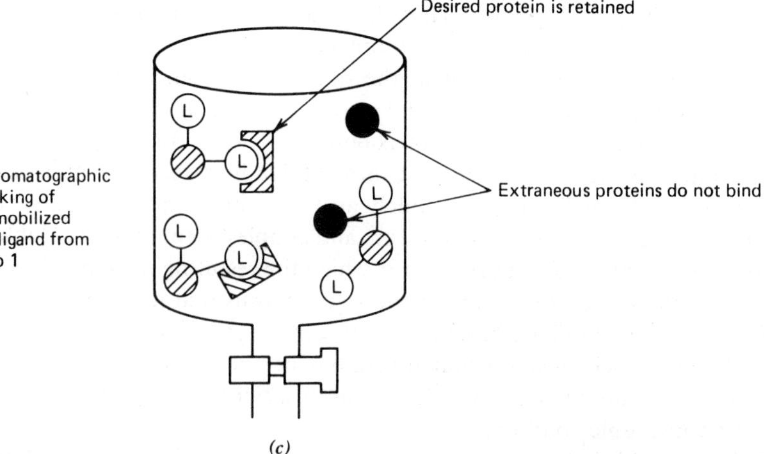

(c)

Fig. 1.1. The steps of affinity chromatography. (*a*) Step 1: Immobilize a bioligand. (*b*) Step 2: Prepare crude extract and free it from any endogenous substrate. (*c*) Step 3: Apply the substrate-free extract to a column of the bioselective adsorbent. (*d*) Step 4: Wash away unwanted proteins. (*e*) Step 5: The desired protein is eluted, possibly with a soluble bioligand. Reprinted with permission From *Affinity Chromatography: Bioselective Adsorption on Inert Matrices*, by W. H. Scouten, Copyright © John Wiley and Sons, Inc. (1981).

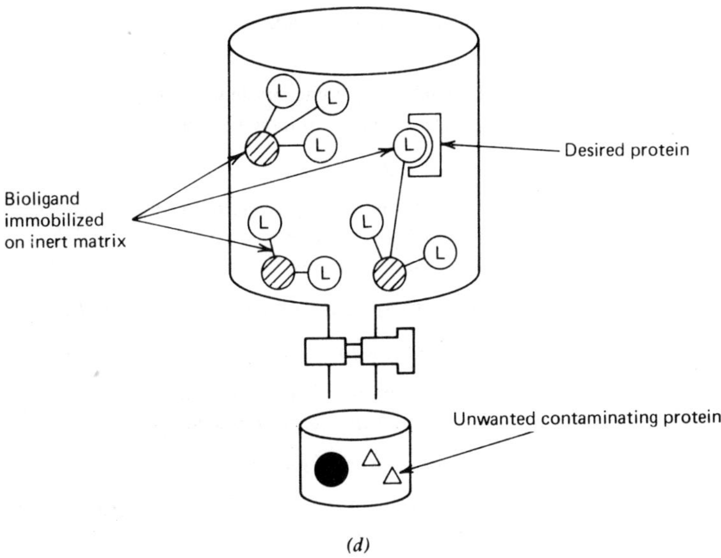

(d)

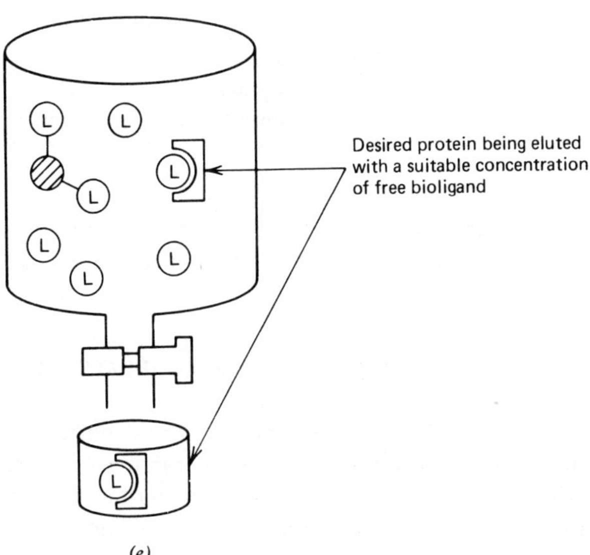

(e)

Fig. 1.1. (*Continued*)

Table 1.1. Types of Affinity Chromatography

I. Bioselective Adsorption: Affinity Based on Biologically Relevant Binding

A. Group specific ligands

1. Lectins
2. Nucleotide cofactors (e.g., NAD, AMP)
3. Immunobiotin-labeled proteins

B. Specific ligands

1. Rarely employed cofactors (e.g., vitamin B_{12})
2. Receptor proteins
3. Immunosorbants

II. Chemiselective Adsorption: Affinity Based on Chemically Defined Interactions

A. No modification involved

1. Hydrophobic chromatography
2. Ion exchange chromatography

B. Transient protein modification

1. Covalent chromatography (active thiols, Hg^{++}, etc.)
2. Borate complexes
3. Arsine oxide complexes
4. Active serines

C. Immobilized protein modifiers (IMP's)

1. Reducing agents
2. Oxidizing agents
3. Reactive site modifiers (e.g., lysine, cysteine)

field. Likewise, we introduce in this chapter some of the problems associated with solid phase biochemistry that will be defined more closely in the later chapters.

1.1. AFFINITY CHROMATOGRAPHY AND RELATED TECHNIQUES

Affinity chromatography is the term applied to the purification of enzymes and similar bimolecules based upon the interaction of an enzyme or biomolecule with an immobilized ligand in the form of an enzyme substrate inhibitor, antibody, or similar substance (1–10; see Chapter 2 for a full discussion). A column of the immobilized ligand is prepared and a relatively crude extract is applied to it. The protein desired binds to the immobilized ligand while extraneous proteins pass through the column. The enzyme–ligand matrix is then washed with buffer and the desired enzyme is eluted with buffer containing the free ligand. Alternatively, elution of the adsorbed enzyme can be effected by changing the pH or ionic strength or using other techniques to lower the enzyme–ligand affinity sufficiently to elute the enzymes (see Figure 1.1).

The term "affinity chromatography" is somewhat of a misnomer for this method because the purification is based upon a biologically defined interaction between a biological ligand or its analogue and a protein, nucleic acid, or similar biomolecule, whereas "affinity chromatography" would seem to be a naturally broader term. Actually, "affinity chromatography and related techniques" is a common symposium title, and it usually includes hydrophobic chromatography (chromatography on immobilized lipids or other hydrophobic materials), covalent chromatography, affinity electrophoresis, and so on. I suggest dividing "affinity chromatography" into two sections: *bioselective* adsorption, as described above, and *chemiselective* adsorption, that is, chromatography of proteins and similar biomolecules on the basis of their chemical structure (usually surface composition) rather than biological activity. Thus chemiselective adsorption would include hydrophobic chromatography, covalent chromatography, borate chromatography, and even, in the broadest sense, ion exchange chromatography. The term bioselective adsorption would be reserved for affinity chromatography based on biologically determined interactions between proteins and immobilized ligands. This potential division is shown in Table 1.1.

The success of affinity chromatography and bioselective adsorption is manifold. Although a few scattered papers in the area appeared before 1970, the first thorough, exciting, and stimulating account of affinity chromatography as a general technique was that published by Cuatrecases et al. in 1970 (11). Since then, hundreds of papers dealing with this method have appeared, as well as several books (1–2, 5, 6, 8). The success of affinity chromatography, however, cannot be measured in terms of quantity alone. Some affinity chromatographic procedures have proved to be the key to the many other advances in todays biotechnological revolution. Consider, for example, the production of urokinase via genetic engineering (12). The DNA polymerase and other DNA-related enzymes used may have been isolated by DNA–cellulose affinity chromatography. The RNA from the total RNA of human fetal kidney cells was enriched for poly-A-containing messenger RNA via affinity chromatography on poly(U)-Sephadex. The urokinase messenger RNA was isolated from the total mRNA via classical ultracentrifugation methods and inserted into *E. coli* plasmid pBR322. This was then cloned in an appropriate host, *E. coli* χ-1776. Urokinase was isolated from the transformed *E. coli* via affinity chromatography on benzamidine–agarose to obtain a urokinase enzyme molecule indistinguishable from the uro-

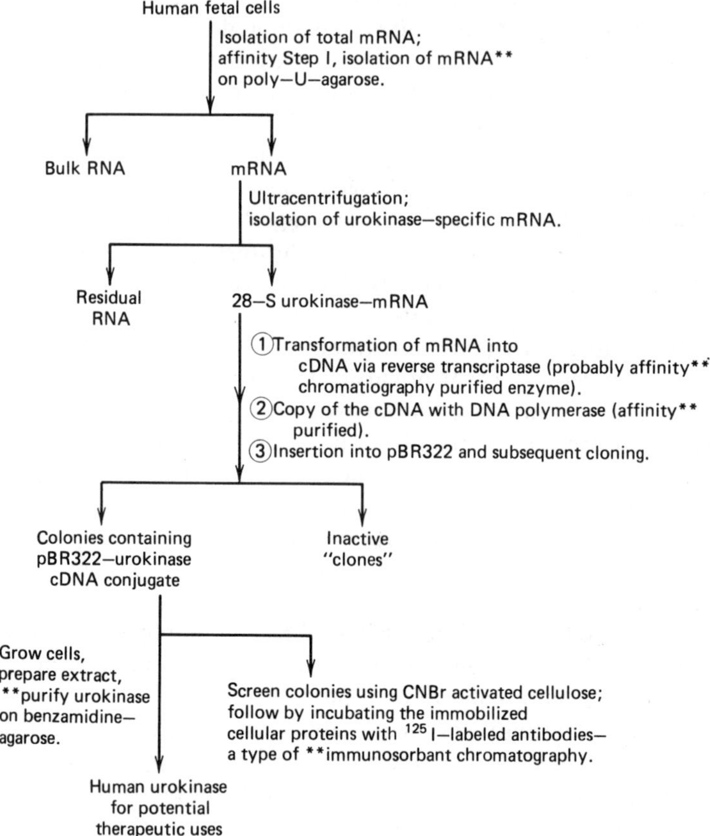

Fig. 1.2. Flow chart of a typical genetic engineering problem and its essential affinity chromatography steps.

kinase isolated from human tissue-culture preparations. Thus the preparation via genetic engineering of this important human protein (one that may well be used in treating embolism and phlebitis and in dissolving other blood clots, thus saving many lives) involved at least three distinct affinity chromatography steps and possibly more. Without the development of affinity chromatography, progress in any aspect of genetic engineering would have been extremely slow, if not impossible. In fact, genetic engineering has developed upon the foundation of bioselective adsorption. (See Figure 1.2.)

Applications of affinity chromatography to analytical techniques are as important as its contribution to genetic engineering. Virtually hundreds of

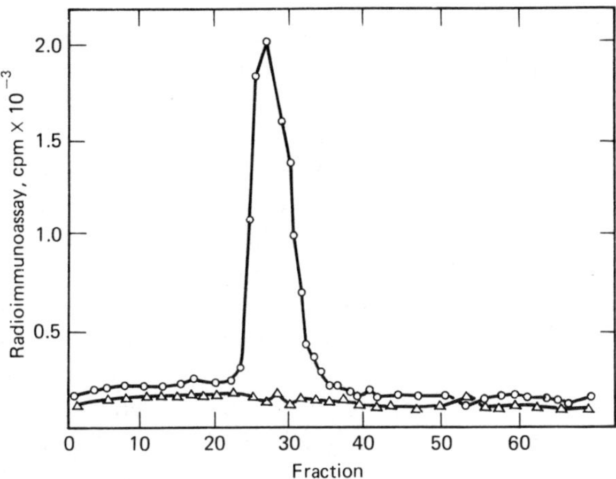

Fig. 1.3. Benzamidine affinity column chromatography of urokinaselike material. One-liter cultures of *E. coli* χ1776 containing either pABB26 or pBR322 were grown overnight, and the cells were collected by centrifugation and lysed. The lysate was dialyzed against starting buffer (0.1 *M* sodium phosphate, pH 7.0/0.4 *M* NaCl). After dialysis, the lysate was loaded on a benzamidine-Sepharose column, washed thoroughly with the starting buffer, and then eluted with 0.1 *M* sodium acetate, pH 4.0/0.4 *M* NaCl. Fractions were collected and aliquots were taken for radioimmunoassay in plastic-well microtiter plates. ●, χ1776-pABB26; △, χ1776-pBR322. From Ratzin et al. (12), with permission.

analytical tools have been devised employing the concepts of bioselective adsorption, from solid phase radioimmunoassay (13) to liquid chromatography (14). Indeed, in the past few years HPLC employing affinity matrices has become very popular and promises to yield even more advances in analytical procedures.

1.2. IMMOBILIZED ENZYMES

The use of immobilized enzymes in analytical and synthetic methods is thoroughly discussed by two experts in these areas, John F. Kennedy (Chapter 7) and J. Tramper (Chapter 8) respectively. Several other excellent treatments are available, including the monograph by P. W. Carr and L. D. Bowers (volume 56 in this series) entitled *Immobilized Enzymes in Analytical and Clinical Chemistry* (15). Indeed, the area of immobilized biocatalysts, including immobilized cells, organelles, and similar biomaterials has received more attention in terms of the number of direct publications (see Chapter 8) than has affinity chromatography or any other area of solid phase biochemistry. Moreover, several industrial procedures and standard clinical assays have been developed which contribute to the commercial importance of the immobilized enzyme field. Coupled with recent developments in genetic engineering (which will yield new, better, or cheaper enzymes), immobilized biocatalysts can be expected to occupy a major role in industrial chemistry in the next decade. Immobilized biocatalysts, however, will always be in competition with classical solution methodology, and fermentation will probably never be displaced from its position in industrial biotechnology as the leading biocatalyst process. Immobilized catalysts would have to be superior to fermentation methods in cost, quality, or novelty of the product before solid phase biocatalysts, with their natural complexity, could displace fermentation and other solution methods, most of which are relatively simple. Immobilized whole cells (see Chapter 15) offer considerable advantage over immobilized enzymes in terms of simplicity and may well become the more widely employed of the two systems for bulk chemical preparation. Immobilized enzymes, on the other hand, will probably fare very well in fine chemicals, pharmaceuticals, and diagnostics, where quality of the product is of prime importance and its cost, and often the raw materials used in its manufacture, is relatively high. Prime candidates for production through solid phase catalysts are research-quality biochemicals where quality is essential and cost is often extreme; coenzyme A, for example, costs about $10,000 an ounce based on commercial catalogue prices! Indeed the cost of such biochemicals where they are used as cofactors in other immobilized enzyme processes, has been one of the prime obstacles in the use of solid phase biocatalysts, since these cofactors must be added, usually in soluble form, to the reaction stream. Substantial progress has, however, been made in the regeneration of cofactors (see Chapter 7) which will mean wider use of immobilized enzymes in the future.

1.3. OTHER SYNTHETIC METHODS EMPLOYING SOLID PHASE BIOCHEMISTRY

Immobilized enzymes as described in the previous section are probably the most widely employed components in the synthetic methods that use solid phase reagents and catalysts. Nonetheless, solid phase protein synthesis, as led by Nobel Laureate R. B. Merrifield, and solid phase DNA synthesis, as developed by Carruthers and Itakura, have had significant impact on many areas of biochemistry. In Chapter 10, John M. Stewart, an early colleague of Merrifield and a prolific researcher in his own right, describes many recent advances in solid phase protein synthesis, in both application and methodology. Constant improvements have made this technique increasingly likely to meet the goal of synthetically produced enzyme analogues with altered activity. However, low yields, racemization of amino acids during synthesis, and the problem of enzymes folding to their native conformation have all inhibited the practical use of this method (16, 17).

Indeed, practical large-scale synthesis of hard-to-obtain enzymes and their analogues appear most likely to result from genetic engineering technology. And genetic engineering, as we have seen, is based upon solid phase biochemical methods. For example, many "promotor" sequences, essential for the genetic engineering of many proteins, are synthesized rapidly and easily by automated solid phase DNA synthesizers, often termed "gene machines." In another application of solid phase DNA synthesis, short DNA sequences, corresponding to all known codons for a known protein sequence of a protein to be purified, have been prepared (18). The resulting DNA mixture is immobilized to form an affinity matrix for the purification of the mRNA (or its complimentary DNA) that codes for the protein desired. A crude mRNA (or cDNA) factor is applied to the immobilized synthetic DNA mixture under annealing conditions (conditions where complimentary single-stranded DNA chains hydrogen-bond to form double-stranded DNA), then the bound DNA is removed using denaturing (hydrogen-bond-breaking) conditions. This yields a nucleic acid fraction enriched in the gene coding for the desired protein. Using the DNA isolated in this manner in a recombinant DNA (genetic engineering) system, a bacterial strain can be created that will synthesize large quantities of the desired protein, assuming all other factors are favorable.

This and other applications of solid phase DNA synthesis are discussed by R. Bruce Wallace and Keiichi Itakura in Chapter 13. A thorough discussion of the most-used methods and the significant applications are given, along with such references as are needed to give the beginner a place to start investigating—if not actually using—solid phase DNA synthesis.

1.4. ANALYTICAL TECHNIQUES BASED ON SOLID PHASE BIOCHEMISTRY

Analytical techniques using solid phase systems can be divided into two general classes: (i) Those systems in which analytical separations are performed using solid phase systems, such as affinity electrophoresis (Chapter 6) or affinity chromatography (Chapter 3); (ii) those systems in which solid phase methods are involved in sample preparation, such as solid phase protein sequencing or such sample detection systems as solid phase radioimmunoassay (13, 19) enzyme electrodes (Chapter 9), enzyme thermistors (20), and transistors (21).

1.4.1. Analytical Separation Based on Solid Phase Biochemical Methods

Most early applications of affinity chromatography were utilized for the purification of specific, usually protein, macromolecules. Very few of these were analytical in purpose but had synthetic aims as steps in a synthetic or natural product isolation. Only recently have good analytical procedures been developed based on affinity methods. For example, Amicon (Danvers, Mass.) has developed an assay for glucosylated hemoglobin, which is elevated in diabetes, based on the "chemiadsorbant" affinity between boryl-agarose and the glucosylated hemoglobin (22). This method, originally developed by Peter Dean and coworkers at Liverpool (23, 24), has been extended by Glad and Mosbach (25, 26) to the use of borylated porous glass in HPLC separation of sugars and glycoproteins. Chapter 4 details some of the applications of borate chromatography in both analytical and preparative procedures.

HPLC applications of affinity chromatography have greatly increased (14, 27, 28) in the past five years, as has "reversed phase" or "hydrophobic chromatography" HPLC separations of proteins and peptides (29). Possibly the most promising application of HPLC-affinity chromatography is in the area of "dye-affinity" chromatography (see Chapter 3). Many organic dyes have surface features similar to NAD, ATP, and other nucleotide coenzymes. These dyes are inhibitors of a wide variety of enzymes, and when coupled to a suitable matrix they can be used to separate many dehydrogenases, kinases, and similar substances. Besides the intrinsic appeal of their own color, these dyes have many advantages for application in affinity chromatography; they are, as dyes used in colorfast textiles, cheap, plentiful, numerous, and readily attached to hydroxyl or amino matrices (30–32). When applied to HPLC (33) they provide an excellent separation medium for enzymes and even isoenzymes.

1.4.2. Solid Phase Biochemical Sensors in Analytical Techniques

Numerous sensing devices have been attached to immobilized enzymes, substrates, antibodies, and other solid phase biochemical reagents in an attempt to form more sensitive, more selective, faster, or more accurate analytical tools. By far the most widely employed of these sensors are enzyme electrodes, that is, electrodes coated in some fashion with an immobilized enzyme. This immobilized enzyme causes the presence of the substance being analyzed to make a change in the parameter (e.g., pH or ion concentration) that is being monitored by the electrode. In this way the enzyme serves as an interface between the sensing electrode and the compound whose presence is being analyzed. George Guilbault details in Chapter 9 the development and application of these devices.

1.5. MATRICES

This book contains two chapters dealing specifically with matrix materials used in solid phase biochemistry: One, Chapter 12, details the problems with using cyanogen-bromide-activated agarose; the other, Chapter 11, deals with the underutilized inorganic matrices including controlled pore glass. Agarose seems like an ideal matrix for many purposes—hydrophilic, noninteractive with proteins, inert, reasonably stable—yet when activated with cyanogen bromide it forms materials with significant problems (although they are minor in many applications), including ion exchange capacity for adsorbing protein nonspecifically, lack of stability, and excess chemical reactivity. Even so, cyanogen-bromide-activated agarose is one of the most widely utilized of all matrices in many areas of solid phase biochemistry. Thus a thorough knowledge of the problems inherent in this technique is essential. M. Wilchek and J. Kohn (Chapter 12) have investigated in detail both the mechanism of the reaction of cyanogen bromide with polysaccharides such as agarose and the problems in ligand leakage from cyanogen-bromide-agarose-prepared affinity matrices. (I have had excellent results with an alternative method of activation of agarose, tosyl chloride activation; it yields matrices with a high ligand loading, preparation is easy, there are no dangerous reagents—cyanogen bromide is extremely toxic—and one achieves a stable ligand–matrix bond. This method is described in Section 1.5.2.)

Unlike agarose, porous glass would seem to be a poor matrix since proteins are readily and tightly adsorbed to untreated ("virgin") porous glass. Moreover, glass is not chemically stable but dissolves in alkaline buffers. (This is not noticed with ordinary glass bottles since the rate of dissolution is very slow and the surface area is low, but a high-surface-area glass kept in aqueous solution at pH 10 to 14 with gentle warming can be dissolved overnight.) In contrast to agarose, however, derivatized porous glass does not adsorb proteins, has no undesired ion exchange properties, and is much more stable than underivatized porous glass. Wolfgang Haller, who developed the process for preparing controlled pore glass, describes the use of porous glass in Chapter 11. He also gives perhaps the first complete description of the process for preparing this matrix. This may be useful for biochemists, not to produce controlled pore glass themselves (it requires a great deal of investment in time and equipment) but to understand its nature and construction.

Many other matrices have been employed in solid phase biochemistry. They are rather thoroughly discussed in volume 59 of this series.

1.5.1. "Mosbach-Gribnau Activation"—Tosylated Agarose (34, 35)

No method of preparing a reactive matrix for ligand or enzyme immobilization lacks some drawback, but the method of sulfonyl chloride activation ("Mosbach-Gribnau activation") of agarose is almost trouble-free. I have used it to make very highly substituted amino agarose. The procedure that follows is an adaptation of that originally employed by Nilsson and Mosbach (34).

Sepharose CL-6B is washed with mixtures of water and acetone with constantly increasing acetone concentration in order to suspend the beads finally in pure acetone (dried over $CaSO_4$). For example, 25 mL of Sepharose CL-6B (settled volume) is washed with 75 mL each of water, water and acetone (3:1), water and acetone (1:3), and finally acetone. The beads are then mixed with 7 grams of tosyl chloride dissolved in ~10 mL of acetone. The beads are stirred with a magnetic stirrer and pyridine (7 mL) is added slowly, drop by drop. The mixture is then left for 2 hr at room temperature. The resulting tosylated agarose is washed thoroughly with acetone, then the acetone is exchanged with water by the reverse of the previous acetone—and—water wash sequence.

The resulting tosylated agarose is washed with 1.5 *M* ethylene diamine in 0.5 *M* $NaHCO_3$, final pH 10.7, and the washed beads are suspended in 50 mL of the ethylenediamine-$NaHCO_3$ solution. The suspension is tumbled (Kendall–Fischer mixer) or shaken at 40°C for 16 to 48 hr, after which the beads are washed with 500 mL each of water, 1 m*M* HCl, 1 *M* $NaHCO_3$ and finally water. Alternatively, ligands containing nucleophiles can be coupled directly to the tosylated agarose.

1.5.2. Immobilized Modification Reagents for Proteins—"IMPS"

A rather novel application of solid phase biochemistry is the immobilization of protein modification reagents such that the immobilized reagent is sterically restricted to the surface of proteins or membranes, thus yielding a protein or membrane surface topology probe (see Figure 1.4). This is discussed in detail in Chapter 14, but as a simple example the immobilization of radiolabeled cysteamine (see Figure 1.5) and its derivatives would form the basis of many potential such reagents. ^{14}C-bis-iodoacetamidyl cysteamine (36) would react with surface thiols, immobilizing the protein/membrane. The

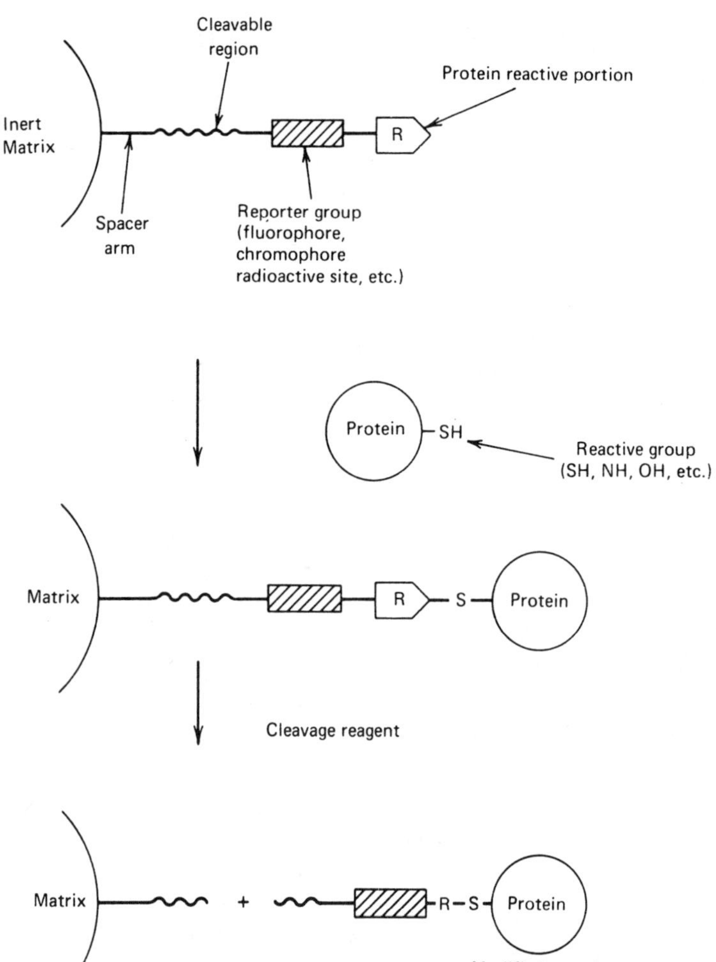

Fig. 1.4. Generalized scheme for the application of immobilized protein modification reagents. Reprinted with permission from *Affinity Chromatography: Bioselective Adsorption on Inert Matrices*, by W. H. Scouten, Copyright © John Wiley and Sons, Inc. (1981).

immobilized protein/membrane could be readily washed free of excess unreacted material, then the immobilized protein isolated by eluting with a reducing agent such as dithiothreitol. The *surface* thiols would now be radioactively labeled while *buried* ones would remain unlabeled. (Buried proteins in a membrane would likewise remain unlabeled.) Subsequent hydrolysis could show which cysteine was on the surface and which was buried.

$$H_2N—CH_2CH_2—{}^{35}S—{}^{35}S—CH_2CH_2—NH_2$$

(A)

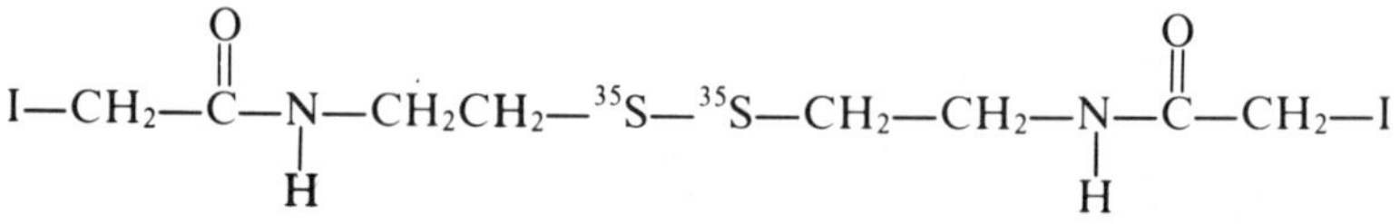

$$I—CH_2—\overset{\overset{\displaystyle O}{\|}}{C}—\underset{\underset{\displaystyle H}{|}}{N}—CH_2CH_2—{}^{35}S—{}^{35}S—CH_2—CH_2—\underset{\underset{\displaystyle H}{|}}{N}—\overset{\overset{\displaystyle O}{\|}}{C}—CH_2—I$$

(B)

Fig. 1.5. The structures of ^{35}S labeled cysteamine (A) and bis-iodoacetamidyl cysteamine (B).

1.6. CONCLUSION

The examples of solid phase biochemistry given in this book are only a sampling of the many potential uses for solid phase biochemical systems. Many more that have not yet even been thought of will surely be developed in the future. Without any doubt, solid phase methods will provide one of the foundations for exploring and applying biochemistry and for translating the vast biochemical knowledge acquired in the past few decades into practical ways to better the life and expand the knowledge of mankind.

REFERENCES

1. W. H. Scouten, *Affinity Chromatography,* Wiley-Interscience, New York (1981).
2. W. B. Jakoby and M. Wilchek, *Methods in Enzymology* **34,** Academic Press, New York (1974).
3. J. Turkova, *Affinity Chromatography,* Elsevier, Amsterdam (1978).
4. W. H. Scouten, *Amer. Lab.* **6,** 23 (1974).
5. C. R. Lowe, *Introduction to Affinity Chromatography,* Elsevier, Amsterdam (1979).
6. P. H. Reiner and A. Walsh, *Chromatographia* **4,** 578 (1971).
7. C. R. Lowe and P. D. G. Dean, *Affinity Chromatography,* Wiley, London (1974).
8. R. B. Dunlop, Ed., *Immobilized Biochemicals and Affinity Chromatography,* Plenum Press, New York (1974).
9. W. H. Scouten, in *Encyclopedia of Polymer Science and Technology,* Supplement 2 (N. Bikales, Ed.), p. 19, John Wiley and Sons, New York (1977).
10. J. Porath, *Biochemie* **55,** 943 (1973).
11. P. Cuatrecasas, *J. Biol. Chem.* **245,** 3059 (1970).

12. R. Ratzin, S. G. Lee, W. J. Schrenk, R. Roychoudhury, M. Chen, M. A. Hamilton, and P. P. Hung, *Proc. Nat. Acad. Sci. U. S.* **78,** 3313 (1981).
13. F. W. Spierto and B. Smarr, *Clin. Chem.* **20,** 631 (1974).
14. S. Ohlson, L. Hansson, P.-O. Larsson, and K. Mosbach, *FEBS Lett.* **93,** 5 (1978).
15. P. W. Carr and L. D. Bowers, *Immobilized Enzymes in Analytical and Clinical Chemistry,* Wiley-Interscience, New York (1980).
16. J. M. Stewart and J. D. Young, *Solid Phase Peptide Synthesis,* Freeman, San Francisco (1969).
17. J. D. Young and J. M. Stewart, *Solid Phase Peptide Synthesis,* Pierce Chemical Co., Rockford, Ill. (1982).
18. D. V. Goeddel, H. M. Shepard, E. Velverton, D. Leung, R. Crea, A. Sloma, and S. Pestka, *Nucleic Acids Res.* **8,** 4057 (1981).
19. K. Catt and G. W. Tregar, *Biochem. J.* **100,** 31c (1966); *Science* **158,** 1570 (1967).
20. B. Danielsson, L. Buelow, C. R. Lowe, I. Satoh, and K. Mosbach, *Anal. Biochem.* **117,** 84 (1981).
21. B. Danielsson, I. Lundström, K. Mosbach, and L. Stiblert, *Anal. Lett.* **12,** 1189 (1979).
22. P. J. Brown, P. D. G. Dean, and V. Bouriotis (Amicon Co.), British in Patent Application 2,024,829, through *Chem. Abstr.* **93,** 65,405 (1980).
23. V. Bouriotis, I. J. Galpin, and P. D. G. Dean, *J. Chromatogr.* **210,** 269 (1981).
24. A. K. Mallia, G. T. Hermanson, R. I. Krohn, E. K. Fujimoto, and R. K. Smith, *Anal. Lett.* **14,** 649 (1981).
25. M. Glad, S. Ohlson, L. Hansson, M.-O. Månsson, and K. Mosbach, *J. Chromatogr.* **200,** 254 (1980).
26. M. Glad, S. Ohlson, P.-O. Larsson, and K. Mosbach, *Abstracts, 4th Int. Affinity Chromatography Conference,* Vedhoven, The Netherlands (1980).
27. V. Kasche, K. Buchholz, and B. Galunsky, *J. Chromatogr.* **216,** 169 (1981).
28. C. R. Lowe, M. Glad, P.-O. Larsson, S. Ohlson, D. A. P. Small, T. Atkinson, and K. Mosbach, *J. Chromatogr.* **215,** 303 (1981).
29. A. H. Nishikawa, S. K. Roy, and R. Puchalski, in *Affinity Chromatography and Related Techniques* (T. C. J. Gribnau, J. Visser, and R. J. F. Nivard, Eds.), Elsevier, Amsterdam (1982) p. 471.
30. P. D. G. Dean and D. H. Watson, *J. Chromatogr.* **165** (1979) 301.
31. J. M. Egly and E. Boschetti in *Affinity Chromatography and Related Techniques* (T. C. J. Gribnau, J. Visser, and R. J. F. Nivard, Eds.), Elsevier, Amsterdam (1982) p. 445.
32. C. R. Lowe, Y. D. Clonis, M. J. Goldfinch, D. A. P. Small, and A. Atkinson, in *Affinity Chromatography and Related Techniques* (T. C. J. Gribnau, J. Visser, and R. J. F. Nivard, Eds.) Elsevier, Amsterdam (1982) p. 389.
33. D. A. P. Small, T. Atkinson, and C. R. Lowe, *J. Chromatogr.* **216,** 175 (1981).
34. K. Nilsson and K. Mosbach, *Biochem. Biophys. Res. Commun.* **102,** 449 (1981).
35. T. C. J. Gribnau, Ph.D. Thesis, University of Nijmegen, Nijmegen, The Netherlands (1977).
36. R. F. Ludena, M. C. Roach, P. P. Trcka, and S. Weintraub, *Anal. Biochem.* **117,** 76 (1981).

CHAPTER

2

INTRODUCTION TO AFFINITY PURIFICATION OF BIOPOLYMERS

A. HIROTOSHI NISHIKAWA

Biopolymer Research Department
Hoffman-LaRoche, Inc.
Nutley, NJ 07110

Nature weaves a seemingly infinite variety into the structure of bipolymers. Biochemists long ago assumed that any compendium of methods and recipes for the purification of these complex substances would surely be an immense if not impossible undertaking. Thus, few tracts have been devoted solely to the subject of the purification of biological macromolecules except for an encyclopedia like the *Methods in Enzymology.* This chapter, by focusing only on the use of the *affinity principle* in the isolation of a biopolymer, hopes to serve as an entree to the purification of all major classes of biological macromolecules. An enormous number of papers which include the use of this concept, is precluding any attempt at complete coverage of the literature. Thus attention here is directed to general principles. While the emphasis mainly centers on the ubiquitous proteins, attention is also directed to nucleic acids and polysaccharides.

2.1. HISTORICAL BACKGROUND

The basic phenomenon, that gives rise to the many applications covered in several chapters of this book, is *adsorption,* one substance sticking to another. Of the many purification techniques used by biochemists, adsorption has an old and venerable history. Since about 1900, the use of adsorption techniques has undergone a gradual evolution from being rather nonspecific to being highly specific. This development was determined as much by the growth in understanding of the highly selective properties of biological macromolecules as by the invention or availability of highly selective sorbents. While both have accelerated rapidly in recent years, it is the making and using of sorbents that exhibit high bioselectivity that has brought adsorption technology into such widespread use. The possibility has emerged that almost any type of bipolymer can be purified by this approach.

2.1.1. Enzymes

In the purification of enzymes, one can place the crystallization of urease by Sumner more than 50 years ago at the dawn of modern enzymology (1). Although this accomplishment was made without the use of any adsorption techniques, its significance was, of course, Sumner's assertion that an enzyme was a *crystallizable protein.* Willstätter, who had pursued adsorption methodology to a fine art through much of the 1920s (2, 3), did not share Sumner's belief. Although the protein nature of enzymes had yet to be settled, Willstätter did attempt to make use of affinity phenomena to purify an enzyme. The example was lipase, which was long observed to have an avidity for lipids. While tristearin was a good (affinity) sorbent for lipase, the recovery of bound enzyme was very difficult (4). Hence the use of aluminum hydroxide gels or kaolin, which also bound lipase but yet afforded ease of recovery, eclipsed the use of sorbents comprised of lipids. Thus the general notion of using affinity adsorption in enzyme purification did not progress at that time.

Yet many examples of selectivity in adsorption of enzymes had been reported since 1907, when Hedin observed that charcoal could bind both alpha- and beta-protease from ox spleen whereas kieselguhr could bind the alpha- but not the beta-enzyme (5). Michaelis and Ehrenreich surveyed a variety of inorganic sorbents and found specific conditions for binding various proteins to them (6). Starkenstein in 1910 observed biospecific adsorption (but perhaps did not realize it) while he studied the effect of chloride ion on amylase when bound to starch particles or not bound to them (7). The removal of amylase from solution by starch particles was observed to be a selective process by Ambard (8). And while a convenient means for analyzing the enzyme and studying its mechanism of action was afforded by the starch-binding phenomenon, it was not used as a basis for purifying the enzyme.

Perhaps uncertainty about the larger question of the chemical nature of enzymes diverted attention from attempting an affinity purification of these entities. For while Bayliss (9) and others marshaled evidence on the colloidal nature of enzymes, Willstätter had shown that enzymes could be adsorbed to inorganic materials such as kieselguhr or bentonite with remarkable selectivity (2) ***and*** then could be purified to the point where little protein was detectable (by methods available then). Thus Willstätter and colleagues held that enzymes were small catalytically active substances to which colloidal proteins were only incidentally, although often importantly, associated (10; see also Sumner, 11).

In 1930 Northrup reported the crystallization of another enzyme, pepsin (12). This enzyme, too, was purified without employing any adsorption methods even though Northrup himself ten years earlier had observed a somewhat specific precipitation (binding) of the protease onto insoluble protein (13). While Sumner's notion that enzymes are crystallizable proteins received reinforcement from the pepsin example, uncertainties persisted, as evident in a standard textbook of the era by Haldane (14). Sumner himself pressed on to show that the catalytic activity of urease (obtained by crystallization) could be destroyed by active pepsin at the same rate as that of its (protein) digestion (15). Waldschmidt-Leitz (a Willstätter colleague) questioned Sumner's idea even into 1934 (16), at which time Northrup was able to demonstrate that edestin (a crystalline, nonenzymic protein) and pepsin combine to form an insoluble complex from which the enzyme could be recovered and recrystallized (17). Thus, by using methods of affinity adsorption (he did not call it that), Northrup unequivocally demonstrated that the crystalline protein of pepsin is identical with the enzyme activity.

That enzymes could adsorb in a specific fashion to substrates or potential substrates continued to be demonstrated in the 1930s. Holmbergh had noted that alpha-amylase could be adsorbed to starch particles in aqueous ethanol suspension while beta-amylase could not (18). Similarly, Tokuoka found that amylase binds to steamed rice in the presence of alcohol (19). But efforts were still not directed to purifying enzymes by this approach.

In 1944 Hockenhull and Herbert (20) used the findings of Holmbergh (18) and Tokuoka (19) to purify an amylase from *Clostridium acetobutylicum* some 300-fold over the crude. While this may not have been the very first instance, it appears to be one of the earliest examples of deliberately using biospecific adsorption to purify an enzyme. By this example the possibility emerged that if an insoluble polymeric substrate were available, a biospecific affinity purification method could be devised for the corresponding enzyme. Of course, the approach was limited mainly to the enzymes that act on proteins, polysaccharides, nucleic acids or other naturally occurring substances that might be obtained in insoluble form. For example, Lineweaver et al. (21) reported the partial purification of fungal polygalacturonase by adsorption and desorption from alginic acid. The adsorption of malt alpha-amylase onto wheat starch was found to be an essential step by Schwimmer and Balls (22) in the isolation and crystallization of the enzyme. Other purification examples were reported throughout the 1950s. Their variety can be seen from these selected cases: Amylase from *Pseudomonas sacchrophila* was isolated on a column packed with potato starch and diatomaceous earth (23); porcine elastase was purified using powdered elastin (24, 25); collagenase from *Clostridium histolyticum* was recovered from ichthyocol particles (26); and neuraminidase was purified from *Vibrio cholerae* using cell walls from ovine or bovine erythrocytes (27).

The early 1950s saw the introduction of ion exchange resins in the purification of a protein, cytochrome c by Paleus and Neilands (28). It was, however, the development of ion exchange celluloses (29), found to be more widely compatible with proteins than synthetic ion exchange resins, that gave rise to a powerful general methodology for protein isolation (30) that is popular to this day. Even so, a review by Zittle in 1953 on the use of adsorbents in protein isolation noted that while specificity could be found with adsorbents like kaolin, hydroxylapatite, or bentonite, there was a need for selective adsorbents (31):

> . . . it would be desirable to have an adsorbent of absolute specificity such that a single adsorption would suffice to remove the wanted protein from a mixture. . . . The binding of substrate and enzyme which is highly specific might provide the desired specific forces. It would be necessary to have a solid resemble the substrate sufficiently so that it could be the site of enzyme–substrate bonding and yet usable under conditions that it would not be altered by the enzyme.

This stimulating review included a citation of a now classical paper by Campbell et al. (32) on antibody purification using a protein antigen immobilized on cellulose particles. In another review of protein isolation published in 1953, Taylor reflected on the purification of amylase on starch powder by Schwimmer and Balls (22) and wrote of the need (33):

> . . . to test the attractive suggestion that the adsorption is related to the same forces which operate in the attraction of the enzyme for its substrate. Further extension of this work would be highly desirable. Insoluble polymers, of known composition, should prove to be particularly suitable for the separation of proteins with specific "affinities."

Contemporaneous with these reviews was the appearance of the remarkable paper by Lerman on mushroom tyrosinase purification (34). Having earlier worked with Campbell on applying the affinity concept to the purification of antibody molecules, Lerman now demonstrated that a substrate analog covalently attached to cellulose could be used to specifically remove the desired enzyme from solution and effect its purification. This appears to be the first case where the enzyme-specific sorbent was prepared by deliberate chemical synthesis.

Ingenious as Lerman's example was, there appeared no great rush in the literature to apply his method to the purification of other enzymes. Perhaps the high resolving power of ion exchange chromatography on cellulosic sorbents more readily commanded the attention of enzymologists, as suggested by the growing avalanche of papers describing this methodology in the 1950s and 1960s. Among the few studies using the affinity principle was a brief account by Erlanger (35) of attempts to purify chymotrypsin by adsorbing it to an artificial substrate: *N*-carbobenzoxy-L-leucyl-D-phenylalanine benzyl ester

(CLPB). Due to their insolubility in water, particles of CLPB functioned conveniently as a specific adsorbent. Whereas affinity with chymotrypsin was evident, the sorbent also bound a number of other proteins in varying degrees, hence it was not attractive for further pursuit.

Of the reports describing the use of naturally occurring polymeric sorbents that continued to emerge throughout the 1950s and 1960s, few referred to Lerman's work. One was De la Haba, who in 1962 reported the crystallization of glycogen phosphorylase from rabbit muscle after adsorption to potato starch powder in a preliminary purification step. In his discussion, De la Haba noted Lerman's work and suggested that very selective methods for enzyme purification could be devised by using insoluble substrates as sorbents (36). It was in the laboratory of McCormick that Lerman's baton was picked up again. With Arsenis in 1964, McCormick reported the preparation of a flavin–cellulose sorbent that was used to purify flavokinase from rat liver (37). A little later Arsenis and McCormick (38) produced a more extensive study involving the preparation of flavin–phosphate cellulose to purify flavin mononucleotide-dependent enzymes.

From his earlier association with McCormick, Baker noted similarities in the problem of designing active-site-directed irreversible enzyme inhibitors and that of preparing enzyme-specific adsorbents. Thus it seemed to him appropriate to review and summarize these studies in his 1967 book on enzyme inhibitors (39). To succeed in purification using "enzyme-specific columns," Baker pointed to the following: (i) the polymer carrier should be inert and nonionic, (ii) more should be known about the locus on an inhibitor molecule where the carrier could be attached without interfering with the enzyme binding, (iii) studies should be made of optimum spacing of the immobilized inhibitors in the carrier so that efficient binding could be attained, (iv) efficient means of deriving the carrier polymer should be found. The last point was prompted by the observation that cellulose derivation in aqueous medium proceeded slowly. Baker's review did not pretend to be exhaustive but rather focused attention on Lerman's principle. It did not for instance include the work of Fritz et al. (40) who used the methods developed by Katchalski and coworkers (Levin et al., 41) to make trypsin insoluble and then use it to purify the trypsin inhibitor from soybean.

2.1.2. Nucleic Acids

The biospecific purification of nucleic acids apparently had to wait for the insights revealed by Watson and Crick (42) on the nature of base-pairing and the double-stranded structure of DNA. As in the case of enzymes, application of affinity adsorption methods developed in conjunction with or slightly after the understanding of biospecific mechanisms. For instance, in

1962, there were reported a number of applications exploiting the base-pair matching tendencies of appropriate nucleotide sequences in different nucleic acid polymers. For example, Bautz and Hall (43) described the purification of T4 (phage)-specific RNA using a cellulose sorbent that contained chemically bonded DNA extracted from the T4 phage. A general method for the purification of RNA complementary to DNA was reported by Bolton and McCarthy (44). Their technique was to immobilize DNA by entrapment in cellulose acetate or agar gels and then hybridize the appropriate RNA molecules to these sorbents.

Nucleic acid chemists like Gilham (45) showed that thymidine polynucleotides bound to cellulose by phosphodiester linkages could be used to purify the complementary deoxyadenosine oligonucleotides. Adler and Rich (46) extended the application of acetylated phosphocellulose developed by Bautz and Hall (43) to immobilize synthetic polyribonucleotides, which were then used to purify other homopolymers. By 1965 the base-pairing concept had been elegantly applied to the isolation of transfer-RNAs (called sRNA at that time). Erhan et al. (47) purified a particular tRNA on the basis that its anticodon was able to recognize and bind to the appropriate triplet codon in the polynucleotide immobilized to a cellulose carrier. Their method simultaneously permitted the isolation of a tRNA specific for a given amino acid and the identification of the triplet codon and anticodon. From McCormick's laboratory came also the report that thymidylate–cellulose could be used to purify adenine-rich polynucleotides (Sander et al., 48).

2.1.3. Antibodies

As in the case of biochemists with enzymes, the affinity purification of antibodies by immunochemists did not make much headway until some basic understanding of these macromolecules had been obtained. The ability of antibodies to recognize and bind specific chemical structures (haptens) was first shown by Landsteiner in 1920 (49). But it was not yet established that antibodies were proteins; the possibility still existed that the activity was only *associated* with proteins. Furthermore, the multiplicity of biological phenomena associated with antibodies was confusing. A "unitarian hypothesis" was put forth by Zinsser (50) to clarify this situation. He made the far-reaching suggestion that the many phenomenologically different reactions observed with antigens—such as agglutination, precipitation, complement fixation, bacteriolysis, or opsonization—were all due to one class of substances, namely antibodies.

By the 1930s accumulating evidence was pointing strongly to the protein nature of antibody molecules (51). Silber and Demidova (52) compared curves for heat inactivation of antibodies and concomitant protein denaturation and concluded that antibodies were proteins. Studies along similar lines by Gerlough and White (53) on tetanus antitoxin also supported this notion. From his extensive studies on antipneumococcus material, which included digestibility by proteases, Felton (54) concluded that antibodies were proteins. He furthermore reported that antibodies were responsible for all the biological activities proposed by Zinsser in 1921.

Amongst the more elegant studies establishing the protein nature of antibodies was that of Kirk and Sumner (55), who showed that selectively purified antiurease was destroyed by papain and pepsin and that, furthermore, injection of this antibody into test animals protected them from urease poisoning. The final physical characterization came at the hands of Tiselius, who with his electrophoresis method clearly showed that gamma globulins in plasma are identical with antibody molecules (56).

The early approaches to purification of antibodies included the nonbioselective methods of adsorption in the manner of Willstätter. But relatively few successes were seen. Biospecific methods of purification, wherein the antibody was recovered from precipitates formed with specific antigens, appeared more promising. The approach taken by Marrack and Smith (57, 58) in their isolation of diphtheria antitoxin was followed by examples from other laboratories. In time, improvements were made in the recovery of antibody from immunoprecipitates (e.g., Heidelberger and Kabat, 59). The work of Kirk and Sumner cited above (55) elegantly showed the potential of biospecific precipitation methods. They precipitated antiurease from serum with the enzyme, washed the precipitate with dilute salt solution, and then destroyed the enzyme with dilute acid. The residue was resolubilized and brought to the isoelectric point of urease whereby the inactive enzyme precipitated and left behind a solution of highly purified antiurease.

One of the earliest studies on the road to manmade immunosorbents was that by Bleyer, who observed that immunoglobulins (then called immunoagglutinins) could bind well to fine colloids and then be selectively desorbed by bacterial preparations that had been used to raise the antisera (60). A notable improvement was made by D'Alessandro and Sofia (61), who isolated antibodies specific for syphilis and tuberculosis by using adsorbents comprised of corresponding antigens coated on particles of kaolin or charcoal. Meyer and Pic (62) followed this work with a detailed study of factors affecting recovery of antibody from immunosorbents made of antigen-coated kaolin.

Having earlier bonded haptens covalently to proteins to obtain immunogens for use in antibody production in vivo, Landsteiner's move to preparing insoluble carrier-bound haptens seemed natural. In the classic paper by Landsteiner and Van der Scheer (63), various haptens were diazo-coupled onto chicken erythrocyte stroma to obtain immunosorbents which were then used to selectively remove certain subpopulations within a mixture of cross-reactive antibodies.

Given the limitations of erythrocyte stroma, the use of a more durable carrier like cellulose was a significant development. This was introduced in 1951 by Campbell and colleagues, who used *p*-aminobenzylcellulose as a medium for immobilizing serum albumin. The insoluble conjugate was then used to bind antialbumin from the sera of rabbits immunized with bovine serum albumin (32). This general approach to preparing insoluble antigen adsorbents was followed by Lerman's method for preparing insoluble hapten adsorbents (64).

Soon, Isliker (65) extended the development of synthetic immunosorbents by using cation exchange resins as the carrier. Blood group antigens as well as human serum albumin were covalently bonded to the carboxyl groups of these resins, which afforded sorbents capable of selectively binding the corresponding antibodies. Manecke and Gillert (66) followed by showing that polyaminostyrene is a useful carrier for either antigen or antibody. This resin had been shown by Grubhofer and Schleith to be useful in preparing immobilized enzymes (67).

Campbell's aminobenzylcellulose method was applied in 1954 by Talmage et al. (68) to purify I^{131}-labeled antihuman IgG from rabbit serum. Yagi et al. (69) carried out quantitative studies of detecting I^{131}-labeled antibodies adsorbed to serum albumin or to ovalbumin bonded to polyaminostyrene. Gyenes et al. (70) coupled various proteins, including ragweed pollen, to polyaminostyrene particles, which were then used to selectively remove corresponding antibodies from rabbit antisera. By 1958 Campbell's laboratory had also explored antigen-coated glass beads as immunosorbents (Sutherland and Campbell, 71).

These developments naturally gave rise to reviews, which included the uses of synthetic immunosorbents. One of the earliest was that of Isliker (72). This was followed by those of Kabat and Mayer (73), Manecke (74), Sehon (75), Weliky and Weetall (76,76a), and Silman and Katchalski (77). This last review included immobilized enzymes as well as antibodies and antigens and served to underscore the potential of immunosorbents in immunochemistry as well as in other biochemical applications.

2.1.4. Miscellany

The literature on *receptors, lectins,* and *viruses,* although not of comparable volume to that on enzymes or antibodies, also reveals an interesting historical perspective on the application of affinity principles to the purification of these substances.

The concept of *receptors* for specific compounds was considered and invoked long before the physical or chemical demonstration of their existence. Indirect support for their presence was suggested by the association of radiolabeled compounds (e.g., steroids) with selectively precipitated fractions from cell extracts (78). One of the first attempts to use covalently immobilized ligands to selectively adsorb receptors was reported by Jensen and colleagues in the mid-1960s (79). They coupled estradiol via the diazo reaction to *p*-aminobenzyl cellulose and used the resulting sorbent to isolate estrogen receptors from extracts of calf uterus. Alternative methods for ligand immobilization were reported by von der Haar and Mueller (80), who coupled estradiol via the 17 alpha side-chain to *p*-aminostyrene. The estrogen receptor was found to bind quite well to the sorbent but it was very difficult to recover (resolubilize) the adsorbed receptor. Thus, this approach needed some refinement to make it practical.

Lectins are a class of proteins that are capable of recognizing and binding specific sugar moieties either in oligomeric or polymeric form (see Section 2.4.2a). The best studied of these, concanavalin, was first isolated in 1916 (81). Using conventional means Sumner crystallized concanavalin A (or con A) as well as the B form a few years later (82). It was not until 1936, however, that con A was shown to be identical with a hemagglutinin that also reacts with glycogen (83).

The use of affinity concepts in the isolation of con A came some three decades later, when in 1967 Olson and Liener (84) reported its purification by adsorption on Sephadex and recovery with low pH buffer. In the same year, Agrawal and Goldstein (85) described a similar approach, but in addition used selective elution with glucose or fructose to recover the lectin. Since then a variety of lectins have been isolated by such affinity adsorption methods, using as sorbents naturally occurring polymers, modified biopolymers, or synthetic polymer-bound materials.

Another binding protein of historical interest to affinity purification is avidin, a protein in chicken egg white that has a great avidity to bind biotin. In the mid-1960s, McCormick devised a preparation of biotin–cellulose to purify the binding protein (86).

If, for purposes of discussion we can consider *viruses* as macromolecules, it is interesting that some years ago investigators attempted to use selective adsorption methods on these entities. A particular example is influenza A strain PR 8, purified by Knight (87), who took advantage of the observation by Hirst (88, 89) of the agglutination of chicken erythrocytes by this virus. In isolating the virus from murine lung tissue, Knight noted that only the virus bound to the erythrocytes and that this selectivity led to an efficient isolation.

Despite this early application of selective adsorption, the course of virus isolation technology traveled a more "physical" route. Taking advantage of the large mass associated with a virus particle, new and sophisticated centrifugation instruments and techniques provided convenient means for isolating these biomacromolecules; as a consequence, selective adsorption techniques were largely set aside until the early 1970s.

2.1.5. Recent Events

Upon perusing the various pathways of bioscience history, it seems that immunochemists were drawn to biospecific purification methods somewhat earlier than enzymologists. Perhaps this resulted from the great difficulties confronting the immunochemists in isolating antigen-specific immunoglobulins. Even if these proteins could be separated from the other serum constituents, available physical methods were inadequate for isolating the antigen-specific IgG subpopulation from the nonspecific one. Enzymes, by contrast, offered more variety, and most distinct species could be isolated eventually by using a combination of methods that dealt with some physical property such as isoelectric point, molecular weight, or solubility. Perhaps due to these successes, the development of biospecific purification methods for enzymes drew little attention until recently.

It was perhaps a unique constellation of factors that produced the current interest in affinity methods, especially as applied to enzymes. In 1968, from a laboratory well known for its contributions to enzyme chemistry, came what appeared to be a new technique: *affinity chromatography* (90). In addition to introducing this mnemonically useful term, Cuatrecasas, Wilchek, and Anfinsen presented a simple procedure for preparing biospecific adsorbents. By using agarose gel beads and the cyanogen bromide coupling reaction developed a few years previously by Axen and Porath (91, 92), these workers solved a key problem pointed out by Baker in his 1967 book (39). This ease of sorbent preparation no doubt accounted for the quick and widespread application of the affinity methods which continues to this day. The simplicity of the affinity method as originally reported was, as might be expected, not universal. Subsequent developments (which will be discussed later in this and other chapters) suggested that skill and care must be exercised to obtain sorbents that exhibit a minimum of nonspecific adsorption by contaminants (see O'Carra, 93).

2.2. REVIEW OF REVIEWS

The enormous quantity of accumulated literature on affinity purification methods makes it, of course, impossible to thoroughly review the material here.

But a survey of review articles, chapters, and books may serve as a useful introduction to the reader. Amassed in Table 2.1 in chronological order are a variety of review articles (varying from lengthy to brief) as well as books that deal with the use of affinity binding in biopolymer purification. Where the title does not adequately reflect the material contained, comments are appended, and where possible, the scope is indicated.

2.3. MODALITIES OF AFFINITY PURIFICATION

Before venturing into the many pathways of affinity purification methods, it is useful to establish some basic definitions and working concepts, as depicted in Figure 2.1.

Affinity binding describes the selective process whereby a macromolecule of interest interacts with or binds to a target species that is specifically recognized by virtue of the latter's chemical structure.

Ligate is the designation for the macromolecule of interest, that which is being sought after. Usually the ligate is a solute in the *mobile phase.*

The *ligand* is the target species or lure to which the ligate is attracted. Frequently the ligand is attached to an insoluble *support* or *matrix* or *carrier.* In such cases, the ligand–support comprises the *immobile phase.* In instances where the support entity is also soluble, the ligand–support would comprise a phase distinct from that of the ligate.

The ligand may be attached to the support via a *leash* or tether of various dimensions. The leash may be short, composed mainlyof the reagent needed to covalently couple the ligand to the support, or it may be made deliberately long to facilitate binding by the ligate.

As suggested in the historical examples, the use of affinity binding in the purification of biopolymers has given rise to a variety of modalities. The most common by far involves an affinity sorbent, which permits the capture of a soluble ligate onto an insoluble phase. *Chromatography* in its various modalities encompasses these methods. In addition, ligands attached to soluble polymers may be used in selective precipitations (*affinity precipitation*) or in selective extractions in a different mobile phase (*affinity partitioning*). These modalities are discussed below.

Another modality, that of using immobilized ligands in an electric field, is called *affinity electrophoresis;* the subject is extensively treated by Bøg-Hansen in Chapter 6 and thus will not be further explored here.

2.3.1. Adsorption and Chromatography

During the course of a historical review of affinity purification of enzymes, it was found that a considerable number were purified by adsorption onto an

Table 2.1. **List of Review Articles and Books on Affinity Purification Methods**

Author	Year	Title	Reference	Comments
Kabat	1943	Immunochemistry of Proteins	94	Includes section on selective purification of antibodies.
Isliker	1957	The Chemical Nature of Antibodies	72	Includes affinity methods of antibody purification
Kabat and Mayer	1961	Purification of Antibodies, Specific Methods	73	Book chapter on immunosorbents; immunoprecipitants.
Manecke	1962	Reactive Polymers and Their Use for the Preparation of Antibody and Enzyme Resins	74	Short review of immunosorbents; emphasis is on author's works.
Sehon	1963	Physicochemical and Immunochemical Methods for the Isolation and Characterization of Antibodies	75	Extensive review of immunosorbents.
Weliky and Weetall	1965	The Chemistry and Use of Cellulose Derivatives for the Study of Biological Systems	76a	Extensive review of chemistry to couple a wide variety of substances to cellulose; immunosorbents.
Silman and Katchalski	1966	Water-Insoluble Derivatives of Enzymes, Antigens, and Antibodies	77	Includes immunosorbents; also a few enzyme and inhibitor purifications.
Baker	1967	Enzyme-Specific Columns	39	Short book chapter; lays down principles and guidelines.
Campbell and Weliky	1967	Immunoadsorbents: Preparation and Use of Cellulose Derivatives	95	Immunoadsorbents using cellulose; antibody isolation.
Goldstein and Katchalski	1968	Use of Water-Soluble Enzyme Derivatives in Biochemical Analysis and Separation	96	Includes some examples of immobilized protease to purify inhibitors and vice versa.

Kato	1969	Affinity Chromatography	97	In Japanese; mostly examples from Anfinsens' laboratory.
Rothfus	1970	Newer Techniques in Protein Isolation and Characterization	98	Includes some immunosorbent examples.
Cuatrecasas and Anfinsen	1971	Affinity Chromatography	99	A review of methods and procedures.
Cuatrecasas and Anfinsen	1971	Affinity Chromatography	100	A more extensive review; 106 references.
Cuatrecasas	1971	Functional Purification of Proteins and Peptides by Affinity Chromatography	101	Short review; emphasis on own work.
Cuatrecasas	1971	Selective Adsorbents Based on Biochemical Specificity	102	Similar to preceding but longer.
Feinstein	1971	Affinity Chromatography of Biological Macromolecules	103	Explores maleic–anhydride polymer carriers as well as agarose.
Friedberg	1971	Affinity Chromatography and Insoluble Enzymes	104	Rather short coverage of purification methods.
Mosbach	1971	Enzymes Bound to Artificial Matrixes	105	Brief diagrammatic presentation of affinity method included.
Porath	1971	Some Recently Developed Fractionation Methods	106	In Japanese; mainly own work.
Reiner and Walch	1971	Affinity Chromatography: Specific Separation of Proteins	107	1969 to spring 1971 papers covered; some theory attempted.
Cuatrecasas	1972	Affinity Chromatography of Macromolecules	108	Extension of previous reviews; theory attempted.
Orth and Brümmer	1972	Carrier-Bound Biologically Active Substances and Their Applications	109	Affinity purification of biopolymers presented as an extension of enzyme engineering.
Porath and Sundberg	1972	Specific Extraction of Enzymes Using Solid Surfaces	110	Many examples of naturally occurring polymers as sorbent; 65 references.

Table 2.1. (*Continued*)

Author	Year	Title	Reference	Comments
Guilford	1973	Chemical Aspects of Affinity Chromatography	111	Deals with variety of chemical reactions used in sorbent preparations
Huggins	1973	Affinity Chromatography: Enzyme-Inhibitor Systems	112	Short explanation of method.
Cromwell	1973	Protein Purification by Immunosorption	112a	Concepts and methods; 33 references.
Munier	1973	Chromatography of Macromolecules of Biological Origin	113	Variety of methods and examples; affinity examples are pre-1968.
Porath	1973	Conditions for Biospecific Adsorption	114	Short review; considers operational problems.
Weetall	1973	Affinity Chromatography	115	Covers papers to 1972; 92 references.
Guilford	1974	Affinity Chromatography	116	Concepts and examples; 59 references.
Kennedy	1974	Chemically Reactive Derivatives of Polysaccharides	117	Extensive review of chemical reactions on polysaccharides; immunosorbents; enzyme affinity systems; nucleic acids.
Lowe and Dean	1974	*Affinity Chromatography*	118	Book; references to mid-1973.
May and Zaborsky	1974	Ligand-Specific Chromatography	119	Papers through 1972, some examples from 1973; 138 references.
O'Carra	1974	Affinity Chromatography of Enzymes	93	Short review; emphasis on operational problems.
Royer and Meyers	1974	Support Materials for Immobilized Enzymes and Affinity Chromatography	127	Chemistry of support materials; 36 references.

Scouten	1974	Affinity Chromatography: Bioselective Adsorption on Inert Matrices	120	Short review; emphasizes porous glass matrices.
Turkova	1974	Affinity Chromatography	121	Extensive reviews; emphasis on 1972 and 1973 papers; 187 references.
Williams	1974	Developments in Affinity Chromatography	122	Short appraisal of problems.
Baum and Wrobel	1975	Affinity Chromatography	123	Book chapter; 74 references.
Dean and Harvey	1975	Applications of Affinity Chromatography	124	Listing of NAD(P) affinity examples.
Nishikawa	1975	Affinity Purification of Enzymes	125	Short review on concept and practice.
Parikh and Cuatrecasas	1975	Affinity Chromatography: Principles, and Applications	126	Book chapter.
Porath and Kristiansen	1975	Biospecific Affinity Chromatography and Related Methods	128	General review; references to 1972.
Turkova	1975	Affinity Chromatography Theory; Affinity Chromatography Practice	129	Short book chapters.
Morris and Morris	1976	Affinity Chromatography	130	Book subchapter.
Wilchek and Hexter	1976	The Purification of Biologically Active Compounds by Affinity Chromatography	131	Book chapter, citations to 1973.
Whitesides and Nishikawa	1976	Affinity Chromatography	145	Sorbent design; affinity concepts on small molecules; 98 references.
Lowe	1977	Affinity Chromatography: Current Status	132	An assessment of general problems and solutions.
Olsen	1977	Affinity Chromatography	133	Brief explanation of principles.
Parikh and Cuatrecasas	1977	Affinity Chromatography in Immunology	134	Antigens and antibodies; 212 references.

Table 2.1. (*Continued*)

Author	Year	Title	Reference	Comments
Boguslaski and Smith	1977	Isolation of Substances from Urine by Affinity Chromatography	135	Variety of ligates from one source; 50 references.
Rode	1977	Purification of Enzymes and Other Biologically Active Proteins by Means of Chromatography on Low Molecular Substances Immobilized on Solid Supports	136	In Polish; 89 references.
Rutishauser and Edelman	1977	Fractionation and Manipulation of Cells with Chemically Modified Fibers and Surfaces	136a	26 references; some practical concerns in cell separation.
Burgett and Greenley	1977	Cibacron Blue F3GA Affinity Chromatography	137	Blue dye as general ligand; 39 references.
Walter	1977	Partition of Cells in Two-Polymer Aqueous Phases: A Surface Affinity Method for Cell Separation	137a	53 references; affinity partitioning of cells.
Turkova	1978	*Affinity Chromatography*	138	Extensive treatise; 405 pages.
Trayer and Winstanley	1978	Immobilized Nucleotides and Their Use in Affinity Chromatography	139	Emphasis on chemistry; 72 references.
Hudson	1978	Lymphocytes, Receptors and Affinity Chromatography	140	Short review and critique on cell fractionation.
Potuzak and Dean	1978	Affinity Chromatography on Columns Containing Nucleic Acids	141	Very brief review on nucleic acid purification.
Nishikawa	**1978**	**Affinity Chromatography of Plasma Proteins: An Overview**	**142**	**71 references.**

Chaiken	1979	Quantitative Uses of Affinity Chromatography	143	Mostly ligand binding constants. 32 references.
Lotan and Nicolson	1979	Purification of Cell Membrane Glycoproteins by Lectin Affinity Chromatography	143a	Extensive review; 178 references.
Nishikawa	1979	Affinity Chromatography	144	Includes biomedical applications.
Dean and Watson	1979	Protein Purification using Immobilized Triazine Dyes	144a	Covers a variety of dyes; 171 references.
Hubert and Della-Cherie	1980	Uses of Water-Soluble Biospecific Polymers for Purification of Proteins	146	Affinity partitioning.
Gray	1980	Affinity Chromatography	147	General review of recent chemical methods.
Farooqui	1980	Purification of Enzymes by Heparin-Sepharose Affinity Chromatography	148	Heparin as general affinity ligand; 87 references.
Sharma and Mahendroo	1980	Affinity Chromatography of Cells and Cell Membranes	148a	Extensive review of methods and problems; 165 references.
Scouten	1981	*Affinity Chromatography: Bioselective Adsorption on Inert Matrices*	148b	Textbook; many examples of methods

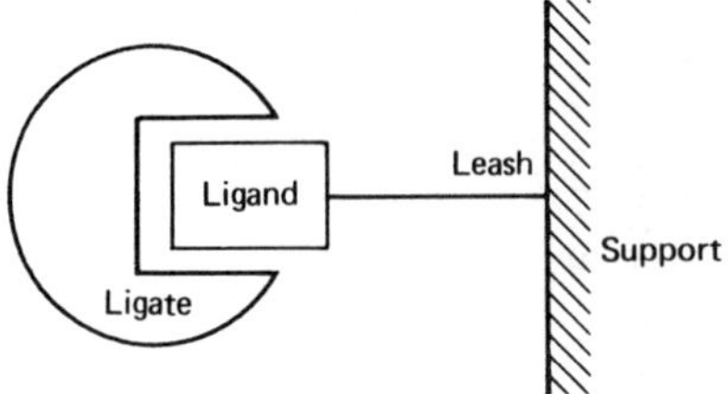

Fig. 2.1. Elements of affinity binding system.

insoluble polymeric substrate. Nonetheless, the examples were primarily limited to enzymes where a suitable naturally occurring sorbent could be obtained.

With the advent of synthetic affinity sorbents, there arose a need to consider a variety of factors so that an optimally functioning sorbent could be obtained. In addition to containing a ligand of suitable selectivity and avidity, the ideal sorbent was expected to be stable to handling and flow of solvents, to be resistant to microbial degradation, to be free of nonspecific binding by unwanted contaminants, to possess a high capacity for ligand (and ligate); and to be convenient for the attachment of ligand (128).

2.3.1a. Supports

The perfectly ideal support material for affinity sorbent preparation does not exist. However, a variety of useful materials have found their way into practice. Table 2.2 lists some of the more common support matrices ranked somewhat in the order of their frequency of use.

Historically, Cellulose saw earliest use as a general support matrix, primarily by immunochemists, but after the late 1960s the availability and greater ease of use of agarose gel made it the most popular. Dextran gel (Sephadex) does not possess adequate rigidity at molecular porosities comparable to that of agarose gel. Hence it has been less used as a support. High molecular porosity is related to high capacity for macromolecular ligates.

Polyacrylamide gels of desirably high porosity are soft and do not form column beds with high flow throughput. But since they are completely man-made they show more resistance to biodegradation. Poly HEMA [poly(hydroxyethylmethacrylate)] gels are available in rigid spherical beads, but are in somewhat limited supply. They tend to show more nonspecific hydrophobic adsorption of proteins than agarose (157).

Silica gels are available as irregular granules and now also as fine spherical beads. They must be carefully coated with suitable organosilanes to diminish nonspecific adsorption of proteins as well as to introduce suitable linkage points for attaching ligands. The highest flow rates are attained through sorbent columns prepared from silica gels.

Table 2.2. Some Support Materials for Affinity Sorbents

Material	Unit Structure	Physical Form	Suppliers (Trade Names)	References on Usage
Agarose	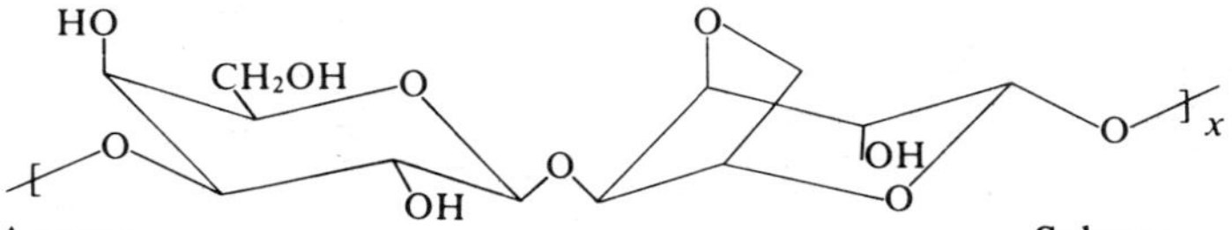	Spheres	Bio-Rad Labs (Bio-Gel A) Pharmacia (Sepharose)	149, 150, 152, 153
Dextran	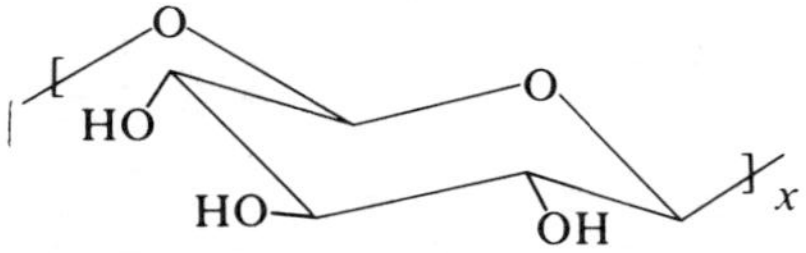	Spheres	Pharmacia (Sephadex)	91, 151
Cellulose	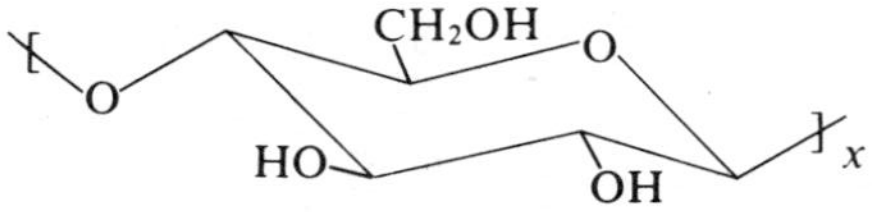	Granules; spheres	Whatman	76a, 117
Polyacrylamide	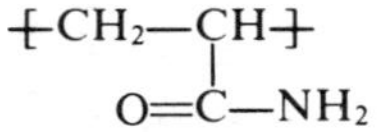 $-[CH_2-CH(C(=O)NH_2)]-$	Spheres	Bio-Rad Labs (Bio-Gel P)	154
Poly(HEMA)[a]	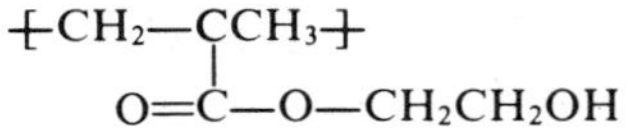 $-[CH_2-CCH_3(C(=O)-O-CH_2CH_2OH)]-$	Spheres	Lachema (Spheron)	155, 155a
Glass/silica gel	$[SiO_2]_x$	Granules; spheres	E. Merck (LiChrosorb, LiChrosphere)	156

[a] poly(hydroxyethylmethacrylate).

The support materials listed in Table 2.2 do not constitute an exhaustive list. They are adequate however, as an introduction to the field. The interested reader is directed to Turkova (138, Chap. 8) for further amplification.

2.3.1b. Coupling Chemistry

Space does not permit a detailed review here of the wide variety of chemical coupling reactions employed in attaching ligands to supports. But, for purposes of orientation it is useful to examine some of the primary coupling reactions onto support matrices as listed in Table 2.3.

Because of the abundance of hydroxyl groups in the more popular support materials, *O*-alkylation has been a convenient route to attaching ligands. Cyanogen bromide has been relatively easy to use (158). But its linkage product (a substituted isourea) has been found unstable under certain conditions (158a) and its obnoxious property as a very toxic lacrymator leave something to be desired. (See Chapter 12.) By comparison, *N,N'*-carbonyldiimidazole is more congenial to handling and yields a more stable urethane (carbamate) linkage (159). Epichlorohydrin (160, 161) and its bromo homolog (150) as well as bifunctional epoxides (bisoxiranes [163]) can afford nonionic linkages that are as chemically stable as the glycosidic ones that make up the backbone in polysaccharide gels. Halo acetates have been used for some time to make carboxymethyl cellulose ion exchangers (29), but more recently their adducts with agarose gels have appeared. A new activation reaction for agarose gels involving tosyl chloride has recently been described by Nilsson and Mosbach (164).

The reaction of cyanuric chloride (trichloro-*s*-triazine) with alcohol groups has features of both acylation and alkylation. The multifunctionality of the reagent, rapid reactivity, and stability of the resulting linkage are attractive features which have led a few laboratories to study its use on agarose gels (165, 166, 162).

The oxidation of vicinal diols by periodate can be used in certain polysaccharides possessing these structures (e.g., cellulose and dextran gels). The resulting dialdehydes can be reacted with alkylamines and sodium borohydride to obtain very stable linkages (167, 168).

Owing to the abundant presence of the relatively unreactive amide groups in poly(acrylamide) gels, Inman and Dintzis found it practical to substitute a portion of these with bifunctional amines. The resulting gel derivatives could then be coupled in a variety of ways to other leash moieties or ligands (154).

Silica gels can be functionalized for covalent coupling of ligands by coating with 3-aminopropyltriethyoxysilane or 3-glycidoxypropyltriethoxysilane. The amino-coated silica gel can be treated with glutaraldehyde and then directly reacted with a protein ligand. Alternatively, one can succinylate the

Table 2.3. Primary Coupling Reactions Onto Support Matrices

Rxn Type/Reagent	Support	1st Derivative[a]	Other Derivative[a]	References
Acylation/BrCN	Agarose; dextran	S(—O—)(—O—)C=NH (cyclic imidocarbonate)	S—O—C(=$^{+}NH_2$)—NH—R	151, 152, 158a
Acylation/CD Imdz	Agarose	S—O—C(=O)—N(imidazole)	S—O—C(=O)—NH—R	159
Alkylation/Epihalohydrin	Agarose	S—O—CH_2CH—CH_2 (epoxide, —O—)	S—O—$CH_2CH(OH)CH_2NH$—R	150, 160, 161
Alkylation/Halo acetate	Cellulose; agarose	S—O—CH_2COOH	S—O—$CH_2C(=O)$—NHR	29
Alkylation/Cyanuric chloride	Cellulose; agarose	S—O—(triazine ring)(Cl)(Cl)	S—O—(triazine ring)(NHR)(NHR′)	165, 166
Oxidation/IO_4^-	Cellulose; dextran	S—CHO	S—CH_2NH—R	167, 168
Substitution/H_2N—R	Poly(acrylamide)	S—$CONHCH_2CH_2NH_2$	S—$CONHCH_2CH_2NHC(=O)R$	154
Acylation/Tosyl Cl	Agarose	S—O—$S(=O)_2$—C_6H_4—CH_3	S—NH—R	164

[a] S = support, R = leash or ligand.

amino groups and follow with carboxyl-activating agents such as those used in peptide chemistry (169). The glycidoxyl-coated silica gels can be reacted directly with leash groups or ligands bearing amino groups.

These, then, constitute some of the more popular ways to introduce ligand and leash functionalities into supports so as to arrive at a selective sorbant with the desired characteristics. More details and examples can be found in extended treatises like that of Turkova (138) or Lowe and Dean (118).

2.3.1c. Ligands

In preparing a synthetic sorbent one can have a variety of ligands to consider and from which to choose depending on the nature of the ligate being sought. In Table 2.4 are some of the more prominent examples of ligand–ligate pairs reported in the literature. In the case of enzymes (Type I) we have already cited in Section 2.1.1 a variety of historical examples of using insoluble polymeric substrates as selective sorbents. In some modern examples, soluble polymeric molecules (e.g., hemoglobin) were immobilized onto gels to obtain selective sorbents (e.g., for proteases). Examples of low molecular weight compounds that are bifunctional (i.e., that contain a site for coupling to the leash or support as well as a binding site for the enzyme) include a number of inhibitors or substrate analogs. These make possible a large variety of affinity sorbent preparations. Although some are monospecific (they will bind only one of several related species of ligates), many are not; this is discussed in Section 2.4.2 on group-specific ligands. Immobilized cofactors and prosthetic groups usually afford group-specific sorbents to which related ligates (e.g., dehydrogenases to immobilized NAD^+) can bind with different avidities.

It is possible to reverse roles and use immobilized enzymes (Type II) to isolate corresponding inhibitors, usually macromolecular, as in the example of immobilized trypsin being used to purify soybean trypsin inhibitor (40, 170).

With immobilized nucleic acid (Type III) ligands, the ligate of interest has frequently been the mRNA complement to the DNA genome moiety (171, 172). Nucleases have also been isolated with such sorbents (173, 174). The use of immobilized oligonucleotides to isolate specific tRNA has already been mentioned in the historical review (Sect. 2.1.2; see also ref. 47). Another example of interest has been the isolation of antibodies (IgG) to nucleic acids on immobilized nucleic acids (175, 176).

The use of immobilized haptens (Type IV) to isolate corresponding antibodies has much similarity to the system (Type VI) where immobilized monosaccharides are used to bind lectins or where small target molecules serve as ligands for binding proteins and receptors (Type VIII). In all of these cases

Table 2.4. Some Ligand–Ligate Combinations in Affinity Systems

Type	Ligand (Immobile Entity)	Ligate (Soluble Entity)	References
I	Inhibitor, cofactor, prosthetic group	Enzyme; apoenzymes	100, 110
II	Enzyme	Polymeric inhibitors	40, 170
III	Nucleic acid, single strand	Nucleic acid, complementary strand	171, 172
IV	Hapten; antigen	Antibody	75, 134
V	Antibody (IgG)	Proteins; viruses; receptors	177, 178, 178a
VI	Monosaccharides; oligosaccharides	Lectins, receptors; antibodies	179, 180, 180a
VII	Lectin	Glycoproteins; receptors; cells	183, 184
VIII	Small target compounds	Binding proteins	187, 188, 189
IX	Binding protein	Small target compounds	190, 191

the ligates have a distinctive function to recognize and bind the structural entities represented by the ligands. An immobilized hexose ligand can bind not only its corresponding lectin but also an antibody raised against it, or even a particular cell-surface receptor.

Antigens differ from haptens generally by being of more complex (usually macromolecular) structure. They possess many epitopes (distinct recognition sites) for binding by antibodies. Since the usual method for eliciting antibodies involves injecting responsive animals with the whole antigen, the resulting antisera are somewhat heterogeneous (even if high titers are attained). With these antisera, an immobilized antigen sorbent yields a population of antibody molecules that recognize several available epitopes. With monoclonal antibodies, the hybridoma selection process yields cell lines that produce an antibody capable of recognizing only one epitope, hence these antibodies are homogeneous with regard to ligand recognition. When monoclonal antibodies are in turn immobilized to produce immunosorbents for the purpose of isolating corresponding antigens, one can expect high specificity towards the ligate. This has been recently demonstrated by Secher and Burke (181) and by Staehelin et al. (182), who have prepared high purity interferon by the use of hybridoma-produced monoclonal antibodies.

The use of immobilized lectins (Type VII) to bind and isolate a variety of oligosaccharide-containing ligates has been pioneered by Sharon and coworkers (185). These ligands have been useful in the isolation and characterization of glycoproteins. More recently, subpopulations of lymphocytes bearing distinct markers (presumably oligosaccharides) have been selectively bound to gel particles bearing immobilized lectins (186). This versatility places lectins in a unique class of group-selective ligands.

The immobilization of small target compounds (Type VIII) to attract and isolate binding proteins is quite analogous to the inhibitor–enzyme (Type I) system or the hapten–antibody (Type V) situation. The reciprocal system of immobilized binding protein to bind and isolate small ligates has been reported here and there (145). The molar capacity of immobilized proteins relative to ligate bindings is generally unfavorable (192).

2.3.2. Affinity Precipitation

Much of what has been discussed with affinity sorbents has application here, except that the immobilized ligand is attached to or is a part of a soluble entity. As shown schematically in Figure 2.2, the affinity reagent is soluble until it contacts the specific ligate and forms an insoluble complex, which can then be separated from the other components by centrifugation or filtration.

As shown in the figure, the ligand is attached in multiple to the soluble carrier so that one immobilized ligand–carrier entity can interact with several ligate molecules simultaneously. To result in precipitation the ligand–ligate interaction must produce an entity of low intrinsic solubility. Usually this is achieved by extensive multiple interactions that result in very high molecular weight ensembles that can no longer stay in solution. It may be expected that the intramolecular spacing of ligands in the affinity precipitation, as well as the molecular size and polyvalency of the ligate species, will have influence on the nature (e.g., granular or flocculent) and extent of the precipitation.

2.3.2a. Immunoprecipitation

The earliest examples of affinity precipitation were seen in reactions of antibody (IgG) with antigen. In this case, the ligate (or ligand if antibody is being sought) was composed of the antigen, which presented several epitopes per molecule that could be bound by the corresponding IgG molecule. Because IgG has two binding sites, it would form bridges between two antigen molecules. And since the antibody preparation was usually heterogeneous, bridges could be formed at several sites on the antigen molecule. These multiple interactions resulted in a tight three-dimensional lattice that precipitated out. Unlike other precipitation phenomena (e.g., salting out) immunoprecipitation usually exhibited an optimum composition where maximum precipitation was obtained. If, compared with the antigen content in the reaction mixture, the antibody titer was in great excess or in marked deficiency, little

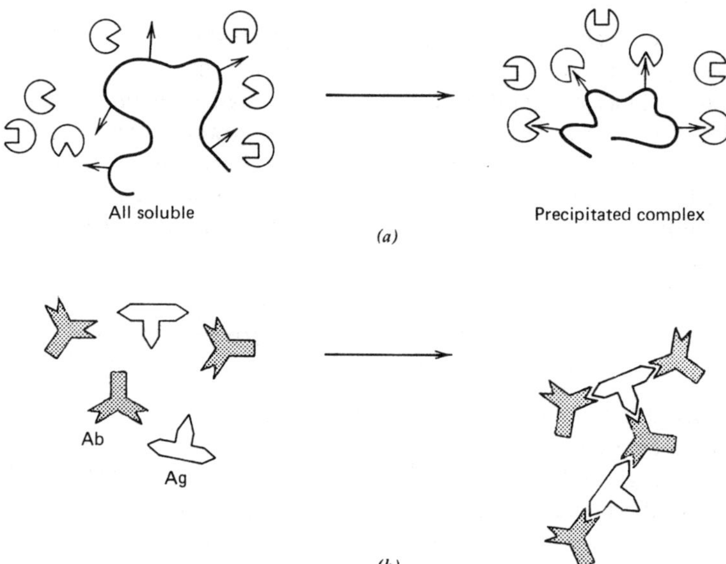

Fig. 2.2. Affinity precipitation systems: (*a*) Enzyme and inhibitor; (*b*) Antigen and antibody.

or no immunoprecipitation was observed. This required a careful study of conditions so that the maximum recovery of a desired ligate can be attained.

Having separated the immunoprecipitate from the mixture, one still has the task of separating antibody from antigen. The antigen–antibody complex can be disrupted and solubilized by a variety of solvent conditions including high pH, low pH, or inclusions of chaotropic agents such as urea, guanidine hydrochloride, magnesium chloride, and sodium iodide (193). The resulting solution must be further treated to recover the antigen and the antibody separately and to be rid of chaotropic agents (if they are used). Size-exclusion chromatography is a convenient procedure for this task, especially if the molecular weight of the antigen is greater than 250,000 or smaller than 100,000 daltons (194). This is because gels are available that would permit the separation of such antigens from the 160,000 dalton IgG molecule. Ion exchange chromatography in the presence of concentrated (4–8 *M*) urea may also be useful. An interesting approach devised by Gough and Adams was to use inactivated *Staphylococcus aureus* preparations (which contain, bound to the cell surface, protein A that binds selectively to immunoglobulin) to isolate antibody-polysome complexes (195).

There may be situations where the immune complexes remain soluble along with soluble excess antigen. Gel filtration has been used to fractionate such mixtures.

Because of technical inconveniences attending the isolation of the immune complex as well as the separation of antibody from antigen, immunoprecipitation has not gained acceptance compared with immunosorbents (see Section 2.4.2c). There may be situations, however, where one could benefit from the examples of Ralt et al. (197) on the immunoprecipitation of ATPase from *E. coli.* Likewise the work on ATP synthetase from beef heart mitochondria by Ludwig and Capaldi (198) and glutamate dehydrogenase by Yeung et al. (199) merit examination.

2.3.2b. Naturally Occurring Affinity Precipitants

In the early 1960s, investigators studying the interaction of glycogen with alpha-amylase noted that this enzyme could be selectively precipitated by limit dextrins of high molecular weight (7000 daltons or higher) (200). Much like antigen–antibody reactions, the precipitation was inhibited by an excess of dextrins or an excess of enzyme (201). Similar observations were made in the binding of muscle phosphorylase to glycogen (202).

2.3.2c. Synthetic Affinity Precipitants

An effort at preparing enzyme-selective precipitants was made by Dennis, who bonded *N*′-oxalyl-*p*-phenylenediamine to bovine serum albumin (203). This agent, BSA-1, complexed with lactate dehydrogenase (LDH) in the presence of NAD^+. The complex was precipitable by ammonium sulfate solutions from 45 to 60% saturation, permitting separation from free LDH. However, size-exclusion chromatography through polyacrylamide gel was needed to separate the complex from BSA-1. Of relevant interest to this work was the preparation of soluble high molecular weight NAD^+ by Wykes et al. (204), who did not evaluate their material as a precipitant.

Copolymerization of acrylamide and allyl glycosides of various sugars yielded a number of *O*-glycosyl derivatives of polyacrylamide copolymers for Horejsi et al. (205). In studying the precipitation of various lectins with these polyacrylamide-glycosides, concanavalin A was found to precipitate strongly and rapidly with *O*-mannosyl and *O*-glucosyl polymers. However, the *Ricinus communis* lectin did not precipitate completely with its corresponding lactose-bearing polymer. Poor results were also seen with the *Glycine soja* lectin with the *N*-acetyl-galactosamine-bearing polymer.

A nonpolymer, bifunctional affinity precipitation agent was developed by Larsson and Mosbach (206):

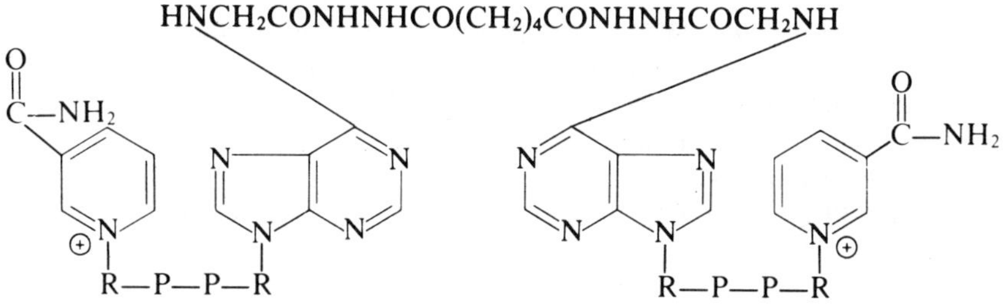

$$HNCH_2CONHNHCO(CH_2)_4CONHNHCOCH_2NH$$

(Bis-NAD structure: two NAD units, each R—P—P—R with nicotinamide and adenine rings, linked through adenine)

This "Bis-NAD" agent efficiently precipitated lactate dehydrogenase in the presence of sodium pyruvate. Gel filtration was used to separate the enzyme from Bis-NAD. While the Bis-NAD is a general class agent capable of precipitating other NAD-binding enzymes, studies with alcohol dehydrogenase (and pyrazole) suggested that a longer connecting leash between the two cofactor groups is needed for some enzymes.

The work of Hubert et al. introduced 3-amino-2-hydroxy-propyl dextran as a water-soluble ligand carrier (207). They coupled estradiol-7α-butyric acid via carbodiimide reaction to the amino-polymer. The resulting dextran–estradiol could be complexed with uterine estrogen receptor and the complex so obtained was separated from unwanted proteins by gel filtration. Highly purified receptor was subsequently released from the complex by exchange with [^{3}H]-estradiol.

Affinity precipitation may offer advantages in selectivity, rapid kinetics, and low contamination. Its advantages in a given situation will have to be empirically determined. The fractionation steps following precipitation are less convenient than is the case with insoluble sorbents.

2.3.3. Liquid–Liquid Partitioning

As with precipitation, affinity partitioning involves the use of soluble polymer carriers bearing selective ligands. It differs, however, by involving two liquid phases instead of a solid phase forming from the liquid phase. The idea of fractionating mixtures of biopolymers by partitioning between two different aqueous phases was pioneered by Albertsson (208). In this approach, two aqueous phases were defined by solutions of polymers (e.g., phase *a* by dextran and phase *b* by polyethylene glycol). Ionic strength, pH, and polymer compositions were adjusted so that the solubility of one protein in a mixture would be enhanced in one phase while the rest would remain in the other. Affinity partitioning evolved when a selective ligand was bonded to one of the phase-defining polymers. Then the avidity to bind the ligand drove the ligate into that phase. The concept is shown schematically in Figure 2.3. A mixture of ligates (lower phase) is placed in contact with a separate phase containing a selective ligand (upper phase). After intermixing, the two phases are allowed to separate again, and one finds a particular ligate species enriched in the upper phase.

An early example was the trypsin partition system reported by Takerkart et al. (209). PEG 9000 (9000 dalton polyethylene glycol) was converted into a diamidino-α, ω-diphenylcarbamyl derivative (PEGPAB):

$$H_2N{-}\overset{\overset{\oplus}{N}H_2}{\overset{\|}{C}}{-}C_6H_4{-}NH\overset{\overset{O}{\|}}{C}O{-}(CH_2CH_2O)_x{-}\overset{\overset{O}{\|}}{C}NH{-}C_6H_4{-}\overset{\overset{\oplus}{N}H_2}{\overset{\|}{C}}{-}NH_2$$

The phenylamidine moiety is known as a potent inhibitor of trypsin. When trypsin was added to the basic partition system of PEG phase (upper) and dextran phase (lower), about 40% of the activity was found in the upper phase. In the system including PEGPAB, in the upper phase 92% of the enzyme was found in the PEG phase. The specificity of the partition system was shown by comparison with the behavior of chymotrypsin, which assayed 20% in the PEG-only phase and 28% in the PEG + PEGPAB phase.

A variety of affinity partition systems are listed on Table 2.5. While a diversity of ligands and ligates are evident in the samples, the liquid phase couples have been limited thus far to PEG–dextran. No doubt one reason for this is the convenience of controlled derivatization of PEG, which has at most only two sites of covalent ligand bonding.

Since affinity partitioning is somewhat affected by the solvent properties of the two immiscible phases, it is not surprising that there are ligates where the PEG–dextran system is not optimal. One approach to augment affinity partitioning has been described by Chaabouni and Della-Cherie (217). They

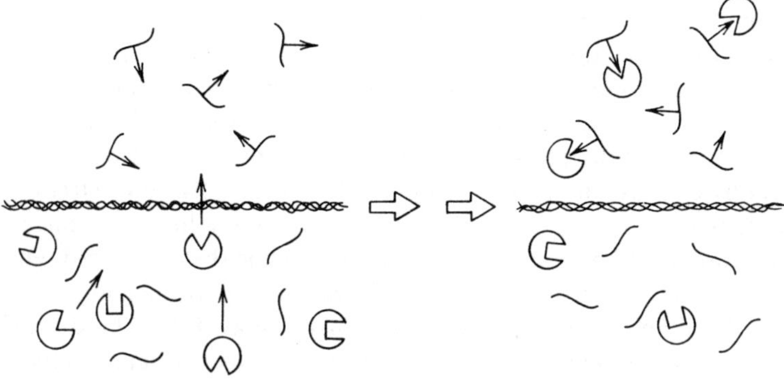

Fig. 2.3. Affinity partitioning (liquid–liquid) system.

Table 2.5. Examples of Affinity Partition Systems

Ligate	Ligand	Ligand Phase	Nonligand Phase	Reference
Bovine serum albumin	Palmitate	PEG	Dextran	210
Erythrocyte vesicles	Palmitate	PEG	Dextran	211
Histones	Palmitate; stearate	PEG	Dextran	212
Erythrocytes	Palmitate; oleate; linoleate; linolenate; deoxycholate	PEG	Dextran	213
Chloroplasts	Deoxycholate	PEG	Dextran	214
	Deoxycholate	Dextran	PEG	
	Palmitate	PEG	Dextran	
Human serum	Adipic; suberic; sebacic; dodecanedioic; tetradecanedioic; hexadecanedioic, octadecandioic	PEG	Dextran	215
Δ_{5-4} 3-Oxosteroid isomerase	Estradiol	PEG	Dextran	216
Δ_{5-4} 3-Oxosteroid isomerase	Estradiol	PEG	Triethylpropylammonium dextran; carboxymethyldextran	217
Δ_{5-4} 3-Oxosteroid isomerase	3-Oxo-4-androsten-17β-yl carboxyl	PEG	Dextran	218
	3-[3,17 β-Dihydroxy-1,3,5(10)-estratrien-7 α-yl] butyroyl	PEG	Dextran	
Concanavalin A	Dextran	Dextran	PEG	219
Myeloma protein	Dinitrophenyl	PEG	Dextran	219
Acetylcholine receptor	4-Trimethylammoniumphenylamino-	PEG	Dextran	220, 221, 222
	trimethylammoniumethyl-	PEG	Dextran	

prepared charged dextran phases by carboxymethylation with chloroacetate or by reaction with epichlorohydrin and triethylamine. In a multistage extraction of oxosteroid isomerase, the inclusion of cationic dextran in the phase opposite the estradiol-bearing PEG resulted in a marked augmentation of the enzyme binding to the ligand phase.

A drawback in affinity partitioning, one long associated with liquid–liquid partitioning involving polymer-defined phases, is the need to separate the ligate from the phase-defining polymer. Gel filtration and ultrafiltration have been used (146).

There is a limitation in the extent of derivatization of the polymer phase. Depending on the solubility properties of the ligand, a high incorporation into one of the phase-defining polymers could result in a marked (and adverse) change in the phase definitions of the polymers involved. This suggests that the avidity of a ligate to the polymer-bound ligand must be quite high.

The advantages of affinity partitioning include very mild interface conditions for handling sensitive ligates such as membrane-bound receptors and whole cells. Where binding constants are high, low concentrations of ligate species can be extracted efficiently. Since the ligand-bearing polymer is low in concentration, there is a minimum of nonspecific adsorption by contaminant protein species (146).

2.4 LIGANDS: STRUCTURE AND FUNCTION

Now that we have briefly surveyed ligand–ligate combinations that address the major groups of biopolymers of interest, in this section we consider in more detail certain aspects of ligand selection and usage.

2.4.1. Selectivity of Ligands

It is implicitly understood in the affinity purification of biopolymers that the ligands employed must possess a high degree of selectivity. But what exactly is high selectivity? We will attempt to arrive at a practical understanding of this concept by examining a variety of ligands.

We shall use the term *selectivity* to describe the unique aspects of a ligand structure that causes it to be *recognized* and *bound* by a ligate. Since the ligate is usually a macromolecule, we shall describe its capability to recognize and bind a ligand as *specificity*. An analogy is made to the "lock and key," where the ligate is the lock and the ligand the key. Some common examples are enzyme and inhibitor, antibody and hapten, and lectin and sugar. Since previously we defined the ligand simply as an *immobile phase entity* and the ligate as a *mobile phase entity*, the present definition of selectivity can be confusing if the ligand is a macromolecule, thus having specificity, and the ligate is a small molecule that is selected and bound by the ligand. Further confusion may arise if both the ligand and ligate are macromolecules, as in the case of immobilized DNA adsorbing to (or being adsorbed by?) complementary mRNA. With enzymes and corresponding polymeric inhibitors or with antibody and corresponding antigen, the accepted mechanisms of binding suggest that specificity should be ascribed to enzymes and antibodies, respectively, whether they are functioning as ligands or ligates.

Selectivity in ligands can vary from narrow to wide. For example, specific antibodies (especially monoclonal) can be directed toward a unique epitope (recognition site) on a ligate protein and recognize little else. By contrast, wide selectivity is observed with immobilized enzyme cofactors such as NAD^+, which can be recognized and bound by a large number of enzymes in certain catalytic categories (139). Furthermore, there are other ligands, such as heparin, which seem to selectively bind a wide array of functionally and structurally unrelated ligates (148). Then, too, there are the immobilized synthetic dyes that have been found useful as general affinity ligands for a wide variety of ligates (144a).

We can derive a general understanding of the key *elements of selectivity* by ligands from a detailed look at a few enzyme–inhibitor systems as models. For convenience, we can designate *primary, secondary,* and *tertiary* elements of selectivity as relating in a general way to increasing goodness-of-fit of the ligand into the binding site of the ligate.

With enzymes, the *primary* elements may include the array of atoms that are acted upon by the catalytic site. For example, —NHOH, which is isoelectronic with HOOH, can be recognized by peroxidases (223). And carbonic anhydrase, which reacts with the structure $\mathrm{HO{-}C({=}O){-}OH}$, can also be inhibited by $\mathrm{R{-}S({=}O)_2{-}NH_2}$ (224, 225). The *secondary* elements may include structural groupings that are adjacent to the primary elements and that enhance binding and improve discrimination.

A comparison of two proteases is instructive. The primary element of specificity for both chymotrypsin and trypsin is the recognition of amide bonds (as well as ester bonds). The secondary elements of specificity for chymotrypsin involve hydrophobic side-chains of amino acids such as the indole moiety of tryptophan, while for trypsin a strong cationic side-chain (guanidino group of arginine) is important. Thus chymotrypsin binds avidly to a ligand composed of tryptophan methyl ester, while trypsin binds better to a ligand of arginine methyl ester.

The avidity of ligate binding to primary or secondary elements can vary considerably. For example, *m*- or *p*-aminobenzamidines are synthetic inhibitors of trypsin and comprise only the secondary elements of specificity. Yet they bind the enzyme better (226) than a ligand made up of glycyl-glycyl-*L*-arginine, which more closely resembles the natural substrate (233).

The *tertiary* elements of a ligand may include structural entities that are further removed from the primary and secondary elements. They may include groups that affect steric accessibility by the binding site (such as that enabled by a leash) or ionic groups that reinforce binding of the other elements. The spatial relationship of tertiary elements to the primary or secondary element is often critical. A favorable relationship can improve ligate binding by tenfold to a hundredfold, while an unfavorable one can diminish binding by comparable orders of magnitude.

When considering the contributions of the various elements to ligand selectivity, one expects that ligands possessing more elements and in distinct array will be more selective towards a unique ligate. Less selectivity is expected with ligands of simpler structure. This is illustrated in Table 2.6.

Very simple ligands such as the *p*-chlorobenzylamino (L-1) or phenylacetyl (L-2) appear to be bound by an assortment of functionally unrelated enzymes such as beta-galactosidase and trypsin. More complex ligands such as the *N*-carbobenzoxyphenylalanyl (L-3) are able to show more discrimination in binding by proteolytic ligates. For example, chymotrypsin is bound but trypsin is not. This is in marked contrast to L-1 and is clearly the effect of multiple structural elements that narrow the selectivity. Some paradoxes can be noted with simple ligands, however. With immobilized *p*-amino-phenylpropionate (L-4), which in its soluble form is a known inhibitor of both carboxypeptidase A and chymotrypsin, only the exopeptidase was bound. The endopeptidase, trypsin, was bound nonspecifically owing to electrostatic attraction.

2.4.1a. Concerns of Structure

Paradoxes such as the one cited above should serve as warnings that the choice of a ligand and the use of the solvent containing the ligand must be made with care. Certain combinations of hydrophobic structural elements with ionic groups may produce strongly binding sorbents of low selectivity akin to ion exchange resins.

Among the early reports calling attention to this problem was that by O'Carra et al. (234). They analyzed the affinity system for beta-galactosidase by Steers et al. (235) and found that the weak beta-thiogalactoside ligand showed unexpectedly strong avidity for the enzyme if a hydrophobic spacer-arm (leash) was used. Indeed, a gel containing the leash moiety only, composed of diaminohexane coupled at one terminus to *n*-phenylglycine and

Table 2.6. The Binding of Proteins to Some Simple Ligands

	Ligand	Binding Ligate	Ref.
L-1	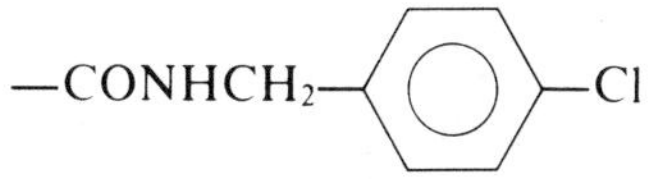 $-CONHCH_2-C_6H_4-Cl$	Trypsin	227
		Chymotrypsin	227
		Lysozyme	227
		Thrombin	228
L-2	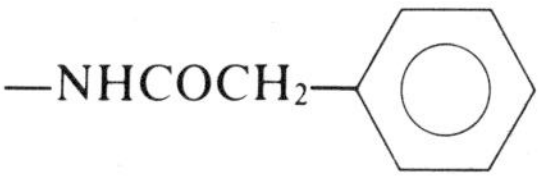 $-NHCOCH_2-C_6H_5$	Chymotrypsin A	229
		Lysozyme	229
		β-Galactosidase	230
L-3	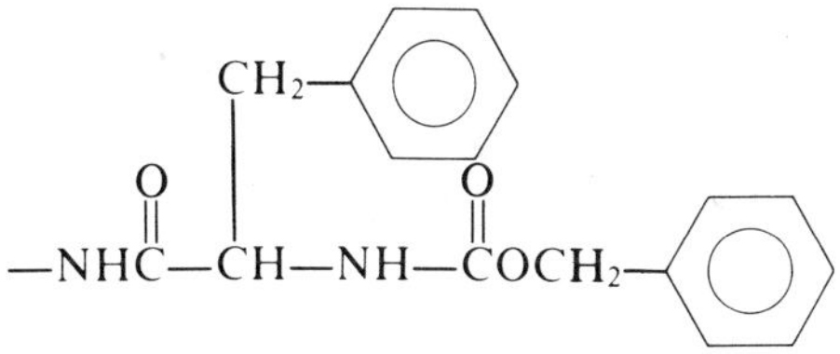 $-NHC(=O)-CH(CH_2C_6H_5)-NH-C(=O)OCH_2-C_6H_5$	Chymotrypsin	231
		Subtilisin	231
		Pepsin	231
		Thermolysin	231
L-4	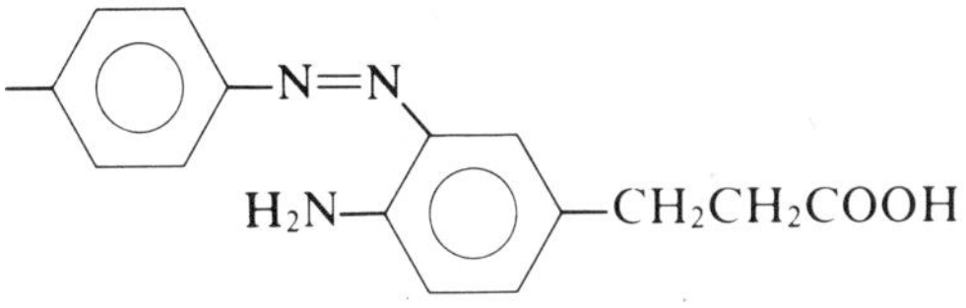 $-C_6H_4-N{=}N-C_6H_2(NH_2)-CH_2CH_2COOH$	Carboxypeptidase A	232
		Trypsin	232

lacking a galactoside group, nevertheless was found to bind the enzyme (234). Related to this were the observations by Hofstee that a number of enzymes could be immobilized through noncovalent binding to substituted agarose gels (236). Nishikawa and Bailon pointed out that in addition to hydrophobic leashes and ligands, the cationic linkage products resulting from cyanogen bromide coupling chemistry gave rise to significant nonspecific binding of proteins (230).

The purification of sialidase on immobilized *N*-(*p*-aminophenyl)oxamic acid (237) was found by Rood and Wilkinson to pose problems of selectivity. They noted that hemagglutinin, hemolysin and phospholipase C were bound to the sorbent along with sialidase (238). Huang and Aminoff, in a detailed look at the leash structure used to immobilize the inhibitor, noted that the tyrosyl leash per se sufficed to bind sialidase (239). This binding in the absence of a ligand pointed to strong nonselective interactions between enzyme and hydrophobic leash. The hydrophobic contributions of the leash structure in sorbents for acetylcholinesterase were noted by Massoulie and Bon (240). The study of chymotrypsin binding to soluble polymers containing D-tryptophan by Blumberg and Katchalski-Katzir provided useful insight into the effects of steric and electrostatic factors in affinity binding (241). Sharma and Hopkins concluded that positive charges on the chymotrypsin molecule may enhance binding to a low avidity ligand like caproyl-D-tryptophan methyl ester. This was demonstrated with acetylated chymotrypsin (242).

2.4.1b. Coupling Chemistry Problems

Many of the problems of ionic contributions to nonspecific adsorption of ligates can be traced to the coupling chemistry used in preparing the sorbents. The popular cyanogen bromide coupling (see Chapter 12) is frequently the source of ionic nonspecific adsorption problems due to the formation of isourea groups (153, 158a):

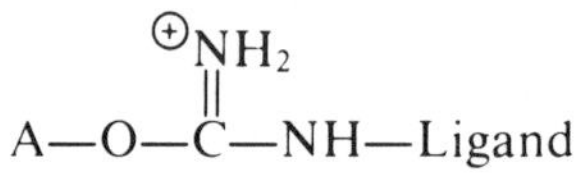

$$\mathrm{A{-}O{-}\overset{\overset{\oplus}{N}H_2}{\overset{\|}{C}}{-}NH{-}Ligand}$$

This can be avoided by using the carbonyl diimidazole method (159) which affords a nonionic (and more chemically stable) urethane linkage:

$$\mathrm{A{-}O{-}\overset{\overset{O}{\|}}{C}{-}NH{-}Ligand}$$

The use of bifunctional oxiranes (see 160–162) can also afford stable, nonionic leash structures, and the resulting sorbents show significantly lower nonspecific protein binding (243). The ether linkage to polysaccharides that results from oxirane coupling is more stable than urethane and is destroyed only under conditions that would destroy the gel.

The matter of linkage stability was the focus of some attention in the mid-1970s, involving particular investigators studying cell surface receptors and hormones. It is instructive to read the accounts of Katzen and Vlahakes (244), Cuatrecasas (245), Davidson and Van Herle (246), Lefkowitz (247), Venter and Kaplan (248), and Yong and Richardson (249). Owing to the very low concentrations of the ligates involved as well as the high affinity constants, the outcomes of these investigations were (or could be) affected by ligand leakage. A number of mathematical analyses of ligand leakage were put forth (250, 251, 252) as well as a thorough chemical analysis by Tesser et al. (253). This latter work helped explain the leakage problems observed by

Schwartz et al. (254), Davidson et al. (255), and Vanquelin et al. (256).

A solution to the leakage problem was put forth by Wilchek (257), who used polylysine as the leash moiety. The multipoint attachment of the polymer leash lowered the possibility of ligand loss. All things considered, the use of oxiranes affords the best approach to avoiding the leakage problem.

2.4.1c. Ligand (Leash) Density

Although often neglected, the density of ligands (and therefore leashes) in the support matrix (μeq/mL of support bed) is of great practical importance. Often a poor coupling yield leading to low ligand density will result in an ineffective sorbent. Too high a density, on the other hand, might result in extensive nonspecific adsorption of unwanted substances. The theoretical assessment of ligand density requires more space than is practical here (see ref. 138), thus comments are directed to some practical aspects.

The maximum possible density of ligand in the support matrix is a practical concern, because it sets an upper limit on capacity for ligate binding. Furthermore, with low molecular weight ligands the effectively accessible concentration of ligands is only a small fraction (it can be as low as 1%) of that actually present (225, 227). This does not seem to be the case (182) with immobilized antibodies. The ligand capacities of agarose gels activated by epibromohydrin (150) are shown in Table 2.7. The values are comparable to those obtained with cyanogen-bromide-activated agarose (227). The observed ligand density limits indicate that for effective affinity adsorption, the dissociation constant K_D between ligate and ligand must be on the order of 2×10^{-5} *M* if one is to employ 2% agarose gel effectively. The figure for 6% agarose gel is 2×10^{-4} *M*. These numbers serve as guidelines for designing affinity sorbents. If a given preparation containing a modest ligand density (e.g., 1 μeq/mL) is found to adsorb a ligate with a weak binding (or corresponding dissociation) constant, this is presumptive evidence that the binding is not specific.

Table 2.7. Ligand (Leash) Densities of Agarose Gels Derivatized with Epibromohydrin (150)

Agarose Density[a] (%)	MW Exclusion Limits (10^6 daltons)	Ligand Concentration (μeq/mL)
2	50	~8
4	15	~24
6	5	~40

[a] The gel "density" is the concentration of dry agarose polymer per 100 mL of gel.

With very tightly binding ligates ($K_D \sim 10^{-10}$), a frequent problem is to prepare sorbents with low (but well-defined) ligand densities. Ligate recovery is often difficult, as in the case of removing antigens from immobilized antibodies. Extreme shifts in pH (high or low) or chaotropic eluents are often necessary to effect recovery (see Section 2.4.2c). An interesting approach to recovering the tightly binding cobalamin-binding proteins is the use of photodegradable leashes for ligand immobilization (258).

2.4.2. General Ligands

From preceding discussions it would appear that relatively few ligands are so narrow in their selectivity as to be bound by only one ligate. In varying degrees most ligands will bind more than one ligate species. But for practical reasons we shall consider as *general ligands* those entities which by design and practice address classes or several classes of ligates. The description may be arbitrary in that the general utility of a number of ligands only awaits demonstration by an application different from that originally reported.

2.4.2a. Class-Selective Ligands

Among the early examples of major class-selective ligands were the immobilized nucleotide enzyme cofactors such as NAD^+. The examples are so numerous as to merit special reviews (e.g., Trayer and Winstanley, 139). The attraction of immobilized cofactors is the possibility of isolating several functionally related enzymes with one column (see Section 2.5). Examples of interest are Clonis and Lowe (259), Lee and Johansson (260), Brodelius et al. (261), and Kaplan and coworkers (262).

Synthetic Systems. Many other ligands of the inhibitor type are used in class-selective affinity sorbents, but of special interest are the synthetic general-affinity ligand systems. Here, a low molecular weight marker is covalently attached to a solute species of interest. This marker-conjugate (which corresponds to the ligate) is selectively removed from solution by an immobilized capture agent (which corresponds to the ligand). As can be seen in Table 2.8, the marker converts the macromolecular solute into a new, selectively retrievable species. This methodology has potential for wide application. However, it appears useful primarily for known systems, that is, where the species to be retrieved has been once purified and conjugated with marker. Thus, as suggested by Horowitz and Whitesides (267), the method affords a convenient means of removing enzymes from a reaction mixture. If, however, the marker can be selectively attached to a desired component, one then has at hand a powerful isolation system.

Table 2.8. Synthetic General-Affinity Systems

Marker Coupled to Solute	Solute Examples	Immobilized Capture Agent	Reference
Dinitrophenyl group	Peptides	Antibody to DNP	257
Biotinyl group	Peptides; proteins	Avidin	263
Biotinyl group	Thymocytes	Avidin	264, 265
Biotinyl group	DNA-RNA hybrid	Avidin	266
p-Sulfonamidobenzoyl group	Glucose-6-phosphate dehydrogenase	Carbonic anhydrase	267
	Hexokinase	Carbonic anhydrase	267
	Lysozyme	Carbonic anhydrase	267

Table 2.9. Some Plasma Proteins Purified on Heparin–Agarose

Protein	Reference
Human antithrombin III	269
Canine antithrombin III	270
Murine antithrombin	271
Thrombin	272
Factor IX	273, 274
Factor XI	274
Factor VII	275
Very-low-density lipoproteins; low-density lipoproteins	277
High-density lipoproteins	278

The most widely exploited combination is biotin–avidin. For examples in detection of cell surface antigens, immunoassay separations, and so on, the review by Boyer and Wilchek should be consulted (268).

Heparin. As a macromolecular ligand, heparin presents a unique situation in affinity purification methodology. A somewhat diverse collection of ligates have been purified with it. A recent review by Farooqui includes a number of enzymes purified on heparin–Sepharose (148). In view of its clinical applications as an anticoagulant, it is not surprising that a number of plasma proteins have been isolated with heparin in immobilized form. Some examples are listed in Table 2.9.

Heparin is a complex polysaccharide containing amino-, carboxy-, and sulfate ester moieties (276):

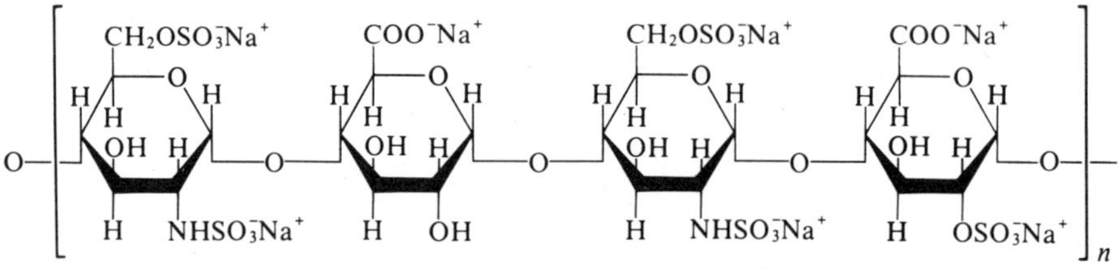

Its effectiveness in an affinity sorbent appears to be dependent on the method used to couple it to the support. For example, heparin bonded via its carboxyl groups to aminohexyl–agarose adsorbed less thrombin than when heparin was bonded via its amino groups (which result from partial desulfatization) in cyanogen–bromide coupling (272). A more recent method of immobilization using trichloro-*s*-triazine appears to afford a stable, effective sorbent (271). A titrimetric method for determining heparin content in sorbents has been reported (279). The recent study of heparinases and heparitinase isolation by affinity chromatography on glycosaminoglycan-agarose is useful to consult (280).

Lectins. While the exact biological functions of lectins are not yet understood, these proteins constitute a fascinating class of specific sugar-binding agents which have found much utility in histochemical and biochemical research (281). Their extensive uses in the purification of all membrane glycoproteins have been recently reviewed by Lotan and Nicolson (143a). General approaches to the purification of lectins have been described by Lis et al. (282) and Vretblad (283).

Owing to their high specificity towards particular sugar residues in complex polysaccharides and glycoproteins, lectins have become a favorite ligand in purifying these types of biopolymers. The best known lectin is concanavalin A (con A), a tetrameric protein capable of binding alpha-D-mannopyranosyl, alpha-D-glucopyranosyl, or beta-D-fructofuranosyl residues. A partial list of what can be isolated with immobilized con A is given in Table 2.10. For the best extent of purification, the ligates have been recovered by eluting the lectin–gel with methyl-alpha-D-glucopyranoside. In addition to con A, there are a number of other lectins with different selectivities. Some are listed in Table 2.11. The list is not comprehensive and is only intended to serve as a starting point and to suggest the wide diversity of glycoproteins to be addressed. Indeed, some researchers, such as Uhlenbruck et al. (304), screened a wide variety of serum glycoproteins (32 in all) against several different lectins as a means of characterizing the serum components. Borate gels have also been used as affinity chromatography materials to purify glycoproteins in a similar fashion. This is discussed in Chapter 4.

Table 2.10. Some Glycoproteins Purified on Immobilized Conconavalin A

Ligate Function	Source	Reference
Blood group substance	Pig	284
Glycoprotein receptor	Pig lymphocyte membrane	285
Glycopeptides	Rat fibroblast trypsinate	183
Structural protein	Bovine achilles tendon	286
Renin substrate	Human plasma	287
Testosterone-binding protein	Human serum	288
Various glycopeptides	Miscellaneous	289
Lectin	Red kidney bean	290
Carcinoembryonic antigen	Human liver tumor	291
Membrane glycoprotein	Semliki Forest virus	292
Gonadotropins	Human urine	293
Egg antigens	*Schistosoma japonicum*	294

Table 2.11. Purification of Glycoproteins with Various Lectins

Ligate	Lectin Source	Binding Residue	Reference
Trypanosome coat protein	Lentil (*Lens culinaris*)	Glucose	295
Basophile protein	Lentil (*Lens culinaris*)	Mannose	296
Fetuin; ceruloplasmin	*Ricinus communis*	Lactose	297
Platelet membrane glycoproteins	*Lens culinaris;* wheat germ	Mannose	298
		N-Acetyl-glucosamine	
Blood group A substance	*Vicia cracca*	*N*-Acetyl-galactosamine	299
Immunoglobulins; alpha-2-macroglobulin	*Vicia faba*	Glucose	300
Bone marrow cells	Wheat germ	*N*-Acetyl-glucosamine	301
Erythropoietin	Wheat germ	*N*-Acetyl-glucosamine	302
	PHA; red kidney bean	*N*-Acetyl-galactosamine	
Gladiolus-style protein	*Tridacna maxima*	Galactose	303

2.4.2b. Artificial Ligands—Dyes

As a general class-selective ligand, synthetic dyes are a serendipic boon. By accident, there were early observations that Blue-Dextran 2000, a commercially available dyed polymer for determining void volumes of gel columns, could inhibit or both inhibit and bind, or bind enzymes, apparently with some selectivity (305). Because handling dyed dextran is cumbersome, Easterday and Easterday developed an agarose gel containing Cibacron F3GA, the triazine-chloride-linked blue dye, directly bonded to the support (306). The class specificity towards dehydrogenases, for example, as elucidated by Stellwagen and coworkers (307) led to the widespread use of blue-dye-bound gels as well as the development of others (144a, 308). An extended treatment of this topic is presented by Dean in Chapter 3.

2.4.2c. Antibodies

In contrast to the other general ligands, IgG's share common physical properties but very different ligate specificities. Immunoglobulins also differ from the preceding ligands in two distinct ways: (i) they are of high molecular weight (160,000 daltons); (ii) they are produced specifically against a known ligate. These distinctions are of sufficient importance for the preparation and use of immunosorbents to require a few special considerations.

Capacity. The high molecular weight of IgG sets a practical limit in its immobilization to a carrier. An agarose gel of 4% polymer density may be able to hold 30 mg of IgG per mL of gel. This corresponds to 0.375 μeq of ligand (in terms of two binding sites per molecule) per mL of gel. By contrast, a low molecular weight ligand such as *p*-aminobenzamidine can be bonded as high as 20–22 μeq per mL of gel. A compensating factor with immobilized antibodies is the usually high binding constant to the ligate ($K_b \sim 10^8$–10^{11}), whereas with usual systems involving enzymes and low molecular weight inhibitors the corresponding constants are typically 10^4–10^6. Of course the capacity of an immunosorbent in terms of weight quantity of ligate depends on the latter's molecular weight. As a practical matter it would not seem too productive to isolate a low molecular weight ligate by immunosorbents, especially at large preparatory scale. However, the isolation of macromolecular ligates may be tractable. For example, if only a third of the binding groups in the immunosorbent (30 mg IgG/mL) are active (e.g., 0.125 μeq binding sites/mL) and the ligate is 20,000 daltons in molecular weight, then the weight capacity will be 2.5 mg/mL gel. Such a yield of ligate is not unwelcome.

Another aspect of capacity unique to macromolecular ligands involves the space (volume) they occupy in the support matrix. A detailed study by Comoglio et al. is particularly instructive on this matter (309). In a series of agarose immunosorbent preparations that ranged from 0.48 mg IgG/mL gel to 48.96 mg IgG/mL, they observed that the residual immunoreactivity declined from 92% (at low IgG density) to 15.6% (at high IgG density). This could be explained as due to "crowding" of IgG molecules in the agarose support and diminution of access by ligate species because of steric hindrance by the protein mass of antibody at high loadings. It remains to be seen if this crowding phenomenon occurs in silica gel or cellulose supports. In a study of immobilized anti-human serum albumin (HSA), Eveleigh has also observed a decline in the binding ratio of HSA/IgG as the antibody loading in agarose gel is increased (310).

Ligand Preparation. Compared with the low molecular weight ligands mostly obtained by chemical synthesis, the preparation of ligate-specific IgG is much more biological. It requires the use of live animals properly inoculated and maintained. There is much art involved in getting "good animal response," that is, high specificity and titer. And patience is required, since several weeks or even months may pass before an adequate production level is attained for a good antibody harvest. General treatises that include various aspects of antibody work should be consulted (311, 312).

A recent innovation that makes possible the production of large amounts of a given antibody is the use of hybridomas obtained by in vitro fusion of myeloma (cancer) cells with splenic antibody-producing lymphocytes. Although general principles and procedures have been established, this approach involves great methodological skill, much practice, and luck. A review by Goding examines many of the fine points of this technique (313).

Compared with other ligand–ligate systems, a distinctive requirement in preparing specific antibodies is the necessity to first isolate the ligate of interest and obtain it at high purity. The high purity is important for obtaining antibodies of high specificity to the ligate of interest. Thus the effort involved in first obtaining some high purity ligate (usually by classical biochemical methods, but recently also involving various other affinity methods) and then undertaking antibody production is usually only justified when large amounts of the ligate are ultimately desired and when its isolation is very difficult by nonimmunosorbent means. The immunosorbent approach is a major undertaking even to reproduce or verify a published procedure. Obtaining a ligate sample from the original investigator is of considerable advantage.

The great attraction of the antibody approach is, of course, the prospect of obtaining an affinity system of high selectivity towards the ligate. This can afford great leverage in ligate purification (over a thousandfold purification can be attained in one step). However, the high selectivity is often accompanied by high avidity of ligate binding. This tightness of binding can pose special problems in ligate recovery, which is usually accomplished by finding conditions under which antigen binding to antibody is weakened. The usual approaches are to elute the adsorbed system with solutions of very high (>9) or low (<5) pH, possibly in combination with concentrated salt solutions (76a). Concentrated solutions of chaotropic agents (e.g., potassium thiocyanate, guanidinium hydrochloride, urea, halo acetates) also have been used to recover ligates (314). Electrophoretic methods, described by Bog-Hansen in Chapter 6, have been employed with mixed results. In general, the more tightly adsorbed ligates require stronger elution conditions, which sometimes require a combination of effects.

When strong solvent conditions are needed to recover the ligate, it is important to rather quickly neutralize the pH from either high or low extremes or dilute out or dialyze away the chaotropic agents to minimize damage to either ligand or ligate. Automated systems have been developed to accomplish this efficiently.

With the high selectivity and avidity exhibited by IgG ligands, it would seem that they are without fault. However, the large molecular size of antibody molecules are also accompanied by a proportionately large surface area, which in practice seems to increase the risk of nonspecific adsorption by unwanted proteins. Some useful strategies in removing nonspecific proteins from immobilized IgG columns were explored by Zoller and Matzku (315). These include intermediate washes of the alpha-foeto protein (AFP) adsorbed to anti-AFP column with a variety of buffers containing high salts (e.g., 2 *M* NaCl, 3 *M* NaSCN, pH 10 buffer, etc.). Intermediate washes with detergent-containing buffer solutions have been found useful by Smith et al. (316). The finding by Dimitriadis that a variety of nonionic detergents are well tolerated in antigen–antibody interactions (317) is reassuring to this approach.

A final point concerning the problem of nonspecific adsorption to immunosorbents deals with the method of ligand immobilization to agarose gel. In a detailed comparative study Murphy et al. (243) showed that ligands immobilized via cyanogen bromide coupling gave rise to immunosorbents that exhibited much more nonspecific adsorption of contaminants than when a bifunctional oxirane was used. Thus, the coupling methods described in refs. 160, 161, and 162 would be preferred for immunosorbent preparation.

Application Examples. It is instructive to examine studies reported in the literature of immobilized antibodies in the purification of various ligates, including enzymes, viruses, and miscellaneous proteins. Table 2.12 lists such studies. Perusal of the list reveals that rabbits are a convenient laboratory

Table 2.12. Some Examples of Ligate Isolation with Immobilized Specific Antibodies

Ligate	Ligand	Support/Coupling	Recovered Eluant	Reference
Hydroxyindole *O*-methyl transferase	Rabbit Ab	Sepharose 4B/BrCN	pH 2.4 buffer 0.5 *M* NaCl	318
Elastase	Rabbit Ab	Sepharose 4B/BrCN	pH 2.8 buffer 0.5 *M* NaCl	319
Dipeptidyl peptidase IV	Rabbit Ab	Sepharose 4B/BrCN	0.002 *M* Tris pH 8, 0.1% Triton X-100	320
Carboxypeptidase N	Rabbit Ab	Sepharose 4B/BrCN	0.1 *M* Tris 2 *M* NaCl pH 8	177
Terminal deoxynucleotidyl transferase	Rabbit Ab	Sepharose 4B/BrCN	0.6 *M* KCl, 0.2 *M* NH_4OH, pH 10.6	321
Angiotensin-1-converting enzyme	Rabbit Ab	Sepharose 4B/BrCN	2 *M* $MgCl_2$ pH 5.8	322
Alpha-amylase inhibitor	Rabbit Ab Enzyme	Sepharose 4B/BrCN Affi-Gel 10	0.05 *M* glycine pH 2.8	323
Lymphocyte antigen	Guinea pig Ab	Sepharose 4B/BrCN	0.2 *M* glycine pH 2.8	324
Astroglial protein	Rabbit Ab	Sepharose 4B/BrCN	1 *M* acetic acid, 5 *M* urea, 0.8 *M* NaCl, pH 2.5	325
Paramecium-i antigen	Rabbit Ab	Sepharose 4B/BrCN	0.05 *M* glycine pH 10.4	326
Alpha-2-macroglobulin	Rabbit Ab	Sepharose 4B/BrCN	0.1 *N* acetate, 0.5 *M* NaCl, pH 2.5	327

Alpha-fetoprotein	Rabbit Ab	Polyacrylamide (entrapped)	0.1 *M* glycine, 0.5 *M* NaCl, pH 2.5	328
Beta$_2$-microglobulin	Rabbit Ab Monoclonal mouse	Sepharose 6B/BrCN	0.05 *M* glycine pH 3, 0.5% Nonidet P40	329
Poliovirus	Guinea pig Ab	Sepharose 4B/BrCN	3 *M* KSCN	330
Acetylcholine receptors	Monoclonal mouse	Sepharose 4B/BrCN	0.2 *M* glycine, 0.5 *M* NaCl, 1% sodium cholate pH 10	331
Interferon	Monoclonal mouse	Affi-Gel-10	0.2 *N* acetate, 0.15 *M* NaCl, 0.1% Triton X-100, pH 2.5	181 182

animal for antibody production, but this may give way increasingly to murine sources as monoclonal antibodies become more commonplace. The convenience of agarose activated by cyanogen bromide has been evident until now. But as performance requirements get more stringent, alternatives that do not use cyanogen bromide should increase.

2.4.3. Ligate Concerns

While much attention up to now has been placed on ligand selectivity, one should not overlook the importance of how ligate specificity may affect success in the affinity purification process. Furthermore, it is useful to examine the milieu of the impure ligate so as to arrive at the best strategy for its affinity purification.

2.4.3a. Ligate Specificity

From the discussions on ligand selectivity it is apparent that one can devise a variety of sorbents, each containing a different ligand, to attempt the purification of a desired ligate species. In a few instances investigators have gone through the effort of screening several affinity sorbents to find the best one for their ligate of interest. More frequently, the situation is one where an affinity sorbent already documented to work, for example, on enzyme X from calf liver, is used (or at least tried out) on enzyme X from snail-gut juice. To assure success in this process it is important to determine certain specificity parameters of the snail-gut juice enzyme. In particular, the pH-profile and avidity towards the ligand should be determined. If the snail enzyme has a weaker binding constant than the one from calf liver, then the sorbent must be prepared with higher ligand density. Likewise, if the pH profile is different in the two enzymes, adjustments need to be made for the snail enzyme to be successfully purified on the affinity sorbent. Furthermore, there may be other variables, for example, the need for the presence of a divalent metal ion in the case of the snail enzyme. It is good practice to know as much as possible about the ligate of interest so that its purification can be successful.

2.4.3b. Ligate milieu

Sometimes the ligate of interest is one of several related species all of which are present in the crude mixture. It is important to determine which of the affinity parameters would allow its separation from its congeners. For example, both NAD^+-dependent and $NADP^+$-dependent dehydrogenases may bind to the same AMP-containing sorbent, but by selective elution with the different coenzymes the two dehydrogenases may be recovered separately. When working with sorbents containing simple ligands, it may be useful to assay the crude mixture for several different activities so that a better assessment of the affinity purification process can be made. When it becomes evident that a given ligand has wide selectivity and that a number of species in the crude mixture are capable of binding to it, the prudent strategy may be to design a more selective sorbent. A final point on the importance of knowing the ligate milieu: contaminants may interfere with the proper ligate–ligand binding. When the presence of such an interfering substance is noted, a prefractionation step may be necessary (see, e.g., ref. 332). In some instances it may be practical to redesign the sorbent.

2.5. OPERATION MODES IN CHROMATOGRAPHY

Given the complexity of biopolymer ligate molecules and the significant difficulties in securing their purification, it is no wonder that a variety of operational modes for using affinity sorbents have been developed. The simple packed column has succeeded in the majority of applications. However, the desire to separate several related ligates (or even unrelated ones) from the same crude mixture has led to interesting modifications of the simple approach. And automation has evolved to make the affinity separation process more efficient.

2.5.1. Single Step/Single Ligate

This, of course, is the most commonly performed in a simple packed column. For efficient operation, however, a network such as that shown in Figure 2.4 is desirable. It is an arrangement amenable to automation. By controlling Valve A, the sample is applied to the affinity sorbent and followed with wash buffer. During this phase Valve B is set to direct the effluent to drain. At the conclusion of the wash, Valve A directs the flow of desorbing agent to the affinity sorbent and Valve B is set to direct the affinity effluent to the molecular sieving gel to remove the desorbing agent. The ligate effluent from the molecular sieving gel passes through the monitor and into the fraction collector.

In some affinity sorbents (e.g., those with macromolecular ligands) it may be more effective to use the sorbent in a batch suspension in the ligate solution. This may result in more even binding of ligate species throughout the sorbent sample. In turn, this may afford better recovery of the ligate in the recovery step, which can be performed by dropping the ligate-bound sorbent

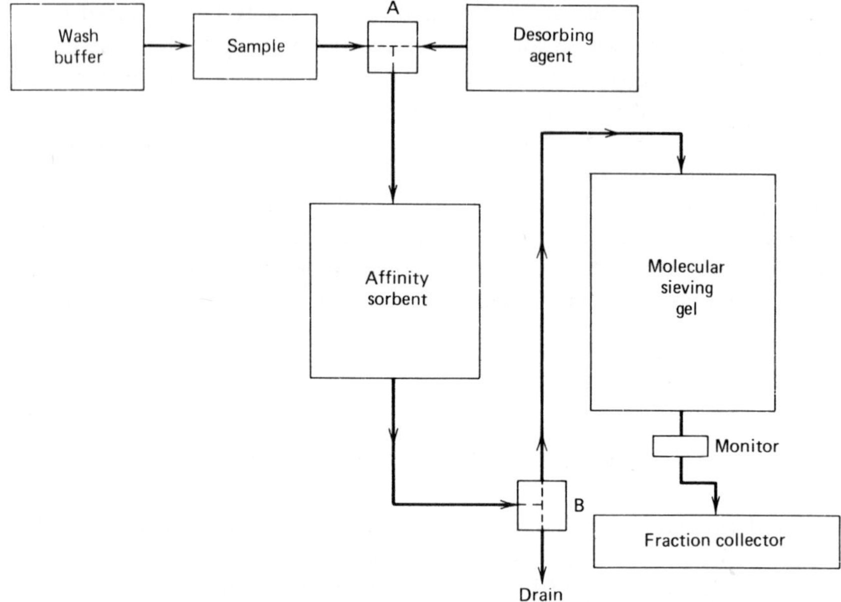

Fig. 2.4. Flow chart for affinity separation.

into a desorbing agent solution. Alternatively, the batch-adsorbed affinity sorbent can be packed into a column and eluted with the desorbing agent solution.

2.5.2. Multiple Ligate/Gradient Elution

There are a number of instances where it is of interest to separate several related ligate species from the same crude mixture. One can employ a sor-

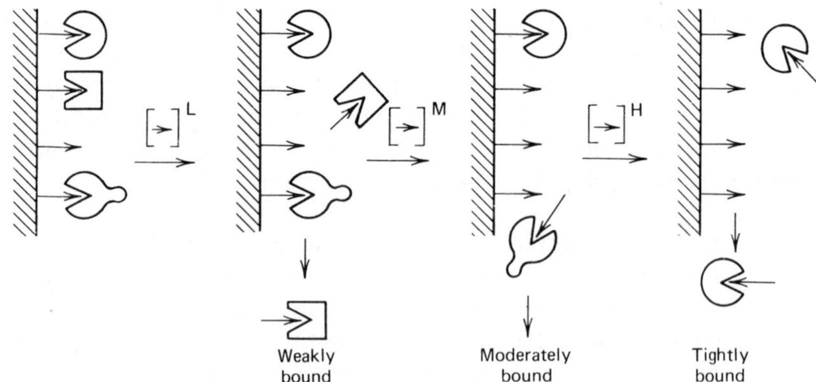

Fig. 2.5. Stepwise (or gradient) elution from general ligand sorbent.

bent containing a general ligand, to which all of ligates will bind in varying avidities. As depicted schematically in Figure 2.5, the different ligates can be recovered by stepwise elution with desorbing agent at different concentrations. At low concentration, the weakly bound ligates are obtained. At the highest concentrations of desorbing agent, the most avidly bound ligates are recovered. Examples of this approach may be found in the work of Easterday and Easterday (306) and Brodelius et al. (333). The gradient may be stepwise or continuous.

2.5.3. Simultaneous Multiple Ligate Binding

When a crude extract contains several ligates of rather different specificities and it is of interest to recover each of them pure, the simultaneous adsorption method is of interest. Selective sorbents for each of the ligate species are enclosed individually in a fine mesh bag (hence the description "tea bag method"). These bags are then simultaneously contacted with the crude-solution-containing ligate species (see Figure 2.6). At the appropriate time the bags are withdrawn and washed in a suitable manner. The ligates are

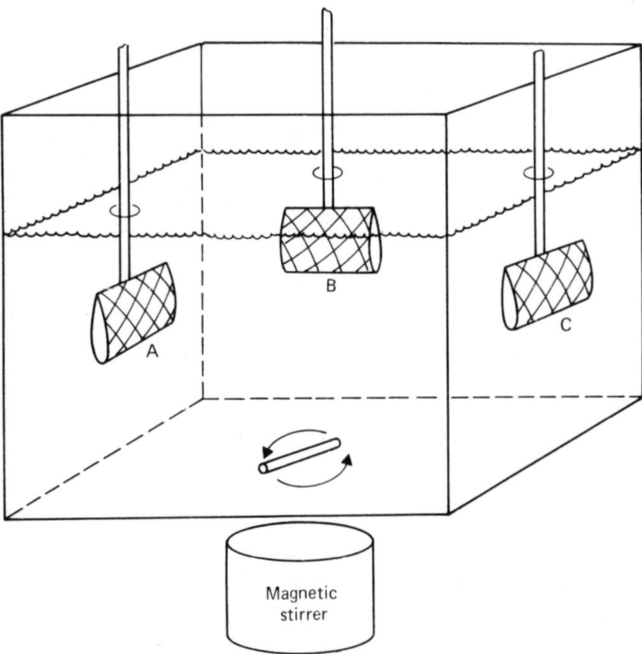

Fig. 2.6. "Tea bag" processing of several biopolymers on three different sorbents (A, B, and C) simultaneously. From Sundberg, Porath, and Aspberg (334), with permission, courtesy of *Biochim. Biophys. Acta*.

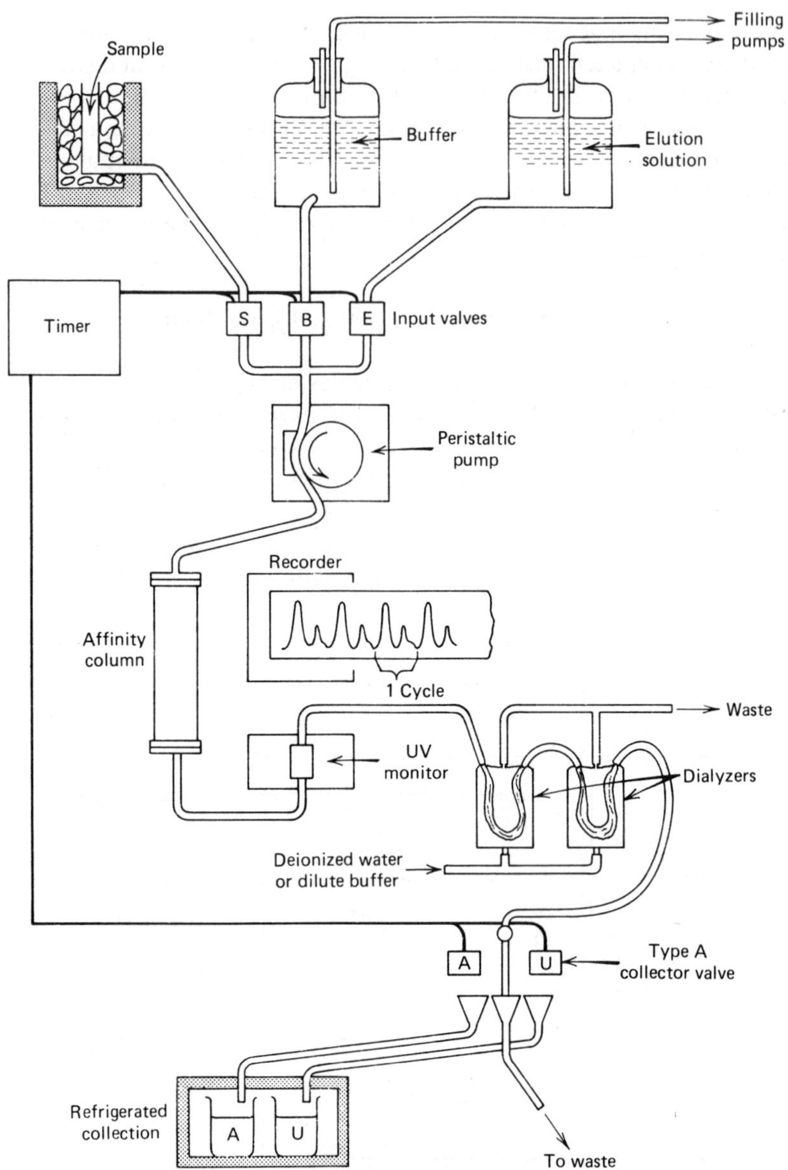

Fig. 2.7. The Cyclum automated affinity purification system. The overall process (sample loading, washing, etc.) is controlled by a master timer. The desired ligate is accumulated by repeating the process cycle, and the monitor's recorder makes tracings of each batch cycle processed. *S*, sample; *B*, buffer; *E*, eluent; *A*, absorbed material; *U*, unbound material. From Anderson et al. (335), with permission, courtesy of *Anal. Biochem.*

recovered by placing the affinity sorbent bag into the appropriate desorbing agent solution. This technique has been described by Sundberg et al. (334).

2.5.4. Single Step/Multisequential

An alternative to the tea bag method is to use several affinity columns in sequence, so that all of the ligates of interest in a given liquid sample may be recovered with a minimum of waste. There may be certain limitations on the number of columns that can be used in a series. Principally these would be a consequence of band spreading due to dilution of the sample volume as it passed from column bed to column bed. Clementson et al. (298) were successful in purifying blood platelet glycoproteins by a sequential passage through a column of immobilized wheat germ agglutinin followed by immobilized *Lens culinaris* lectin. This sequential multicolumn method is well suited for process automation.

2.5.5. Repetitive Single Step/Automatic Recycling

To improve reproducibility in analytical chromatography processes and to remove the tedium (and the errors it generates), the practice today is to automate. One of the earliest examples was the Cyclum developed by Anderson et al. (335) at the Oak Ridge National Laboratories. Owing to the extra complexity of immunosorbent systems (e.g., the need to remove the harsh desorbing agent quickly from contact with ligate), automation was a logical and productive development. Its scheme is presented in Figure 2.7. A more complex system has been reported by Folkersen et al. (336). Pahud and Schwarz (337) have described a system using the LKB Ultragrad gradient maker at the heart of the automation network.Robinson et al. (338) devised an electromechanical timing device to facilitate the large-scale production of enzymes.

REFERENCES

1. J. B. Sumner, *J. Biol. Chem.* **69,** 435–441 (1926).
2. R. Willstatter, *Ber.* **55,** 3601–3623 (1922).
3. R. Willstatter, *J. Chem. Soc.,* 1359–1381 (1927).
4. R. Willstatter, E. Waldschmidt-Leitz, and F. Memmen, *Z. Physiol. Chem.* **126,** 93–131 (1923).
5. S. G. Hedin, *Biochem. J.* **2,** 112–116 (1907).
6. L. Michaelis and M. Ehrenreich, *Biochem. Z.* **10,** 283–299 (1908).
7. E. Starkenstein, *Biochem. Z.* **24,** 210–218 (1910).
8. L. Ambard, *Bull. Soc. Chim. Biol.* **3,** 51–65 (1921).

9. W. M. Bayliss, *The Nature of Enzyme Action,* Longmans, Green, London (1925), Chap. 3.
10. E. Waldschmidt-Leitz, *Science* **78,** 189–190 (1933).
11. J. B. Sumner, *J. Chem. Educ.* **14,** 255–259 (1937).
12. J. H. Northrup, *J. Gen. Physiol.* **13,** 739–766 (1930).
13. J. H. Northrup, *J. Gen. Physiol.* **2,** 113–122 (1920).
14. J. B. S. Haldane, *Enzymes,* Longmans, Green, London (1930), p. 174–176.
15. J. B. Sumner, J. S. Kirk, and S. F. Howell, *J. Biol. Chem.* **98,** 543–553 (1932).
16. E. Waldschmidt-Leitz, *Ann. Rev. Biochem.* **3,** 39–58 (1934).
17. J. H. Northrup, *J. Gen. Physiol.* **17,** 165–194 (1934).
18. O. Holmbergh, *Biochem. Z.* **258,** 134–140 (1933).
19. I. Tokuoka, *J. Agr. Chem. Soc. Jpn.* **12,** 1189–1194 (1936).
20. D. J. D. Hockenhull, and D. Herbert, *Biochem. J.* **39,** 102–106 (1944).
21. H. Lineweaver, R. Jang, and E. F. Jansen, *Arch. Biochem. Biophys.* **20,** 137–152 (1949).
22. S. Schwimmer and A. K. Balls, *J. Biol. Chem.* **179,** 1063–1074 (1949).
23. P. S. Thayer, *J. Bacteriol.* **66,** 656–663 (1953).
24. N. H. Grant and K. C. Robbins, *Arch. Biochem. Biophys.* **66,** 396–403 (1957).
25. D. A. Hall, *Arch. Biochem. Biophys.* **67,** 366–377 (1957).
26. P. M. Gallop, S. Seifter, and E. Meilman, *J. Biol. Chem.* **227,** 891–906 (1957).
27. G. Schramm and E. Mohr, *Nature* **183,** 1677–1678 (1959).
28. S. Paleus and J. B. Neilands, *Acta Chem. Scand.* **4,** 1024–1030 (1950).
29. H. A. Sober and E. A. Peterson, *J. Amer. Chem. Soc.* **76,** 1711–1712 (1954).
30. S. Moore and W. H. Stein, *Advan. Protein Chem.* **11,** 191–236 (1956).
31. C. A. Zittle, *Advan. Enzymol.* **14,** 319–374 (1953).
32. D. H. Campbell, E. Luescher, and L. S. Lerman, *Proc. Nat. Acad. Sci. U. S.* **37,** 575–578 (1951).
33. J. F. Taylor, in *The Proteins,* Vol. 1, Part A. (H. Neurath and K. Bailey, Eds.), Academic Press, New York (1953), 1–85.
34. L. S. Lerman, *Proc. Nat. Acad. Sci. U. S.* **39,** 232–236 (1953).
35. B. F. Erlanger, *Biochim. Biophys. Acta* **27,** 646–647 (1958).
36. G. de la Haba, *Biochim. Biophys. Acta* **59,** 672–680 (1962).
37. C. Arsenis and D. B. McCormick, *J. Biol. Chem.* **239,** 3093–3097 (1964).
38. C. Arsenis and D. B. McCormick, *J. Biol. Chem.* **241,** 330–334 (1966).
39. B. R. Baker, *Design of Active-Site-Directed Irreversible Enzyme Inhibitors,* Wiley, New York (1967), pp. 301–308.
40. H. Fritz, H. Schultz, M. Neudecker, and E. Werle, *Angew. Chem.* **78,** 775 (1966).
41. Y. Levin, M. Pecht, L. Goldstein, and E. Katchalski, *Biochemistry* **3,** 1905–1913 (1964).
42. J. D. Watson and F. H. C. Crick, *Nature* **171,** 737–738 (1953).
43. E. K. F. Bautz and B. D. Hall, *Proc. Nat. Acad. Sci. U. S.* **48,** 400–408 (1962).
44. E. T. Bolton and B. J. McCarthy, *Proc. Nat. Acad. Sci. U. S.* **48,** 1390–1397 (1962).
45. P. T. Gilham, *J. Amer. Chem. Soc.* **84,** 1311–1312 (1962).
46. A. J. Adler and A. Rich, *J. Amer. Chem. Soc.* **84,** 3977–3979 (1962).

47. S. Erhan, L. G. Northrup, and F. R. Leach, *Proc. Nat. Acad. Sci. U. S.* **53,** 646–652 (1965).
48. E. G. Sander, D. B. McCormick, and L. D. Wright, *J. Chromatogr.* **21,** 419–423 (1966).
49. K. Landsteiner, *Biochem. Z.* **104,** 280–299 (1920).
50. H. Zinsser, *J. Immunol.* **6,** 289–299 (1921).
51. M. Heidelberger, *Ann. Rev. Biochem.* **1,** 655–674 (1932).
52. L. A. Silber and M. W. Demidova, *Z. Immunitatsforsch.* **77,** 504–513 (1932).
53. T. D. Gerlough and W. White, *J. Immunol.* **27,** 367–377 (1934).
54. L. D. Felton, *Science* **79,** 277–279 (1934).
55. J. S. Kirk and J. B. Sumner, *J. Immunol.* **26,** 495–504 (1934).
56. A. Tiselius, *Biochem. J.* **31,** 1464–1477 (1937).
57. J. R. Marrack and F. C. Smith, *Proc. Roy. Soc. B* **106,** 1–19 (1931).
58. J. R. Marrack and F. C. Smith, *Brit. J. Exptl. Path.* **13,** 394–402 (1932).
59. M. Heidelberger and E. A. Kabat, *J. Exper. Med.* **67,** 181–198 (1938).
60. L. Bleyer, *Z. Immunitats. Exper. Therap.* **33,** 478–503 (1922).
61. G. d'Alessandro and F. Sofia, *Z. Immunitatsforsch. Exper. Therap.* **84,** 237–250 (1935).
62. K. Meyer and A. Pic, *Ann. Inst. Pasteur* **56,** 401–412 (1936).
63. K. Landsteiner and J. van der Scheer, *J. Exp. Med.* **63,** 325–339 (1936).
64. L. S. Lerman, *Nature* **172,** 635–636 (1953).
65. H. Isliker, *Ann. N.Y. Acad. Sci.* **57,** 225–238 (1953).
66. G. Manecke and K.-E. Gillert, *Naturwissenschaften* **42,** 212–213 (1955).
67. N. Grubhofer and L. Schleith, *Hoppe-Seyler's Z. Physiol. Chem.* **97,** 108–112 (1954).
68. D. W. Talmage, H. R. Baker, and W. Akeson, *J. Infect. Dis.* **94,** 199–212 (1954).
69. Y. Yagi, K. Engel, and D. Pressman, *J. Immunol.* **85,** 375–386 (1960).
70. L. Gyenes, B. Rose, and A. H. Sehon, *Nature* **181,** 1465 (1958).
71. G. B. Sutherland and D. H. Campbell, *J. Immunol.* **80,** 294–298 (1958).
72. H. C. Isliker, *Adv. Prot. Chem.* **12,** 387–463 (1957).
73. E. A. Kabat and M. M. Mayer, *Experimental Immunochemistry,* 2nd ed., Chas. C. Thomas, Springfield, Ill. (1961), pp. 781–797.
74. G. Manecke, *Pure Appl. Chem.* **4,** 507–520 (1962).
75. A. H. Sehon, *Brit. Med. Bull,* **19,** 183–191 (1963).
76. N. Weliky, H. H. Weetall, R. V. Gilden, and D. H. Campbell, *Immunochemistry* **1,** 219–229 (1964).
76a. N. Weliky and H. H. Weetall, *Immunochemistry* **2,** 293–322 (1965).
77. I. H. Silman and E. Katchalski, *Ann. Rev. Biochem.* **35,** 873–908 (1966).
78. G. P. Talwar, S. J. Segal, A. Evans, and O. W. Davidson, *Proc. Nat. Acad. Sci. U. S.* **52,** 1059–1066 (1964).
79. E. V. Jensen, E. R. DeSombre, and P. W. Jungblut, *2nd Int. Congr. Hormonal Steroids, Milan, 1966,* Excerpta Medica, Amsterdam (1967), pp. 492–500.
80. B. von der Haar and G. C. Mueller, *Biochim. Biophys. Acta* **176,** 626–631 (1969).
81. D. B. Jones and C. O. Johns, *J. Biol. Chem.* **28,** 67–75 (1916).

82. J. B. Sumner, *J. Biol. Chem.* **37,** 137–141 (1919).
83. J. B. Sumner and S. F. Howell, *J. Bacteriol.* **32,** 227–237 (1936).
84. M. Olson and I. E. Liener, *Biochemistry* **6,** 105–111 (1967).
85. B. B. L. Agrawal and I. J. Goldstein, *Biochim. Biophys. Acta* **147,** 262–271 (1967).
86. D. B. McCormick, *Anal. Biochem.* **13,** 194–198 (1965).
87. C. A. Knight, *J. Exp. Med.* **83,** 11–24 (1946).
88. G. K. Hirst, *Science 94,* 22–23 (1941).
89. G. K. Hirst, *J. Exp. Med.* **76,** 195–209 (1942).
90. P. Cuatrecasas, M. Wilchek, and C. B. Anfinsen, *Proc. Nat. Acad. Sci. U. S.* **61,** 636–643 (1968).
91. R. Axen and J. Porath, *Nature* **210** 367–369 (1966).
92. J. Porath, R. Axen, and S. Ernback, *Nature* **215,** 1491–1492 (1967).
93. P. O'Carra, in B. Spencer, Ed., *Industrial Aspects of Biochemistry,* Fed. Europe. Biochem. Soc., North-Holland, Amsterdam (1974), pp. 107–134.
94. E. A. Kabat, *J. Immunol.* **47,** 513–587 (1943).
95. D. H. Campbell and N. Weliky, in *Methods in Immunology and Immunochemistry,* Vol. 1, (C. A. Williams and M. W. Chase, Eds.), Academic Press, New York (1967), pp. 365–385.
96. L. Goldstein and E. Katchalski, *Z. Anal. Chem.* **243,** 375–396 (1968).
97. I. Kato, *Tampakushita Kakusan Koso, Tokyo* **14,** 1143–1151 (1969).
98. J. A. Rothfus, *J. Amer. Oil Chem. Soc.* **47,** 316–325 (1970).
99. P. Cuatrecasas and C. B. Anfinsen, in *Methods in Enzymology,* Vol. 22 (W. B. Jakoby, Ed.), Academic Press, New York (1971), pp. 345–378.
100. P. Cuatrecasas and C. B. Anfinsen, *Ann. Rev. Biochem.* **40,** 259–277 (1971).
101. P. Cuatrecasas, *J. Agr. Food Chem.* **19,** 600–604 (1971).
102. P. Cuatrecasas, in *Biochemical Aspects of Reactions on Solid Supports* (G. R. Stark, Ed.), Academic Press, New York (1971), pp. 79–109.
103. G. Feinstein, *Naturwissenschaften* **58,** 389–396 (1971).
104. F. Friedberg, *Chromatogr. Rev.* **14,** 121–131 (1971).
105. K. Mosbach, *Sci. Amer.* **224**(3), 26–33 (1971).
106. J. Porath, *Tampakushitsu Kakusan Koso* **16,** 313–323 (1971).
107. R. H. Reiner and A. Walch, *Chromatographia* **4,** 578–587 (1971).
108. P. Cuatrecasas, *Advan. Enzymol.* **36,** 29–89 (1972).
109. H. D. Orth and W. Brümmer, *Angew. Chem. Int. Ed.* **11,** 249–346 (1972).
110. J. Porath and L. Sundberg, in *Chemistry of Biosurfaces,* Vol. 2 (M. L. Hair, Ed.), Marcel Dekker, New York (1972), pp. 633–661.
111. H. Guilford, *Chem. Soc. Rev.* **2,** 249–270 (1973).
112. K. G. Huggins in *Methodological Developments in Biochemistry,* Vol. 2. *Preparative Techniques* (E. Reid, Ed.), Longmans, Green, London (1973), pp. 109–112.
112a. O. Cromwell in *Methodological Developments in Biochemistry,* Vol. 2. *Preparative Techniques* (E. Reid, Ed.), Longmans, Green, London, 1973, pp. 113–126.
113. R. L. Munier in *Experimental Methods in Biophysical Chemistry* (C. Nicolau, Ed.), Wiley, New York (1973), Chap. 6.

114. J. Porath, *Biochimie* **55,** 943–951 (1973).
115. H. H. Weetall, *Separation Purification Meth.* **2,** 199–229 (1973).
116. H. Guilford, *Ann. Rep. Prog. Chem.* **71,** 56–74 (1974).
117. J. F. Kennedy, *Advan. Carbohyd. Chem. Biochem.* **29,** 305–405 (1974).
118. C. R. Lowe and P. D. G. Dean, *Affinity Chromatography,* Wiley-Interscience, New York, 1974.
119. S. W. May and O. R. Zaborsky, *Separation Purification Meth.* **3,** 1–86 (1974).
120. W. H. Scouten, *Amer. Lab.* **6**(8), 23–39 (1974).
121. J. Turkova, *J. Chromatogr.* **91,** 267–291 (1974).
122. K. W. Williams, *Lab. Prac.* **23,** 521–524 (1974).
123. G. Baum and S. J. Wrobel, in *Immobilized Enzymes, Antigens, Antibodies, and Peptides: Preparation and Characterization* (H. H. Weetall, Ed.), Marcel Dekker, New York (1975), pp. 419–496.
124. P. D. G. Dean and M. J. Harvey, *Process Biochem.* **10,** (9) 5–10 (1975).
125. A. H. Nishikawa, *Chem. Tech.* **5,** 564–571 (1975).
126. I. Parikh and P. Cuatrecasas, in *Methods of Protein Separation,* Vol. 1 (N. Catsimpoolas, Ed.), Plenum, New York (1975), pp. 255–276.
127. G. P. Royer and W. E. Meyers, in *Bonded Stationary Phases in Chromatography* (E. Grushka, Ed.), Ann Arbor Science Pub., Ann Arbor, Mich. (1974), pp. 93–112.
128. J. Porath and T. Kristiansen, in *The Proteins,* Vol. 1, 3rd ed. (H. Neurath and R. L. Hill, Eds.), Academic Press, New York, 1975, pp. 95–178.
129. J. Turkova, in *Liquid Column Chromatography: A Survey of Modern Techniques and Applications,* (Z. Deyl, K. Macek, and J. Janek, Eds.), Elsevier, Amsterdam (1975), pp. 89–100, 369–376.
130. C. J. O. R. Morris and P. Morris, *Separation Methods in Biochemistry,* Wiley, New York (1976), pp. 223–233.
131. M. Wilchek and C. Hexter, in *Methods of Biochemical Analysis,* Vol. 23 (D. Glick, Ed.), Interscience, New York (1976), pp. 347–385.
132. C. R. Lowe, *Int. J. Biochem.* **8,** 177–181 (1977).
133. K. W. Olsen, *Amer. Lab.* **9**(8), 21–28 (1977).
134. I. Parikh and P. Cuatrecasas in *Immunochemistry of Proteins* (M. Z. Attassi), Plenum Press, New York (1977), pp. 1–44.
135. R. C. Boguslaski and R. S. Smith, *J. Chromatogr.* **141,** 3–12 (1977).
136. W. Rode, *Postepy Biochem.* **23,** 113–127 (1977).
136a. U. S. Rutishauser and G. M. Edelman, in *Methods of Cell Separation* (N. Catsimpoolas, Ed.), Plenum Press, New York, 1977, Chap. 5.
137. M. W. Burgett and L. V. Greenley, *Amer. Lab.* **9,** 74–83 (1977).
137a. H. Walter, in *Methods of Cell Separation* (N. Catsimpoolas, Ed.), Plenum, New York (1977), Chap. 8.
138. J. Turkova, *Affinity Chromatography,* Vol. 12, *Journal of Chromatography Library,* Elsevier, Amsterdam (1978).
139. I. P. Trayer and M. A. Winstanley, *Int. J. Biochem.* **9,** 449–456 (1978).
140. L. Hudson, *J. Chromatogr.* **159,** 123–128 (1978).
141. H. Potuzak and P. D. G. Dean, *FEBS Lett.* **88,** 161–166 (1978).
142. A. H. Nishikawa, in *Proceedings of the International Workshop on Technology*

for Protein Separation and Improvement of Blood Plasma Fractionation (H. E. Sandberg, Ed.), DHEW Publ. No. (NIH) 78-1422 (1978), pp. 422–435.

143. I. M. Chaiken, *Anal. Biochem.* **97,** 1–10 (1979).

143a. R. Lotan and G. L. Nicholson, *Biochim. Biophys. Acta* **559,** 329–376 (1979).

144. A. H. Nishikawa, in *Encyclopedia of Chemical Technology,* Vol. 6, 3rd ed., Wiley, New York (1979), pp. 34–54.

144a. P. D. G. Dean and D. H. Watson, *J. Chromatogr.* **165,** 301–319 (1979).

145. G. M. Whitesides and A. H. Nishikawa in *Applications of Biochemical Systems in Preparative Organic Chemistry,* Vol. 10, *Technique of Organic Chemistry,* 3rd ed. (J. B. Jones, D. Perlman, and C. J. Sih, Eds.), Interscience, New York (1976), pp. 929–968.

146. P. Hubert and S. E. Della-Cherie, *J. Chromatogr.* **184,** 325–333 (1980).

147. G. R. Gray, *Anal. Chem.* **52,** 9R–15R (1980).

148. A. A. Farooqui, *J. Chromatogr.* **184,** 335–345 (1980).

148a. S. K. Sharma and P. P. Mahendroo, *J. Chromatogr.* **184,** 471–499 (1980).

148b. W. H. Scouten, *Affinity Chromatography: Bioselective Adsorption on Inert Matrices,* Wiley-Interscience, New York (1981).

149. P. Cuatrecasas and I. Parikh, *Biochemistry* **11,** 2291–2299 (1972).

150. A. H. Nishikawa and P. Bailon, *J. Solid-Phase Biochem.* **1,** 33–49 (1976).

151. R. Axen and S. Ernback, *Eur. J. Biochem.* **18,** 351–360 (1971).

152. J. Porath, R. Axen, and S. Ernback, *Nature* **215,** 1491–1492 (1967).

153. M. Wilchek, *FEBS Lett.* **33,** 70–72 (1973).

154. J. K. Inman and H. M. Dintzis, *Biochemistry* **8,** 4074–4082 (1969).

155. J. Turkova and A. Seifertova, *J. Chromatogr.* **148,** 293–297 (1978).

155a. J. Coupek, M. Krivakova, and S. Pokorny, *J. Polym. Sci. Polym. Symp.* **42,** 185–190 (1973).

156. H. Engelhardt and D. Mathes, *J. Chromatogr.* **142,** 311–320 (1977).

157. P. Strop, F. Mikes, and Z. Chytilova, *J. Chromatogr.* **156,** 239–254 (1978).

158. J. F. Kennedy, J. A. Barnes, and J. B. Matthews, *J. Chromatogr.* **196,** 379–389 (1980).

158a. M. Wilchek, T. Oka, and Y. J. Topper, *Proc. Nat. Acad. Sci. U. S.* **72,** 1055–1058 (1975).

159. G. S. Bethell, J. S. Ayers, W. S. Hancock, and M. T. W. Hearn, *J. Biol. Chem.* **254,** 2572–2574 (1979).

160. J. Ellingboe, B. Alme, and J. Sjovall, *Acta Chem. Scand.* **24,** 463–467 (1970).

161. J. Porath and L. Sundberg, *Nature* **238,** 261–262 (1972).

162. L. T. Hodgins and M. Levy, *J. Chromatogr.* **202,** 381–390 (1980).

163. L. Sundberg and J. Porath, *J. Chromatogr.* **90,** 87–98 (1974).

164. K. Nilsson and K. Mosbach, *Eur. J. Biochem.* **112,** 397–402 (1980).

165. T. Lang, C. J. Suckling, and H. C. S. Wood, *J. Chem. Soc. Perkins Trans.* **1,** 2189–2194 (1977).

166. T. H. Finlay, V. Troll, M. Levy, A. J. Johnson, and L. T. Hodgins, *Anal. Biochem.* **87,** 77–90 (1978).

167. R. D. Guthrie, *Adv. Carbohyd. Res.* **16,** 105–158 (1961).

168. C. J. Sanderson and D. V. Wilson, *Immunology* **20,** 1061–1065 (1971).

169. S. Ohlson, L. Hansson, P.-O. Larsson, and K. Mosbach, *FEBS Lett.* **93,** 5–9 (1978).

170. H. Fritz, I. Trautschold, H. Haendle, and G. Werle, *Ann. N.Y. Acad. Sci.* **146,** 400–412 (1968).

171. G. R. Stark and J. G. Williams, *Nucleic Acids Res.* **6,** 195–203 (1978).

172. D. F. Smith, P. F. Searle, J. G. Williams, *Nucleic Acids Res.* **6,** 487–506 (1978).

173. H. Schaller, C. Nüsslein, F. J. Bonhoeffer, C. Kurz, and I. Nietzschmann, *Eur. J. Biochem.* **26,** 474–481 (1972).

174. H. Slok, *Nucleic Acids Res.* **2,** 587–593 (1975).

175. D. S. Terman, I. Stewart, J. Robinette, R. Carr, and R. Harbeck, *Clin. Exp. Immunol.* **24,** 231–237 (1976).

176. R. C. Manak and E. W. Voss Jr. *Immunochemistry,* **15,** 643–651 (1978).

177. E. Simonianova and M. Petakova, *Coll. Czech. Chem. Commun.* **44,** 626–630 (1979).

178. P. D. W. Areson, S. E. Charm, and B. L. Wong, *Biotechnol. Bioeng.* **22,** 2207–2217 (1980).

178a. J. Heinrich, P. F. Pilch, and M. P. Czech, *J. Biol. Chem.* **255,** 57–60 (1980).

179. R. J. Banes and G. R. Gray, *J. Biol. Chem.* **252,** 57–60 (1977).

180. M. Horisberger, *Carbohydr. Res.* **53,** 231–237 (1977).

180a. J. H. Pazur, K. L. Dreher, and L. S. Forsberg, *J. Biol. Chem.* **253,** 1832–1837 (1978).

181. D. S. Secher and D. C. Burke, *Nature* **285,** 446–450 (1980).

182. T. Staehelin, H.-F. Kung, D. Hobbs, and S. Pestka, *J. Biol. Chem.,* **256,** 9750–9754 (1981).

183. S. Ogata, T. Muramatsu, and A. Kobata, *J. Biochem.* **78,** 687–696 (1975).

184. K. O. Lloyd, *Arch. Biochem. Biophys.* **137,** 460–468 (1970).

185. H. Lis and N. Sharon, *Ann. Rev. Biochem.* **42,** 541–575 (1973).

186. D. H. Boldt and R. D. Lyons, *J. Immunol.* **123,** 808–816 (1979).

187. W. Rosner and H. L. Bradlow, *J. Clin. Endocrinol. Metab.* **33,** 193–198 (1971).

188. A. Matsuura, A. Iwashima, and Y. Nose, *Biochem. Biophys. Res. Commun.* **51,** 241–246 (1973).

189. B. A. Sage and J. D. O'Conner, *Anal. Biochem.* **73,** 240–246 (1976).

190. A. D. Landman and N. N. Landman, *J. Chem. Educ.* **53,** 571–592 (1976).

191. S. G. Mayhew and M. J. J. Strating, *Eur. J. Biochem.* **59,** 539–544 (1975).

192. K. K. Stewart and R. F. Doherty, *Proc. Nat. Acad. Sci. U. S.* **70,** 2850–2852 (1973).

193. S. Avrameas and T. Ternynck, *Biochem. J.* **102,** 37–39 (1967).

194. J. Folkersen, B. Teisner, P. Svendsen, and S.-E. Svehag, *J. Immunol. Method.* **23,** 127–135 (1978).

195. N. M. Gough and J. M. Adams, *Biochemistry* **17,** 5560–5566 (1978).

196. J. Menzel, *Z. Immunol. Forsch.* **149,** 9–19 (1975).

197. D. Ralt, N. Nelson, and D. Gutnick, *FEBS Lett.* **91,** 85–89 (1978).

198. B. Ludwig and R. A. Capaldi, *Biochem. Biophys. Res. Commun.* **87,** 1159–1167 (1979).

199. A. T. Yeung, K. J. Turner, N. F. Bascomb, and R. R. Schmidt, *Anal. Biochem.* **110,** 216–228 (1981).
200. A. Levitzki and M. Schramm, *Bull. Res. Council of Israel* **11A,** 258 (1963).
201. A. Levitzki, J. Heller and M. Schramm, *Biochem. Biophys. Acta* **81,** 101–107 (1964).
202. Z. Selinger and M. Schramm, *Biochem. Biophys. Res. Commun.* **12,** 208–214 (1963).
203. D. Dennis, *Anal. Biochem.* **24,** 541–544 (1968).
204. J. R. Wykes, P. Dunnill and M. D. Lilly, *Biochim. Biophys. Acta* **286,** 260–268 (1972).
205. V. Horejsi, P. Smolek, and J. Kocourek, *Biochim. Biophys. Acta* **538,** 293–298 (1978).
206. P.-O. Larsson and K. Mosbach, *FEBS Lett.* **98,** 333–338 (1979).
207. P. Hubert, J. Mester, E. Della-Cherie, J. Neel, and E.-E. Baulieu, *Proc. Nat. Acad. Sci. U. S.* **75,** 3143–3147 (1978).
208. P. A. Albertsson, *Partition of Cell Paticles and Macromolecules,* 2nd ed., Wiley-Interscience, New York, 1971.
209. G. Takerkart, E. Segard and M. Monsigny, *FEBS Lett.* **42,** 218–220 (1974).
210. V. P. Shanbhag and G. Johansson, *Biochem. Biophys. Res. Commun.* **61,** 1141–1146 (1974).
211. H. Walter and E. J. Krob, *FEBS Lett.* **61,** 290–293 (1976).
212. C.-G. Axelsson and V. P. Shanbhag, *Eur. J. Biochem.* **71,** 419–423 (1976).
213. E. Eriksson, P.-A. Albertsson, and G. Johansson, *Molec. Cell Biochem.* **10,** 123–128 (1976).
214. H. Westrin, P.-A. Albertsson, and G. Johansson, *Biochim. Biophys. Acta* **436,** 696–706 (1976).
215. V. P. Shanbhag and G. Johansson, *Eur. J. Biochem.* **93,** 363–367 (1979).
216. P. Hubert, E. Della-Cherie, J. Neel, and E.-E. Baulieu, *FEBS Lett.* **65,** 169–174 (1976).
217. A. Chaabouni and E. Della-Cherie, *J. Chromatogr.* **171,** 135–143 (1979).
218. A. Chaabouni, P. Hubert, E. Della-Cherie and J. Neel, *Makromol. Chem.* **179,** 1135–1144 (1978).
219. S. D. Flanagan and S. H. Barondes, *J. Biol. Chem.* **250,** 1484–1489 (1975).
220. S. D. Flanagan, P. Taylor, and S. H. Barondes, *Nature* **254,** 441–443 (1975).
221. S. D. Flanagan, S. H. Barondes, and P. Taylor, *J. Biol. Chem.* **251,** 858–865 (1976).
222. S. D. Flanagan, P. Taylor, and S. H. Barondes, *Croat. Chem. Acta* **47,** 449–457 (1975).
223. L. Reimann and G. R. Schonbaum, in *Methods in Enzymology* (S. Fleischer and L. Packer, Eds.), Academic Press, New York (1978), pp. 514–521.
224. S. O. Falkbring, P. O. Göthe, P. O. Nyman, L. Sundberg, and J. Porath, *FEBS Lett.* **24,** 229–235 (1972).
225. P. L. Whitney, *Anal. Biochem.* **57,** 467–476 (1974).
226. H. F. Hixson and A. H. Nishikawa, *Arch. Biochem. Biophys.* **154,** 501–509 (1973).

227. A. H. Nishikawa, P. Bailon, and A. H. Ramel, *J. Macromol. Sci. Chem.* **A10,** 149–190 (1976).
228. A. R. Thompson and E. W. Davie, *Biochim. Biophys. Acta* **250,** 210–215 (1971).
229. P. Bailon and A. H. Nishikawa, *Prep. Biochem.* **7,** 61–87 (1977).
230. A. H. Nishikawa and P. Bailon, *Arch. Biochem. Biophys.* **168,** 576–584 (1975).
231. K. Fujiwara and D. Tsuru, *Int. J. Peptide Protein Res.* **9,** 18–26 (1977).
232. G. Oshima and K. Nagasawa, *J. Biochem.* **81,** 1285–1291 (1977).
233. T. Kumazaki, K. Kasai, and S. Ishii, *J. Biochem.* **79,** 749–755 (1976).
234. P. O'Carra, S. Barry, and T. Griffin, *Biochem. Soc. Trans.* **1,** 289–290 (1973).
235. E. Steers Jr., P. Cuatrecasas, and H. Pollard, *J. Biol. Chem.* **246,** 196–200 (1971).
236. B. H. J. Hofstee, *Biochem. Biophys. Res. Commun.* **53,** 1137–1144 (1973).
237. P. Cuatrecasas and G. Illiano, *Biochem. Biophys. Res. Commun.* **44,** 178–183 (1971).
238. J. I. Rood and R. G. Wilkinson, *Biochim. Biophys. Acta* **334,** 168–178 (1974).
239. C. C. Huang and D. Aminoff, *Biochim. Biophys. Acta* **371,** 462–469 (1974).
240. J. Massoulie and S. Bon, *Eur. J. Biochem.* **68,** 531–539 (1976).
241. S. Blumberg and E. Katchalski-Katzir, *Biochemistry* **18,** 2126–2133 (1979).
242. S. K. Sharma and T. R. Hopkins, *J. Chromatogr.* **110,** 321–326 (1975).
243. R. F. Murphy, J. M. Conlon, A. Imam, and G. J. C. Kelly, *J. Chromatogr.* **135,** 427–433 (1977).
244. H. M. Katzen and G. J. Vlahakes, *Science* **179,** 1142–1143 (1973).
245. P. Cuatrecasas, *Science* **179,** 1143–1144 (1973).
246. M. B. Davidson and A. J. Van Herle, *New Engl. J. Med.* **289,** 695–696 (1973).
247. R. J. Lefkowitz, *New Engl. J. Med.* **289,** 696 (1973).
248. J. C. Venter and N. O. Kaplan, *Science* **185,** 459–460 (1974).
249. M. S. Yong and J. B. Richardson, *Science* **185,** 460–461 (1974).
250. T. C. J. Gribnau and G. I. Tesser, *Experientia* **30,** 1228–1230 (1974).
251. H. Thoni, *Experientia* **31,** 251 (1975).
252. J. Lasch, *Experientia* **31,** 1125–1126 (1975).
253. G. I. Tesser, H.-U. Fasch, and R. Schuyzer, *Helv. Chim. Acta* **57,** 1718–1730 (1974).
254. J. Schwartz, D. F. Nutting, H. M. Goodman, J. L. Kostyo, and R. E. Fellows, *Endocrinology* **92,** 439–445 (1973).
255. M. B. Davidson, A. J. Van Herle, and L. E. Gerschenson, *Endocrinology* **92,** 1442–1446 (1973).
256. G. Vanquelin, M.-L. Lacombe, J. Hanoune, and A. D. Strosberg, *Biochem. Biophys. Res. Commun.* **64,** 1076–1082 (1975).
257. M. Wilchek in *Immobilized Biochemicals and Affinity Chromatography,* Vol. 42, *Advances in Experimental Medicine and Biology,* (R. B. Dunlop, Ed.), Plenum Press, New York (1974), pp. 15–31.
258. D. W. Jacobsen, Y. O. Montejano, and F. M. Huennekens, *Anal. Biochem.* **113,** 164–171 (1981).

259. Y. D. Clonis and C. R. Lowe, *Eur. J. Biochem.* **110,** 279–288 (1980).
260. C.-Y. Lee and C.-J. Johansson, *Anal. Biochem.* **77,** 90–102 (1977).
261. P. Brodelius, P.-O. Larsson, and K. Mosbach, *Eur. J. Biochem.* **47,** 81–89 (1974).
262. C.-Y. Lee, L. H. Lazarus, D. S. Kabakoff, P. J. Russell Jr., M. Laver, and N. O. Kaplan, *Arch. Biochem. Biophys.* **178,** 8–18 (1977).
263. A. Bodansky and M. Bodansky, *Experientia* **26,** 327 (1970).
264. K. Hofmann, F. M. Finn, and Y. Kiso, *J. Amer. Chem. Soc.* **100,** 3585–3590 (1978).
265. M. L. Jasiewicz, D. R. Schoenberg, and G. C. Mueller, *Exp. Cell Res.* **100,** 213–217 (1976).
266. J. Manning, M. Pellegrini and N. Davidson, *Biochemistry* **16,** 1364–1370 (1977).
267. R. Horowitz and G. M. Whitesides, *J. Amer. Chem. Soc.* **100,** 4632–4633 (1978).
268. E. A. Boyer and M. Wilchek, *Trends Biochem. Sci.* **3,** N257–N259 (1978).
269. M. Miller-Anderson, H. Borg, and L.-O. Anderson, *Thromb. Res.* **5,** 439–452 (1974).
270. P. S. Damus and G. A. Wallace, *Biochem. Biophys. Res. Commun.* **61,** 1147 (1974).
271. T. H. Finlay, V. Troll, and L. T. Hodgins, *Anal. Biochem.* **108,** 354–359 (1980).
272. I. Danishefsky, F. Tzeng, M. Ahrens, and S. Klein, *Thromb. Res.* **8,** 131–140 (1976).
273. K. Fujiwara, A. R. Thompson, M. E. Legas, G. G. Meyer, and E. W. Davie, *Biochemistry* **12,** 4938–4945 (1973).
274. P. W. Gentry and B. Alexander, *Biochem. Biophys. Res. Commun.* **50,** 500–509 (1973).
275. W. Kisiel and E. W. Davie, *Biochemistry* **14,** 4928–4934 (1975).
276. S. S. Stivala and J. Ehrlich, *Polymer* **15,** 197–203 (1974).
277. P.-H. Iverius, *J. Biol. Chem.* **247,** 2607–2613 (1972).
278. S. R. Srinivasan, B. Radhakrishnamurthy, and G. S. Berenson, *Arch. Biochem. Biophys.* **170,** 334–340 (1975).
279. M. Ya. Varshavskaya, A. L. Klibanou, V. S. Goldmacher, and V. P. Torchilin, *Anal. Biochem.* **95,** 449–451 (1979).
280. N. Ototani, M. Kikuchi, and Z. Yosizawa, *Carbohydr. Res.* **88,** 291–303 (1981).
281. I. J. Goldstein, L. A. Murphy, and S. Ebisu, *Pure Appl. Chem.* **49,** 1095–1103 (1977).
282. H. Lis, R. Lotan, and N. Sharon, *Ann. N.Y. Acad. Sci.* **234,** 232–238 (1974).
283. P. Vretblad, *Biochim. Biophys. Acta* **434,** 169–176 (1976).
284. K. O. Lloyd, *Arch. Biochem. Biophys.* **137,** 460–468 (1970).
285. D. Allen, J. Auger, and M. J. Crumpton, *Nature* **236,** 23–25 (1972).
286. J. C. Anderson, *Biochim. Biophys. Acta* **379,** 444–455 (1975).
287. K. Hiwada, H. Tanaka, K. Nishimura, and T. Kokubu, *Clin. Chim. Acta* **74,** 203–206 (1977).

288. B. C. Nisula, D. L. Loriaux, and Y. A. Wilson, *Steroids* **31,** 681–690 (1978).
289. S. Narasimhan, J. R. Wilson, E. Martin, and H. Schachter, *Can. J. Biochem.* **57,** 83–96 (1979).
290. T. K. Datta and M. K. Ray, *Indian J. Exp. Biol.* **17,** 323–324 (1979).
291. J. E. Coligan and H. S. Slayter, *Molec. Immunochem.* **16,** 129–135 (1979).
292. K. Mattila, *Biochim. Biophys. Acta* **579,** 62–72 (1979).
293. S. Matsuura and H.-C. Chen, *Anal. Biochem.* **106,** 402–410 (1980).
294. C. E. Carter and D. G. Colley, *Molec. Immunol.* **18,** 219–225 (1981).
295. J. E. Strickler, P. E. Mancini, and C. L. Patton, *Exp. Parasitology* **46,** 262–276 (1978).
296. R. M. Helm, D. H. Conrad, and A. Froese, *Int. Arch. Allergy Appl. Immun.* **58,** 90–98 (1979).
297. A. Surolia, A. Ahmad and B. K. Bachhawat, *Biochim. Biophys. Acta* **404,** 83–92 (1975).
298. K. J. Clemetson, S. L. Pfueller, E. F. Luescher, and C. S. P. Jenkins, *Biochim. Biophys. Acta* **464,** 493–508 (1977).
299. T. Kristiansen, *Biochim. Biophys. Acta* **338,** 246–253 (1974).
300. P. Ziska and J. Mohr, *Acta Biol. Med. Ger.* **36,** 1197–1198 (1977).
301. N. A. Nicola, A. W. Burgess, D. Metcalf, and F. L. Battye, *Aust. J. Exp. Biol. Med. Sci.* **56,** 663–679 (1978).
302. J. L. Spivak, D. Small, J. H. Shaper, and M. D. Hollenberg, *Blood* **52,** 1178–1188 (1978).
303. P. A. Gleeson, M. A. Jermyn, and A. E. Clarke, *Anal. Biochem.* **92,** 41–45 (1979).
304. G. Uhlenbruck, R. Newman, G. Steinhausen, and H. G. Schwick, *Z. Immunitatsforsch.* **153,** 183–187 (1977).
305. G. Kopperschlager, R. Freyer, W. Diezel, and E. Hoffmann, *FEBS Lett.* **1,** 137 (1968).
306. R. L. Easterday and I. M. Easterday, in *Immobilized Biochemicals and Affinity Chromatography* (*Advances in Experimental Medicine and Biology,* Vol. 42), (R. B. Dunlap, ed.), Plenum Press, New York (1974), 123–133.
307. E. Stellwagen, *Accts. Chem. Res.* **10,** 92–98 (1977).
308. A. R. Ashton and G. M. Polya, *Biochem. J.* **175,** 501–506 (1978).
309. S. Comoglio, A. Massaglia, E. Rolleri, and U. Rosa, *Biochim. Biophys. Acta* **420,** 246–257 (1976).
310. J. W. Eveleigh, *Characteristics and Practical Application of Agarose Based Immunosorbents,* U. S. Energy Research and Development Administration Report ORNL-TM-4962, (1975).
311. D. M. Weir, Ed., *Handbook of Experimental Immunology,* Vol. 1, *Immunochemistry,* 3rd ed., Blackwell Scientific, Oxford (1978).
312. H. Van Vunakis and J. J. Langone, *Immunochemical Techniques,* Part A (*Methods in Enzymology,* Vol. 70), Academic Press, New York (1980).
313. J. W. Goding, *J. Immunol. Method.* **39,** 285–308 (1980).
314. S. Avrameas and T. Ternynck, *J. Biol. Chem.* **242,** 1651–1659 (1967).
315. M. Zoller and S. Matzku, *J. Immunol. Method.* **11,** 287–295 (1976).

316. J. A. Smith, J. G. R. Hurrell, and S. J. Leach, *Anal. Biochem.* **87,** 299–305 (1978).
317. G. J. Dimitriadis, *Anal. Biochem.* **98,** 445–451 (1979).
318. R. Kuwano, Y. Yoshida, and Y. Takahashi, *J. Neuro. Chem.* **31,** 815–824 (1978).
319. T. Ooyama, J. Nakao, and K. Katayoma, *Biochem. Med.* **19,** 231–235 (1978).
320. B. Svensson, M. Danielsen, M. Staum, L. Jeppesen, O. Noren, and H. Sjostrom, *Eur. J. Biochem.* **90,** 489–498 (1978).
321. B. I. S. Srivastava, J. Y. H. Chan, and F. A. Siddiqui, *J. Biochem. Biophys. Method.* **2,** 1–9 (1980).
322. J. J. Lanzillo, R. Polsky-Cynkin, and B. L. Fanburg, *Anal. Biochem.* **103,** 400–407 (1980).
323. K.-H. Pick and G. Wober, *Prep. Biochem.* **9,** 293–302 (1979).
324. L. T. Clement, A. M. Kask, and E. M. Shevach, *Immunochemistry* **15,** 393–399 (1978).
325. D. C. Rueger, D. Dahl, and A. Bignami, *Anal. Biochem.* **89,** 360–371 (1978).
326. R. H. Davis Jr., and E. Steers Jr., *Immunochemistry*, **15,** 371–378 (1978).
327. J. E. McEntire, *J. Immunol. Method.* **24,** 39–44 (1978).
328. G. J. Mizejewski, R. Simon, and M. Vonnegut, *J. Immunol. Method.* **31,** 333–339 (1979).
329. S. Pahlman, I. Ljungsted-Pahlman, A. Sanderson, P. J. Ward, A. Grant, and J. Hermon-Taylor, *Brit. J. Cancer* **40,** 701–709 (1979).
330. F. Brown, B. O. Underwood, and K. H. Fantes, *J. Med. Virol.* **4,** 315–319 (1979).
331. V. A. Lennon, M. Thompson, and J. Chen, *J. Biol. Chem.* **225,** 4395–4398 (1980).
332. S. K. Roy and A. H. Nishikawa, *Biotech. Bioeng.* **21,** 775–785 (1979).
333. P. Brodelius, P.-O. Larsson, and K. Mosbach, *Eur. J. Biochem.* **47,** 81–89 (1974).
334. L. Sundberg, J. Porath, and K. Aspberg, *Biochim. Biophys. Acta* **221,** 394–395 (1970).
335. N. G. Anderson, D. D. Willis, D. W. Holladay, J. E. Caton, J. W. Holleman, J. W. Eveleigh, J. E. Attrill, F. L. Ball, and N. L. Anderson, *Anal. Biochem.* **66,** 159–174 (1975).
336. J. Folkersen, B. Teisner, J. Westergaard, and S.-E. Svehag, *J. Immunol. Method.* **23,** 137–147 (1978).
337. J. J. Pahud and K. Schwarz, *Science Tools* **23,** 40–42 (1976).
338. P. J. Robinson, M. A. Wheatley, J.-C. Janson, P. Dunnill, and M. D. Lilly, *Biotech. Bioeng.* **16,** 1103–1112 (1974).

CHAPTER

3

AFFINITY CHROMATOGRAPHY ON IMMOBILIZED DYES

PETER D. G. DEAN and FIRDAUSI QADRI

Department of Biochemistry
University of Liverpool
Liverpool L69 3BX, United Kingdom

The concept of group-specific affinity supports in protein separation is now well established (Lowe and Dean, 1). Immobilized triazine dyes are excellent examples of group-specific affinity matrices; see Table 3.1. The advantages of these immobilized ligands over many of the adsorbents mentioned in the table are numerous (see Section 3.3.21), but the ease of preparation of immobilized dye matrices and the ubiquitous nature of their interactions with macromolecules are fundamental reasons why these dyes have attracted so much attention. Furthermore, triazine chemistry is being used increasingly in affinity chromatography (Table 3.2). In this chapter, we examine the background, mechanism, and applications of immobilized dyes, especially in protein purification. In screening the literature for this review, we have looked for applications of dyes covered by the following trade names: Procion (ICI), Cibacron (Ciba-Geigy), Lewafix (Bayer AG), Remazol (Hoechst). We have not attempted to exhaustively review the applications of noncovalently bound dyes.

3.1. GENERAL NATURE OF THE TRIAZINE DYES

The triazine dyes are so called because they are based upon the chemistry of cyanuric chloride (1,3,5-trichloro triazine).

The reactive triazine dyes introduced by ICI in 1954 were originally designed for textile applications. the Procion dyes, which contain reactive chlorotriazine groups, consist of a variety of aromatic chromophores linked to the triazine ring via NH bridges. Many of the early dyes, which are referred to as Procion MX dyes, contain dichloro triazinyl groups. This series of dyes reacts extremely rapidly in basic solutions with a wide variety of different materials. A later series produced from the same company used monochlorotriazines, called Procion H dyes, which, because they were more

Table 3.1. Some Group-Specific Affinity Ligands

Ligand	Macromolecule to be Purified	Reference
1. NAD, ATP, and other nucleotides	Nucleotide-dependent proteins	Lowe & Dean, 1
2. Dyes	Very varied	This chapter
3. Phenylboronates	Diols; nucleotides; sugars; glycoproteins	Bouriotis, Galpin, & Dean, 3
4. Heparin	Various	Farooqui, 4
5. Lectins	Sugars; glycoproteins	Lis & Sharon, 5; Spivak et al., 6
6. Antinormal IgG's	Pregnancy and other "invaded host" proteins	Sutcliffe et al., 7; Heinzel et al., 7a
7. Protein A and C1q	IgG and immune complexes	McFarthing & Dean, 8; Langone et al., 9

stable in the cold, were very useful additions to the dying and printing field. The Cibacron dyes are structurally extremely similar to the Procion series and contain the same general arrays of triazine rings and aromatic chromophores. An independent series, the Lewafix dyes, use the diazine ring system as an alternative to triazine but have very similar aromatic chromophores to the Procion–Cibacron series. The chemistry of Remazol Blue is somewhat different from the triazine-based dyes (see Land and Byfield, 16; Mislovicova, 17).

The most important property of these dyes is their ability to become covalently attached to polymer surfaces. They may be used to permanently dye materials such as cellulose, wools, and cloths. A very complete range of dye shades has been produced in both dichloro- and monochloro-triazine series by varying the chromophoric residues. Examples of some of the dye variants are shown in Figure 3.1. Phthalocyanines produce turquoise shades; an-

Table 3.2. Applications of Triazine Chemistry in Affinity Chromatography

Ligand	Macromolecule	Reference
1. Oxamate	LDH	Lang et al., 10
2. Nucleic acid	Nucleoproteins	Biagioni et al., 11; Dean et al., 12
3. Various small and large ligands	Various	Finlay et al., 13
4. Triazine dyes	Mainly enzymes	This chapter
5. Glucose-6-phosphate DH (immobilization)	—	Smith & Lenhoff, 14
6. Spacers and nucleophiles	Model enzymes	Hodkins & Levy, 15

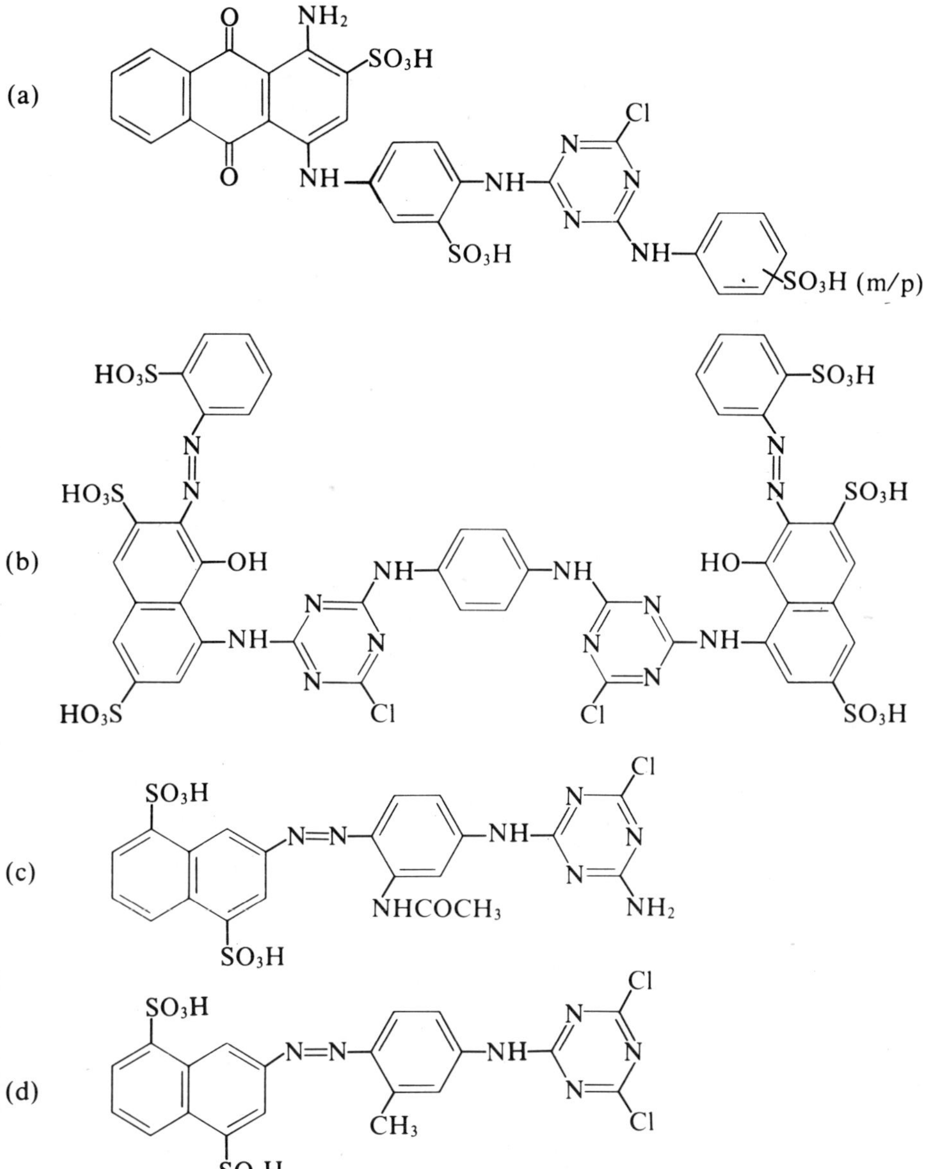

Fig. 3.1. The structures of four typical triazine dyes: (a) Cibacron Blue F3G-A (Procion Blue H-B); (b) Procion Red HE-3B; (c) Procion Yellow H-A; (d) Procion Yellow MX-R. From Lowe, Small, and Atkinson (18), with permission.

thraquinone chromophores are usually blue. Green dyes are usually associated with anthraquinones, and phthalocyanines are coupled to azo or stilbene groups. Reds and yellows are often azo dyes, whereas violets, browns, and blacks are associated with metal complexes (copper, etc.) of dihydroxyazo systems. Only a few dyes are chemically well defined in the literature, however, a reflection of the complex patent situation associated with their manufacture. As pointed out by Lowe, Small, and Atkinson (18), it is inappropriate to ascribe standard IUPAC nomenclature for these complex substances, and they recommend the use of the color index number and the standard commercial name (e.g., Cibacron Blue 3G-A or Procion Red HE-3B). Table 3.3 lists a series of names and color index numbers of some commercial dyes.

3.1.1. Routes of Synthesis

The synthetic routes to triazine dyes centers on the relative reactivity of the chlorine atoms in the triazine ring. The rates of reaction of the three chlorine atoms in *s*-trichlorotriazine (cyanuric chloride) are markedly different (see Lang, Suckling, and Wood, 10; Beech, 19). Displacement of the first chlorine by amines or hydroxyl compounds proceeds very rapidly even in the cold ($t_{1/2}$ = 30 sec). The reaction of the second chlorine is comparatively slow at room temperature but is rather rapid at about 40°C ($t_{1/2}$ = 30 min), whilst the last chlorine is displaced more slowly and requires prolonged exposure at elevated temperatures before displacement is achieved by hydroxyl compounds such as cellulose ($t_{1/2}$ = 24 hr). The dyes are prepared by sequential replacement of these chlorine atoms by a reactive oxygen or amino-containing chromophore. Other routes of synthesis involve (i) the formation of complex chromophoric amines followed by coupling to the triazine, or (ii) a

Table 3.3. Color Index Numbers and Names of Some "Reactive" Dyes

Commercial Name		Color Index No.
Reactive Black 1	BGA Cibacron Black	17916
Reactive Blue	3G-A Cibacron Blue	61211
Reactive Brilliant Blue	BRP	61210
Reactive Brilliant Red	3BA	18105
Reactive Brilliant Red	3A	18156
Reactive Brilliant Yellow	3GP	18972
Reactive Rubine	RE	17912
Reactive Turquoise Blue	GE	74460
Reactive Turquoise Blue	GFP	74459
Reactive Violet	2RP	18157
Remazol Blue		61200

stepwise addition of amines to the triazine followed by further derivatization such as diazotization (i.e., treatment of the resulting molecule with a diazotized amine). Thus Procion Red HE-3B might be prepared by condensation of two molecules of cyanuric chloride with one of *p*-phenylenediamine followed by reaction of the tetrachloro(*bis*)triazine with 8-amino naphthol 3-6-disulfonic acid. This molecule may then be coupled to diazotized orsonilic acid (*o*-amino phenylsulfonic acid).

3.1.2. Some Chromophoric Groups Used in Dyes

Some examples of typical group chromophores are given in Figure 3.2. A detailed account of the different chromophores and general approaches to the synthesis of triazine dyes may be found in Beech (19).

3.1.3. Identification of Triazine Dyes

The following techniques have been used in the identification and verification of specific dye structures: Thin layer chromatography on silica gel,

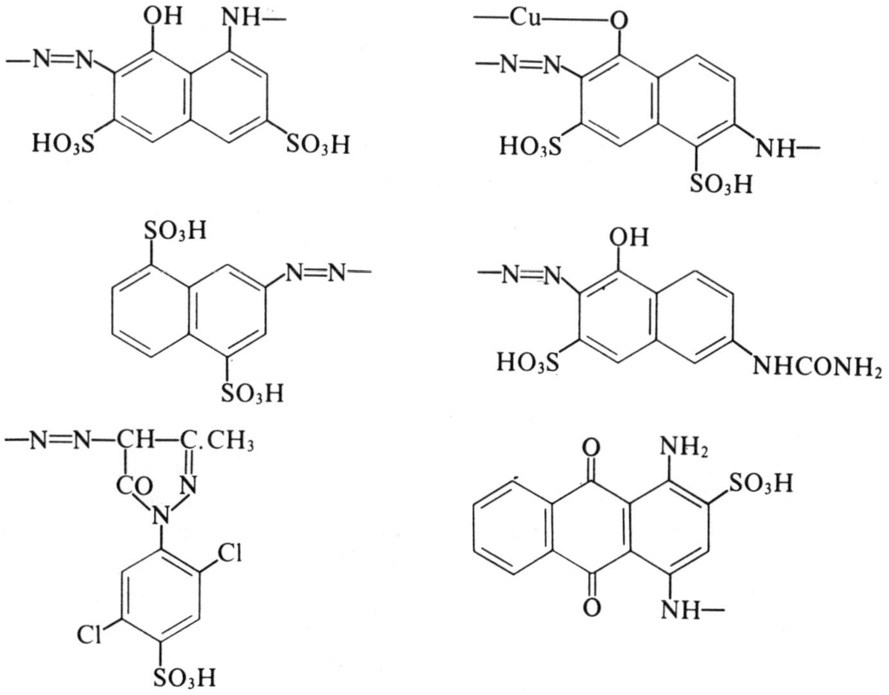

Fig. 3.2. Some chromophores commonly encountered in triazine dyes.

paper chromatography, ultraviolet and infrared spectroscopy (KBr disk), NMR spectroscopy (particularly useful in the aromatic region when identifying structures), mass spectroscopy (of limited use without suitable derivatization).

3.1.4. Purification of Triazine Dyes

Many of the dye preparations contain several contaminants. Surfactants and buffers (see Lowe, Small, and Atkinson, 18) are often added as aids to solubility or as stabilizing reagents to prevent self-hydrolysis of dyes. Salts are used as material fillers (see Klein and Foreman, 20; Ewen, 21). Several authors warn of the dangers of not purifying the dyes before use, particularly in structural studies (see Weber et al., 22; Lowe, Small, and Atkinson, 18). Thin layer chromatography is a particularly useful way of removing small quantities of these contaminants (see Weber et al., 22). Other workers have used paper chromatography (Edwards and Woody, 23; Wu, 24; Beissner and Rudolph, 25, 26). Witt and Roskoski (27) describe three solvent systems suitable for use in the chromatographic purification of dyes. Other workers (Qadri and Dean, unpublished observations) have used isoelectric focusing in dye purification. In general, silica or cellulose chromatography appear to be the most generally used methods of dye purification, especially to remove partially completed chromophoric contaminants.

3.2. MECHANISMS FOR BINDING PROTEINS TO DYE COLUMNS

The literature contains a large number of papers that discuss the mechanism of dye binding and the structural features of enzymes that bind to dyes. As is common with many papers on affinity chromatography, considerable confusion arises from deductions made from the ability of certain ligands to desorb an enzyme from an affinity support. A typical example is found in Kulbe and Schuer (28). The implication is that the mechanism of an enzyme's adsorption is necessarily directly connected to its desorption from the solid phase. In our opinion, these two features of chromatography, adsorption and desorption, should be considered separately. The above confusion is countered by kinetic studies, which give considerable insight into the way these dyes behave. Clonis and Lowe (29) have shown how the kinetics of inhibition by various dyes elegantly fit in with the properties of the enzyme on dyed gels. Useful general discussions may be found in Fulton (30), Yon (31), and Beissner, Quiocho, and Rudolph (32). In this context, Beissner and coworkers say, "These dye columns are not probes for the dinucleotide fold nor affinity columns in the strictest sense, but localized cation exchangers having some hydrophobic interactions as well. This does not preclude the dye from interacting at other sites on a protein by hydrophobic forces, electrostatic forces, or a combination of both."

Generalizations concerning the mechanisms of dye–protein interactions are difficult to make. Forces associated with hydrophobic, ionic, van der Waals, and even charge-transfer interactions may be advocated to explain individual data. It is probable that no single type of interaction is adequate to describe any one protein's binding to a dye molecule. It is also probable that no single type of interaction predominates throughout the protein field.

In an attempt to understand the mechanism of interaction of the dyes with enzymes, a number of different approaches have been used. They are described in the following sections.

3.2.1. Equilibrium Dialysis

Thompson and Stellwagen (33) have used equilibrium dialysis to estimate the dissociation constant and the number of binding sites of Cibacron Blue 3G-A for lactate dehydrogenase (rabbit muscle). They observed that each subunit of the enzyme bound one dye molecule with a dissociation constant for the protein subunit:dye binary complex of 0.71 μM.

One disadvantage of equilibrium dialysis for studying triazine dye–protein interactions has been observed, however, (Qadri and Dean, unpublished observations); some triazine dyes (Cibacron Blue 3G-A and Procion Red HE-3B) bind to Visking membranes, making calculations of dye binding difficult. Thus, although equilibrium dialysis is used routinely for studying binding of proteins to drugs, it is less suitable for the measurement of binding of proteins to charged, large molecular weight triazine dyes.

3.2.2. Frontal Analysis

Frontal analysis has been used (Lowe and Dean, 1) to measure the capacity of proteins for and their affinity for immobilized ligands. This technique has been further extended (Nichol et al., 34) and simplified by Dean and Watson (35) to measure the affinity of proteins with immobilized ligands in terms of an apparent association constant K_A (app). In this procedure a constant concentration of the protein under consideration is applied to a column until the concentration of protein in the eluate reaches that of the sample being applied. The apparent association constant is measured using the equation:

$$K_A\,(\text{app}) = \frac{Ve - Ve^*}{Vs \times L - E(Ve - Ve^*)} \tag{1}$$

where Ve = elution volume at half saturation of the adsorbent
Ve^* = volume at half saturation of the enzyme in a control experiment with unsubstituted matrix
Vs = total volume of the gel available to the enzyme
L = ligand concentration of the adsorbent gel
E = concentration of enzyme in the mobile phase

Dean and Watson (36) observed that the capacities of the immobilized dyes Cibacron Blue 3G-A and Procion Red HE-3B for enzymes are an order of magnitude higher than their capacities for immobilized nucleotides. The K_A (app) of any particular enzyme was found to be dependent upon the purity of the extract applied (Watson, Harvey, and Dean, 37; Qadri, 38). This effect has been attributed to competition from other contaminating proteins for the binding sites which results in lower capacities for the enzyme being studied (Watson, Harvey, and Dean, 37).

Frontal analysis is critically dependent on the flow rate of the column. Dean and Watson (35) have shown that the capacity of columns decreases with an increase of flow rate for agarose and pellicular supports (Figure 3.3). Lowe et al. (39) have made similar observations and have stressed the impor-

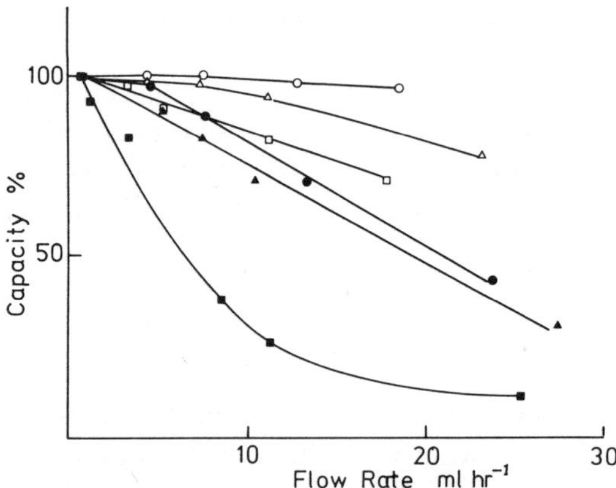

Fig. 3.3. The variation of capacity of NADP$^+$ matrices for various enzymes with increasing flow rate. Capacities were measured by frontal analysis. Closed symbols refer to Sepharose-based ligands, open symbols refer to Matrex 201R-based ligands. Circles, dihydrofolate reductase (*L. casei*); triangles, isocitrate dehydrogenase (pig heart); squares, L-glutamate dehydrogenase (*N. crassa*). From Dean and Watson (158), with permission.

tance of contact time between sample and triazine dye columns. We have observed (Qadri, 38) that the flow rate of triazine dye columns should not exceed 15 $mL/cm^2/hr$. On increasing the flow rate beyond 30 $mL/cm^2/hr$, a wide range of enzymes that otherwise bound to the immobilized dye column at the reduced flow rate were eluted in the wash. The effects of ligand concentration on frontal analysis are particularly important (see Section 3.3.8).

3.2.3. Affinity Electrophoresis

Affinity electrophoresis, first developed by Takeo and Nakamura (40), has recently been exploited to study protein and immobilized dye interactions (Johnson, Metcalf, and Dean, 41; Ticha, Horejsi, and Barthova, 42). In this technique, the effects of both biospecific interactions and electrophoretic mobilities are combined to measure the dissociation constants of particular protein–ligand interactions such as that between an enzyme and its substrate or antibody and antigen. A relationship for the dissociation constant of enzyme and immobilized ligand has been established using affinity electrophoresis data for lectins and immobilized sugars (Horejsi, Ticha, and Kocourek, 43). The method is particularly useful because only very small amounts of protein samples are required and, providing a suitable detection system is available, the protein samples applied need not be pure. Furthermore, a wide range of affinity interactions from weak to moderately strong (10^{-3} to 10^{-6} M K_{diss}) can be studied by this method.

The value of K_{diss} can be obtained from measuring the relative mobility of the enzyme in the absence (R_{mo}) and the presence of ligand (R_{mi}) in the gel:

$$\frac{R_{mo}}{R_{mi}} = 1 + \frac{c}{K} \tag{2}$$

Values of R_{mo}/R_{mi} are plotted against the concentration of immobilized ligand c; a straight line should be obtained, the abscissa intercept of which equals $-K$.

Using equation (2), dissociation constants have been calculated between bovine serum albumin and immobilized Cibacron Blue 3G-A (Johnson, Metcalf, and Dean, 41) (Figure 3.4). Similarly, the retardation of the enzyme 6-phosphogluconate dehydrogenase on polyacrylamide gels containing immobilized Cibacron Blue 3G-A and Procion Red HE-3B has been used (Metcalf, Crow, and Dean, 44) to measure dissociation constants between enzyme and dye (Figure 3.5). Affinity electrophoresis using triazine dyes as the ligand is a useful, rapid (2–3 hr) method for determining dissociation constants.

A word of caution: Since the dyes absorb intensely and since the measurement of the dissociation constant using equation (2) is highly dependent

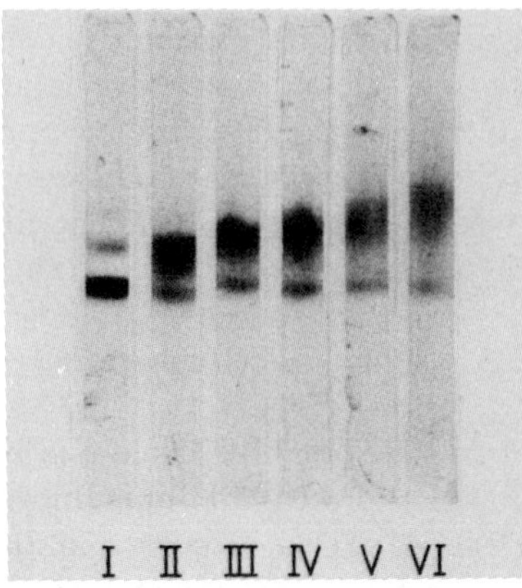

Fig. 3.4. The electrophoretic mobility of defatted bovine serum albumin on immobilized Cibacron Blue F3G A–agarose–polyacrylamide and control gels. I, Control gel; II–IV, increasing concentrations of immobilized Cibacron Blue F3G A. 20, 34, 46, 60, and 90 μM dye. From Johnson et al. (41), with permission.

on the ligand concentration of the gels, additional care must be taken to determine the dye concentration accurately. We have found that the determination of ligand concentration is most accurately carried out by measuring the absorbance of the polyacrylamide-agarose-dye gel directly in the spectrophotometer (Johnson, Metcalf, and Dean, 41; Metcalf, Crow, and Dean, 44).

When using affinity electrophoresis to study protein–immobilized-dye interaction, the values obtained may differ from those obtained using free solution inhibition studies or from those obtained with dye–agarose matrices. Several factors may give rise to different results; a few are (i) accessibility of

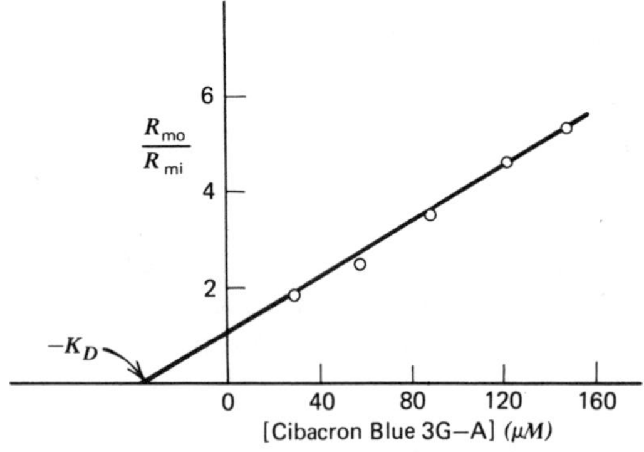

Fig. 3.5. Determination of K_D by affinity electrophoresis. From Metcalf et al. (44), with permission.

ligand, (ii) ion exchange properties of the dye, (iii) working temperature, (iv) pH.

3.2.4. Inhibition Studies

Reports in the literature on the nature of inhibition of enzymes by Cibacron Blue 3G-A are varied. In keeping with Wilson's view that the dye is a universal ligand (Wilson, 45), it has been suggested that interactions take place at the nucleotide (Böhme et al., 46; Kopperschläger et al., 47, 48; Wilson, 45; Thompson and Stellwagen, 33), the polynucleotide (Moe and Piszkiewicz, 49), the dihydrofolate (Wilson, 45; Chambers and Dunlap, 50), the tetrahydrofolate (Ramesh and Rao, 51, 52) and the phospholipid (Barden et al., 53) and fatty acid binding sites of various proteins (Metcalf, Crow and Dean, 44). For some enzymes the dye was found to be a competitive inhibitor with respect to nucleotides, coenzymes, and nucleic acids (Wilson, 45; Thompson et al., 54; Moe and Piszkiewicz, 49; Seelig and Colman, 55; Morrill, Thompson, and Stellwagen, 56, Morrill and Thompson, 57); for other enzymes the dye was found to be either a mixed inhibitor (Wilson, 45; Ramesh and Rao, 51, 52; Moe and Piszkiewicz, 49; Pompon and Lederer, 58; Schuber and Pascal, 59) or noncompetitive for others (Ahmad, Surolia, and Bachnawat, 60; Kumar and Krakow, 61; Beissner and Rudolph, 62; Kapoor and O'Brien, 63) or uncompetitive for dikinase (Redman and Hornemann, 64) (but the enzyme will not bind to the immobilized dye!). For pyruvate kinase, Kapoor and O'Brien (63) found the dye Cibacron Blue 3G-A to be a noncompetitive inhibitor with respect to both substrates ADP and phosphoenolpyruvate but competitive with respect to fructose 1,6-diphosphate, an allosteric effector of the enzyme. For 6-phosphogluconate dehydrogenase from *Bacillus stearothermophilus,* Cibacron Blue 3G-A was found to be a noncompetitive inhibitor with respect to both 6-phosphogluconate and $NADP^+$, which suggested that the dye bound at a site or sites on the enzyme different from the $NADP^+$ and 6-phosphogluconate binding sites (Qadri, 38). Procion Red HE-3B showed competitive inhibition of NADH binding to aldehyde reductase (Ki $=1.8 \times 10^{-7}$ M) (Turner and Hryszko, 65).

In summary, the dyes exhibit too wide a range of inhibitor characteristics for generalizations to be drawn.

3.2.5. Relationship Between the Binding of Enzymes to Immobilized Dyes and Free Solution Inhibition

Qadri and Dean (66) have measured the concentration of some related triazine dyes that resulted in 50% inhibition (IC_{50}) of the activity of the 6-phosphogluconate dehydrogenase. A parallel study was also carried out to measure the binding of the enzyme to the immobilized triazine dyes (Table 3.4). The dyes that bound the enzyme well in the immobilized state showed IC_{50} values that were comparatively low (less than ~250 μM). The study suggests there is a relationship between the binding of the enzyme to the dye columns and the dyes' capacity to inhibit the enzyme in free solution. A similar observation has been made by Stockton et al. (67) on the binding and inhibition of enzymes by Procion Red HE-3B and Cibacron Blue 3G-A. These authors observed that enzymes that bound to the immobilized dye matrices were also significantly inhibited by the dye in free solution.

3.2.6. Affinity Labelling Studies

Reports in the literature suggest that Cibacron Blue 3G-A may inhibit enzymes covalently by displacing the reactive chlorine atom of the triazine ring (Apps and Gleed, 68; Moe and Piszkiewicz, 49; Witt and Roskoski, 27; Weber et al., 22; Witt and Roskoski, 69). It is therefore worth considering hydrolysis of the dye (Moe and Piszkiewicz, 49) to remove the reactive chlorine atom and use the resulting alcohol to clarify inhibition studies. Irreversible inhibition of enzymes (Witt and Roskoski, 27; Clonis and Lowe, 70) by Cibacron Blue 3G-A has been observed in time-dependent inhibition experiments. Irreversible inhibition of enzymes is characterized by a progressive increase of inhibition with time, ultimately reaching complete inhibition even in the presence of dilute concentrations of the inhibitor (Dixon and Webb, 71). Witt and Roskoski have used such time-dependent inhibition studies to detect irreversible inhibition of the catalytic subunit of protein kinase with Cibacron Blue 3G-A, and they have observed that within 50 min almost all the activity of the enzyme was irreversibly lost. Similar studies (Qadri, 38) using Cibacron Blue 3G-A and Procion Red HE-3B failed to detect irreversible inhibition by the dyes of the enzymes malate dehydrogenase, aspartate aminotransferase, and 6-phosphogluconate dehydrogenase. No enzyme activity was lost after incubation with either dye over a period of 2 hr.

3.2.7. Spectral Shifts

One of the first methods used to study the interaction of Cibacron Blue 3G-A with proteins was difference spectral titration (Thompson, Cass, and Stellwagen, 54; Thomspon and Stellwagen, 33; Ashton and Polya, 72, 73). The visible absorption bands of dyes are well separated from the region where most proteins absorb, and the perturbation of these bands on binding to proteins provides a convenient monitor of the dye–protein interaction. Cibacron Blue 3G-A exhibits a red shift in the presence of several enzymes; in free solution the dye has an absorbance maximum at 610 nm, whereas in the pres-

Table 3.4. Binding of *Bacillus Stearothermophilus* 6-Phosphogluconate Dehydrogenase to Free and Immobilized Triazine Dyes[a]

	Free Solution [Concn. Required to give 50% inhibition (μM)]	Column	Concn. of KCl Coincident with Peak of Enzyme Activity (M)
Procion Red HE-3B	3	Binds	See text
Procion Red H-3B	222	Binds	0.22
Procion Red MX-2B	260	—	—
Procion Scarlet MX-G	453	—	—
Cibacron Blue F3G-A	27	Binds	0.23
Procion Blue MX-3G	168	Binds weakly (leaks)	0.12
Procion Blue MX-R	265	—	—
Procion Brown H-2G	150	Binds	0.21
Procion Yellow MX-R	220	—	—

[a] From Qadri and Dean (66), with permission.

ence of some enzymes a maximum at 660 to 690 nm is observed (Figure 3.6). The effect of substrates, inhibitors, and nucleotides on these different spectra has been used to study dye–protein interaction (Thompson and Stellwagen, 33). These studies were extended to attempt to predict the presence or absence of the nucleotide binding domain known as the dinucleotide fold (Rossmann et al., 76; Rossmann, Moras, and Olsen, 76). It was originally concluded that the difference spectra generated by the protein–Cibacron Blue 3G-A interaction was a consequence of the presence of the dinucleotide fold. This conclusion was based on the observation that enzymes such as lactate dehydrogenase, malate dehydrogenase, and glyceraldehyde 3-phosphate dehydrogenase, which are known to contain dinucleotide binding sites, and phosphoglycerate kinase, which contains a mononucleotide binding site, were able to produce a red shift of the visible absorption spectrum of Cibacron Blue 3G-A (Thompson and Stellwagen, 33).

As more and more protein–dye interactions have been studied, it became clear that the above suggestion was based on too narrow a range of proteins. The dinucleotide fold is constructed by approximately 150 amino acid residues. However, phospholipase A_2 from cobra venom produces a characteristic shift of the dye's spectrum although it is a very small protein comprising only about 120 residues (Barden et al., 53). And many proteins that have been shown not to contain a dinucleotide fold, such as ribonuclease, micrococcal nuclease, chymotrypsin, and serum albumins, produce

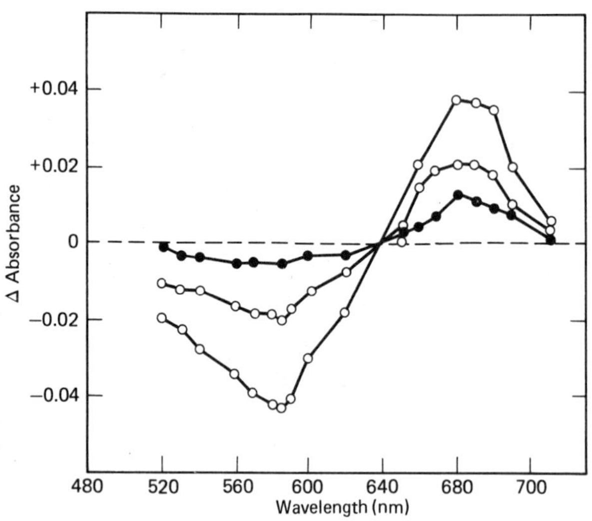

Fig. 3.6. The different concentrations of dye used were 34.4 μM (solid circles), 51.6 μM (unfilled circles), and 77.4 μM (slashed circles) purified Cibacron Blue 3G-A.

similar spectral shifts on interaction with Cibacron Blue 3G-A (Nakagawa, Thompson, and Stellwagen, personal communication). Moreover, addition of sodium chloride or the hydrophobic solvent ethylene glycol to the dye solution, or an increase in the dye concentration, produces spectral shifts, indicating that the red shift is not restricted to the binding of proteins containing the nucleotide binding sites. Stellwagen and his group have re-examined protein interactions with Cibacron Blue 3G-A and suggested that the interaction of cationic groups in proteins with the anionic dye promotes intra- or intermolecular dye stacking, which produces the observed spectral shift. Therefore the phenomenon is not restricted to functional sites created solely by the dinucleotide fold (Stellwagen, 77).

It is now widely accepted that difference spectroscopy is a useful tool for studying dye–protein interactions. The method as applied to these interactions by no means indicates that the dye binds specifically to only one type of structural domain. Grazi et al. (78) showed by difference spectroscopy of aldolase that at least three dye molecules are bound per subunit, which indicates that the NAD binding domain is not necessarily the only structure that binds Cibacron Blue 3G-A. Aromatic dye molecules are thought to bind preferentially to active site regions of globular proteins (Glazer, 74) probably because of a specific arrangement of charged groups, hydrophobic regions, and hydrogen bond donors or acceptors.

3.2.8. Circular Dichroism Studies

The measurement of the induced circular dichroism as a function of wavelength indicates changes in molecular configuration of the dye as a result of binding.

Edwards and Woody (79, 80) have used circular dichroism studies to suggest that although Cibacron Blue 3G-A is not a specific analog of nucleotides and coenzymes, the disposition of the several functional groups of the dye and their conformational freedom allows the dye to bind such that aromatic residues fit into hydrophobic pockets while other parts of the dye can make other favorable interactions. The blue dye was found to be very sensitive to the binding sites of proteins. M_4 and H_4 isozymes of lactate dehydrogenase showed very similar circular dichroism spectra (Edwards and Woody, 23), whereas glucose-6-phosphate, alcohol, and malate dehydrogenases all showed strikingly different circular dichroism spectra. Either NAD or NADH caused the magnitude of the induced CD spectra to decrease. Although the dye Cibacron Blue 3G-A interacts directly with some substrate and cofactor binding sites, circular dichroism studies, like other spectral and inhibition data, offer strong evidence that other nonspecific hydrophobic and ionic interactions also take place.

3.2.9. Nuclear Magnetic Resonance (NMR) Studies

Nuclear magnetic resonance difference spectra (Leatherbarrow and Dean, 81) have been used to analyze the binding of Cibacron Blue 3G-A to human serum albumin. These studies suggest that similar amino acids are involved in the interaction of bilirubin and Cibacron Blue 3G-A with human serum albumin and thus the triazine dye was thought to bind at the bilirubin binding site of human serum albumin. The study indicted that three molecules of Cibacron Blue 3G-A bind to HSA, which is known to have three equivalent binding sites for bilirubin. The NMR data indicated interactions with lysine and possibly arginine groups as well as aromatic amino acids. From other evidence, both lysine and arginine have been thought to be present in the bilirubin binding site of HSA. The NMR method of studying the binding of Cibacron Blue 3G-A with proteins is useful because many of the dyes do not resonate in the aliphatic region. However, the conclusions based on NMR difference spectra are not in agreement with more recent studies using affinity electrophoresis (Metcalf, Crow, and Dean, 44).

3.2.10. Fluorescence Studies

Fluorescence quenching techniques have been used to gain insight into the structure-function relationship of pyruvate kinase (from *N. crassa*) with Cibacron Blue 3G-A. Ligand-induced conformational changes were monitored by the quenching of intrinsic fluorescence of aromatic residues of proteins by Cibacron Blue 3G-A (Kapoor and O'Brien, 63). It was observed that addition of increasing amounts of dye to the enzyme resulted in a progressive decrease in the fluorescence of surface tryptophan residues. The quenching of fluorescence of exposed aromatic groups was subject to reversal on addition of the effector molecule fructose 1,6-disphosphate. On the basis of fluorescence data, difference spectroscopy, free solution inhibition data, and column elution experiments, Kapoor and O'Brien (63) have proposed that Cibacron Blue 3G-A binds to the allosteric binding site of pyruvate kinase and not to the substrate (phosphoenolpyruvate and ADP) binding sites. These workers have also proposed that structural features other than the dinucleotide fold must be recognized by Cibacron Blue 3G-A (cf. Thompson, Cass, and Stellwagen, 54). They suggest several types of binding sites, making up different regions on an enzyme's surface, can accommodate negatively charged phosphate groups in proximity to apolar residues and can recognize Cibacron Blue 3G-A. Among these are the classical dinucleotide fold, the linear and cyclic nucleotide binding sites, and other as yet undescribed structural entities.

3.2.11. X-ray Crystallographic Studies

X-ray crystallographic studies have shown that the meta isomer of Cibacron Blue 3G-A binds to horse liver alcohol dehydrogenase. The anthraquinone ring of the dye structure (Figure 3.7, part A) binds in the hydrophobic pocket of the dinucleotide fold (Biellmann et al., 82). The binding of the dye resembles NAD^+ binding from the adenine site to the adenine diphosphate moiety. But the nicotinamide ribose part of NAD^+ and the terminal sulfoanilinyl group of Cibacron Blue 3G-A (Figure 3.7, part D) bind very differently. It has been suggested (Biellmann et al., 82) that because ADH binds apolar substrates (as opposed to cofactors) it cannot accommodate anionic groups at the substrate site. However, for enzymes that use negatively charged substrates (a property which many enzymes have in common), the binding of the dye may be possible with a combination of substrate and cofactor binding sites.

3.2.12. Analogue Studies

In an effort to understand which particular features of the structure of Cibacron Blue 3G-A are important in the binding (and inhibition) of proteins, some comparisons have been carried out between Cibacron Blue 3G-A, parts of the dye, and other related dye structures. It has been shown that the requirement for particular features of the dye varies from one protein to another (Land and Byfield, 16). In some cases the terminal sulfoanilinyl group (part D in Figure 3.7) does not seem to be required (e.g., in lactate and malate dehydrogenases) whereas others (e.g., glucose 6-phosphate dehy-

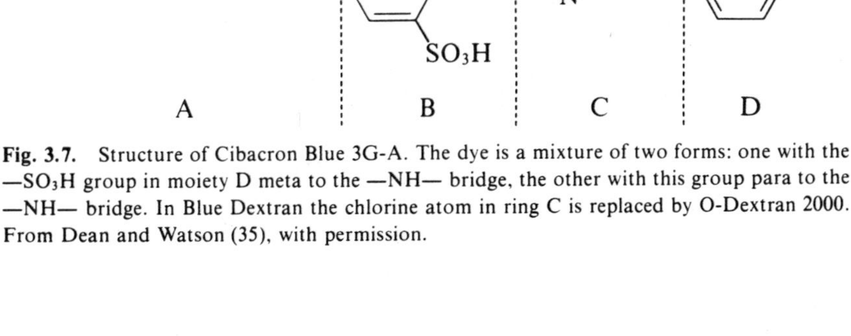

O NH$_2$ SO_3H O NH– –NH– Cl N N N –NH– SO_3H SO_3H

A B C D

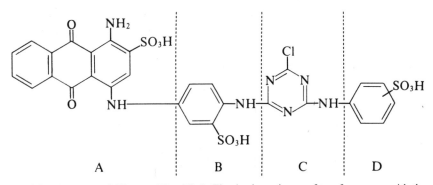

Fig. 3.7. Structure of Cibacron Blue 3G-A. The dye is a mixture of two forms: one with the —SO_3H group in moiety D meta to the —NH— bridge, the other with this group para to the —NH— bridge. In Blue Dextran the chlorine atom in ring C is replaced by O-Dextran 2000. From Dean and Watson (35), with permission.

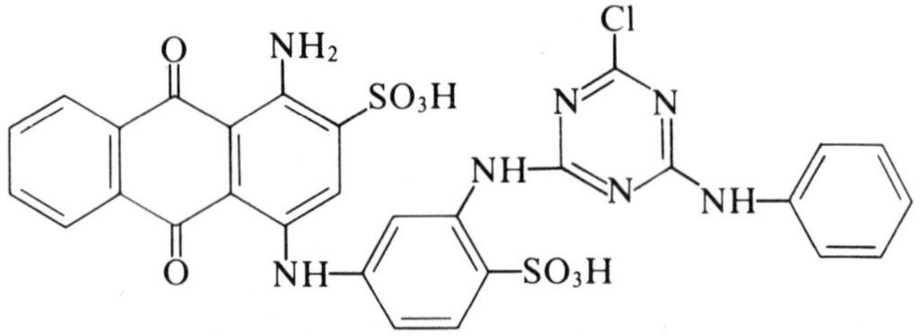

Fig. 3.8. Structure of Cibacron Brilliant Blue BRP.

drogenase) require this moiety for binding. This particular ring also appears to be required for 6-phosphogluconate dehydrogenase from *B. stearothermophilus* to interact with Cibacron Blue 3G-A and with a number of related triazine dye structures (Qadri and Dean, 66). An interchange of the position of attachment of the amino group linking the dye base to the triazine ring with the sulfonic acid residue (ring B of the Cibacron Blue 3G-A structure in Figure 3.7) does not seem to affect all enzymes adversely. Although phosphofructokinase (Böhme et al., 46) interacts with Cibacron Blue 3G-A directly, it does not interact with the structural isomer Cibacron Brilliant Blue BRP (Figure 3.8). A similar effect was observed with 6-phosphogluconate dehydrogenase (Qadri, 38). In addition, human serum albumin not only requires this particular orientation of the amino and sulfonic acid group (part B in Figure 3.7) but also other parts of the Cibacron Blue 3G-A molecule (Leatherbarrow and Dean, 81; Land and Byfield, 16).

Proteins that have been compared for their affinity for different anthraquinone dyes (Beissner and Rudolph, 26) usually bind 1-amino-4-(4′-aminophenyl) anthraquinone-2,3′-disulfonic acid (ASSO) (Figure 3.9), whereas other studies have indicated the anthraquinone ring alone (part A of the CB structure) is the critical feature in the interaction. On the other hand, we

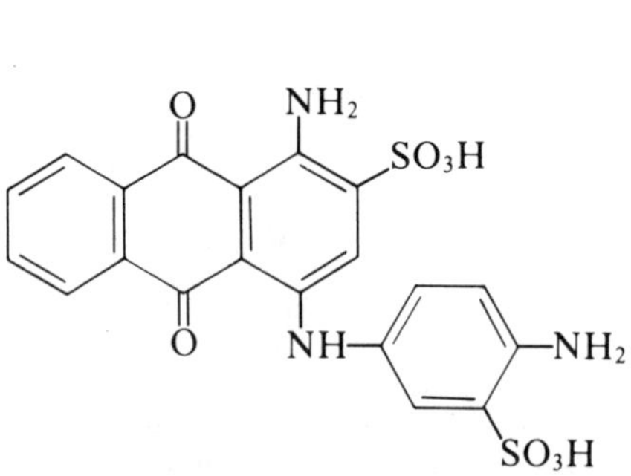

Fig. 3.9. Structure of 1-amino-4-(4′-aminophenyl) anthraquinone-2,3′-disulfonic acid (ASSO).

have observed that the anthraquinone bromamine acid was a very poor inhibitor of 6-phosphogluconate dehydrogenase (Table 3.5).

In a different study it was observed that increasing the hydrophobicity of the ASSO component (Figure 3.9) by substitution of the phenyl ring with alkyl side-chains resulted in a greater inhibition of enzymes compared with Cibacron Blue 3G-A and dyes containing comparatively more polar groups (Bornmann and Hess, 83). Likewise the binding of interferon to immobilized aminoanthracene adsorbents indicates that a hydrophobic contribution is involved in Cibacron Blue binding to some proteins (Jankowski et al., 84). Figure 3.10 shows some of the structural analogs that are available for Cibacron Blue 3G-A.

3.3. CHROMATOGRAPHIC APPLICATIONS OF DYES

3.3.1. Reactivity of Dyes and Spacer Design

Before considering coupling the dyes to a matrix, thought should be given to which functional groups are available on the matrix surface before coupling

Table 3.5. The Binding of 6-Phosphogluconate Dehydrogenase to Free and Immobilized Triazine Dyes[a]

Dye	IC_{50}[b] (μM)	Binding[c] to Column	β Value[d] (mM)	Ligand Concentration (μmol/g wet weight of gel)
Procion Red HE-3B	3	++	230	2.2
Procion Red H-3B	222	++	220	4.9
Procion Red MX-2B	260	−		4.8
Procion Scarlet MX-G	453	−		4.8
Procion Brown H-2G	150	++	210	2.6
Procion Yellow MX-R	220	−		2.3
Procion Blue MX-3G	168	+	120	3.8
Procion Blue MX-R	220	−		3.9
Cibacron Blue 3G-A	27	++	260	2.1
Bromamine acid	1490	NA[e]		NA
Red Component	9938	NA		NA

[a] From Qadri (38), with permission.
[b] IC_{50} = concentration of dye required to give 50% inhibition of enzyme.
[c] ++ denotes strong binding; + denotes weak binding with some leakage; − denotes no binding.
[d] β value is the molarity of KCL coincident with the peak of enzyme activity.
[e] Not applicable. These dyes could not be immobilized to agarose.

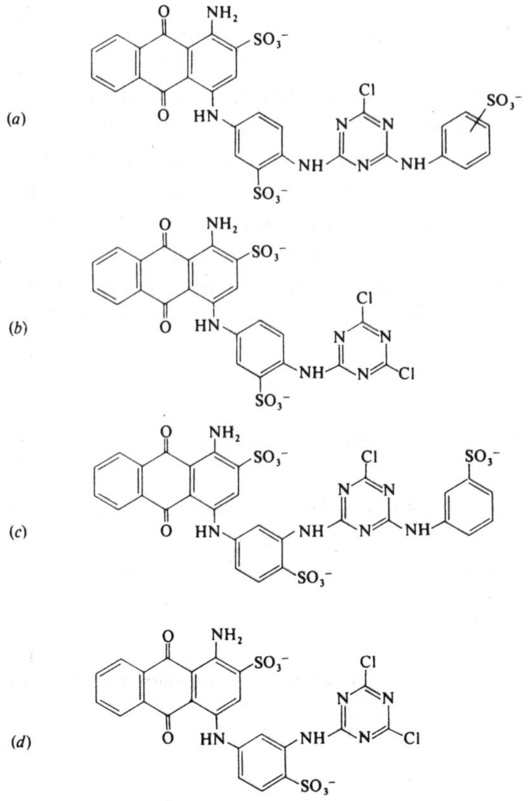

Fig. 3.10. Structures of the triazine dyes used in these experiments: (*a*) Cibacron Blue F3G A, (*b*) Procion Blue MX3G, (*c*) Cibacron Brilliant Blue, (*d*) Procion Blue MXR. From Leatherbarrow and Dean (81), with permission.

(see also Section 3.3.7). Dichlorotriazine dyes (Procion MX dyes) react very readily with the hydroxyl group at relatively low temperatures and rapidly with amines. Hence spacer molecules may be prepared using alcohols as functional groups providing that MX dyes are used. On the other hand, the preparation of immobilized H (monochlorotriazine) dyes to hydroxyl-containing spacers is more difficult to achieve (MX dyes are 50 times more reactive than H dyes) and requires higher temperatures. Spacers are easily designed by using the difference in reactivities of MX and H dyes. Increasing the pH by 1.7 pH units increased the reaction rate of H dyes by about fifty-fold. The reaction rate of an H dye at 70°C is about the same as an MX dye at 20°C. Thus, manipulation of temperature and pH can be used to vary dye binding to suit the particular matrix and desired ligand concentration. (See Sumner, 85; Ingamello, et al., 86; Sumner and Taylor, 87; Shore, 88–90; Lang, Suckling, and Wood, 10; Beech, 19).

3.3.2. Procedures for Coupling Dyes to Matrices

Many procedures have been reported in the literature for coupling dyes (Beech, 19; Baird et al., 91; Heyns and de Moor, 92; Böhme et al., 46; Lowe et al., 18; Dudman and Bishop, 93; Ryan and Vestling, 94). The preferred procedures for MX and H dyes used in our laboratory are outlined in Table 3.6. Some authors use sodium hydroxide instead of sodium carbonate (Beech, 19 and Mislovicova et al., 17). Vigorous mechanical stirring should be avoided to prevent the gels from being damaged by the coupling procedure.

Vessels made of nylon and polyethylene should not be used, as these materials have been reported to take up the dyes.

Table 3.6. Optimized Coupling Procedures for Triazine Dyes[a]

1. *Monochlorotriazine Dyes (e.g., Cibacron Blue 3G-A, Procion Red HE-3B, and Procion Dyes in General)*

Agarose gel (Matrex Gel or Sepharose, 20 g moist wt) is suspended in distilled water (100 mL) and left to stand for 5 min. The water is removed (sintered funnel) and the washing repeated twice. The gel is then suspended in water (70 mL) and a solution of the dye (200 mg) in water (20 mL) is added. The mixture is placed in a screw cap vessel on a rotary mixer (Coulter) for 5 min, after which 20% (w/v) sodium chloride (10 mL) is added. (This "salting in" step is essential in order to take the dye into the matrix; the procedure speeds coupling and results in less hydrolysis of the reactive chlorine groups). The mixing is continued for 30 min at room temperature. A solution of 0.5 mL of 5 *M* NaOH (or 1 *M* Na_2CO_3) is added. NaOH leads to higher ligand concentrations (see Beech, 19); a pH increase of 1.7 increases the coupling rate fiftyfold). Other bases may be used to control coupling pH except those containing nucleophilic groups such as amines (Beech, 19). After addition of the base, the mixture is incubated (Coulter mixer or shaking water bath) for one to three days at 40 to 80°C, depending on the degree of coupling desired. The gel is then filtered and washed on a sintered funnel with distilled water (500 mL), 1*M* NaCl (after standing for 10 min in 1 *M* NaCl), 500 mL, 6 *M* urea (stand in 100 mL for 10 min), and distilled water (500 mL). The gel is stored at 4°C in 0.02 *M* azide (Na).

2. *Dichlorotriazines (e.g., Procion MX series)*

The procedure described above is repeated except that the final incubation is for 1 to 24 hr at room temperature. The washing and storing procedures are otherwise similar.

[a] From Dean and Watson (35), with permission.

3.3.3. Washing and Storing Gels

The storage of dye matrices is rather easier in our experience than other affinity materials such as immobilized nucleotides. They are best stored at 0–4°; freezing has a deleterious effect on the gels as beads are shattered on freezing and thawing. However, the inclusion of buffers such as Tris and phosphate has no adverse effects on dyed gel storage. Inclusion of sodium azide 0.02% (w/v) to 0.2% (w/v) has been advocated. Gels are stable for several years under these conditions and show a remarkable lack of bacterial or fungal invasion. It has been reported in several instances that the dyes are themselves bacteriostatic. No harmful side effects in workers exposed to dyes have been reported. The dyes have also failed to show a positive response in Ames tests. We have attempted limited Ames screening of both Procion Red HE-3B and Cibacron Blue 3G-A and have failed to observe any mutagenicity. Similar reports come from the manufacturers of Procion dyes (Crabtree, H. C., personal communication). On recovery of dye matrices after prolonged storage, the supernatant above the dyed gel very frequently contains appreciable quantities of unbound ligand. While it is important to remove this unbound dye, it appears to affect the properties of the gel only very slightly and should not be taken as a reason for rejecting gel samples that have leached in this way. The removal of unbound dye is clearly an important issue that will be dealt with under the section dealing with regeneration of matrices. Dye columns should not be overexposed to light (Chambers, 95). Used (unwashed) gels very rapidly grow bacterial colonies (P. Brown, unpublished observations) and it is particularly important to regenerate dyed gels before storage. Very few instances have been recorded of dye ligands being removed from gel supports by proteins, but at least one case has been observed by McLennan (personal communication) where nucleic acids (DNA) remove very substantial quantities of dye from a Procion Green H4-G-Sepharose 6B column.

3.3.4. Spectral Properties

3.3.4a. Determination of Dye Concentrations

There is considerable evidence in the literature that the dyes of the Procion and Cibacron type can self-associate in concentrated solutions. NMR studies (Nakagawa and Stellwagen, personal communication; Leatherbarrow and Dean, 81; and others) have shown that the chemical shifts and intensities of individual aromatic protons vary with ligand concentration in a quite dramatic fashion. Similar spectroscopic studies (Hey, Jessup, and Dean, unpublished observations) in the visible region show comparable effects, thus up to concentrations of 40–50 μM (Procion Red HE-3B) it is possible to obtain

Table 3.7. Extinction Coefficients for Some Triazine Dyes

Dye	λ_{max}nm	$M^{-1}cm^{-1}$	Reference
Cibacron Blue 3G-A	610	13,600	Thompson & Stellwagen, 33
	620	13,300	Barden et al., 53; Clonis & Lowe, 149
Procion Red HE-3B	530	30,100	Dean & Watson, 35
	526	30,700	Clonis & Lowe, 149
Procion Red H-3B	530	18,100	Clonis & Lowe, 149
Remazol Blue	590	5,930	Mislovicova et al., 17

linear relationships between observed optical density and ligand concentration, but above these values, departures from Lambert Beer's law are observed. Reisler et al. (96) report no departure from Beer's Law up to a concentration of 7×10^{-7} M for Cibacron Blue 3G-A.

The extinction coefficient for several dyes are given in Table 3.7.

3.3.4b. Determination of Immobilized Dye Concentrations

It will be seen later that the ligand concentration can be all important in determining the nature and degree of protein purification. Ligand concentrations should always be quoted if separations are to be reproduced (intra- and inter-laboratories). Preferably both wet and dry weight values should be given, since many gels have different swelling characteristics. Most workers consider that hydrolysis of the ligand is necessary before determination of the ligand concentration can be made (Chambers, 95; Angal and Dean, 97; Lowe et al., 39). The use of strong acid (HCl) at elevated temperatures has been advocated by several workers (Easterday and Easterday, 98; Beissner and Rudolph, 26; Heyns and de Moor, 92). None of these methods is particularly satisfactory in our view on account of the quite rapid decomposition of many dyes at extremes of pH. An alternative method has been proposed which has not received widespread attention, the use of titration to determine the total number of equivalents of charge introduced into the gel (Neame, personal communication). Several authors advocate drying of the gels to remove all water before determining the weight of gel taken for a ligand concentration determination. Thus, Heyns and de Moor (92) describe (i) washing with methanol to remove water, (ii) washing with ether to remove methanol, and (iii) desiccation *in vacuo* to remove ether. One final method that may be used is direct visible wavelength scanning of melted gels or polyacrylamide-agarose copolymer disc gels (Johnson, Metcalf, and Dean, 41). Typical ligand concentrations for Cibacron Blue gels may be found in Angal and Dean (97) and Mislovicova et al. (17). The latter report concentrations of 65 μmol/g dry wt of cellulose whereas the former report values of 0.2 to 3.2 μmol/mL gel.

3.3.5. Chemical Modification of Dyes

Chemical modification of immobilized ligands can either give insight into the mechanism of a particular dye or can provide useful alternatives to the dye from which they were derived.

Many dye structures are particularly suited to chemical modification. Selective removal of acetoxy residues in some dye structures may be achieved by gentle treatment with base. Alternatively, diazonium linkages may be reductively cleaved using solutions of sodium dithionite (Cuatrecasas, 99).

3.3.6. Chromophoric Impurities of Dyes and Their Effect on Dye–Ligand Chromatography

The definitive work that demonstrates the difference in properties between individual dye structures is undoubtedly that of Biellmann et al. (82). These workers showed that the meta isomer of Cibacron Blue 3G-A (see Figure 3.7) was unable to bind effectively to crystals of alcohol dehydrogenase, whereas the para isomer bound tightly when crystals were soaked in solutions of the dye. Furthermore, they showed that the binding of the para form of the dye closely approximated the structural features of the majority of the NAD molecule in the nucleotide binding site (hence in at least one major dye at least half the dye may not interact with a given protein at all). It has been shown by other workers (Weber et al., 22) that minor impurities in Cibacron Blue preparations are potent inhibitors of certain enzyme activities. Reisler et al. (96) make the point that it is important to check whether or not the different contaminants behave in the same way as the parent dye, hence the need to purify dyes prior to mechanistic studies and presumably prior to immobilization for chromatography. A number of workers claim that some minor components of the dyes bind more effectively than the pure molecules (Bornmann and Hess, 83; Ashton and Polya, 73). On the other hand, instances have been cited (cyclic AMP-dependent protein kinase) where the inhibition by Cibacron Blue is the same whether or not impurities are present (Witt and Roskoski, 27).

3.3.7. Matrix Studies

A very wide variety of different support matrices have been investigated for the purposes of dye ligand chromatography. For a general comparison see

Angal and Dean (97). The importance of the matrix is summed up by Beissner and Rudolph (26), who point out that both the adsorption and elution phases are effected in a very pronounced way by the nature of the matrix. Few detailed comparisons have been made of individual matrices. Some discussion may be found in Burgett and Greenley (99a), Baird et al. (91), and Easterday and Easterday (98). Table 3.8 lists studies in which some different matrices have been compared, and Table 3.9 lists the sources of some of the matrices.

Higher ligand concentrations of Procion Red HE-3B were obtained with the crosslinked agarose Matrex Gel than with Sepharose 6B (Dean et al., 12). Similar results were obtained on comparison of Matrex Gel with the crosslinked agarose, Sepharose CL-6B. It is possible that the particular crosslinking of Matrex Gel allows more dye molecules to come into contact with hydroxyl groups than in the case of Sepharose 6B or Sepharose CL-6B.

3.3.8. Ligand Concentration

Chromatographic techniques for detecting interactions of proteins with immobilized ligands are dependent on another important factor, the ligand concentration of the dye. Very often misleading information appears in the

Table 3.8. Different Support Matrices Examined for Dye Ligand Chromatography

Matrix	References
Sephadex	Böhme et al., 46; Roschlau & Hess, 100; Easterday & Easterday, 98; Baird et al., 91; Khang et al., 101; Jaenicke & Berson, 102; Burgett & Greenley, 99a; Schroeder, 103; Kulbe & Schuer, 28
Agarose	Easterday & Easterday, 98; Heyns & de Moor, 92; Ryan & Vestling, 94; Angal & Dean, 97; Bollin et al., 104; Kulbe & Schuer, 28; Burgett & Greenley, 99a
Sephacryl	Fischer & Witt, 105
Cellulose	Böhme et al., 46; Angal & Dean, 97; Beissner & Rudolph, 26
Beaded cellulose	Mişlovikova et al., 17
Crosslinked agaroses	Bollin et al., 104; Fischer & Witt, 105; Schroeder, 103
Glass (control-pored)	Anderson & Jervis, 106; Jervis, 107
Polyacrylamide	Kopperschläger et al., 108; Meldolesi et al., 109
Polyacrylamide (affinity electrophoresis)	Adinolphi & Hopkinson, 110, 111
Polyacrylamide + agarose (affinity electrophoresis)	Johnson et al., 41; Metcalf, Crowe, & Dean, 44

Table 3.9. Some Support Matrices

Support	Trade Name	Company
Agarose (crosslinked)	Matrex Gel, Dyematrex kit (5 different dyes)	Amicon
Agarose	Affigel Blue	Biorad
Agarose	Sepharose series	Pharmacia
Agarose (crosslinked)	Sepharose Cl series	Pharmacia
Cellulose	—	Whatman
Dextran (crosslinked)	Sephadex series	Pharmacia
Dextran (crosslinked) (Polyacrylamide bridges)	Sephacryl	Pharmacia
Epoxy activated acrylate beads	Eupergit	Rohme & Kaas
Polyacrylamide-agarose copolymer	Ultrogel series	LKB
Beaded cellulose	—	Chemapol (Prague)
Controlled-pore glass	—	Pierce-Corning

literature, associating the failure of some dye columns to retard enzymes with the lack of affinity of the proteins rather than with the low ligand concentration of the matrix. This type of information has often been used to interpret the mechanism of interaction of triazine dye columns. Some workers have apparently overlooked ligand concentration (Pannell et al., 112; Beissner and Rudolph, 26; Jankowski et al., 84; Wilson, 45) when comparing the relative affinity of a protein for different immobilized Cibacron Blue 3G-A columns. These workers concluded that binding was due to the unavailability of one or other functional group of the dye Cibacron Blue 3G-A for interaction with the particular protein. In our view, comparisons should be made over a range of ligand concentrations when investigating the mechanism of binding.

With respect to dye concentrations it has been observed (Qadri, 38) that different triazine dyes give rise to immobilized dye columns of widely varying dye concentrations. Some dyes have better dyeing characteristics than others (Beech, 19) and not all MX dyes give high ligand concentrations. Different support matrices also give widely varying ligand concentrations with the same dye (Dean et al., 12; Qadri and Dean, 66). This is because the affinity of the triazine dyes, and hence the resulting ligand concentration, is related in a complex manner to the chemical structure of the dyes. For these reasons, capacity and affinity of proteins for immobilized dyes should be compared using similar ligand concentration matrices, and the matrices should be prepared accordingly.

3.3.9. Spacer Effects

The relative affinities of proteins with different forms of the same immobilized dye appear to vary with the particular requirements of the protein being studied. Occasionally Blue Dextran-agarose gives better results (Vician and Tishkoff, 113; Thang et al., 114; Bollin et al., 104) than directly linked Cibacron Blue 3G-A-agarose (cf. Apps and Gleed, 68; Leatherbarrow and Dean, 81; Travis et al., 115; Pannell and Newman, 116). This difference has been ascribed to the bulky dextran spacer. However, there are reports of increased capacity when triazine-linked Cibacron Blue 3G-A adsorbents were substituted for Blue Dextran supports (Böhme et al., 46; Apps and Gleed, 68; Travis et al., 115; Reyes and Sandquist, 117). Indeed, Kulbe and Schuer (28) found that Blue Dextran-agarose would not bind phosphoglycerate kinase from yeast but would bind the liver enzyme. These observations may be an effect of the difference in ligand concentration of dye obtained when Blue Dextran is immobilized to supports (Reyes and Sandquist, 117; Qadri, 38).

Cibacron Blue 3G-A linked to supports via the amino group of the anthraquinone ring either directly (Pannell, Johnson, and Travis, 112; Jankowski et al., 84) or by spacer arms (Apps and Gleed, 68; Leatherbarrow and Dean, 81, Beissner and Rudolph, 26) showed a sharp decrease in binding for some proteins compared with the triazine-linked Cibacron Blue 3G-A supports. This suggests that the antraquinone ring of the dye is important for interactions with proteins so that when the latter ring is too close to the support, dye–protein interactions are not allowed. Similar explanations have been made for Blue Dextran supports (Böhme et al., 46). However, comparisons with differently immobilized Cibacron Blue matrices (Qadri, 38) indicate that this effect is also due to dependence of the binding of proteins on the dye ligand concentration. The variations in the binding of proteins to differently immobilized dyes have been attributed to the unavailability of certain groups of the dye for interaction with the protein rather than to differences in dye concentration. We have observed that the binding of enzymes to the triazine-activated supports containing dye molecules attached to spacer arms was weaker than for the corresponding unspaced Procion Red HE-3B and Cibacron Blue supports (Qadri and Dean, unpublished observations). The capacities for total protein were lower although the supports bound more enzyme, and when eluted non-specifically with salt, a greater purification was obtained than that achieved on unspaced triazine dye supports. The effect of the spacer on triazine-activated supports is being studied in our laboratories in more detail in order to understand if the introduction of the spacer actually confers some specificity in dye–protein interaction.

3.3.10. pH Effects

In our laboratory, we studied the variation of the binding of plasma proteins to Cibacron Blue 3G-A as a function of pH and observed that raising the pH of the elution buffer weakened the binding of proteins to Cibacron Blue–Sepharose (Angal and Dean, 97). At pH 5.0, 95% of the plasma proteins bound to the column; at pH 9.0, only 40% were bound (mainly albumin).

The pH values at which the various plasma proteins emerged from the column did not correlate with the isoelectric points of the proteins, suggesting that specific surface sites and not overall net change were involved. In contrast, the binding of 6-phosphogluconate dehydrogenase from *B. stearothermophilus* to Cibacron Blue-Sepharose was found to decrease when the pH was lowered below 6.5 or increased above pH 8.5 (Qadri, 38). It is important to examine the pH dependence of individual proteins over as wide a range as possible in order to optimize separations based on immobilized triazine dyes.

Electrostatic interactions are also important in the binding of enzymes to dye columns, since increasing the ionic strength (effected by increasing the eluting salt concentration) was usually successful in desorbing bound enzymes from dye columns. For some enzymes however, the presence of salt increases the binding, as in the hydrophobic interactions with fibroblast interferon (Jankowski et al., 84), rabbit interferon (Bollin et al., 104), and human heart isocitrate dehydrogenase (Seelig and Colman, 55). In such cases proteins cannot be eluted with salt and eluents such as 50% ethylene glycol are required to desorb proteins from dye columns.

3.3.11. Temperature Effects

A number of workers have studied the effect of temperature on the binding of enzymes to various affinity ligands (Harvey, Lowe, and Dean, 118; Comer et al., 119) and triazine dye adsorbents (Qadri and Dean, 66). The binding of 6-phosphogluconate dehydrogenases from both yeast and *B. stearothermophilus* increased with increasing temperature. The enzyme from *B. stearothermophilus* showed a greater change in its temperature-dependent binding to Procion Red HE-3B Sepharose. For the yeast enzyme, the binding to Cibacron Blue 3G-A column showed a smaller increase with increasing temperature. Since enzymes from both thermophilic and mesophilic sources showed increases in binding with increasing temperature, we have suggested that a hydrophobic parameter is important in the binding of these enzymes to the dye ligands.

3.3.12. Ligand Presaturation Studies

The binding of enzymes and proteins has been reported to be affected by the presence of endogeneous ligands: thus the binding of serum albumins to Cibacron Blue 3G-A is affected by fatty acids (Metcalf, Crow, and Dean, 44). Similarly, horse liver alcohol dehydrogenase from crude extracts (Roy and Nishikawa, 120) did not bind to Cibacron Blue 3G-A columns in the presence of endogeneous ligands. The binding of pig heart malate dehydrogenase to Procion Red HE-3B Sepharose is abolished in the presence of $NADH^+$ or the inhibitor maleic acid (Qadri, 38).

6-Phosphogluconate dehydrogenase from *B. stearothermophilus* is not retarded by Cibacron Blue-Sepharose in the presence of $NADP^+$ (Qadri and Dean, 66). Similar effects for the same enzyme are observed during affinity electrophoresis in Cibacron Blue gels (Metcalf, Crow, and Dean, 44). On removal of $NADP^+$ using activated charcoal (Eby and Kirtley, 121), the enzyme was found to bind to immobilized Cibacron Blue 3G-A both in column experiments and affinity electrophoretic studies (Note that dialysis will not remove all nucleotide).

Although the binding of the enzyme 6-phosphogluconate dehydrogenase to Cibacron Blue 3G-A Sepharose was drastically inhibited in the presence of $NADP^+$, there was very little change in its behavior to Procion Red HE-3B. This difference in the behavior of Cibacron Blue and Procion Red is explained in terms of the absolute value of their affinity constants and the effect of ligands on these parameters. It has been shown by affinity electrophoresis that the K_D for enzyme–Procion Red HE-3B showed only a threefold increase in the presence of $NADP^+$. Ligand presaturation studies suggest that Cibacron Blue 3G-A competes with $NADP^+$ for the nucleotide binding site of the enzyme, since in the presence of $NADP^+$ the binding of the dye to the enzyme is abolished. However, free solution inhibition data (Qadri, 38) showed that the dye was noncompetitive with 6-phosphogluconate dehydrogenase with respect to both substrates $NADP^+$ and 6-phosphogluconate. It is possible in this case that the binding of the enzyme to the free and immobilized dye is different. The charge on the enzyme may be important in the binding of the enzyme to the immobilized dye, so that in the presence of bound $NADP^+$, which alters the overall charge on the protein, the binding is deleteriously affected.

Pompon, Guiard, and Lederer (122) showed recently that apocytochrome b_5 reductase binds much more tightly to Blue Dextran-Sepharose than does the holoenzyme. Other workers have made similar observations; Adinolfi and Hopkinson (110, 111) describe the need to remove presaturating ligand before affinity electrophoresis of isozyme mixtures on Cibacron Blue-polyacrylamide. Redman and Hornemann (64) also mention the need to dialyze a protein prior to chromatography, and they point out that in the enzyme they were studying (pyruvate phosphate dikinase) long columns had the same effect as prior dialysis. Baird et al. (91) found the need to remove pterin from the enzyme carboxypeptidase G prior to chromatography. Barden et al. (53) and Schroeder (103) both discuss the ways in which ligand presaturation can affect chromatography. Schroeder has shown that human serum albumin binds so tightly to anthraquinone-containing dyes similar to Cibacron Blue that not only does the presaturating ligand not affect subsequent chromatography but the adsorbed fatty acids (and bilirubin) may be removed by solvent washes prior to elution of the protein, while the protein is still bound to the gel. Kelleher, Smith, and Pannell (123), on the other hand, claim that charcoal treatment does not improve the binding of rat albumin to immobilized Cibacron Blue, although it is difficult to compare this data with ours (Metcalf, Crow, and Dean, 44) because no ligand concentrations were given by these authors.

3.3.13. Biospecific Elution from Triazine Dye Columns

The ability of nucleotides to elute dehydrogenases and kinases from Blue Dextran columns was originally used as a criterion for the detection of the dinucleotide fold (Thompson, Cass, and Stellwagen, 54). According to Scopes (124), however, biospecific elution from adsorbents can take place when the charge that is responsible for the binding of the enzyme to the adsorbent is neutralized by the eluting ligand leading to the desorption of the enzyme. On the other hand, biospecific elution may also be explained in terms of the ligand's ability to induce a conformational change on the bound enzyme that alters the nature of the protein surface (e.g., charge distribution) and causes elution of the enzyme (Von der Haar, 125).

The criterion of biospecific elution should not be used as an indication of biospecific adsorption, since the two phenomena should be considered separately in adsorption chromatography. In truly reversible chromatography (such as ligand-mediated chromatography; see Myöhänen, Bouriotis, and Dean, 126), elution and adsorption are inextricably associated. In adsorption chromatography this is not necessarily true (Yon, 31). Seelig and Colman (55) have argued that binding should be increased with increasing salt concentrations if the binding is hydrophobic. The majority of dye–protein interactions, however, are weakened by increasing salt concentrations. These data suggest that many mechanisms of binding involve important ionic contributions.

On the other hand, there are instances where it has been shown that biospecific elution failed: Kulbe and Schuer (28) and Stockton et al. (67) note that not all enzymes can be eluted biospecifically. Nevertheless, De la Rosa et al. (127) have shown that FAD and NAD^+ alone will not elute nitrate reductase from immobilized Cibacron Blue. However, the two coenzymes together were successful eluants. Clonis and Lowe (29) have also shown that combinations of nucleotides will elute proteins but single coenzymes are not as successful. Table 3.10 lists some instances of successful use of biospecific elution from triazine dye columns.

Table 3.10. The Use of Biospecific Elution in the Purification of Enzymes from Triazine Columns[a]

Enzyme	Ligand	Reference
Isoleucyl RNA-synthetase	Isoleucine; Mg^{++}; ATP	Moe & Piszkiewicz, 128, 129
3(17)β-Hydroxysteroid dehydrogenase	NADP(H) (but not NAD)	Heyns & de Moor, 92
MDH and LDH	NADP; NAD	Stockton et al., 67
Creatine kinase	Mg^{++}	Fischer & Whitt, 105
MDH, G6PDH, LDH, aldolase, PFK, glycerophosphate DH	ATP: NADH: NADP	Beissner & Rudolph, 25, 26
Interferon	Polynucleotide	de Maeyer-Guignard et al., 131
Carboxypeptidase G	p-Aminobenzoyl glutamate	Baird et al., 91
tRNA synthetases	Substrates	Bruton & Atkinson, 131
tRNA nucleotidyl transferase	tRNA	Deutscher and Masiakowski, 132
LDH, cyclic nucleoside phosphodiesterase, nucleoside diphosphate kinase	(In detergent) NADH + detergent thiocyanate detergent GTP detergent	Robinson et al., 134
IMPDH: SAMP synthetase	IMP vs NAD	Clonis & Lowe, 29, 149
Pyruvate phosphate dikinase	NADH: acetyl CoA CoA	Redman & Hornemann, 64
Alanine dehydrogenase	ATP and guethidine	
PEP Carboxylase		
Pyruvate kinase		
Cyclic AMP protein kinase		
ADH	NAD	Lamkin & King, 135
Cytochrome b_5 reductase	FAD + NAD	Pompon et al., 122
nitrate reductase	FAD + NAD	de la Rosa, 127
13 Examples of biospecific elution		Thompson et al., 54
Orotidylase decarboxylase	UMP, OMP	Reyes & Sandquist, 117

[a] *Key to abbreviations*: MDH = malate dehydrogenase; LDH = lactate dehydrogenase; G6PDH = glucose 6-phosphate dehydrogenase; PFK = phosphofructokinase; IMPDH = inosine monophosphate dehydrogenase; SAMP synthetase = adenylosuccinate synthetase; PEP carboxylase = phosphoenolpyruvate carboxylase; ADH = alcohol dehydrogenase.

3.3.14. Dye Screening

Several authors advocate the preliminary examination of a large number of different dyes before attempting to purify an enzyme using an immobilized triazine dye; see Table 3.11. We have found that this dye screening is extremely useful in (i) determining the general properties of a protein on such supports, and (ii) obtaining additional information on dye columns that work better by "negative" adsorption of other proteins than the more conventional "positive" adsorption columns. We have conducted such dye screenings for more than 20 enzymes with 80 different dye columns over the past five years and have learned from them how best to apply this information for novel applications. The data that may be sought from such screening are capacity, the ability of the column to increase the specific activity or both. Thus in a study of alkaline phosphatase from calf intestine, we (Bouriotis and Dean, 137) have found a number of dyes that will purify the enzyme to a considerable degree but have rather low capacities, among them—Procion Scarlet MX-G. On the other hand, several dyes have very high capacities but purify the enzyme to a lesser degree. In our experience, the separate operation of each column presents few problems and it is unnecessary to apply a gradient of eluate to obtain satisfactory results. In a recent screening with glucose-6-phosphate dehydrogenase (Hey and Dean, unpublished observations), the added purification factor achieved after applying to 1 *M* KCl gradients was only twofold to threefold greater than when the enzyme was eluted with a single pulse of 1 *M* KCl. This screening method also indicates those dyes that are likely to bind the enzyme so tightly that successful recovery would probably be difficult to achieve. In many of these cases the enzyme is still active while attached to the column, which might have applications in enzyme immobilization but this has yet to be demonstrated.

Although screening gives relevant information for those various enzymes with the same catalytic activity that come from similar species, it is by no means certain that the information obtained from an enzyme from one source can be used in selecting the best dye for the purification of the same enzyme from another source. Seelig and Coleman (55) show that pig and human isocitrate dehydrogenase behave very similarly in two separate dye screens. Examples of species differences exist in Kawai and Eguchi (147, 148) and Chambers and Dunlap (50). Clonis and Lowe (149) have shown for two enzymes that red, blue, and green dyes bind much more tightly than yellow or orange dyes. These authors recommend using the latter dyes first.

The information obtained from a dye screen will depend to a large extent upon the choice of buffer molecules, pH of the buffer, the presence or absence of metals such as magnesium, and whether or not the sample has been successfully dialyzed to remove all presaturing ligands. In general, an enzyme's binding can be enhanced (see below) by the appropriate choice of lower buffer pH or higher concentrations of magnesium. Ideally, these should be examined in a brief series of experiments with selected dyes before conducting a full screening operation.

Finally, dye screening has its own problems: (i) As described above, it may not select the best dye for the purification of a partially purified enzyme when the screening was conducted using a crude extract. (ii) The use of crude extracts may lead to extensive column fouling. (iii) Dye screening has to be carried out at a arbitrary ligand concentration.

3.3.15. Tandem Columns

The use of two or more dye columns in series has been advocated by several workers, in particular Lowe et al. (39) and Fulton and Carlson (136). Tandem column arrangements can be set up in several ways. For example, a "negative" column can be placed above a (positive) binding column such that some of the contaminants that would normally bind to the lower column are removed before adsorption to the lower column. Separation of the two columns before elution can be used to increase selectivity. In a further development of this concept Moe and Piszkiewicz (128, 129) showed that by suitable manipulation of the concentration of ATP, magnesium, and isoleucine, successive uses of the same column could produce useful purification factors. In our laboratory, we have used immobilized Cibacron Blue to remove albumin from plasma proteins before using a second dye column to purify β-globulins (McFarthing and Dean, 8). We have also used tandem columns to optimize the use of triazine dyes in the purification of the *B. stearothermophilus* 6-phosphogluconate dehydrogenase (Qadri and Dean, unpublished observations). Several dye column combinations could be used beneficially to remove protein in a tandem application. One further use of tandem columns of this type is particularly relevant to large scale applications. Very often dye columns become contaminated by protein that is not irreversibly bound but is extremely difficult to remove. The use of a protecting tandem column might well provide a disposable means of overcoming this problem.

3.3.16. Negative Adsorption Chromatography

Dao et al. (150) show the successful use of negative adsorption during the purification of glucose 6-phosphate dehydrogenase on immobilized triazine dyes, and Travis et al. (115) describe a similar use for the isolation of IgG. Menter and Burke (151) have used triazine dyes to separate inactive from active isocitrate dehydrogenase of *E. coli.* Thompson, Cass, and Stellwagen

Table 3.11. Dye Screening Programs Using Immobilized Dyes[a]

Enzyme	No. of Dyes Assayed	Flow Rate/Hr	Column Dimension	Running Buffer: Ion	Running Buffer: pH	Running Buffer: Salt or Buffer	Wash	Elution Gradient or Pulse	Reference
IMP DH	10	7.5	0.5 × 5 cm	Pi	7.0	30 m*M*	9 mL	0–1.5 KCl, G*	Lowe et al., 39
SAMP Synthetase	2	7.5	0.45 × 3.2 cm	Tris/Cl + 1.5 m*M* $MgCl_2$ DTT Pi	7.5	50 m*M*	10–12 mL	IMP 0–20 m*M*, G*	Clonis & Lowe, 149
IMPDH	2	7.5	0.45 × 3.2 cm		7.0	50 m*M*	10–12 mL	Nucleotide G*	
LDH MDH	2	NA	1 × 5 cm	Pi	7.5	10 m*M*	"as needed"	1 m*M* NADH	Stockton, 67
Aldehyde reductase	2	10	0.5 × 5 cm	Pi	7.5	10 m*M*	"as needed"	KCl, G*, 0–1 *M*	Turner & Hryszko, 65
Carboxy peptidase	11	NA (15°C)	5 mL	Tris/Ac	6.0	20 m*M*	10 mL	Substrate or 1 *M* NaCl	Baird et al., 91
Human albumin	24	NA	1 mL	Tris/Cl	8.0	50 m*M*	NA "thorough"	0.2 *M* SCN	Schroeder, 103
PYRK LDH	5	NA	2 mL	Tris/Cl	7.5	20 m*M*	10 mL	1.5 *M* KCl	Fulton & Carlson, 136

PGK MDH LDH G6PDH	3	NA	1 mL (7 × 4.0 cm)	Tris EDTA DTT Pi	7.5	10 m*M*	2.5 mL	KCl pulses 0.01, 0.11 *M*	Beissner & Rudolf, 26
tRNA synthetase	32	NA	1 mL	PMSF	6.5	10 m*M*	4 mL	50 m*M* Amino acid or ATP/mg/AA	Bruton & Atkinson, 131
Cyclic nucleotide Phosphatase LDH	25	Inhibition studies		—	—	—	—	—	Ashton & Polya, 72, 73
Dehydrogenases Kinases	10	Inhibition studies		—	—	—	—	—	Bornmann & Hess, 83
ADH MDH PYRK Carnitine acetyl transferase G6PDH 6PGDH HK	53	3–4	0.5 mL	Tricine + 25 m*M* Mg^{++} for kinases	8.0	50 m*M*	5 mL	1.5 *M* KCl or substrates or nucleotides	Qadri, 38

[a] *Key to abbreviations*: MDH = malate dehydrogenase; LDH = lactate dehydrogenase; G6PDH = glucose 6-phosphate dehydrogenase; PFK = phosphofructokinase; IMPDH = inosine monophosphate dehydrogenase; SAMP synthetase = adenylosuccinate synthetase; PEP carboxylase = phosphoenolpyruvate carboxylase; ADH = alcohol dehydrogenase; PYRK = pyruvate kinase; CS = citrate synthetase; 6PGDH = 6-phosphogluconate dehydrogenase; HK = hexokinase; G = gradient (NA = data not available).

(152) describe the separation of apoenzymes from holoenzymes (coenzyme plus enzyme).

The effects of presaturating ligands on enzyme binding can be used to advantage to prescribe the separation of particular proteins by inclusion into the presaturating buffers and enzyme-suitable ligands that will permit a good separation. Redman and Hornemann (64) describe the use of negative dye column steps in the purification of phosphate dikinase. The removal of albumin by negative chromatography has been described by many authors. Koj et al. (153) and Tewksbury et al. (154) have shown in addition that IgG may be effectively removed using immobilized Cibacron Blue by a process of retardation rather than specific adsorption.

Baird et al. (91) claim that dramatic changes in the binding of enzymes to immobilized dyes of very similar structures can be used to design useful purification schemes.

3.3.17. Sample Preparation

Presaturation studies have shown (see Section 3.3.12) that unless presaturating ligands are carefully and reproduceably removed from protein preparations, a protein might not behave as expected on dye ligand chromatography. Therefore, the first and most important consideration in preparing the enzyme sample is to ascertain whether or not presaturating ligands have a deleterious effect on column performance. Exhaustive dialysis is not usually recommended in this situation. Either a rapid exposure to charcoal or ultrafiltration through a hollow fiber system is far more convenient and can handle very large volumes. The latter can also be used to incorporate dialysis into the ultrafiltration step, giving rapid and convenient sample preparation. The pH of the protein extract to be applied to a column is also critical, as is the concentration of the protein. It is not always best to apply protein to the column in the most concentrated form in which the enzyme can be prepared. Some workers advocate prefiltering the extract through a 0.2 μm filter (Ryan and Vestling, 94) to eliminate the possibility of bacterial contamination of the column; others advocate the addition of 0.1% bovine serum albumin to the column wash (Mislovikova et al., 17). The concentration of salts and detergents must also be carefully regulated (see Robinson et al., 134). We have found it convenient to examine the behavior of the enzyme during the dye screening both with and without charcoal pretreatment before application to the column. Column loading is an important factor to consider: If columns are overloaded, a limited purification factor might be obtained. Column loading must be considered in conjunction with the choice of the most suitable ligand concentration (see Fulton, 30).

3.3.18. Large-scale Applications

Several large-scale applications of triazine dye columns have already appeared in the literature. The most obvious, perhaps, is the isolation of albumin from plasma, but other applications have steadily appeared since the work of Kulbe and Schuer (28) who described the preparation of 7 g of protein from 500 mL column. Clonis and Lowe (149) have described a preparation of several milligrams of adenylosuccinate synthetase. Lowe, Small, and Atkinson (18) described the large-scale purifications of β-hydroxybutyrate dehydrogenase and malate dehydrogenase using immobilized triazine dyes, and Bruton et al. (155) described a method for the preparation of 10 g of glycerokinase using a similar system.

3.3.19. Analytical and Laboratory-scale Applications of Triazine Dyes

The list of individual enzymes that have been examined with a variety of triazine dyes is almost too large to be comprehensible. However, Table 3.12 collects together examples of three groups of enzymes and species that have been examined to date. The table shows quite clearly that enzymes are no longer restricted to dehydrogenases and kinases and that almost every class of protein has been investigated with these materials. A number of interesting applications are worthy of comment: The separation of diaphorase isoenzymes using triazine dye columns has been demonstrated by Adinolfi and Hopkinson (110, 111). The same group has shown the separation of these isoenzymes using NAD(H) elution but not by NADPH elution (Edwards, Potter, and Hopkinson, 156, 157). Another interesting application is the ability of triazine dyes in the separation of wild-type from mutant enzymes (for one example see Dean and Watson, 158). Barden et al. (53) have shown that subunits may be prepared from aggregates using immobilized triazines. De Abreu (159) has shown that multienzyme complexes may also be isolated using immobilized dyes. Watson, Harvey, and Dean (37) have shown that the capacity of immobilized Procion Red HE-3B for NADP-dependent dehydrogenases is greater than that of immobilized Cibacron Blue and also that NAD-dependent dehydrogenases bind with greater capacity to immobilized Cibacron Blue than to Procion Red HE-3B. But this generalization is probably not supported by later literature (see Turner and Hryszko, 65; and Qadri and Dean, 66).

3.3.20. Advantages of Dye Columns

Many workers have claimed advantages in chromatographing proteins on immobilized dye columns. The most often cited advantages are (i) the con-

Table 3.12. Enzymes Examined with Triazine Dyes

Enzyme	EC No.	Source	Reference
ADP/CDP reductase	—	Human lymphoblast	Chang & Cheng, 185
Alanine dehydrogenase	1.4.1.1	*Streptomyces verticillatus*	Redman & Hornemann, 64
Alcohol dehydrogenase	1.1.1.1	Horse liver	Roy & Nishikawa, 120; Dean & Watson, 189; Ticha et al., 42; Thompson et al., 152, 186
		Yeast	Watson et al., 37; Easterday & Easterday, 98
		Cotton seeds	Lankin & King, 135
		Rainbow trout	Bauermeister & Sargent, 187
		Human	Adinolfi & Hopkinson,
		Lycopersicon esculentum	Nicolas & Crouzet, 188
Aldehyde dehydrogenase	1.2.1.5	Yeast	Tamaki et al., 190
			Bostian & Betts, 191
		Bovine liver	Hulse & Henderson, 192, 193
Aldehyde reductase	1.1.1.2	*Sacc. cerevisiae*	Tamaki et al., 190
		Pig kidney	Davidson & Flynn, 194, 195
		Rat liver	Stockton et al., 67; Turner & Hryszko, 65
		Ox Brain	Whittle & Turner, 170
		Yeast	
Aldose reductase	1.1.1.21	Pig brain	Boghosian & McGuinness, 196
Cytochrome B5 reductase	1.6.2.2	Rabbit liver	Stellwagen, 143
			Pompon et al., 122
		Human erythrocytes	Yubisui & Takeshita, 197
Dihydrofolate reductase	1.5.1.3	Rat liver; ox brain	Turner et al., 160
		Lactobacillus casei	Dean & Watson, 189
			Watson et al., 37; Chambers & Dunlap, 50
		Chicken liver; *L. casei*	Subramanian & Kaufman, 198
		Carcinoma	Johnson et al., 199
		Yeast	Wu et al., 200
Dihydropteridine reductase	1.6.99.7	Rat liver; brain	Turner et al., 160
			Aksnes & Jones, 201
			Chauvin et al., 452
Flavocytochrome b_2	1.1.2.3	Yeast	Pompon & Lederer, 58
Glucose dehydrogenase	1.1.1.47	*Gluconobacter suboxydans*	Adachi et al., 202

Glucose-6-phosphate dehydrogenase	1.1.1.49	Heterolactic bacteria	Kawai & Eguchi, 203, 204
		Yeast	Wilson, 45; Easterday & Easterday, 98; Land & Byfield, 16; Watson et al., 37; Fulton, 30
		Lactobacillus buchneri	Kawai & Eguchi, 204; Cheng & Domin, 205
		Rat liver cytoplasm	Dao et al., 150
		Leuconostoc mesenteroides	Choi et al., 206
		Bovine liver	Wilson, 45
		B. methylomonas	Steinbach et al., 207
Glutamate dehydrogenase	1.4.1.4	*Aspergillus niger*	Watson et al., 37
		Pig kidney	Khang et al., 208
		Neurospora crassa + enzyme mutants	Watson et al., 37; Dean & Watson, 158
		Bacteroides thetaiotaomicron	Glass & Hylemon, 209
		Beef liver	Dean & Watson, 158
Glutathione reductase	1.6.4.2	Human erythrocyte	Staal et al., 210
		Yeast	Watson et al., 37 Dean & Watson, 158
Glyceraldehyde-3-phosphate dehydrogenase	1.2.1.12	Rabbit muscle	Thompson et al., 54, 152; Fulton, 30; Kalman et al., 211
		Yeast	Easterday & Easterday, 98; Fulton, 30
		Plant	Cerff, 451
Glycerol phosphate dehydrogenase	1.1.1.8	*Escherichia coli*	Edgar & Bell, 212, 213
Homoserine dehydrogenase	1.1.1.3	Cultured maize cells	Walter et al., 214
Hydroxyacyl-CoA dehydrogenase	1.1.1.35		Frevert & Kindl, 215
3-Hydroxy-3-methyl glutaryl-CoA reductase	1.1.1.34	Rat liver	Srikantaiah et al., 220; Tormanen et al., 221; Edwards et al., 222; Ness et al., 224
		Chicken liver	Rogers et al., 456; Beg et al., 225
15-Hydroxyprostaglandin dehydrogenase	1.1.1.141	Human placenta	Westbrook et al., 216; Lin & Jarabak, 217; Mak & Jeffery, 218
		Rabbit kidney	Korff & Jarabak, 219
3(17)β-Hydroxysteroid dehydrogenase	1.1.1.51	Rat erythrocytes	Heyns & de Moor, 92
Inosine monophosphate dehydrogenase	1.2.1.14	*E. coli*	Lowe et al., 18; Gilbert et al., 226; Clonis & Lowe, 149
Isocitrate dehydrogenase	1.1.1.42	*Drosophila melanogaster*	Williamson et al., 227
		Bacillus stearothermophilus	Nagoaka et al., 167
		Human heart; pig heart	Seelig & Colman, 55; Watson et al., 37
		Bovine mammary	Farrell & Warwick, 228
		E. coli	Vasquez & Mongillo, 229

Table 3.12. (*continued*)

Enzyme	EC No.	Source	Reference
			Garnak & Reeves, 230
			Mentner & Burke, 151
2-Ketogluconate reductase	1.1.1.69	*Acetobacter ascendens*	Adachi et al., 232
		Gluconobacter suboxydans	Adachi et al., 231
9-Ketoprostaglandin reductase	1.1.1.141	Rabbit kidney	Korff & Jarabak, 219
		Human placenta	Lin & Jarabak, 217
Lactate dehydrogenase	1.1.1.27	Rat liver	Wroblewski & La Due, 459; Fulton, 30; Mislovicova et al., 17
		Rat hepatoma	Ryan & Vestling, 94; Stellwagen et al., 161; Vestling, 233
		Rabbit muscle	Land & Byfield, 16; Thompson et al., 54; Weininger & Banaszak, 234; Wilson, 45; Easterday & Easterday, 98
		Bacillus thermus aquaticus	Lakatos et al., 235
		Dog; mouse; rat; human, chicken; frog; fish	Nadal-Ginard & Markert, 236
		Fish muscle	Anderson & Jervis, 237; Jervis, 107
		Rat heart	Stockton et al., 67
		Pig heart	Watson et al., 37
		Dogfish muscle	Anderson & Jervis, 237
		Lactobacillus casei	Gordon & Doelle, 238; Thompson et al., 152
		Soybean seedlings	Barthova et al., 239
		Potato plant	Jervis & Schmidt, 240; Rothe et al., 241; Ashton & Polya, 73
			(see also Beissner & Rudolf, 26; Robinson et al., 134)
Lipoamide dehydrogenase	1.6.4.3	*Malbranchea pulchella*	McKay & Stevenson, 242
Malate dehydrogenase	1.1.1.37	Pig heart	Thompson et al., 54; Weininger & Banaszak, 234; Wilson, 45; Land & Byfield, 16
		Yeast	Guerrero & Guitierrez, 243; Watson et al., 37; Hagele et al., 244, 245
		Rat heart	Stockton et al., 67
		Fish muscle	Anderson & Jervis, 237; Jervis, 107
		Rat liver and heart	Kuan et al., 246

		Rabbit muscle	Fulton, 30
		Mouse liver and hepatoma	Felder, 247
		Bacteria	Kay & Down, 248
Malic enzyme (NADP)	1.1.1.40	Maize leaf	Asami et al., 249
5,10-Methylene tetrahydrofolate reductase	1.1.1.68	Bovine liver	Stockton et al., 67
		Ox brain	Turner et al., 160
NADH:Nitrate reductase	1.6.6.1	*Chlorella vulgaris*	Funkhouser et al., 250; Solomonson, 251
		Chlorella fusca	Guerrero et al., 252
		Spinach leaves	Guerrero et al., 252
		Higher plants	Campbell & Smarrelli, 253, 254, 254a
		Hordeum vulgare	Kuo et al., 255
		Wheat leaves	Sherrard & Dalling, 256
NADPH diaphorase	1.6.4.3	Human	Edwards et al., 223, 224
NAD(P)H:Nitrate reductase	1.6.6.2	Spinach leaves	Notton et al., 257
		Rhodotorula glutinus	Guerrero & Guitierrez, 243
		Chlorella variegata	Hipkin et al., 258
		Ankistrodesmus braunii	de la Rosa et al., 127, 259
		Aspergillus niger	Downey & Steiner, 260
NADPH:Nitrate reductase	1.6.6.3	*Neurospora crassa*	Amy et al., 261
		Maize	Campbell, 254
		Nicotiana tabacum	Mendel, 262
NAD(P)H:Nitrate reductase	1.6.6.4	*Neurospora crassa*	Greenbaum et al., 263
2-Oxoaldehyde dehydrogenase	1.2.1.23	Rat liver	Van der Jagt & Davidson, 264
Prolyl hydroxylase	1.14.11.1	Rat dermis	Pannell & Newman, 116
Pyruvate dehydrogenase complex	1.2.2.2	*Azotobacter vinelandii*	de Abreu et al., 265
	(1.2.4.1)	*E. coli*	
3-Phosphoglycerate dehydrogenase	1.1.1.95	Chicken liver	Grant et al., 266
6-Phosphogluconate dehydrogenase	1.1.1.44	Yeast	Thompson et al., 54; Wilson, 45; Watson et al., 37; Fulton, 30
		B. stearothermophilus	Dean et al., 12, 38, 44 (Suppl.) 66
		E. coli	Wolf & Shea, 267
Quinate dehydrogenase	1.1.1.24	*Neurospora crassa*	Barea & Giles, 268
Ribonucleotide reductase	1.17.4.1	*Anabaena*	Gleason & Frick, 269
Thioredoxin reductase	1.6.4.5	Calf thymus	Thelander et al., 270
		Human tumor cells	Cory & Fleischer, 271; Cory et al., 272; Chang & Cheng, 185

Table 3.12. (*continued*)

Enzyme	EC No.	Source	Reference
Semialdehyde dehydrogenase	—	—	Maruyama et al., 273
Shikimate dehydrogenase (3-dehydroquinate dehydratase)	1.1.1.25	*Physcomitrella patens*	Polley, 274
		Neurospora crassa	Barea & Giles, 268
Threonine dehydrogenase	1.1.1.103	*E. coli*	Boylan & Dekker, 275
Adenosine kinase	2.7.1.10	L. 1210 cells	Chang et al., 276
Adenylate kinase	2.7.4.3	Rat muscle & liver	Tamura et al., 277
		Rabbit muscle	Thompson et al., 54
		Human erythrocytes	Cheng & Domin, 205
		Pig heart	Itakura et al., 278
		Bovine heart mitochondria	Tomasselli & Noda, 279, 280
		Rabbit muscle	Fulton, 30
		Thiobacillus novellus	McKellar et al., 281
ATP diphosphate hydrolase		Pig pancreas	LeBel et al., 282
ATP-AMP phosphotransferase	2.7.4.3	Beef heart	Tomasselli & Noda, 279, 280
Carbamyl phosphate synthetase I	2.7.2.5	Frog liver	Mori & Cohen, 168
Carnitine acetyltransferase	2.3.1.7	Pigeon breast muscle	Fulton, 30
Choline acetyltransferase	2.3.1.6	Human brain and placenta	Roskoski et al., 283
CMP kinase	2.7.4.5	Human erythrocytes	Cheng & Domin, 205
			Land & Byfield, 16
Creatine kinase	2.7.3.2	Rabbit muscle	Easterday & Easterday, 98
		Lepomis cyanellus	Fischer & Whitt, 105
		Rabbit muscle	Fulton, 30
Cytidylate kinase		Human erythrocytes	Sears & Beydok, 284
Deoxycytidine kinase	2.7.1.74	*Lactobacillus acidophilus*	Deibel & Ives, 285, 286
		Calf thymus; pig lymph node tumor; pig lymphoma liver, human lymphocytes	Baxter et al., 287
Glucose 1,6-bis phosphate synthetase	—	Bovine liver	Ueda et al., 290
Glucurono kinase	2.7.1.43	*L. longiflorum*	Gillard & Dickinson, 291
Glutamine phosphoribosylpyrophosphate amidotransferase	2.4.2.14	*E. coli*	Messenger & Zalkin, 292
Glutamate synthase	2.6.1.53	Yeast	Masters & Rowe, 293
Glycerokinase	2.7.1.30	*B. stearothermophilus*	Scawen et al., 155

Glycogen phosphorylase kinase	—	Liver	Sakai et al., 288
		Thyroid	Sand et al., 289
GTP-AMP phosphotransferase	2.7.4.10		Tomasselli et al., 294
Hexokinase	2.7.1.1	Yeast	Easterday & Easterday, 98; Fulton, 30; Land & Byfield, 16
		Rat muscle	Farmer & Easterby, 295
		Rat brain	Wilson, 45, 460
Lecithin:cholesterol acyl transferase	2.3.1.43	Human plasma	Chung et al., 296
NAD kinase	2.7.1.23	Pigeon liver	Apps & Gleed, 68
Nucleoside diphosphate kinase	2.7.4.6	Human erythrocytes	Cheng & Domin, 205
Nucleotide kinase	2.7.4.4	Calf thymus	Imizawa & Eckstein, 298
Nucleotide phosphate hydrolase	2.7.1.95	*Asperium nidus*	Grivell & Jackson, 297
Orotate phosphoribosyltransferase	2.4.2.10	Yeast	Reyes & Sandquist, 117, 144
Phosphofructokinase	2.7.1.11	*E. coli*	Thompson et al., 54; Kotlarz & Buc, 299; Babul, 300, 301
		Human platelet	Akkerman et al., 302; Kahn et al., 303, 304
		Saccharomyces carlsbergensis	Tamaki & Hess, 305
		Thermus x-1, *E. coli*	Cass & Stellwagen, 306
		Pig kidney	Khang et al., 208; Mendicino et al., 307
		Rat thyroid	Meldolesi et al., 109
		Lactobacilli	Simon & Hofer, 308;Kawai & Eguchi, 147
		Yeast	Thompson et al., 54; Kopperschläger et al., 47, 48, 108, 309; Diezel et al., 310, Böhme et al., 46
		Human	Cottreau et al., 311
		Leuconostoc dextranum	Kawai & Eguchi, 148
		Rat liver	Kagimoto & Uyeda, 312 (see also Böhme et al., 313)
Phosphoglycerate kinase	2.7.2.3	Yeast	Thompson et al., 54; Kulbe & Schuer, 28 (see Kulbe et al., 314); Stellwagen, 143; Fulton, 30
		Silver beet (*Beta vulgaris*)	Cavell & Scopes, 315
		Horse erythrocytes	Rose & Dube, 316 (see also Beissner & Rudolf, 26, 450)
		Leuconostoc plantarum	Kawai & Eguchi, 147
Phosphoglycerate mutase	2.7.5.3	Rabbit muscle	Thompson et al., 54
		Leuconostoc dextranum	Kawai & Eguchi, 148
Phosphorylase kinase	2.7.1.38		Skuster et al., 317

Table 3.12. (*continued*)

Enzyme	EC No.	Source	Reference
Phosphoserine aminotransferase		Bovine liver	Lund et al., 318
Polynucleotide kinase	2.7.1.78	Bacteriophage T_4	Nichols et al., 319
Protein kinase	2.7.1.37	Pig kidney	Mendicino et al., 307
		Bacteriophage T	Pai et al., 320
		Bovine brain	Witt & Roskoski, 27, 69
		Calf lung	Kobayashi & Fang, 321
		Rabbit muscle (inhibitor)	Demaille et al., 322
		Bovine heart	Armstrong et al., 323
Pyruvate kinase	2.7.1.40	Yeast	Haeckel et al., 324; Roschlau & Hess, 100
		Human erythrocyte	Marie et al., 325; Staal et al., 326; Blume et al., 327; Marie et al., 325
		Rabbit muscle	Easterday & Easterday, 98
		Pig kidney	Zimmerman & Fern, 328
		Human kidney	Harkins et al., 329
		Neurospora crassa	Kapoor & O'Brien, 63
		Human liver	Marie & Kahn, 330
		Mung bean seeds	Babul, 300, 301
		Porcine kidney medulla	Zimmerman & Fern, 328
		Rabbit muscle	Fulton, 30
		Phaseolus aureus	Morelli & Kayne, 331
		Rat liver	Riou et al., 332
Pyruvate phosphate dikinase	2.7.9.1	*Streptomyces verticillatus*	Redman & Hornemann, 64
Rhodopsin kinase		Bovine	Shichi et al., 333; Shichi & Somers, 334
Acyl-acyl carrier protein synthetase		*E. coli*	Rock & Cronan, 335; Rock & Garwin, 336
Acyl-CoA synthetase I	6.2.1.3	*Candida lipolytica*	Hosaka et al., 337
Acid phosphatase	3.1.3.2	Prostate	de Vries et al., 338
Adenylate cyclase	4.6.1.1	Bovine brain	Stellwagen & Baker, 339
		Beef brain	Stellwagen, 143
		Brain	Stellwagen & Baker, 339
			Walseth & Johnson, 340
Adenylosuccinate synthetase	6.3.4.4	*Azotobacter vinlandii*	Markham & Reed, 341
		E. coli	Clonis & Lowe, 149

Albumin		Human serum	Travis et al., 342; Fulton, 30; Travis & Pannell, 343
		Rabbit serum	Burgett & Greenley, 99a
		Human rat; equine; rabbit; ovine; guinea pig; turkey; monkey; bovine	Metcalf et al., 44; Antoni et al., 344 (see Smith et al., 457)
		Human	Angal & Dean, 97, 164; Leatherbarrow & Dean, 81; Schroeder, 103; Metcalf, Crow & Dean, 44; Young & Webb, 345; Land et al., 346, 347, 16; Hanford, 348; Kelleher and Smith, 349; Dean et al., 12
		Bovine	Johnson et al., 41
Aldolase	4.1.2.13	Rabbit muscle	Grazi et al., 78
Aminoacyl tRNA synthetase	6.1.1.20	Yeast	Nikodem et al., 350; Bruton & Atkinson, 131
Angiotensinogen		Rat plasma	Hilgenfeldt et al., 351; Hilgenfeldt & Hackenthal, 352
Alkaline phosphatase	3.1.3.1	Calf intestine	Bouriotis & Dean, 137
Anti-clostridium-toxin antibodies		Horse serum	Worthington & Mulders, 353
Antigen, Barr-Epstein (nuclear)		Human lymphoid	Luka et al., 354
α_1-Antitrypsin		Rabbit plasma	Koj et al., 153; Travis et al., 115
		Human	
Arom multienzyme system		*Neurospora crassa*	Gaertner & Cole, 355
Aryl-sulfate sulfohydrolase (A and B)	3.1.6.1	Rat liver, kidney and brain; human urine; chicken brain	Ahmad et al., 60
		Pseudomonad C-12B	Shaw et al., 356
ATP citrate lyase	4.1.3.8	Rat	Hoffmann et al., 357; Yen & Mach, 358
Bisphosphoglycerate synthase	5.4.2.1	Horse erythrocytes	Rose & Dube, 316
Blood clotting factor X		Human	Vician & Tishoff, 113
Blood clotting factor VIII		Human	Rocha et al., 359
Blood coagulation factors		Human	Swart & Hemker, 360
Calmodulin		Brain	Walseth & Johnson, 340
Carbamyl phosphate synthetase	2.7.2.5	Frog liver	Mori & Cohen, 168, 168a; Mori et al., 361
Carboxypeptidase G	3.4.12.10	*Pseudomonas sp.*	Baird et al., 91
Carnitine acetyl transferase		Pig heart	Qadri, 38; Fulton 30
Chromatin protein	—	—	Kristensen & Holtlund, 362
Citrate lyase, see ATP			
Citrate synthetase	4.1.3.7	Pig heart; chicken heart	Fulton, 30 Wang et al., 458 Stellwagen, 143
Complement factors		Human serum	Gee et al., 363

Table 3.12. (*continued*)

Enzyme	EC No.	Source	Reference
Cyclic nucleotide phosphodiesterase	3.4.1.17	Bovine brain	Morrill et al., 56; Ashton & Polya, 72, 73; Morrill & Thompson, 57
Cytochrome c		Horse heart	Land & Byfield, 16
DNA		Various	Bünemann & Müller, 177
			Koller et al., 179
DNA binding protein		*E. coli*	Meyer et al., 364
DNA polymerase	2.7.7.7	*Artemia salina*	McLennan, 365
		E. coli	Thompson et al., 54
		HeLa cells	Brissac et al., 366
		Drosophila melanogaster	Bank et al., 367
		Phage T4	Lindell et al., 368
		Chicken embryo	Yamaguchi et al., 369
DNA polymerase–regulatory protein		Calf thymus	Steinberg & Grindley, 370
Elastase	3.4.4.7	Human leucocyte	Baugh & Travis, 371
Endonuclease, see Restriction			
Enolase	4.2.1.11	Rat brain	Suzuki et al., 372
Estradiol, see Receptors		Rabbit muscle	Stellwagen et al., 161
Fatty acid CoA ligase	6.2.1.3	Rat liver	Tanaka et al., 373; Philipp & Parsons, 374
Ferritin		Human liver	Kelleher et al., 123, 377
Ferrocytochrome c		Horse heart	Thompson et al., 54
α-Fetoprotein		Human fetal serum	Wu et al., 375; Young & Webb, 345; Gold et al., 376
Follicle stimulating hormone		Human	Bell et al., 142
Follitrophin receptor		Calf testis	Abou–Issa & Reichert, 378
Fructose *bis*-phosphatase	3.1.3.11	Chicken; bumble bee	Leyton et al., 169
		Rabbit muscle	Thompson et al., 54
		Hog kidney	Mendicino et al., 307
		Mouse liver	Tashima et al., 379
			Cruz et al., 380
		Rabbit intestine	Mizunuma et al., 381
Glutamate synthase	2.6.1.53	Soybean	Chiu et al., 453
Glutamine PRPP amidotransferase	2.4.2.14	*E. coli*	Messenger & Zalkin, 292
Glutamine synthetase	6.3.1.2	*Azotobacter vinelandii*	Lepo et al., 382

		Anabaena sp.	Tuli et al., 383
		Yeast	Masters & Rowe, 293
		Pig brain	Jaenicke & Berson, 102
γ-Glutamyl hydrolase	3.4.12.10	Bovine liver	Silink et al., 384
Glycerol-3-phosphate acyltransferase			Larson et al., 454
Glyoxylase I	4.4.1.5	Sheep liver	Uotila & Koivusalo, 385
		Rabbit liver	Elango et al., 386
		Human erythrocytes	Schimandle & Jadt, 387
Hepatitis BE antigen			Howard & Zuckerman, 388
Histidine-rich glycoprotein		Human serum	Morgan, 389
Hydroxybenzoate hydroxylase	1.14.13.2	*P. fluorescens*	Muller et al., 455
Hydroxycinnamoyl transferase			Rhodes et al., 390
Initiation factors 4Aγ4D		Eukaryotic	Van der Mast & Voorma, 391
Interferon		Mammalian	Bollin et al., 104
		Human fibroblast and lymphoid	Erickson and Paucker, 395, 396; Cesario et al., 397; Knight et al., 398
		Human leucocyte	Berg & Heron, 399
		Human immune	Halling et al., 400; Mizrahi et al., 401
		Human fibroblast	Tan et al., 402
		Human fibroblast, human leucocyte	Jankowski et al., 84
		Mouse (Erlich tumor); mouse; rabbit; hamster; horse; human	de Maeyer-Guignard et al., 392; Inglot et al., 393; Kawakita et al., 394; Ashton & Polya, 72, 73
		Human leukocyte and lymphoblast	Erickson & Paucker, 395
		Mouse immune	Wietzerbin et al., 403
β-Lactamase	3.5.2.6	*Streptomyces cellulosae*	Ogawara & Horikawa, 404
Lecithin-cholesterol acyl transferase	2.3.1.43	Human plasma	Chung et al., 296
Lipoproteins		Serum	Wille, 405
α-2-Macroglobulin		Human plasma	Virca et al., 406
Murein transglycosylase	3.2.1	*E. coli*	Kusser & Schwarz, 407
Myosin		Rabbit skeletal muscle	Toste & Cooke, 408
		Bovine thyroid	Kobayashi et al., 409
NAD-glycohydrolase	3.2.2.6	Calf spleen	Schuber & Pascal, 59
Neoantigens		Human cancer tissues	Thompson et al., 410
Nucleotide pyrophosphatase	3.6.1.9	*Phaseolus aureus*	Reddy et al., 411
Octopine synthase	1.5.1a	Crown gall	Hack & Kemp, 412
Orotate phosphoribosyltransferase	2.4.2.10	Yeast	Reyes & Sandquist, 117, 144

Table 3.12. (*continued*)

Enzyme	EC No.	Source	Reference
Orotidylate decarboxylase	4.1.1.23	Yeast	Reyes & Sandquist, 144
		Baker's yeast	Reyes & Sandquist, 117
			(see Levine et al., 413)
Phenol sulfotransferase	2.8.2.1	Rat liver	Sekura & Jakoby, 414
Phosphodiesterase	3.1.4.1	*Kufi* venom	Barmina et al., 415
		Crotalus adamanteus venom	Oka et al., 416
		Bovine pineal	Sankavan & Lovenberg, 417
		Brain	Walseth & Johnson, 340
Phosphoenolpyruvate carboxylase		*Streptomyces verticillatus*	Redman & Hornmann, 64
		E. coli	Scrutton, 418
Phospholipase A_2	3.1.1.4	*Naja naja*	Barden et al., 53
			Darke et al., 419
Plasminogen		Human plasma	Harris & Byfield, 420
			Land & Byfield, 16
Poly-ADP ribose polymerase		Calf thymus	Mandel et al., 421
Polynucleotide phosphorylase	2.7.7.8	*E. coli*	Drocourt et al., 133; Buckingham & Thang, 166
Pregnancy-specific glycoprotein (SP_1)		Human	Angal & Dean, 164
Propionyl-CoA carboxylase	6.4.1.3	Human liver	Kalousek et al., 422
Pyrogen		Human plasma	Critz, 423
Pyruvate dehydrogenase (four-component complex)		*Azotobacter vinlandii*	de Abreu et al., 265
Receptor, estradiol		Mouse uterine	Kumar et al., 146, 428
		Rat mammary tissue	Tenenbaum & Leclercq, 145
			Ratajczak & Hahnel, 429
Receptor, transferrin		Rabbit reticulocyte	Witt & Woodworth, 430
Receptor, vitamin D		Chicken intestine	Pike & Haussler, 431
			McCain et al., 432
Rennin	3.4.4.3	Human plasma	Takii et al., 433
R-enzyme preparation		Sweet corn	Marshall, 424
Restriction endonucleases (Alu I, Bam Hl, Xho I, Hae III, Bgl I and II)		Various	Baksi et al., 425
Restriction endonuclease (Pal I)		*Providencia alcalifaciens*	Baksi & Rushizky, 426; George & Chirikjian, 427

Type II Restriction Endonucleases (Bg I, Xho, Pal, Alu I, Bam HI, Hae III)		Various	Baksi & Rushizky, 426
Rhodopsin kinase		Bovine rods	Shichi et al., 333, 334
Ribonuclease	2.7.7.16	Bovine pancreas	Thompson et al., 54; Land & Byfield, 16
Ribonuclease H	3.1.4.34	Rat liver	Sawai et al., 434, 434a, 434b
Ricin		*Ricinus communis*	Appukuttan & Bachhawat, 435
mRNA guanylyl transferase		HeLa nuclei	Venkatesan et al., 436
tRNA-nucleotidyl transferase		Rabbit liver	Deutscher & Masiakowski, 133
RNA polymerase	2.7.7.6	*Azotobacter vinelandii*	Kumar & Krekow, 61; Thompson, 54
		E. coli	Halling et al., 162, 163
		B. subtilis	
tRNA synthetase	6.1.1.5	*Euglena gracilis*	Imbault, 437; Sarantoglou et al., 438
		E. coli	McDonald et al., 439; Moe & Piszkiewicz, 128; Buckingham & Thang, 166
		B. stearothermophilus	Bruton & Atkinson, 131
Serine hydroxymethyltransferase	2.1.2.1	Monkey liver	Ramash & Rao, 51, 52
Sex hormone binding globulin	—	Human	Iqbal & Johnson, 440
Succinyl CoA transferase	2.8.3.5	Pig heart	White & Jencks, 441
		Sheep kidney	Sharp & Edwards, 442
Terminal deoxynucleotidyl transferase	2.7.7.31	Human lymphoblast	Deibel & Coleman, 443, 444
Thiosulfate sulfur transferase	2.8.1.1	Bovine liver	Horowitz, 445
Thymidine phosphate transferase		*Aslenium nidus*	Grivell & Jackson, 297
Transketolase (does not bind)			Wood & Fletcher, 446
Troponin			Reisler et al., 96
Trypsin	3.4.4.4		Johnson & Travis, 447; Haff & Easterday, 448
Uronosyl C-5 epimerase			Malmstrom et al., 449

siderable saving in expense; (ii) the time saved, (iii) the possibilities of large-scale work; (iv) the group-specific nature of the dyes; (v) the considerable stability of the column to chemical and enzymic degradation (the stability to bacterial degradation is noted by Turner and Hryszko, 65, and by Robinson et al., 134); (vi) the fact that under the right conditions chromatography on the dyes is not interfered with by detergents (Robinson et al., 134). An increasing number of workers use both negative and positive chromatographic steps in enzyme purification with immobilized triazine dyes (Section 3.3.15; Redman and Hornemann, 64). The high capacities of dye columns have been shown by several workers (Lowe et al., 39; Travis et al., 115). Mislovikova et al. (17) have reported the capacity of beaded cellulose dye supports to be in the region of 1 mg/mL, which although not as high as for the corresponding agarose matrices (40–50 mg/mL for human serum albumin—Travis et al., 115), nevertheless could represent a saving in the cost of the matrix. But this point has yet to be proved. The simplicity in preparing immobilized dye columns has been cited as a major advantage. The fact that they are readily reusable has been pointed out by Stockton et al. (67). Their reusability is also cited by Turner, Pearson, and Mason (160) and Lamkin and King (135) (up to 40 times reuse is claimed for a single column). Stellwagen et al. (161) have claimed high yields of protein recovered from triazine dye columns from batch and gradient elution of columns. The dyes are chemically simple to use and avoid the difficulties and hazards of the cyanogen bromide activation that is a feature of many other affinity systems. Many dye columns can be operated at high flow rates, which is a time-saving feature (Reisler et al. 96). Many dye columns can be stored dry (Lamkin and King, 135), another advantage not often seen in affinity systems since most of the agarose gels used have to be stored wet, which encourages bacterial degradation—particularly for ligands like immobilized nucleotides.

With one notable exception (see Section 3.3.1) the triazine dyes are not readily hydrolyzed from agarose supports. This is in distinct contrast to CNBr-immobilized ligands where nucleophilic buffers (e.g., glycine) and high protein concentrations are best avoided. Immobilized triazine dyes may therefore be used in a wide spectrum of biological applications (see Dean and Watson, 35).

3.3.21. Chromatographic Check List

The checklist, Table 3.13, may be useful to those embarking on specific applications of dye ligand chromatography. The 13 items are elaborated here.

1. Ligand concentration. It is important to select the appropriate ligand concentration and also the most suitable matrix for that separa-

Table 3.13. Checklist to Review Before Separating Proteins of Immobilized Dye Columns

1. Ligand concentration; optimum matrix
2. Temperature
3. Pretreatment of column to remove unbound dye and to remove bound protein left behind from previous experiments
4. Preequilibration of column with appropriate buffer, pH, metal ion, salts
5. Sample preparation
6. Flow rate
7. Column geometry
8. Ionic strength of buffer (ionicity)
9. Use of detergents on chaotropic ions
10. Prior dye screening
11. Elution conditions
12. Apoenzyme or holoenzyme
13. Previous purification steps

tion (see Bollin et al., 104). Some idea of whether the ligand concentration is suitable or not may be judged from small-column chromatography of the extract followed by elution with a gradient of salt. Selection can then be made of a dye concentration that allows the enzyme to be eluted at about 0.2 *M* KCl. Low levels of ligand are preferred in certain separations (see Lowe, Small, and Atkinson, 18).

2. Temperature. Optimization of the column temperature is important. Most workers use 0–4° C. It may be recalled that low temperatures discourage hydrophobic interactions.
3. Pretreatment of column. The column may be pretreated with NaOH/urea to remove unbound ligand, particularly if the matrix has not been previously used. The recommended procedures include the following: (i) 2 *M* KCl and 6 *M* guanidine HCl (Reisler et al., 96). (ii) 0.5 *M* NaOH urea (Fulton, 30)—in this case the matrix used was Matrex Gel (some columns of Sepharose will not stand this treatment in our hands). (iii) High salt followed by water (Ryan and Vestling, 94).
4. Pre-equilibration of column. The appropriate buffer and pH must be selected together with metal ions and other salts to optimize the binding and stability of an enzyme during chromatography. Angal and Dean (164) recommend that low pHs should be used during the adsorption phase, increasing higher pHs. This concept is supported by Redman and Hornemann (64). On the other hand, Lamkin and King (135) claim that low pHs were not successful.

5. Sample preparation. The protein concentration of the extract applied to the column must be selected to optimize both separation and purification. Halling et al. (162, 163) recommend pretreatment of the column to remove proteases using immobilized haemoglobin-agarose.
6. Flow rate. It is important to control the column flow rate. Lowe, Small, and Atkinson (18) have shown that the loading phase flow rate can be particularly important in certain instances.
7. Column geometry. Redman and Hornemann (64) recommend the use of long columns and believe that this has the same effect as dialysis prior to application. Some authors advocate the use of batch techniques rather than column application (Iqbal and Johnson, 165).
8. Ionic strength of buffer. Buffers of low ionic strength encourage ionic interactions. Care should be taken to consider ionicity when extrapolating from one buffer system to another. Land and Byfield (16) discuss the ionic nature of the dye–protein interaction. Certain metal ions enhance the binding of some enzymes (Fulton, 30; Drocourt, 133). Electrostatic interactions may be strengthened with divalent cations such as Mg^{2+} and in some cases Mn^{2+}, which enhance the binding of several enzymes to columns of immobilized Cibacron Blue 3G-A (Nagaoka et al., 167; Moe and Piszkiewicz, 129; Mori and Cohen, 168; Fulton, 30) and Procion Red HE-3B (Fulton, 30). Divalent cations probably form electrostatic bridges between the anionic dye and the anionic group in the protein binding sites (Stellwagen, 77), or cations may lower the surface charge on the protein, facilitating access by the dye molecule.
9. Use of detergent or chaotropic ions. Stockton et al. (67) used the nonionic detergent lubrol to chromatograph protein, and Robinson et al. (134) have examined in some detail the relationship between the charge of a detergent and the binding of a protein to immobilized Procion dyes. Following on from the work of Robinson et al., it may be useful to try novel nondenaturing, nonionic detergents (for membrane isolation) of the type described by Hjelmeland (461).

   Chaotropic ions (thiocyanate) have been used principally to desorb albumins while glycerol has been used to elute troponin from immobilized Cibacron Blue (Reisler et al., 96) (see also Seelig and Coleman, 55).
10. Prior dye screening. A previous dye screening has many advantages in selecting the most appropriate dye column for a particular separation. However, when extrapolating from both dye screens and other

people's work on different enzyme sources, complications can arise (see Leyton, 169; Bollin et al., 104).

11. Elution conditions. Many authors claim that KCl is an adequate elution buffer for most systems. In initial experiments KCl is therefore generally recommended (Moe and Piszciewicz, 128; de la Rosa et al., 127; Lowe, Small, and Atkinson, 18). KCl gradients are commonly used (Whittle and Turner, 170) but batch salt elution is also used (Stellwagen et al., 161). Inorganic phosphate has been used to effect elution by Bruton and Atkinson (132), Pompon and Lederer (58) and Land and Byfield (16), while ammonium chloride effectively elutes RNA polymerase (Halling et al., 162).
12. Apoenzyme or holoenzyme. It is important in the case of dissociable ligands to consider whether or not that ligand should be removed prior to chromatography (see Pompon, Guiard, and Lederer, 122). Reyes and Sandquist (117) recommend gel filtration before dye chromatography; other authors recommend charcoal treatment (Metcalf, Crow and Dean, 44).
13. Previous purification steps. Many purification procedures employing immobilized triazine dyes have incorporated protamine sulfate fractionation or ammonium sulfate fractionation before the dye chromatography.

3.3.22. Post-column considerations

Once the enzyme has been removed from the column it is advisable to rapidly regenerate the column and system using combinations of solutions such as urea + NaOH (Moe and Piszkiewicz, 128, 129; Fulton, 30).

3.4. OTHER APPLICATIONS OF DYES

3.4.1. Arguments in Favor of Using Triazine Dyes as Probes

Ashton and Polya have shown (72, 73) that several dyes can be used as probes for proteins containing nucleotide binding sites and that the Cibacron Blue structure is not unique for this purpose. From this and other work it has been suggested that the dyes can be used as mononucleotide as well as dinucleotide site probes. Another observation that seems to be in keeping with the ability of some of the dyes to simulate NAD is shown by the work of Wu (24), who has demonstrated that an NAD-like inhibition of protein synthesis is shown by Cibacron Blue 3G-A. Baksi et al. (138–141) have claimed that Cibacron Blue simulates the phosphodiester backbone of nucleic acids. Witt and Roskoski (69) claim that the ability of Cibacron Blue to bind to the nucleotide binding site is demonstrated by the dye's ability to protect cyclic-AMP-dependent protein kinase. Clonis and Lowe (70) claim that the dyes can discriminate between different nucleotide binding sites.

3.4.2. Arguments Against Using Triazine Dyes as Probes

Beissner et al. (32) show that the *E. coli* arabinose binding protein, known to contain a classical nucleotide binding site, will not bind to immobilized Cibacron Blue. These authors also point out that the cat muscle pyruvate kinase closely resembles triose phosphate isomerase in the folding pattern of the active site, but triose phosphate isomerase does not bind to Cibacron Blue 3G-A whereas the pyruvate kinase does bind. Hence enzymes with very similar structures behave differently in their interactions with the blue dye. It seems clear, at least to us, that the dyes are not specific for any one type of structure, for instance the nucleotide binding site.

Several proteins that do not necessarily contain nucleotide binding regions (Barden et al., 53) bind the dyes. Proteins that bind to Cibacron Blue and which are thought not to include the dinucleotide fold include albumin, follicle stimulating hormone (Bell et al., 142), aldolase (Stellwagen, 143), orotate decarboxylase (Reyes and Sandquist, 144), and troponin T (Reisler et al., 96). Tenenbaum and Leclercq (145) have shown that the rat uterine estrogen receptor binds to Cibacron Blue only in the absence of estrogen; when estradiol is added to the receptor, the complex will no longer bind to the dye. This protein also has no known reason to bind nucleotide or contain a nucleotide-binding site. This is in contrast to observations of Kumar et al. for the mouse protein, which behaves in the opposite fashion (Kumar et al., 146).

3.4.3. Histochemical Stains

Procion Yellow has been used to label central neurones (Purves and McMahon, 171; Kellerth, 172), and Procion Brown, which contains chromium, may be used directly for electron microscopic studies (Christiansen, 173). Levine et al. (174) describe the use of Procion Reds as a histological agent in decalcification, and Goland and Graud (175) have reported improved preservation, fewer misinterpretations, and better imaging of tissues using these dyes. Other ophthalmological applications in the area have been described by Laties and Liebman (176). The chemistry of triazine dye reactions with proteins are discussed by Shore (88–90). Other histochemical applications are reviewed by Lowe, Small, and Atkinson (18).

3.4.4. Affinity Labeling

Monochlorotriazine dyes are poor affinity labels, but a number of workers have shown that dichlorotriazine dyes react rather readily with some proteins (Clonis and Lowe, 70; Witt and Roskoski, 27; Apps and Gleed, 68; Weber et al., 22). This subject is more fully discussed under mechanism of binding of proteins to dye columns (see Section 3.2.6).

3.4.5. Dyes as Probes for Nucleic Acid Structures and the Separation of Nucleic Acids

Bünemann and Müller (177, 178) have developed a polyacrylamide-based gel containing immobilized base-pair-specific dyes. These materials show AT specificity in the case of malachite green and GC specificity in the case of immobilized phenyl neutral red. These authors have shown that the separation of base-specific strands of sheared DNA from bacterial sources is possible, as well as higher molecular weight DNA from calf thymus. These authors have also described the separation of defined fragments of coliphage DNA. The excellent resolving power of this method for nucleic acids is comparable only to DNA fractionation in buoyant-density gradients in the presence of base-pair-specific ligands. The method has the great advantage of being simple to operate and it is chemically stable, allowing repeated re-use of the same column. Other applications of these two gels include the separation of supercoiled DNA from other forms and the isolation of DNA from agarose gels by electrophoretic desorption (Bünemann and Müller, 177, 178; Koller et al., 179.

3.4.6. High Performance Liquid Affinity Chromatography

Developments in high performance liquid chromatography have permitted the application of this powerful technique to protein separation. Recent work has shown that this technique may also be useful in protein separations based on immobilized dyes (Metcalf and Dean, unpublished observations).

3.4.7. Molecular Weight Markers

Blue Dextran has been used for many years as a molecular weight marker for gel filtration chromatography.

The triazine dyes have been suggested as very appropriate protein markers for molecular weight determination using polyacrylamide gel electrophoresis (Lowe, Small, and Atkinson, 18).

3.4.8. Chromogenic Substrates

Triazine dyes have been used for the determination of enzyme activity: by labeling the substrate (in this case dextran) with Cibacron Blue it is possible to quantify the activity of dextranase (and other enzymes) (see Ewen, 21; Klein and Foreman, 20; Mathewson and Pomeranz, 180; Jacobsberg, Kantrowitz, and Lipscomb, 181; Isseroff and Dietz, 182 and Somerville and Quiocho, 183).

3.4.9. Affinity Electrodes

Lowe (184) has described an interesting application of immobilized Cibacron Blue for the determination of albumin in solution. The dye is attached to oxidized titanium wires and when these are placed into solutions containing albumin, it is possible with a reference electrode to demonstrate protein dependent voltages which permit the estimation of albumin concentrations in the range of 1–10 μg per mL.

ACKNOWLEDGMENTS

We are extremely grateful to Dr. E. Metcalf, Y. Hey, K. McFarthing, T. Myöhänen, C. Longstaff, and S. Fulton for their help in preparing the manuscript. We also thank Mr. H. Long for collating the references.

REFERENCES

1. C. R. Lowe and P. D. G. Dean, *Affinity Chromatography*, Wiley, New York (1974).
2. C. R. Lowe and P. D. G. Dean, *FEBS Lett.* **18,** 31 (1971).
3. V. Bouriotis, I. F. Galpin, and P. D. G. Dean, *J. Chromatogr.* **210,** 267 (1981).
4. A. A. Farooqui, *J. Chromatogr.* **184,** 335 (1980).
5. H. Lis and N. Sharon, *Ann. Rev. Biochem.* **42,** 541–4 (1973).
6. J. L. Spivak, D. Small, J. H. Shaper, and M. D. Hollenberg, *Blood* **52,** 1178 (1978).
7. R. G. Sutcliffe, B. M. Kukulska, L. V. B. Nicholson, and W. F. Paterson, (1979) in *Placental Proteins* (A. Klopper and T. Chard, Eds.), Springer-Verlag, Berlin, p 55.

7a. W. Heinzel, I. Rahimi-Laridjani, and H. Grimminger, *J. Immunol. Methods* **9,** 337 (1976).

8. K. McFarthing, S. Angel, and P. D. G. Dean, *Anal. Biochem.* **122,** 186 (1982).
9. J. S. Langone, M. D. P. Boyle, and T. Borsos, *J. Immunol.* **121,** 327 (1978).
10. T. Lang, C. J. Suckling, and H. C. S. Wood, *J. Chem. Soc. Perkin Trans.* **19,** 408 (1977).

10a. C. J. Suckling, J. R. Sweeney, and H. C. S. Wood, *J. Chem. Soc. Chem. Comm*, 173 (1975).
11. S. Biagioni, R. Sisto, A. Ferraro, P. Caiafa, and C. Turano, *Anal. Biochem.* **89,** 616 (1978).
12. P. D. G. Dean, F. Qadri, W. Jessup, V. Bouriotis, S. Angal, H. Potuzak, R. J. Leatherbarrow, T. Miron, E. George, M. R. A. Morgan, in *Affinity Chromatography and Molecular Interactions* Inserm Colloquia, Vol 86, (J. Egly, Ed.) Inserm, Paris (1979), p. 321.
13. T. H. Finlay, V. Troll, M. Levy, A. L. Johnson, and L. T. Hodgkins, *Anal. Biochem.* **87,** 77 (1978).
14. N. L. Smith and H. M. Lenhoff, *Anal. Biochem.* **61,** 392 (1974).
15. L. T. Hodgkins and M. Levy, *J. Chromatogr.* **202,** 381 (1980).
16. M. Land and P. G. H. Byfield, *Int. J. Biol. Macro.* **1,** 223 (1979).
17. D. Mislovicova, P. Gemeiner, L. Kuniak, and J. Zemek, *J. Chromatogr.* **194,** 95 (1980).
18. C. R. Lowe, D. A. P. Small, and A. Atkinson, *Int. J. Biochem.* **13,** 13 (1981).
19. W. F. Beech, in *Fibre Reactive Dyes,* Logos, London, (1970).
20. B. Klein and J. A. Foreman *Clin. Chem.* **26,** 250 (1980).
21. L. M. Ewan, *Clin. Chim. Acta* **47,** 233 (1973).
22. D. Weber, K. Willeford, J. Moe, and D. Piszkiewicz, *Biochem. Biophys. Res. Commun.* **86,** 252 (1979).
23. R. A. Edwards and R. W. Woody, *Biochem. Biophys. Res. Commun.* **79,** 470 (1977).
24. J. M. Wu, *FEBS Lett.* **110,** 297 (1980).
25. R. S. Beissner and F. B. Rudolph, *J. Chromatogr.* **161,** 127 (1978).
26. R. S. Beissner and F. B. Rudolph, *Arch. Biochem. Biophys.* **189,** 76 (1978).
27. J. J. Witt and R. Roskoski, *Biochemistry,* **19,** 143 (1980).
28. K. Kulbe and R. Schuer, *Anal. Biochem.* **93,** 46 (1979).
29. Y. D. Clonis and C. R. Lowe, *Eur. J. Biochem.* **110,** 279 (1980).
30. S. Fulton, in *Dye Ligand Chromatography* Amicon Corp., Lexington, Mass. (1980).
31. R. J. Yon, *Biochem. J.* **161,** 233 (1977).
32. R. S. Beissner, F. A. Quiocho, and F. B. Rudolph, *J. Molec. Biol.* **134,** 847 (1979).
33. S. T. Thompson and E. Stellwagen, *Proc. Nat. Acad. Sci. U. S.* **73,** 361 (1976).
34. L. W. Nichol, A. G. Ogston, D. J. Winzor, and W. H. Sawyer, *Biochem. J.* **143,** 435 (1974).
35. P. D. G. Dean and D. H. Watson, in *Affinity Chromatography* (O. Hofmann-Ostenhoff et al. Eds.), Pergamon Press, Oxford (1978) p. 25.
36. P. D. G. Dean and D. H. Watson *Biochem. Soc. Trans.* **5,** 1099 (1977).
37. D. H. Watson, M. J. Harvey, and P. D. G. Dean, *Biochem. J.* **173,** 591 (1978).
38. F. Qadri, Ph.D. Thesis, University of Liverpool, 1980.
39. C. R. Lowe, M. Hans, N. Spibey, and W. T. Drabble, *Anal. Biochem.* **104,** 23 (1980).
40. K. Takeo and S. Nakamura, *Arch. Biochem. Biophys.* **153,** 1 (1972).
41. S. J. Johnson, E. J. Metcalf, and P. D. G. Dean, *Anal. Biochem.* **109,** 63 (1980).

42. M. Ticha, V. Horejsi, and J. Barthova, *Biochim. Biophys. Acta* **534,** 58 (1978).
43. V. Horejsi, M. Ticha, and J. Kocourek, *Biochim. Biophys. Acta* **499,** 290 (1977).
44. E. J. Metcalf, B. Crow, and P. D. G. Dean, *Biochem. J.* **199,** 465 (1981).
45. J. E. Wilson, *Biochem. Biophys. Res. Commun.* **72,** 816 (1976).
46. H.-J. Böhme, G. Kopperschläger, J. Schulz, and E. Hofmann, *J. Chromatogr.* **69,** 209 (1972).
47. G. Kopperschläger, W. Diezel, R. Freyer, S. Liebe, and E. Hofmann, *Eur. J. Biochem.* **22,** 40 (1971).
48. G. Kopperschläger, H. J. Böhme, W. Diezel, and S. Liebe, *Symp. Chromatogr. Clin. Biochem.* **3** (1971).
49. J. G. Moe and D. Piszkiewicz, *Biochemistry* **13,** 2810 (1979).
50. B. B. Chambers, and R. B. Dunlap, *J. Biol. Chem.* **254,** 6515 (1979).
51. K. S. Ramesh and N. A. Rao, *Biochem. J.* **187,** 249 (1980).
52. K. S. Ramesh and N. A. Rao, *Biochem. J.* **187,** 623 (1980).
53. R. E. Barden, R. A. Deems, E. A. Dennis, and P. L. Darke, *Biochemistry* **19,** 1621 (1980).
54. S. T. Thompson, K. H. Cass, and E. Stellwagen, *Proc. Nat. Acad. Sci. U. S.* **72,** 669 (1975).
55. G. F. Seelig and R. F. Colman, *J. Biol. Chem.* **252,** 3671 (1977).
56. M. E. Morrill, S. T. Thompson, and E. Stellwagen, *J. Biol. Chem.* **254,** 4371 (1979).
57. M. E. Morrill and S. T. Thompson, *Fed. Proc.* **38,** 232 (1979).
58. D. Pompon and F. Lederer, *Eur. J. Biochem.* **90,** 563 (1978).
59. F. Schuber and M. Pascal, *Biochimie* **59,** 735 (1977).
60. A. Ahmad, A. Surolia, B. K. Bachhawat, *Biochim. Biophys. Acta,* **481,** 542 (1977).
61. S. A. Kumar and J. S. Krakow, *J. Biol. Chem.* **252,** 5724 (1977).
62. R. Beissner and F. B. Rudolph, *J. Biol. Chem.* **254,** 6273 (1979).
63. M. Kapoor and M. D. O'Brien, *Can. J. Microbiol.* **26,** 613 (1980).
64. K. L. Redman and U. Hornemann, *J. Antibiot.* **33,** 863 (1980).
65. A. J. Turner and J. Hryszko, *Biochim. Biophys. Acta* **613,** 256 (1980).
66. F. Qadri and P. D. G. Dean, *Biochem. J.* **191,** 53 (1980).
67. J. Stockton, A. G. M. Pearson, L. J. West, and A. J. Turner, *Biochem. Soc. Trans.* **6,** 200 (1978).
68. D. K. Apps and C. D. Gleed, *Biochem. J.* **159,** 441 (1976).
69. J. J. Witt and P. Roskoski, *Biochemistry,* **14,** 4503 (1975).
70. Y. D. Clonis and C. R. Lowe, *Biochem. J.* **191,** 247 (1980).
71. M. Dixon and E. C. Webb, in *Enzymes,* 3rd ed., Longman, London (1979), p. 332.
72. A. R. Ashton and G. M. Polya, *Fed. Proc.* **37,** 1539 (1978).
73. A. R. Ashton, and G. M. Polya, *Biochem. J.* **175,** 501 (1978).
74. A. N. Glazer, *Proc. Nat. Acad. Sci. U. S.* **65,** 1057 (1970).
75. M. G. Rossmann, A. Liljas, C.-I. Branden, and L. J. Banaszak, in *Enzymes,* Vol. 2, Academic, New York (1975), p. 61.
76. M. G. Rossmann, D. Moras, and K. W. Olsen, *Nature* **250,** 194 (1974).

77. E. Stellwagen, in *Inserm Colloquia* **86** (J. M. Egly, ed.) (1979), p. 345.
78. E. Grazi, A. Diiasio, G. Trombetta, and E. Magri, *Arch. Biochem. Biophys.* **190,** 405 (1978).
79. R. A. Edwards and R. W. Woody, *Fed. Proc.* **83,** 839 (1977).
80. R. A. Edwards and R. W. Woody, *Biochemistry* **18,** 5197 (1979).
81. R. J. Leatherbarrow and P. D. G. Dean, *Biochem. J.,* **189,** 27 (1980).
82. J. -F. Biellmann, J.-P. Samama, C.-I. Branden, and H. Eklund, *Eur. J. Biochem.* **102,** 107 (1979).
83. L. Bornmann and B. Hess, *Naturforsch.* **32,** 756 (1977).
84. W. J. Jankowski, W. von Nuenchhausen, E. Sulkowski, and W. A. Carter, *Biochemistry* **15,** 5182 (1976).
85. H. H. Sumner, *J. Soc. Dyers Colourists,* **76,** 672 (1960).
86. W. Ingamello et al. *J. Soc. Dyers Colourists,* **78,** 274 (1962).
87. H. H. Sumner and B. Taylor, *J. Soc. Dyers Colourists,* **83,** 445 (1967).
88. J. Shore, *J. Soc. Dyers Colourists* **84,** 408 (1968).
89. J. Shore, *J. Soc. Dyers Colourists* **84,** 545 (1968).
90. J. Shore, *J. Soc. Dyers Colourists* **85,** 14 (1969).
91. J. K. Baird, R. F. Sherwood, R. J. G. Carr, and A. Atkinson, *FEBS Lett.* **70,** 61 (1976).
92. W. Heyns and P. de Moor, *Biochim. Biophys. Acta* **358,** 1 (1974).
93. W. F. Dudman and C. T. Bishop, *Can. J. Biochem.* **46,** 3079 (1968).
94. L. D. Ryan and C. S. Vestling, *Arch. Biochem. Biophys.* **160,** 279 (1974).
95. G. Chambers, *Anal. Biochem.* **83,** 551 (1977).
96. E. Reisler, J. Liu, M. Mercola, and J. Horwitz, *Biochim. Biophys. Acta* **623,** 243 (1980).
97. S. Angal and P. D. G. Dean, *Biochem. J.* **167,** 301 (1977).
98. R. L. Easterday and I. M. Easterday, *Advan. Exp. Med. Biol.* **42,** 123 (1974).
99. P. Cuatrecasas, *J. Biol. Chem.* **245,** 3059 (1970).
99a. M. W. Burgett and L. V. Greenley, *Amer. Lab.* **9,** 74 (1977).
100. P. Roschlau and B. Hess, *Z. Physiol. Chem.* **353,** 441 (1971).
101. N. Q. Khang, H.-J. Boehme, and E. Hofmann, *Acta Biol. Med. Ger.* **36,** 1019 (1978).
102. L. Jaenicke and W. Berson, *Hoppe Seyler's ZPC,* **358,** 883 (1977).
103. D. D. Schroeder, *Protides Biol. Fluids,* **27,** 809 (1979).
104. E. J. R. Bollin, K. Vastola, D. Oleszek, and E. Sulkowski, *Prep. Biochem.* **8,** 259 (1978).
105. S. E. Fischer and G. S. Whitt, *Anal. Biochem.* **94,** 89 (1979).
106. P. A. Anderson and L. Jervis, *Biochem. Soc. Trans.* **6,** 263 (1978).
107. L. Jervis, in *Chromatography of Synthetic and Biological Polymers* Vol. 2 (R. Epton, ed.), Ellis Horwood, Chichester, p. 231 (1978).
108. G. Kopperschläger, R. Freyer, W. Diezel, and E. Hofmann, *FEBS Lett.* **1,** 137 (1968).
109. M. F. Meldolesi, V. Macchia, and P. Laccetti, *J. Biol. Chem.* **251,** 6244 (1976).
110. A. Adinolfi and D. A. Hopkinson, *Ann. Human Genetics* **41,** 339 (1978).
111. A. Adinolfi and D. A. Hopkinson, *Ann. Human Genetics* **43,** 109 (1979).
112. R. Pannell, D. Johnson, and J. Travis, *Biochemistry* **135,** 5439 (1974).

113. L. Vician and G. H. Tishkoff, *Biochim. Biophys. Acta* **434,** 199 (1976).
114. M.-N., Thang, J. L. Drocourt, M. K. Chalbi-Alix, D.-C. Thang, J. Lubochinsky, A. Ruet, A. Sentenac, J. Gangloff, and G. Dirheimer, *Inserm Colloquia* **86,** (J. M. Egly, Ed.) p. 303 (1979).
115. J. Travis, J. Bowen, D. Tewksbury, D. Johnson, and R. Pannell, *Biochem. J.* **157,** 301 (1976).
116. R. W. Pannell and R. A. Newman, *Fed. Proc.* **37,** 1528 (1978).
117. P. Reyes and R. B. Sandquist, *Anal. Biochem.* **88,** 522 (1978).
118. M. J. Harvey, C. R. Lowe, and P. D. G. Dean, *Eur. J. Biochem.* **41,** 353 (1974).
119. M. J. Comer, D. B. Craven, M. J. Harvey, A. Atkinson, and P. D. G. Dean, *Eur. J. Biochem.* **55,** 201 (1975).
120. S. K. Roy and A. H. Nishikawa, *Biotech. Bioeng.* **21,** 775 (1979).
121. D. Eby and M. E. Kirtley, *Biochemistry* **10,** 2677 (1971).
122. D. Pompon, B. Guiard, and F. Lederer, *Eur. J. Biochem.* **110,** 565 (1980).
123. P. C. Kelleher, C. J. Smith, and R. Pannell, *J. Chromatogr.* **173,** 415 (1979).
124. R. K. Scopes, *Biochem. J.* **161,** 265 (1977).
125. F. Von der Haar, *Eur. J. Biochem.* **34,** 84 (1973).
126. T. Myöhänen, V. Bouriotis, and P. D. G. Dean, **197,** 683 (1981).
127. M. A. de La Rosa, J. Diez, J. M. Vega, and M. Losada, *Eur. J. Biochem.* **106,** 249 (1980).
128. J. G. Moe and D. Piszkiewicz, *Fed. Proc.* **35,** 1467 (1976).
129. J. G. Moe and D. Piszkiewicz, *FEBS Lett.* **72,** 147 (1976).
130. J. de Maeyer-Guignard, M. N. Thang, and E. de Maeyer, *Proc. Nat. Acad. Sci. U. S.* **74,** 3787 (1977).
131. C. J. Bruton and A. Atkinson, *Nucleic Acid Res.* **7,** 1579 (1979).
132. M. P. Deutscher and P. Masiakowski, *Nucleic Acid Res.* **5,** 1947 (1978).
133. J. L. Drocourt, D. C. Thang, and M. N. Thang, *Eur. J. Biochem.* **82,** 355 (1978).
134. J. B. Robinson Jr., D. G. Wick, J. M. Strottmann, and E. Stellwagen, *Fed. Proc.* **39,** Abstr. 1287 (1980).
134a. J. B. Robinson Jr., J. M. Strottmann, D. G. Wick, E. Stellwagen, *Proc. Natl. Acad. Sci.* **77,** 5847 (1980).
135. G. E. Lamkin and E. E. King, *Biochem. Biophys. Res. Commun.* **72,** 560 (1976).
136. S. P. Fulton and E. R. Carlson, *Amer. Lab.* Oct. p. 55.
137. V. Bouriotis and P. D. G. Dean, *J. Chromatogr.* **206,** 521 (1981).
138. K. Baksi and G. W. Rushizky, *Anal. Biochem.* **99,** 207 (1979).
139. K. Baksi and G. W. Rushizky, *Fed. Proc.* **38,** 486 (1979).
140. K. Baksi and G. W. Rushizky, *J. Supramol. Struct.* **8,** 145 (1979).
141. K. Baksi, D. L. Rogerson, and G. W. Rushizky, *Biochemistry* **17,** 4136 (1978).
142. J. Bell, E. Rosenkovich, and D. Rabinowitz, *Proc. Soc. Exp. Biol. Med.* **149,** 565 (1975).
143. E. Stellwagen, *Acc. Chem. Res.* **10,** 92 (1977).
144. P. Reyes and R. B. Sandquist, *Fed. Proc.* **35,** 1752 (1976).

145. A. Tenenbaum and G. Leclercq, *J. Steroid Biochem.* **13,** 829 (1980).
146. S. A. Kumar, T. A. Beach, and H. W. Dickerman, *Proc. Nat. Acad. Sci. U. S.* **76,** 2199 (1979).
147. K. Kawai and V. Eguchi, *J. Ferm. Tech.* **58,** 383 (1980).
148. K. Kawai and V. Eguchi, *J. Biochem.* **88,** 1227 (1980).
149. Y. Clonis and C. R. Lowe, *Biochim. Biophys. Acta.* **659,** 86 (1981).
150. M. L. Dao, J. J. Watson, R. Delaney, and B. C. Johnson, *J. Biol. Chem.* **254,** 9441 (1979).
151. P. Menter and W. Burke, *Fed. Proc.* **38,** 673 (1979).
152. S. T. Thompson, R. Cass, and E. Stellwagen, *Anal. Biochem.* **72,** 293 (1976).
153. A. Koj, M. W. Hatton, K. L. Wong, and L. Regoeczi, *Biochem. J.* **169,** 589 (1978).
154. D. A. Tewksbury, M. R. Premeau, M. L. Dumas, and W. L. Frome, *Circul. Research* **41,** 29 (1977).
155. C. Bruton, P. M. Hammond, and A. Atkinson, *J. Appl. Biochem.* **1,** 289 (1979).
156. Y. H. Edwards, J. Potter, and D. A. Hopkinson, *Biochem. J.* **187,** 429 (1980).
157. Y. H. Edwards, J. Potter, and D. A. Hopkinson, *Ann. Hum. Genet.* **42,** 293 (1979).
158. P. D. G. Dean and D. H. Watson, (1978) in *Affinity Chromatography* (O. Hofmann-Ostenhoff et al., Eds.) Pergamon Press, Oxford, p. 25.
159. R. A. de Abreu, A. de Kok, C. de Graaf-Hess, and C. Veeger, *Eur. J. Biochem.* **81,** 357 (1977).
160. A. J. Turner, A. G. M. Pearson, and R. J. Mason, in *Chemistry and Biology of Pteridines* (G. M. Brown, Ed.), Elsevier, Amsterdam (1979), p. 501.
161. E. Stellwagen, R. H. Cass, S. T. Thompson, and M. Woody, *Nature* **257,** 716 (1975).
162. S. M. Halling, F. J. Sanchez-Anzaldo, R. Fukuda, R. H. Doi, and C. F. Meares, *Biochemistry* **16,** 2880 (1977).
163. S. M. Halling, K. C. Burtis, and R. H. Doi, *J. Biol. Chem.* **252,** 9024 (1977).
164. S. Angal and P. D. G. Dean, *FEBS Lett.* **96,** 346 (1978).
165. M. J. Iqbal and M. W. Johnson, *J. Steroid Biochem.* **8,** 977 (1977).
166. R. U. Buckingham and M. N. Thang, *Nucleic Acid. Res.* **6,** 2919 (1979).
167. T. Nagaoka, A. Hachimori, A. Takeda, and T. Samejima, *J. Biochem.* **81,** 71 (1977).
168. M. Mori and P. Cohen *Fed. Proc.* **37,** 1341 (1978).
168a. M. Mori and P. Cohen, *J. Biol. Chem.* **253,** 8337 (1978).
169. J. F. Leyton, A. M. Chinelatto, H. A. El-Dorry, and M. Bacila, *Arch. Biochem. Biophys.* **202,** 168 (1980).
170. S. R. Whittle and A. J. Turner, *J. Neurochem.* **6,** 1453 (1978).
171. D. Purves and U. McMahon, *J. Cell Biol.* **55,** 205 (1972).
172. J. O. Kellerth, *Brain Res.* **50,** 415 (1973).
173. B. Christiansen, *Science* **182,** 1255 (1973).
174. P. Levin, et al. *Science* **146,** 1676 (1964).
175. P. Goland and N. Graud, *Am. J. Phys. Anthrop.* **29,** 210 (1968).
176. A. Laties and P. Liebman, *Science* **168,** 1475 (1970).

177. H. Bünemann and W. Müller, in *Affinity Chromatography* (O. Hoffmann-Ostenhoff et al., Eds.), Pergamon, Elmsford (1978), p. 353.
178. H. Bünemann and W. Müller, *Nucleic Acids. Res.* **5,** 1059 (1978).
179. B. Koller, H. Delius, H. Bünemann, and W. Müller, *Gene* **4,** 227 (1978).
180. P. R. Mathewson and Y. Pomeranz, *J. Assoc. Offic. Anal. Chem.* **62,** 198 (1979).
181. L. B. Jacobsberg, E. R. Kantrowitz, and W. N. Lipscomb, *J. Biol. Chem.* **250,** 9238 (1975).
182. R. P. Isseroff and G. W. Dietz, Jr., *J. Biochem. Biophys. Methods* **1,** 85 (1979).
183. L. L. Somerville and F. A. Quiocho, *Biochim. Biophys. Acta* **481,** 493 (1977).
184. C. R. Lowe, *FEBS Lett.* **106,** 405 (1979).
185. C.-H. Chang and Y.-C. Cheng, *Cancer Res.* **39,** 436 (1979).
186. S. Thompson, K. Cass, and E. Stellwagen, *Fed. Proc.* **34,** 567 (1975).
187. A. Bauermeister and J. Sargent, *Biochem. Soc. Trans.* **6,** 222 (1978).
188. M. Nicolas and J. Crouzet, *Phytochemistry* **19,** 15 (1980).
189. P. D. G. Dean and D. H. Watson, *J. Chromatogr.* **165,** 301 (1979).
190. N. Tamaki, M. Nakamura, K. Kimuar, and T. Hama, *J. Biochem.* **82,** 73 (1977).
191. R. A. Bostian and G. J. Betts, *Biochem. J.* **173,** 773 (1978).
192. J. Hulse and L. Henderson, *J. Biol. Chem.* **255,** 1146 (1980).
193. J. Hulse and L. Henderson, *Fed. Proc.* **38,** 676 (1979).
194. W. W. Davidson and T. G. Flynn, *J. Biol. Chem.* **254,** 3724 (1979).
195. W. W. Davidson and T. G. Flynn, *Biochem. J.* **177,** 595 (1979).
196. R. A. Boghosian and E. T. McGuinness, *Biochim. Biophys. Acta* **567,** 278 (1979).
197. T. Yubisui and M. Takeshita, *J. Biol. Chem.* **255,** 2454 (1980).
198. S. Subramanian and B. T. Kaufman, *J. Biol. Chem.* **255,** 587 (1980).
199. S. J. Johnson, K. J. Stevenson, and V. S. Gupta, *Can. J. Biochem.* **58,** 1252 (1980).
200. J. Wu, J. R. Florance, and K. Hoogsteen, *Fed. Proc.* **39,** 1771 (1980).
201. A. Aksnes and L. T. Jones, *Arch. Biochem. Biophys.* **202,** 342 (1980).
202. O. Adachi, K. Matsushita, E. Shinagawa, and M. Ameyama, *Agric. Biol. Chem.* **44,** 301 (1980).
203. K. Kawai and Y. Eguchi, *J. Ferment. Technol.* (*Japan*) **54,** 609 (1976).
204. K. Kawai and Y. Eguchi, *J. Ferment. Technol.* (*Japan*) **54,** 128 (1976).
205. Y. C. Cheng and B. Domin, *Anal. Biochem.* **85,** 425 (1978).
206. Y. D. Choi, M. H. Han, and S. M. Byun, *Korean Biochem. J.* **11,** 12 (1978).
207. R. A. Steinbach, H. Sahm, and H. Schuette, *Eur. J. Biochem.* **87,** 409 (1978).
208. N. Q. Khang, H.-J. Bohme, and E. Hoffmann, *Acta Biol. Med. Ger.* **35,** 1425 (1976).
209. T. L. Glass and P. B. Hylemon, *J. Bacteriol.* **141,** 1320 (1980).
210. G. E. Staal, J. Visser, and C. Veeger, *Biochim. Biophys. Acta* **185,** 39 (1969).
211. M. Kalman, M. Nuridsany, and J. Ovadi, *Biochim. Biophys. Acta* **614,** 285 (1980).
212. J. R. Edgar and R. M. Bell, *Fed. Proc.* **36,** 857 (1977).
213. J. R. Edgar and R. M. Bell, *J. Biol. Chem.* **253,** 6348 (1978).

214. T. J. Walter, J. A. Connelly, B. G. Gengenbach, and F. Wold, *J. Biol. Chem.* **254,** 1349 (1979).
215. J. Frevert and H. Kindl, *Eur. J. Biochem.* **107,** 79 (1980).
216. C. Westbrook, Y.-M. Lin, and J. Jarabak, *Biochem. Biophys. Res. Commun.* **76,** 943 (1977).
217. Y.-M. Lin and J. Jarabak, *Biochem. Biophys. Res. Commun.* **81,** 1227 (1978).
218. C. T. Mak and J. Jeffery, *Biochem. Soc. Trans.* **6,** 1165 (1978).
219. J. Korff and J. Jarabak, *Prostaglandins* **20,** 111 (1980).
220. M. V. Srikantaiah, C. D. Tomanen, W. L. Redd, J. E. Hardgrave, and T. J. Scallen, *J. Biol. Chem.* **252,** 6145 (1977).
221. C. D. Tormanen, W. L. Redd, M. V. Srikantaiah, and T. J. Scallen, *Biochem. Biophys. Res. Commun.* **68,** 754 (1976).
222. P. Edwards, D. Lemongello, and A. Fogelman, *Fed. Proc.* **37,** 1524 (1978).
223. P. Edwards, D. Lemongello, and A. Fogelman, *Biochim. Biophys. Acta* **574,** 123 (1979).
224. G. Ness, C. Spindler, and M. Moffler, *Fed. Proc.* **38,** 672 (1979).
225. Z. H. Beg, J. A. Stonik, and H. B. Brewer, *FEBS Lett.* **80,** 123 (1977).
226. M. J. Gilbert, C. R. Lowe, and W. T. Drabble, *Biochem. J.* **183,** 481 (1979).
227. J. H. Williamson, D. Krochko, and M. M. Bentley, *Comp. Biochem. Physiol.* **65,** 339 (1980).
228. H. Farrell and P. Warwick, *Fed. Proc.* **37,** 1780 (1978).
229. B. Vasquez and V. Mongillo, *Fed. Proc.* **38,** 841 (1979).
230. M. Garnak and H. Reeves, *J. Biol. Chem.* **254,** 7915 (1979).
231. O. Adachi, T. Chiyonobu, E. Shinagawa, K. Matsushita, and M. Ameyana, *Agr. Biol. Chem.* (*Tokyo*) **42,** 2057 (1978).
232. O. Adachi, E. Shinagawa, K. Matsushita, and M. Ameyama, *Agr. Biol. Chem.* (*Tokyo*) **43,** 75 (1979).
233. C. S. Vestling, in *Isozymes II* (C. L. Markert, Ed.), Academic Press, London, p. 87 (1975).
234. M. S. Weininger and L. J. Banaszak, *J. Mol. Biol.* **119,** 443 (1978).
235. G. Lakatos, G. Haiarz, and P. Zavodszky, *Biochem. Trans.* **6,** 1195 (1978).
236. B. Nadal-Ginard and C. L. Markert, in *Isozymes II* (C. L. Markert, Ed.), Academic Press, London, p. 45 (1975).
237. P. A. Anderson and L. Jervis, *Biochem. Soc. Trans.* **5,** 728 (1977).
238. G. Gordon and H. Doelle, *Eur. J. Biochem.* **67,** 543 (1976).
239. J. Barthova, M. Pecka, and S. Leblova, *Collect. Czech. Chem. Comm.* **42,** 3705 (1977).
240. I. Jervis and C. N. G. Schmidt, *Biochem. Soc. Trans.* **5,** 1767 (1977).
241. G. M. Rothe, A. Corper, and W. Bacher, *Phytochemistry* **17,** 1701 (1978).
242. D. J. McKay and K. J. Stevenson, *Biochemistry* **18,** 4702 (1979).
243. M. G. Guerrero and M. Guitierrez, *Biochim. Biophys. Acta* **482,** 272 (1977).
244. E. Hägele, J. Neeff, and D. Mecke, *Physiol. Chem.* **158,** 243 (1977).
245. E. Hägele, J. Neeff, and D. Mecke, *Eur. J. Biochem.* **83,** 67 (1978).
246. K. N. Kuan, G. L. Jones, and C. S. Vestling, *Biochemistry* **18,** 4366 (1979).
247. M. R. Felder, *Arch. Biochem. Biophys.* **190,** 647 (1978).
248. W. Kay and J. Down, *Current Microbiology* **1,** 293 (1978).

249. S. Asami, K. Inque, K. Matsomoto, A. Murachi, and T. Akazawa, *Arch. Biochem. Biophys.* **194,** 503 (1979).

250. E. A. Funkhouser, T.-C. Shen, and R. Ackerman, *Plant Physiol.* (Bethesda) **65,** 939 (1980).

251. L. P. Solomonson, *Plant Physiol.* **56,** 853 (1975).

252. M. C. Guerrero, K. Jetschmann, and W. Volker, *Biochim. Biophys. Acta* **482,** 19 (1977).

253. W. H. Campbell and J. Smarrelli Jr., *Plant Physiol.* **61,** 1978 (1978).

254. W. H. Campbell, *Z. Pflanzenphysiol.* **88,** 357 (1978).

254a. J. Smarrelli Jr. and W. H. Campbell, *Plant Sci. Lett.* **16,** 2 (1979).

255. T. Kuo, A. Kleinhofs, and R. L. Warner, *Plant Sci. Lett.* **17,** 371 (1980).

256. J. H. Sherrard and M. J. Dalling, *Plant Physiol.* **63,** 346 (1979).

257. B. A. Notton, R. J. Fido, and E. J. Hewitt, *Plant Sci. Lett.* **8,** 165 (1977).

258. C. R. Hipkin, B. A. Al-Bassam, and P. J. Syrett, *Planta* **144,** 137 (1979).

259. M. A. de la Rosa, J. Diez, and J. M. Vega, *Rev. Espan. Fisiol.* **36,** 177 (1980).

260. R. J. Downey and F. X. Steiner, *J. Bacteriol.* **137,** 105 (1979).

261. N. K. Amy, R. H. Garrett, and B. M. Anderson, *Biochim. Biophys. Acta* **480,** 83 (1977).

262. R. R. Mendel, *Biochem. Physiol. Pflanzen* **175,** 216 (1980).

263. P. Greenbaum, K. N. Prodouz, and R. H. Garrett, *Biochim. Biophys. Acta* **526,** 52 (1978).

264. D. L. Vander-Jagt and L. M. Davison, *Biochim. Biophys. Acta* **484,** 260 (1977).

265. R. A. de Abreu, A. de Kok, and C. Veeger, *FEBS Lett.* **82,** 89 (1977).

266. G. A. Grant, L. M. Keefer, and R. A. Bradshaw, *J. Biol. Chem.* **253,** 2724 (1978).

267. R. E. Wolf Jr. and F. M. Shea, *J. Bacteriol.* **138,** 171 (1979).

268. J. L. Barea and N. H. Giles, *Biochim. Biophys. Acta* **524,** 1 (1978).

269. F. K. Gleason and T. D. Frick, *J. Biol. Chem.* **255,** 7728 (1980).

270. L. Thelander, S. Eriksson, and M. Akerman. *J. Biol. Chem.* **255,** 7426 (1980).

271. J. G. Cory and A. E. Fleischer, *Cancer Res.* **39,** 4600 (1979).

272. J. G. Cory, A. E. Fleischer, and J. B. Munro, *J. Biol. Chem.* **253,** 2898 (1978).

273. K. Maruyama, N. Ariga, M. Tsuda, and K. Deguchi, *J. Biochem.* (Tokyo) **83,** 1125 (1978).

274. L. D. Polley, *Biochim. Biophys. Acta* **526,** 259 (1978).

275. S. A. Boylan and E. E. Dekker, *Biochem. Biophys. Res. Commun.* **85,** 190 (1978).

276. C.-H. Chang, R. W. Brockman, and L. L. Bennett Jr., *J. Biol. Chem.* **255,** 2366 (1980).

277. T. Tamura, H. Shiraki, and H. Nakagawa, *Biochim. Biophys. Acta* **612,** 56 (1980).

278. T. Itakura, K. Watanabe, H. Shiokawa, and S. Kubo, *Eur. J. Biochem.* **82,** 431 (1978).

279. A. G. Tomasselli and L. H. Noda, *Fed. Proc.* **38,** 671 (1979).

280. A. G. Tomasselli and L. H. Noda, *Eur. J. Biochem.* **103,** 481 (1980).

281. I. R. C. McKellar, A. M. Charles, and B. J. Butler, *Arch. Microbiol.* **124,** 275 (1980).

282. D. Le Bel, G. G. Poirer, S. Phaneuf, P. St. Jean, J. F. Laliberté, and A. R. Beaudoin, *J. Biol. Chem.* **255,** 1227 (1980).
283. R. Roskoski, C.-T. Tim, and L. M. Roskoski, *Biochemistry* **14,** 5105 (1975).
284. D. W. Sears and S. Beydok, in *Physical Principles and Techniques of Chemistry*, Part C (S. Leech, Ed.), Academic Press, London (1973), p. 445.
285. M. R. Deibel and D. H. Ives, *J. Biol. Chem.* **252,** 8235 (1977).
286. M. R. Deibel and D. H. Ives, *Fed. Proc.* **36,** 857 (1977).
287. A. Baxter, L. M. Currie, and J. P. Durham, *Biochem. J.* **173,** 1005 (1978).
288. K. Sakai, S. Matsumura, Y. Okimura, H. Yamamura, and Y. Nishizuka, *J. Biol. Chem.* **254,** 6631 (1979).
289. G. Sand, H. Brocas, and C. Erneux, *Acta Endocrinol.* (*Copenhagen*) **82:**S204 (1976).
290. M. Ueda, M. Hirose, R. Sasaki, and H. Chiba, *J. Biochem.* (Tokyo) **83,** 1721 (1978).
291. D. F. Gillard and D. B. Dickinson, *Plant Physiol.* **62,** 706 (1978).
292. L. Messenger and H. Zalkin, *J. Biol. Chem.* **254,** 3382 (1979).
293. D. Masters and B. Rowe, *Fed. Proc.* **38,** 724 (1979).
294. A. G. Tomasselli, R. H. Schirmer, and L. H. Noda, *Eur. J. Biochem.* **93,** 257 (1979).
295. E. Farmer and J. Easterby, 1980, personal communication.
296. J. Chung, D. Abano, G. Fless, and A. Scanu, *J. Biol. Chem.* **254,** 7456 (1979).
297. A. R. Grivell and J. F. Jackson, *Biochem. J.* **155,** 571 (1976).
298. M. Imizawa and F. Eckstein, *Biochim. Biophys. Acta* **570,** 284 (1979).
299. D. Kotlarz and H. Buc, *Biochim. Biophys. Acta* **484,** 35 (1977).
300. J. Babul, *Fed. Proc.* **36,** 723 (1977).
301. J. Babul, *J. Biol. Chem.* **253,** 4350 (1978).
302. J. W. Akkerman, G. Gorter, J. J. Sixma, and G. E. J. Staal, *Biochim. Biophys. Acta* **370,** 102 (1974).
303. A. Kahn, D. Cottreau, and M. C. Meienhofer, *Biochim. Biophys. Acta* **611,** 114 (1980).
304. A. Kahn, M. C. Meienhofer, D. Cottreau, J.-L. Lagrange, and J. C. Dreyfus *Hum. Genet.* **48,** 93 (1979).
305. N. Tamaki and R. Hess, *Hoppe-Seyler's Z. Physiol. Chem.* **356,** 399 (1975).
306. K. H. Cass and E. Stellwagen, *Arch. Biochem. Biophys.* **171,** 682 (1975).
307. J. Mendicino, F. Leibach, and S. Reddy, *Biochemistry* **17,** 4662 (1978).
308. W. A. Simon and H. W. Hofer, *Biochim. Biophys. Acta* **481,** 450 (1977).
309. G. Kopperschläger, Reports of the 5th Annual Meeting of the Biochemical Society, G. D. R., 1968.
310. W. Diezel, H.-J. Böhme, K. Nissler, R. Freyer, W. Hielmann, G. Kopperschläger, and E. Mann, *Eur. J. Biochem.* **38,** 479 (1973).
311. D. Cottreau, M. J. Levin, and A. Kahn, *Biochim. Biophys. Acta* **568,** 183 (1979).
312. T. Kagimoto and K. Uyeda, *J. Biol. Chem.* **254,** 5584 (1979).
313. H. J. Böhme, R. Freyer, P. Retterrath, W. Schellenberger, and E. Hofmann, *Acta Biol. Med. Ger.* **37,** 173 (1978).
314. K. D. Kulbe, M. Bojanovski, H. Foellmer, J. Fuchs, and R. Schuer, "Affinity Chromatography" *Inserm Colloquia* **86** (J. Egly, Ed.), 444 (1979).

315. S. Cavell and R. K. Scopes, *Eur. J. Biochem.* **63,** 483 (1976).
316. Z. B. Rose and S. Dube, *Arch. Biochem. Biophys.* **177,** 284 (1976).
317. J. R. Skuster, K. F. J. Chan, and D. J. Graves, *J. Biol. Chem.* **255,** 2203 (1980).
318. K. Lund, D. Merrill, and R. Guynn, *Fed. Proc.* **38,** 355 (1979).
319. B. P. Nichols, T. D. Lindell, E. Stellwagen, and J. E. Donelson, *Biochim. Biophys. Acta* **526,** 410 (1978).
320. E. Pai, H. Ponta, H. Rahmsdorf, M. Hirsch-Kauffmann, P. Herrlich, and M. Schweiger, *Eur. J. Biochem.* **55,** 299 (1975).
321. R. Kobayashi and V. S. Fang, *Biochem. Biophys. Res. Comm.* **69,** 1080 (1976).
322. J. Demaille, K. Peters, and E. Fischer, *Biochemistry* **16,** 3080 (1977).
323. R. Armstrong, H. Kondo, J. Garnot, E. Kaiser, and A. Mildvan, *Biochemistry* **18,** 1230 (1979).
324. R. Haeckel, B. Hess, W. Lauterborn, and K.-H. Wüster. *Hoppe-Seyler's Z. Physiol. Chem.* **349,** 699 (1968).
325. J. Marie, A. Kahn, and P. Boivin, *Biochim. Biophys. Acta* **481,** 96 (1977).
326. G. E. J. Staal, J. F. Koster, H. Kamp, L. van Milligen-Boersma, and C. Veeger, *Biochim. Biophys. Acta* **227,** 86 (1971).
327. K. G. Blume, R. W. Hoffbauer, D. Busch, H. Arnold, and G. W. Lohr, *Biochim. Biophys. Acta* **227,** 364 (1971).
328. J. Zimmerman and M. Fern, *Fed. Proc.* **38,** 673 (1979).
329. R. N. Harkins, J. A. Black, and M. B. Rittenberg, *Biochemistry* **16,** 3831 (1977).
330. J. Marie and A. Kahn, *Enzyme* **22,** 407 (1977).
331. J. Morelli and F. J. Kayne, *Fed. Proc.* **36,** 718 (1977).
332. J. P. Riou, T. H. Claus, and S. J. Pilkis, *J. Biol. Chem.* **253,** 656 (1978).
333. H. Shichi, R. Somers, and T. Williams, *Fed. Proc.* **38,** 332 (1979).
334. H. Shichi and R. L. Somers, *J. Biol. Chem.* **253,** 7040 (1978).
335. C. O. Rock and J. R. Cronan, *J. Biol. Chem.* **254,** 7116 (1979).
336. C. O. Rock and J. R. Garwin, *J. Biol. Chem.* **254,** 7123 (1979).
337. K. Hosaka, M. Mishina, T. Tanaka, T. Kamiryo, and S. Numa, *Eur. J. Biochem.* **93,** 197 (1979).
338. G. P. De Vries, A. W. Slob, A. C. Jobsis, A. E. F. H. Meijer, and G. T. B. Sanders, *Am. J. Clin. Pathol.* **72,** 944 (1979).
339. E. Stellwagen and B. Baker, *Nature* **261,** 719 (1976).
340. T. F. Walseth and R. A. Johnson, *Fed. Proc.* **39,** Abstract 127 (1980).
341. G. D. Markham and G. H. Reed, *Arch. Biochem. Biophys.* **184,** 24 (1977).
342. J. Travis, D. Garner, and J. Bowen, *Biochemistry* **17,** 5647 (1978).
343. J. Travis and R. Pannell, *Clin. Chim. Acta* **49,** 49 (1973).
344. G. Antoni, R. Botti, M. C. Casagli, and P. Neri, *Boll. Soc. Ital. Biol. Sper.* **54,** 1913 (1979).
345. J. L. Young and B. A. Webb, *Anal. Biochem.* **88,** 619 (1978).
346. P. G. H. Byfield, M. Land, D. G. Williams, and J. M. Rideout, *Biochem. Soc. Trans.* **6,** 666 (1978).
347. P. G. H. Byfield, M. Land, D. G. Williams, and J. M. Rideout, *Clin. Chim. Acta.* **87,** 253 (1978).

348. R. Hanford, W. d'A Maycock, and L. Vallet, (1976) in *Chromatography of Synthetic and Biological Polymers* (R. Epton, Ed.), Vol. 2, Ellis Horwood, Chichester, p. 288.
349. P. C. Kelleher and C. J. Smith, *Biochim. Biophys. Acta,* **317,** 231 (1973).
350. V. M. Nikodem, R. C. Johnson, and J. R. Fresco, *Fed. Proc.* **36,** 822 (1977).
351. U. Hilgenfeldt, N. Blandfort, and E. Hackenthal, in *Enzymatic Release of Vasoactive Peptides* (F. Gross, Ed.), Raven Press, New York, quoted in ref. 352.
352. U. Hilgenfeldt and E. Hackenthal, *Biochim. Biophys. Acta* **579,** 375 (1979).
353. R. W. Worthington and M. S. G. Mulders, *S. Afr. Onderstepoort J. Vet. Res.* **46,** 121 (1979).
354. J. Luka, T. Lindahl, and G. Klein, *J. Virol.* **27,** 604 (1978).
355. F. H. Gaertner and K. W. Cole, *Arch. Biochem. Biophys.* **177,** 566 (1976).
356. D. J. Shaw, K. S. Dodgson, and G. F. White, *Biochem J.* **187,** 181 (1980).
357. G. E. Hoffmann, J. Schiessl, and L. Weiss, *Hoppe-Seyler's Z. Physiol. Chem.* **360,** 1445 (1979).
358. C. S. Yen and D. O. Mack *Nutr. Rep. Intern.* **22,** 245 (1980).
359. E. Rocha, J. M. Solana, J. Fernandez, M. J. Narvaiza, B. Cuesta, and M. Hernandez, *Sangre*, **24,** 567 (1979).
360. A. C. W. Swart and H. C. Hemker, *Biochim. Biophys. Acta.* **222,** 692 (1970).
361. M. Mori, S. M. Morris Jr., and P. P. Cohen, *Proc. Nat. Acad. Sci. U. S.* **76,** 3179 (1979).
362. T. Kristensen and J. Holtlund, *J. Chromatogr.* **192,** 494 (1980).
363. A. P. Gee, T. Borsos, and M. D. P. Boyle, *J. Immunol. Method.* **30,** 119 (1979).
364. R. R. Meyer, J. V. Scott, A. Kornberg, and J. Glassberg, *J. Biol. Chem.* **255,** 2897 (1980).
365. A. G. McLennan, 1980 personal communication.
366. C. Brissac, M. Rucheton, C. Brunel, and P. Jeanteur, *FEBS Lett.* **61,** 38 (1976).
367. G. R. Bank, J. A. Boezi, and I. R. Lehman, *J. Biol. Chem.* **254,** 9886 (1979).
368. T. D. Lindell, B. P. Nichols, J. E. Donelson, and E. Stellwagen, *Biochim. Biophys. Acta.* **562,** 231 (1979).
369. M. Yamaguchi, K. Tanabe, Y. N. Taguchi, M. Nishizawa, T. Takahashi, and A. Matsukage, *J. Biochem.* **255,** 9942 (1980).
370. J. Steinberg and G. Grindey, *Fed. Proc.* **38,** 484 (1979).
371. R. J. Baugh and J. Travis, *Biochemistry* **15,** 836 (1976).
372. F. Suzuki, Y. Umeda, and K. Kato, *J. Biochem.* (Tokyo) **87,** 1587 (1980).
373. T. Tanaka, K. Hosaka, M. Hoshimaru, and S. Numa, *Eur. J. Biochem.* **98,** 165 (1979).
374. D. P. Philipp and P. Parsons, *J. Biol. Chem.* **254,** 10,776 (1979).
375. J. T. Wu, L. H. Wu, and A. C. Madsen, *Biochem. Med.* **23,** 336 (1980).
376. P. Gold, A. Labitan, H. C. G. Wong, S. O. Freedman, J. Krupey, and J. Shuster, *Cancer Res.* **38,** 6 (1978).
377. C. J. Smith and P. C. Kelleher, in *α-Foeto protein* Inserm Colloquia **28** (R. Masseyeff, Ed.), p. 86 (1974).
378. H. Abou-Issa and L. E. Reichert Jr., *Biochim. Biophys. Acta.* **631,** 97 (1980).

379. Y. Tashima, H. Mizunuma, and M. Hasegawa, *J. Biochem.* (Tokyo) **86,** 1089 (1979).
380. Z. M. Cruz, M. M. Tanizaki, H. A. El-Dorry, and M. Bacila, *Arch. Biochem. Biophys.* **198,** 424 (1979).
381. H. Mizunuma, M. Hasegawa, and Y. Tashima, *Arch. Biochem. Biophys.* **201,** 296 (1980).
382. L. E. Lepo, G. Stacey, O. Wyss, and F. R. Tabita, *Biochim. Biophys. Acta* **568,** 428 (1979).
383. R. Tuli, N. Jawali, and J. Thomas, *Indian J. Exp. Biol.* **17,** 1239 (1979).
384. M. Silink, R. Reddel, M. Bethel, and P. B. Rowe, *J. Biol. Chem.* **250,** 5982 (1975).
385. L. Uotila and M. Koivusalo, *Eur. J. Biochem.* **52,** 493 (1975).
386. N. Elango, S. Janaki, and A. R. Rao, *Biochem. Biophys. Res. Commun.* **83,** 1388 (1978).
387. C. M. Schimandle and D. L. V. Jagt, *Arch. Biochem. Biophys.* **195,** 261 (1979).
388. C. G. Howard and A. J. Zuckerman, *J. Med. Virol.* **4,** 303 (1980).
389. W. T. Morgan, *Biochim. Biophys. Acta.* **535,** 319 (1978).
390. M. J. C. Rhodes, L. S. C. Wooltorton, and E. J. Lourençq, *Phytochemistry* **18,** 1125 (1979).
391. C. Van der Mast and H. O. Voorma, *Biochim. Biophys. Acta.* **607,** 512 (1980).
392. J. de Maeyer-Guignard and E. de Maeyer, *C. R. Acad. Sci. Paris, Ser. D.* **283,** 709 (1976).
393. A. D. Inglot, B. Kisielow, and O. Inglot, *Arch. Virol.* **60,** 43 (1979).
394. M. Kawakita, B. Cabrer, H. Taira, M. Rebello, E. Suttery, H. Weideli, and P. Lengyel, *J. Biol. Chem.* **253,** 598 (1978).
395. J. S. Erickson and K. Paucker, *Anal. Biochem.* **98,** 214 (1979).
396. J. S. Erickson, B. J. Dalton, and K. Paucker, *Arch. Virol.* **63,** 253 (1980).
397. T. C. Cesario, P. Schryer, A. Mandel, and J. G. Tilles, *Proc. Soc. Exp. Biol. Med.* **153,** 486 (1976).
398. E. Knight Jr., M. W. Hunkapiller, B. D. Korant, R. W. F. Hardy, and L. E. Hood, *Science* **207,** 525 (1980).
399. K. Berg and I. Heron, *Scand. J. Immunol.* **11,** 489 (1980).
400. S. Halling, F. Sanchez-Anzaldo, R. Fukuda, R. Doi and C. Meares, *Fed. Proc.* **36,** 883 (1977).
401. A. Mizrahi, J. A. O'Malley, W. A. Carter, A. Takatsuki, G. Tamura, and E. Sulkowski, *J. Biol. Chem.* **253,** 7612 (1978).
402. Y. H. Tan, F. Barakat, W. Berthold, H. Smith-Johannsen, and C. Tan, *J. Biol. Chem.* **254,** 8067 (1979).
403. J. Wietzerbin, S. Stefanos, M. Lucero, E. Falcoff, J. A. O'Malley, and E. Sulkowski, *J. Gen. Virol.* **44,** 773 (1979).
404. H. Ogawara and S. Horikawa, (1979): *J. Antibiot.* (Tokyo) **32,** 1328 (1979).
405. L. E. Wille, *Clin. Chem. Acta.* **71,** 355 (1976).
406. G. D. Virca, J. Travis, P. K. Hall, and R. C. Roberts, *Anal. Biochem.* **89,** 274 (1978).
407. W. Kusser and U. Schwarz, *Eur. J. Biochem.* **103,** 277 (1980).

408. A. P. Toste and R. Cooke, *Anal. Biochem.* **95,** 317 (1979).
409. R. Kobayashi, R. Goldman, D. Hartshorne, and J. Field, *J. Biol. Chem.* **252,** 8285 (1977).
410. D. M. P. Thompson, D. N. Tataryn, and R. Schwartz, *Brit. J. Cancer* **41,** 86 (1980).
411. A. R. V. Reddy, V. S. Ananthanarayanan, and N. A. Rao, *Arch. Biochem. Biophys.* **198,** 89 (1979).
412. E. Hack and J. D. Kemp, *Plant Physiol.* **65,** 949 (1980).
413. H. L. Levine, R. S. Brady, and F. H. Westheimer, *Biochemistry* **19,** 4993 (1980).
414. R. D. Sekura and W. B. Jakoby, *J. Biol. Chem.* **254,** 5658 (1979).
415. L. V. Barmina, S. F. Oreshkova, and V. K. Starostina, *Prikl. Biokhim. Mikrobiol.* **15,** 269 (1979).
416. J. Oka, K. Ueda, and O. Hayaishi, *Biochem. Biophys. Res. Commun.* **80,** 841 (1978).
417. K. Sankavan and W. Lovenberg, *Fed. Proc.* **38,** 232 (1979).
418. M. C. Scrutton, *Biochem. Soc. Trans.* **6,** 182 (1978).
419. P. L. Darke, R. E. Barden, R. A. Deems, and E. A. Dennis, *Fed. Proc.* **39,** 1993 (1980).
420. N. D. Harris and P. G. Byfield, *FEBS Lett.* **103,** 162 (1979).
421. P. Mandel, H. Okazaki, and C. Niedergang, *FEBS Lett.* **84,** 331 (1977).
422. F. Kalousek, M. D. Darigo, and L. E. Rosenberg, *J. Biol. Chem.* **225,** 60 (1980).
423. W. J. Critz, 64th Annual Meeting of the Federation of American Societies for Experimental Biology, Anaheim, Calif., USA. *Fed. Proc.* **39,** Abstr. 3805 (1980).
424. J. J. Marshall, *J. Chromatogr.* **53,** 379 (1970).
425. K. Baksi, D. L. Rogerson, and G. W. Rushizky, *Biochemistry 17,* 4136 (1978).
426. K. Baski and G. W. Rushizky, *Fed. Proc.* **37,** 1414 (1978).
427. J. George and J. G. Chirikjian, *Nucleic Acids Res.* **5,** 2223 (1978).
428. S. A. Kumar, T. A. Beach, and H. W. Dickerman, *Proc. Nat. Acad. Sci.* **77,** 3341 (1980).
429. T. Ratajczak and R. Hahnel, *J. Steroid. Biochem.* **13,** 439 (1980).
430. D. P. Witt and R. C. Woodworth, *Biochemistry* **17,** 3913 (1978).
431. J. W. Pike and M. R. Haussler, *Proc. Nat. Acad. Sci.* **76,** 5485 (1979).
432. T. A. McCain, M. R. Haussler, D. Okrent, and M. R. Hughes, *FEBS Lett.* **86,** 65 (1978).
433. Y. Takii, N. Takahashi, T. Inagami, and N. Yokosawa, *Life Sci.* **26,** 347 (1980).
434. Y. Sawai, M. Yanokura, and K. Tsukada, *J. Biochem.* (Tokyo) **86,** 757 (1979).
434a. Y. Sawai, S. U. J. Saito, N. Sagano, and K. Tsukada, *J. Biochem.* (Tokyo) **85,** 1301 (1978).
434b. Y. Sawai, M. Unno, and K. Tsukada, *Biochem. Biophys. Res. Commun.* **84,** 313 (1978).
435. P. S. Appukuttan and B. K. Bachhawat, *Biochim. Biophys. Acta* **580,** 10 (1979).
436. S. Venkatesan, A. Gershowitz, and B. Moss, *J. Biol. Chem.* **255,** 2829 (1980).

437. P. Imbault, V. Sarantoglou, and J. H. Weil, *Biochem. Biophys. Res. Commun.* **88,** 75 (1979).
438. V. Sarantoglou, P. Imbault, and J. H. Weil, *Biochem. Biophys. Res. Commun.* **93,** 134 (1980).
439. T. McDonald, L. Breite, K. L. W. Pangburn, S. Hom, J. Manser, and G. M. Nagel, *Biochemistry* **19,** 1402 (1980).
440. M. J. Iqbal and M. W. Johnson, *J. Steroid. Biochem.* **10,** 535 (1979).
441. H. White and W. P. Jencks, *J. Biol. Chem.* **251,** 1708 (1976).
442. J. A. Sharp and M. R. Edwards, *Biochem. J.* **173,** 759 (1978).
443. M. Deibel and M. Coleman, *Fed. Proc.* **38,** 487 (1979).
444. M. Deibel and M. Coleman, *J. Biol. Chem.* **254,** 8634 (1979).
445. P. M. Horowitz, *Anal. Biochem.* **86,** 751 (1978).
446. T. Wood and S. Fletcher, *Biochim. Biophys. Acta.* **527,** 249 (1978).
447. D. A. Johnson and J. Travis, *Anal. Biochem.* **72,** 573 (1976).
448. L. A. Haff and R. L. Easterday in *Theory and Practice in Affinity Chromatography* (P. V. Sundaram and F. Eckstein, Eds.), Academic Press, New York (1978), p. 21.
449. A. Malmstrom, L. Roden, D. S. Feingold, I. Jacobsson, G. Backstrom, and U. Lindahl, *J. Biol. Chem.* **255,** 3878 (1980).
450. R. Beissner, *Fed. Proc.* **38,** 646 (1979).
451. R. Cerff, *Eur. J. Biochem.* **94,** 243 (1979).
452. M. M. Chauvin, K. K. Korri, A. Tirpak, R. C. Simpson, and K. G. Scrimgeour, *Canad. J. Biochem.* **57,** 178 (1979).
453. J. Y. Chiu and P. D. Shargool, *Plant Physiol.* **63,** 409 (1979).
454. T. J. Larson, V. A. Lightner, P. R. Green, P. Modrich, and R. M. Bell, *J. Biol. Chem.* **256,** 9421 (1980).
455. F. Muller, G. Voordouw, W. J. H. Van Berkel, P. J. Steennis, S. Visser, and P. J. Van Rooijen, *Eur. J. Biochem.* **101,** 235 (1979).
456. D. H. Rogers, S. R. Panini, and H. Rudney, *Anal. Biochem.* **101,** 107 (1980).
457. K. T. Smith, M. L. Failla, and R. J. Cousins, *Biochem. J.* **184,** 627 (1979).
458. A. H.-J. Wang, M. L. Sherman, and A. Rich, *Biochem. Biophys. Res. Commun.* **82,** 150 (1978).
459. F. Wroblewski and J. S. La Due, *Proc. Soc. Exp. Biol. Med.* **90,** 210 (1955).
460. J. E. Wilson, *Biochem. Biophys. Res. Commun.* **82,** 745 (1978).
461. L. M. Hjelmland, *Proc. Nat. Acad. Sci. U. S.* **77,** 6368 (1980).

CHAPTER

4

BORATE CHROMATOGRAPHY

ALAN BERGOLD and WILLIAM H. SCOUTEN

Chemistry Department
Bucknell University
Lewisburg, PA 17837

Since Biot (1) reported in 1842 the interaction between borate and sugars, complexes of the borate type have been widely used in a number of applications. These applications range from ion exchange chromatography of anionic borate–sugar complexes to the use of organoboronates as protecting groups in synthetic reactions employing carbohydrates. An area that has proved especially useful to the biochemist and which has received increasing attention over the past decade is the use of chromatographic materials that contain immobilized boronic acid groups.

Boronates react readily and under mild conditions with a wide range of polar functional groups that are oriented in the proper geometry. Although complexes of 1.2-diols, 1,3-diols, and the catechol type compounds are the most studied, a variety of other functional groups such as 1,2-hydroxy acids and 1,2-hydroxylamines will also react with boronates to form complexes. Many substances possessing these types of functional groups are biologically important. Some of these compounds are listed in Table 4.1.

In a sense, boronates can be considered a "general lectin." Although the specificity for a given carbohydrate is less than for a true lectin, boronates have several advantages over lectins. First, the boronates are much more stable than the lectins. Second, as a consequence of the greater stability and the varying degree of interaction of boronates with different substances, the limits one must work in to obtain a separation are more flexible. Finally, the low cost of boronates makes boronate affinity chromatography an attractive tool for biochemical investigations.

4.1. INTERACTION OF BORONATES AND SUBSTANCES OF BIOLOGICAL INTEREST

The reaction between boric acid and polyhydroxy compounds has been the subject of several reviews (2–4). The chemistry of this interaction is compli-

Table 4.1. Biological Substances Expected to Form Boronate Complexes

Class	Examples
Carbohydrates	Mannose; ribose; fructose
Glycoproteins	Glycosylated hemoglobin
Catecholamines	Epinephrine
Nucleosides	Adenine; cytosine
α-Hydroxy acids	Lactic acid; 6-phosphogluconate
Dihydroxy compounds	Steroids; prostaglandins; cardiac glycosides
Aromatic α-hydroxy acids and amides	Salicylic acid; salicylamide

cated by the formation of a variety of species in solution. In addition to forming the 1:1 borate:diol complex, borate can also form a dimeric 1:2 borate:diol complex with borate as the central linkage. The use of boronic acids as the complexing agent simplifies the chemistry since the formation of 1:2 borate:diol complexes are essentially precluded. Boronic acids also offer several other advantages which are discussed later. Because phenylboronic acid has been widely studied for use in liquid chromatography, and because derivatives of the phenyl ring provide a convenient means of introducing boronate ligands into solid supports, the interaction of boronates with polyhydroxy compounds is discussed largely in terms of phenylboronic acid as the ligand.

Phenylboronic acid (PBA), like boric acid, is a Lewis acid and it ionizes not by direct deprotonation but by hydration and subsequent ionization to give the tetrahedral phenylboronate anion (1). PBA (pK_a 8.86) is three times as strong an acid as boric acid and approximately 10 times as strong as *n*-butaneboronic acid (5). As would be expected, the pK_a for the equilibrium depicted in Scheme 1 can be altered by substituent groups on the ring. Sub-

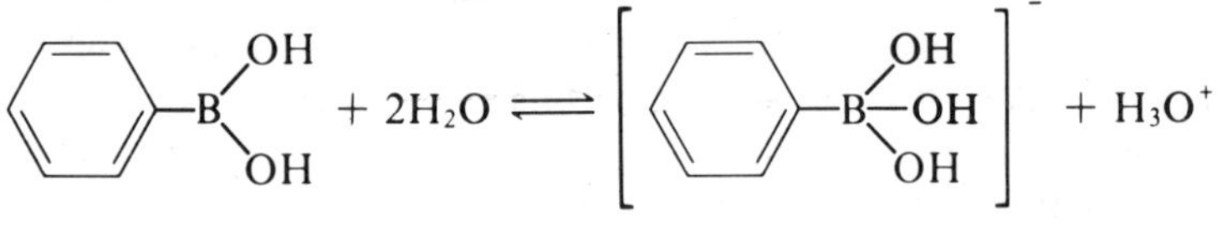

$$C_6H_5B(OH)_2 + 2H_2O \rightleftharpoons [C_6H_5B(OH)_3]^- + H_3O^+$$

Scheme 1

stitution of an alkyl or phenyl group in the ortho position decreases the acid strength, while meta and para substitution only slightly decrease it (6). Substitution of electron withdrawing groups can considerably increase the acid strength, as is evident for 3-nitrophenylboronic acid (pK_a 7.30) (7). However, substitution in the ortho position by a nitro group of a bulky substituent can lower the acidity of the boronic acid. Ortho nitro groups have been shown to form a chelate with the dihydroboryl group (8) while bulky substituents lower the acceptor strength of the boron atom and therefore affect the ability of boron to hydrate prior to ionization (9).

Ionization of the boronic acid is an important factor in the complexation of polyhydroxy compounds by boronic acids. Upon ionization, boron changes from trigonal coordination to the tetrahedrally coordinated anion. The bond angles and bond lengths associated with these two forms are depicted in Scheme 2 (10).

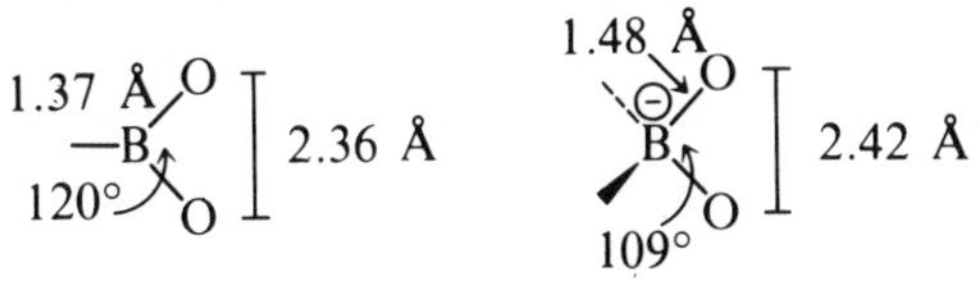

1.37 Å, 120°, 2.36 Å; 1.48 Å, 109°, 2.42 Å

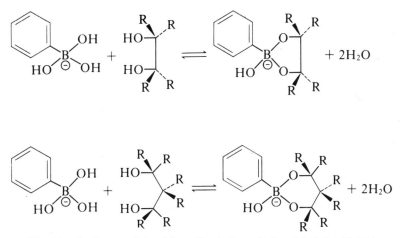

Scheme 2

Compounds that contain hydroxylic functional groups can react with PBA in two ways. The first type of interaction that may occur is the formation of a cyclic ester complex, as illustrated in Figure 4.1. These complexes can form with 1,2- or 1,3-diol-containing compounds when the hydroxyl groups are oriented in the proper geometry. The second type of interaction that may occur is that exhibited by monofunctional compounds, such as alcohols, to form the dialkyl boronate derivatives. In the context of boronate affinity techniques, we are mainly concerned with the formation of the cyclic complexes. Although the dialkyl derivatives are hydrolytically unstable, they

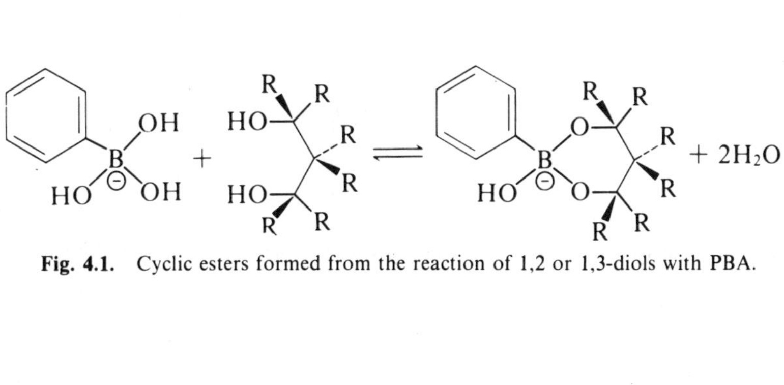

+ $2H_2O$

+ $2H_2O$

Fig. 4.1. Cyclic esters formed from the reaction of 1,2 or 1,3-diols with PBA.

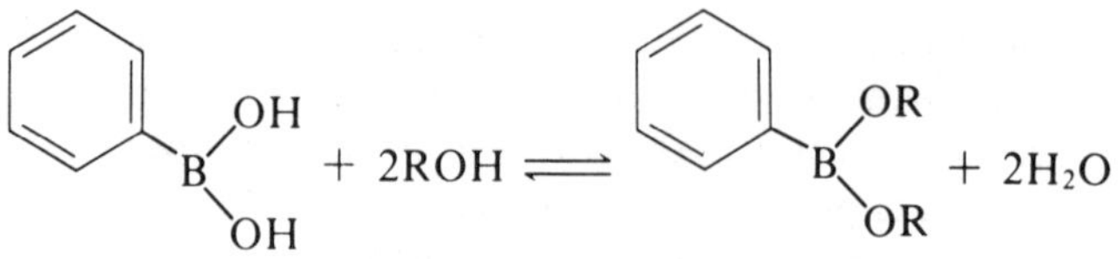

$$C_6H_5B(OH)_2 + 2ROH \rightleftharpoons C_6H_5B(OR)_2 + 2H_2O$$

Scheme 3

may account for small interactions exhibited between polyhydroxy compounds and boronates when no complex formation is predicted.

Figure 4.1 depicts the anionic form of PBA reacting with the diol to form the complex as postulated by Lorand and Edwards (11). Another possible scheme for this reaction is that proposed by Barker and co-workers in Figure 4.2 (12). In this scheme, the complex is formed by sequential nucleophilic attack of the diol oxygen atoms on the boron atom of the free acid. The result is either the anionic or neutral complex. Several important points are evident when examining these two mechanisms. First, the two schemes are indistinguishable at equilibrium. Second, if the anionic complex is formed, it should be stabilized by electron withdrawing groups on the aromatic ring. And finally, the formation of the anionic complex is favored in alkaline solution.

The extent to which neutral anionic complexes are formed depends on both the type of diol being complexed and the conditions under which the complex is formed. As was already mentioned, the formation of the anionic complex is favored at alkaline pH. At neutral or acidic pH values, the possibility of a neutral complex existing is highly dependent on the diol. Neutral complexes involve the trigonally coordinated form of the boronate. Al-

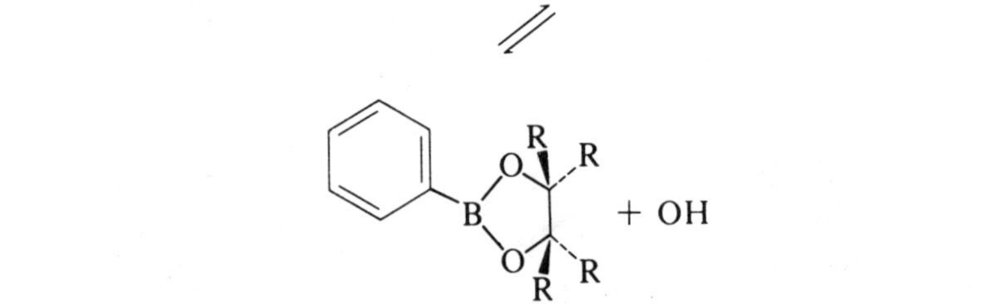

$$C_6H_5B(OH)_2 + \text{HO–CR}_2\text{–CR}_2\text{–OH} \rightleftharpoons \text{anionic complex (HO–B}^{\ominus}\text{ with O–CR}_2\text{–CR}_2\text{–O ring)} + H_2O + H^+$$

$$\rightleftharpoons \text{neutral cyclic boronate ester} + OH$$

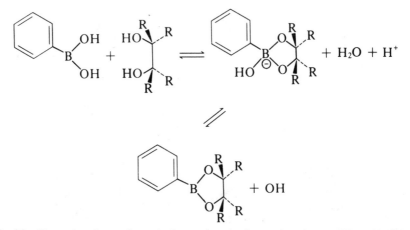

Fig. 4.2. Formation of neutral or anionic complexes by the reaction of neutral PBA with diols.

though six-membered ring complexes from 1,3-diols can incorporate trigonally coordinated boron, the five-membered ring complexes from 1,2-diols and trigonal boronates are strained and are hydrolytically unstable, as demonstrated by Bowie and Musgrave (13). The tetrahedral boronates, however, can be incorporated into strainless five- and six-membered rings. Alteration of solution pH provides a method of adjusting complex stability and therefore a means of binding and eluting substrates of interest. Under the conditions commonly employed in boronate affinity chromatography, the anionic complex is the predominant species in solution. The neutral complexes are normally stable only in dry organic solvents and thus are important to the synthetic carbohydrate chemists who use boronates as protective groups (10).

4.1.1. Requirements for Formation of Boronate Complexes

It was alluded to previously that the formation of boronate complexes is dependent on two factors. The first of these is the coordination geometry of the boron atom in the boronate. The second factor of importance is the geometry of the diol itself and its ability to assume the conformation necessary for complexation to occur. Since the geometry of PBA is invariant under a given set of conditions, it is the conformation of the polyhydroxy compounds that governs complexation and the ability to separate them on an immobilized PBA phase.

The requirements for the formation of a cyclic PBA : diol complex are usually formulated in terms of the diol hydroxyls being *cis* and coplanar. Although PBA readily forms complexes with *cis* and coplanar diols, a variety of diols react with PBA and do not satisfy the above requirements. A broader set of spatial requirements has been proposed (14). For vicinal diols to react, the dihedral angle between the hydroxyl groups must be between 0 and 60°. A more general requirement that allows for the reaction of PBA with 1,3-diols is that the oxygen–oxygen distance must be between 2.2 and 2.5 Å.

The intramolecular interactions that result when a carbohydrate undergoes a conformational change are pivotal in the ability to form complexes with PBA. An excellent example of these interactions occurs in a six-membered ring containing vicinal hydroxyl groups. When the hydroxyls are *cis* (axial-equatorial) they are capable of undergoing complexation with PBA. *Trans* hydroxyls (diequatorial), however, do not undergo complexation. Examination of a molecular model presents an anomaly in that the oxygen–oxygen separation distance appears the same in both cases. When the model is distorted to bring the hydroxyls into an eclipsed arrangement for the *cis* conformation, it is evident that interactions between axial substituents are reduced. Repeating the process for the *trans* model results in an increase in axial–axial interactions. Thus, in the *trans* case the driving force for complexation is not sufficient to overcome the resulting interactions. The role of intramolecular interaction is also important in compounds with acyclic diols, as will be seen later.

Once formed, the question of diols–boronate stabilities and thus the magnitude of the formation constant is very complex. The ease of hydrolysis of the cyclic esters is dependent not only on ring size but also on the organic moiety of the boronate. Arylboronates are more stable to hydrolysis than their alkyl counterparts (15). The stability of the complexes is also dependent upon the type of substituent groups present on the diol itself (13). Groups that can inhibit the coordination of water with boron stabilize the ester toward hydrolysis. This is the case with some carbohydrates, such as ribose, which has a third hydroxyl in the proper position to act as a third ligand. Compounds containing these groups exhibit a greater degree of interaction with PBA and consequently a larger interaction with an immobilized boronate stationary phase.

4.1.2. Boronate–Monosaccharide Complexes

The formation of a boronate: monosaccharide complex is contingent upon the requirements outlined in the previous section. Namely, the monosaccharide must have at least one diol group with the necessary orientation. Many monosaccharides contain more than one such group and can therefore form several types of boronate complexes. Glucose, for example, can form complexes involving the 1,2- and 4,6-hydroxyl groups. Information on the actual structures of boronate: monosaccharide complexes in aqueous solution is severely lacking. The major source of structural information is drawn from structural analysis on the complexes isolated from dry organic solvents and from electrophoretic data on the monosaccharides and their derivatives.

Electrophoretic data on boronate complexes are commonly indexed according to an M_G value defined as follows:

$$M_G = \frac{\text{true distance of migration of the substance}}{\text{true distance of migration of D-glucose}}$$

The true distances of migration are obtained by correcting for movement due to electroendoosmotic flow. Table 4.2 contains a partial list of M_G values for some monosaccharides and their derivatives along with R_F values from paper chromatography in the presence and absence of PBA. Also included in the table are the elution volumes for the monosaccharides on a boronate affinity matrix.

The data in Table 4.2 illustrate the differing ability of pyranosidic and fu-

Table 4.2. Comparison of Electrophoretic and Chromatographic Enhancements with Retention Volume for Some Selected Monosaccharides and Their Derivatives

Compound	$M_G{}^a$ (16)	$R_F{}^b$	$R_F{}^c$ (17)	Elution Volume (ml)[d] (18)
Erythritol	0.1	0.23	0.31	51
L-Arabinitol	0.6	—	—	76
Xylitol	0.9	0.14	0.45	93
D-Mannitol	1.0[e]	0.08	0.43	109
myo-Inositol	0.0	0.02	0.02	56
epi-Inositol	1.8	0.01	0.04	—
D-Inositol	0.0	—	—	52
Sorbitol	1.3	0.08	0.45	182
Sucrose	—	—	—	52
D-Glucose	1.0	0.08	0.08	52
D-Ribose	4.7	0.25	0.50	85
D-Fructose	9.3	0.11	0.12	128
D-Xylose	1.8	0.15	0.15	—
D-Galactose	1.8	0.06	0.08	—
Methyl-D-glucofuranoside	2.0	—	—	—
Methyl-D-glucopyranoside	0.0	0.20	0.21	—
Methyl-D-mannofuranoside	16.0	—	—	—
Methyl-D-mannopyranoside	0.0	—	—	—
2-*O*-Methyl-D-glucose	0.0	—	—	—
4-*O*-Methyl-D-glucose	0.0	—	—	—
6-*O*-Methyl-D-glucose	0.5	—	—	—

[a] Electrophoresis was carried out in a 0.05 *M* aqueous solution of sulfonated phenylboronic acid buffered with monopotassium phosphate at pH 6.5.
[b] R_F values are for descending chromatograms on Whatman No. 1 paper using ethyl-acetate–acetic-acid–water (9:2:2 v/v).
[c] Same solvent with 0.55% PBA added.
[d] Eluting buffer was 1 *M* sodium acetate 0.05 *M* *N*-methyl-morpholinium-Cl, pH 7.5 on a 55 × 1 cm column of *N*-(*m*-dihydroxyborylphenyl)-carbonylmethyl cellulose.
[e] For the glycitols, mannitol is the standard against which migration distances are compared.

ranosidic sugars to form boronate complexes. The zero M_G values for the methyl pyranosides indicate that vicinal, *trans* hydroxyl groups do not react with sulfonated phenyl boronic acid. Böeseken and Foster made similar observations using boric acid as the complexing agent (2, 3).

The very high mobility of the methyl-α-D-mannofuranoside shows the importance of planar *cis* 1,2-diols in complex formation. In mannofuranside the hydroxyls at C_2, C_3, and C_5 may be situated for possible formation of a tridentate complex with PBA and may explain the much greater mobility of mannofuranoside compared with the glucofuranoside (16). The zero M_G

value for 2-*O*-methyl-D-glucose, in which the C_3, C_5, C_6 hydroxyls are free, suggests that the furanose form of this sugar is not involved in complex formation. It is also unlikely that these free hydroxyl groups can form a tridentate complex as in the case of mannofuranoside.

Tridentate conformation can also occur for six-membered ring compounds. These complexes occur on rings that contain *cis* 1,3-axial hydroxyl groups and are stabilized towards hydrolysis by an intervening equatorial hydroxyl group (see Scheme 4). In the pentose series, only ribose has such a grouping. For the aldohexoses, allose, gulose, and talose show enhancement.

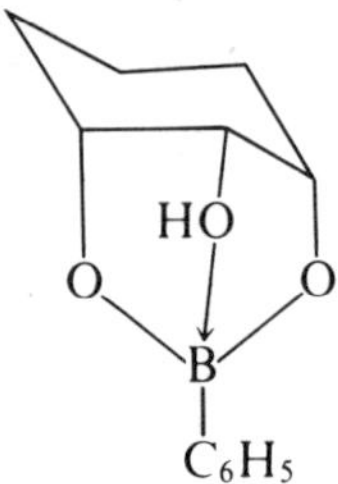

Scheme 4

Oxygen atoms in the pyranoid ring may also result in some stabilization provided they are at the proper distance, as is the case for methyl-3,6-anhydro-D-glucopyranosides (10). The inositols are also capable of forming tridentate complexes. Most notable are *epi*- and *cis*-inositol, which have formation constants with borate of 7000 and 1.1×10^6 respectively.

Paper chromatography in the presence of PBA also shows mobility enhancement due to complex formation. The alditols all exhibited enhancement, as did ribose. A comparison of these results with electrophoretic data reveals some marked differences. The most notable difference was for fructose, which showed no enhancement during chromatography but underwent strong complexing with sulfonated PBA. Despite the differences exhibited, both methods appear to be selective for the *cis,cis*-triol group, since ribose was enhanced in both systems. The differences that occur may be the result of instability of a given complex under the conditions used. Differences also exist between sulfonated PBA and borate as complexing agents (16). Sulfonated PBA does not form a 4,6 derivative with 2-*O*-methyl-D-glucose whereas borate does. Since the sulfonated PBA electrophoresis was carried out at pH 6.5 while that for borate was done at pH 10.0, this discrepancy may also be due to differences in stability for the given conditions. When developing a boronate affinity method, it is more important to experimentally examine the stability of a given complex than to make predictions from published data using different conditions.

4.1.3. Extension of Boronate–Monosaccharide Complex Formation to Other Biological Compounds

Boronate affinity techniques are applicable to the isolation of biological macromolecules largely because the condensation between these macromolecules and carbohydrates readily form boronate complexes. For example, RNA and nucleosides form boronate complexes because ribose is present in its furanose form. The manner in which the moiety undergoing complexation is attached to the macromolecule can have a profound effect on its interaction with PBA.

The most pronounced effect on complexation occurs when the groups involved in the interaction are the point of attachment to another molecule. Such is the case for the disaccharides of D-glucose with (1 → 2) or (1 → 4) linkages, which exhibit lower M_G values than those disaccharides with (1 → 3) linkages (3). The complexation obviously will be influenced to a much smaller extent by attachment at a position that is remote from the complexation site. In order for complexation to occur, the carbohydrate must also be sterically accessible to the boronate.

The discussion of boronate complexes up to this point has been solely in terms of carbohydrates. There are many other biological compounds that form complexes, and these have been pointed out in Table 4.1. In general, the complexes formed by these molecules are less stable than most carbohydrates. The catecholamines are an exception, however. Catechol–boronate complexes owe their stability to the favorable position of the two aromatic hydroxyls.

As in the case of monosaccharide complexes, it is important to determine the stability of a macromolecule complex under the conditions desired or needed. Although prediction can help one decide whether to try a boronate affinity separation, the final decision must be withheld until the experimental evidence is in.

4.2. PREPARATION AND USE OF BORONATE AFFINITY ADSORBENTS

The first report of a chromatographic stationary phase containing immobilized boronic acid groups was by Solms and Deul (20). The material was prepared by acid-catalyzed condensation polymerization of *M*-aminobenzene boronic acid, *M*-diaminobenzene, and formaldehyde. Although the resin was suitable for the separation of simple mixtures of sugars, it was chromatographically inefficient and contained residual ion exchange sites. Letsinger and Hamilton (21) were the first to prepare the so-called popcorn polymers using *p*-vinylbenzeneboronic acid, styrene, and diallylmalleate. However, the boronate materials were not tested for their chromatographic properties.

The movement toward applying boronate affinity matrices to biochemical problems was started in 1970 when Weith et al. described the synthesis of cellulose derivatives containing dihydroboryl groups (18). The advantage of these derivatives was the large pores in the matrix, allowing facile diffusion of the macromolecule to the boronate group.

4.2.1. Matrices

Matrices currently employed in boronate affinity methods can be divided into three groups. The first group consists of the matrices most commonly employed in affinity chromatography, such as cellulose, agarose, polyacrylamide, and glass. These matrices all require some type of activation in order to couple the boronate ligand to the surface of the matrix. Polymers that contain PBA as an integral part of the polymer backbone constitute the second class. These polymers are usually synthesized by using a derivative of PBA that contains a vinyl group and polymerizing with some type of crosslinking agent. Polymers such as polystyrene, which are functionalized to bind PBA, are included in the first category. The third group is unique and is new to the affinity field. To date, only one matrix fits into this category and it is polychorotrifluoroethylene. The affinity matrix forms, with a PBA derivative, what the authors term a reversed-phase boronate support (22).

In boronate affinity adsorbents employing polysaccharides as the matrix, there is the potential for interaction between *cis*-1,2-diols and the ligand. This interaction would compete with other ligandophiles and effectively lower the capacity of the adsorbent. This was suggested by Hageman and Kuehn, who observed that cellulose and Sephadex boronate gels did not bind ATP as well as polyacrylamide gel (23). For adsorbents having similar boron contents but exhibiting different capacity toward a ligandophile, internal complexation may be the cause. The possibility of this type of interaction has resulted in a move away from the glucose-based polysaccharide supports toward other matrices such as polyacrylamide and agarose.

Currently, five commercial boronate affinity adsorbents are available on the market. Two of these are based upon a polyacrylamide matrix and are available from Bio-Rad (Richmond, Calif.) and Pierce Chemical (Rockford, Ill.). Aldrich Chemical (Milwaukee, Wis.) markets a material called boric acid gel. This gel is a methacrylic acid polymer with *m*-aminobenzeneboronic acid incorporated into the polymer backbone. Amicon (Danvers, Mass.) supplies an agarose-based adsorbent which has 5% crosslinking and is available in three ligand concentrations ranging from 10 to 100 μmol of boron per mL. of gel. A cellulose-based gel is also available from Collaborative Research (Waltham, Mass.).

4.2.2. Attachment of the Boronate Ligand

The majority of synthetic schemes for immobilizing the boronate ligand are based upon *m*-aminobenzeneboronic acid. The reasons for using this precursor are the commercial availability, the wide variety of reactions that the arylamine group can undergo, and the ability to alter the pK_a of the boronate group by adding substituent groups on the phenyl ring. Lowering of the boronate pK_a enables one to use less alkaline conditions to form the boronate complex and allows compounds that are unstable in alkaline media to be separated by this method.

The cellulose derivatives reported by Weith et al. (18) were synthesized by converting carboxymethylcellulose to the azide form and then coupling with *m*-aminobenzeneboronic acid to give *N*-(*m*-dihydroxyborylbenzene)carbamylmethylcellulose (DBCM-cellulose). The resulting derivative was found to contain about 0.2 mmol of boronate groups per gram of dry cellulose. Approximately 33% of carboxyl groups in the original cellulose remained underivatized. As a result, the derivative is anionic above pH 5.0. Since boronate matrices are inherently negative above the pK_a of the immobilized boronate, other contributions to the total negative charge on the matrix are undesirable. To circumvent this problem, the authors tried an alternative procedure. Aminoethylcellulose was condensed with *N*-(*m*-dihydroxyborylbenzene)-succinamic acid using *N*-cyclohexyl-*N*′-*β*-(4-methylmorpholinium)ethylcarbodiimide *p*-toluene-sulfonate as the activating agent. The *N*-[*N*′-(*m*-dihydroxyborylbenzene)succinamyl]aminoethylcellulose(DBAE-cellulose) (see Scheme 5) contained approximately 0.6 mmol of boronate groups per gram

$(HO)_2B$–C$_6$H$_4$–NH_2 + succinic anhydride (CH_2–CO–O–CO–CH_2) ⟶ $(HO)_2B$–C$_6$H$_4$–$NHCOCH_2CH_2COOH$

(1)

–$OCH_2CH_2NH_2$ + (**1**) ⟶ –$OCH_2CH_2NHCOCH_2CH_2CONH$–C$_6H_4$–$B(OH)_2$

DBAE—Cellulose

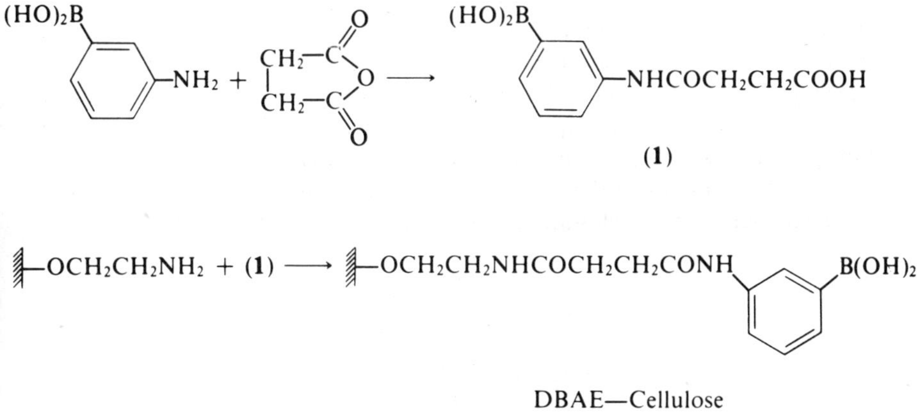

Scheme 5

of dry cellulose corresponding to a substitution of 60% of the amino groups in the starting material. At neutral pH this derivative exhibits the properties of a polycation due to residual aminoethyl groups. McCutchan et al. (24) have reported a method for removing these charge sites by acetylation with acetic anhydride. Acetylated DBAE-cellulose can also be obtained from Collaborative Research.

Yurkovich and coworkers (25) have taken an interesting approach towards stabilizing the arylboronic acid anion by synthesizing bipolar derivatives. The affinity matrices they synthesized were based on dextran and contained bipolar 2-{[(4-boronophenyl)methyl]ethylammonio}ethyl and diethylammonio}ethyl groups. DEAE-Sephadex A-25, DEAE-Sephadex LH-20, and O-[2-(ethylamino)-ethyl]cellulose were reacted with *tris*[4-(bromomethyl) phenyl]boroxin in *N,N*-dimethylformamide (see Scheme 6). The resulting derivatives had capacities ranging from 1.52 mmol of boron per g in the case of DEBAE-Sephadex A-25 to 0.2 mmol boron per gram for the cellulose. The extent of modification was 60 to 70%.

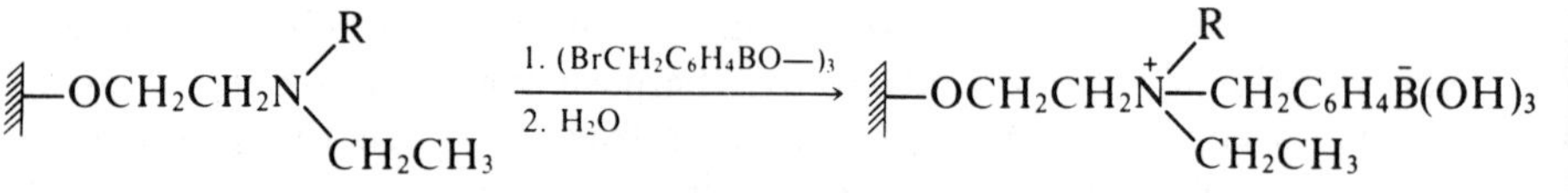

$$\text{—OCH}_2\text{CH}_2\text{N}(\text{R})(\text{CH}_2\text{CH}_3) \xrightarrow[\text{2. H}_2\text{O}]{\text{1. (BrCH}_2\text{C}_6\text{H}_4\text{BO—)}_3} \text{—OCH}_2\text{CH}_2\overset{+}{\text{N}}(\text{R})(\text{CH}_2\text{CH}_3)\text{—CH}_2\text{C}_6\text{H}_4\overset{-}{\text{B}}(\text{OH})_3$$

$R = H$ or CH_2CH_3

Scheme 6

The synthesis of boronate agarose supports has been reported by several groups (22, 26–29). Two of these matrices were synthesized with the aim of applying them to the isolation of enzymes and made use of hydrophobic spacer arms. Akparov and Stepanov (26) derivatized CH-Sepharose (a Sepharose derivative containing ε-aminocaproic acid residues with free carboxyl groups) with *p*-methylaminebenzeboronic acid. This gel contained 5–10 μmol of boronate groups per mL of swollen Sepharose and was used to purify several serine proteinases. Bouriotis et al. (28) have prepared a gel which is very similar to that described above. In this case, the authors synthesized 6-aminocaprayl-3-amino-phenylboronic acid (see Scheme 7) and

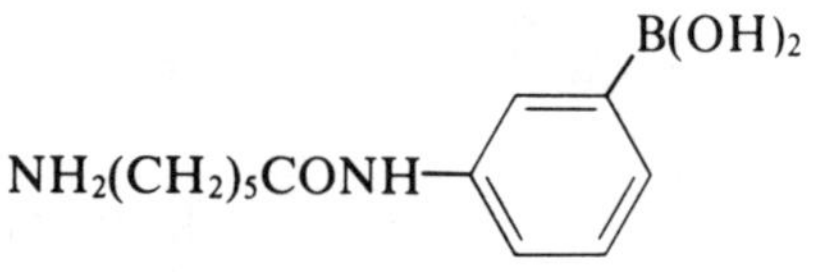

$$\text{NH}_2(\text{CH}_2)_5\text{CONH—C}_6\text{H}_4\text{—B(OH)}_2$$

Scheme 7

reacted it with Sepharose 6B, which had been cyanogen-bromide activated. Also reported in this paper was a method for coupling *M*-aminophenylboronic acid(APBA) to Sepharose activated by CNBr or by epoxy activation. The capacities reported ranged from 34 μmol/mL to 146 μmol/mL, epoxy activation with APBA giving the highest yield. Boronate–agarose derivatives have also been prepared by using commercially available supports such as CM-Bio-Gel A and Affi-Gel 10 (Bio-Rad) (22, 27) or Reacti-Gel (Pierce Chemical Co.) (29).

The preparation of polyacrylamide matrices containing immobilized boronate groups has followed two paths, using either the acyl azide procedure of Inman and Dintzis (30) or a succinylated polyacrylyl-hydrazine derivative as described by Uziel et al. (31). Uziel's procedure involves coupling APBA to succinylated polyacrylylhydrazide in the presence of 1-ethyl-3(3-dimethylaminopropyl)carbodiimide(EDAC) as depicted in Scheme 8. The derivatized gel contained 4.9 mmol of boronate groups per g of the starting hydrazide gel. The synthesis of this gel was improved by another group of workers (32).

$$\text{—}CONH_2 \xrightarrow{NH_2NH_2} \text{—}CONHNH_2 \xrightarrow{\text{succinic anhydride}} \text{—}CONHNHCO(CH_2)_2COOH$$

$$\xrightarrow{H_2N\text{-}C_6H_4\text{-}B(OH)_2\ +\ EDAC} \text{—}CONHNHCO(CH_2)_2CONH\text{—}C_6H_4\text{—}B(OH)_2$$

Scheme 8

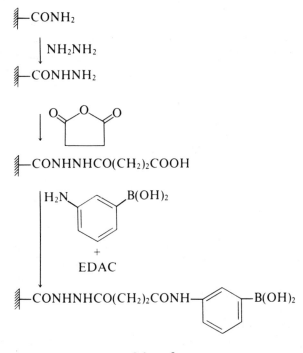

Compared with Uziel's procedure, a finer mesh (200–400) polyacrylyl-hydrazide gel with a lower concentration of hydrazide groups (1.2 mmol/g) was used, resulting in a gel that exhibited decreased shrinkage when either the ionic strength or pH of the eluent was changed. This procedure also resulted in a higher percentage of reactive groups in the starting material being derivatized (100% vs. 83%).

Recently, Olsson (34) published a comparison of the derivatives formed by the acylazide technique and those described by Davis et al. (32). The two derivatives perform equally with respect to retention and recovery of some purine nucleosides. The method reported by Olsson, however, took less time to prepare and cost less. Most of the cost differential was due to the use of hydrazide beads for the procedure described by Uziel and Davis. A method for producing boronate polyacrylamide by a modification of the method of Weith et al. (18) has also been described (23).

Duncan and Gilham (33) were the first to report the use of glass as a support for boronate affinity chromatography. The reported procedure was similar to that used for synthesizing DBAE-cellulose. Amino-substituted porous glass was reacted with *N*-(*m*-dihydroxyborylphenyl)succinamic acid in the presence of water soluble carbodiimide. The resulting material had a low ligand coverage of 0.06 mmol/g. Glad and coworkers (35) have recently developed a microparticulate silica boronic acid support suitable for high performance chromatography. Epoxy-substituted silica was obtained by reacting porous silica with γ-glycidoxypropyltrimethoxysilane in anhydrous toluene. This procedure results in a thin molecular coating of epoxy groups, which the authors state is necessary to prevent slow solute diffusion and low chromatographic efficiency. A reference column of the epoxy-substituted silica did not show any retardation of the nucleosides, nucleotides, or carbohydrates studied. In addition, the recovery of human serum albumin was reported to be 98% for a 10 μL sample. No mention of the conditions employed was made. The epoxy-substituted silica was reacted with *m*-aminobenzene boronic acid in water adjusted to pH 8.5 to yield the boronate silica support. Unfortunately, the authors do not mention the stability of their support. Since the elution conditions were all in the range of pH 7.5–8.5 it is possible the column could degrade with time due to dissolution of the silica matrix. A support with monolayer coverage would be less stable than a support with a polymeric coating under these conditions. The use of PBA derivatives with lower pK_a's, such as the nitro derivative, would allow conditions more favorable to silica stability.

The successful synthesis of 4-vinylbenzeneboronic acid was the factor that led to the boronate-containing polymers of Letsinger and Hamilton. Since that time, there have been numerous reports involving the use of this monomer to incorporate phenyl boronate groups into a polymer chain. Barker et al. (36) polymerized the iminodiethyl derivative of 4-vinylbenzeneboronic acid with ethylvinylbenzene, using azobisisobutyronitrile as the catalyst, to form a lightly crosslinked gel. As a result of the low crosslinking, the polymer swells in alkaline aqueous solution allowing monosaccharides access to the internal volume of the beads. The boronic acid gel was used to separate mixtures of monosaccharides and to study the effects of pH and temperature on retention factors; 4-vinylbenzene has also been copolymerized with styrene, *N,N'*-methylenebis(acrylamide) and with a mixture of acrylamide and *N,N'*-methylenebis(acrylamide) (37). These authors also describe a method for preparing poly(*p*-vinylbenzeneboronic acid)-coated polystyrene beads by interstitial homopolymerization. Of the four methods tested, the homopolymerization yielded a product that was more satisfactory for the separation of L-dopa from L-tyrosine. Seymour and Frechet (38) have described the covalent modification of polystyrene (Amberlite XE 305) to yield a boronate polymer containing 1–2 mmol of boronate per gram of polymer. This procedure results in a polymer with a capacity 10 to 20 times greater than boronate polymer formed by interstitial homopolymerization. The authors found that the highest capacity was suitable for the solid phase synthesis of glycosides such as methyl 2,3-di-*O*-benzoyl-α-D-galactopyranoside.

Schott (39) has described the synthesis of a dihydroxyboryl substituted methacrylic acid polymer with the structure shown in Scheme 9. In this

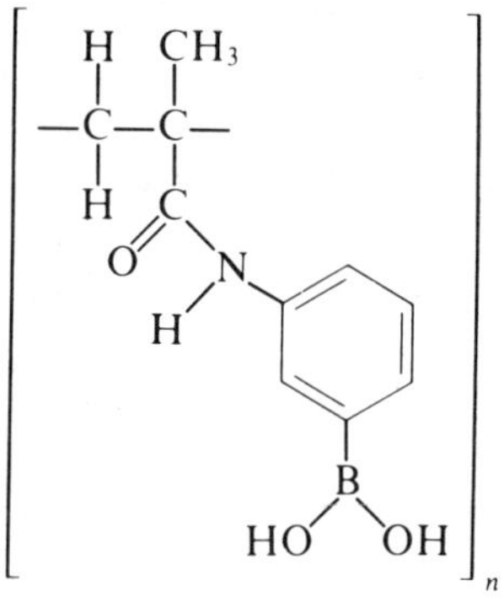

$$\left[-CH_2-C(CH_3)(C(=O)NH-C_6H_4-B(OH)_2)- \right]_n$$

Scheme 9

study, the boron content of the polymers ranged from 0.11 to 0.58%. It is interesting that the boronate content and exclusion limit could be varied simply by altering the relative amounts of boronate monomer and crosslinking reagent in the reaction mixture. The capacity of the gel seems to depend on the nature, particularly the chain length, of the oligonucleotides. This observation most probably results from the inability of large molecules to reach boronate groups on the interior of the gel.

Wulff and coworkers (40, 41) have described a procedure for preparing a chromatographic sorbent with a chiral cavity for the resolution of racemates.

The sorbents contain a boronic acid group locked into a fixed position in a polymer cavity. 4-Nitrophenyl-α-D-mannoside-2,3; 4,6-di-*O*-(4-vinylphenylboronate) was polymerized with divinylbenzene in the presence of an inert solvent to obtain a macroporous polymer. A high degree of crosslinking ensured a rigid polymer with boronic acid groups locked into a fixed stereochemical arrangement upon removal of the template (4-nitrophenyl-α-D-mannoside). Although the polymer exhibited high separation factors ($\alpha =$ 1.05–2.32) in an HPLC analysis of the optical isomers of the template, the columns were inefficient as measured by their plate height and they exhibited excessive tailing. The high specificity of the sorbent was demonstrated by a lack of resolution of racemates other than the template. The sorbents only displayed their specificity for the template under conditions in which the trigonal boronate complex with the template could exist. Apparently, formation of a tetrahedral boronate changed the stereochemical arrangement enough so that the sorbent could not distinguish between the isomers of the template.

Singhal and coworkers (22) have reported the synthesis of a reversed-phase boronate stationary phase in their attempt to separate aminoacyl-tRNAs. The authors reacted *m*-aminophenylboronic acid with decanoyl chloride in dioxane to form *m*(decanoylamino)phenylboronic acid. About 1 g of this compound was dissolved in chloroform and mixed with 10 g of 10-μm-diameter polychlorotrifluoroethylene beads. After evaporation of the chloroform, the beads were washed with acetate buffer to remove any poorly absorbed affinity groups. Efficiency of the column reportedly did not change even after 30 separations of adenosine and 2-*O*-methyladenosine.

4.2.3. Characterization of the Adsorbent

Characterization of boronate affinity adsorbents in terms of the quantity of boronate ligand bound to the support can be readily accomplished in one of several ways. The most common method used is spectrophotometric determination of the amount of uncoupled *m*-aminophenylboronic acid remaining in the filtrate and washes of the gel. Table 4.3 lists some of the common boronate derivatives used in preparing affinity adsorbents and their λ_{max} and ϵ_{max} values. The values listed in this table are for specific conditions. When analyzing for uncoupled boronate derivatives under different conditions, care should be taken to establish a new calibration curve since many of the boronate derivatives exhibit substantial changes in λ_{max} and ϵ_{max} when changing pH, ionic strength, and solvent.

Elemental analysis is readily applicable to the characterization of boronate adsorbents due to the presence of boron. Provided the underivatized support is free of boron, this data gives direct information on the boronate

Table 4.3. Spectrophotometric Characteristics of Some Boronate Derivatives Used in Preparing Affinity Adsorbents

Boronate Derivative	λ_{max}	ϵ_{max}	Reference
m-Aminophenylboronic acid	280 nm	1230[a]	18
	293 nm	1560[b]	18
	293 nm	1610[b]	34
N-(*m*-Dihydroxyborylphenyl)-succinamic acid	243 nm	10,800	18
6-Aminocaproyl-3-aminophenylboronic acid	242 nm[c]	—	28
m-(Decanoylamino)phenylboronic acid	219[d]	—	22
	246[d]	—	22

[a] In 0.2 *M* ammonia.
[b] In 1 *M* sodium phosphate, pH 7.0.
[c] In 0.1 *M* sodium phosphate, pH 7.0.
[d] In ethanol.

content of the gel. Glass matrices, however, may contain some boron, which would give a high indication of the boronate content. Hageman and Kuehn (23) have determined the degree of coupling by oxidative cleavage of the boronate groups from the acrylamide beads followed by colorimetric quantitation of borate in the neutralized filtrate. Cleavage was accomplished by using the acidic hydrogen peroxide procedure of Letsinger and Nazy (43). The borate concentration of the filtrate was determined by complexation with azomethylene-H according to the method of Wolf (44).

The acid–base properties of the boronate group provide another convenient means of evaluating the adsorbent. This type of analysis can supply two types of information, the boronate content of the gel and the pK_a of the immobilized boronate group. Figure 4.3 shows the results obtained from a titration of a polyacrylamide adsorbent (42). The average pK_a from three titrations was found to be 9.2 and agrees well with the value of 8.86 for underivatized benzeneboronic acid. The slightly higher pK_a obtained for the immobilized boronate is most likely due to the electron donating capacity of the amino substituent at the meta-position of the benzene ring. Unfortunately, few reports of pK_a values for immobilized boronate groups have appeared in the literature to date (28, 42). More investigations of this type would be useful for two reasons. First, a comparison of the boronate pK_a before and after immobilization would provide information on the effects of the synthetic method and the type of support used. Supports containing residual charge sites would probably have a noticeable effect on the pK_a. Secondly, the titration data reveals the pH range in which the boronate is ionized and therefore the conditions needed for maximum binding of a species to occur.

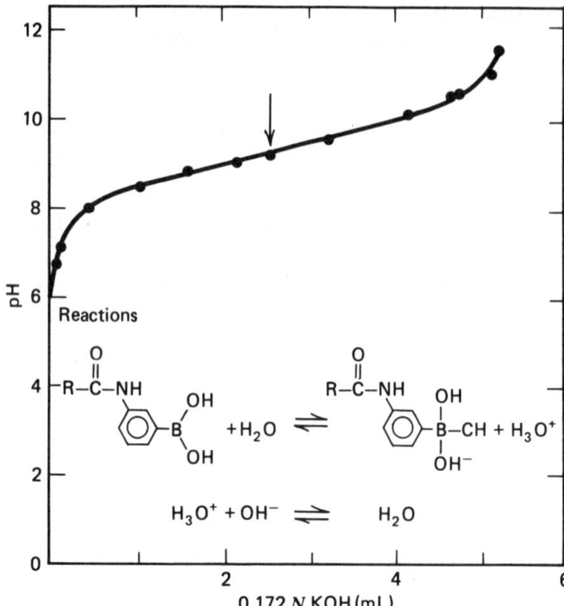

Fig. 4.3. Titration of P-150 boronate gel. A sample containing 0.93 gm dry weight of derivatized gel was suspended in 50 mL of distilled water. This suspension was stirred and titrated with 0.172 *N* potassium hydroxide. From this titration curve it was calculated that 0.961 meq of boronate groups was bound per gram dry weight of derivatized gel. The average pK_a from the titration trials was calculated to be 9.2 (arrow). From Maestas et al. (42, p. 228), with permission.

All of the methods previously discussed give the investigator some idea of the efficiency obtained in coupling the boronate ligand to the support. The results from these methods are not, however, always a good indication of the gel capacity for a specific biomolecule. For this reason Maestas et al. (42) have examined the maximum binding capacity of several boronate gels for a variety of nucleotides and biomolecules of use in the affinity purification of enzymes. The results of their study are contained in Table 4.4. Although none of biomolecules tested exhibited a 100% saturation of available binding sites, it should be kept in mind that these studies were done in a pH range (8.4–8.5) where only 16 to 20% of the boronate groups are ionized. Only two substances, NAD and lactic acid, are bound to a greater extent. As was previously mentioned, boronate gels may also exhibit a reduced capacity towards macromolecules due to exclusion effects. These results indicate that the maximum binding capacity of the adsorbent is strongly dependent upon the substance being considered and the conditions under which the experiment is carried out.

Table 4.4. Maximum Binding Capacities of P-150-Boronic Acid Beads for Several Biomolecules

Biomolecule	Amount Bound (μmol/ml wet gel)	Estimated Percent Saturation of Boronate Groups[a]	Amount Recovered in Borate Wash (μmol/ml wet gel)
NAD^+	52[b]	39	37
NADH	23	17	20
FAD	7	5	6
Pyridoxal	2.5	2	0.76
Epinephrine	16	12	16
Citric acid	21	16	1.6
Lactic acid	83	61	0

[a] The percentage was calculated assuming 0.135 meq boronate/ml wet-packed gel.
[b] From Maestas et al. (42).

4.2.4. Binding and Elution Conditions

When choosing conditions for a boronate affinity separation, two considerations are of importance. Consideration of the moiety involved in complexing with the boronate will provide some guidelines in choosing the initial conditions in the separation. Second, the stability of the biomolecule to the above conditions will ultimately determine the feasibility of the separation.

In general, the retention volume of a particular biomolecule depends on (i) the availability and configuration of the diol in the compound, (ii) the pH of the eluent, (iii) the ionic strength and the nature of the cations in the eluent, and, in the case of some compounds such as the nucleosides, (iv) the nature of other groups in the molecule. The factors in (i) above have already been discussed in Sections 4.1.1 and 4.1.2. The effect of pH, ionic strength, and cations on retention volume is exemplified by some work done by Bouriotis et al. (28) and is shown in Table 4.5. It is seen from this data that the increase in pH in going from pH 7.0 phosphate buffer to pH 8.45 Hepes buffer increased the retention volume of FAD. The addition of either 1 *M* NaCl or 0.1 *M* $MgCl_2$ produced another sixfold to tenfold increase in retention volume.

The effect of change of pH of the eluting solvent can be understood in terms of an increase in the overall concentration of the anionic form of the dihydroxyboryl group that is active in complex formation. The change in retention volumes caused by a change in salt concentration can also be accounted for by considering the anionic form of the covalently bound dihydroxylboryl groups. It can be postulated that field effects due to the proximity of neighboring negatively charged groups would tend to raise the

Table 4.5. Effect of Varying the pH and Salt Concentration on the Interaction of FAD with PBA-Agarose Columns

	Reaction Volume (× V_0)
0.1 *M* potassium phosphate, pH 7.0	2.5
0.1 *M* potassium phosphate + 1 *M* NaCl	4.6
50 m*M* Hepes, pH 8.4	4.0
50 m*M* Hepes + 1 *M* NaCl	26
50 m*M* Hepes + 0.1 *M* $MgCl_2$	25

NOTE: Columns containing PBA-Sepharose 6B(CNBr-activated, 1 mL) were equilibrated with the appropriate buffer before application of a pulse of 1 m*M* FAD (0.2 ml) at 4°C. Flow rate 0.7 mL/hr. Retention volume is defined as the lowest volume after which the nucleotide was detected and is expressed as a multiple of the column volume. From reference 28.

p*K* of the ionization of the surface-bound boronate groups above that for unbound boronate groups. An increase in the ionic strength of the eluent, however, should decrease the p*K* of the ionization due to the shielding of these charges on the surface of the support. Thus, at a fixed pH, an increase in salt concentration increases the concentration of boronate anions with a resulting increase in retention volume. The use of a divalent cation in the eluent is apparently more efficient in shielding these charge interactions since a magnesium chloride concentration only one-tenth that of sodium chloride is enough to evoke nearly the same increase in retention. Divalent cations also play an important role in the boronate chromatography of the charged nucleotides, as will be seen in a later section.

The choice of a buffer is solely dependent upon the desired pH range one desires to work in and the stability of the biomolecule in that media. Most of the buffers used in biochemical research have been used, and all work well with one exception. Tris buffers have a tendency to decrease the capacity of boronate gels. Hageman and Kuehn (23) have observed a 20-fold decrease in the capacity of a polyacrylamide-boronate gel towards ATP when Tris-Cl buffers were used. Likewise, Goitein and Parson (46) found that Tris buffers resulted in decreased retention of IMP and PRibATP. In our laboratory the use of Tris buffers resulted in glycosylated homoglobin being eluted in the void volume of a boric acid gel column. These results are most likely the result of a complexation between boronate ligands and the free hydroxyl groups of Tris.

The effect of temperature on the binding of diols to boronate adsorbents has largely been ignored. Barker et al. (45) studied the effect of temperature on the retention of glucose and fructose. They found that an increase in temperature resulted in an increase in retention. The retention factor for glucose increased from 0.73 to 0.83 on going from 20 to 50°C, while the retention factor for fructose nearly doubled (1.34–2.6). Our results on the separation of glycohemoglobins also exhibit an increase in retention when the temperature increases. Our system also exhibited a decreased recovery of glycohemoglobin from the column when temperature increased. At this time it is not known whether this is a result of increased boronate–diol interaction or a nonspecific interaction. More work on temperature effects is certainly needed, especially in cases where the boronate adsorbent is saturated with a biomolecule and used for affinity separation of enzymes.

Reversal of the boronate–diol complex is readily accomplished under mild conditions and is one of the attractive features of boronate affinity methods. The conditions chosen for solution must be made with the stability and the desired composition of the final media in mind. Most commonly, elution is achieved by lowering the pH and ionic strength of the eluent. This can be accomplished by using a buffer of lower pH or by using water as the eluent. Using water has the advantage of not having to dialyze the collected fractions to remove buffer salts.

Elution can also be carried out by adding to the eluent an agent that competes with the boronate–diol complex. The most commonly used agents are sorbitol and borate. Pace and Pace (27) have carried out a fairly detailed examination of elution conditions for some different RNAs. The results of their study indicate that sorbitol works well for the elution of some types of RNA and not for others. In order for either borate or sorbitol to effectively compete with the boronate–diol complex it must have access to the complexation site. In the chromatography of some large macromolecules this site may be shielded by the bulk of the molecule itself, rendering the competitor ineffective. The most effective use of the competitive agent occurs when the complexation equilibria with the competitor is more favorable than the equilibria between the immobilized boronate ligand and the diol on the biomolecule of interest.

Before leaving the subject of binding and conditions of elution, something must be said about nonspecific interactions, since very few systems do not exhibit some nonspecific effects. As was already mentioned, charge interactions can be largely eliminated by adjusting the ionic strength of the eluting medium and by elminating extraneous charge sites on the matrix by acetylation or other means. Nonspecific interactions can also be reduced to a minimum by saturating the nonspecific sites with the biomolecule and then washing the column to remove excess biomolecule, as was done by Pace and Pace (27). Occasionally, it is necessary to reduce interactions between the biomolecules themselves in order to achieve binding to the matrix. This was observed by Rosenberg and Gilham (47), who added 20% by volume DMSO to the eluent in order to achieve binding of tRNA to boronate cellulose column.

Okayama et al. (48) reduced interactions between ADP-ribosylated nuclear proteins with 6 *M* guanidine-HCl. Other instances exist in the literature where organic modifiers have been added to the eluent to eliminate nonspecific interactions. A judicial choice of modifier can only be made by the experimenter after considering the application at hand and information gleaned from published data.

4.3. APPLICATION OF BORONATE AFFINITY CHROMATOGRAPHY TO THE ISOLATION AND DETERMINATION OF BIOLOGICAL SUBSTANCES

4.3.1. Carbohydrates

Weith and coworkers (18) were the first group to apply boronate to affinity adsorbents to the separation of carbohydrates. The separation was carried out at pH 7.5 on a cellulose–boronate matrix. A summary of their results is contained in Table 4.2. Relative retention volumes of the various sugars are determined purely by their content of *cis*-glycol groups possessing the coplanar conformation. Thus, sucrose and the two inositols that are incapable of locating any two of their hydroxyls in this configuration pass through the column with little interaction and elute close to the column void volume. Secondary interactions, such as those exhibited in the separation of nucleotides, are apparently not active or are insignificant.

Goitein and Parsons (46) have used a cellulose–boronate matrix to separate ribose-5-phosphate from its metabolic derivative 5-phospho-α-D-ribose-1-diphosphate (PRibPP). PRibPP is separated from its metabolic derivative not by virtue of its ability to complex with boronate but rather by its inability to form a complex. The presence of the negatively charged 1-pyrophosphate moiety in configuration results in electrostatic repulsion with the boronate anion. Using a pH 8.5 glycylglycine buffer, PRibPP was found to pass unretarded through a small boronate column whereas inosinic acid (IMP) and N^1-(5′-phospho-β-ribosyl)-adenosine-triphosphate (PRibATP) were strongly bound. With a pH 8.5 Tris elution buffer, PRibPP again passes unretarded through the column while its metabolic precursor ribose-5-phosphate and the derivatives IMP and PRibATP are partially retarded.

Glad et al. (35) have demonstrated the utility of their silica-based boronate matrix in the separation of some common monosaccharides. The separation was achieved using a 10 m*M* pyrophosphate, pH 8.25 eluent at a flow rate of 1 ml/min. Glucose and sucrose, which do not have their hydroxyls groups oriented most favorably for complexation, elute close to the column void volume, whereas those compounds containing vicinal *cis*-diols such as sorbitol are more strongly retained. Capacity factors for the carbohydrates ranged from 0.14 for glucose to 4.14 for sorbitol.

It is doubtful that boronate affinity matrices will replace some of the more commonly used methods for the separation of monosaccharides. The great utility of boronate chromatography lies in an area that to date has not been examined. This area is the separation of large polysaccharides and oligosaccharides. Boronates are known from the work of Burnett et al. (49) to complex carbohydrates on cell surfaces. These workers have synthesized a fluorescent boronate that binds reversibly to *Bacillus subtilis* cell walls. The use of boronate adsorbents could prove very useful in the isolation of cell surface carbohydrates and polysaccharides that are not well separated by ion exchange and gel filtration.

4.3.2. Nucleosides and Nucleotides

Nucleosides and nucleotides complex readily with boronates by virtue of the 2′,3′-diol on the ribose moiety. Gilham and coworkers (18, 50) have made an extensive study of nucleosides and nucleotides on boronate–cellulose matrices. Table 4.6 contains some of the results of their study. These experiments

Table 4.6. Retention Volumes of Nucleosides and Nucleotides on Columns of DBCM-Cellulose[a]

	Volume (mL)			
Compound	Buffer	Buffer + 1 *M* NaCl	Buffer + 0.1 *M* $MgCl_2$	Buffer + 1 *M* NaCl and 0.1 *M* $MgCl_2$
Cytidine	15.3	36.4	40.0	
Uridine	15.7	42.0	41.0	44.0
Adenosine	31.0	136.0	121.0	
Uridine 3′,5′-diphosphate	8.3	10.1	10.5	
Uridine 5′-phosphate	8.4	18.0	23.8	
Uridine 5′-diphosphate	8.3	15.2	28.1	
Uridine 5′-triphosphate	8.3	13.3	24.3	25.0
Adenosine 5′-phosphate		32.1	49.0	
Adenosine 5′-diphosphate		24.4	59.3	
Adenosine 5′-triphosphate		19.2	44.6	

[a] The columns had dimensions 40 × 0.6 cm, and the elutions were carried out at 20° and at a flow rate of about 3 ml/hr. Each elution solvent contained the buffer, 0.05 *M* *N*-methylmorpholinium chloride, and the final pH of each solvent was adjusted to 7.5 at 20°. The retention volumes were taken as the volumes of each solvent required to elute the sample and were measured from the point of addition of the sample at the top of the column to the point of its peak concentration eluting from the bottom. From Rosenberg, Wiebers, and Gilham (50).

indicate that the four factors previously mentioned, which are operative in determining the retention volume of a biomolecule, are in effect here. In addition, the nature of the nucleoside base plays a role in determining retention volume. The role of the base is probably attributable to the superimposition of other binding forces, such as H-bonding. Gilham et al. also demonstrated the pH and ionic strength were operative in determining retention only if the nucleoside contained the ribose moiety. For example, while the ribonucleotides were bound more firmly when the pH was increased, the deoxyribonucleotides exhibit little or no change in retention volume when the pH or ionic strength was increased.

The data in Table 4.6 reveal that the total charge on a nucleotide is very important in determining the elution position. In these experiments uridine 3′,5′-diphosphate has been used as a control to detect any interaction with the cellulose derivative other than complexation with boronate since it does not have a *cis*-diol group. Under conditions of low ionic strength the nucleotides are not retained and are eluted with the control. The addition of 1 *M* sodium chloride results in a shielding of the charge interaction between the negatively charged boronate anions and the negatively charged nucleotides. The effect of divalent cations is well illustrated in these results. Interestingly, magnesium ions interact with the nucleotide diphosphates in a manner that results in a greater shielding than occurs with the monophosphates and triphosphates. Magnesium and sodium ions exert their effect at the same site, since the addition of 1 *M* sodium chloride to the magnesium-containing buffer does not alter the retention volume of uridine-5′-triphosphate. Yurkevich et al. (25) have also investigated the chromatography of the adenosine phosphates on their bipolar boronate adsorbents. Their results are consistent with those of Gilham except that they were able to achieve binding at low pH values (2.5–3.8) owing to the stabilizing nature of the bipolar structure.

A boronate–cellulose derivative has been utilized in the assay of ribonucleoside reductases (51). The boronate cellulose column is used for a preliminary group separation of deoxynucleotides from the radioactively labeled substrate ribonucleotides. Total time for the entire assay is less than 5 hr. Gilham (50) also examined the retention behavior of 15 dinucleotide phosphates. The results indicate that the type of nucleotide base occurring in both the 3′ and 5′ position is important in determining the retention volume of the molecule. From studies on the mononucleosides it was found that cytidine has the smallest retention volume and exhibits the least amount of secondary interaction with the support. By using CpC as the reference, the authors were able to estimate the magnitude of the secondary binding effects of any nucleoside in the 3′ or 5′ position relative to cytidine in the corresponding position.

The recent development of high efficiency boronate affinity supports has

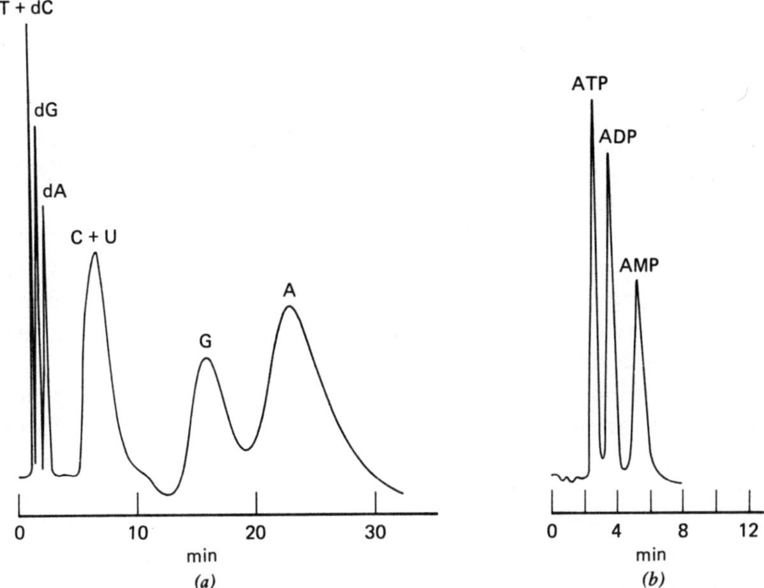

Fig. 4.4. (*a*) Separation of nucleotides with boronic acid silica. (*b*) Separation of nucleotides with boronic acid silica. Both separations employ 0.1 *M* sodium phosphate pH 7.5 as the eluent. From Glad et al. (35), with permission.

resulted in a new method for the high-speed analysis of nucleosides and nucleotides. Figure 4.4 illustrates the types of separation that can be achieved with the silica boronic acid support developed by Glad et al. (35). The amount of time needed to complete the separation in Figure 4.4*a* is of prime importance. While this separation took only 30 min, the separation of the same mixture on DBAE-cellulose using the same eluent took 30 hr (18). The separation of the mono-, di-, and triphosphates of adenosine follows the elution order expected on the basis of their total charge. It is also possible to separate AMP from 3′,5′-cyclic adenosine monophosphate with the same system.

The reversed-phase boronate matrix synthesized by Singhal and coworkers (22) works quite well for the separation of 2′-*O*-methylnucleosides from the unmethylated nucleosides. Recoveries of the nucleosides ranged from 92 to 96%. The separation described by the authors is not an isocratic elution as was the case with Glad's separation. Elution is accomplished by change to a pH 4.5 ammonium acetate buffer. Although resolution of the unmethylated ribonucleosides is not good, the authors state that resolution could probably be imporved by modifying the elution conditions. A similar separation was carried by this group on DBCM-cellulose with a total separation time of 176 min, compared with 46 min for the reversed-phase matrix.

A particularly attractive use for boronate affinity matrices has been in the area of sample cleanup prior to HPLC or G.C. A considerable number of references have appeared on the subject of isolating nucleosides and nucleotides from biological matrices by means of a boronate support (31, 32, 34, 52–55). Uziel et al. (31) were the first group to report the use of a boronate adsorbent for the removal of modified nucleosides from urine. Nucleosides were removed from urine by applying urine directly to a polyacrylamide-boronate column and washing with pH 8.8 ammonium acetate buffer. The nucleosides were then eluted with water followed by 1 *M* acetic acid. Recovery of the nucleosides was 100 ± 5%.

Current methods for isolating nucleosides from biological fluids generally employ some variation of Uziel's original procedure. A major variation is the use of a finer mesh polyacrylamide with a lower initial concentration of hydrazide groups. This results in fewer problems with shrinkage when the pH and ionic strength are changed to elute the nucleosides. Davis and coworkers (32) report analytical recoveries of greater than 90% at concentrations comparable to those found in normal urine, while eight nucleosides exhibited recoveries of greater than 85% at the 1 μg/ml level. Some of the nucleosides isolated by this method include pseudouridine, cytidine, inosine, 1-methyladenosine, N^2-methyl-guanosine, and N^2,N^2-dimethylguanosine (32, 52). Boric acid gel has also been used to isolate 5-fluorouridine from urine prior to capillary gas chromatography.

4.3.3. RNA and Oligoribonucleotides

The information that has resulted from studies of the interaction between nucleosides, nucleotides and polyols with boronate-containing matrices has permitted the extension of the technique to the isolation of RNA, oligonucleotides and polynucleotides. The initial work, as in the case of nucleosides, was carried out by Gilham and coworkers (50). In their studies on polynucleotides they found that raising the solvent pH favors retention, with the maximum effect occurring at about pH 8.5. Magnesium chloride is also required for the same reasons that it is needed in the chromatography of nucleotides. This group also discovered that an increase in chain length results in a decrease in retention. This effect is probably due to increased steric exclusion from the boronate ligand and an increase in total charge on the molecules.

Experiments on the retention of tRNA on the same cellulose supports revealed that DMSO was required to achieve binding. Addition of 20% DMSO to the buffer containing sodium-magnesium chloride resulted in a substantial increase in binding. The percentage of tRNA bound depended on whether or not the tRNA fraction was purified and on the temperature at which the experiment was carried out. Maximum binding occurred with purified tRNA at 0°C. The function of the denaturant in these binding studies probably involves a loosening of the tRNA structure near the 3′ terminal end, thereby alleviating some of the steric restriction to the approach of the 2′,3-diol to the boronate groups on the cellulose.

Boronate-cellulose adsorbents have been found useful in a number of applications. Rosenberg and Gilham (47) have used boronate-cellulose for the isolation of 3′ terminal polynucleotides obtained by ribonuclease T_1 digestion of the RNAs of bacteriophages f_2, Q_1B and GA as well as the isolation of rRNA and tRNA. Phenylalamine rRNA has been separated from phenylalanine tRNA with 2′-deoxyadenosine or 3′-deoxyadenosine incorporated into the 3′ terminal position (56). McCutchan and Gilham (57) have utilized DBAE–cellulose in a preliminary separation of synthetic oligonucleotides pdT-dT_n-dT and pdT-dT_n-A. The two fractions collected from boronate chromatography are then subjected to anion-exchange chromatography to isolate the individual members of each series. DNA can also be easily removed from RNA by employing acetylated DBAE-cellulose (58).

Pace and Pace (27) investigated the use of supports other than cellulose for boronate affinity chromatography. Their main goal was to find a support suitable for the separation of large biopolymers, and which did not contain the residual charge sites characteristic of boronate cellulose. Their results on polyacrylamide and agarose matrices indicate that although boronate polyacrylamide specifically binds *cis*-diol containing mononucleotides or short oligonucleotides, it does not retain polyribonucleotides with sufficient tenacity for preparative use. On the other hand, boronate agarose proved suitable for the fractionation of trace amounts of macromolecular RNA but not for mononucleotides. Presumably, the majority of the boronate groups within the pores of the polyacrylamide P-G were inaccessible to RNA molecules.

From the experiments carried out by Pace and Pace it is evident that the tenacity of the interaction with the boronate support is highly dependent on the RNA. For example, the tRNA used in Figure 4.5*a* binds rather weakly in comparison to 23S RNA. As a consequence, some of the tRNA bleeds from the column during the initial washing of the column. For cases such as this, longer columns should be used and prolonged washing avoided. These workers also found that elution with a competing *cis*-diol is not always effective; like the graded interaction with the boronate support, it depends on the type of bound RNA. The tRNA in Figure 4.5*a* is almost completely eluted when the initial buffer containing 0.1 *M* sorbitol is added at point S. However, mouse L-cell 5S rRNA or *B. brevis* 23S rRNA bind very strongly to the boronate agarose. None of the 5S rRNA is removed by washing or by eluting

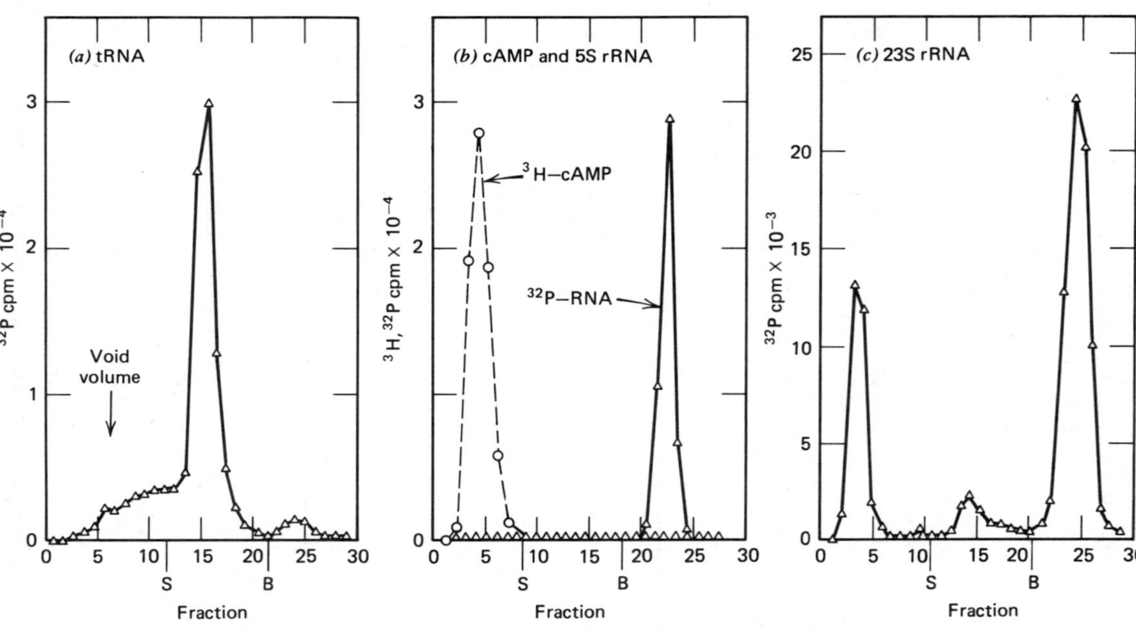

Fig. 4.5. Sorbitol and acid elution of tRNA, 5S rRNA and 23S rRNA from boronate agarose. From Pace and Pace (27), with permission.

with the wash buffer plus sorbitol. Only a pH 5.0 sodium acetate buffer removes the RNA. The 23S rRNA can also be eluted most efficiently with the pH 5.0 acetate buffer. In this case, however, 26% of the RNA flows through the column with the initial buffer and 5% is removed with the initial buffer containing sorbitol. Either the initial conditions to be modified to achieve complete binding or the capacity of the support towards this RNA needs to be examined to make sure the capacity has not been exceeded.

The rapid separation of aminoacyl-tRNA from unacylated tRNA can be readily accomplished by boronate affinity chromatography. The boronate matrix can interact with the unacylated tRNAs by complexing with the 2′,3′-diol of the terminal nucleotide while the aminoacyl-tRNAs pass through the column unable to complex due to substitution at the 2′ or 3′ position. The alkaline conditions normally used to achieve binding to the affinity matrix cause hydrolysis of the aminoacyl-tRNA bond. The high concentration of NaCl and $MgCl_2$ needed to eliminate ionic interactions contribute to a low solubility of the RNAs in the buffer and hence a low binding capacity of the cellulose.

Gilham et al. (24) have tried to optimize the separation conditions for aminoacyl-tRNA on acetylated DBAE-cellulose. Acetylation of the residual aminoethyl groups on the cellulose reduces the strong ionic interactions at neutral pH and allows the use of low-ionic-strength buffers. The conditions chosen for fractionation of *E. coli* aminoacyl-tRNA was pH 7.2–7.7/0.05 *M* 4-methyl-morpholine buffer containing 0.2 *M* NaCl, 0.01 *M* $MgCl_2$, and 20% ethanol (v/v). These conditions produced no significant deacylation during the time needed for chromatography. Recovery from the column was at least 95%. Elution was accomplished using a pH 5.0 0.05 *M* sodium acetate buffer containing 0.2 *M* NaCl. Gilham has also reported separations of yeast aminoacyl-tRNA on unacetylated DBAE-cellulose (33).

Singhal et al. (22) has compared several boronate matrices for the separation of aminoacyl-tRNA. The most interesting support they tested was a reversed-phase boronate support. A chromatogram of the separation achieved on this support is shown in Figure 4.6. The authors tried purification of a number of tRNAs. Recovery of the nonretained aminoacyl-tRNAs was about 80%. However, recovery of tRNAs containing the modified base queuine was extremely poor (6–17%). The purity of the recovered queuine tRNAs was satisfactory. The most useful aspect is the low pH (4.5–6.8) at which the separation can be carried out. This method may be very useful for those aminoacyl-tRNAs which are very susceptible to hydrolysis.

Boronate affinity chromatography can also be applied to nucleic acid substances not having a 2′,3′-diol. Rosenberg et al. (50) have developed a means of incorporating sorbitol into compounds containing a terminal phosphate group. Aqueous solutions of the nucleotides or other terminal-

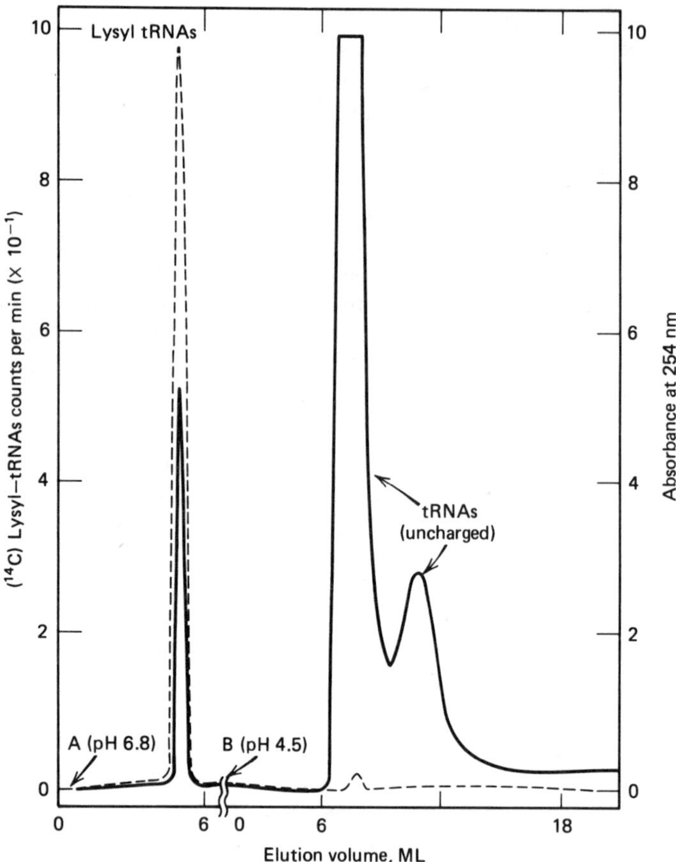

Fig. 4.6. Separation of aminoacyl–tRNA from uncharged tRNAs by reversed-phase boronate chromatography. About 3 mg of rat liver tRNAs containing (^{3}H)lysyl–tRNA (≃4 nmol) in 0.2 mL of buffer *A* (50 m*M* sodium cacodylate, 10 m*M* $MgCl_2$, 0.5 *M* (NaCl, pH 6.8) was applied to an RPB column of 23 × 0.63 cm dimensions. After desorbing the ultraviolet-absorbing material with buffer *A*, the column was eluted with buffer *B* (50 m*M* sodium acetate, 10 m*M* $MgCl_2$, 1.0 *M* NaCl, pH 4.5) in a batchwise manner. The column was eluted with a flow rate of 0.3 mL/min using a Beckman 'AccuFlo' pump, at 4°C. Fractionation of the uncharged tRNAs on this column can be achieved by eluting the column with a linear gradient of chloride ion (0.2 *M* vs 1.0 *M* NaCl) at pH 4.5. From Singhal et al. (22), with permission.

phosphate-containing compounds are converted to the diester by activation with *n*-cyclohexyl-*N*′-*β*-(4-methylmorpholinium)ethylcarbodiimide in the presence of relatively high concentrations of sorbitol. The authors have been able to bind tRNA containing a 5′-terminal phosphate group to boronate cellulose at pH 8.5 using this procedure. This RNA is not normally bound at pH 8.5 in the absence of DMSO. *N*-Methylglucamine can also be incorporated into the terminal phosphate in the same manner as sorbitol. The methylglucamine group can easily be removed from the biopolymers by mild acid treatment, and it binds quite well to boronate cellulose.

4.3.4. Catecholamines

Boronate gels have proved as useful in the removal of catecholamines from biological fluids as in the case of nucleosides. Higa et al. (59) first reported the isolation of epinephrine and norepinephrine from urine using boric acid gel (Aldrich). The recovery of catecholamines from the gel was very dependent on pH, as expected. At or below pH 5.0, most of the catecholamines could be eluted from the column with water, resulting in a loss during the stage of washing unwanted urine component off the column. The recovery values rose with pH to a maximum value of 86% at pH 6.5 for norepinephrine and 96% at pH 7.5 for epinephrine. Their methodology involved adjusting the urine to pH 7.5 with 1 *M* phosphate buffer and either NaOH or HCl depending on the initial pH of the urine. The sample was applied to the column and was then washed with H_2O. Elution of the adsorbed catecholamines was carried out with .023 *M* HCl. The calibration curves were linear over the range of 0.05 to 1.6 μg. Recoveries were 79% for NE and 94% for E. Hansson et al. (60) have also used boric acid gel for the isolation of catecholic amino acids and catecholamines. Since L-dopa was only bound at pH values of 8.0 or greater, the authors could separate it from 5-*S*-cysteinyldopa and dopamine, which are strongly retained at a loading pH of 6.8. 5-*S*-Cysteinyldopa and dopamine could also be separated by first eluting 5-*S*-cysteinyldopa with triply distilled water at pH 5.6 and then eluting dopamine at pH 1.75 with a small volume of mineral acid. Recovery of the catecholic amino acids and catecholamines was nearly complete. A method for separating L-tyrosine from L-dopa has also been published (37).

4.3.5. Glycoprotein, Enzymes, and Peptides

In our laboratory we have been particularly interested in the application of immobilized boronates to the separation of glycoproteins. In particular, the separation of glycosylated hemoglobin from unglycosylated hemoglobin was studied. Glycosylated hemoglobin (HbA_{1C}) is formed by a nonenzymatic glycosylation of the *N*-terminal valines of the β chains in hemoglobin. The initial adduct between valine and glucose is a Schiff base. The Schiff base then undergoes an Amadori rearrangment to the more stable form 1-deoxy-1-(*N*-valyl)-fructose. It is possible that the same type of reaction occurs with amine groups on lysine side-chains.

Figure 4.7*a* illustrates a separation of hemolyzed whole blood on a boric acid gel column. Figure 4.7*b* is the separation of a mixture containing synthetic glucosyl-valyl-histidine and valyl-histidine. In a separate experiment valyl-histidine was not retained on the column. Fractions 26 and 32 in Figure 4.7*b* were subjected to hydrolysis and analyzed by TLC. The results indicate that the bound material is glucosyl-valyl-histidine and is homogeneous despite its appearance. Clearly, the glucose moiety is involved in complexation with the boronate ligand. Recovery of bound hemoglobin for the conditions indicated in Figure 4.7 is typically 93% or better. The boric acid gel is very stable; two months of everyday use resulted in no deterioration of the separation efficiency (62).

Mallia et al. (29) have reported the separation of HbA_{1C} on a boronate agarose affinity column. Their results are almost identical with ours. Separation conditions were carried out at pH 8.5 with 0.25 *M* ammonium acetate buffer containing 0.05 *M* $MgCl_2$. Bound fractions were eluted using 0.2 *M*

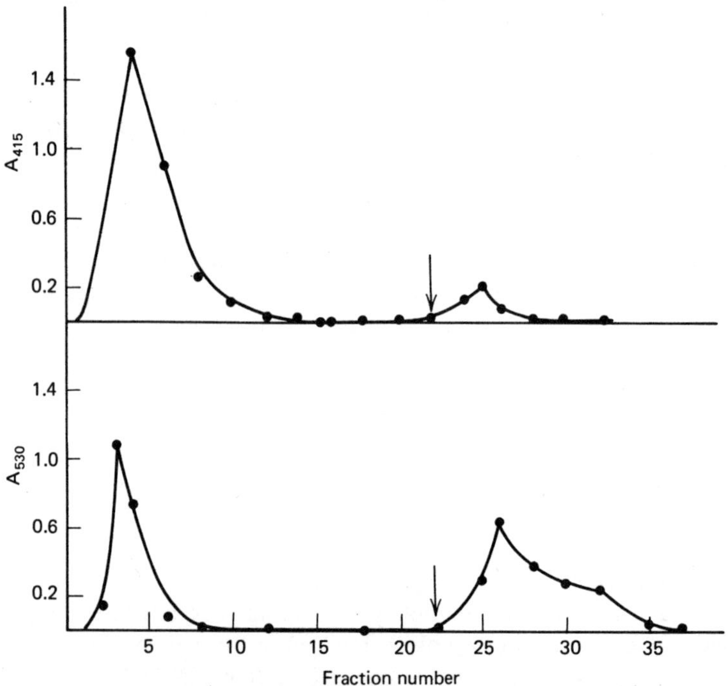

Fig. 4.7. Separation of (*a*) glycohemoglobin from whole human blood and (*b*) glycosyl-valyl-histidine from valyl-histidine. Separations were carried out on a 1 × 4.5 cm column of boric acid gel (Aldrich Chemical). Initial buffer was 0.05 *M* ammonium-acetate pH 9.0. Elution was achieved using distilled water. Fractions of 1.0 mL were collected at a flow rate of 1.0 mL/min. Histidine-containing fractions were detected by the Pauly reaction.

sorbitol in 0.1 *M* Tris (pH 8.5) containing 50 m*M* EDTA. These procedures are a vast improvement over the cation exchange methods currently used for the estimation of HbA_{1C}. Separation times are short (15–30 min), highly reproducible, and are not subject to errors resulting from small pH changes (±0.1) or temperature fluctuations of ±8°C.

The application of boronate adsorbents to the fractionation of glycoproteins has been largely ignored. However, a recent report shows that the technique might have considerable potential in this area (61). The report concerned the solubilization, by Triton X-100, of asialoglycoproteins. Apparently the binding of borate to these glycoproteins restores their solubility. The ability of borate to bind to the glycoproteins indicates that boronate affinity matrices may be useful in their isolation and purification prior to characterization.

ADP-ribosyalted nuclear proteins are readily isolated on dihydroxyboryl polyacrylamide (48). Fractionation was achieved at pH 8.2 using a morpholine · HCl buffer containing 6 *M* quanidine · HCl. When quanidine · HCl was omitted from the buffer, severe nonspecific interactions occurred among the proteins and also between the proteins and the gel. Nonribosylated protein passed through the column without interaction. The ribosylated proteins were then eluted with 150 m*M* potassium phosphate (pH 6.0). Total time for the separation was about 5 hr.

Annamalai et al. (63) have described a novel method for isolating affinity labeled nucleotide regulatory sites. Rabbit muscle pyruvate kinase was labeled with 5′-*p*-fluorosulfonylbenzoyl adenosine and bovine liver glutamate dehydrogenase was modified with 5′-*p*-fluorosulfonylbenzoyl guanosine. Tryptic digests of both enzymes were then subjected to boronate chromatography. At pH 7.8 on DBAE-polyacrylamide the 5′-SO_2B_2-ade peptides are selectively retained while all other peptides elute in the void volume. The nucleosidyl peptide of glutamate was isolated in the same way, providing a quick and simple one-step purification.

Recently, Rose et al. (64) have published a procedure for isolating arginine-containing peptides. Their procedure involves the well-known condensation reaction between the guanidino group of arginine and cyclohexane 1,2-dione (Figure 4.8). The pH at which this reaction is carried out has a dramatic effect on the nature and number of produces formed. At pH 7–9, however, only a single product, DHCH-arginine, is formed. DHCH-arginine is remarkably stable in borate buffer at pH 8–9, with no arginine liberated over a 24 hr period at 37°C. At high pH in the absence of borate, cyclohexane 1,2-dione and arginine are spontaneously regenerated. This method provides a convenient means of isolating arginine-peptides and regenerating the native peptide. Rose and coworkers applied the technique to the isolation and analysis of an arginine-containing peptide from insulin.

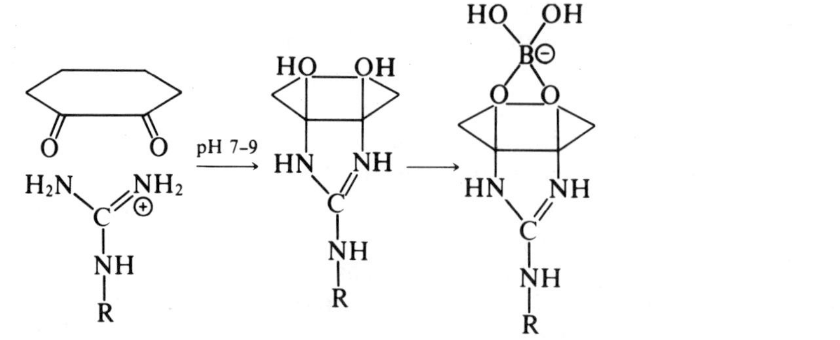

DHCH-arginine DHCH-arginine borate complex

Fig. 4.8. Reaction of the guanidino group of arginine with cyclohexane-1,2-dione to form a *cis*-diol-containing adduct.

Up to this point, all discussions concerning the interaction of biomolecules with boronate were in terms of a covalent complex formed between the boronate ligand and a diol moiety in the biomolecule. Boronic acid derivatives are potent inhibitors of the serine proteinases. Substituted boronic acids exert their inhibitory effect by reproducing the transition state and by forming a covalent bond with the serine residue at the active site. This configuration is depicted in Figure 4.9. One of the hydroxy groups of the tetrahedral boronate forms a covalent bond with the serine residue while the other two hydroxyls serve as hydrogen bond acceptors from donor groups in the "oxyanion hole" of the enzyme. Formation of a phenylboronic acid–glycerol complex stabilizes the tetrahedral state of the boron atom and improves the stability of the enzyme ligand complex. The pH optimum for interaction of

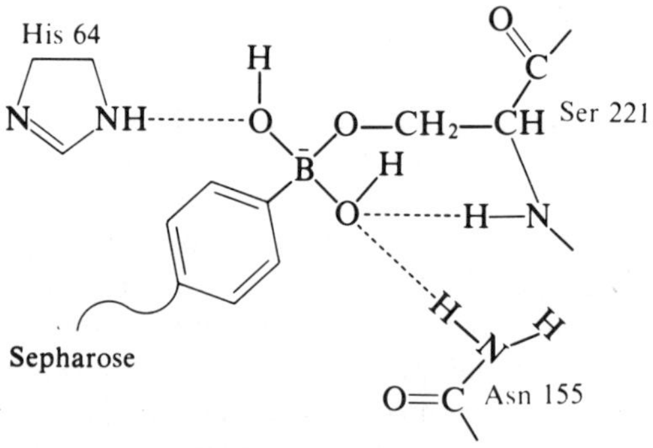

Fig. 4.9. Hypothetical structure of CHPB–Sepharose complex with the active site of a serine proteinase (amino acid residue numbering is given for subtilisin). Glycerol presumably binds with both free hydroxyl groups. From Akparov and Stepanov (26).

subtilisin BPN′ occurs at pH 7.5 (26). The binding properties of boronate-Sepharose in glycerol-containing solutions was used for the direct isolation of subtilisin from *Bacillus subtilis* A-50 culture filtrate (26). In a single step by elution through a boronate-Sepharose column, equilibrated with 0.5 *M* glycerol in 0.05 *M* phosphate buffer (pH 7.5), a subtilisin preparation was obtained with a specific activity of 1.8 units/mg, which can be compared with a commercial preparation of subtilisin with a specific activity of 0.94 units/mg. Hence phenylboronate supports may be useful in the isolation of a variety of serine proteinases. In addition, it would be interesting to examine the effect of substituents on the phenyl ring on retention.

A recent interest in the use of boronate adsorbents saturated with biomolecules has emerged because of their potential use in the affinity purification of enzymes. There are several attractive reasons for applying these supports in this manner. First, the supports are capable of binding a large variety of potential ligands. Second, the binding of these ligands is readily accomplished. Third, the ligands can be removed and replaced with another ligand. This last advantage, however, is sometimes a decided disadvantage.

Maestas et al. (42) have described the use of polyacrylamide-boronate beads saturated with UTP. The purification of UDPG pyrophosphorylase from slime mold using the UTP saturated column is shown in Figure 4.10.

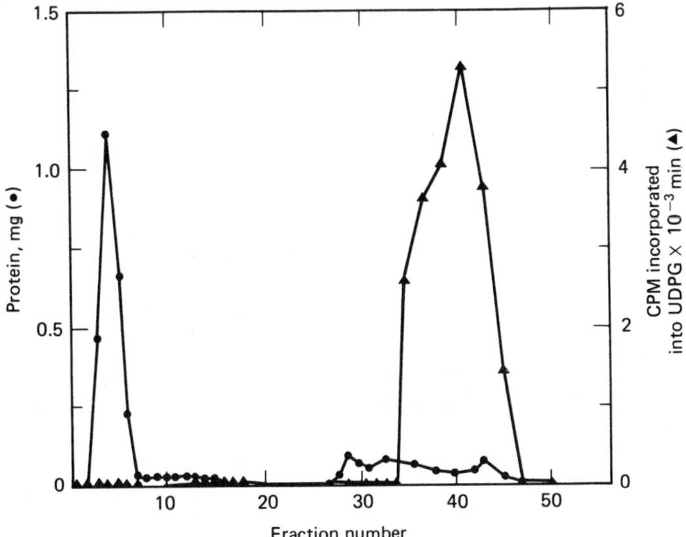

Fig. 4.10. Affinity chromatography of UDPG pyrophosphorylase from the slime mold *Physarum polycephalum* on a P-150 boronate acrylamide column saturated with UTP. From Maestas et al. (42), with permission.

The enzyme was not eluted until UTP was displaced by washing the column with pH 7.0 phosphate buffer containing glucose. On the basis of enzymatic assay, a purification factor of 1060-fold was achieved. Some of this increase may be superficial, however, since UDPG-pyrophosphorylase shows marked activation upon partial purification. On the basis of protein recovered, a 9.3-fold purification was realized.

Bouriotis and coworkers (28) are also involved in the use of biomolecule-saturated boronate adsorbents for affinity chromatography. The work they published is a comparison of an agarose-boronate support, which the authors prepared, and Matrix gel PBA (Amicon). Eleven nucleotides, including NAD^+, FAD, NADH, and $NADP^+$ were tested for retention on the agarose PBA matrix. Of these, NAD^+, $NADP^+$, and ATP were tested for suitability as a saturating ligand in affinity chromatography. Figure 4.11 is the result obtained from chromatographing G-6PDH from yeast protein on Matrix gel PBA saturated with $NADP^+$. A purification factor of twelvefold and a yield of 69% were achieved. A curious result is that G-6PDH had to be eluted with 2 m*M* $NADP^+$ from this column while the agarose column saturated with $NADP^+$ only retarded G-6PDH slightly. The authors feel that this difference may be the result of differing capacities (41 μmol/mL for Matrix gel vs 34 μmol/mL for agarose PBA) or the crosslinking present in the Matrix gel.

The work of Maestas and Bouriotis reveal several problems with these

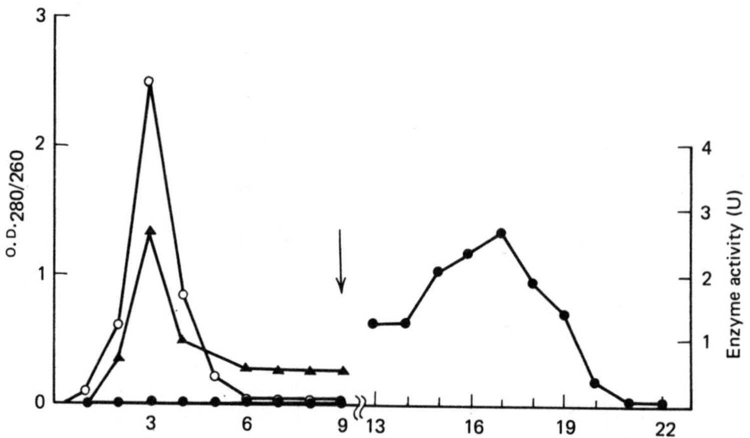

Fig. 4.11. Chromatography of yeast G-6-PDH by $NADP^+$-presaturated Matrex gel-PBA. Yeast enzyme concentrate (7.5 mg in 0.5 mL) in buffer (ligand concentration, 41 mol/mL) was applied to a column of Matrex gel PBA (1 mL). The concentrate contained 19.4U of G-6-PDH. Fractions (1 mL) were collected and assayed for enzyme activity (●), O.D. 260 (▲), and O.D. 280 (O). Elution was effected with 2 m*M* $NADP^+$ (*arrow*). From Bouriotis, Galpin, and Dean (28), with permission.

types of systems. First, the pK of the immobilized boronate is too high. As a result, chromatography at a pH in excess of 8.0 is often necessary to achieve sufficient binding of the ligand. Second, the utility of the methods is limited by the capacity of the gels, particularly the agarose supports. Despite these problems, the technique shows great promise, especially in the area of being able to immobilize more than one affinity ligand at a time. Needed now are boronic acid derivatives with lower pK's and supports with fairly high capacities and high exclusion limits.

4.4. CONCLUSIONS

Despite the rather stringent requirements a molecule must meet in order to form boronate complexes, boronate affinity chromatography is applicable to a wide range of biomolecules. This range will be extended even further with more intensive investigations into the use of boronate supports saturated with biomolecules in affinity separations.

Probably the area of research which is most in need of attention is the synthesis of new boronic acid derivatives with a lower pK. Breakthroughs in this area would open new avenues of investigation. Although all of the supports commonly used in affinity separation have been applied to boronate adsorbents, the ideal support is yet to be found. This is true in all types of affinity separations. The search should not stop, however, for all chromatographic techniques would benefit from a truly inert, derivatized support.

Boronate affinity chromatography will never outclass the time-honored techniques of ion exchange chromatography and gel permeation. It does, however, have its own distinct advantages and belongs in the kit of techniques all biochemists require.

REFERENCES

1. M. Biot, *Compt. Rend., Acad. Sci. (Paris)* (1842).
2. J. Böeseken, *Adv. Carbohydr. Chem.* **4,** 189 (1949).
3. A. B. Foster, *Adv. Carbohydr. Chem.* **12,** 81 (1957).
4. C. A. Zittle, *Adv. Enzymol.* **12,** 493 (1951).
5. D. L. Yabroff, G. E. K. Branch, and H. J. Almquist, *J. Amer. Chem. Soc.* **55,** 2935 (1933).
6. D. L. Yabroff, G. E. K. Branch, and B. Bettman, *J. Amer. Chem. Soc.* **56,** 1850 (1934).
7. K. Torssell, J. H. McClendon, G. F. Somers, *Acta. Chem. Scand.* **12,** 1373 (1958).

8. E. L. Muetterties, *The Chemistry of Boron and its Compounds,* Wiley, New York (1967) p, 498.
9. M. F. Lappert, *Angew, Chem.* **69,** 684 (1957).
10. R. J. Ferrier, *Adv. Carbohydr. Chem.* **35,** 31 (1978).
11. J. P. Lorand and J. O. Edwards, *J. Org. Chem.* **24,** 769 (1959).
12. S. A. Barker, A. K. Chopra, B. W. Hatt, and P. J. Somers, *Carbohydr. Res.* **26,** 33 (1973).
13. R. A. Bowie and O. C. Musgrave, *J. Chem. Soc.* 3945 (1963).
14. T. E. Acree, *Adv. Chem. Ser.* **117,** 208 (1973).
15. A. Finch, P. J. Gardner, P. M. McNamara, and G. R. Wellum, *J. Chem. Soc., A* 3339 (1970).
16. P. J. Garegg and B. Lindberg, *Acta Chem. Scand.* **15,** 1913 (1961).
17. E. J. Bourne, E. M. Lees, and H. Weigel, *J. Chromatogr.* **11,** 253 (1963).
18. H. L. Weith, J. L. Wiebers, and P. T. Gilham, *Biochemistry* **9,** 4396 (1970).
19. S. J. Angyal, and D. J. Mchugh, *J. Chem. Soc.* 423 (1957).
20. J. Solms and H. Deul, *Chimia* **11,** 311 (1957).
21. R. L. Letsinger and S. B. Hamilton, *J. Amer. Chem. Soc.* **81,** 3009 (1959).
22. R. P. Singhal, R. K. Bajaj, C. M. Buess, D. B. Smoll, and V. N. Vakharia, *Anal. Biochem.* **109,** 1 (1980).
23. J. H. Hageman and G. D. Kuehn, *Anal. Biochem.* **80,** 547 (1977).
24. T. F. McCutchan, P. T. Gilham, and D. Soll, *Nucleic Acids Res.* **2,** 853 (1975).
25. A. M. Yurkevich, I. I. Kolodkina, E. A. Ivanova, and E. I. Pichuzhkina, *Carbohydr. Res.* **43,** 215 (1975).
26. V. H. Akparov and V. M. Stepanov, *J. Chromatogr.* **155,** 329 (1978).
27. B. Pace and N. Pace, *Anal. Biochem.* **107,** 128 (1980).
28. V. Bouriotis, I. J. Galpin, and P. D. G. Dean, *J. Chromatogr.* **210,** 267 (1981).
29. A. K. Mallia, G. T. Hermanson, R. I. Krohn, E. K. Fujimoto, and P. K. Smith, *Anal. Lett.* **14**(B8), 649 (1981).
30. J. K. Inman, and H. M. Dintzis, *Biochemistry* **8,** 4074 (1969).
31. M. Uziel, L. H. Smith, and S. A. Taylor, *Clin. Chem.* **22,** 1451 (1976).
32. G. E. Davis, R. D. Suits, K. C. Kuo, C. W. Gehrke, T. P. Waalkes, and E. Borek, *Clin. Chem.* **23,** 1427 (1977).
33. R. E. Duncan, and P. T. Gilham, *Anal. Biochem.* **66,** 532 (1975).
34. R. A. Olsson, *J. Chromatogr.* **176,** 239 (1979).
35. M. Glad, S. Ohlson, L. Hansson, M. O. Mansson, and K. Mosbach, *J. Chromatoger.* **200,** 254 (1980).
36. S. A. Barker, B. W. Hatt, P. J. Sommers, and R. R. Woodbury, *Carbohyr. Res.* **26,** 55 (1973).
37. C. A. Elliger, B. G. Chan, and W. L. Stanley, *J. Chromatogr.* **104,** 57 (1975).
38. E. Seymour, and J. M. M. Frechet, *Tetrahed. Lett.* **1149** (1976).
39. H. Schott, E. Rudloff, P. Schmidt, R. Roychoudhury, and H. Kossel, *Biochemistry* **12,** 932 (1973).
40. G. Wulff, W. Vesper, R. Grobe-Einsler, and A. Sarhan, *Makromol. Chem.* **178,** 2799 (1977).
41. G. Wulff, and W. Vesper, *J. Chromatogr.* **167,** 171 (1978).

42. R. R. Maestas, J. R. Prieto, G. D. Kuehn, and J. H. Hageman, *J. Chromatogr.* **189,** 225 (1980).
43. R. L. Letsinger, and J. R. Nazy, *J. Amer. Chem. Soc.* **81,** 3013 (1959).
44. B. Wolf, *Soil Sci. Plant Anal.* **5,** 39 (1974).
45. S. A. Barker, B. W. Hatt, P. J. Somers, and R. R. Woodbury, *Carbohydr. Res.* **26,** 55 (1973).
46. R. K. Goitein, and S. M. Parsons, *Anal. Biochem.* **83,** 641 (1978).
47. M. Rosenberg, and P. T. Gilham, *Biochim., Biophys. Acta* **246,** 337 (1971).
48. H. Okayama, K. Veda, and O. Hayaishi, *Proc. Nat. Acad. Sci. U. S.* **75,** 1111 (1978).
49. T. J. Burnett, H. C. Peebles, and J. H. Hageman, *Biochem. Biophys. Res. Commun.* **96,** 157 (1980).
50. M. Rosenberg, J. L. Wiebers, and P. T. Gilham, *Biochemistry* **11,** 3623 (1972).
51. E. C. Moore, D. Peterson, L. Y. Yang, C. Y. Yeung, and N. F. Neff, *Biochemistry* **13,** 2904 (1974).
52. K. C. Kuo, C. W. Gehrke, and R. A. McCune, *J. Chromatogr.* **145,** 383 (1978).
53. C. W. Gehrke, K. C. Kuo, G. E. Davis, R. D. Suits, T. P. Waalkes, and E. Borek. *J. Chromatogr.* **150,** 455 (1978).
54. A. P. DeLeenheer and C. F. Gelijkens, *J. Chromatogr. Sci.* **16,** 552 (1978).
55. E. H. Pfadenhauer and S. D. Tong, *J. Chromatogr.* **162,** 585 (1979).
56. M. Sprinzl, K. H. Scheit, H. Sternbach, F. ven der Haar, and F. Cramer, *Biochem. Biophys. Res. Commun.* **51,** 881 (1973).
57. T. F. McCutchan and P. T. Gilham, *Biochemistry* **12,** 4840 (1973).
58. S. Ackerman, B. Cool, and J. J. Furth, *Anal. Biochem.* **100,** 174 (1979).
59. S. Higa, T. Suzuki, A. Hayashi, I. Tsuge, and Y. Yamamura, *Anal. Biochem.* **77,** 18 (1977).
60. C. Hansson, G. Agrup, H. Rersman, A. M. Rosengren, and E. Rosengren, *J. Chromatogr.* **161,** 352 (1978).
61. R. S. Pratt and G. M. W. Cook, *Biochem. J.* **179,** 299 (1979).
62. A. F. Bergold and W. H. Scouten, unpublished results (1979).
63. A. E. Annamalai, P. K. Pan, and R. F. Colman, *Anal. Biochem.* **99,** 85 (1979).
64. K. Rose, J. D. Priddle, and R. E. Offord, *J. Chromatogr.* **210** 301 (1981).

CHAPTER

5

NUCLEIC ACID AFFINITY CHROMATOGRAPHY

THOMAS O. FOX and CHARALAMBOS SAVAKIS*

Department of Neuroscience
Children's Hospital Medical Center and Harvard Medical School
Boston, MA 02115

Biopolymers that have specific or fortuitous affinities for nucleic acids can be chromatographed on matrices to which nucleic acids have been immobilized. Two methods used extensively for this purpose are DNA-cellulose affinity chromatography for proteins and oligo[dT]-cellulose and poly[U]-Sepharose chromatography for separating different classes of RNA. Under appropriate conditions, enzymatic and regulatory proteins, including bacterial DNA (1, 2) and RNA (1) polymerases and procaryotic repressors (3), adhere to columns or batches of DNA-cellulose as a consequence of their physiologically functional binding sites for nucleic acids. Other proteins (4) also bind to these columns and can be fractionated empirically. In some cases, a possible association in vivo of the protein with DNA might be established by a combination of other analyses, whereas other proteins can bind to DNA-cellulose columns strictly as a result of electrostatic interactions that are without functional significance. Whether this method is used to analyze specific DNA-interacting proteins, proteins with possible in vivo DNA-binding properties, or proteins that only bind to DNA under experimental conditions, chromatography with these analytical matrices can yield very effective resolutions of the DNA-binding class of proteins from an organism (1) and purifications of a wide range of selective proteins (5).

Two types of information can be obtained with nucleic acid affinity chromatography: characteristics of the adhering macromolecules, independent of the affinity matrix, and properties of the affinity moiety that determine or influence the binding. Usually this method has been used to separate proteins or polynucleotides that bind to these matrices, to fractionate them from interacting molecules other then nucleic acid, or to resolve heterogene-

*Biological Laboratories, Harvard University, Cambridge, Massachusetts. Current address: Department of Genetics, University of Cambridge, Cambridge, England.

ous species. For such analyses of proteins the use of DNA may be favorably indicated, though binding to it is not indicative itself of in vivo function. It is possible, however, to use these methods analytically to probe physiologically relevant protein–nucleic acid interactions and functions and thereby attempt to elucidate the nature and parameters of the nucleic acid moiety that is attached to the matrix. Several reports (6, 3) provide reviews of these issues and conditions for approaching questions of affinity and specificity of polynucleotide binding sites for specified proteins.

One area of biological and biochemical research that has benefited particularly from nucleic acid affinity chromatography–both to fractionate and analyze specific proteins and to prepare separated messenger RNA–is the investigation of regulatory mechanisms for steroid hormone action. Several models of putative steroid receptors (7, 8, 9) propose that these complexes act at the cellular nucleus, where they are observed to accumulate. The apparent affinities of many steroid receptors with DNA suggest that in cells they may adhere to DNA either directly or indirectly. Taking advantage of this ability to adhere to DNA-cellulose, many investigators utilize this method for probing these mechanisms. Since DNA-cellulose chromatography of steroid receptors has not previously been reviewed selectively, this chapter draws upon these studies heavily, though not exclusively, to summarize and contrast certain steroid receptor properties while illustrating several aspects of nucleic acid affinity chromatography. Additional facets of steroid receptors and their interactions with nucleic acids and non-nucleic acid components in the cell are beyond the scope of this chapter; the discussions contained here represent only examples of the investigations into steroid receptor mechanisms.

Methods for preparing and utilizing DNA-cellulose to chromatograph proteins are available (1, 4). The most noted and used of these treatises (4) is quite adequate and reliable for initiating and applying this method. Also available for reference are another useful review of DNA-binding proteins that have been examined with DNA-cellulose chromatography (10) and descriptions of related methods with DNA attached to Sephadex and Sepharose supports (11), agarose and acrylamide gels (10), and additional covalent attachment methods (12). Further variations on DNA-cellulose chromatography include the use of bromodeoxyuridine-substituted DNA (13) and chromatin (14) attached to cellulose. An alternative approach, with the advantages of more defined matrices, is to use oligo(dT)-cellulose (15, 16), other oligodeoxynucleotide–cellulose complexes (17), and ATP-Sepharose (18) to chromatograph proteins that also bind to DNA. These alternative methods have been used to chromatograph steroid receptors as well as other DNA-binding proteins, and they provide unique information which complements that obtained with DNA-cellulose. A recent, thorough summary and outline of the general approaches available for chromatography with immobilized nucleic acids is contained in (5).

This chapter examines several practical aspects of the preparation and use of DNA-cellulose chromatography for analyses of proteins, and it discusses the use of oligo[dT]-cellulose and poly[U]-Sepharose chromatography for obtaining separated classes of messenger RNA molecules. Particular attention is directed at the appropriateness of specified procedures for different applications, pitfalls that have been encountered, and potentials and suggestions for future studies.

5.1. PREPARATION OF DNA-CELLULOSE COLUMNS FOR CHROMATOGRAPHY OF PROTEINS

Cellulose and DNA solutions can be combined to prepare two types of DNA-cellulose, native and denatured. It is advised to begin chromatographing proteins with denatured DNA-cellulose, since proteins that bind to native DNA-cellulose also bind to denatured DNA-cellulose, along with a large number of additional proteins (19, 20). This selectivity was used (20) in a protocol for removing those proteins that adhere to both native and denatured DNA-cellulose to obtain the proteins that are specific for denatured DNA-cellulose binding; first samples were run through a column of native DNA-cellulose and then the nonadhering proteins were rechromatographed on a column of denatured DNA-cellulose. Probably the binding to denatured DNA-cellulose of two types of proteins—those that adhere to single-stranded DNA and those that adhere to double-stranded DNA—is a consequence of the presence of both single-stranded and local double-stranded DNA regions on DNA-cellulose prepared with denatured DNA. Native DNA-cellulose can also be used to select and chromatograph double-strand specific DNA-binding proteins.

We (19, 21, 22, 23, 24) have followed closely the methods of Alberts and Herrick (4; see also refs. 1 and 2) to prepare DNA-cellulose. It is neither necessary nor usually beneficial to use homologous DNA, since virtually all DNA-binding proteins that have been examined exhibit low affinity with DNAs of nonspecific nucleotide base sequence, whether or not they also can be shown to exhibit high affinity for specific base sequences. As a highly reliable and reproducible source of DNA that is both suitable and adequate, we use highly polymerized commercial calf thymus DNA (Worthington Biochemical Corporation, Piscataway, N. J.). Both native and denatured DNA-cellulose are prepared by drying the appropriate DNA solution onto Munktell 410 cellulose (Cellex 410, Bio-Rad Laboratories, Richmond, Calif.) (4).

Cellulose is washed by suspending it in 95% ethanol (7 times, 700 mL/

500 g cellulose each time) to remove residual pyridine. Solvent is removed by vacuum filtration through a large sintered-glass Buchner funnel (Pyrex, medium porosity). The cellulose is then washed in glass-distilled water (1 L for 500 g), suspended in 0.1 *N* NaOH for 5 min, washed again with water, suspended in 0.1 *N* HCl for 2 hr, and finally washed 6 times with water. After verifying that the pH of the last wash is near neutrality, the cellulose is air-dried, and sometimes dried more thoroughly under vacuum, before storage and use. To prepare native DNA-cellulose, the dried and powdered cellulose is added to a solution of native DNA and then processed as described (4, 19).

Some variations can enhance the preparation and use of DNA-cellulose. Using sterile dissecting forceps and scissors, we cut 0.5 × 5 mm pieces of native calf thymus DNA and tease it apart. This increases its rate of solution and yields a more disperse solution with minimal clumps, thus facilitating even adsorption to cellulose. The DNA is dissolved (2 mg/mL) in 10 m*M* K_2HPO_4, 1 m*M* Na_3EDTA, with gentle overnight mixing with a magnet stirrer set for low rotation, at 2°C. This step can be aided by evacuating the DNA-buffer suspension; this removes bubbles and speeds wetting and sinking of the DNA. Heating at this stage also results in faster solution and can be tolerated when all reagents are clean (G. Herrick, personal communications).

The DNA is then denatured by dispensing the solution into 15 mL fractions in tempered glass test tubes and immersing these tubes in 100°C water for 15 min. The tubes are removed one at a time, immediately immersed in an ice-water slurry for 5 min, and twirled frequently to insure rapid thermal equilibration. Denaturation is monitored by measuring hyperchromicity; we typically obtain an apparent increase in O.D. at 254 nm of 25 to 39%. The denatured solution is then poured into a shallow glass pan, and 1 *M* Tris-HCl pH 7.4 (21°C) is added to a final concentration of 20 m*M*.

Dry cellulose powder is added to the buffered denatured DNA solution at a proportion of approximately 30–35 g per 100 mL of solution; this produces a thick paste without standing puddles when spread approximately 5 mm thick in a shallow Pyrex dish. The dish is covered with gauze, and room-temperature air is blown over it for 12–15 hr. The DNA-cellulose, now with a cakey consistency, is transferred to a large beaker. It is left loose, without tight packing, to permit further drying. The beaker containing DNA-cellulose is placed in a large desiccator jar containing Drierite and P_2O_5 and is dried for 2–3 days with a vacuum pump connected through a trap with dry ice in 2-methoxy-ethanol. This is continued until either no additional water uptake is detected in the desiccant or no additional weight is lost from the DNA-cellulose. Rigorous drying at this stage results in the highest yield of adhered DNA (greater than 90% of input DNA). The dried DNA-cellulose is then resuspended and allowed to settle 4 times in five-fold excess of 10 m*M*

Tris (pH7.4 at 21°C), 1 m*M* EDTA with a minimum of 6 hr between each buffer change. This removes fine cellulose particles and unbound DNA. It is finally resuspended as a 1:1 (buffer volume:packed volume of DNA-cellulose) slurry in the same buffer with a final concentration of 1 *M* NaCl and frozen at −20°C for storage until use.

Two methods have been used to measure the efficiency of DNA adherence to cellulose. DNA can be removed from the DNA-cellulose by hydrolysis or boiling and measured by UV absorbance (4, 10). We find, however, that with continued storage at −20°C the DNA remains tightly bound to the cellulose matrix (19). Accordingly, to determine the DNA content accurately, we find routinely that it is convenient and reliable to assay the DNA-cellulose directly by the diphenylamine color test for deoxyribose (25). A diluted slurry is mixed with the reagent as usual and is centrifuged to remove cellulose prior to detection; a parallel slurry of cellulose provides a blank with background absorption. As mentioned above, we routinely obtain about 90% input DNA bound to denatured DNA-cellulose. This is almost double the yield that is usually reported for dried DNA-cellulose preparations (10, 12). As do others, we obtain a lower yield (about 50%) when we have prepared native DNA-cellulose (1). This is very similar to the result reported for binding of calf thymus DNA (50% for native; 80–90% for denatured) to Whatman No. 1 paper disks (26).

DNA-cellulose is quite stable to frozen storage and multiple reuse. Similar chromatographic patterns are observed as long as several years after initial preparation. DNA-cellulose also can be reused many times. Generally, we find no differences in chromatographic patterns between the first and subsequent uses of a column. Routinely, we use a given column only one or two times and then remove and refreeze the slurry, so that reused DNA-cellulose is usually repacked and poured prior to reuse. The DNA-cellulose is used continually, with dozens of freeze–thaw cycles, until either adsorption efficiencies of known DNA-binding proteins decrease or the DNA content of the preparation drops significantly (less than 50% of the original). Since some extracts and chromatographic protocols will degrade the matrix relatively quickly, the longevity described above certainly will not be attained in every application.

To examine steroid receptors, we commonly chromatograph cytosol samples equivalent to 75 mg of tissue per mL of denatured calf thymus DNA-cellulose. Receptor yields are optimal over approximately a fivefold range of tissue concentrations, with the maximum we load being 200 mg tissue weight per mL of DNA-cellulose. Much below this range, yields are sometimes variable, presumably because of nonspecific loss of proteins, and above this range, yields of steroid receptors are reduced by overloading of column sites. For proteins that are single-strand-specific and bind selectively to denatured

DNA-cellulose (20), a concentration more than six times higher than the above was chromatographed on DNA-cellulose; that separation (extract from 500 g frozen calf thymus on 400 mL packed volume of denatured DNA-cellulose) was close to maximal, without loss of major species of DNA-binding proteins (20). Whether an extract is cytosolic or of nuclear origin, is crude or partially fractionated, and whether a protein of interest is minor or major, or single-strand-specific or not, will determine the capacity achievable with DNA-cellulose chromatography. With these guidelines for a start, capacities can be determined empirically by rechromatographing the nonadhering fraction from an initial column to learn if some proteins remain that will bind to a second column.

Typically, to analyze estrogen or androgen receptors, we use columns with 2 to 5 mL packed volume of DNA-cellulose. Samples are incubated with the appropriate concentration of radioactive hormone for 1 hr or more before application to the DNA-cellulose columns. Ionic strength of extracts and buffers is monitored by measuring the conductivity of 10 mL aliquants in deionized H_2O with a Radiometer CDM3 instrument (Copenhagen, Denmark). The ionic strength of the extract is reduced, when necessary, by addition of buffer containing no added NaCl, to be equivalent to extract buffer with 50 m*M* NaCl. This, or some procedure for desalting, is essential for extracts of tissues that contain appreciable salts, which interfere with protein binding. Samples are loaded by gravity onto DNA-cellulose columns that have been equilibrated with buffer containing 50 m*M* NaCl. Binding of some proteins is enhanced by exposure of the column containing the loaded sample to an elevated temperature (for example, 21°C) for up to one hour and then cooling it to 2°C again prior to washing and eluting.

Columns are washed with a minimum of 12 column-volumes using a flow rate regulated by a peristaltic pump at 1 column-volume/hr. In earlier experiments, 0.2 mg/mL of bovine serum albumin (crystallized bovine fraction V, Pentex, Miles Laboratories, Elkhart, Ind.) was included in elution buffers used for chromatography. For studies of androgen and estrogen receptors in kidney and brain extracts, equally efficient or better results were obtained without carrier protein (23), so it was eliminated in subsequent experiments, as has been reported for sufficiently high cell extract concentrations (10). Radioactivity in the wash-through fractions is monitored until a low, constant background is reached. Samples are then eluted either with a step in ionic strength or by pumping buffer with a linear gradient of NaCl (for example, 50–300 m*M*) through each column. For determining radioactivity only, samples are collected in 50 × 16 mm polypropylene vials (W. Sarstedt, Inc., Princeton, N. J.). However, radioactivity irregularly adheres to these vials, so when aliquants must be removed and counted first—to select samples for subsequent assays—elution is into polystyrene tubes.

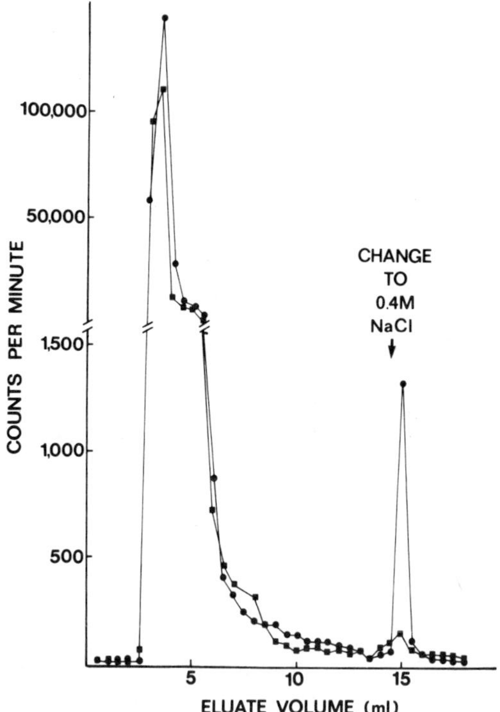

Fig. 5.1. DNA-cellulose chromatography to distinguish and separate the estradiol-binding proteins: mouse alpha-fetoprotein and estrogen receptor. An extract of hypothalamus from 1-week-old mice was prepared in buffer (0.01 *M* Tris-HCl, pH 8.1, 21°C; 1 m*M* Na_3 EDTA [ethylenediamine tetraacetate]; 1 m*M* mercaptoethanol; 0.15 *M* NaCl; and 10% [v/v] glycerol). After centrifugation, aliquots were incubated with 4.8 n*M* [^{3}H]estradiol alone (●) or [^{3}H]estradiol combined with 320 n*M* diethylstilbestrol (DES) (■). The labeled extracts were chromatographed on DNA-cellulose columns (1.5 mL packed volume). Alpha-fetoprotein eluted in the wash, and estradiol receptors eluted with buffer containing 0.4 *M* NaCl. [Data from T. O. Fox, *Nature* **258,** 441, (1975).]

Figure 5.1 demonstrates the separation of two estradiol-binding proteins from perinatal mouse hypothalamus (27). The first, perinatal estradiol-binding protein, or alpha-fetoprotein, does not adhere to DNA-cellulose and is not significantly inhibited by the synthetic estrogen, diethylstilbestrol (DES), from binding radioactive estradiol. Accordingly, the alpha-fetoprotein is seen in the wash and is relatively unaffected by DES competition, whereas estradiol receptor binds to DNA-cellulose, is inhibited by DES, and elutes when the ionic strength is increased with a step of NaCl. Using this separation, another protocol was designed to allow detection of estradiol receptors from embryonic mouse hypothalamus (Fig. 5.2; ref. 28). The extract was

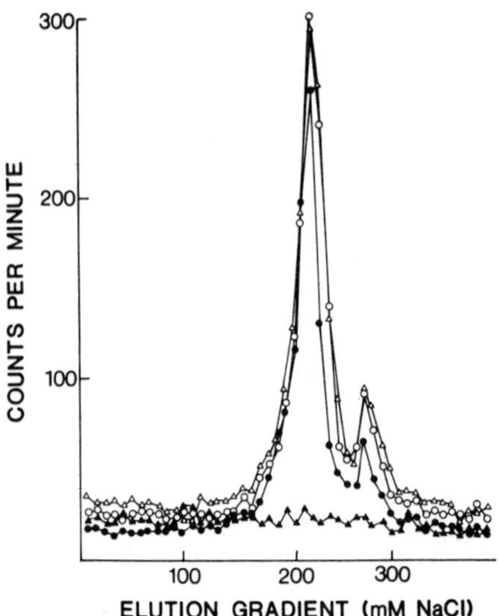

Fig. 5.2. DNA-cellulose chromatography of estradiol receptor from brain cytosol extracts of embryonic mice. Hypothalami from mouse embryos 3 days prior to birth were dissected and placed in buffer containing 10 m*M* Tris-HCl, pH 8.1 (21°C); 1 m*M* Na_3 EDTA; 1 m*M* mercaptoethanol; 150 m*M* NaCl; and 10% (v/v) glycerol. This buffer and all subsequent steps were at 2°C. The tissue was homogenized in a Kontes Teflon glass homogenizer. The homogenate was combined with two rinses and was centrifuged at 40,000 rpm for 60 min in a Beckman type 40 rotor. The cytosol fraction obtained was adjusted, using conductivity measurement, to 50 m*M* NaCl. Portions of the extract were chromatographed directly on 2 mL DNA-cellulose columns, equilibrated with buffer containing 50 m*M* NaCl and 1.0 mg of bovine serum albumin per mL. After 2.5 hours of washing to remove alpha-fetoprotein, 2 mL of buffer containing [^{3}H]estradiol at the specified concentrations was layered on and allowed to completely enter the colunms. These columns were incubated for 1 hr and then washed with 12.5 volumes of buffer over 16 hr, a procedure that removes the unbound radioactive hormone. The [^{3}H]estradiol-bound contents of the columns were eluted with a 24-mL linear gradient of NaCl (50 to 400 m*M*). Radioactivity of 0.5 mL fractions was determined in a two-phase system with Omnifluor (New England Nuclear) toluene scintillation counting fluid. *Symbols:* 3.5 n*M* [^{3}H]estradiol, ●; 8.4 n*M* [^{3}H]estradiol, △; 13.0 n*M* [^{3}H]estradiol, ○; 8.0 n*M* [^{3}H]estradiol plus a hundredfold excess nonradioactive DES, ▲. [Data from C. C. Vito and T. O. Fox, *Science* **204,** 517 (1979).]

chromatographed on DNA-cellulose, and alpha-fetoprotein was washed away. Then radioactive estradiol was added to the column, the receptor that had adhered became labeled, the excess unbound radioactivity was washed away, and the labeled receptor was then eluted with a gradient of increasing NaCl concentration (Fig. 5.2). Both qualitative characteristics of the receptor's elution and quantitative aspects of hormone binding were detected simultaneously.

Many agents besides salts have been used to elute proteins from DNA-cellulose, for example, various soluble RNAs and DNAs (4) and sodium dextran sulfate (20). For steroid receptors, different DNAs (29), synthetic oligo- or polynucleotides (29, 30), pyridoxal phosphate (31), and dyes (32) have provided very effective elutions from DNA-cellulose and oligodeoxynucleotide-celluloses while also providing valuable information. Alternatively, dyes have been used to block binding of steroid receptors to DNA-cellulose (33). Further elaborations and combinations of these ligands should provide a wealth of differential methods for binding and eluting specific proteins of interest.

As for some other methods, DNA-cellulose chromatography does not always work the first time. Among the problems that have been encountered are tissue extracts that cause the DNA to elute from the column with the flow-through. We have never encountered this problem, but it might be caused selectively by certain proteins or result from inadequate prewashing of the DNA-cellulose after preparation and prior to chromatography, such that tightly binding proteins would strip poorly adhered DNA from the cellulose matrix. As a precaution against such a possibility, we generally chromatograph a standard tissue extract through new, unused DNA-cellulose to test that it is functioning properly. This suggested pretreatment was reported to be effective in a study of androgen receptors from rat brain (34). Methods for preparing extracts to chromatograph on DNA-cellulose, including the removal of endogenous DNA which can complete with the DNA on the column, controls with columns or removable plugs of cellulose, and other initials considerations have been discussed at length (4).

5.2. APPROPRIATENESS OF DNA-CELLULOSE CHROMATOGRAPHY FOR THE SEPARATION OF PROTEINS

As an alternative among many methods in protein chemistry, DNA-cellulose chromatography can provide particularly useful resolutions and separations of proteins. It has, for example, distinguished androgen receptors of wild-type mice from putative residual androgen receptors of an androgen-resistant mutant when other methods had not given a similar resolution (35, 22). As stated above, however, it is not necessary that binding to DNA by respective proteins either occurs physiologically or that such an in vivo interaction is inferred. What is most important is the extent to which the method provides convenient separations of proteins or unique information.

To minimize the binding of proteins that have no specific affinity with DNA but that would adhere electrostatically to anionic residues of the DNA-cellulose, it is important to perform DNA-cellulose chromatography under conditions of sufficient ionic strength. This can be achieved with concentrations of salts and buffers that equal or exceed formulations for physiological saline (equivalent to approximately 0.15 *M* NaCl or KCl), as have been used to examine estrogen receptors (36, 21). For some proteins, however, a lower ionic strength is required for sufficient stability of binding, and buffers with 0.05 *M* NaCl are used when loading these samples (1, 20, 35). Since proteins with adequate regions of basic amino acid residues still might adhere by ionic bonding, chromatographic conditions of moderate or high ionic strength do not assure that proteins that adhere to DNA-cellulose do so specifically. On the other hand, it remains possible that proteins which chromatograph on DNA-cellulose entirely under conditions of low ionic strength—for example, some glucocorticoid receptors—do so by binding sites that probably exhibit affinities for DNA as a result of physiological functions (37, 38, 39).

Proteins that probably do not interact with DNA in the cell can adhere to DNA-cellulose and chromatograph quite reproducibly under conditions of low ionic strength (40), presumably by ionic interactions. In one of the earliest reports of DNA-cellulose analysis of mammalian cell proteins, newly synthesized proteins in growing and nongrowing cell cultures were compared (41). Different peaks of DNA-binding proteins were observed that were either more prevalent in serum-deprived nondividing mouse cells or were more prevalent in serum-supported dividing cultures. It was shown that the former class of proteins comprised underhydroxylated precursors of collagen; their detection by DNA-cellulose chromatography was subsequently prevented by culturing with sufficient serum or added ascorbic acid (40). These proteins adhered to DNA-cellulose in buffers containing 50 m*M* NaCl and were eluted with buffer containing only 150 m*M* NaCl. No evidence suggests that binding of these proteins to DNA is physiologically meaningful, and such binding might not even be possible in an intact cell.

In contrast to the low-salt-eluting collagen precursors, the other proteins detected in this study could be eluted from DNA-cellulose with higher ionic strength buffers (42). In addition, these proteins exhibited much higher affinity for single-stranded than for double-stranded DNA and thereby resemble other DNA-binding proteins that have been analyzed extensively with these methods (20, 43). Like the gene-32 protein of T4 bacteriophage (44), these proteins can be shown to elicit DNA unwinding and to enhance DNA polymerization activity (45, 46). Thus a bacteriophage mutation and in vitro studies suggest that these DNA-cellulose binding proteins have important DNA interacting functions in the cell. While some of these proteins can be chromatographed on DNA-cellulose in buffers containing 50–150 m*M* NaCl (20, 19), additional proteins with selectivity for single-stranded DNA are eluted with 200 m*M*—2.0 *M* NaCl (42, 20, 19). Although low ionic strength increases the probability of electrostatic interactions unrelated to specific DNA interactions during DNA-cellulose chromatography, proteins that might bind to DNA in vivo also appear to adhere under these conditions. In either case, for proteins that bind to DNA-cellulose without a likely intracellular DNA interaction (40), and for proteins that might exhibit both in vivo and in vitro binding (20, 43), as does the T4 gene 32 protein (1), the method can yield well-fractionated protein preparations quite suitable for further purification.

An assumption frequently made from elution characteristics is that proteins that have higher affinity with DNA elute from DNA-cellulose with higher ionic strength. While this assumption is sometimes presented cautiously (37), it must be proved rigorously if conclusions about affinity are desired, or these must be determined under appropriate conditions and protocols (6). For two forms of the estradiol receptor, the order of their respective affinities with DNA, as measured by an alternative method under conditions that approach equilibrium (47), was as expected from their order of elution from DNA-cellulose. Likewise, a correlation was reported for single-strand-specific DNA-binding proteins, with the higher-salt-eluting fractions (20) being more effective in stimulating melting of DNA (45) and in stimulating DNA polymerase activity (46). Therefore, elution can offer hypotheses about relative affinities of proteins with DNA, but these must be confirmed or challenged independently, including determinations by other methods (47, 20, 45, 46, 48). For the *lac* repressor, which has an affinity with specific operator DNA about 10^6-fold higher than with nonoperator DNA (49), binding to these sequences was distinguished by DNA-cellulose chromatography (3). Gradient elution from non-operator-containing DNA-cellulose was obtained with about 250 mM salt; elution from operator-containing DNA-cellulose was delayed and occurred with an average salt concentration of 1.0 M.

5.3. SOME PITFALLS ENCOUNTERED WITH DNA-CELLULOSE CHROMATOGRAPHY

Many attempts have been made to identify proteins that bind specifically to DNA from a selected source; for example, species-specific binding of proteins to homologous DNA. Since many physical and technical factors can influence the chromatography of proteins on DNA-cellulose, criteria for a successful demonstration of specificity must be met. If proteins from any source bind differently to alternate preparations of DNA-cellulose, several properties of the DNA used to prepare the respective DNA-celluloses, and of the resultant bound DNA, should be examined. These include the size and degree of homogeneity of DNA strands, base composition, the extent of methylation or other modifications, and the amount of single-stranded lengths and double-stranded or looped "hairpin" segments on the matrix. Since these factors vary among DNA sources and methods of preparation, and they affect DNA-cellulose chromatography, they could result in differential binding of proteins, which could be attributed to species-specificity. Therefore, such factors must be eliminated systematically before a conclusion of sequence specificity is presented.

Early analysis of rat proteins yielded results, from which it was concluded that homologous binding to species-specific DNA occurs (50, 51). Proteins were prepared from rat liver nuclei and were labeled with ^{131}I. In contrast to studies of bacterial DNA-binding proteins (1) and mammalian proteins (4, 41, 19), these proteins did not bind reproducibly to DNA-cellulose in buffers with 50–140 m*M* NaCl (50). To avoid this result, the proteins were mixed with DNA-cellulose in high salt (0.6 *M* NaCl) and with 5 *M* urea, and the mixture was dialyzed to a low ionic strength. Subsequently, fractions were eluted with sequential steps of 0.14, 0.6, and 2.0 *M* NaCl. By these procedures, binding of the rat proteins appeared greater when DNA-cellulose prepared with rat DNA was employed than when salmon sperm or *E. coli* DNA was used. Similar results were obtained with phosphorylated nonhistone chromatin proteins (52).

The requirement of a dialysis step in the above procedure in order to achieve the binding of proteins to DNA-cellulose in the previous experiments (50, 52) may have resulted from the fact that the proteins had been iodinated in a prior step (50). Tyrosines are derivatized in this reaction, and aromatic amino acids are important for the binding of some proteins to nucleic acids (53). For example, several tyrosine residues are important in DNA binding by the *lac* repressor of *E. coli* (54). The *N* terminus of the *lac* repressor contains its DNA-binding capacity. This region is rich in tyrosines (55), and, when isolated, this portion exhibits low affinity binding with nonoperator DNA sequences (56). Perhaps most telling, the DNA-binding CAP protein (cAMP receptor) was reported to lose its characteristic cAMP-dependent binding to DNA after iodination by either the chloramine T or lactoperoxidase methods (57). Therefore, tyrosines that are important for DNA binding might be found in many proteins; they could even be probed by selective iodinations. This might explain the reduction of binding to DNA-cellulose encountered earlier (50).

It is especially difficult to interpret the significance of elutions of proteins that had been adsorbed to DNA-cellulose by dialysis to reduce ionic strength. Trapping and coprecipitation must be considered as possible occurrences, and the physical features of different DNA preparations could influence the results obtained. Where possible, extracts should be cleared of particulate matter prior to loading on DNA-cellulose, and the DNA used in preparing the matrix should be clean. Experiments that define binding of

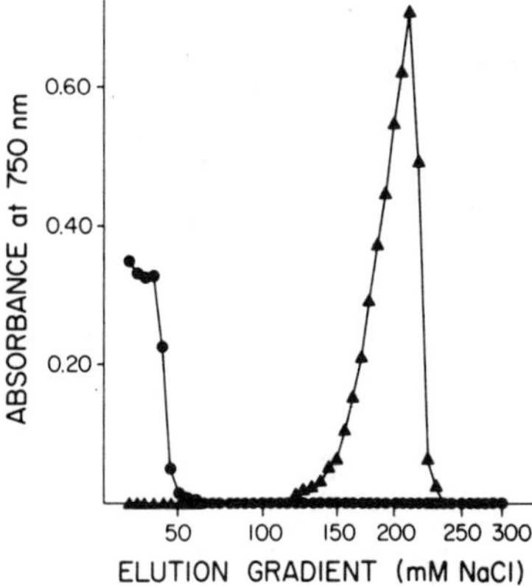

Fig. 5.3. Interaction of lysozyme and albumin with DNA-cellulose. Two columns (2.5 mL packed volume, each) of DNA-cellulose were loaded with extract buffer containing 50 m*M* NaCl and either 0.1 mg/mL of egg white lysozyme (▲) or 0.2 mg/mL of bovine serum albumin (●). After washing with a total of 60 mL of buffer containing carrier protein, the columns were eluted with buffer lacking protein, but containing a linear (50 to 300 m*M* NaCl) gradient. From the 0.5-mL samples, aliquants were taken for determination of protein [after the method of O. H. Lowry, N. J. Rosebrough, A. L. Farr, and R. J. Randall, *J. Biol. Chem.* **193,** 265 (1951)]. For lysozyme, 20 μL/sample was assayed, and for bovine serum albumin, 200 μL/sample. [Data from T. O. Fox, S. E. Bates, C. C. Vito, and S. J. Wieland, *J. Biol. Chem.* **254,** 4963 (1979).]

proteins to DNA or DNA-cellulose by cosedimentation are subject to artifacts caused by aggregation (47).

Another consideration for interpreting DNA-cellulose chromatography is that the elution of a given protein might be altered by other proteins that are present in the extract or on the column. For example, the very basic protein, lysozyme, binds to DNA at low ionic strength (58, 59), and can interact with estrogen (23), androgen (23), and glucocorticoid (60) receptors. Figure 5.3 depicts the chromatography of lysozyme and of albumin—which did not adhere—on DNA-cellulose. When androgen and estrogen receptors from mouse brain were chromatographed on DNA-cellulose (Fig. 5.4), the presence of lysozyme grossly altered the elution of androgen receptors but not estrogen receptors. This differential effect apparently resulted from the observed fact that estrogen receptors from this tissue fortuitously elute with the same salt dependence as lysozyme; therefore, only a slight alteration of estrogen receptor elution was observed when lysozyme was present. On the one hand, this model experiment indicates that extraneous factors, such as basic proteins, can introduce artifacts during DNA-cellulose chromatography. On the other hand, such effects could be used to good advantage technically and might even occur physiologically. Interactions of histones

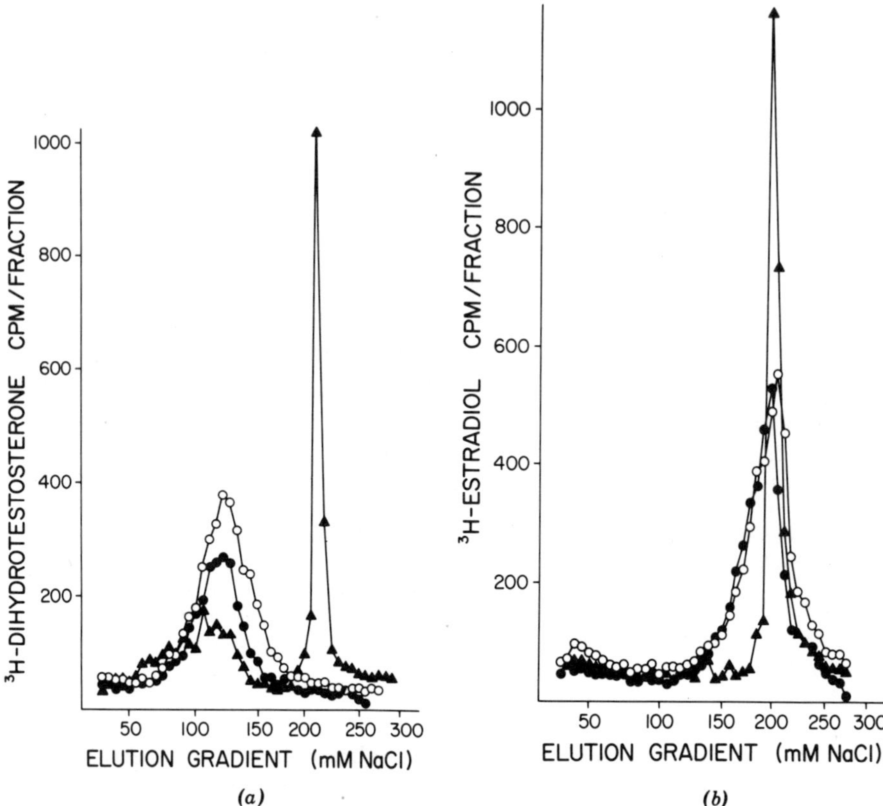

Fig. 5.4. DNA-cellulose chromatography of putative androgen and estrogen receptors with buffers containing different putative carrier proteins. Cytosol extracts of mouse brain were prepared in buffers with 50 m*M* NaCl, labelled with [^{3}H]dihydrotestosterone (a) or [^{3}H]estradiol (b) and chromatographed on DNA—cellulose columns, which previously had been equilibrated with buffers containing 50 m*M* NaCl and, respectively, no protein (○), 0.1 mg/mL of egg white lysozyme (▲), or 0.2 mg/mL of bovine serum albumin (●). Receptors were eluted with a linear gradient of 50 to 300 m*M* NaCl in the respective buffers. [Data from T. O. Fox, S. E. Bates, C. C. Vito, and S. J. Wieland, *J. Biol. Chem.* **254,** 4963 (1979)].

with estradiol receptors and their effects on binding of these receptors to oligo(dT)-cellulose and DNA-cellulose are being investigated (61, 62).

Proteins that bind selectively to highly repetitious DNA could be picked up in a comparison of different DNAs. With a nitrocellulose filter assay, a protein from embryos of *Drosophila melanogaster* was found to bind highly repeated satellite DNA from the same species (63). Another *Drosophila* protein, with specificity for a short sequence of cloned heat-shock-region DNA (64), was detected by the method of protein blotting (59). Perhaps apparent species-specific proteins detected with DNA-cellulose chromatography (50, 65) recognize repetitious sequences that might be sufficiently prevalent in noncloned DNA.

5.4. CHROMATOGRAPHY AND ANALYSES OF STEROID RECEPTORS

In the absence of high steroid concentrations, putative receptors for steroid hormones tend to be found in the soluble, cytosol, fraction of tissue or cell homogenates. After exposure to moderate or high steroid concentrations, a greater proportion of the receptor—now occupied with steroid—is found in the homogenate pellet, more tightly bound to nuclear sites. Thus the binding of ligand appears to shift an equilibrium toward increased nuclear occupation (66). The possibility that nuclear retention is mediated by the binding of occupied steroid receptors to DNA has led to a large volume of work (67, 36, 68) in which DNA-cellulose chromatography is used to examine putative steroid receptors. Historically, these experiments have almost exclusively involved first labeling receptors in crude or partially fractionated extracts with radioactive steroids and then following the fate of the bound radioactivity to monitor separations of the binding proteins and to deduce parameters of their binding to DNA. For a variety of extracts and labeled steroids, binding of the hormone–protein complexes to DNA-cellulose can be demonstrated.

The technical requirements for binding to DNA-cellulose vary considerably for the different steroid receptors and for receptors with similar ligand specificities but different tissue sources. Consequently, the extent to which these interactions with DNA reflect in vivo binding to DNA may differ greatly; evidence to date for correlations of DNA-cellulose binding and nuclear binding of steroid receptors occur for only some such receptors (37, 39). For these reasons it is quite important to distinguish what is known about individual receptor types with regard to their parameters of DNA-cellulose chromatography. Generalizations drawn from data on a given receptor has been shown in several cases not to apply to steroid receptors generally or even, in some cases, to any but a given steroid receptor type.

Some distinctive characteristics of receptors for glucocorticoids, estrogens, protestins, and androgens are briefly summarized in Table 5.1. Just as it is important to distinguish the behaviors of the receptor types and avoid unjustified premises for a given steroid receptor based on properties of another steroid receptor, it is also important to recognize that these characteristics represent typical but not invariable differences.

Glucocorticoid receptors require "activation" for binding to DNA (68), which is usually achieved by warming the solution (20°C) (68) or by elevating the ionic strength of the extract (68, 77). This is generally not essential for

Table 5.1. Differences in DNA-cellulose Chromatography of Receptors for Various Steroids

	Receptor	Distinctive DNA-binding Properties	Elution (m*M* salt)[a]	References
GR	Glucocorticoid	Elution altered in some mutants; activation required for all DNA-binding	130–170	37, 69
ER	Estrogen	Binding and elution at higher salt concentrations than for other steroid receptors; 210 m*M* eluting form transformed to the 250 m*M* eluting form	210, 250	36, 70, 28
PR	Progestin	Very weak binding of one fraction	60, 180	71, 72, 73
AR	Androgen	Elution pattern for one mutant distinguished from wild-type and from other mutants	140, 180	35, 74, 75, 76

[a] These approximate values were obtained with several different ionic and buffer compositions; they represent examples from a range of possible results, and in some cases were determined by interpolation from the published data. Values obtained depend empirically upon buffer strength and composition, pH, the degree of saturation of column sites, and the species, tissue source, and purity of the extract used.

DNA-cellulose binding by other steroid receptors, although it is often included in protocols and in some cases can be shown to enhance apparent binding. Very interestingly, a study of binding of mouse steroid receptors to oligo(dT)-cellulose and other oligodeoxynucleotide-celluloses indicates that only the glucocorticoid receptor, and not estrogen or androgen receptors, requires "activation" to yield appreciable binding (S. R. Gross, S. A. Kumar, and H. W. Dickerman, manuscript in preparation). DNA-binding is also reported to increase after dilution or gel filtration of certain extracts (78) in agreement with cell-free nuclear binding experiments (79). These experiments have been interpreted to suggest that extracts may contain DNA-binding inhibitors that are either large (60) or relatively small (78). The existence of a low molecular weight inhibitor of nuclear binding by estrogen receptors was suggested by dialysis and gel filtration (80). Only for glucocorticoid receptors, however, does the literature consistently indicate a requirement of receptor activation for DNA-binding.

An important consequence of this property is that two-step chromatography on DNA-cellulose can yield a significant increment in purification of glucocorticoid receptors (81, 82, 83). Prior to heat activation, extracts are chromatographed on DNA-cellulose to adsorb many DNA-binding proteins. Then the nonbinding portion of the extract is heat-activated with bound glucocorticoid to protect the receptor, and this extract is again chromatographed. Efforts to utilize this approach have been much less successful for other steroid receptors (e.g., estrogen receptors) since they show little requirement for a bound ligand, or for an activation step, in order for them to bind to DNA-cellulose (28) other than a limited dependence (under specified conditions) on relatively high ionic strength (36).

An additional feature of the glucocorticoid receptor system is that mutant cell lines processing altered receptor forms have provided genetic correlations between nuclear retention and binding to DNA-cellulose (37, 38). In mouse lymphoma cell lines selected (84) for their resistance to the killing action of the synthetic glucocorticoid dexamethasone, several classes of receptor mutations were identified (85). In two classes of these variants, receptors are present and demonstrable by hormone binding and exhibit either decreased or enhanced nuclear retention, respectively. In both classes, altered binding to DNA-cellulose is detected (37). For the receptor mutants that possess decreased nuclear retention, the receptors adhere to DNA-cellulose but elute at lower ionic strength than is normal; for the mutants possessing enhanced nuclear retention, elution from DNA-cellulose occurs at a higher than normal ionic strength. Thus the DNA binding of these mutants correlates with nuclear parameters and supports the hypothesis that these receptors can adhere to DNA in the cell. Comparable evidence for a correlation between nuclear and DNA binding is not available for other steroid receptors.

Glucocorticoid-resistant variants of hepatoma cells with receptor defects have also been obtained by fluorescence-activated cell sorting, and they provide further genetic correlations of receptor function (86). Using a system in which glucocorticoid induction of mouse mammary tumor virus normally occurs (87), evidence has been obtained (39) to support models of sequence-specific DNA-binding by glucocorticoid receptors (88) and by other eucaryotic regulators in general (89, 49).

For estrogen receptors, ligand occupation and temperature activation are not required for binding at low ionic strengths (less than 200 mM NaCl) (36, 28), below which the other steroid receptors elute during DNA-cellulose chromatography (Table 5.1). Thus "activation" of estrogen receptors is not directly associated with the *ability* to bind DNA under typical conditions. As a consequence, a two-step purification of estrogen receptors by DNA-cellulose chromatography has not been very effective, in contrast with successful fractionations of glucocorticoid receptors (81, 82, 83).

The estrogen receptor is unique among steroid receptors in that a larger, "transformed" species of the receptor is detected in cell nuclei of estradiol-primed animals, and this "nuclear" species can form during DNA-cellulose chromatography (36, 90, 21, 91). The estrogen receptor that predominates in cytosol fractions of tissue homogenates sediments as a 4S species, whereas the predominant "nuclear" form sediments as a 5S species that also has been shown by permeation and sedimentation criteria to have a higher molecular weight (36, 90, 92). The 5S receptor is "activated" relative to the 4S receptor in that it has higher affinity with DNA in solution (47) and it elutes from DNA-cellulose at a higher ionic strength (240–260 m*M* NaCl) than does the 4S receptor (200–220 m*M* NaCl) (36, 90, 70, 23, 28). Moreover, it has acquired an additional polypeptide. Since the 5S receptor has a higher affinity with DNA in solution and elutes from DNA-cellulose at higher ionic strength, it may be that activation of the estrogen receptor is simply formation of the stable 5S receptor with the result that DNA binding is enhanced. Thus for this receptor, activation causes in *increased* affinity with DNA and not an absolute dependence on any binding at all; both forms of the estradiol receptor adhere to DNA-cellulose in low ionic strength buffers without a requirement for activation. Accordingly, in examining "activation" of estrogen receptor binding to DNA-cellulose, one must distinguish the elution of 4S and 5S forms. If conditions that permit both to bind and elute together are used (93), then no effect of "activation" should be detected; conclusions about estrogen receptor activation are misleading if they are drawn from such experiments.

In evaluating the binding of proteins to DNA-cellulose, a question arises whether the proteins of interest contact DNA directly or through mediation by other proteins or macromolecules. As an example, two types of indirect binding have been proposed for steroid receptors. The two characteristics that emphatically distinguish the estrogen receptors from other putative steroid receptors—complexing with another protein termed "Subunit X" (90), and elution from DNA-cellulose at higher salt concentrations—might be interrelated, as discussed in the above paragraph. We consider this "activation" or "transformation" of the estradiol receptor in the context of models and evidence for chromatin "acceptor" proteins (94, 95, 96, 97, 9, 98). Acceptor proteins are believed to mediate receptor–nucleus interactions by binding to DNA and thereby providing an indirect site of adherence for steroid receptors. This hypothesis for nuclear binding has been less often proposed for the estrogen receptor system. Perhaps this model has not been necessitated for estrogen receptors, since they have apparent high affinity with DNA-cellulose compared with other steroid receptors and therefore direct binding to DNA more readily has been envisioned. A possible model, though, suggests how common features of the various steroid receptors may be related by differential utilization of similar proteins; opposite extremes in their properties may be manifest as "transformed nuclear" estrogen receptors or progesterone receptor "acceptor proteins," respectively. This model is depicted in Figure 5.5.

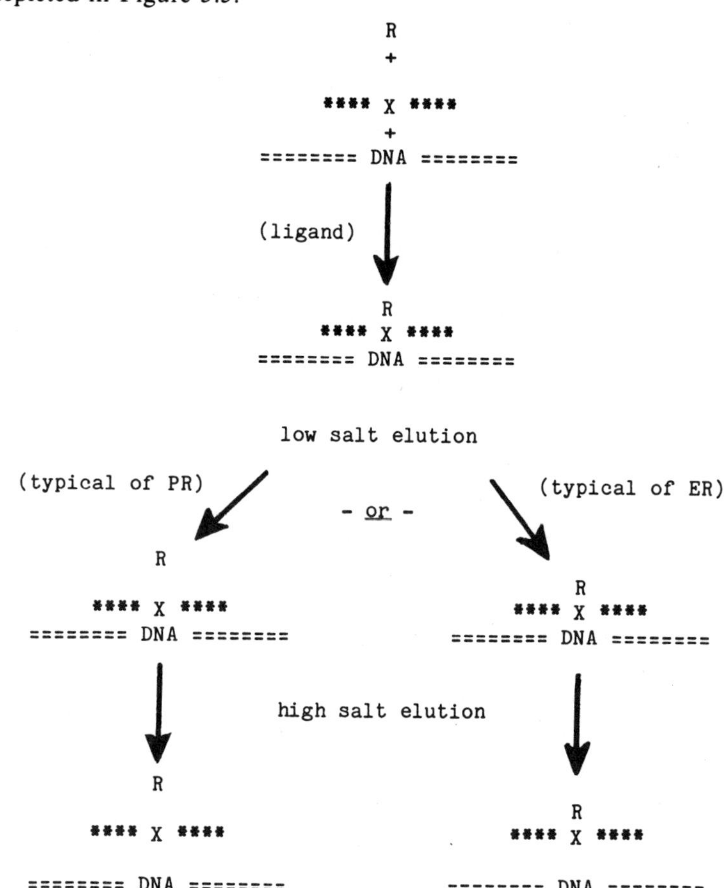

Fig. 5.5. Models for "acceptor" proteins. PR: progesterone (progestin) receptor; ER: estradiol (estrogen) receptor. This model depicts binding of steroid receptors (R) to an acceptor protein, *X*, (*****X*****) which itself binds to DNA (========DNA========). Direct binding of receptors to DNA is not considered or depicted in this figure. The "acceptor" protein may serve (left-side pattern) as a bridge between steroid receptors—for example, the progesterone receptor—and binding sites on DNA. With elution, this bridge protein dissociates from the receptor and then from the DNA. The same bridge may occur for estradiol receptor (right-side pattern), except that the order of elution is reversed: *X* dissociates from the DNA first, resulting in the layer "5S, nuclear" estradiol receptor form. Therefore, these two models, the "acceptor protein hypothesis" (94, 95) and transformation of the "cytoplasmic" 4S-sedimenting estradiol receptor to the "nuclear" 5S form (36, 90, 92), might involve similar proteins—receptor and *X*—distinguished only by relative stability in salts.

It may be that all steroid receptors can bind to DNA by adhering to proteins or to other factors—for the purpose of this hypothesis, called *X*—that bind to DNA. This binding would be in addition to binding directly to DNA with either low or high affinity; dynamically, all three types of interaction might occur to different extents. For progestins, as an example, the receptors could interact with *X*, which itself adheres to DNA; *X* may exhibit greater salt extractability from the receptor than from DNA. Consequently, these receptors would usually elute from DNA-cellulose at relatively low ionic strengths (left-side pattern). In contrast, the estrogen receptor could have the ability to form a stable complex with *X*, and might generally elute when *X* elutes from the DNA-cellulose (right-side pattern). As depicted in Figure 5.5, this model would provide that progestin, and possibly other steroid receptors, elute from DNA with lower ionic strength and without an increase in molecular weight, whereas the estrogen receptor would elute later in an increasing salt gradient and would exhibit sedimentation and permeation properties indicative of a higher molecular weight. Of course, some proportion of estrogen receptors would not form stable complexes and would exhibit the left-side pattern of elution (Figure 5.5). The purpose of this model is to indicate how "acceptor" proteins, if they exist, might interact in different ways with distinct steroid receptors. As a consequence, the proposed property of estrogen receptor to complex with a protein *X* may be a simple variant of the theoretical "acceptor" model. With methods to examine receptors in intact cells, without the artificial changes of ionic strength encountered in cell-free studies, greater similarities among different steroid receptors might be apparent.

A striking feature of the progestin receptors is that they exist as two species that display quite different affinities with DNA-cellulose. Present in roughly equal amounts, the A and B progesterone receptors are similar in hormone binding characteristics but differ in size and can be separated by chromatography on ion exchange columns (DEAE-cellulose) and on DNA-cellulose (72, 99, 73). When loaded on DNA-cellulose in 50 m*M* KCl, only the protein A binds, and it elutes with 180–190 m*M* KCl (72, 73). Protein B can adhere to DNA-cellulose below 30 m*M* KCl and elute with 60 m*M* KCl (99). While a salt step at 200 m*M* KCl elutes protein A from DNA-cellulose (71), a unitary ammonium sulfate fraction of progesterone receptor in another study eluted from DNA-cellulose with 0.05—0.15 *M* KCl (100). Thus both forms of the progesterone receptor can bind to DNA-cellulose, but one typically (99), and apparently both (100) can elute at ionic strengths that are distinctively low for putative steroid receptors and for "DNA-binding" proteins in general. However, only as much as half the amount of estrogen receptors from the hypothalamic/preoptic area of mice adhered to DNA-cellulose (21), in contrast with estrogen receptors from mouse or rat uterus, which can virtually all bind to DNA-cellulose. For all steroid receptors, evidence is incomplete with regard to whether binding to DNA is direct or is mediated by other factors. The suggestion that such binding is all artifactitious can be considered (100), but the especially weak binding that is typically demonstrated for one fraction of progesterone receptor, and generally for the receptors reported in this study (100), makes the meaning and reliability of this binding suspect and makes generalizations from it to other steroid receptors unjustified. Nevertheless, DNA-cellulose chromatography of progestin receptors remains useful, especially for receptor A (72), and the distinction between the A and B proteins provides one of the intriguing model systems for speculations on the mechanisms of steroid hormone action (98).

Androgen receptors bind to DNA-cellulose in low to moderate salt concentrations (35, 74, 22, 76) and were among the first steroid receptors to be analyzed (67) and fractionated (101) by virtue of their interactions with DNA-cellulose. This method has been effective for comparing different species of putative androgen receptors. DNA-cellulose chromatography results in a distinction between normal androgen receptors from wild-type mice and the reduced level of residual receptors in androgen-resistant mice with the mutation of testicular feminization (*Tfm*) (35, 74, 22). Figure 5.6 illustrates several variables of androgen receptors that can be analyzed by DNA-cellulose chromatography. Receptor differences as detected with different radioactive ligands, among different tissues of the same animal, and between wild-type and mutant animals are distinguished by their characteristic elution profiles. The distinguishable elution patterns detected for androgen receptors from wild-type mice and androgen-resistant mutant mice were intrinsic to the hormone binding entities, since mixed extracts (22) from wild-type and mutant mice yielded additive patterns (Figure 5.7), as had been shown earlier for mutant glucocorticoid receptors (37). DNA-cellulose chromatography also permitted comparisons between this mouse mutation and a similar, though phenotypically distinguishable, androgen-resistant rat mutation (76, 102, 103). DNA-cellulose is being utilized to compare putative androgen receptors from cultured human skin cells (102, 104); these appear similar in gradient elution profiles (102) to those from rodents, though for the human cells the receptors elute with lower ionic strength (102). Defects have been distinguished by DNA-cellulose chromatography (T. O. Fox and D. Blank, unpublished data) among putative androgen receptors from patients with syndromes of androgen resistance. These exhibit similarities to the patterns for androgen-resistance mutations of mice (35, 74, 22) and rats (76, 102, 103).

Although the conditions for binding to DNA-cellulose and the parameters of elution differ in important ways for the four major classes of putative

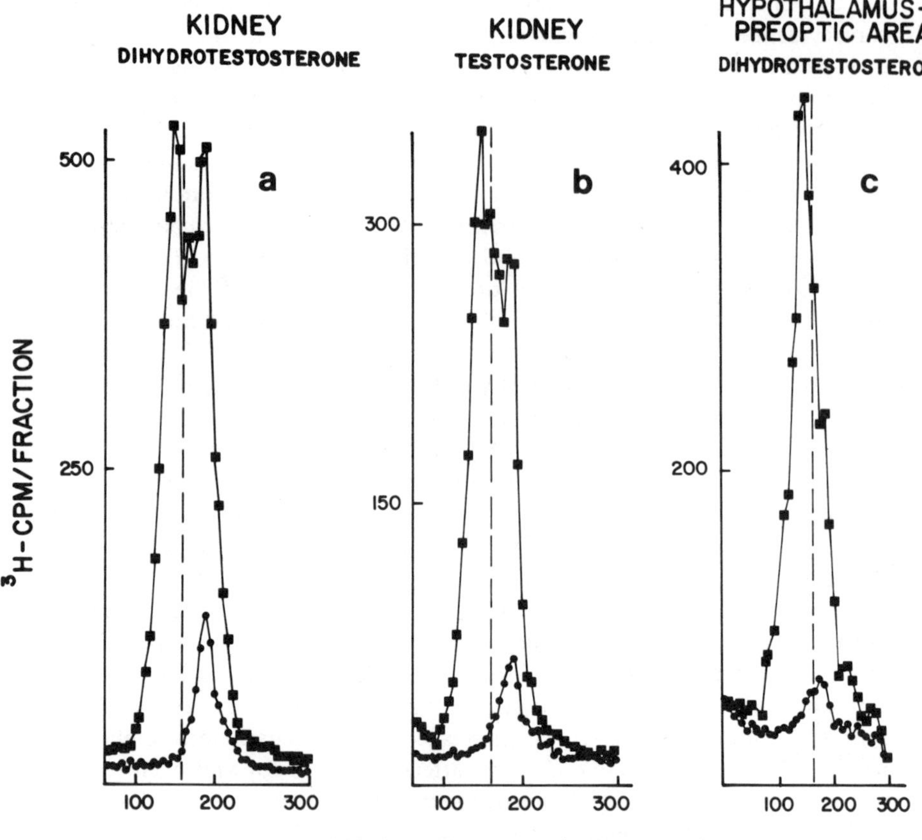

Fig. 5.6. Variations with tissue, ligand, and an androgen-resistant mutation of putative androgen receptor elution from DNA-cellulose. Cytosol extracts of kidneys of androgen-resistant mouse mutants with testicular feminization (*Tfm*) (circles) and of the wild-type (squares) were incubated with either 10 n*M* [^{3}H]dihydrotestosterone (*a*) or 10 n*M* [^{3}H]testosterone (*b*). For (*c*), cytosol extracts of tissue blocks of hypothalamus of *Tfm* mutants (circles) or wild-type (squares) were incubated with 10 n*M* [^{3}H]dihydrotestosterone. Samples were chromatographed on 2 mL DNA-cellulose columns in buffer with 50 m*M* NaCl, and receptors were eluted with a linear gradient of 50 to 400 m*M* NaCl. The dashed line in all three panels is arbitrarily placed at 160 m*M* NaCl to emphasize the classes of lower-salt-eluting and higher-salt-eluting receptors that are detected. [Data from S. J. Wieland and T. O. Fox, ,*Cell* **17**, 781 (1979).]

steroid receptors detailed above, several qualitative features emerge as possible similarities. For example, evidence exists for all four receptor types to suggest that in each case two major forms of receptor can be distinguished on the basis of DNA binding and eluting peaks. This is well established for estrogen receptors (36, 70), has been claimed for androgen receptors (22),

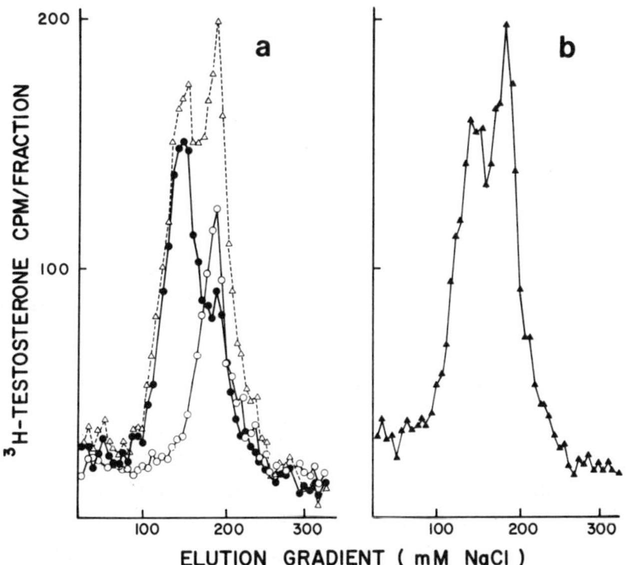

Fig. 5.7. Chromatography of mixed cytosol extracts of wild-type and *Tfm*/Y mutant kidneys. Each sample was incubated with 10 n*M* [^{3}H]testosterone, loaded at 50 m*M* NaCl onto a 5 mL DNA-cellulose column, and then washed with the same buffer. The sample contents were eluted with a linear gradient of 50 to 400 m*M* NaCl. (*a*) Cytosol of 4.5 *Tfm*/Y kidneys (○); cytosol of 0.5 wild-type kidney (●); summation of *Tfm*/Y plus wild-type (after correction for background). (▲). (*b*) Cytosol of 4.5 *Tfm*/Y kidneys mixed with cytosol of 0.5 normal kidney. After 1 hr incubation, this mixture was labeled with [^{3}H]testosterone and then chromatographed, as in (*a*). [Data from S. J. Wieland and T. O. Fox, *Cell* **17,** 781 (1979).]

has been shown for progesterone receptors (99), and is suggested by data for glucocorticoid receptors (69). In the last case, evidence indicates that two forms that differ in DNA-cellulose binding and in size are related as precursor and proteolyzed forms; the higher-salt-eluting form of glucocorticoid receptor is smaller than the lower-salt-eluting form and is derived from it by proteolytic digestion (69). Two additional facts are interesting in this regard. First, the glucocorticoid receptors of nt^i lymphoma-cell-line mutants, which are resistant to killing by dexamethasone, exhibit higher apparent affinities with both nuclei and DNA-cellulose and exhibit slower sedimentation through sucrose density gradients (37) than glucocorticoid receptors from the wild-type. Perhaps they are like the proteolyzed form that was observed (69). Second, in parallel fashion, a proteolytically derived fragment of bacteriophage T4 gene 32 protein—lacking about 60 C-terminal amino acids—has enhanced DNA affinity and DNA destabilizing activity (105).

The higher-salt-binding form of the progesterone receptor is also smaller than the other progesterone receptor, but experiments do not indicate that the smaller of these is a proteolytic product of partial digestion of the larger form (99). In contrast, the higher-salt-eluting form of estrogen receptor is larger than the lower-salt-eluting form and is derived from it by complexing with a second protein (36, 90, 92), whereas the androgen receptor forms that elute with lower and higher salt concentrations are similar in relative sedimentation rates (22, 106). In this case, each androgen receptor peak rechromatographs respectively with lower salt or higher salt (22, 107). This indicates that these particular peaks probably result from intrinsic heterogeneity of the receptors rather than heterogeneity of binding sites on the DNA-cellulose. The above similarities among receptor types are intriguing, but it may be only coincidental that several steroid receptors chromatograph on DNA-cellulose with double peaks, and the causes of these dichotomies might prove to be unrelated to one another.

To evaluate the possible physiological significance of binding to DNA and to explain receptor mediation of the actions of bound ligands, repressor proteins of procaryotes have been considered as a model for steroid receptors. This model has been used to assess the observed low affinity interactions of steroid receptors with DNA and the theoretical possibility that binding with high affinity to specific base sequences could be detected (88). The *E. coli* lactose repressor protein has high affinity for the specific *lac* operator DNA sequence (108, 109) and has lower, but significant, affinity for nonspecific DNA (110, 89, 49). The latter binding masks the specific operator binding owing to the much higher concentration of nonspecific DNA. It was proposed that if both low and high affinity binding occur for steroid receptors, then the low affinity binding to nonspecific DNA would prevent detection of specific high affinity receptor binding (88). Accordingly, it was predicted that specific steroid receptor binding to DNA will be detectable when the specific sequences of DNA are obtained sufficiently devoid of nonspecific DNA. Evidence consistent with this prediction is now available (39).

Not only is the low affinity binding of *lac* repressor important technically in that it competes in vitro with the high affinity binding to the operator sequence, but it should function in the cell to influence the effective concentration of available repressor (89, 49). As expected from such calculations, alterations in the binding of repressor for low affinity sites can alter the repression state in cells. This has been examined in the *E. coli* X86 mutation, which has an altered *lac* repressor (111). It was demonstrated (112) that the X86 *lac* repressor has about fiftyfold lower dissociation constant both for *lac* operator DNA and for nonspecific DNA, and therefore the X86 phenotype can be explained by a lower probability of available repressor binding to the operator (112). The significance of this mutation and phenotype is twofold. First, it reaffirms the importance of low affinity binding sites in the modulation of functional high affinity binding to DNA. Second, it indicates that a mutation which enhances DNA binding generally, by decreasing the dissociation rate, can result in reduced availability for limited high affinity sites.

Presumably in eucaryotes, if steroid receptors bind with high affinity to specific DNA sites, then the low affinity of the type and strength that has been detected may be an unavoidable property of steroid receptors, and it may be important physiologically. A consequence of this proposal is that steroid receptor mutants should be found which exhibit abnormally high binding to nuclei and to DNA with resultant *lack* or reduction of activity. Among mutant mouse lymphoma cell lines, as shown above, the nt^i phenotype is resistant to cell killing by dexamethasone, the nt^i glucocorticoid receptor displays greater nuclear adherence, and its elution from DNA-cellulose is delayed, requiring a higher salt concentration (37). Although the mechanism of nt^i resistance is not known, DNA-cellulose may be useful in determining mutations in those steroid receptors, which have altered affinity with DNA and whose altered DNA-binding constants cause abnormal or paradoxical physiological effects of induction or suppression.

5.5. OLIGO(dT)-CELLULOSE AND POLY(U)-SEPHAROSE AFFINITY CHROMATOGRAPHY OF RNA

Affinity chromatography on immobilized oligo-deoxythymidylate [oligo(dT)] or poly-uridylate [poly(U)] has been used extensively during the past ten years in studies of eucaryotic messenger RNAs (mRNAs). Procedures for oligo(dT)-cellulose and poly(U)-Sepharose synthesis and various chromatographic applications of these matrices have been reviewed recently (5). This short review is an attempt (i) to point out some important practical aspects of oligo(dT)-cellulose and poly(U)-Sepharose chromatography, (ii) to compare the two methods, and (iii) to examine the practical limitations of these methods in separating different RNA classes.

The existence of mRNA molecules with a stretch of poly(A) at the 3′ terminus has been documented in all eucaryotes studied, from mammals (113, 114) to yeast (115). The poly(A) is added to newly synthesized RNA in the cell nucleus, and its function remains a matter of conjecture. Its existence, however, makes possible the chromatographic separation of polyadenylated RNA from ribosomal RNA (rRNA), which constitutes as much as 99% of the total mass of polysomal RNA.

Both oligo(dT)-cellulose and poly(U)-Sepharose are relatively easy to prepare; they are also commercially available. Oligo(dT)-cellulose can be synthesized using the *N,N′*-dicyclohexyl-carbodiimide reaction for the polymerization of thymidine-5′-monophosphate on cellulose (116). Commercially available oligo$(dT)_{12-18}$-cellulose is widely used. Poly(U)-Sepharose can be prepared by coupling poly(U) to CNBr-activated agarose beads (Sepharose) (117, 118).

Binding of polyadenylated RNA to oligo(dT)-cellulose or to poly(U)-Sepharose is performed at conditions that stabilize double-stranded hybrids between the poly(A) on the RNA molecules and the immobilized oligo(dT) or poly(U): high ionic strength (0.5 *M* NaCl is generally used), near-neutral pH, and relatively low temperatures (room temperature to 4°C). Bound RNA is eluted by lowering the ionic strength or increasing the temperature or both, or by using a denaturing agent like formamide.

In oligo(dT)-cellulose chromatography, the length of the AT hybrids is presumably limited by the chain length of the oligo(dT), which is usually smaller than 20. Binding can be performed at room temperature, although binding at 4°C is advisable (see below); elution is generally performed at room temperature with H_2O or with a buffer of low ionic strength, such as 10 m*M* Tris-HCl pH 7.5 (119, 120). In poly(U)-Sepharose chromatography the hybrids formed between poly(U) and poly(A) are longer. The length of the hybridized regions is mainly determined by the length of the poly(A), which in most cases is longer than 100 nucleotides. Elution from poly(U)-Sepharose must therefore be performed at higher temperatures (121) or in the presence of formamide (117) in order to break the greater number of base pairs.

Poly(U)-Sepharose-bound RNA can be fractionated according to poly(A) length by applying a formamide concentration gradient (122) or by increasing the temperature (121). With glass-immobilized poly(U), there is a linear relationship between the logarithm of poly(A) length and the elution temperature of pure poly(A) with lengths between 20 and 200 nucleotides. However, poly(A) attached to RNA has a weaker interaction with poly(U) than does pure poly(A) (123). A similar destabilizing effect of the non-poly(A) moiety of the RNA molecule is also suggested by the data of Vournakis et al. (124) in the case of oligo(dT)-cellulose chromatography. For elution with buffers containing formamide, high-purity formamide should be used, especially if the eluted RNA is to be translated in vitro.

Oligo(dT)-cellulose chromatography has been the method of choice for large-scale preparation of polyadenylated RNA because it is simple and the columns have a large capacity for poly(A) and are reusable for very long times. Detailed chromatographic procedures have been described by Aviv and Leder (119), by Kates (125), and by Efstratiadis and Kafatos (120). The binding capacity of oligo(dT)-cellulose varies widely among different commercial preparations and between batches, thus the capacity of any single preparation should be determined before use if large quantities of RNA are to be processed (see 120 for a procedure). High capacity batches of oligo(dT) cellulose can bind up to 1 mg poly(A) per g resin. Theoretically, 1 g of such resin should bind all the polyadenylated RNA from 250 to 1000 mg of RNA, assuming the polyadenylated RNA comprises 1–2% of polysomal RNA and that the poly(A) represents 10–20% of the polyadenylated RNA mass. Poly(U)-Sepharose exhibits a limited capacity for very large RNA molecules, presumably because such molecules are excluded from the agarose matrix and can interact only to poly(U) on the surface of the beads (126).

To ensure proper performance, the following precautions should be taken when applying the RNA: First, RNA should be heat-treated before application, to disrupt aggregates that can result in artifactual binding of nonpolyadenylated RNAs to the column. Aggregates form when RNA is ethanol-precipitated or stored frozen in solution. Heating the RNA in buffer of low ionic strength for 10 minutes at 65°C followed by rapid cooling at 0°C is recommended (127). Second, it is advisable to apply the RNA at concentrations lower than 2–3 mg/mL. Although the performance of oligo(dT)-cellulose seems to be concentration-independent, certain RNAs, like reticulocyte polysomal RNA, tend to aggregate at high concentrations (120). Contamination of the bound RNA with varying amounts of rRNA is a common result, especially with oligo(dT)-cellulose (127, 128). Heat-treatment of the eluted RNA followed by a second round of chromatography removes most of the contaminating rRNA.

The efficiency of binding of polyadenylated RNA by oligo(dT)-cellulose is high. Zimmerman et al. (129) reported that after passing *Drosophila* polysomal RNA twice through an oligo(dT)-cellulose column, the bound material contained more than 95% of the polyadenylated molecules present in the starting RNA preparation. Comparable or better retention of polyadenylated RNA has been obtained by other workers using RNAs from different sources (130, 131). RNA free of any poly(A) detectable by hybridization with tritiated poly(U) can be obtained by multiple passages through oligo(dT)-cellulose (132). Quantitative data of the same kind are not available for poly(U)-Sepharose, because measurements of poly(A) content are based on hybridization with labeled poly(U), and the RNA eluted from poly(U)-Sepharose is usually contaminated with small amounts of poly(U). Nevertheless, in studies where the performance of oligo(dT)-cellulose and of poly(U)-Sepharose were compared using the same RNA preparation, the yields of polyadenylated RNA were identical (131, 129).

Direct determination of the size of poly(A) tracts from RNA purified by affinity chromatography suggest that oligo(dT)-cellulose is as efficient as poly(U)-Sepharose in retaining RNA molecules with short poly(A) tracts. Poly(U)-Sepharose retains RNAs with poly(A) as short as 25 nucleotides—as determined by electrophoresis after digestion with ribonucleases A and T1 (122). Using the same method to determine poly(A) lengths, Vournakis et al. (124) showed that at 2°C, oligo(dT)-cellulose retains mRNA molecules with poly(A) tracts of at least 20 nucleotides. The data of Vournakis and coworkers are in good agreement with an indirect determination of the minimum poly(A) length required for binding of globin mRNA to oligo(dT)-cellulose, which yielded minima of approximately 30 and 16 nucleotides for binding at 22°C and at 4°C, respectively (133).

Taken together, the results outlined above strongly suggest that oligo(dT)-cellulose and poly(U)-Sepharose do not differ in their specificities for polyadenylated RNA. However, there have been at least two reports in which a specific mRNA binds better to poly(U)-Sepharose than to oligo(dT)-cellulose (134, 135).

It has been assumed that with a few exceptions all eucaryotic mRNAs are polyadenylated (136, 137), and occasionally the terms polyadenylated RNA and mRNA are used indiscriminately. There is convincing evidence, however, that many mRNA species are non-polyadenylated, at least by the criterion of binding to oligo(dT)-cellulose or poly(U)-Sepharose. In HeLa cells and sea urchin embryos, histone mRNAs appear to lack poly(A) (138, 139). The relatively abundant mRNAs encoding globin, protamine, and actin can exist in two forms in the same tissue, one polyadenylated and one apparently nonpolyadenylated (140, 141, 142, 135). Nonpolyadenylated mRNAs have been detected in a variety of species and cell types, including slime molds (143), HeLa cells (130), *Xenopus* oocytes (144), mouse brain (131), and *Drosophila* (129). This class of unidentified mRNAs constitutes a large fraction of the total mRNA in certain cell types: It can represent 30 to 80% of the mass of the newly synthesized mRNA (130, 144) and 30% of the steady-state mRNA levels (130). Moreover, molecular hybridization studies suggest that polyadenylated and nonpolyadenylated mRNAs represent two largely non-overlapping populations, *i.e.*, that they are transcripts of different genes: In *Drosophila* larvae and in mouse brain, approximately 70% and 45%, respectively, of the expressed genes may be transcribed into nonpolyadenylated mRNAs (129, 131). It is not known whether the so-called nonpolyadenylated mRNAs lack poly(A) completely or contain poly(A) tracts that are too short to bind to oligo(dT)-cellulose or to poly (U)-Sepharose. Whatever the case, their study may provide some clues about the functions of poly(A) in eucaryotic mRNA.

In addition to oligo(dT)-cellulose and poly(U)-Sepharose, other affinity matrices have been used for analysis or preparation of polyadenylated RNA. Binding to cellulose nitrate membrane-filters (Millipore filters) and to poly(U)–glass fiber filters can be used as rapid assay methods for poly(A)-containing molecules in RNA populations (145, 124, 123, 125). Millipore filters have certain disadvantages: Binding may not be efficient when the poly(A) is short (146; see, however, 124), and they also bind other polynucleotides, such as poly(U) and (G) (125). Poly(U)–fiber glass filters are specific for poly(A) and have high capacities for RNAs with poly(A) tracts longer than 15–20 nucleotides; they can be prepared simply by immobilizing poly(U) to glass fiber filters by UV irradiation (123, 125).

Poly(U)-cellulose, prepared by attaching poly(U) to cellulose by UV irradiation, can be used for preparative purposes (123). Poly(U)-cellulose columns have the disadvantage that poly(U) is leached out from the column and contaminates the RNA (125). However, its simple preparation, high capacity, and low cost, may make poly(U)-cellulose attractive for certain applications.

ACKNOWLEDGMENTS

We thank Glenn Herrick for numerous conceptual and factual suggestions. Portions of our research reported here were supported generously by the March of Dimes Birth Defects Foundation, the National Institutes of Health, and the Mental Retardation Research Center at Children's Hospital Medical Center.

REFERENCES

1. B. M. Alberts, F. J. Amodio, M. Jenkins, E. D. Gutmann, and F. L. Ferris, *Cold Spring Harbor Symp. Quant. Biol.* **33,** 289 (1968).
2. R. M. Litman, *J. Biol. Chem.,* **243,** 6222 (1968).
3. G. Herrick, *Nucleic Acids Res.,* **8,** 3721 (1980).
4. B. Alberts and G. Herrick, in *Methods in Enzymology* Vol. 21D (L. Grossman and K. Moldave, Eds.), Academic Press, New York (1971), pp. 198–217.
5. W. H. Scouten, *Affinity Chromatography,* Wiley-Interscience, New York, 1981.
6. P. L. deHaseth, C. A. Gross, R. R. Burgess, and M. T. Record Jr., *Biochemistry* **16,** 4777 (1977).
7. E. V. Jensen, T. Suzuki, T. Kawashima, W. E. Stumpf, P. W. Jungblut, and E. R. DeSombre, *Proc. Natl. Acad. Sci. USA* **59,** 632 (1968).
8. K. R. Yamamoto and B. M. Alberts, *Ann. Rev. Biochem.* **45,** 721 (1976).
9. J. Gorski and F. Gannon, *Ann. Rev. Physiol.* **38,** 425 (1976).
10. G. H. Stein, in *Methods in Cell Biology,* Vol. 17 (G. Stein, J. Stein, and L. J. Kleinsmith, Eds.), Academic Press, New York (1978), pp. 271–283.
11. V. G. Allfrey and A. Inoue, in *Methods in Cell Biology,* Vol. 17 (G. Stein, J. Stein, and L. J. Kleinsmith, Eds.), Academic Press, New York, 1978, pp. 253–270.
12. H. Potuzak and P. D. G. Dean, *FEBS Lett.* **88,** 161 (1978).
13. J. Kallos, T. M. Fasy, V. P. Hollander, and M. D. Bick, *Proc. Nat. Acad. Sci. USA* **75,** 4896 (1978).
14. T. C. Spelsberg, E. Stake, and D. Witzke, in *Methods in Cell Biology,* Vol. 17

(G. Stein, J. Stein, and L. J. Kleinsmith, Eds.), Academic Press, New York (1978), pp. 303–324.
15. S. Thrower, C. Hall, L. Lim, and A. N. Davison, *Biochem. J.* **160,** 271 (1976).
16. K. H. Thanki, T. A. Beach, and H. W. Dickerman, *J. Biol. Chem.* **253,** 7744 (1978).
17. S. A. Kumar, T. A. Beach, and H. W. Dickerman, *Proc. Nat. Acad. Sci. USA* **77,** 3341 (1980).
18. V. K. Moudgil and D. O. Toft, *Proc. Nat. Acad. Sci. USA* **72,** 901 (1975).
19. T. O. Fox and A. B. Pardee, *J. Biol. Chem.* **246,** 6159 (1971).
20. G. Herrick and B. Alberts, *J. Biol. Chem.* **251,** 2124 (1976a).
21. T. O. Fox and C. Johnston, *Brain Res.* **77,** 330 (1974).
22. S. J. Wieland and T. O. Fox, *Cell* **17,** 781 (1979).
23. T. O. Fox, S. E. Bates, C. C. Vito, and S. J. Wieland, *J. Biol. Chem.* **254,** 4963 (1979).
24. C. C. Vito and T. O. Fox, *Develop. Brain Res.,* **2,** 97 (1982).
25. K. Burton, *Biochem. J.* **62,** 315 (1956).
26. R. J. B. King and J. Gordon, *Nature New Biol.* **240,** 185 (1972).
27. T. O. Fox, *Nature* **258,** 441 (1975).
28. C. C. Vito and T. O. Fox, *Science* **204,** 517 (1979).
29. J. Kallos and V. P. Hollander, *Nature* **272,** 177 (1978).
30. S. Liao, S. Smythe, J. L. Tymoczko, G. P. Rossino, C. Chen, and R. A. Hipakka, *J. Biol. Chem.* **255,** 5545 (1980).
31. M. H. Cake, D. M. DiSorbo, and G. Litwack, *J. Biol. Chem.* **253,** 4886 (1978).
32. S. A. Kumar, T. A. Beach, and H. W. Dickerman, *Proc. Nat. Acad. Sci. USA* **76** 2199 (1979).
33. J. Andre, A. Pfeiffer, and H. Rochefort, *Biochemistry* **15,** 2964 (1976).
34. I. Lieberburg, N. MacLusky, and B. S. McEwen, *Brain Res.* **196,** 125 (1980).
35. S. J. Wieland, T. O. Fox, and C. Savakis, *Brain Res.* **140,** 159 (1978).
36. K. R. Yamamoto and B. M. Alberts, *Proc. Nat. Acad. Sci. USA* **69,** 2105 (1972).
37. K. R. Yamamoto, M. R. Stampfer, and G. M. Tomkins, *Proc. Nat. Acad. Sci. USA* **71,** 3901 (1974).
38. K. R. Yamamoto, U. Gehring, M. R. Stampfer, and C. H. Sibley, *Rec. Prog. Horm. Res.* **32,** 3 (1976).
39. F. Payvar, O. Wrange, J. Carlstedt-Duke, S. Okret, J.-A. Gustafsson, and K. R. Yamamoto, *Proc. Nat. Acad. Sci. USA* **78,** 6628 (1981).
40. R. L. Tsai and H. Green, *Nature New Biol.* **237,** 171 (1972).
41. J. Salas and H. Green, *Nature New Biol.* **229,** 165 (1971).
42. R. L. Tsai and H. Green, *J. Mol. Biol.* **73,** 307 (1973).
43. S. R. Planck and S. H. Wilson, *J. Biol. Chem.* **255,** 11547 (1980).
44. B. M. Alberts, *Fed. Proc.* **29,** 1154 (1970).
45. G. Herrick and B. Alberts, *J. Biol. Chem.* **251,** 2133 (1976b).
46. G. Herrick, H. Delius, and B. Alberts, *J. Biol. Chem.* **251,** 2142 (1976).
47. K. R. Yamamoto and B. Alberts, *J. Biol. Chem.* **219,** 7076 (1974).
48. A. Revzin and P. H. von Hippel, *Biochemistry* **16,** 4769 (1977).
49. S.-Y. Lin and A. D. Riggs, *Cell* **4,** 107 (1975).

50. L. J. Kleinsmith, J. Heidema, and A. Carroll, *Nature* **226,** 1025 (1970).
51. G. S. Stein, J. S. Stein, and L. J. Kleinsmith, *Sci. Amer.* **232,** 46 (1975).
52. L. J. Kleinsmith, *J. Biol. Chem.* **218,** 5648 (1973).
53. C. Helene, *Nature New Biol.* **234,** 120 (1971).
54. J. H. Miller, in *The Operon* (J. H. Miller, and W. S. Reznikoff, Eds.), Cold Spring Harbor Laboratory (1978), pp. 31–88.
55. K. Adler, K. Beyreuther, E. Fanning, N. Geisler, B. Gronenborn, A. Klemm, B. Muller-Hill, M. Pfahl, and A. Schmitz, *Nature* **237,** 322 (1972).
56. N. Geisler and K. Weber, *Biochemistry* **16,** 938 (1977).
57. P. Nissley, W. B. Anderson, M. Gallo, I. Pastan, and R. L. Perlman, *J. Biol. Chem.* **247,** 4264 (1972).
58. D. Cattan and D. Bourgoin, *Biochim. Biophys, Acta* **161,** 56 (1968).
59. B. Bowen, J. Steinberg, U. K. Laemmli, and H. Weintraub, *Nucleic Acids Res.* **8,** 1 (1980).
60. S. S. Simons, Jr., H. M. Martinez, R. L. Garcea, J. D. Baxter, and G. M. Tomkins, *J. Biol. Chem.* **251,** 334 (1976).
61. K. H. Thanki, T. A. Beach, A. I. Bass, and H. W. Dickerman, *Nucleic Acids Res.* **6,** 3859 (1979).
62. J. Kallos, T. M. Fasy, and V. P. Hollander, *Proc. Nat. Acad. Sci. USA* **78,** 2874 (1981).
63. T.-S. Hsieh and D. L. Brutlag, *Proc. Nat. Acad. Sci. USA* **76,** 726 (1979).
64. R. S. Jack, W. J. Gehring, and C. Brack, *Cell* **24,** 321 (1981).
65. A. Sen and G. J. Todaro, *Proc. Nat. Acad. Sci. USA* **75,** 1647 (1978).
66. D. Williams and J. Gorski, *Proc. Nat. Acad. Sci. USA* **69,** 3464 (1972).
67. W. I. P. Mainwaring and F. R. Mangan, *Advan. Biosci.* **7,** 165 (1971).
68. J. D. Baxter, G. G. Rousseau, M. C. Benson, R. L. Garcea, J. Ito, and G. M. Tomkins, *Proc. Nat. Acad. Sci. USA* **69,** 1892 (1972).
69. O. Wrange and J.-A. Gustafsson, *J. Biol. Chem.* **253,** 856 (1978).
70. T. O. Fox, in *Proceedings of the 2nd Annual* (*1976*) *Maine Biomedical Science Symposium,* University of Maine Press, Orono (1977a), pp. 544–572.
71. R. W. Kuhn, W. T. Schrader, W. A. Coty, P. M. Conn, and B. W. O'Malley, *J. Biol. Chem.* **252,** 308 (1977).
72. W. A. Coty, W. T. Schrader, and B. W. O'Malley, *J. Steroid Biochem.* **10,** 1 (1979).
73. W. V. Vedeckis, W. T. Schrader, and B. W. O'Malley, *Biochemistry* **19,** 343 (1980).
74. T. O. Fox, C. C. Vito, and S. J. Wieland, *Amer. Zool.* **18,** 525 (1978).
75. C. C. Vito, S. J. Wieland, and T. O. Fox, *Nature* **282,** 308 (1979).
76. S. J. Wieland and T. O. Fox, *J. Steroid Biochem.* **14,** 409 (1981).
77. E. Milgrom, M. Atger, and E.-E. Baulieu, *Biochemistry* **12,** 5198 (1973).
78. J. A. Goidl, M. H. Cake, K. P. Dolan, L. G. Parchman, and G. Litwack, *Biochemistry* **16,** 2125 (1977).
79. S. J. Higgins, G. G. Rousseau, J. D. Baxter, and G. M. Tomkins, *J. Biol. Chem.* **248,** 5866 (1973).
80. B. Sato, R. A. Huseby, and L. T. Samuels, *Endocrinology* **102,** 545 (1978).
81. H. J. Eisen and W. H. Glinsmann, *Biochem. J.* **171,** 177 (1978).

82. O. Wrange, J. Carlstedt-Duke, and J.-A. Gustafsson, *J. Biol. Chem.* **254,** 9284 (1979).
83. H. J. Eisen, *Proc. Nat. Acad. Sci. USA* **77,** 3893 (1980).
84. C. H. Sibley and G. M. Tomkins, *Cell* **2,** 213 (1974a).
85. C. H. Sibley and G. M. Tomkins, *Cell* **2,** 221 (1974b).
86. J. R. Grove, B. S. Dieckmann, T. A. Schroer, and G. M. Ringold, *Cell* **21,** 47 (1980).
87. G. M. Ringold, K. R. Yamamoto, G. M. Tomkins, J. M. Bishop, and H. E. Varmus, *Cell* **6,** 299 (1975).
88. K. R. Yamamoto and B. Alberts, *Cell* **4,** 301 (1975).
89. P. H. von Hippel, A. Revzin, C. A. Gross, and A. C. Wang, *Proc. Nat. Acad. Sci. USA* **71,** 4808 (1974).
90. K. R. Yamamoto, *J. Biol Chem.* **249,** 7068 (1974).
91. T. O. Fox, *Brain Res.* **120,** 580 (1977b).
92. A. C. Notides and S. Nielsen, *J. Biol. Chem.* **249,** 1866 (1974).
93. A. Bailly, B. Le Fevre, J.-F. Savouret, and E. Milgrom, *J. Biol. Chem.* **255,** 2729 (1980).
94. B. W. O'Malley, D. O. Toft, and M. R. Sherman, *J. Biol. Chem.* **246,** 1117 (1971).
95. T. C. Spelsberg, A. W. Steggles, and B. W. O'Malley, *J. Biol. Chem.* **246,** 4188 (1971).
96. T. C. Spelsberg, G. M. Pikler, and R. A. Webster, *Science* **194,** 197 (1976a).
97. T. C. Spelsberg, R. A. Webster, and G. M. Pikler, *Nature* **262,** 65 (1976b).
98. W. T. Schrader and B. W. O'Malley, in *Steroids and Their Mechanism of Action in Nonmammalian Vertebrates* (G. Delrio and J. Brachet, Eds.), Raven Press, New York (1980), pp. 179–187.
99. W. V. Vedeckis, W. T. Schrader, and B. W. O'Malley, in *Steroid Hormone Receptor Systems* (W. W. Leavitt and J. H. Clark, Eds.), Plenum, New York (1979), pp. 309–327.
100. C. L. Thrall and T. C. Spelsberg, *Biochemistry* **19,** 4130 (1980).
101. W. I. P. Mainwaring and R. Irving, *Biochem. J.* **134,** 113 (1973).
102. T. O. Fox, D. Blank, K. L. Olsen, D. E. Vaccaro, and S. J. Wieland, *Abstracts,* The Endocrine Society Annual Meeting, 1981, No. 716.
103. K. L. Olsen and T. O. Fox, *Abstracts,* Society for Neuroscience Annual Meeting, 1981.
104. S. W. Rothwell, T. R. Brown, and C. J. Migeon, *Abstracts,* The Endocrine Society Annual Meeting, 1981, No. 717.
105. R. L. Burke, B. M. Alberts, and J. Hosoda, *J. Biol. Chem.* **255,** 11,484 (1980).
106. T. O. Fox and S. J. Wieland, *Endocrinology* **109,** 790 (1981).
107. S. J. Wieland, Ph.D. dissertation, Harvard University, 1979.
108. W. Gilbert and B. Muller-Hill, *Proc. Nat. Acad. Sci. USA* **58,** 2415 (1967).
109. A. D. Riggs, S. Bourgeois, and M. Cohn, *J. Mol. Biol.* **53,** 401 (1970).
110. S. Lin and A. D. Riggs, *J. Mol. Biol.* **72,** 671 (1972).
111. G. C. Chamness and C. D. Willson, *J. Mol. Biol.* **53,** 561 (1970).
112. M. Pfahl, *J. Mol. Biol.* **106,** 857 (1976).
113. J. E. Darnell, R. Wall, R. J. Tushinski, *Proc. Nat. Acad. Sci. USA* **68,** 1321 (1971).

114. J. R. Greenberg and R. P. Perry, *J. Mol. Biol.* **72,** 91 (1972).
115. C. S. McLaughlin, J. R. Warner, J. Edmonds, H. Nakazato, and M. Vaughan, *J. Biol. Chem.* **248,** 1466 (1973).
116. P. T. Gilham in *Methods in Enzymology,* Vol. 21, Part D (L. Grossman and K. Moldowe, Eds.), Academic Press, New York (1971), pp. 191–197.
117. U. Lindberg and T. Persson, *Eur. J. Biochem.* **31,** 246 (1972).
118. U. Lindberg and T. Persson, in *Methods in Enzymology* Vol. 34 (W. B. Jakoby and M. Wichek, Eds.), Academic Press, New York (1974), p. 496.
119. H. Aviv and P. Leder, *Proc. Nat. Acad. Sci. USA* **69,** 1408 (1972).
120. A. Efstratiadis and F. C. Kafatos, in *Methods in Molecular Biology,* Vol. 18 (J. Last, Ed.), Dekker, New York (1976), pp. 1–124.
121. J. N. Ihle, K.-L. Lee, and F. T. Kenney, *J. Biol. Chem.* **249,** 38 (1974).
122. R. A. Firtel, K. Kindle, and M. P. Huxley, *Fed. Proc.* **35,** 13 (1976).
123. R. Sheldon, C. Jurale, and J. Kates, *Proc. Nat. Acad. Sci. USA* **69,** 417 (1972).
124. J. N. Vournakis, R. E. Gelinas, and F. C. Kafatos, *Cell* **3,** 265 (1974).
125. J. Kates, in *Methods in Cell Biology,* Vol. 7 (D. M. Prescott, Ed.), Academic Press, New York (1973), pp. 53–65.
126. R. G. Deeley, J. I. Gordon, A. T. Burns, K. P. Mullinix, M. Binastein, and R. F. Goldberger, *J. Biol. Chem.* **252,** 8310 (1977).
127. M. E. Haines, M. H. Carey, and R. D. Palmiter, *Eur. J. Biochem.* **43,** 549 (1974).
128. D. J. Shapiro and R. T. Schimke, *J. Biol. Chem.* **250,** 1759 (1975).
129. J. L. Zimmerman, D. L. Fouts, and J. E. Manning, *Genetics* **95,** 673 (1980).
130. C. Milcarek, R. Price, and S. Penman, *Cell* **3,** 1 (1974).
131. J. Van Ness, I. H. Maxwell, and W. E. Hahn, *Cell* **18,** 1341 (1979).
132. Y. Kaufmann, C. Milcarek, H. Berissi, and S. Penman, *Proc. Nat. Acad. Sci. USA* **74,** 4801 (1977).
133. U. Nudel, H. Soreq, U. Z. Littauer, G. Marbaix, G. Huez, M. Leclerq, E. Hubert, and H. Chantrenne, *Eur. J. Biochem.* **64,** 115 (1976).
134. W. Wetekam, K. P. Mullinix, R. G. Deeley, H. M. Kronenberg, J. D. Eldridge, M. Meyers, and R. F. Goldberger, *Proc. Nat. Acad. Sci. USA* **72,** 3364 (1975).
135. T. Hunter and J. I. Garrels, *Cell* **12,** 767 (1977).
136. G. Brawerman, *Progr. Nucleic Acid Res. Mol. Biol.* **17,** 118 (1976).
137. J. E. Darnell, *Progr. Nucleic Acid Res. Mol. Biol.* **22,** 327 (1979).
138. M. Adesnic and J. E. Darnell, *J. Mol. Biol.* **67,** 397 (1972).
139. M. Grunstein and P. Schnell, *J. Mol. Biol.* **104,** 323 (1976).
140. L. M. Houdebine, *FEBS Lett.* **66,** 110 (1976).
141. A. Cann, R. Gambino, J. Banks, and A. Bank, *J. Biol. Chem.* **249,** 7536 (1974).
142. L. Gedamu, K. Iatrou, and G. H. Dixon, *Cell* **10,** 443 (1977).
143. H. F. Lodish, A. Jacobson, R. Firtel, T. Alton, and J. Tuchman, *Proc. Nat. Acad. Sci. USA* **71,** 5103 (1974).
144. L. Miller, *Dev. Biol.* **64,** 118 (1978).
145. G. Brawerman, J. Mendecki, and S. Y. Lee, *Biochemistry* **11,** 637 (1972).
146. G. Brawerman, in *Methods in Cell Biology,* Vol. 7 (D. M. Prescott, Ed.), Academic Press, New York (1973), pp. 1–52.

CHAPTER

6

AFFINITY ELECTROPHORESIS OF GLYCOPROTEINS

T. C. BØG-HANSEN

The Protein Laboratory, University of Copenhagen
Sigurdsgade 34, DK-2200
Copenhagen N, Denmark

Affinity electrophoresis is the term commonly used to describe an electrophoretic system in which interacting components are allowed to react during electrophoresis (1). To date, this principle has been mainly used for analytical purposes. An early systematic study of electrophoresis of interacting systems was conducted by Nakamura (2). The reaction of antigen and antibody during electrophoresis is a well-known example of the principle of interaction of components during electrophoresis, such as in counter immunoelectrophoresis, in rocket immunoelectrophoresis and in crossed immunoelectrophoresis—collectively referred to as electroimmunoassays, quantitative immunoelectrophoresis, or gel electro-immunoprecipitation methods (3). A short introduction to analytical affinity electrophoresis was given in reference (4). For preparative purposes, affinity electrophoresis has mainly been tried for the separation of bound material from affinity matrices after an ordinary initial column chromatographic adsorption step, that is, as an alternative elution with a specific displacer. Examples will be given below.

The combination of quantitative immunoelectrophoresis and affinity electrophoresis with lectins was developed for identifying, quantifying, and characterizing glycoproteins. It has also been used for predicting preparative separations. The approach for the characterization of glycoproteins with lectins by affinity electrophoresis has clearly been analytical. The basic principles may be applied to other interacting systems.

Three avenues may be explored when studying interacting components by analytical electrophoresis: the interaction can take place before, during, or after electrophoresis. This chapter considers the systems in which interaction takes place during electrophoresis, as the most information about the nature of the interaction can be obtained in this way. Moreover, the characterization of electrophoretically separated glycoproteins and glycopeptides by the binding of labeled lectins has been described elsewhere (5). Several types of label can be used, including flourescein isothiocyanate labels, radioactive labels, enzyme labels, and antibody labels. The reaction of glycoconjugates with lectins after electrophoresis, in a system analogous to the classical immunoelectrophoretic analysis of Grabar, has been studied by others (6–10) and is mentioned below.

In this laboratory, electroimmunoprecipitation is the preferred reference method, as proteins remain in their native state during an electroimmunoprecipitation experiment. Thus they retain their biological activity and protein–ligand interactions are not impaired. The formation of a protein–ligand complex before or during electrophoresis tends to change the electrophoretic and antigenic behaviour of the protein. Both aspects may readily be studied by crossed immunoelectrophoresis, which can be adapted to include macromolecular ligands in the agarose gel.

The agarose gel matrix is especially well suited to the study of native macromolecules. One important feature of these gels is the large pore size, which allows free migration of even very large macromolecular complexes (11).

Crossed immunoelectrophoresis has several unique features that can be used to advantage in the analysis and characterization of multicomponent protein mixtures. The use of antibodies in the reference gel offers the possibility of specific identification of proteins, and the size of the immunoprecipitation peaks depends on the antibody-antigen ratio, which can be adapted so that minor components need not be obscured by major components. The agarose gel immunoprecipitation methods and the preparation of antibodies has been described elsewhere (12, 13). However, immunoprecipitation limits the analytical uses of lectins somewhat due to the glycoprotein nature of the antibodies and the necessity for nondenaturing conditions. The technical details of the compound methods of lectin affinity immunoelectrophoresis are readily available (1, 14–30).

Many proteins have been characterized and purified by affinity chromatography (31), including group-specific affinity chromatography with lectins, as reviewed by Dulaney (32). It is the object of the first part of this chapter to describe a general "table top" approach to facilitate:

1. Identification of ligand-binding proteins.
2. Characterization of the reactive site or sites.
3. Quantification of ligand-binding proteins.
4. Prediction of preparative affinity separations.

Some examples of preparative affinity separations are also reviewed and discussed.

6.1. DESCRIPTION OF THE "COPENHAGEN APPROACH"

6.1.1. Method A: Reactions with Lectin Before Electrophoresis

The types of crossed immunoelectrophoretic precipitation patterns obtained after incubation of glycoprotein with lectin are shown schematically in Figure 6.1. The control pattern is shown in part *a*. Part *b* shows the electrophoretic pattern after reaction of glycoproteins with immobilized lectin. The incubation was performed prior to electrophoresis, by mixing immobilized lectin with the protein sample. The supernatant fluid can be analyzed by crossed immunoelectrophoresis. If a protein is known not to bind to the lectin, its precipitate will be unchanged and it may be used as an internal reference (Protein 1 in Figure 6.1). The most common type of result is that precipitates disappear from the pattern or are reduced in area. The disappearance may either be complete, as seen for Protein 2, or partial, as seen for Protein 3. The reduction in the immunoprecipitation area of Protein 3 indicates that only a fraction of this protein is bound. Other patterns could also occur in theory. The precipitate could split into several lines with the same first-dimension mobility (Figure 6.1*a*, Protein 4 → 4*a* + 4*b*) or with different first-dimension mobility (Protein 5 → 5 + *X*). This could be accompanied by a change of precipitate morphology (Protein 4) or profile (Protein 5).

Where free lectin is used instead of immobilized lectin, some proteins appear as lectin-complexes in the supernatant fluid and thus are seen in electrophoresis (Figure 6.1*c*). This results in a pattern of partial identity (19) mediated by lectin crosslinking of different protein species; this is discussed below.

At low lectin-to-glycoprotein ratios, the pattern is affected by the lectin concentration, but above the "saturation point" the pattern is little changed by further additions of more lectin. The loss of an immunoprecipitate in the reference gel of crossed immunoelectrophoresis is interpreted as binding of the missing protein by the lectin under study.

Other electrophoretic systems may be employed instead of crossed immunoelectrophoresis, and the reactions described here for crossed immunoelectrophoresis have their counterparts in the other systems.

6.1.2. Method B: Lectin in an Intermediate Gel

The principle of incorporating lectin into an intermediate gel in crossed immunoelectrophoresis was introduced under the term "crossed immunoaffinoelectrophoresis" (1). This method is analogous to the use of specific antibodies in the intermediate gel, which is a highly specific and sensitive

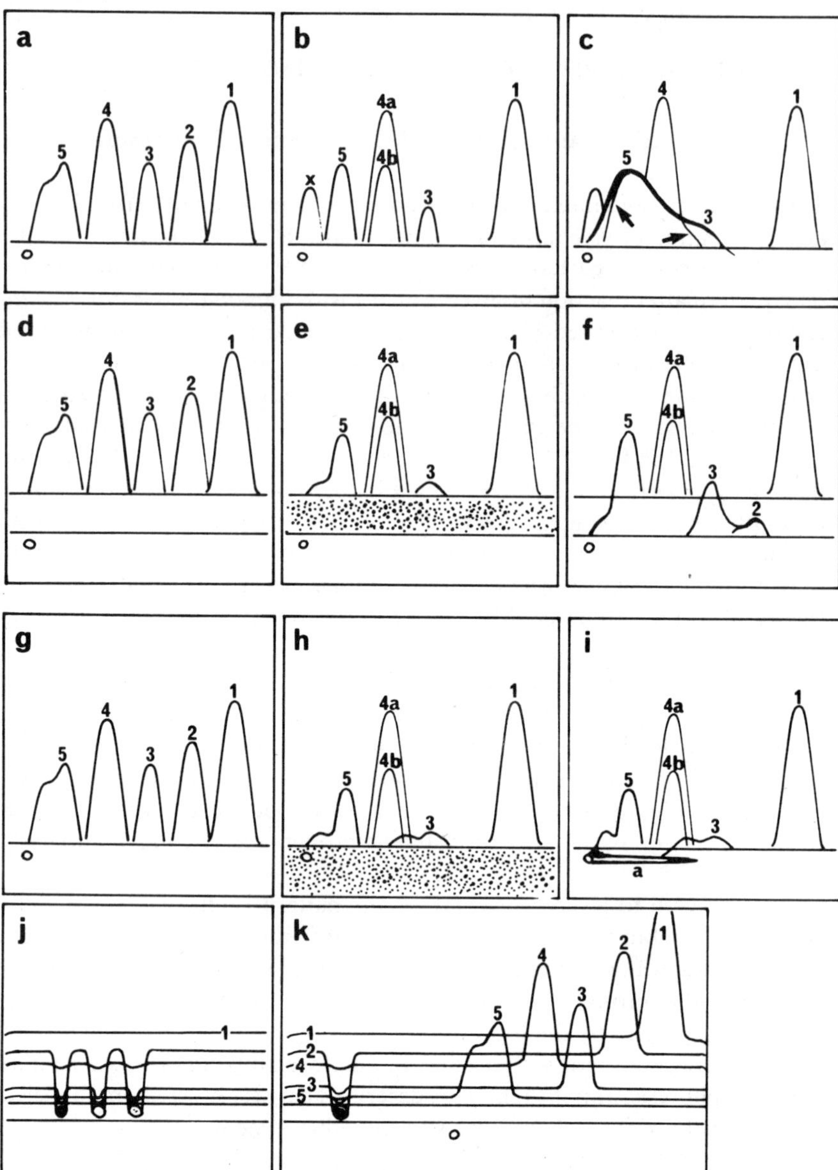

Fig. 6.1. Schematic representation of the reactions of glycoproteins with lectins. The methods are based upon crossed immunoelectrophoresis, and specific antibodies are applied in the reference gels to give the numbered immunoprecipitates. *Method A*: Reaction with lectin before electrophoresis: *a*, the control pattern without lectin; *b*, the pattern after reaction with immobilized lectin; *c*, the pattern after reaction with free lectin. *Method B*: *d*, the control pattern without lectin; *e*, the pattern with immobilized lectin in the intermediate gel; *f*, the pattern with free lectin in the intermediate gel. *Method C*: *g*, the control pattern without lectin; *h*, the pattern with immobilized lectin in the first dimension gel; *i*, the pattern with free lectin in the first dimension gel. *Method D*: *j*, the pattern in line immunoelectrophoresis with *in situ* absorption with lectins; *k*, the combined pattern of line-crossed immunoelectrophoresis and *in situ* absorption with lectins.

method for the detection and identification of interacting antigens and antibodies (33). Since its introduction, crossed immunoelectrophoresis has been used to identify many glycoproteins and to predict separation experiments with lectin affinity chromatography of glycoproteins (1, 15, 16, 19, 34–42).

Immobilized lectin was added to the intermediate gel in the original experiments. However, only a limited number of reactions are possible. Characteristic reactions are the disappearance or diminution of precipitates (see Figure 6.1, parts *d* and *e*). This is characteristic of simple glycoproteins. Theoretically it is reasonable to expect that more complex glycoproteins, such as macromolecular complexes containing glycoproteins, would give more complex patterns after reaction with an immobilized lectin. Splitting of precipitates may occur after reaction in the intermediate gel, as shown in Figure 6.1, part *e* (Protein 4 → 4*a* + 4*b* of the same first-dimension mobility). The change in precipitate profile seen for Protein 5 part *e* is characteristic for complex glycoproteins and is seen, for example, with human serum complement C3. Any of these reactions may be accompanied by changes in precipitate morphology. Changes in the precipitate patterns, including observations with both free and immobilized lectins, are shown in Table 6.1.

Figure 6.1*f* shows the reactions with free lectin in the intermediate gel. In addition to the precipitate changes mentioned with immobilized lectin, some notable features appear with free lectins: the formation of affinity precipitates and coprecipitation of individual glycoproteins (discussed below).

6.1.3. Method C. Lectin in the First-Dimension Gel

The principle of incorporating lectin into the first-dimension gel was introduced in reference (17) and is analogous to the use of specific antibodies in the first-dimension gel. This has proven to be a highly specific and sensitive method for the detection and identification of interacting antigens and antibodies (43). Since its introduction, the method with lectins in the first-dimension gel has been used for identification, characterization, and quantification of glycoproteins. It has also been used to predict which lectins may be suitable for use on affinity matrices for the preparation of glycoproteins (19, 22–26, 35, 38, 39, 44–71).

Figure 6.1, parts *h* and *i* shows typical reactions when proteins are electrophoresed through a gel containing immobilized and free lectin respectively. The pattern is highly dependent on the lectin in the range of concentrations used. Protein 1 does not bind to the lectin and is used as an internal reference of unchanged position. Table 6.1 gives some examples of reactions frequently observed.

Alterations in immunoprecipitation patterns can be seen reflecting the interaction of the protein with lectin, and sometimes an affinity precipitate can

Table 6.1. Changes in the Immunoprecipitation Pattern in Crossed Immunoelectrophoresis with Lectins[a]

No.	Reaction with Lectin in One-Dimension Gel	Reaction with Lectin in Intermediate Gel	Interpretation
1.	No reaction	No reaction	No molecules have affinity for the lectin.
2.	Disappearance of precipitate	Disappearance of precipitate	All molecules have affinity for the ligand. The binding to immobilized lectin may be observed with glycoprotein enzymes. An affinity precipitate may appear, indicating that some or all molecules contain two or more binding sites for the ligand.
3.	Shift of position in electrophoretic pattern.		Binding to ligand.
4.	Appearance of a multipeak precipitate ("camel" precipitate)		Reveals various molecular forms with different affinity to the ligand.
5.	Decrease of precipitate size	Decrease of precipitate size	Only part of the molecules have binding sites, or the binding is weak.
6.	Increase of precipitate size	Increase of precipitate size	Binding of ligand leads to stearic hindrance of antibody binding.
7.	Change in precipitate profile	Change in precipitate profile	As 4.
8.	Change in precipitate morphology A. Splitting of precipitate B. Change to diffuse precipitate	Change in precipitate morphology A. Splitting of precipitate B. Change to diffuse precipitate	As 6, or interrupted precipitation.
9.	Reactions of "partial identity"	Reactions of "partial identity	As 4 and 6, or coupling of different proteins through one molecule of ligand (the ligand mediates the "partial identity").

[a] Reactions seen for various glycoproteins with con A. With immobilized con A–Sepharose reactions, 6 and 9 have not been seen. Modified from Bøg-Hansen (22).

be seen in the first-dimension gel. Changed profiles can also be seen. In Figure 6.1, part *i*, Protein 2 is lost from the pattern. Protein 3 is shifted cathodically to appear as a double-peak precipitate, each peak with a characteristic retardation that is dependent on the lectin concentration. A number of human serum glycoproteins show multipeak precipitates in this system. Alpha-1-acid glycoprotein (AGP or orosomucoid), for example, shows this pattern when con A and WGA are incorporated into the first-dimension gel (Figure 6.2). Generally, lectins with a low mobility at pH 8.6 (such as con A from Pharmacia and Pharmindustrie) induce a cathodic shift of the binding glycoproteins.

6.1.4. Method D. A Small Amount of Lectin in a Well

A modification of the immunoelectrophoretic technique that has the added advantage of using very conservative amounts of lectin was introduced by Krøll and Andersen (72). They suggested placing a small amount of lectin in a so-called line immunoelectrophoresis (see Figure 6.1, part *j*). In this variant method, the antigen (the protein mixture) is melted into the lower gel, and after electrophoresis into the antibody-containing gel, each protein component precipitates as a line across the plate. The lectin well is placed in a blank intermediate gel between the antigen gel and the antibody gel. If there is an interaction between a protein and the lectin, a change of the pattern may be observed as a drop in the line-precipitate. Identification of the line is usually performed in a combined "crossed-line immunoelectrophoresis" (Figure 6.1*k*). In addition to its use in the identification of glycoproteins, this method was suggested for the evaluation of binding properties of chro-

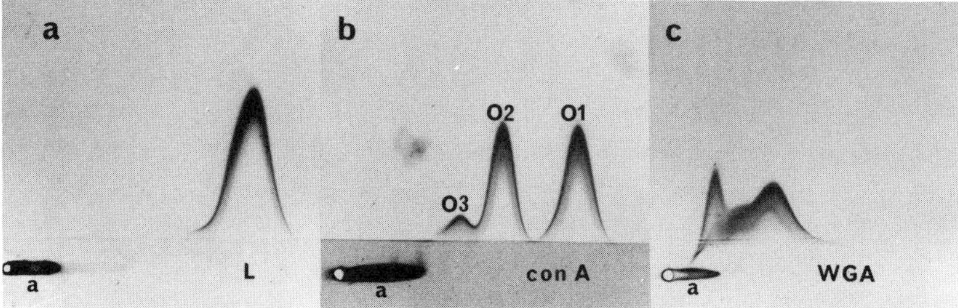

Fig. 6.2. Reaction of orosomucoid from human serum with lectins in the first dimension. Total serum was analyzed against a specific antibody against orosomucoid. *a*, the pattern of orosomucoid with lentil lectin in the first dimension gel (only slightly deviated from the normal reference pattern); *b*, the pattern with con A in the first dimension gel (01, 02, 03 denote subpopulations of orosomucoid); *c*, the pattern with WGA in the first dimension gel.

matographic media (72). A systematic application of this method is described by Brogren et al. in their test of the reaction of chicken lymphocyte MHC-alloantigens with 16 different lectins (73).

6.2. INTERPRETATION OF THE PATTERNS

6.2.1. Visualization of Proteins Bound to Immobilized Lectins

Glycoproteins bound to immobilized lectins cannot be specifically stained by Coomassie Brilliant Blue or other protein stains because the immobilized lectins are themselves proteins and thus will also stain. Only when a bound protein can be detected by a specific method can binding be visualized. Such methods include autoradiography of radioactively labeled glycoproteins or histochemical staining of glycoprotein enzymes. An example is shown in Figure 6.3. As in many other esterases, human serum cholinesterase can be stained by the histochemical stain for esterase directly after immunoelectrophoresis (74). Figure 6.3 shows that this enzyme binds to con A particles (Figure 6.3*b*) and precipitates with free con A in an affinity precipitate (Figure 6.3*c*).

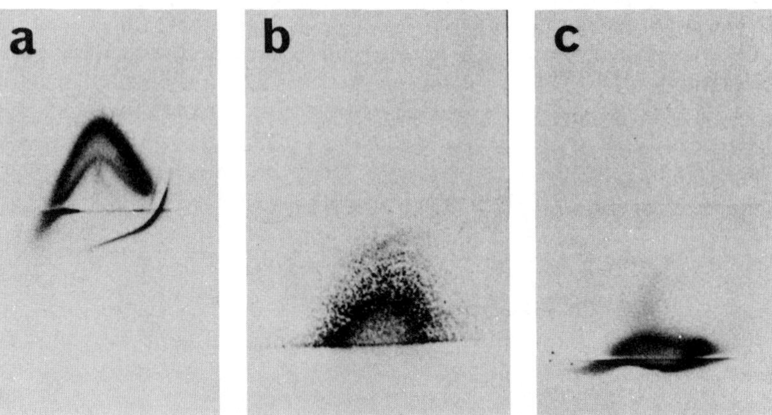

Fig. 6.3. Specific detection of lectin-bound protein. Total human serum was analyzed with multispecific antibodies in the reference gel. After immunoelectrophoresis, the plates were stained for esterase to reveal the cholinesterase. *a*, the pattern without lectin; *b*, the pattern with immobilized con A in the intermediate gel; *c*, the pattern with free con A in the intermediate gel. Adapted from Bøg-Hansen and Brogren (16).

An approach which utilizes the binding of radioactively labeled lectins in the identification of glycoproteins is described in ref. 75.

6.2.2. Appearance of Affinity Precipitate

The appearance of lectin-precipitated glycoprotein in one or more discrete "affinity precipitates" is most remarkable after electrophoresis with free lectin. Such precipitates are seen when human serum proteins are electrophoresed through a first dimension gel or through an intermediate gel containing free con A (as in Figures 6.1*f* and 6.1*i*, and Figure 6.2). A prerequisite for the formation of an affinity precipitate is that both the lectin and glycoprotein must have at least two binding sites per molecule. This is found to be the case for most lectins, with very few exceptions. Therefore we may deduce that the majority of the glycoproteins in the affinity precipitate have at least two binding sites. On the other hand, it cannot be excluded that some glycoproteins with only one site will bind to an already existing affinity precipitate. The intermediate-gel method is so discriminative that it was possible to distinguish between the reactions of two human glycoprotein enzymes, urine acid phosphatase (prostate) and serum cholinesterase (16). Both enzymes bound to con A-Sepharose but only cholinesterase gave an affinity precipitate with free con A (see Figure 6.3*c*), indicating two or more binding sites per molecule for cholinesterase (glycoprotein Type 2; cf. below, quantification by the intermediate gel technique, Method B) and only one binding site for urine acid phosphatase (glycoprotein Type 1, cf. below).

6.2.3. Reactions of "Partial Identity"

When mixtures of glycoproteins are analyzed with free lectin (cf. Figure 6.1*c*, *f*, and *i*), the common carbohydrate moieties will mediate reactions of partial identity and the lectin will crosslink all glycoproteins having binding capacity into a common precipitate. Lectins could be considered "promiscuous." The precipitation pattern of partially identical proteins in quantitative immunoelectrophoresis has been described in detail by Bock and Axelsen (76).

6.3. DETERMINATION OF AFFINITY

The magnitude of the electrophoretic shift during the first-dimension electrophoresis in a gel with lectin is an expression of the affinity between the glycoprotein and the lectin. Commercially available lectins often have a low mobility at pH 8.6, the pH of the crossed immunoelectrophoresis, and therefore the most common shift is a retardation of the faster migrating glycoproteins. The conditions in the electrophoresis gel could be compared with conditions in an affinity chromatography column in that higher affinity means stronger binding resulting in a larger electrophoretic shift, usually a retardation.

6.3.1. The Retardation Coefficient

A simple way to express the relative affinity of proteins for a particular ligand is to calculate the retardation coefficients,

$$R = \frac{l_0}{l_r} - 1$$

where l_r and l_0 are the migration distances in the first dimension electrophoresis with and without ligand respectively (19). The retardation coefficient may be calculated from a single experiment with ligand. In conditions where an excess of ligand is present, the retardation coefficient is independent of the ligand concentration, assuming influences resulting from differences in molecular weights, charge densities, and so on, can be disregarded. The advantage of this method is that only one experiment needs to be performed with lectin incorporated in the gel.

6.3.2. The Concentration-dependent Retardation

The concentration-dependent retardation can be examined using increasing concentration of lectin in the first dimension gel. This is shown for human

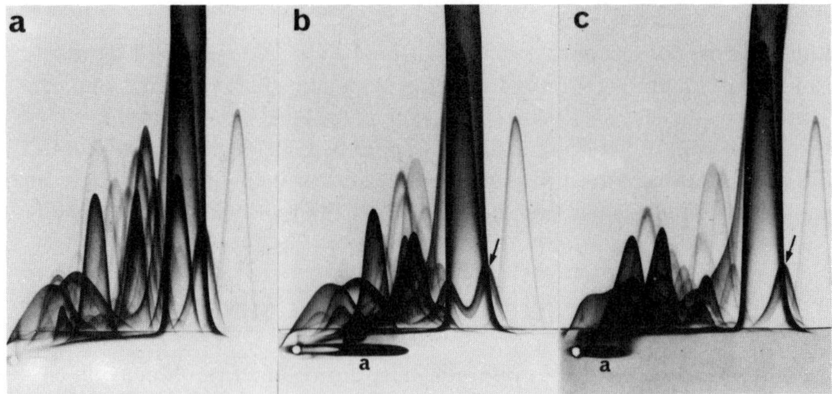

Fig. 6.4. The effect of increasing the amount of lectin in the first dimension gel. Total human serum was analyzed with multispecific antibodies in the reference gel. *a*, the control pattern without lectin; *b*, the pattern with 2×10^{-5} *M* con A in the first-dimension gel; *c*, the pattern with 8.5×10^{-5} *M* con A in the first-dimension gel.

serum in Figure 6.4. When free con A is incorporated into the first dimension gel, some proteins show a characteristic multipeak precipitate. For some con A–binding glycoproteins like orosomucoid, only certain fractions of the protein bind, whereas one fraction does not bind (arrow, Figure 6.4; see also Figure 6.2*b*) (17, 23, 67, 68, 77). This was verified by affinity column chromatography (58).

Below a certain limit, the retardation is totally independent of the amount of glycoprotein. A protein load of 50 ng per analysis did not change the equilibrium conditions within the normally used lectin concentration interval (see below, 6.3.4).

Proteins can exhibit a rather complex microheterogeneity pattern when analyzed in this way; murine AFP (alpha-fetoprotein) shows four microheterogeneity forms upon reaction with con A. These forms were designated 0, 1, 2, and 3 according to increasing con A affinity, form 0 being nonreactive to con A (27).

6.3.3. The General Takeo–Nakamura Plot

The theoretical background of affinity electrophoresis was worked out by Takeo and Nakamura (78) in their original experiments with enzymes and substrates. Using their theory, it is possible to calculate the affinity between a retarded protein and a lectin as the dissociation constant from a Takeo-Nakamura plot, which is a simple relation between the relative migration velocity and the lectin concentration:

$$\frac{1}{R_{mi}} = \frac{1}{R_{mo}}\left(1 + \frac{c}{K}\right) \tag{1}$$

Where K is the dissociation constant of the lectin–glycoprotein complex; c is the concentration of lectin expressed as the concentration of binding sites, the "normality"; R_{mo} is the mobility of the glycoprotein without lectin; R_{mi} is the mobility of the glycoprotein in presence of lectin and R_{mc} is the mobility of the lectin–glycoprotein complex in relation to an internal standard such as bromophenol-blue-marked albumin, prealbumin, or the nonreacting component of orosomucoid. Using this method the dissociation constants for con A complexes of a number of glycoproteins were calculated (23).

The Takeo-Nakamura plot is implicitly confined to the situation where the complex between the interacting components is electrophoretically immobile, a condition that is not met a priori with free macromolecular ligands as lectins but is always found with immobilized lectins. It is often preferable, however, to work with free lectins because it saves the effort involved in preparing the immobilized lectins and eliminates the difficulty of estimating the exact amount of bound lectin. Complexes between glycoproteins and free lectins would be expected to have an eigenmobility. Nonlinear Takeo-Nakamura plots for kidney phosphorylase in complex with glycoprotein (78) and human serum glycoproteins in complex with con A have been described (23). Therefore the general equation for electrophoretic determination of dissociation constants was derived:

$$\frac{1}{R_{\text{mo}} - R_{\text{mi}}} = \frac{K}{R_{\text{mo}} - R_{\text{mc}}}\frac{1}{c} + \frac{1}{R_{\text{mo}} - R_{\text{mc}}} \tag{2}$$

where R_{mc} is the mobility of the con A–glycoprotein complex, taken in relation to the internal standard [the other symbols are as those used in equation (1)]. This equation represents a straight line when (lectin concentration)$^{-1}$ is plotted against $(R_{\text{mo}} - R_{\text{mi}})^{-1}$. The intercept on the $(R_{\text{mo}} - R_{\text{mi}})^{-1}$ axis is $(R_{\text{mo}} - R_{\text{mc}})^{-1}$ and the intercept on the c^{-1} is $-K^{-1}$. The slope of the line is $K/(R_{\text{mo}} - R_{\text{mc}})$. The equation gives the dissociation constant K as well as the mobility of the con A–glycoprotein complex, R_{mc} (see Figure 6.5). The values of K and R_{mc} were calculated by the least-squares method. The equation and

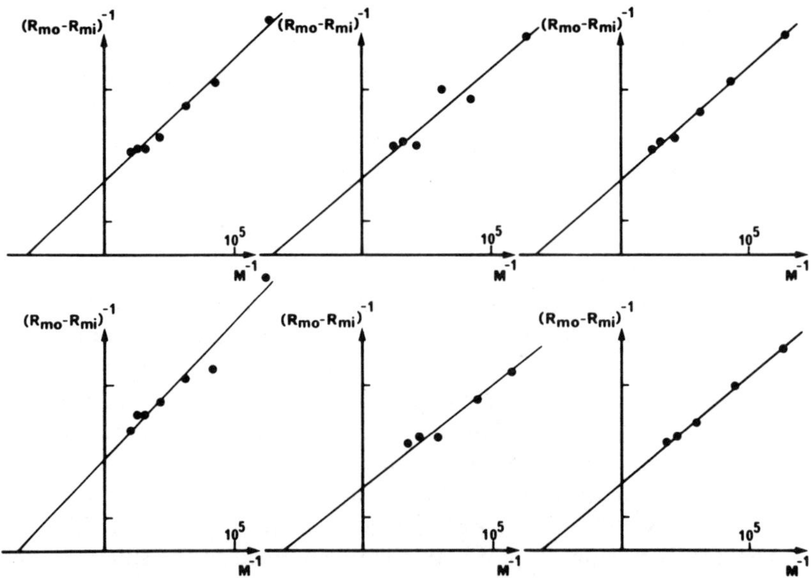

Fig. 6.5. The general plot for calculation of dissociation constants. Six different human serums were analyzed with variable amounts of con A in the first-dimension gel as shown in Figure 6.4. The retardation of a fraction of orosomucoid, 02 (see Figure 6.3*b*) was calculated and plotted according to the general equation (2); see text for further details.

its derivative were subsequently verified by Horejsi, who used other symbols (79, 80).

In the simple case where the mobility of the con A–glycoprotein complex is zero ($R_{mc} = 0$), equation (2) becomes equation (1).

No simple alternative method for determination of the affinity between macromolecules such as glycoproteins and lectins has been published to date.

A prerequisite for exact determination of the dissociation constant is that the mobility of the complex is not influenced by the gel matrix. The agarose gel is especially suited for work with free macromolecular ligands forming large complexes, because the pores allow penetration of macromolecular aggregates of up to 2×10^7 daltons (11).

One practical problem present in the accurate determination of the dissociation constant of a lectin–glycoprotein complex arises from the need to accurately determine the concentration of available lectin. This is particularly apparent in cases where immobilized lectin is used as the volumetric determination of the gel and where the determination of bound lectin per unit matrix is difficult. Factors that complicate determination of the latter are the change in binding properties, and the number of available binding sites, after immobilization. The concentration of ligand is calculated as the monomer concentration, and it was not possible to estimate the effect of the multimeric nature of con A.

The last and greatest obstacle to general usage of this method for diagnostic purposes is the nonreproducibility of commercial lectin preparations. Therefore, international collaboration has been initiated to test lectins and to draw up a set of guidelines for lectin producers (81, 82).

An apparent advantage of the multivalent approach with antibodies is that all binding proteins in the extract or homogenate can be determined. From one set of experiments, a set of data could be obtained for each glycoprotein in human serum reacting with con A (23).

When individual glycoproteins from individual human sera were compared, a considerable and unsystematic variation in the measured dissociation constants for all but a few glycoproteins was found (23). Among these, orosomucoid was surprisingly constant, both with respect to its appearance as three components with a constant ratio (67, 68) and with respect to dissociation constants (see Figure 6.5). When serum from individuals low in alpha-1 antitrypsin was tested in this way and compared with normal serum, we observed the same pattern of a major component and a minor component in both, with equal dissociation constants to con A, even though the antitrypsin concentration in pathological serum was sometimes as low as 1% of the normal concentration (23).

6.3.4. The Electrophoretic Conditions and Control Experiments

The electrophoretic parameters were varied with the conclusion that the best conditions are those that are optimal for electroimmunoprecipitation. In a way, this is a limitation to the general use of the methodology, as the variation of the dissociation constant as a function of pH and of temperature can only be studied within narrow limits. Routinely, electroimmunoprecipitation is performed in the pH range of 8.5 to 8.9 and in the temperature interval of 16 to 18°C; exceptionally, the pH can be chosen as low as 5.5 and as high as 9.2, and the temperature may be from 10 to 30°C. Only small changes in the values of dissociation constants were observed in the narrow intervals.

Immobilized lectin (Sepharose-bound con A) is found to influence the electrophoretic properties of the first-dimension gel. Therefore, several control experiments must be performed to see whether alterations in the immunoprecipitation pattern are due to lectin binding or to other nonspecific properties of the lectin gel. One type of control experiment utilizes specific displacers, the low molecular weight sugar inhibitors of lectin binding. Other types of control experiments are discussed in ref. 25, such as the variation of lectin concentration and the denaturation of the lectin.

The influence of competing glycoproteins can also be tested. The pattern of a glycoprotein enzyme was studied with and without competing glycoproteins. The same multipeak precipitate was observed in both cases, which indicated that the influence of competing glycoproteins on the lectin-binding is negligible under our standard conditions, with a glycoprotein amount of about 50 ng per analysis (26). Under some conditions, however, a less detailed pattern was observed in the presence of competing glycoprotein, thus showing that the pattern observed with total serum does indeed reflect a microheterogeneity of the serum glycoproteins.

6.3.5. Evaluation of the Dissociation Constants

The method described here for determination of dissociation constants is a two-dimensional method that allows analysis of more complex systems as components are specifically identified by their immunological characteristics. The absolute requirement of a number of analyses performed at well-determined ligand concentrations could appear to be a drawback when comparing this system with frontal or elution analysis affinity chromatography. On the other hand, the electrophoreses are simple to perform and require only a minimum of material for the determination. Furthermore, the ligand need not be immobilized (see Chapter 2).

The dissociation constants of glycoproteins for con A determined by our method are in the order of 10^{-5} M. This concurs with measurements of dissociation constants for two human serum glycoproteins obtained by different means, to be 0.4 10^{-5} M for purified transferrin, 0.7 10^{-5} M for transferrin glycopeptide, and 10^{-6} M for IgM by a precipitation inhibition test (83). A summary of results obtained with affinity electrophoretic determination of binding constants between a number of enzymes and their substrates was published in reference (23).

The development of an easy method for determining the affinity of antibodies has high priority. Thus Caron et al. (84) studied possibilities for assessing antialbumin affinity in agarose affinity electrophoresis by determining the "partitioning coefficient" K_p between immunosorbent and soluble antigen or antibody. A more direct approach for antibodies was used by Takeo and Kabat (85) in an elegant study in analogy with the original work on enzymes. They measured association constants for mouse myeloma proteins with antidextran and anti-isomaltose oligosaccharide activity by electrophoresis in polyacrylamide gels containing dextran ($K_a = 3 \times 10^2$ to 6×10^4 mL/g) or isomaltose oligosaccharides ($K_a = 8.3 \times 10^4$ M^{-1}). They concluded that affinity electrophoresis is useful for obtaining binding constants ranging from 10^2 to 10^6 M^{-1}.

In a similar system of one-dimensional affinity electrophoresis, Johnson et al. (86) have studied binding of albumins to immobilized Cibacron Blue F3G-A, and have listed prerequisites for successful use of this system. They found that affinity electrophoresis in pure agarose gels containing immobilized Cibacron Blue is seriously affected by dehydration, as initially observed by Caron et al. (87), and they overcame this problem by inclusion of polyacrylamide into the dye–agarose gels. The first prerequisite was that the dye–agarose gels (in beaded form) should be extensively washed so that no unbound dye leaches out of the gel. The presence of any such free ligand will complicate the determination of dissociation constants. Second, gels should be pre-electrophoresed for at least 30 min to remove persulfate and free ligands. Third, the dye–agarose beads should be thoroughly melted so that the agarose does not solidify as clumps within the polyacrylamide. Fourth, reproducible ligand concentrations in the agarose–polyacrylamide gels are more easily achieved by adding undyed agarose to dyed agarose rather than by decreasing the amount of dye or the incubation time during the preparation of the dyed agarose. In this way they could study the dissociation constants for the interaction of a microbial 6-phosphogluconate dehydrogenase with Procion Red HE3B and with Cibacron Blue F3G-A, and the effects of lipid presaturation on the dissociation constants of albumins from different species (86). Needless to say that the homogeneity of the gel matrix was essential.

6.4. QUANTITATION OF MICROHETEROGENEITY FORMS

The heterogeneity of glycoproteins in their reaction with lectins, as observed with these methods and with other methods, may now be ascribed to the existence of populations of microheterogeneity forms. Each population may have a characteristic carbohydrate structure that possibly reflects different metabolic stages related to the function of the protein. The present methods may be used to assess the quantitative changes of these microheterogeneity forms that take place in vivo.

6.4.1. Quantitation by the Intermediate Gel Technique, Method B

Precipitation with free lectin *vs.* binding to immobilized lectin forms a basis for a classification of the heterogeneity forms or the microheterogeneity forms of a glycoprotein into three classes: Type 0, molecules without binding sites; Type 1, molecules with one binding site; and Type 2, molecules with two or more binding sites.

The experimental basis for quantification of each of these three types of heterogeneity forms or microheterogeneity forms requires three different experiments. The first experiment is without lectin in the intermediate gel, the second experiment includes immobilized lectin in the intermediate gel, and the third experiment includes free lectin in the intermediate gel. By careful planimetry of the enclosed area of each precipitate, a relative estimate can be obtained for the total amount of each glycoprotein (the first experiment), the amount of glycoprotein of Type 1 plus Type 2 (the second experiment), and the amount of glycoprotein of Type 2 (the third experiment). A simple calculation will give the relative distribution between the heterogeneity classes or the microheterogeneity classes (27).

6.4.2. Quantitation by the First Dimension Gel Technique, Method C

It is possible to distinguish between microheterogeneity forms, differing in their affinity to the lectin, by the first-dimension technique. The amount of each distinct microheterogeneous component is determined by careful planimetry as above. The amount of glycoprotein in the affinity precipitate can also be found in an experiment that includes an intermediate gel containing an inhibitory sugar (a specific displacer), which is carried into the first-dimension gel by the electroendosmosis to liberate the bound proteins. Examples of the calculation of relative amounts of individual components of serum glycoproteins may be found in the work of several authors (49–52, 67, 68, 70, 88). These studies strongly indicate the relationship between the carbohydrate part of glycoconjugates and the function and metabolism of the glycoconjugate; the three forms of orosomucoid present are normally in a well-defined ratio, but this is shifted during inflammation (77), during pregnancy, and during treatment with steroids (67, 68).

The number of detected microheterogeneity forms depends upon the lectin preparation. The reason for this inconsistency between lectin preparations is not clear but may lie in differences in the binding specificity of various commercial lectins prepared from different lines or cultivars of the same plant; such differences are known to occur (29, 90).

6.4.3. Preparative Use of Lectin Affinity Immunoelectrophoresis

Crossed immunoelectrophoresis and line immunoelectrophoresis, though principally analytical methods, may be used for semipreparative purposes when only small amounts of pure protein are required. For preparation of immunogens for antibody production, crossed immunoelectrophoresis represents an easy shortcut to problems encountered in conventional protein separations. The technology shown here may be applied to secure optimal separation of lectin-binding glycoproteins from nonbinding proteins, and it may also be used to prepare small amounts of individual microheterogeneity forms of single glycoproteins.

6.5. OTHER ANALYTICAL ELECTROPHORETIC METHODS

At the beginning of this chapter, we briefly mentioned the use of labeled lectins for the identification of glycoprotein bands after electrophoresis. This is a very sensitive way for identification of glycoprotein bands after acrylamide-gel electrophoresis and is a good supplement to the mentioned technology or for preliminary screening for glycoproteins. The methodology and references are given in ref. 5. The blotting technique (89) may add increased sensitivity to the identification.

6.5.1. Selection Electrophoresis

It is not always necessary, or even desirable, to perform crossed immunoelectrophoresis as the reference method, although antibodies may be available. "Selection electrophoresis" is a simpler method: a combination of one-dimensional electrophoresis in agarose with subsequent diffusion of lectin from a trough, analogous to the classical immunoelectrophoretic analysis of Graber (8–10). The method has hardly been used to any extent. However, the great potential of glycoprotein analysis using lectins was shown by

Osunkoya and Williams (7) in their studies on glycoproteins from various pathological sera.

6.5.2. One-Dimensional Affinity Electrophoresis

To determine that protein–lectin binding occurs, or for the identification of proteins that precipitate with lectins, one-dimensional electrophoresis may be performed into a gel with either immobilized lectin or free lectin, respectively. The use of one-dimensional affinity electrophoresis for the study of enzymes from barley has been reported (15), and the method was also used for quantitation of native glycoproteins (Figure 6.6) as well as for denatured glycoproteins. Protein denaturation does not interfere with the structure of the binding site, and glycoproteins denatured to various degrees all gave affinity precipitates of the same height (20).

One-dimensional affinity electrophoresis in which the components are reversed may be a powerful screening method for glycoprotein-precipitating lectins (14, 27). The gel contains glycoproteins (for instance human serum), and extracts of various plant sources can be electrophoresed into the glycoprotein-containing gel, resulting in the formation of affinity precipitates if lectin is present in the extracts. In addition, this type of experiment allows lectin quantification (14).

6.5.3. Affinity Electrophoresis in a Preformed *p*H Gradient

Protein–ligand interactions may be tested over a pH span in the same experiment. By combining the electrophoretic titration technique of Rosengren et

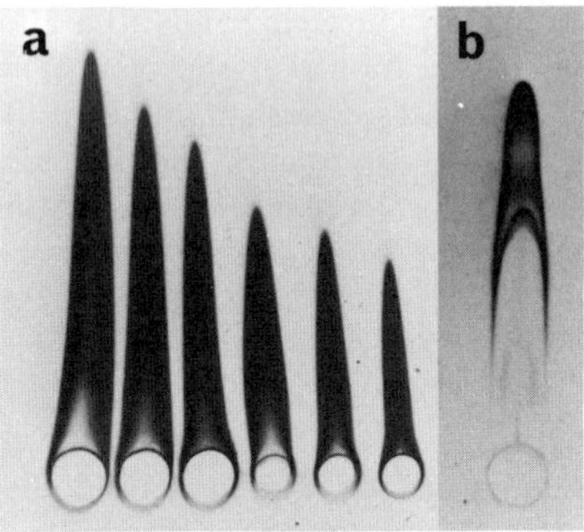

Fig. 6.6. One-dimensional rocket affinity electrophoresis of glycoproteins into gel-containing lectin. *a*, the pattern of alpha-2-macroglobulin (900, 675, 540, 450, 360, and 225 μg) into con A (40 $\mu g/cm^2$); *b*, the pattern of 12 μg of purified human erythrocyte *M,N*-glycoprotein (glycophorin) into WGA (50 $\mu g/cm^2$). Adapted from Bøg-Hansen, Bjerrum, and Brogren (20).

al. (109) with affinity electrophoresis, Ek and Righetti (110) were able to measure the pH dependence of the dissociation constant. The electrophoretic titration technique uses a flat-bed electrophoresis in polyacrylamide, where the first dimension of the electrophoresis is without sample but with ampholytes to form a pH gradient. Before electrophoresis in the second dimension, the sample is applied in a slit across the pH gradient. The second dimension will give the mobility of the protein at various pH values in a continuous manner (109). In the affinity experiments of Ek and Righetti (110), glycogen was incorporated at various concentrations in the polyacrylamide gel as substrate for rabbit muscle phosphorylases to give the dissociation constants for the enzyme-substrate complexes from pH 5 to pH 9. This technique has some similarity to the traditional "crossing diagrams" of Nakamura (2) and seems to be very promising for screening purposes.

6.5.4. Compound Gel Method: Polyacrylamide and Agarose

SDS-polyacrylamide gel electrophoresis of denatured glycoproteins can be combined with lectin affinity electrophoresis (24). After separation of the proteins in the SDS gel, lanes can be cut and soaked for 15 min in the agarose gel buffer or stored frozen. The second dimension of the electrophoresis is performed in agarose containing a nonionic detergent such as Triton X-100, Lubrol, or Berol in high concentration. This method may be used quantitatively for determining the amount of glycoprotein applied or the amount of lectin-binding glycoprotein in each band, since the precipitate formation is dependent upon the amount of glycoprotein and the concentration of lectin (24).

It is remarkable that the SDS in the acrylamide gel does not disturb the binding of glycoproteins to the lectin. This is presumably due, in part, to the neutralizing effect of the nonionic detergent, by including SDS into the nonionic detergent-micelles, so that the lectin is not exposed to free SDS. Also some lectins are known to be stable and reactive in the presence of detergent (91). Furthermore, denaturation of the glycoprotein has little effect on its binding to lectin (con A). Whether denaturation is performed with detergent, heat, or acid treatment, the precipitation in affinity precipitates is quantitative irrespective of the degree of denaturation (20).

6.6. PREPARATIVE AFFINITY ELECTROPHORESIS

The development of affinity chromatography has been the biggest breakthrough in the purification of macromolecules (cf. chapter 2). In the design of affinity matrices, attention has been directed towards the characteristics of the matrix itself and towards the chemistry of coupling ligands with and without spacer arms and so on, in order to obtain optimal binding of a protein to the affinity matrix. Often a high degree of specificity is sought without having the accompanying problem occur, desorption of the specifically bound material. The normal way for desorption to occur is by using a more or less specific displacer.

A few early exploratory experiments into a possible preparative use of the principle of affinity electrophoresis were unsuccessful. One preparative system that was explored is isotachophoresis in acrylamide gel containing con A. This was not pursued because the potential for investigating the analytical aspects appeared more interesting (18). To my knowledge, applications of affinity electrophoresis for the preparation of proteins have not been reported.

6.6.1 Electrophoretic Desorption of Affinity Gel Matrices

One frequent difficulty in preparative affinity chromatography is that the adsorbed component sticks too tightly to the immobilized ligand. Therefore, one of the most significant parameters of affinity systems is the method of desorption. The outcome of the purification can depend totally on the choice of the method. In certain critical instances, the lack of a suitable elution procedure has meant that the powerful tool of affinity chromatography cannot be applied (92). This is particularly true in the case of interaction of small haptens with antibodies and in cases where the material to be eluted is unstable under the conventional conditions of elution (92). In order to overcome this problem, Dean and coworkers have used electrophoretic desorption of the affinity column after loading and washing (cf. 92). Electrophoretic desorption was developed as a mild, nonchaotropic technique for the removal of material from affinity matrices and immunoadsorbents (93–99). It appears particularly suited to the desorption of antibodies from the high-association complexes with immobilized antigen (94, 96–99).

A technical solution to enable electrophoretic desorption to occur was found, and this enabled a systematic investigation of optimal conditions (92). Low current, buffer of low ionic strength, and thin layers of matrix are desirable. The temperature can be varied according to the nature of the particular interaction and the stability of the material to be desorbed. The instrument that Dean et al. use can desorb 200 mg or more of human serum albumin from 20 mL immobilized Cibacron Blue at a satisfactory rate (92). Much more work, however, is needed in order to show the general applicability of this approach.

6.6.2. Desorption of Affinity Matrices by Electrofocusing into Granulated Gel

Preparative and analytical electrofocusing (isoelectric focusing) is a very discriminative method, resolving proteins by their isoelectric points. One great advantage of electrofocusing over ordinary electrophoresis is that protein bands are not diluted during an experiment. By using this electrophoretic method for desorption after the initial chromatographic adsorption step as above, Rautenberg et al. have shown an elegant way of purifying mg amounts of glycoprotein (100, 101). A total homogenate of Trypanosoma cells containing the "variant surface glycoprotein" was adsorbed onto con A–Sepharose, and unbound proteins were washed from the affinity column in an ordinary way. To liberate the bound proteins, the matrix with the bound glycoproteins was mixed with granulated gel (Ultrodex) with 2% Ampholines and focused in a flat bed of granulated gel. The focusing liberated the variant surface glycoproteins from the con A–Sepharose to appear as distinct bands, which could be detected for subsequent elution. Detailed analysis showed a correlation between content of sialic acids and position in the pH gradient as well as a relation to the ability to form a well-defined affinity precipitate in one-dimensional affinity electrophoresis with con A.

An important factor for an uncomplicated focusing was found to be the pretreatment of con A–Sepharose. The con A matrix was pretreated with 6 *M* urea in order to remove any con A subunits that were not covalently bound to the matrix. When this pretreatment was omitted, the con A subunits were liberated during electrofocusing to band with the glycoproteins.

Following this protocol, the recovery of the variant surface glycoprotein after electrofocusing desorption, dialysis, and lyophilization was found to be 75% compared with only 5% obtained in the corresponding conventional affinity chromatographic procedure with con A–Sepharose. These authors also found it possible to perform the desorption step in Ultrodex-Ampholine-containing urea up to a concentration of 8 *M* without impairing the pH gradient formation and the separation (101).

6.6.3. The Use of Our Analytical Approach to Predict the Results of Preparative Experiments

In initial experiments with con A–Sepharose and serum proteins, a strict correlation was observed between the intermediate gel technique (Method B) and ordinary column affinity chromatography, even though there were differences in various conditions—pH and temperature for instance, not to mention hydrostatic flow and electrophoretic forces. Therefore, the analytical electrophoresis experiments were suggested as a prediction method for preparative separation experiments (1). By careful analysis of experiments as outlined here, we were able to predict the results of preparative experiments with ease and with minimal use of protein (19, 58). A remarkable correlation was also found for the three con A fractions of AFP (alpha-fetoprotein) (44).

In another study, we compared different lectins for their fractionation potential for human serum proteins, and we could demonstrate that there is a strict correlation between electrophoresis and column separation chromatography only when glycoprotein is bound strongly in the affinity precipitate. When glycoprotein was not bound in the affinity precipitate but was retarded in the first dimension, we could not predict the result of affinity chromatography. Thus for one lectin, retardation in electrophoresis corresponds to binding on a lectin column, but this may not necessarily be the case for another (58). Parameters such as binding capacity and elution conditions cannot be determined directly from the electrophoretic experiments.

6.7. BIOMEDICAL APPLICATIONS

The analytical electrophoretic approach has been proposed as an alternative to other preparative methods, one that may be used when only a small amount of biological specimen is available. It might be used routinely for diagnosis and screening of human diseases. The pattern of a number of major serum proteins and their variation is reported in (88).

6.7.1. Diagnosis of Malignant Diseases and Neural Tube Defects

Several important marker proteins in human diseases are glycoproteins, and their interaction with lectins may give important clues to the state of the disease involved. Alpha fetoprotein (AFP) is one such protein. In normal pregnancies, amniotic fluid AFP examined between 15 and 33.5 weeks of pregnancy consists of 12 to 45% of the con A nonreactive form (102), whereas fetal serum has been reported to contain a significantly lower proportion of con A-nonreactive variants (2–10%) (102, 103). This has led to the demonstration that the pattern of con A-affinity variants in amniotic fluid is potentially valuable in diagnosing fetal abnormalities (104). In fetal abnormalities characterized by a change in AFP compartmentalization, such as transudation of fetal serum across exposed fetal membranes in the presence of neural tube defects, a shift has been observed (102, 104) in the pattern of amniotic fluid AFP con A affinity variants to resemble the pattern of fetal serum. The measurements of con A affinity variants in amniotic fluid has recently been performed, using affinity electrophoresis with con A included in the first-dimension gel (46, 53, 59, 65, 66, 105). This method has been found to be simpler and more reliable than the previous chromatographic techniques (105).

The affinity electrophoresis system has also been used in the comparison of fetal and hepatoma AFP lectin-affinity variant patterns (46, 54). The proportion of the con A-affinity variants seems to be similar in hepatoma sera and similar to the proportion in fetal sera. The LCA-affinity patterns vary among hepatoma sera, but generally the percentage of the LCA-reactive variant is higher in hepatoma sera than in fetal sera (102). A one-dimensional variant of the affinity immunoelectrophoretic method with lectin in the intermediate gel (Method B) is being tested as a routine diagnostic tool for the identification and quantification of nonbinding and binding AFP components (81).

6.7.2. Activation of Glycoprotein Enzymes

An interesting phenomenon was observed when some plant acid phosphatases were tested in crossed immunoelectrophoresis with con A and lentil lectin incorporated in the first-dimension gel (Method C). In addition to binding, we observed a remarkable increase in the staining intensity. The increase with con A and lentil lectin was also found by spectrophotometric analysis of the enzyme activity, and it led us to speculate that one of the biological roles of a lectin is to act as a modulator of enzyme activities during biological processes (55).

6.7.3. Generalizations: Other Interacting Systems

Also using the first-dimension technique, Method C, but with hydrophobic matrices, Bjerrum showed how to identify hydrophobic proteins (106). This is especially useful in distinguishing integral membrane proteins from proteins loosely associated with membranes.

By performing crossed immunoelectrophoresis with heparin in the first dimension gel, Bleyl and Peichl (107) were able to separate active antithrombin from the inactive complex and measure it quantitatively in patients under heparin therapy.

This example shows that it is possible to distinguish between various functional forms of active macromolecules. Experiments of this type also offer an opportunity to distinguish between activator control (control by inhibitors and activators and by activation of precursor forms and inactivation by partial degradation) from control by *de novo* synthesis and total degradation.

Horejsi (79, 80) has given some theoretical consideration to the general quantitative use of affinity electrophoresis, and recently Bjerrum et al. (108)

discussed some generalizations of the present approach for a description of receptor functions of membrane proteins and Ramlau and Bock (41) showed some other interacting systems. Horejsi has recently reviewed affinity electrophoresis, including the preparation of affinity gels, qualitatively and quantitatively, and has summarized the theory and future prospects of this technique (111).

6.9. CONCLUDING REMARKS

In general, any macromolecular or particle-bound ligand or interacting substance may be included during electrophoresis to give a ligand-induced reaction, such as a change in the electrophoretic mobility or a change in the morphology or profile of the protein. As seen from some of the examples mentioned above, the ligand need not be charged.

The main advantages of the analytical approach to electrophoresis appear to be:

1. It can separate macromolecules that interact with a specific ligand from those that do not.
2. It can be used for studies of interacting macromolecules.
3. Interacting components need not be purified.
4. A multitude of proteins reacting with the same ligand can be studied simultaneously.
5. It can be generalized to interactions other than those between lectins and glycoproteins.
6. It can be used for the prediction of preparative experiments.

ACKNOWLEDGMENTS

I thank Ms. Pia Jensen for her skillful assistance and Dr. Jean Pigott for her help and patience as my English teacher. The studies were supported by the Danish Medical Research Council.

REFERENCES

1. T. C. Bog-Hansen, *Anal. Biochem.* **56,** 480 (1973).
2. S. Nakamura, *Cross Electrophoresis:* Its Principle and Applications. Igaku Shoin, Tokyo, and Elsevier, Amsterdam (1966).
3. N. H. Axelsen, Ed., *Immunoprecipitation Techniques in Gel, Scand. J. Immunol.* Suppl. 10 (1982).
4. V. Horejsi, M. Ticha, and J. Kocourek, *TIBS* **4,** 1, N6 (1979).

5. H. Bittiger and H. P. Schnebli, *Concanavalin A As a Tool,* Wiley, London (1976).
6. S. Murakawa and S. Nakamura, *Bull. Yamaguchi Med. Sch.* **10,** 11 (1963).
7. B. O. Osunkoya and A. I. O. Williams, *Clin. Exp. Immunol.* **8,** 205 (1971).
8. G. A. Spengler and R. M. Weber, *J. Immunol. Methods* **32,** 71 (1980).
9. G. A. Spengler and R. M. Weber, in *Lectins: Biology, Biochemistry, and Clinical Biochemistry,* Vol. 1 (T. C. Bog-Hansen, Ed.), W. de Gruyter, Berlin (1981), p. 231.
10. M. Harboe, E. Saltved, O. Closs, and S. Olsnes, *Scand. J. Immunol.* **4,** Suppl. 2, 125 (1975).
11. K. Pluzek, in *10th International Seaweed Symposium* (T. Levring Ed.) W. de Gruyter, Berlin (1981) p. 711.
12. P. J. Svendsen in *Electrophoresis: A Survey of Techniques and Applications,* Part A, *Techniques* (Z. Deyl, F. M. Everaerts, Z. Prusik, and P. J. Svendsen, Eds.), *Journal of Chromatography Library,* Vol. 18A, Elsevier Amsterdam (1979), p. 133.
13. T. C. Bøg-Hansen, I. Lorenc-Kubis, and O. J. Bjerrum, in *Electrophoresis '79, Advanced Methods: Biochemical and Clinical Applications* (B. J. Radola, Ed.), W. de Gruyter, Berlin (1980), p. 173.
14. T. C. Bøg-Hansen and M. Nord, *J. Biol. Educ.* **8,** 167 (1974).
15. T. C. Bøg-Hansen, C. H. Brogren, and I. McMurrough, *J. Inst. Brewing* **80,** 443 (1974).
16. T. C. Bøg-Hansen and C. H. Brogren, *Scand. J. Immunol.* **4,** Suppl. 2, 135 (1975).
17. T. C. Bøg-Hansen, O. J. Bjerrum, and J. Ramlau, *Scand. J. Immunol.* **4,** Suppl. 2, 141 (1975).
18. T. C. Bøg-Hansen, P. J. Svendsen, and O. J. Bjerrum, in *Progress in Isoelectric Focusing and Isotachophoresis* (P. G. Righetti Ed.), North-Holland Publishing Company, Amsterdam (1975), p. 347.
19. T. C. Bøg-Hansen, P. Prahl, and H. Løwenstein, *J. Immunol. Methods* **22,** 293 (1978).
20. T. C. Bøg-Hansen, O. J. Bjerrum, and C. H. Brogren, *Anal. Biochem.* **81,** 78 (1978).
21. T. C. Bøg-Hansen, in *Affinity Chromatography* (J. M. Egly, Ed.), *Les Colloques de l'Inserm,* 86, Inserm, Paris (1979), p. 399.
22. T. C. Bøg-Hansen, in *Protides of the Biological Fluids,* Vol. 27 (H. Peeters, Ed.), *Proceedings of the 27th Colloquium,* Pergamon Press, Oxford (1980), p. 659.
23. T. C. Bøg-Hansen and K. Takeo, *Electrophoresis* **1,** 67 (1980).
24. T. C. Bøg-Hansen, in *Electrophoresis '79, Advanced Methods: Biochemical and Clinical Applications* (B. J. Radola, Ed.) W. de Gruyter, Berlin (1980), p. 193.
25. T. C. Bøg-Hansen, *Scand. J. Immunol.* **17,** Suppl. 10 (1983).
26. T. C. Bøg-Hansen, P. Jensen, F. Hinnerfeldt, and K. Takeo, in *Lectins: Biology, Biochemistry, and Clinical Biochemistry* (T. C. Bøg-Hansen, Ed.), W. de Gruyter, Berlin (1981), p. 241.
27. T. C. Bøg-Hansen and J. Hau, in *Electrophoresis: A Survey of Techniques and Applications,* Part B, *Applications, Journal of Chromatography Library,* Vol 18B (Z. Deyl Ed.), Elsevier, Amsterdam (1982), pp. 219–252 (1982).

28. T. C. Bøg-Hansen and J. Hau, *Acta Histochem.* **71,** 47 (1982).
29. T. C. Bøg-Hansen, Ed., *Lectins: Biology, Biochemistry, and Clinical Biochemistry,* Vol. 1, W. de Gruyter, Berlin (1981).
30. T. C. Bøg-Hansen, Ed., *Lectins: Biology, Biochemistry, and Clinical Biochemistry,* Vol. 2, W. de Gruyter, Berlin (1982).
31. J. Turkova, *Affinity Chromatography, Journal of Chromatography Library,* Vol. 12, Elsevier, Amsterdam (1978).
32. J. T. Dulaney, *Mol. Cell. Biochem.* **21,** 43 (1979).
33. P. J. Svendsen and N. H. Axelsen, *J. Immunol. Methods* **1,** 169 (1979).
34. S. Bisati, L. Mikkelsen, and C. H. Brogren, in *Lectins: Biology, Biochemistry, and Clinical Biochemistry* (T. C. Bøg-Hansen Ed.), W. de Gruyter, Berlin (1981), p. 387.
35. O. J. Bjerrum and T. C. Bøg-Hansen, in *Biochemical Analysis of Membranes* (A. H. Maddy Ed.), Chapman and Hall, London (1981), p. 378.
36. O. J. Bjerrum, *Biochim. Biophys. Acta* **472,** 135 (1977).
37. S. Bisati, C. H. Brogren, and M. Simonsen in *Protides of the Biological Fluids,* Vol. 27 (H. Peeters, Ed.), *Proceedings of the 27th Colloquium,* Pergamon Press, Oxford (1980) pp. 471.
38. J. Gerlach, O. J. Bjerrum, H. C. Rank, and T. C. Bog-Hansen, in *Protides of the Biological Fluids,* Vol. 27, (H. Peeters, Ed.), *Proceedings of the 27th Colloquium,* Pergamon Press, Oxford (1982), p. 479.
39. I. Hagen and G. Gogstad, in *Lectins: Biology, Biochemistry, and Clinical Biochemistry,* Vol. 1 (T. C. Bog-Hansen, Ed.), W. de Gruyter, Berlin (1981), p. 347.
40. H. Løwenstein, T. C. Bøg-Hansen, and B. Weeke, in *Protides of the Biological Fluids,* Vol. 27 (H. Peeters, Ed.) *Proceedings of the 27th Colloquium,* Pergamon Press, Oxford (1980), p. 611.
41. J. Ramlau and E. Bock, in *Affinity Chromatography* (J. M. Egly, Ed.), *Les Colloques de l'Inserm,* **86,** Inserm, Paris (1979), p. 147.
42. B. F. Vestergaard and T. C. Bog-Hansen, *Scand. J. Immunol.* **4,** Suppl. 2, 211 (1975).
43. H. S. Platt, B. M. Sewell, T. Feldman, and R. L. Souhami, *Clin. Chim. Acta* **46,** 419 (1973).
44. B. Bayard and J. P. Kerckaert, *Biochem. Biophys. Res. Commun.* **95,** 777 (1980).
45. C. Fournier, J. P. Kerckaert, B. Bayard, M. Collyn, and G. Biserte, in *Protides of the Biological Fluids,* Vol. 27 (H. Peeters, Ed.), *Proceedings of the 27th Colloquium,* Pergamon Press, Oxford (1980), p. 623.
46. J. Breborowics and A. Mackiewicz, in *Lectins: Biology, Biochemistry, and Clinical Biochemistry,* Vol. 1 (T. C. Bøg-Hansen, Ed.), W. de Gruyter, Berlin (1981), p. 303.
47. P. Guldager and T. C. Bøg-Hansen, in *Protides of the Biological Fluids,* Vol. 27 (H. Peeters, Ed.) *Proceedings of the 27th Colloquium,* Pergamon Press, Oxford (1979), p. 401.
48. I. Hagen and O. J. Bjerrum, in *Protides of the Biological Fluids,* Vol. 27, (H. Peeters, Ed.), *Proceedings of the 27th Colloquium,* Pergamon Press, Oxford (1979) p. 875.

49. J. Raynes, *Biomedicine* **36,** 77 (1982).
50. J. Hau, P. Svendsen, B. Teisner, and J. Brandt, *Biol. Reprod.* **124,** 163 (1981).
51. J. Hau, P. Svendsen, B. Teisner, and G. Thomsen Pedersen, in *Lectins: Biology, Biochemistry, and Clinical Biochemistry,* Vol. 1 (T. C. Bog-Hansen, Ed.), W. de Gruyter, Berlin (1981), p. 327.
52. J. Hau, P. Svendsen, B. Teisner, G. Thomsen Pedersen, and B. Kristiansen, *Biol. Reprod.* **24,** 683 (1981).
53. J. P. Kerckaert and B. Bayard, in *Lectins: Biology, Biochemistry, and Clinical Biochemistry,* Vol. 1, (T. C. Bøg-Hansen, Ed.) W. de Gruyter, Berlin (1981), p. 271.
54. J. P. Kerckaret, B. Bayard, and G. Biserte, *Biochim. Biophys. Acta* **576,** 99 (1979).
55. I. Lorenc-Kubis and T. C. Bøg-Hansen in *Lectins: Biology, Biochemistry and Clinical Biochemistry,* Vol. 1 (T. C. Bøg-Hansen, Ed.), W. de Gruyter, Berlin (1981), p. 157.
56. A. Mackiewicz and J. Breborowicz, in *Lectins: Biology, Biochemistry, and Clinical Biochemistry,* Vol. 1 (T. C. Bøg-Hansen, Ed.), W. de Gruyter, Berlin (1981), p. 315.
57. C. S. Nielsen and O. J. Bjerrum, *Biochim. Biophys. Acta* **466,** 496 (1977).
58. M. Nilsson and T. C. Bog-Hansen, *Protides of the Biological Fluids,* Vol. 27 (H. Peeters, Ed.), *Proceedings of the 27th Colloquium,* Pergamon Press, Oxford (1979), p. 509.
59. B. Nørgaard-Pedersen, K. Toftager-Larsen, J. Philip, and P. Hindersson, *Clin. Gen.* **17,** 355 (1980).
60. J. D. Oppenheim, P. Owen, M. S. Nachbar, K. Colledge, and H. S. Kapian, *Immunol. Commun.* **6,** 167 (1977).
61. R. L. Ory, R. R. Mod, and T. C. Bøg-Hansen, in *Protides of the Biological Fluids* Vol. 27 (H. Peeters, Ed.), *Proceedings of the 27th Colloquium,* Pergamon Press, Oxford (1980), p. 387.
62. R. L. Ory, T. C. Bøg-Hansen, and R. R. Mod, in *Antinutrients and Natural Toxicants in Foods* (R. L. Ory, Ed.), Food and Nutrition Press, Westport, Conn. (1981), p. 159.
63. I. R. Pedersen and C. H. Mordhorst, in *Lectins: Biology, Biochemistry, and Clinical Biochemistry,* Vol. 1 (T. C. Bøg-Hansen, Ed.), W. de Gruyter, Berlin (1981), p. 395.
64. T. Plesner, O. J. Bjerrum, and M. Wilken, in *Lectins: Biology, Biochemistry, and Clinical Biochemistry,* Vol. 1 (T. C. Bøg-Hansen, Ed.), W. de Gruyter, Berlin (1981), p. 363.
65. K. Toftager-Larsen, and B. Nørgaard-Pedersen, in *Lectins: Biology, Biochemistry, and Clinical Biochemistry,* Vol. 1 (T. C. Bøg-Hansen, Ed.), W. de Gruyter, Berlin, (1981), p. 293.
66. K. Toftager-Larsen, P. Lund Petersen, and B. Nørgaard-Pedersen, in *Lectins: Biology, Biochemistry and Clinical Biochemistry,* Vol. 1 (T. C. Bøg-Hansen, Ed.), W. de Gruyter, Berlin (1981), p. 283.
67. C. Wells, E. Cooper, R. M. Glass, and T. C. Bøg-Hansen, *Clin. Chem. Acta* **109,** 59 (1981).

68. C. Wells, E. Cooper, and T. C. Bøg-Hansen, in *Lectins: Biology, Biochemistry, and Clinical Biochemistry,* Vol. 1 (T. C. Bøg-Hansen, Ed.), W. de Gruyter, Berlin (1981), p. 339.
69. M. M. Andersen, J. Hau. and T. C. Bøg-Hansen, in *Lectins: Biology, Biochemistry, and Clinical Biochemistry,* Vol. 2 (T. C. Bøg-Hansen, Ed.), W. de Gruyter, Berlin (1982), p. 779.
70. M. Bowen, J. Raynes, and E. H. Cooper, in *Lectins: Biology, Biochemistry, and Clinical Biochemistry,* Vol. 2 (T. C. Bøg-Hansen, Ed.), W. de Gruyter, Berlin, (1982), p. 403.
71. L. Faye and J. P. Salier, in *Lectins: Biology, Biochemistry, and Clinical Biochemistry,* Vol. 2 (T. C. Bøg-Hansen, Ed.), W. de Gruyter, Berlin (1982), p. 605.
72. J. Krøll and M. M. Andersen, *J. Immunol. Methods* **9,** 141 (1975).
73. C. H. Brogren and S. Bisatti, in *Lectins: Biology, Biochemistry, and Clinical Biochemistry,* Vol. 1 (T. C. Bøg-Hansen, Ed.), W. de Gruyter, Berlin (1981), p. 375.
74. C. H. Brogren and T. C. Bøg-Hansen *Scand. J. Immunol.* **4,** Suppl. 2, 37 (1975).
75. O. J. Bjerrum, T. C. Bøg-Hansen, T. Plesner, and M. Wilken, in *Lectins: Biology, Biochemistry, and Clinical Biochemistry,* Vol. 1 (T. C. Bøg-Hansen, Ed.), W. de Gruyter, Berlin (1981), p. 259.
76. E. Bock and N. H. Axelsen, *Scand. J. Immunol.* Suppl. 1, 95 (1973).
77. I. Nicollet, J. P. Lebreton, M. Fontaine, and M. Hiron, *Biochim. Biophys. Acta* **668,** 235 (1981).
78. K. Takeo and S. Nakamura, *Arch. Biochem. Biophys.* **153,** 1 (1972).
79. V. Horejsi, in *Affinity Chromatography* (J. M. Egly, Ed.) *Les Colloques de l'Inserm,* 86, Inserm, Paris (1979), p. 391.
80. V. Horejsi, *J. Chromatography* **178,** 1 (1979b).
81. T. C. Bøg-Hansen, J. Breborowicz and H. Franz, in *Lectins: Biology, Biochemistry, and Clinical Biochemistry,* Vol. 2 (T. C. Bøg-Hansen, Ed.), W. de Gruyter, Berlin, (1982), p. 791.
82. T. C. Bøg-Hansen, in *Lectins Biology, Biochemistry, and Clinical Biochemistry,* Vol. 3 (T. C. Bøg-Hansen and G. A. Spengler, Eds.), W. de Gruyter, Berlin, (1982), in press.
83. N. M. Young and M. A. Leon, *Biochim. Biophys. Acta* **365,** 418 (1974).
84. M. Caron, A. Faure, R. G. Keros, and P. Camillot, *Biochim. Biophys. Acta* **491,** 558 (1977).
85. K. Takeo and A. Kabat, *J. Immunol.* **121,** 2305 (1978).
86. S. J. Johnson, E. C. Metcalf, and P. D. G. Dean, *Anal. Biochem.* **109,** 63 (1980).
87. M. Caron, A. Faure, and P. Cornillot, *J. Chromatogr.* **103,** 160 (1975).
88. F. Hinnerfeldt, J. Albrechtsen, and T. C. Bøg-Hansen, in *Electrophoresis '82* (D. Statachos, Ed.), W. de Gruyter, Berlin, (1982), in press.
89. W. N. Burnette, *Anal. Biochem.* **112,** 195 (1981).
90. K. Taketa, E. Toguchi, and M. Izumi, *Electrophoresis '82* (D. Statachos, Ed.), W. de Gruyter, Berlin, in press (1982).
91. R. Lotan, G. Beattle, W. Hubbell, and G. L. Nicolson, *Biochemistry* **16,** 1786 (1977).

92. P. D. G. Dean, F. Qadri, W. Jessup, V. Bouriotis, S. Angal, H. Potuzak, R. J. Leatherbarrow, T. Miron, E. George, and M. R. A. Morgan, in *Affinity Chromatography* (J. M. Egly, Ed.), *Les Colloques de l'Inserm,* 86, Inserm, Paris (1979), p. 321.
93. P. J. Brown, M. J. Leyland, J. P. Keenan, and P. D. G. Dean, *FEBS Lett.* **83,** 256 (1977).
94. D. Grenot and C. Cuilleron, *Biochem. Biophys. Res. Commun.* **79,** 274 (1977).
95. M. J. Iqbal, P. Ford, and M. W. Johnson, *FEBS Lett.* **87,** 235 (1978).
96. M. R. A. Morgan, P. J. Brown, M. J. Leyland, and P. D. G. Dean, *FEBS Lett.* **87,** 239 (1978).
97. M. R. A. Morgan, P. M. Johnson, and P. D. G. Dean, *J. Immunol. Methods* **23,** 381 (1978).
98. M. R. A. Morgan, E. J. Kerr, and P. D. G. Dean, *J. Steroid Biochem.* **9,** 767 (1978).
99. M. R. A. Morgan, N. A. Slater, and P. D. G. Dean, *Anal. Biochem.* **92,** 144 (1979).
100. E. Reinwald, P. Rautenberg, and H. J. Risse, *Biochim. Biophys. Acta* **668,** 119 (1981).
101. P. Rautenberg, E. Reinwald, and H. J. Risse, in *Lectins: Biology, Biochemistry, and Clinical Biochemistry,* Vol. 2 (T. C. Bøg-Hansen, Ed.), W. de Gruyter, Berlin, (1982), p. 619.
102. C. J. Smith and P. C. Kelleher, *Biochem. Biophys. Acta* **605,** 1 (1980).
103. E. Ruoslahti, E. Engvall, A. Pekkala, and M. Sippala, *Int. J. Cancer* **22,** 515 (1978).
104. C. J. Smith, P. C. Kelleher, L. Belanger, and L. Dallaire, *Brit. Med. J.* **1,** 920 (1979).
105. P. Hinderson, K. Toftager-Larsen, and B. Nørgaard-Pedersen, *Lancet,* Oct. 27, 906 (1979).
106. O. J. Bjerrum, *Anal. Biochem.* **89,** 331 (1978).
107. H. Bley and H. Peichl, in *Electrophoresis '79, Advanced Methods: Biochemical and Clinical Applications* (B. J. Randola Ed.), W. de Gruyter, Berlin (1980), p. 733.
108. O. J. Bjerrum, J. Ramlau, E. Bock, and T. C. Bøg-Hansen, in *Techniques for Membrane Receptor Characterization and Purification* (S. J. Jacobs and P. Cuatrecasas, Eds.), Chapman and Hall, London (1980), p. 115.
109. A. Rosengren, B. Bjellquist, and V. Gasparic, in *Electrofocusing and Isotachophoresis* (B. J. Radola and D. Graesslin, Eds.), W. de Gruyter, Berlin (1977), p. 165.
110. K. Ek and P. G. Righetti, *Electrophoresis* **1,** 137 (1980).
111. V. Horejsi, *Anal. Biochem.* **112,** 1 (1981).

CHAPTER

7

IMMOBILIZED ENZYMES

JOHN F. KENNEDY

Research Laboratory for the Chemistry of Bioactive Carbohydrates and Proteins
Department of Chemistry
University of Birmingham
Birmingham, B15 2TT
ENGLAND

and

JOAQUIM M. S. CABRAL

Laboratório de Engenharia Bioquímica
Instituto Superior Técnico
Universidade Técnica de Lisboa
1000 LISBOA
PORTUGAL

Enzymes are biological catalysts consisting of protein or glycoproteins; they participate in many chemical reactions that occur in living systems. Enzymes have been exploited by human beings since ancient times, well before their nature and function or even the animals, plants, and microorganisms from which they were derived were known or understood.

Not until the beginning of this century were enzymes shown to be the agents responsible for all fermentation processes, and not until 1926, when Sumner (1) crystallized the first enzyme, urease, were they shown to be distinct chemical compounds with well-defined physical and chemical properties.

With the understanding of the nature of enzymes and their catalytic potential, the use of enzymes has gradually been extended in a variety of fields, such as brewing, food production, textiles, pharmaceuticals, and medicine.

Unlike ordinary chemical catalysts, enzymes have the ability to catalyze reactions under very mild conditions in neutral aqueous solutions at normal temperature and pressure, reducing the possibility of damage to heat-sensitive substrates and also reducing the energy requirements and corrosion effects of the process. And with their high degree of substrate specificity, enzymes produce almost no undesirable by-products, thus decreasing not only materials costs but also downstream environmental burdens.

In spite of these advantages, the use of enzymes in industrial applications has been limited because most enzymes are relatively unstable, the cost of enzyme isolation and purification is still high, and it is technically very difficult and expensive to recover active enzyme from the reaction mixture after completion of the catalytic process. This restricts the use of soluble enzymes to essentially batch operations, followed by procedures to isolate the product from the reaction mixture, which involves pH or heat treatment to remove the enzyme and other contaminating proteins by denaturation. This is uneconomical, as active enzyme is lost after each batch reaction.

To eliminate some of these deleterious effects, two possible approaches have been investigated. One is the synthetic approach, using recent developed techniques of organic chemistry to synthesize catalysts that have enzyme-like activities. These catalysts are sometimes called "synzymes". The other approach is to attach enzymes onto water-insoluble solid support materials. Such attachment immobilizes the enzyme molecules and makes them insoluble in aqueous media.

If active and stable immobilized enzymes having appropriate substrate specificities are prepared, most of the above disadvantages are eliminated and it becomes possible to use enzymes conveniently in the same way as ordinary solid catalysts used in synthetic chemical reactions. Some of the key advantages of immobilized enzymes over their soluble counterparts are summarized in Table 7.1.

7.1. HISTORICAL PERSPECTIVE

It is difficult to discuss the current status of immobilized enzyme technology without an appreciation of its roots. Goldstein and Katchalski-Katzir (2) have provided us with an excellent survey of the field from its rather diffuse and unspectacular beginnings to its current status as a sophisticated technology with the potential for dramatically affecting our future life-styles.

The first immobilized enzyme was probably prepared by Nelson and Griffin (3) in 1916. They reported that β-D-fructofuranosidose (invertase) ex-

Table 7.1. Advantages of Immobilized Enzymes over Soluble Enzymes

1. Enzymes can be reused.
2. Processes can be operated continuously and can be readily controlled.
3. Products are easily separated.
4. Effluent problems and materials handling are minimized.
5. In some cases, enzyme properties (activity and stability) can be altered favorably by immobilization.

tracted from yeast was adsorbed on charcoal, and the adsorbed enzyme showed the same activity on native enzyme.

The first group of scientists to make use of this phenomenon was the immunologists; in the 1930s work on the application of adsorbed antigens for the isolation of specific antibodies (for review see Isliker (4)) was reported. The unpredictable behavior of these systems and the inability to obtain clean separations led the early investigators to the realization that fixation by forces stronger than adsorption was necessary. It is thus not surprising that the initial attempts at covalent fixation onto water-insoluble supports were carried out by the immunologists.

In 1936, Landsteiner and Van der Scheer (5) described the coupling of diazotized haptens to blood cell stroma, and the utilization of the insoluble preparations for the isolation of the corresponding antibodies. Their work was followed, after the interruption of World War II, by the first experiments on covalent binding of a variety of antigens to chemically well defined water-insoluble polymeric supports. In 1949, Michael and Evers (6) described the covalent binding of physiologically active protein to carboxymethyl cellulose. Further, in 1951, Campbell et al. (7) prepared immobilized antigen by binding ovalbumin to diazotized 4-aminobenzyl cellulose, and the isolation of ovalbumin antibodies on the immunoadsorbent thus obtained. Subsequently, a number of reports on the preparation of immobilized antigens and antibodies of polymers by other methods appeared (8–10).

Although it has been known for a long time that enzymes in water-insoluble form show catalytic activity, the first attempt to immobilize an enzyme to improve its properties for a particular application was not made until 1953, when Grubhofer and Schleith (11) immobilized pepsin and other enzymes such as carboxypeptidase, diastase, and ribonuclease by presumably covalently binding them to diazotized polyaminostyrene resin.

Following Grubhofer's investigation, Manecke and coworkers (9, 10, 12) and Brandenberger (13–15) used poly(aminostyrene) and poly(isocyanatostyrene) to immobilize enzymes. The amounts of bound proteins and the enzymic activities retained in the immobilized preparations obtained by these methods were, however, relatively poor, presumably owing to the hydrophobicity of the supports. Shortly thereafter, in the early 1960s, Katchalski-Katzir and his coworkers (16–19) at the Weizmann Institute of Science, Israel, and Manecke (20) carried out extensive studies on new immobilization techniques using a number of different carriers with varying degrees of hidrophilicity, which produced conjugates having higher levels of bound protein and much improved activity and stability characteristics.

These various early workers, for the most part biochemists, biophysicists, and other basic scientists, began to appreciate that immobilized enzymes could be used not only as highly specific water-insoluble reagents, but also as models for bound enzymes in living systems, and as a new class of highly specific and efficient industrial catalysts.

From the mid-1960s on, in Japan, Israel, Europe, and the United States, research on the immobilization of enzymes has proceeded along at least three parallel paths. It was clear by this time that the chemical and mechanical nature of the support material was extremely important in determining the characteristics of the immobilized enzyme, and some workers began attempts to redesign carriers in order to achieve optimal binding and enzyme stability. Simultaneously, others sought to develop milder, more general immobilization techniques as an alternative to covalent methods (21–23). The third area of development was the study of immobilized enzymes in continuous reactors (24, 25). This marked the initial contributions of chemical engineers, who could bring to enzyme research their knowledge and experience in the design and analysis of continuous heterogeneous catalytic reactors.

By the end of the 1960s, many of the fundamental principles of immobilized enzyme technology had been developed, and emphasis began to shift towards practical applications. In 1969, Chibata and coworkers succeeded in industrializing a continuous process for the optical resolution of DL-amino acids using immobilized aminoacylase. This was the world's first industrial application of an immobilized enzyme.

By 1970, there had accumulated perhaps 300 papers in the literature concerning enzyme immobilization. In August 1971 the first Enzyme Engineering Conference was held at Henniker, New Hampshire, for the purpose of bringing together researchers who shared the interests and goals of developing immobilized enzyme technology, to discuss the status of the field, and to present guidelines for its future. As will be described below, a definition and a classification of immobilized enzymes were proposed at the Conference. This conference is now biennial, and the main topics have continued to be immobilized enzymes, with publication of the proceedings under the heading Enzyme Engineering (26–30).

By 1975, the number of reports had increased to approximately 1000. Besides these reports, many reviews and books (26–65) have also been published, as has a new journal that includes review articles, research papers, patent reports, a quick-reference literature survey of immobilized enzymes, and bioaffinant reports and current views, all relevant to the area of biotechnology (66).

Although enzymes are produced by all living organisms, enzymes from microbial sources are most suitable for industrial purposes owing to low-cost production, a short time for production, possible mass production, and the conditions for production not being restricted by location or season.

To avoid the need to extract intracellular enzymes from microbial cells or to utilize multienzyme systems, direct immobilization of whole microbial cells has recently been attempted.

7.2. CLASSIFICATION OF IMMOBILIZED ENZYMES

Immobilized enzymes, as drafted at the first Enzyme Engineering Conference, are defined as "enzymes which are physically confined or localized in a certain defined region of space with retention of their catalytic activities, and which can be used repeatedly and continuously." Before that time, various terms such as "water-insoluble enzyme," "trapped enzyme," "fixed enzyme," "matrix-supported enzyme," and "insolubilized enzyme" were used.

The term immobilized enzyme includes: (i) enzymes modified to a water-insoluble form by suitable techniques, (ii) soluble enzymes used in reactors equipped with a semipermeable ultrafiltration membrane, allowing the passage of reaction products resulting from the hydrolysis of high molecular weight substrates but retaining the enzyme molecules inside the reactor, and (iii) enzymes the mobility of which has been restricted by attachment to another macromolecule, with the composite resultant molecule being water soluble.

There are several ways of classifying the various types of immobilized enzymes as defined above. Zaborsky (31) proposed a classification based on the nature of the interaction responsible for immobilization: the immobilization can be achieved through either chemical means, which "include any that involve the formation of at least one covalent (or partially covalent) bond between residues of an enzyme and a functionalized water-insoluble polymer or between two or more enzyme molecules," or physical methods, which "include any that involve localizing an enzyme in any manner whatsoever which is not dependent on covalent bond formation."

At the first Enzyme Engineering Conference (26), Sundaram and Pye recommended classifying immobilized enzymes as either entrapped or bound, based on the nature of the resulting complex. Classifications based on the nature of support and type of reaction are also possible and used. From a practical point of view, the method of classification makes little difference, since generally the same groups are recognized. Also, many of the more successful techniques have evolved as combinations of the basic methods.

The classification that is presented here is not fundamentally different from that recommended by Sundaram and Pye and by Chibata (57); it attempts to combine the nature of interaction responsible for immobilization and the nature of the support. Figure 7.1 gives this classification system and lists the various individual methods that are discussed in detail in subsequent sections of this chapter.

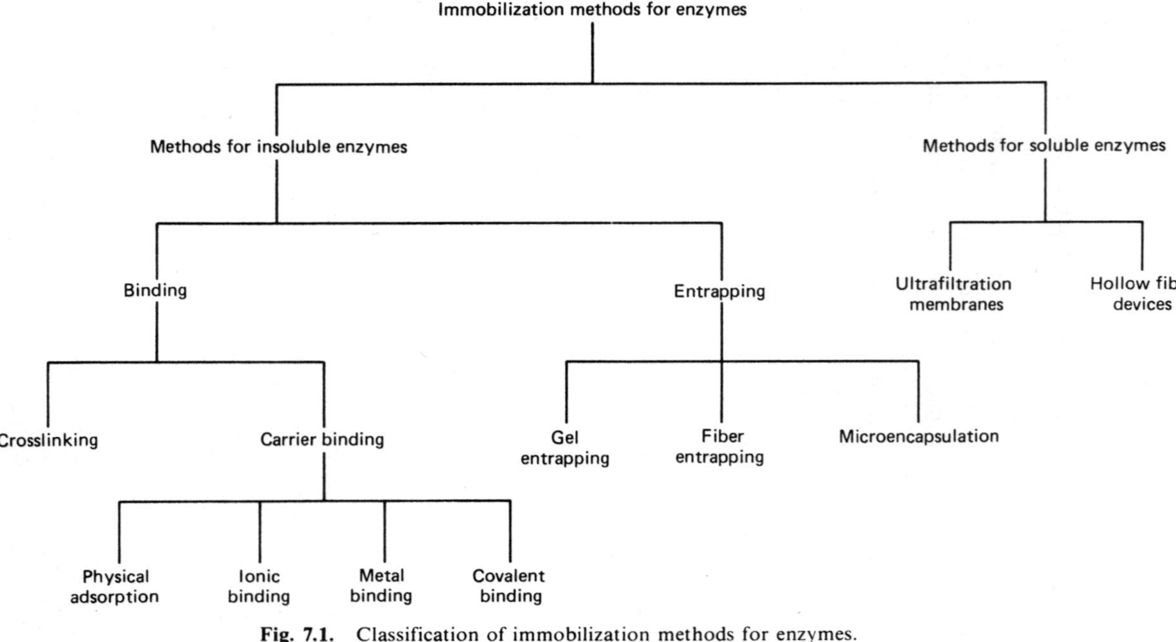

Fig. 7.1. Classification of immobilization methods for enzymes.

7.3. TECHNIQUES OF ENZYME IMMOBILIZATION

Over the last few years a large number of methods have been developed to immobilize enzymes on solid matrices. Several recent reviews (40, 42, 45, 57–67) discuss these methods in detail. The aim in this chapter is to review the more recently published data on the preparation of immobilized enzymes, according to the classification suggested.

7.3.1. Crosslinking Method of Immobilization

The crosslinking immobilization method is based on the formation of covalent bonds between the enzyme molecules, by means of bifunctional or multifunctional reagents, leading to three-dimensional, crosslinked enzyme aggregates that are completely insoluble in water, without the use of water-insoluble carriers.

This method involves the addition of an appropriate amount of crosslinking agent to an enzyme solution at conditions that give the desired insoluble derivative. Optimum conditions for achieving good insolubility while retaining considerable enzymic activity must unfortunately be determined for each case by trial and error until enough is known of the enzyme's primary, secondary, and tertiary structures to allow prediction of the best immobilization chemistry that will exhibit minimal distortion of the enzyme and its derivatives.

Some of the reagents that have been used for the preparation of immobilized enzymes are given in Table 7.2. These reagents possess two identical functional groups (homobifunctional reagents). Heterobifunctional or multifunctional agents have been more often used for the binding of enzymes to carriers than for intermolecular crosslinking.

The main reagents are carbonyl-containing compounds that react with L-lysyl residues of the enzyme (Schiff's base); diazo compounds with L-lysyl, L-histidyl, L-tyrosyl, L-arginyl or L-cysteinyl residues (diazo coupling); isocyanates with primary amino groups (peptide bond); alkyl iodides with nucleophilic groups (alkylation); and iodoacetamides with L-cysteinyl residues (alkylation).

Among the considerable number of crosslinking agents described in the literature (31, 37, 38, 40), glutaraldehyde has found widespread use for enzyme immobilization and is by far the most commonly used crosslinking reagent. Quiocho and Richards (68) in 1964 obtained carboxypeptidase A (EC 3.4.17.1) immobilized by intermolecular crosslinkage with glutaraldehyde, as shown in Scheme 1. This may be the first report of this immobilization method.

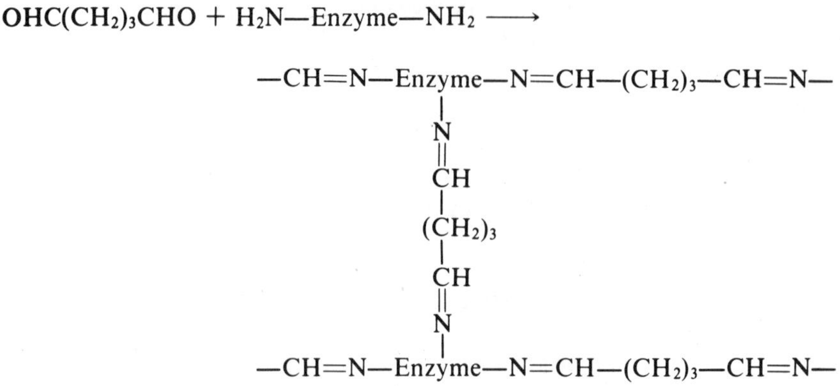

$$OHC(CH_2)_3CHO + H_2N\text{—Enzyme—}NH_2 \longrightarrow$$

$$\begin{array}{c}
\text{—CH=N—Enzyme—N=CH—}(CH_2)_3\text{—CH=N—} \\
| \\
\text{N} \\
\| \\
\text{CH} \\
| \\
(CH_2)_3 \\
| \\
\text{CH} \\
\| \\
\text{N} \\
| \\
\text{—CH=N—Enzyme—N=CH—}(CH_2)_3\text{—CH=N—}
\end{array}$$

Scheme 1

The linkages formed between the enzyme and glutaraldehyde are irreversible and survive extremes of pH and temperature. However, the straightforward participation of the reagent's aldehydo groups to form an aldimine (Schiff's base) bond with protein amino groups has been questioned. It has been suggested that the glutaraldehyde reaction most probably involves conjugate addition of enzyme amino groups to ethylenic double bonds of α, β-unsaturated oligomers, contained in the commercial aqueous glutaraldehyde solutions usually used (Scheme 2).

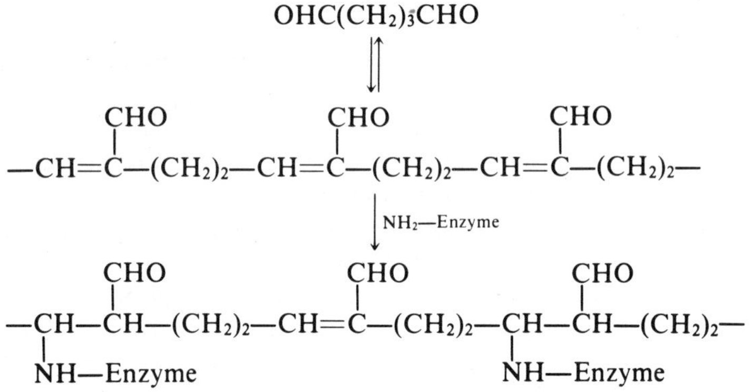

$$OHC(CH_2)_3CHO$$

$$\Updownarrow$$

$$\begin{array}{c}
\text{CHO} \qquad\qquad \text{CHO} \qquad\qquad \text{CHO} \\
\text{—CH=}\overset{|}{\text{C}}\text{—}(CH_2)_2\text{—CH=}\overset{|}{\text{C}}\text{—}(CH_2)_2\text{—CH=}\overset{|}{\text{C}}\text{—}(CH_2)_2\text{—}
\end{array}$$

$$\Big\downarrow \; NH_2\text{—Enzyme}$$

$$\begin{array}{c}
\text{CHO} \qquad\qquad\qquad \text{CHO} \qquad\qquad\qquad \text{CHO} \\
\text{—}\underset{\text{NH—Enzyme}}{\underset{|}{\text{CH}}}\text{—}\overset{|}{\text{CH}}\text{—}(CH_2)_2\text{—CH=}\overset{|}{\text{C}}\text{—}(CH_2)_2\text{—}\underset{\text{NH—Enzyme}}{\underset{|}{\text{CH}}}\text{—}\overset{|}{\text{CH}}\text{—}(CH_2)_2\text{—}
\end{array}$$

Scheme 2

According to Richards and Knowles (69) this mechanism explains the stability of the bond, which cannot be due to a simple Schiff's base formation, and it also explains the lower reactivity on proteins, including enzymes, of solutions of freshly distilled glutaraldehyde.

Studies (70–73) of the preparation of immobilized enzymes by this method, with glutaraldehyde, have shown that insolubilization of protein is critically

Table 7.2. Multifunctional Reagents for the Preparation of Immobilized Enzymes by Intermolecular Crosslinking

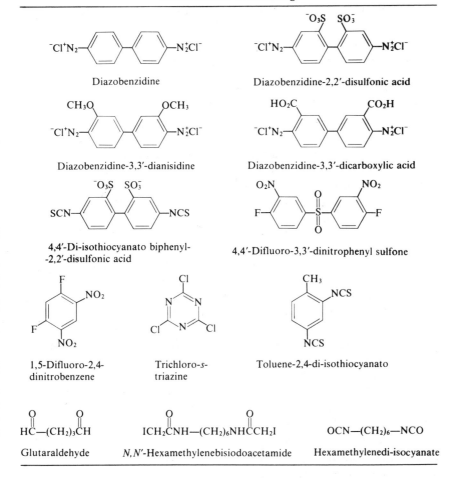

$^{-}Cl^{+}N_2$–C$_6$H$_4$–C$_6$H$_4$–$N_2^{+}Cl^{-}$ Diazobenzidine	$^{-}O_3S$, SO_3^{-}; $^{-}Cl^{+}N_2$–C$_6$H$_3$–C$_6$H$_3$–$N_2^{+}Cl^{-}$ Diazobenzidine-2,2′-disulfonic acid
CH_3O, OCH_3; $^{-}Cl^{+}N_2$–C$_6$H$_3$–C$_6$H$_3$–$N_2^{+}Cl^{-}$ Diazobenzidine-3,3′-dianisidine	HO_2C, CO_2H; $^{-}Cl^{+}N_2$–C$_6$H$_3$–C$_6$H$_3$–$N_2^{+}Cl^{-}$ Diazobenzidine-3,3′-dicarboxylic acid
$^{-}O_3S$, SO_3^{-}; SCN–C$_6$H$_3$–C$_6$H$_3$–NCS 4,4′-Di-isothiocyanato biphenyl--2,2′-disulfonic acid	O_2N, NO_2; F–C$_6$H$_3$–SO$_2$–C$_6$H$_3$–F 4,4′-Difluoro-3,3′-dinitrophenyl sulfone
F, NO_2, F, NO_2 1,5-Difluoro-2,4-dinitrobenzene	Cl, N, N, Cl, N, Cl Trichloro-*s*-triazine
CH_3, NCS, NCS Toluene-2,4-di-isothiocyanato	
$HC(=O)$—$(CH_2)_3$$C(=O)H$ Glutaraldehyde	$ICH_2C(=O)NH$—$(CH_2)_6NHC(=O)CH_2I$ *N,N′*-Hexamethylenebisiodoacetamide
OCN—$(CH_2)_6$—NCO Hexamethylenedi-isocyanate	

dependent on a delicate balance of factors such as the concentrations of the enzyme and crosslinking reagent, the pH and ionic strength of the solution, the temperature, and the time of the reaction. Thus, when applied to solutions of low protein concentrations, glutaraldehyde gives rise initially to water soluble oligomers, but with more concentrated solutions, high molecular weight insoluble polymers are obtained. Protein concentrations of 50–200 mg/mL are optimal for homogeneous immobilization of proteins (73). The nature of the enzyme also has an influence on its insolubilization; for example, L-lysine-rich proteins are readily insolubilized. With increasing glutaraldehyde concentration in the mixture, the yield of immobilized activity first increases and then decreases, giving rise to a maximum. The optimal ratio of glutaraldehyde-to-protein concentration is usually in the range of 10% (w/w), and the best activity yields are generally obtained with concentrations ranging from 0.3 to 0.6% glutaraldehyde.

Jansen, Tomimatsu, and Olson (71) observed that the pH values for the most rapid insolubilization of bovine serum albumin, soybean trypsin inhibitor, lysozyme (EC 3.2.1.17), and papain (EC 3.4.22.2) were found to be nearly the same as the isoelectric points of these proteins, but the formation of insoluble and active chymotrypsin (EC 3.4.21.1) was most rapid at pH 6.2 (isoelectric point, pH 8.6) and chymotrypsinogen A at pH 8.2 (isoelectric point, pH 9.5). They also observed that the degree of insolubilization with chymotrypsin was greater at lower ionic strengths. The reaction is also temperature-dependent and time-dependent; at room temperature the time necessary for crosslinking is 1–3 hr, and generally after 10–30 min, depending on the enzyme, a gel appears. The crosslinking reaction also works at lower temperatures (4°C), which are preferred for more labile molecules, though immobilization then requires more time.

The most important advantage of the crosslinking method is that one bi- or multi-functional reagent can be utilized to prepare, in a simpler manner, immobilized enzymes, that are almost pure protein (31).

However, immobilization of enzymes solely by their intermolecular crosslinking into large aggregates has found rather limited application because of the difficulties encountered in controlling such reactions, the need for a large amount of enzyme, the often unavoidable inactivation of enzyme caused by the participation of the active center on the crosslinkage, and the mechanical properties of the gelatinous nature of these enzyme derivatives. Carrying out the crosslinking reactions on precipitates, obtained by the addition of salts or an organic solvent, has permitted in a few cases a certain degree of control on the properties of the final product. This immobilization method has been largely employed in conjunction with other methods, to overcome the shortcomings pointed out above.

A somewhat different process has been described (74) in which the enzyme forms a complex with a water-soluble agent. This water-insoluble complex is prepared by adding a solution of polyphenol, (for example tannic acid, which is a polymeric glycoside derivative of gallic acid) to a solution of the enzyme. β-D-Fructofuranosidase (EC 3.2.1.21) (invertase) and β-D-galactosidase (EC 3.2.1.23) (lactase) (74, 75) were immobilized by this procedure, but at low pH values dissociation of the enzyme–tannic acid complex occurs and the complexes have inadequate mechanical characteristics for use in continuous reactors; for enhancing these properties the complex has generally been crosslinked with glutaraldehyde.

Table 7.3. Some Principal Examples of Immobilized Enzymes Prepared by Intermolecular Crosslinking

Crosslinking Agent	Enzyme	Reference
Glutaraldehyde	Alcohol dehydrogenase (EC 1.1.1.1)	76
	Glutamate dehydrogenase (EC 1.4.1.2)	77
	Penicillin amidase (EC 3.5.1.11)	78
	Catechol 1,2-dioxygenase (EC 1.13.11.1)	79
	Phenol 2-monooxygenase (EC 1.14.13.7)	80
	Phosphatase (EC 3.1.3.1/2)	81
	Carboxypeptidase A (EC 3.4.17.1)	68, 82
	Ribonuclease A (EC 3.1.26/27.x)	84
1,5-Difluoro-2,4-dinitrobenzene	Carboxypeptidase A (EC 3.4.17.1)	82
	Ribonuclease A (EC 3.1.26/27.x)	83
Diazobenzidine	Carboxypeptidase A (EC 3.4.17.1)	82
Tannic acid	Pullulanase (EC 3.2.1.41)	85

Some immobilized enzymes prepared by the crosslinking method are listed in Table 7.3.

7.3.2. Carrier Binding Method of Immobilization

The carrier binding method, which consists of the binding of enzymes to solid supports, is the oldest (3) and the most prevalent method of immobilizing enzymes. Much has been written on various aspects of the method, and it is the one that commonly comes to mind when referring to immobilized enzymes. The carrier binding method can be further divided into four categories according to the binding mode of the enzyme: physical adsorption, ionic binding, metal binding, and covalent binding.

When enzymes are immobilized in this way, care is required in the selection of carriers as well as the binding procedures. A carrier judiciously chosen can enhance the operational stability of the immobilized system. Although it is recognized that there is no universal carrier, a number of desirable characteristics should be common to any material considered for immobilizing enzymes.

Messing (86) indicates a method for selecting appropriate carriers based on their properties. The chosen carrier should be stable in solution and should not deteriorate under operational solution conditions of composition and pH. It should also be mechanically rigid and show little compaction in high flow-rate continuous operations using fixed bed reactors.

Of the many carriers that have been suggested for immobilizing enzymes, organic or inorganic, natural or synthetic, it appears that the inorganic supports can best fulfill the requirements for industrial use because of their stabilities against physical, chemical, thermal, and microbial degradation, their mechanical strength, and their structural stability (which helps avoid compaction and large pressure drops during flow processes) and their ability to regenerate by a relatively simple pyrolysis process.

In the case of organic supports, the regenerability of the catalyst, that is, the reversibility of attachment between enzyme and carrier, is usually practically possible only with adsorption or ionic binding techniques. In fact the regenerability was a key factor in choosing the ionic binding of aminoacylase on DEAE-Sephadex over covalent or entrapment methods in the design of a commercial process for continuous production of L-aminoacids (21). In special instances reversible covalent chemistry has been designed to allow reversible covalent attachment of the enzyme—for example, via disulfide linkages (87).

The most important advantage of organic supports is the ease with which enzymes can be attached, namely, in the covalent binding method, yielding enzymic preparations with good activity and operational stability. However, according to Weetall (88), these derivatives are susceptible to microbial attack, and also, when used in columns the structure of the matrix changes with varying pH and solvent conditions.

Carriers can be classified according to their composition, as organics and inorganics. Organic carriers can be divided into natural and synthetic polymers. Some examples of natural polymers that have been used as matrices for immobilization of enzymes are polysaccharides such as cellulose, starch, agar, agarose, dextran, chitin and chitosan, proteins such as collagen, silk, albumin and gelatin, and activated carbon. Examples of synthetic polymers are acrylamide-based polymers, copolymers of maleic anhydride, polystyrene, copolymers of acrylic and methacrylic acid, polyvinylalcohol, nylon, and polypeptides.

Inorganic carriers used are metal oxides such as titania, zirconia, alumina, magnetite, and other minerals such as bentonite, kaolinite, clays, hydroxyapatite, calcium phosphate, silica gel, pumice stone, sand, celite, diatomaceous earth, and hydrous metal oxides.

Although this classification into organic and inorganic carriers has been used to categorize the nature of the support, according to Messing (86) this description is not adequate for the full characterization of very pertinent carrier parameters such as surface area and pore diameter, both of which in turn affect the loading of the enzyme. Therefore, the carriers can be classified, based on morphology, as nonporous and porous carriers.

Here, although there are some advantages of this last classification, the former classification based on the chemical nature of matrices will be used for each specific carrier binding method and any relevant morphology will be considered within such division.

7.3.2a. Physical Adsorption

The physical adsorption method is historically the earliest method of enzyme immobilization. By this method, Nelson and Griffin (3) in 1916 immobilized β-D-fructofuranosidase (invertase) by adsorption onto aluminum hydroxide. It is based, like its name suggests, on the physical adsorption of enzyme protein on the surface of solid matrices. This method is also the easiest way of preparing immobilized enzymes. It consists of bringing an aqueous solution of the enzyme into contact with the adsorbent surface.

There are four distinct techniques for obtaining the immobilized enzyme preparation (89) by physical adsorption: (i) the static procedure; (ii) electrodeposition; (iii) the reactor loading process; (iv) mixing or shaking bath loading. Of these four techniques, the most frequently employed for laboratory preparation is the mixing or shaking bath loading, in which the carrier is placed in the enzyme solution and mixed with a stirrer or continuously agitated in a shaking water bath, resulting in a rather uniform loading of enzyme. The prime candidate for commercial purposes is the reactor loading process, in which the support is loaded into the reactor that will be used for the processing system and the enzyme is then added to the reactor and recirculated or agitated through the reactor. In the electrodeposition procedure the carrier is placed next to one of the electrodes in an enzyme bath, the current is turned on, and the enzyme migrates to the carrier and is deposited on the surface. The static procedure is the most inefficient of these four techniques and requires more time. In this method no agitation or stirring is used, resulting in a nonuniform and rather low enzyme loading unless the carrier is exposed to the enzyme for many days.

Because no reactive species are involved, there are little or no conformational changes in the enzyme on immobilization. On this account, one may obtain a derivative with a specific activity similar to that of the soluble counterpart. The adsorption is dependent on the experimental variables such as pH, nature of the solvent, ionic strength, quantity of enzyme and adsorbent, time, and temperature. Good discussions of these variables have been presented by several authors (90–95). A close control of these variables is required owing to the relatively weak binding forces between protein and adsorbent (hydrogen bonds, van der Waals forces, hydrophobic interactions, etc.).

A major influence on the quantity of enzyme adsorbed to a solid support is enzyme concentration exposed to the unit surface of carrier during the

Table 7.4. Some Principal Examples of Immobilized Enzymes Prepared by Physical Adsorption

Carrier	Enzyme	Reference
Organic Supports		
Phenoxyacetyl-cellulose	Alkaline phosphatase (EC 3.1.3.1)	96
	Glyceraldehyde-3-phosphate dehydrogenase (EC 1.2.1.12)	96
	β-D-Xylosidase (EC 3.2.1.27)	97
Palmitoyl-cellulose	Triacylglycerol lipase (EC 3.1.1.3)	98
Tannin-aminohexyl-cellulose	Aminoacylase (EC 3.5.1.14)	99
	Naringinase	100
Tannin-TEAE-cellulose	Pullulanase (EC 3.2.1.41)	101
Concanavalin A-agarose	Phosphodiesterase 1 (EC 3.1.4.1)	102
	β-D-Fructofuranosidase (EC 3.2.1.26)	103
	Acid phosphatase (EC 3.1.3.2)	104, 105
Lecithin-agarose	D-Glucose oxidase (EC 1.1.3.4)	106
	Peroxidase (EC 1.11.1.7)	106
	Catalase (EC 1.11.1.6)	106
	α-D-Glucosidase (EC 3.2.1.20)	106
Heptyl-agarose	Chlorophyllase (EC 3.1.1.14)	107
Hexyl-agarose	Guanylate cyclase (EC 4.6.1.2)	108
Octyl-agarose	Guanylate cyclase (EC 4.6.1.2)	108
	Xanthine dehydrogenase (EC 1.2.1.37)	109
Decyl-agarose	Guanylate cyclase (EC 4.6.1.2)	108
Amino hexadecyl-agarose	Chlorophyllase (EC 3.1.1.14)	107
Amino hexaundecyl-agarose	Chlorophyllase (EC 3.1.1.14)	107
Activated carbon	Glucoamylase (EC 3.2.1.3)	110
Glassy carbon	D-Glucose oxidase (EC 1.1.3.4)	111
Inorganic Supports		
Molecular sieve 4A	Trypsin (EC 3.4.21.4)	112

Table 7.4. (*continued*)

Carrier	Enzyme	Reference
Zeolite	D-Glucose oxidase (EC 1.1.3.4)	113
Silica gel	Trypsin (EC 3.4.21.4)	114
	Alcohol dehydrogenase (EC 1.1.1.1)	115
Octadecyl silica	Adenosine deaminase (EC 3.5.4.4)	116
Methyl aerosil	Chlorophyllase (EC 3.1.1.14)	117
Aluminum hydroxide	Urease (EC 3.5.1.5)	118

immobilization process. The activity increases with increasing enzyme concentration, approaching a saturation value asymptotically at higher enzyme concentrations. Both time and temperature are important parameters in immobilization by adsorption, particularly with porous carriers, since diffusion is an important factor for immobilizing enzymes in such carriers.

Countering the advantages already described, a strong disadvantage is desorption of the protein from the carrier, which occurs during utilization, owing to the weakness of the involved binding forces, with subsequent loss of catalytic activity and contamination of products. High concentrations of salt or substrate have been shown to enhance the rate of desorption of the enzyme. This disadvantage can render some enzymes active for only a short period of time, and thus adsorption techniques are of limited reliability when absolute immobilization of an enzyme is desired.

Some immobilized enzymes that have been prepared by physical adsorption are listed in Table 7.4.

7.3.2b. Ionic Binding

The ionic binding immobilization method is based mainly on ionic binding of enzyme protein to solid supports containing ion-exchange residues. In some cases physical adsorption may also take place in the process. The main difference between physical adsorption and ionic binding is the strength of the linkage of the enzyme to the carrier.

As in the physical adsorption process, the immobilization of enzyme is easily carried out, using the same procedures. The resultant binding forces (ion–ion interactions) are stronger than those in the case of the physical adsorption, although less strong that in the covalent binding. Nevertheless, leakage of enzyme from the carrier may occur in some situations, as with

Table 7.5. Some Principal Examples of Immobilized Enzymes Prepared by Ionic Binding

Carrier	Enzyme	Reference
Anion Exchanges		
DEAE-cellulose	Glucoamylase (EC 3.2.1.3)	123
	Phosphodiesterase 1 (EC 3.1.4.1)	124
	D-Glucose isomerase (EC 5.3.1.18)	125
	Methanol oxidase	126
	Dextransucrase (EC 2.4.1.5)	127
	Choridazon dihydrodial dehydrogenases A and B	128
AE-cellulose	D-Glucose isomerase (EC 5.3.1.18)	125
TEAE-cellulose	Pullulanase (EC 3.2.1.41)	85
DEAE-Sephadex®	D-Glucose isomerase (EC 5.3.1.18)	125
	Phenol 2-monooxygenase (EC 1.14.13.7)	80, 129
	Dextransucrase (EC 2.4.1.5)	127
	β-D-Fructofuranosidase (EC 3.2.1.26)	103
QAE-Sephadex®	Hexokinase (EC 2.7.1.1)	130
	Creatine kinase (EC 2.7.3.2)	130
	Carbamoyl phosphate synthetase (EC 6.3.4.16/6.3.5.5)	130
DEAE-Bio-Gel® A	Phenol 2-monooxygenase (EC 1.14.13.7)	80, 129
Amberlite® IRA 93	D-Glucose oxidase (EC 1.1.3.4)	131
Amberlite® IRA 94	β-D-Fructofuranosidase (EC 3.2.1.26)	132
Amberlite® IRA 910	D-Glucose oxidase (EC 1.1.3.4)	131
Amberlite® IRA 938	D-Glucose oxidase (EC 1.1.3.4)	131
Cation Exchanges		
CM-cellulose	Penicillin amidase (EC 3.5.1.11)	78
Cellulose phosphate	Aminoacyl-tRNA synthetases	136
CM-Sephadex®	β-D-Fructofuranosidase (EC 3.2.1.26)	103
SP-Sephadex®	Dextransucrase (EC 2.4.1.5)	127
	Glucoamylase (EC 3.2.1.3)	133
Dextran sulfate	Lactate dehydrogenase (EC 1.1.1.27)	134
Amberlite® IRC-50	Cholesterol oxidase (EC 1.1.3.6)	135
	β-D-Fructofuranosidase (EC 3.2.1.26)	132
Amberlite® IRC-200	Rennet	137

physical adsorption, with variation of pH or when high substrate or high ionic strength solutions are used.

Owing to the ionic character of the linkage between the enzyme and the support and the mild conditions for immobilization, little conformational changes occur with the enzyme, again leading to immobilized forms with high enzymic activities.

The first enzyme that was immobilized by this method was catalase (EC 1.11.1.6) in 1956 by Mitz (119), with DEAE-cellulose as the carrier. The first industrial application of immobilization of enzymes used aminoacylase (EC 3.5.1.14) immobilized on DEAE-Sephadex (21) in order to produce L-amino-acids from racemic mixtures of *N*-acetyl-DL-aminoacids.

Ion exchangers are used as the carriers for ionic binding. Most frequently they are organic supports with ion exchange residues, although there are also inorganic forms (especially silica) with the same or similar ion exchange residues. The organic polymers are derivatives of polysaccharides, mainly cellulose and dextran, or synthetic polymers are mainly polystyrene based.

According to the type of ion-exchange residues, these carriers may be classified as anion or cation exchangers, respectively based on the properties of exchanging the anions Cl^-, OH^-, etc. of carrier with anions of the solution (anionic residues of enzyme) and the cations H^+, Na^+, etc. with cations of the solution (cationic residues of enzyme).

Suitable substances that are used with the carrier base to form ion exchange materials include compounds containing amino, alkylamino, guanido, and quaternary amino groups for the preparation of anion exchangers (DEAE-, TEAE-, ECTEOLA-, etc.) and compounds containing sulfate, phosphate, and carboxyl groups for cation exchangers (SO_4^{2-}, PO_4^{3-}, CM-).

Ionic bindings of enzymes to these supports have been reported (120–122) with success in their subsequent applications, mainly after increasing the charge on the enzyme protein by covalently linking the enzyme to a charged polymer, for example chymotrypsin (EC 3.4.21.1) (120) and trypsin (EC 3.4.21.4) (121) to a water-soluble copolymer of acrylic acid and maleic anhydride, and glucoamylase (EC 3.2.1.3) (122) to a water-soluble copolymer of ethylene and maleic acid. These polyanionic enzyme derivatives were strongly bound ionically and practically irreversibly bound to anionic exchangers, such as DEAE-cellulose and DEAE-Sephadex, to give much more active and stable complexes than the immobilized preparation obtained using untreated enzyme (122).

Some immobilized enzymes that have been prepared by ionic binding are listed in Table 7.5.

7.3.2c. Chelation, or Metal Binding

Active immobilized enzymes have been prepared using transition metal compounds for the activation of the surface of organic and inorganic carriers (or using hydrous oxide, precipitated from a metal salt, as carrier) by the formation of chelates between the enzyme and the activated carrier.

This technique of immobilization was initially used by Novais (138, 139) when cellulose was steeped in a solution of titanium(IV) chloride followed by filtration or drying or both. The solid was washed with buffer to remove excess salt solution, and the enzyme coupling was achieved by contacting the enzyme with the activated support, usually in a buffer solution at the optimum pH of the enzyme. It was postulated that the titanium compound is chelated with cellulose and that this cellulose titanate can bind enzymes and other bioactive compounds by chelation too (138, 140), via carboxyl, hydroxyl, amino, and thiol groups. The transition metal salts used on this activation process have been $TiCl_3$, $TiCl_4$, $Ti_2(SO_4)_3$, $FeCl_3$, $FeCl_2$, $FeSO_4$, $ZrCl_4$, $SnCl_2$, $SnCl_4$ and VCl_3.

The same technique of drying a mixture of the carrier and the metal salt solution has been used to activate not only organic supports such as filter paper, sawdust, chitin, or alginic acid, but also inorganic supports such as celite (141), glass (138, 142), horneblend (143), or glass wool (142).

It was thought by the original workers that after activation and washing, metal–chloride bonds remained for the chloride ligand to be replaced by chelating enzyme ligand. It is more certain that this is not the case (see below). Some studies (144) on the activation of inorganic materials by this method suggest that it apparently involves the formation of an adsorption layer of hydrated oxide, precipitated on the surface. Kent et al. (144) have noted that while titanium(IV)-activated organic carriers are brown-purple products, the inorganic matrices such as glass, celite, and ceramics remain white, suggesting that the activation of these two types of carriers is different. However, this again is a questionable or incorrect conclusion.

Owing to the variable results obtained with the former technique on application to inorganic materials, other procedures have been used (145–147). These include the activation of the inorganic carrier (such as nonporous glass) by coating it with a film or layer of imperfectly crystallized titanium (IV) oxide obtained by refluxing an acidic titanium(IV) chloride solution in the presence of the support (145). (Owing to relatively poor operational stability, other compounds have been used to coat the oxide, such as tin(II) chloride, 5-amino-salycylic acid, or tannic acid, and better operational stability has been obtained in some cases.) In another very similar procedure, the support is contacted with a solution of titanium(IV) chloride which is then heated to approximately 80°C and held at that temperature for a short period, usually 1 hr (146, 147). In these two variations of the former technique, refluxing and heating the titanium(IV) chloride solution cause the formation of a metal oxide and/or a hydrous metal oxide that adheres intimately, tenaciously, and evenly upon particles of the support, thereby forming a composite material having improved properties. The superior ability of the hydrous metal oxide to bind biologically active, nitrogen-containing organic substances can be attributed to chelation.

The hydrous metal oxide was originally prepared as a matrix independently from the titanium(IV) chloride treatment method and has been used as the only support for the immobilization of aminoacids, peptides, proteins, and antibiotics by Kennedy et al. (148–152). The preparation of these hydrous oxide gels, particularly those of Ti(IV) and Zr(IV), is accomplished by neutralizing the solution of metal chloride with ammonia and chelating the enzyme to the resulting precipitate. This is a very simple and viable method. It is also possible to form the hydrous oxide gel in the presence of the enzyme, resulting in an active enzyme form (149). Preparation of this hydrous oxide gel under controlled conditions in the presence of a matrix leads to the formation of a coating on this matrix (153). This has been employed as a convenient method for the immobilization of α-amylase (EC 3.2.1.1) on magnetic iron oxide (153).

It was from his independent work on the hydrous oxides that Kennedy demonstrated that the parallel use of treatment of supports with metal chlorides led ultimately at working stages to hydrolysis of the impregnate to yield the hydrous metal oxide. His mechanism for the immobilization on hydrous metal oxides involves hydroxyl ligand replacement by amino, carboxyl, hydroxyl, and similar groups of the enzyme (148). Messing (154) has employed a stannous bridge for the immobilization of urease (EC 3.5.1.5) to controlled pore titania, obtaining a yellow-cream-colored support, according to the mechanism shown in Scheme 3. The results show that urease bound to porous titania activated in this way had better storage stability than urease on

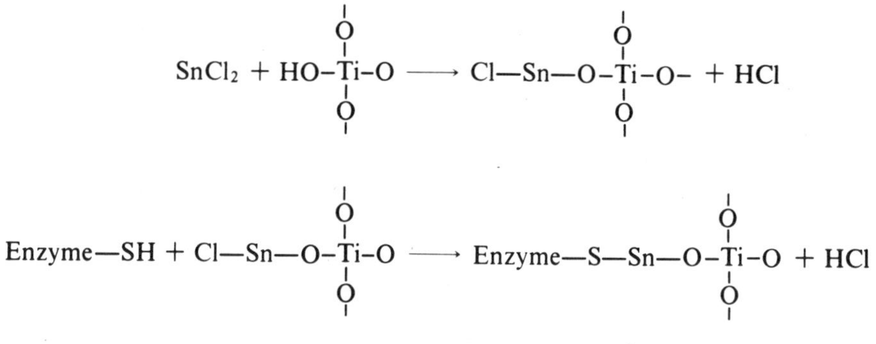

$$\mathrm{SnCl_2} + \mathrm{HO{-}}\underset{\underset{|}{\mathrm{O}}}{\overset{\overset{|}{\mathrm{O}}}{\mathrm{Ti}}}\mathrm{{-}O} \longrightarrow \mathrm{Cl{-}Sn{-}O{-}}\underset{\underset{|}{\mathrm{O}}}{\overset{\overset{|}{\mathrm{O}}}{\mathrm{Ti}}}\mathrm{{-}O{-}} + \mathrm{HCl}$$

$$\mathrm{Enzyme{-}SH} + \mathrm{Cl{-}Sn{-}O{-}}\underset{\underset{|}{\mathrm{O}}}{\overset{\overset{|}{\mathrm{O}}}{\mathrm{Ti}}}\mathrm{{-}O} \longrightarrow \mathrm{Enzyme{-}S{-}Sn{-}O{-}}\underset{\underset{|}{\mathrm{O}}}{\overset{\overset{|}{\mathrm{O}}}{\mathrm{Ti}}}\mathrm{{-}O} + \mathrm{HCl}$$

Scheme 3

untreated titania, probably because of advantageous interactions between the sulfydryl groups of the enzyme protein and the stannous ions on the support, involving more than a single ion–ion interaction and with perhaps some covalent character. According to Messing, this procedure is rather simple and offers additional versatility with respect to surface contributions for the purpose of coupling. This technique is perhaps second only to direct adsorption insofar as economic considerations and ease of handling are concerned.

Kennedy and Kay (155) refer to the possibility of improving the binding capacity of the hydrous oxides by activation with polymeric 1,3-diaminobenzene. Cabral et al. (156, 157), using the former experimental method of Novais on the activation of controlled pore glass, obtained immobilized enzyme preparations with very high activity (tenfold superior to those obtained with the same carrier by the refluxing technique). Their results were reproducible with glucoamylase, unlike those of other researchers who had obtained some variable results by application of the original procedure. But in spite of the high activity preparations obtained, the operational stability was

Table 7.6. Principal Examples of Immobilized Enzymes Prepared by Chelation (Metal Binding)

Carrier	Enzyme	Reference
Organic Supports		
Cellulose	Nuclease P_1	160
	β-D-Fructofuranosidase (EC 3.2.1.26)	103
DEAE-cellulose	Pectin lyase (EC 4.2.2.10)	161
Glass fiber paper	Papain (EC 3.4.22.2)	162, 163
Polypropylene	Papain (EC 3.4.22.2)	162
Nylon	D-Glucose dehydrogenase (EC 1.1.1.47)	170
Duolite® A-7	Triacylglycerol lipase (EC 3.1.1.3)	164
Polystyrene sulfonate	D-Glucose isomerase (EC 5.3.1.5.18)	165
Inorganic Supports		
Alumina	Glucoamylase (EC 3.2.1.3)	147
	β-D-Xylosidase (EC 3.2.1.37)	97
Stainless steel	β-D-Xylosidase (EC 3.2.1.37)	97
Horneblend	Glucoamylase (EC 3.2.1.3)	146, 158
Enzacryl®-TiO	Glucoamylase (EC 3.2.1.3)	158
Controlled pore glass	Pectin lyase (EC 4.2.2.10)	161
	Glucoamylase (EC 3.2.1.3)	156
Controlled pore silica	Glucoamylase (EC 3.2.1.3)	166
Lead glass	Glucoamylase (EC 3.2.1.3)	145, 166, 167
Silica gel	Trypsin (EC 3.4.21.4)	168
Silochrome	Pronase (see EC 3.4.24.4)	169
Hydrous titanium(IV) oxide	Papain (EC 3.4.22.2)	140, 171

very poor, owing not only to leakage of enzyme to solution but also to a strong deactivation of the bound enzyme. Desorption of protein was verified to be more temperature-dependent than substrate-dependent.

Although, as noted, this method is very simple in its application to immobilize enzymes—only one step for the activation of the support and another for the binding of the enzyme—the operational stabilities obtained with high molecular substrates are poor owing to the influence of the titanium(IV)-activated glass on the parameters, which control the deactivation of enzyme (156, 158). Tannic acid was used with relative success for the attenuation of this effect.

Although Messing refers to this method as partially covalent, and although Kennedy (140–142), Emery et al. (139), and Coughlin et al. (159) consider it as chelation of enzyme on hydrous metal oxide, it is "true" that—as with many immobilizations, even covalent ones—desorption can occur under operation or long-term storage, showing that weak interactions between the enzyme and the activated support are an important feature of this technique, so that some authors (144, 145) refer to it as an adsorption method rather than as partially covalent or chelation.

Another disadvantage of this method is its application to organic matrices; these may be degraded by the low pH of solutions when acidic titanous and titanic chlorides solutions are used. However, many matrices used in other methods can suffer similarly. More recent work shows that good success and operational stability can be achieved (157).

Examples of immobilized enzymes that have been prepared by this method are listed in Table 7.6.

7.3.2d. Covalent Binding

The covalent binding method is based on the covalent attachment of enzymes to water-insoluble matrices. This method has been the most widespread and one of the most thoroughly investigated approaches to enzyme immobilization.

The selection of conditions for immobilization by covalent binding is more difficult than in the other three carrier binding methods. The reaction conditions required are relatively complicated and not usually mild.

As with crosslinking, covalent binding is strong, so that stable immobilized enzyme preparations have been obtained that do not lose enzyme into the solution, even in presence of substrate or high ionic strength solutions. The immobilization of an enzyme by covalent attachment to a support matrix should involve only functional groups of the enzyme that are not essential for its catalytic action, and thus the active center of the enzyme must be unaffected by the various reagents that are used.

To achieve higher activities in the resulting immobilized enzyme preparations by preventing inactivation reactions with the essential aminoacid residues of the active site, several attempts (31) have been made: (i) covalent attachment of the enzyme in the presence of a competitive inhibitor or substrate, (ii) a reversible covalently linked enzyme–inhibitor complex, (iii) a chemically modified soluble enzyme whose covalent linkage to matrix is achieved by newly incorporated residues, and (iv) a zymogen precursor.

The wide variety of binding reactions and of matrices with functional groups capable of covalent coupling, or susceptible to being activated to give such groups, makes this a generally applicable method of immobilization. Nevertheless, the compositional and structural complexity of proteins has not allowed, except in a very limited number of cases, the application of general rules by means of which the method best suited for a specific task could be predicted. It would be most satisfactory if the enzyme tertiary, primary, and active site structures were to be known so that a potential linkage point least likely to be involved in activity could be selected for a covalent derivatization.

Three main factors have to be taken into account for covalent immobilization of enzymes by a specific method: (i) the functional group of proteins suitable for covalent binding under mild conditions, (ii) the coupling reactions between the enzyme and the support, (iii) the functionalized supports suitable for enzyme immobilization. These three factors are discussed in the following sections.

Functional Groups of Enzymes for Covalent Binding. Enzymes are heteropolymers built up from more than twenty types of monomer units of amino acid residues. Table 7.7 summarizes the residues of amino acids that have functional groups in the side-chain suitable for linking to a support matrix.

From this list, the amino acids with amide groups (L-glutamine and L-asparagine) and those with hydrocarbon side-chains (L-alanine, L-leucine, L-isoleucine, L-valine, L-phenylalanine, and L-proline) and glycine are absent owing to their relatively low concentration on the exposed protein surfaces and, even more important, owing to their nonreactive hydrophobic natures (Table 7.7). Increased hydrophobicity tends to increase the chances of a residue being buried inside the protein. Table 7.8 shows the average amino acid composition of a number of proteins, from the point of view of the above-mentioned relative reactivity of residues. Means and Feeney (172) made an interesting comparison that summarizes the number of reactions in which each amino acid side-chain can participate: L-Cys – 31, L-Lys – 27, L-Tyr – 16, L-His – 13, L-Met – 7, L-Trp – 7, L-Arg – 6, L-Glu – 4, L-Asp – 4, L-Ser – O, L-Thr – O.

Most of the reactions described below, which are the coupling reactions involving the active side chains, are classified as carbonyl-type reactions with

Table 7.7. Reactive Residues of Proteins[a]

$-NH_2$	ε-Amino of L-lysine (L-Lys) and N-terminus amino group
$-SH$	Thiol of L-cysteine (L-Cys)
$-COOH$	Carboxyl of L-aspartate (L-Asp) and L-glutamate (L-Glu) and C-terminus carboxyl group
$-C_6H_4-OH$	Phenolic of L-tyrosine (L-Tyr)
$-N(H)-C(=NH)-NH_2$	Guanidino of L-arginine (L-Arg)
(imidazole ring, N, N)	Imidazole of L-histidine (L-His)
$-S-S-$	Disulfide of L-cystine
(indole ring, N)	Indole of L-tryptophan (L-Trp)
CH_3-S-	Thioether of L-methionine (L-Met)
$-CH_2OH$	Hydroxyl of L-serine (L-Ser) and L-threonine (L-Thr)

[a] From P. A. Srere and K. Uyeda, in *Methods in Enzymology,* Vol. 44, (K. Mosbach, Ed.), Academic Press, New York (1976), p. 11.

Table 7.8. Average Composition of Proteins (Reactive Residues Only)[a]

Residue	Percent
Ser	7.8
Lys	7.0
Thr	6.5
Asp	4.8
Glu	4.8
Arg	3.8
Tyr	3.4
Cys	3.4
His	2.2
Met	1.6
Trp	1.2

[a] From M. Dayhoff and L. T. Hunt, "Atlas of Protein Sequence and Structure," National Biomedical Research Foundation, Washington, D.C., 1972.

the nucleophilic groups of the protein, -NH_2, -SH and -OH. In terms of nucleophilic reactivity, the sulfur type of anion is one or two orders of magnitude larger than that of most nitrogen and oxygen compounds of comparable basicity. However, the thioesters formed are much less stable than the esters, and these in turn are less stable than the substituted amines that are formed.

Comparing the data in Tables 7.7 and 7.8, the taking into account these considerations, it appears that the most convenient residues for immobilization are the L-lysyl residues, followed by L-cysteine, L-tyrosine, L-histidine, L-aspartic acid, L-glutamic acid, L-arginine, L-tryptophan, L-serine, L-threonine and L-methionine.

General Coupling Reactions for Covalent Binding. The coupling of enzyme molecules to solid supports involves mild reactions between amino acid residues of the enzyme, listed below, and several groups of functionalized carriers. The functional reaction groups of carriers for direct coupling with enzymes, the major reacting groups of enzymes, and the coupling reactions are listed in Table 7.9.

The simplest procedure for producing water-insoluble, surface-bonded, enzyme-support conjugates consists of contacting an enzyme solution with a preformed reactive carrier. However, only a few supports contain these reactive groups for direct coupling of enzymes. Examples of such preformed reactive carrier synthetic polymers are maleic anhydride based copolymers, methacrylic acid anhydride based copolymers, nitrated fluoracryl methacrylic copolymers, and iodoalkylmethacrylates.

The support materials most commonly used, however, do not possess these reactive groups but rather hydroxyl, amino, amido, and carboxyl groups, which have to be activated for immobilization of enzyme. The various potential methods of activation of these supports are described below.

As can be seen from Table 7.9, several types of coupling reactions can be used, and most of the common covalent coupling reactions involve coupling through protein amino groups, thiol groups, carboxyl groups, or aromatic rings of L-tyrosine and L-histidine. The major classes of coupling reactions used for the immobilization of enzymes are:

1. Diazotization
2. Amide (peptide) bond formation
3. Alkylation and arylation
4. Schiff's base formation
5. Ugi reaction
6. Amidination reactions

Table 7.9. Immobilization Methods of Enzymes by Covalent Binding, Showing the Match Between Carrier and Enzyme Functional Groups

Reactive Group of Carrier	Reacting Group of Enzyme	Coupling Reactions
$—C_6H_4—N_2^+Cl^-$ (Diazonium salt)	$—NH_2$ $—SH$ $—C_6H_4—OH$	Diazo linkage
$—C(=O)—O—C(=O)—$ (Acid anhydride)	$—NH_2$	Peptide bond formation
$—CH_2CON_3$ (Acyl azide)	$—NH_2$ $—SH$ $—C_6H_4—OH$	Peptide bond formation
$(—O)_2C=NH$ (Imidocarbonate)	$—NH_2$	Peptide bond formation
$—R—NCS$ (Isothiocyanate)	$—NH_2$	Peptide bond formation
$—C_6H_4—NCO$ (Isocyanate)	$—NH_2$	Peptide bond formation
$—CH_2COCl$ (Acyl chloride)	$—NH_2$	Peptide bond formation
$(—O)_2C=O$ (Cyclic carbonate)	$—NH_2$	Peptide bond formation
$—COO—C(—NH—R')=N^+H—R''$ (*O*-Acylisourea)	$—NH_2$	Peptide bond formation

Table 7.9. *(continued)*

Reactive Group of Carrier	Reacting Group of Enzyme	Coupling Reactions
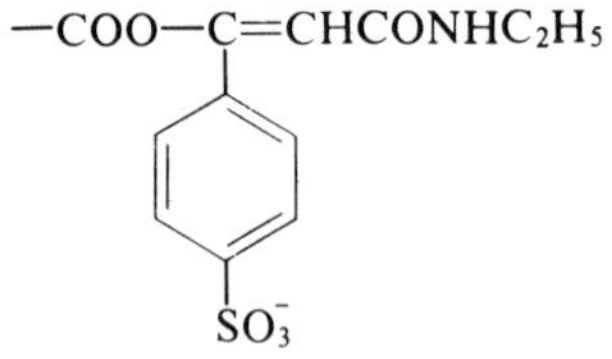 $-COO-C(=CHCONHC_2H_5)-C_6H_4-SO_3^-$ (Woodward's reagent K derivative)	$-NH_2$	Peptide bond formation
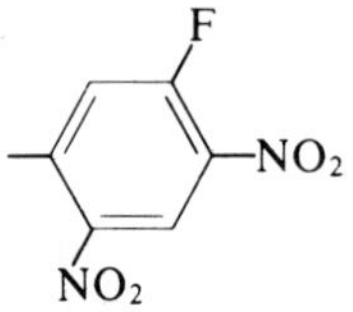 $-C_6H_2(F)(NO_2)_2$ (*m*-Fluorodinitroanilide)	$-NH_2$	Arylation
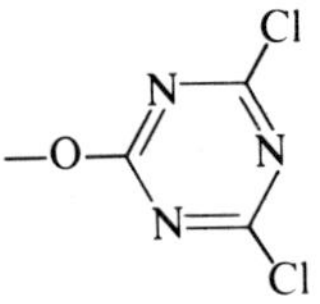 $-O-C_3N_3Cl_2$ (Triazinyl)	$-NH_2$	Arylation
$-O-CH-CH_2$ (ring closed by X); X = >NH, >O, >S (e.g. Oxirane)	$-NH_2$ $-OH$ $-SH$	Alkylation
$-O-CH_2-CH_2-SO_2-CH = CH_2$ (Vinylsulfonyl)	$-NH_2$ $-SH$ $-OH$	Alkylation
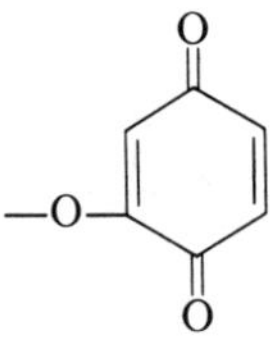 $-O-C_6H_3O_2$ (*p*-benzoquinone ring: O=, =O) (Vinyl keto)	$-NH_2$ $-SH$ $-OH$	Arylation
$-CHO$ (Aldehyde)	$-NH_2$	Schiff base formation
$-\overset{+}{N}H=C(R_1)(R_2)$	$-CO_2H$ $-NH_2$	Ugi reaction

Table 7.9. *(continued)*

Reactive Group of Carrier	Reacting Group of Enzyme	Coupling Reactions
$-C(=NH)-OC_2H_5$ (Imidoester)	$-NH_2$	Amidination reaction
$-CN$ (Cyanide)	$-NH_2$	Amidination reaction
$-S-S-$(2-pyridyl) (Disulfide residue)	$-SH$	Thiol–disulfide interchange
$-C_6H_4-HgCl$ (Mercury derivative)	$-SH$	Mercury–enzyme interaction
$M^{\cdot}$ (Matrix radical)	$E^{\cdot}$ (Enzyme radical)	γ-irradiated induced coupling
$-NH_2$ (Amine)	$-NH_2$ $-CO_2H$	Peptide bond formation (in presence of condensing reagents)
$-CONHNH_2$ (Acylhydrazide)	$-NH_2$ $-CO_2H$	Peptide bond formation (in presence of condensing reagents)

7. Thiol–disulfide interchange reactions
8. Mercury–enzyme interactions
9. γ-Irradiation-induced coupling

Diazotization. The diazo coupling method is one of the most commonly used for coupling enzyme to carrier. Historically, it is among the oldest described in the literature (7, 11, 16–18). It is based on the diazo linkage between enzyme proteins and aryldiazonium electrophilic groups of the carrier.

These diazonium carriers are prepared by treating the support containing aromatic amino groups with nitrite in an acidic medium (Scheme 4). The

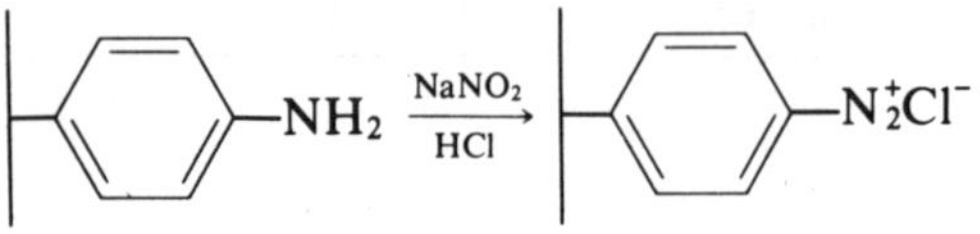

$$\text{—}C_6H_4\text{—}NH_2 \xrightarrow[HCl]{NaNO_2} \text{—}C_6H_4\text{—}N_2^+Cl^-$$

Scheme 4

functional groups on the enzyme participating in diazo coupling are mainly activated aromatic rings such as phenol (L-tyrosine) and imidazole (L-histidine), to form the corresponding azo derivatives.

In addition to these groups, several other amino acids in proteins react under similar conditions with the diazonium salt to give the azo derivatives I and II or the disubstituted bisazo derivative of primary amines (III), as shown in Scheme 5. The guanido group (L-arginine) and indolyl group (L-tryptophan) can also be attacked by the electrophilic aryldiazonium ion.

Examples of immobilized enzymes that have been prepared covalently by the diazo coupling method are listed in Table 7.10.

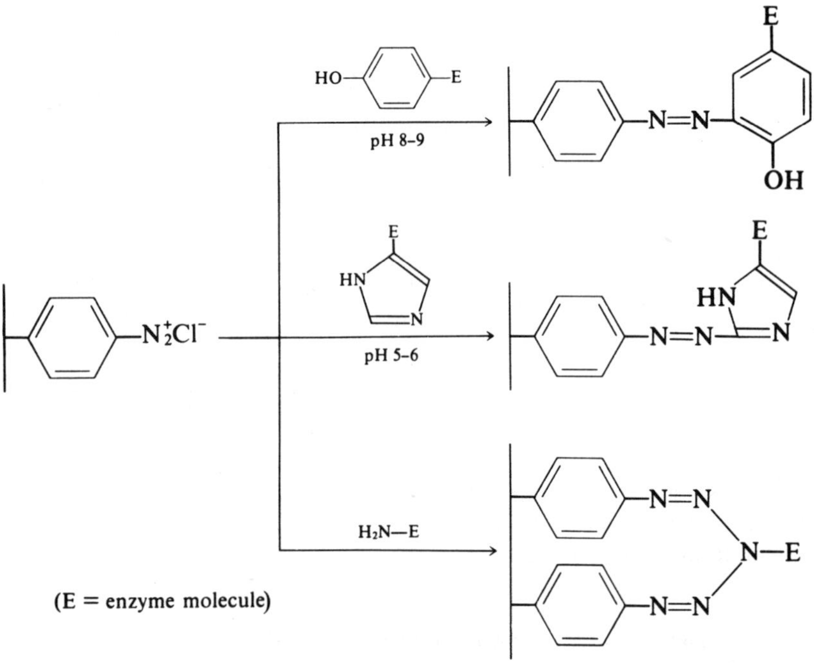

HO—C₆H₄—E, pH 8–9 → —C₆H₄—N=N—(aryl with E, OH) (I)

imidazole-E, pH 5–6 → —C₆H₄—N=N—(imidazole with HN, N, E) (II)

H_2N—E → (—C₆H₄—N=N)₂N—E (III)

—C₆H₄—$N_2^+Cl^-$

(E = enzyme molecule)

Scheme 5

Table 7.10. Principal Examples of Immobilized Enzymes Prepared by Diazotization Coupling

Carrier	Enzyme	Reference
Organic Supports		
Cellulose	β-D-Galactosidase (EC 3.2.1.23)	173
	D-Glucose oxidase (EC 1.1.3.4)	173
	Trypsin (EC 3.4.21.4)	173
	Papain (EC 3.4.22.2)	173
	Pepsin (EC 3.4.23.1)	173
	β-Amylase (EC 3.2.1.2)	101
Enzacryl® AA	β-D-Xylosidase (EC 3.2.1.37)	97
	Dextransucrase (EC 2.4.1.5)	127
	Tyrosinase (EC 1.10.3.1/1.14.18.1)	174
Synthetic pulp	Trypsin (EC 3.4.21.4)	175
	Papain (EC 3.4.22.2)	175
	Aminoacylase (EC 3.5.1.14)	175
	Urease (EC 3.5.1.5)	175
Agar	Protein kinase (EC 2.7.1.37)	176
Styrene/maleic anhyd.	Chymotrypsin (EC 3.4.21.1)	177
Glycidyl methacrylate/ ethylene dimethacrylate	Penicillin amidase (EC 3.5.1.11)	178
Polyacrylamide/nylon	Urease (EC 3.5.1.5)	179
Polythylene terephtalate	Trypsin (EC 3.4.21.4)	180
Inorganic Supports		
Controlled pore glass	Aldehyde dehydrogenase (EC 1.1.1.3)	181
	Phospholipase A_2 (EC 3.1.1.4)	182
Controlled pore silica	Hydrogenase (EC 1.18.3.1)	183

Amide Bond Formation. The amide binding method is based on the formation of amide (peptide) bonds, or similar bonds, between the enzyme protein and a water-insoluble carrier by the application of the peptide synthesis technique. Several procedures are available depending on the different derivatized solid matrices (see Table 7.9).

The common mechanistic feature in the peptide bond formation method is the attack of the nucleophilic groups (amino, hydroxyl, thiol) of enzyme at an activated functional group on the carrier. These nucleophiles are most effective in their unprotonated forms (—NH_2, —C_6H_4—O^-, —S^-) at pH values above their pK_a values. Owing to the irreversible denaturation of the protein at high pH, however, these reactions are commonly carried out at intermediate pH values (7.5–8.5) and at low temperature (4°C).

Table 7.11. Principal Examples of Immobilized Enzymes Prepared by Amide Bond Formation

Carrier	Enzyme	Reference
Acid Anhydride Derivatives		
Ethylene/maleic anhydride	Alkaline phosphatase (EC 3.1.3.1)	184
	Naringinase	185
Butanediol divinylether/anhydride	Lactate dehydrogenase (EC 1.1.1.27)	186
	Trypsin (EC 3.4.21.4)	186
	Chymotrypsin (EC 3.4.21.1)	186
	Papain (EC 3.4.22.2)	186
	Ficin (EC 3.4.22.3)	186
	Bromelain (EC 3.4.22.4)	186
	Subtilisin (EC 3.4.21.14)	186
	Subtilopeptidase B	186
	Pronase (see EC 3.4.24.4)	186
Methylvinylether/maleic anhydride	Alkaline phosphatase (EC 3.1.3.1)	184
	Naringinase	185
Isobutylvinylether/maleic anhydride	Naringinase	185
Acyl Azide Derivatives		
Enzacryl® AH	β-D-Xylosidase (EC 3.2.1.37)	97
	Tyrosinase (EC 1.10.3.1/1.14.18.1)	174
	Dextransucrase (EC 2.4.1.5)	128
Glycidyl methacrylate/ethylene dimethacrylate	Penicillin amidase (EC 3.5.1.11)	178
Poly(acryloylmorpholine)	Carbonate dehydratase (EC 4.2.1.1)	187
Acrylic acid/isothiocyanatostyrene	Papain (EC 3.4.22.2)	188
Polyethylene terephtalate	Trypsin (EC 3.4.21.4)	180
Poly(vinylalcohol)	β-D-Glucosidase (EC 3.2.1.21)	189
Nylon/polyacrylamide	β-D-Galactosidase (EC 3.2.1.23)	190
	Papain (EC 3.4.22.2)	190

Collagen	L-Iditol dehydrogenase (EC 1.1.1.14)	191
	D-Glucose oxidase (EC 1.1.3.4)	192
	Aspartate aminotransferase (EC 2.6.1.1)	193, 194
Controlled pore glass	Parathion hydrolase	195
Controlled pore silica	Parathion hydrolase	195
Cyclic Imidocarbonate Derivatives		
Agarose	Citrate synthase (EC 4.1.3.7/28)	196
	Tyrosinase (EC 1.10.3.1/1.14.18.1)	174
	β-D-Galactosidase (EC 3.2.1.23)	197
	Xanthine dehydrogenase (EC 1.2.1.37)	109
	Phenol 2-monooxygenase (EC 1.14.13.7)	80
	Hydroxysteroid dehydrogenase (EC 1.1.1.50, etc.)	198
	Xanthine oxidase (EC 1.2.3.2)	199, 200
	Superoxide dismutase (EC 1.15.1.1)	199
	Catalase (EC 1.11.1.6)	199
	Catechol-1,2-dioxygenase (EC 1.13.11.1)	79
	D-Fructose bisphosphate aldolase (EC 4.1.2.13)	209
	Fumarate hydratase (EC 4.2.1.2)	201
	L-Malate dehydrogenase (EC 1.1.1.37)	201
	Trypsin (EC 3.4.21.4)	202, 206
	Colipase	203
	Triacylglycerol lipase (EC 3.1.1.3)	203
Cellulose	Xanthine oxidase (EC 1.2.3.2)	200
	Nuclease P_1	204
	L-Asparaginase (EC 3.5.1.1)	205
Dextran	Trypsin (EC 3.4.21.4)	206
2-Hydroxyethylmethacrylate	Polygalacturonase (EC 3.2.1.15)	207
Polythiol/4-vinylpyridine	Trypsin (EC 3.4.21.4)	208
Hornblende	Xanthine oxidase (EC 1.2.3.2)	200
Glass beads	Chymotrypsin (EC 3.4.21.1)	209

Table 7.11. (*continued*)

Carrier	Enzyme	Reference
Isocyanate Isothiocyanate Derivatives		
Cellulose	β-Amylase (EC 3.2.1.2)	210
Enzacryl® AA	Tyrosinase (EC 1.10.3.1/1.14.18.1)	211
Polyurethane	β-D-Fructofuranosidase (EC 3.2.1.26)	212
Polypropylene glycol	Trypsin (EC 3.4.21.4)	213
Polyethylene terephtalate	Trypsin (EC 3.4.21.4)	180
Synthetic pulp	Trypsin (EC 3.4.21.4)	175
	Papain (EC 3.4.22.2)	175
	Aminoacylase (EC 3.5.1.14)	175
	Urease (EC 3.5.1.5)	175
Glycidyl methacrylate/ethylene dimethacrylate	Penicillin amidase (EC 3.5.1.11)	178
Activated carbon	D-Glucose oxidase (EC 1.1.3.4)	110
Aerosil	α-Amylase (EC 3.2.1.1)	214
Silochrome	Amylosubtilin	215
	Pronase (see EC 3.4.24.4)	169
Acyl Chloride Derivatives		
Amberlite® IRC-50	Catalase (EC 1.11.1.6)	13, 216
	Triacylglycerol lipase (EC 3.1.1.3)	217
Amberlite® XE-64	L-Glutamate dehydrogenase (EC 1.4.1.1)	218
Cyclic Carbonate Derivative		
Cellulose *trans*-2,3-carbonate	Dextranase (EC 3.2.1.11)	219
	β-D-Glucosidase (EC 3.2.1.21)	220, 221
	Trypsin (EC 3.4.21.4)	222

Carbodiimide Mediated Derivatives		
Agarose	Riboflavin kinase (EC 2.7.1.26)	224
	Phenol 2-monooxygenase (EC 1.14.13.7)	80, 129
	β-D-Xylosidase (EC 3.2.1.37)	97
	Cholesterol oxidase (EC 1.1.3.6)	135
Cellulose	Papain (EC 3.4.22.2)	225
	Chymotrypsin (EC 3.4.21.1)	225
	α-Amylose (EC 3.2.1.1)	225, 226
	Glucoamylase (EC 3.2.1.3)	225, 226
	Poly-D-galacturonase (EC 3.2.1.15)	225
Nylon	Lactate dehydrogenase (EC 1.1.1.27)	227
	Phenol 2-monooxygenase (EC 1.14.13.7)	129
Nylon/acrilonitrile	β-D-Fructofuranosidase (EC 3.2.1.26)	228
	Pepsin A (EC 3.4.23.1)	228
	Acid phosphatase (EC 3.1.3.2)	228
	Alkaline phosphatase (EC 3.1.3.1)	228
Acrylamide/acrylic acid	Chymotrypsin (EC 3.4.21.1)	229
Glycidyl methacrylate/ethylene dimethacrylate	Penicillin amidase (EC 3.5.1.11)	178
Amberlite® IRC-50	β-D-Fructofuranosidase (EC 3.2.1.26)	132
Glassy carbon	D-Glucose oxidase (EC 1.1.3.4)	111
Activated carbon	D-Glucose oxidase (EC 1.1.3.4)	110, 230
	Glucoamylase (EC 3.2.1.3)	230
	D-Gluconolactonase (EC 3.1.1.17)	230
Aerosil	α-Amylase (EC 3.2.1.1)	214
Glass	Trypsin (EC 3.4.21.4)	231
Controlled pore glass	Aldehyde dehydrogenase (EC 1.2.1.3)	181

Table 7.11 lists some immobilized enzymes that have been prepared by the amide bond formation method. Individual reactions are now described.

ACID ANHYDRIDE DERIVATIVES. Acid anhydrides react with amino groups of proteins as shown in Scheme 6. As can be seen in this scheme, carboxylic

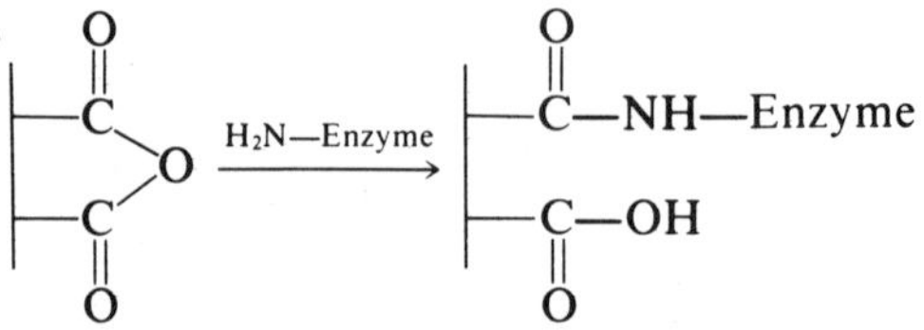

$$\left|\begin{matrix} -C(=O) \\ -C(=O) \end{matrix}\right\rangle O \xrightarrow{H_2N\text{—Enzyme}} \left|\begin{matrix} -C(=O)\text{—NH—Enzyme} \\ -C(=O)\text{—OH} \end{matrix}\right.$$

Scheme 6

acid groups are formed (in addition to the amide linkages) by cleavage of anhydride rings; these do not react with functional groups of enzyme but are ionized in the aqueous solutions spontaneously by OH^- ions, that is, at higher pH values, generating free carboxyl groups with a highly negative charge. The negative charge gives undesirable ionogenic properties to the preparation and may unfavorably affect the activity and the stability of the enzyme. To overcome this disadvantage it is usual to neutralize some of the negatively charged species by the addition of diamines during the immobilization procedure.

ACYL AZIDE DERIVATIVES. The azide method is one of the most frequently employed procedures in the field of peptide synthesis and is one of

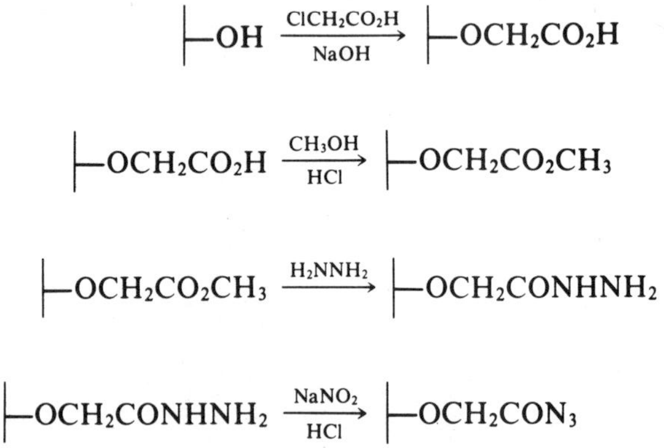

$$\vdash OH \xrightarrow[NaOH]{ClCH_2CO_2H} \vdash OCH_2CO_2H$$

$$\vdash OCH_2CO_2H \xrightarrow[HCl]{CH_3OH} \vdash OCH_2CO_2CH_3$$

$$\vdash OCH_2CO_2CH_3 \xrightarrow{H_2NNH_2} \vdash OCH_2CONHNH_2$$

$$\vdash OCH_2CONHNH_2 \xrightarrow[HCl]{NaNO_2} \vdash OCH_2CON_3$$

Scheme 7

the oldest methods of immobilizing enzymes. The azide carriers are prepared from hydroxyl or carboxyl supports, as indicated in Scheme 7.

The acyl azide derivative reacts with enzyme protein involving predominantly the primary amino groups of enzyme to form the amide linkage (Scheme 8).

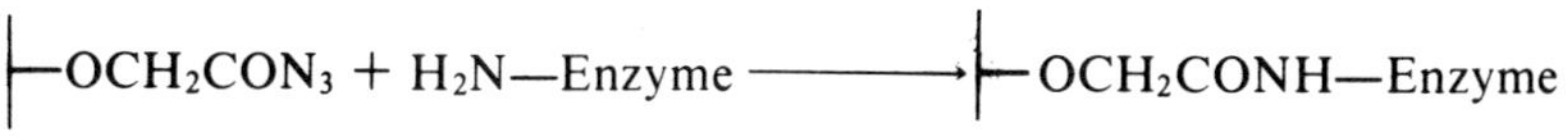

$$\vdash OCH_2CON_3 + H_2N\text{—Enzyme} \longrightarrow \vdash OCH_2CONH\text{—Enzyme}$$

Scheme 8

Side reactions with aliphatic or aromatic hydroxyl groups or thiol groups occur under the same conditions.

CYCLIC IMIDOCARBONATE DERIVATIVES. The imidocarbonate method is probably one of the most commonly used for laboratory preparations of immobilized enzyme derivatives as well as insoluble adsorbents for affinity chromatography. It was first developed by Axèn et al. (232).

The method consists in the reaction of cyanogen halide (mostly CNBr) with hydroxyl groups of the carrier at high pH (10–11.5) to give an inert carbamate (II) and reactive imidocarbonate (III) through the intermediate cyanide (I), which is very labile, as indicated in Scheme 9.

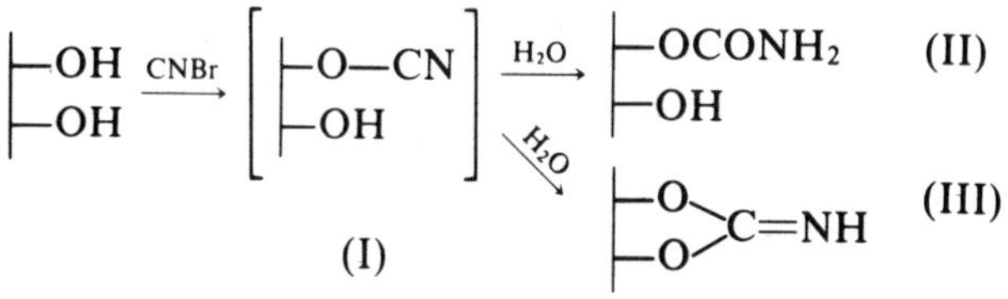

$$\begin{matrix}\vdash OH \\ \vdash OH\end{matrix} \xrightarrow{CNBr} \left[\begin{matrix}\vdash O\text{—}CN \\ \vdash OH\end{matrix}\right] \xrightarrow{H_2O} \begin{matrix}\vdash OCONH_2 \\ \vdash OH\end{matrix} \quad (II)$$

$$(I) \qquad \xrightarrow{H_2O} \begin{matrix}\vdash O \\ \vdash O\end{matrix}\!\!>\!C\!=\!NH \quad (III)$$

Scheme 9

Immobilization is then carried out in a basic medium (pH 9–10) and results in the formation of the different types of structures: *N*-substituted isoureas (IV), *N*-substituted imidocarbonates (V), and *N*-substituted carbamates (VI), as shown in Scheme 10. However, it seems from additional evidence accumulated in the past few years that major reaction products of the imidocarbonates with amines are probably the substituted isourea structures (see 233–236 and see Chapter 12).

Although this immobilization procedure is simple, there are some disadvantages such as the high toxicity of cyanogen bromide, the high pH of the derivatizing and coupling reactions, and also the presence in the products of charged groups, which cause interfering and unreliable nonspecific adsorptions.

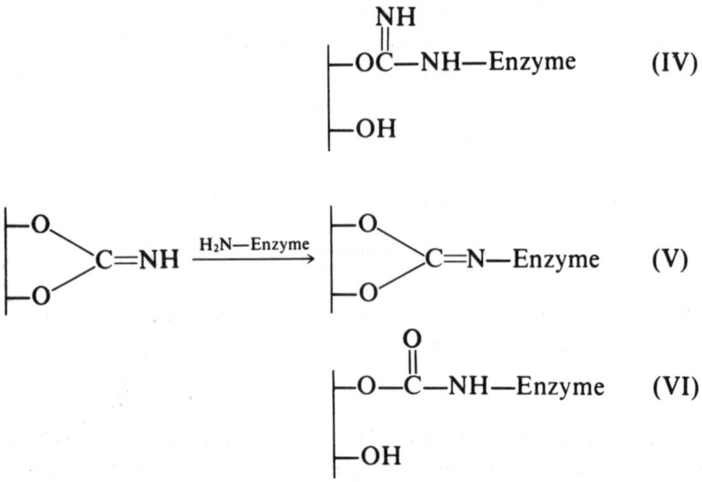

$$\begin{array}{l} \quad\;\; NH \\ \quad\;\; \| \\ \vdash OC{-}NH{-}Enzyme \qquad (IV) \\ | \\ \vdash OH \end{array}$$

$$\left[\begin{array}{l}{-}O\diagdown \\ \qquad C{=}NH \\ {-}O\diagup\end{array}\right. \xrightarrow{H_2N{-}Enzyme} \left[\begin{array}{l}{-}O\diagdown \\ \qquad C{=}N{-}Enzyme \\ {-}O\diagup\end{array}\right. \qquad (V)$$

$$\begin{array}{l} \qquad\;\; O \\ \qquad\;\; \| \\ \vdash O{-}C{-}NH{-}Enzyme \qquad (VI) \\ | \\ \vdash OH \end{array}$$

Scheme 10

An analogous active product is obtained by reaction of hydroxyl groups of polysaccharides with alkyl or aryl cyanates (Scheme 11). The results obtained are similar to those from the cyanogen activation, but the cyanate method has not become widely used, possibly because the preparation of the

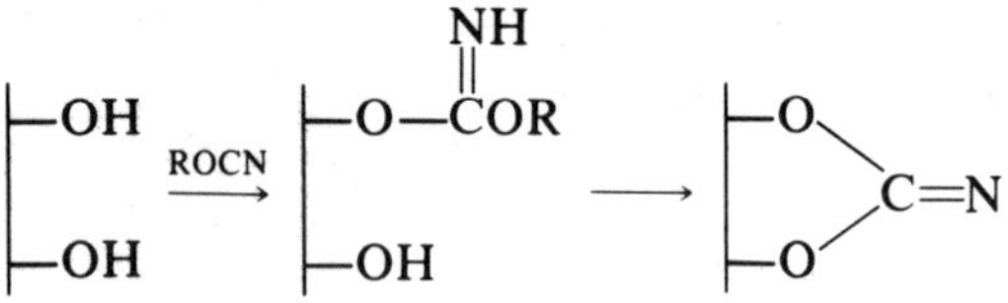

$$\left[\begin{array}{l}{-}OH \\ \\ {-}OH\end{array}\right. \xrightarrow{ROCN} \left[\begin{array}{l}{-}O{-}\overset{NH}{\overset{\|}{C}}OR \\ \\ {-}OH\end{array}\right. \longrightarrow \left[\begin{array}{l}{-}O\diagdown \\ \qquad C{=}N \\ {-}O\diagup\end{array}\right.$$

Scheme 11

alcohol cyanate involves the reaction of cyanogen halide with alcohol or phenol and thus is not only one reaction step longer but requires continued handling of toxic compounds. Furthermore cyanogen bromide activated polysaccharide matrices are available commercially.

ISOCYANATE AND ISOTHIOCYANATE DERIVATIVES. Carriers containing aromatic amino and acyl azide groups can be modified at alkaline pH to the corresponding isocyanate or isothiocyanate derivatives, by reaction respectively with phosgene or thiophosgene. These isocyanate or isothiocyanate functional groups have been used as electrophiles. They react with free amino groups of the enzyme to form an amide or thioamide linkage of the substituted urea or thiourea (Scheme 12).

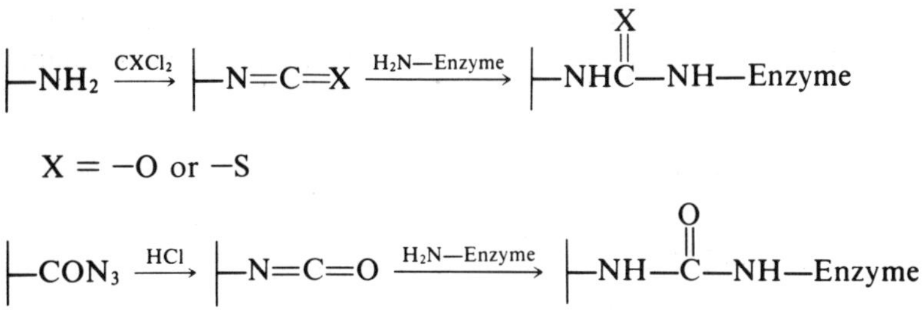

$$\vdash NH_2 \xrightarrow{CXCl_2} \vdash N{=}C{=}X \xrightarrow{H_2N\text{—Enzyme}} \vdash NH\overset{\overset{\displaystyle X}{\|}}{C}\text{—NH—Enzyme}$$

$X = -O$ or $-S$

$$\vdash CON_3 \xrightarrow{HCl} \vdash N{=}C{=}O \xrightarrow{H_2N\text{—Enzyme}} \vdash NH\text{—}\overset{\overset{\displaystyle O}{\|}}{C}\text{—NH—Enzyme}$$

Scheme 12

The reaction with other nucleophiles of the enzyme, such as thiol, imidazole, or aromatic hydroxyl and carboxyl groups, leads to relatively unstable derivatives that decompose at mildly alkaline pH values or in the presence of nucleophiles such as hydrazine (172, 237, 238).

The isothiocyanate derivative has a higher stability and for this reason it has been used more frequently than the isocyanate derivative. The isothiocyanate derivative can also be obtained from aliphatic amino-type carriers.

ACYL CHLORIDE DERIVATIVES. Carboxy functional supports can be activated by refluxing in a solution containing thionyl chloride; an acyl chloride derivative is obtained (Scheme 13). The acyl chloride derivative can be

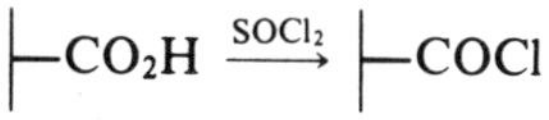

$$\vdash CO_2H \xrightarrow{SOCl_2} \vdash COCl$$

Scheme 13

reacted with enzymes at low temperatures for immobilization according to Scheme 14.

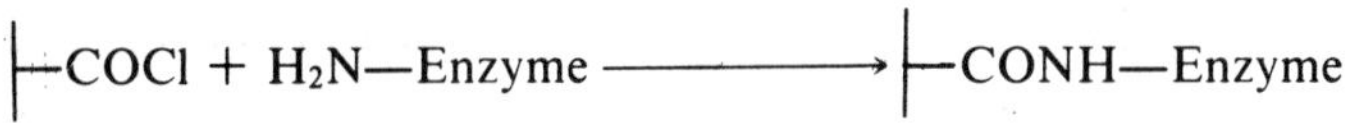

$$\vdash COCl + H_2N\text{—Enzyme} \longrightarrow \vdash CONH\text{—Enzyme}$$

Scheme 14

This method of immobilization was used by Brandenberger (217) in 1956 for the immobilization of triacylglycerol lipase (EC 3.1.1.3) on the carboxychloride derivative of a cation exchanger, Amberlite IRC-50, and this report is considered to be the first on the amide binding methods.

CYCLIC CARBONATE DERIVATIVES. A simple method to attach to the carrier under very mild conditions is described by Barker et al. (222, 223). The activated support is obtained by the reaction of the hydroxyl groups of the support with ethylchloroformate in a dimethylsulfoxide medium with the addition of dioxane and triethylamine. It is completed within a few minutes (Scheme 15).

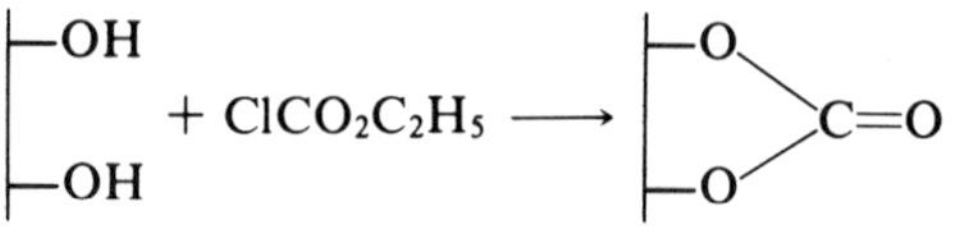

$$\left[\begin{array}{l}—OH\\ \\ —OH\end{array}\right. + ClCO_2C_2H_5 \longrightarrow \left[\begin{array}{l}—O\\ \\ —O\end{array}\right\rangle C{=}O$$

Scheme 15

Immobilized enzyme preparations can be obtained by reaction of nucleophilic groups of the enzyme at pH 7–8 with various modes of binding, obtaining *N*-substituted imidocarbonate (I) and *N*-substituted carbonate (II) (Scheme 16), as with the cyclic imidocarbonate method.

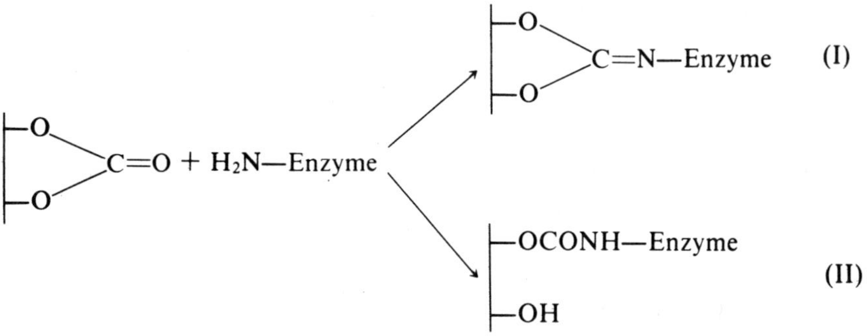

$$\left[\begin{array}{l}—O\\ \\ —O\end{array}\right\rangle C{=}O + H_2N\text{—Enzyme} \begin{array}{l}\nearrow \left[\begin{array}{l}—O\\ \\ —O\end{array}\right\rangle C{=}N\text{—Enzyme} \quad \text{(I)}\\ \\ \searrow \left[\begin{array}{l}—OCONH\text{—Enzyme}\\ \\ —OH\end{array}\right. \quad \text{(II)}\end{array}$$

Scheme 16

By comparison with the cyclic imidocarbonate reaction method using cyanogen bromide, the work with ethyl chloroformate or phosgene is much cheaper, simpler, and in the case of ethyl chloroformate, less toxic. However, it has not been used much.

Condensing Reagents. The most general methods for activation of supports containing carboxy groups involve the use of carbodiimides and similar reagents. As shown in Scheme 17, carbodiimides react with carboxyl groups at slightly acid pH values to give *O*-acylisourea derivatives (I), and the activated carboxyl groups subsequently can rearrange to an acyl urea (II) or react with amino groups of the enzyme to yield the corresponding amides (III). The intermediate can also react with other nucleophiles, e.g. hydroxyl and thiol groups, to give different carboxylate derivatives but at much lower reaction rates.

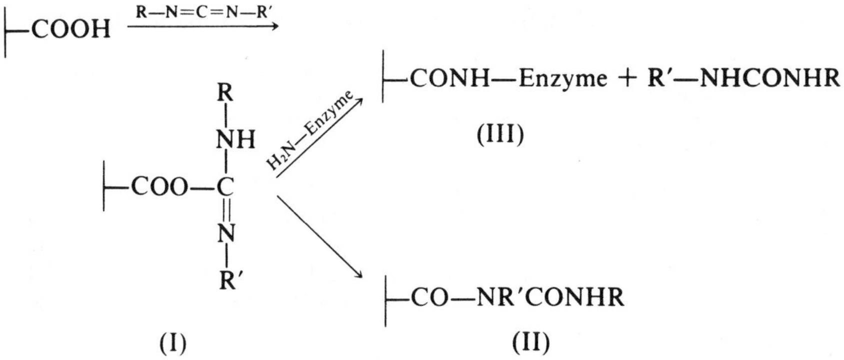

$$\vdash COOH \xrightarrow{R-N=C=N-R'} \vdash COO-C(=N-R')-NH-R \ \text{(I)}$$

$$\text{(I)} \xrightarrow{H_2N-Enzyme} \vdash CONH-Enzyme + R'-NHCONHR \ \text{(III)}$$

$$\text{(I)} \longrightarrow \vdash CO-NR'CONHR \ \text{(II)}$$

Scheme 17

Similar carbodiimide reagents, such as Woodward's reagent K, can be used in the same way, as shown in Scheme 18.

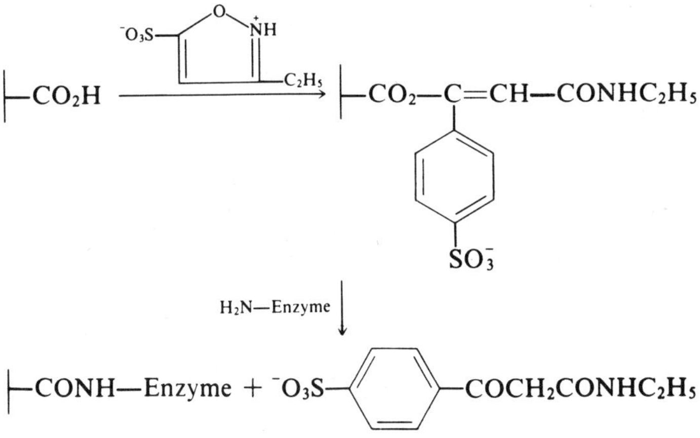

$$\vdash CO_2H \xrightarrow{\text{Woodward's reagent K } (^-O_3S\text{-isoxazolium}, N\text{-}C_2H_5)} \vdash CO_2-C(C_6H_4SO_3^-)=CH-CONHC_2H_5$$

$$\xrightarrow{H_2N-Enzyme} \vdash CONH-Enzyme + {}^-O_3S-C_6H_4-COCH_2CONHC_2H_5$$

Scheme 18

These types of condensing reagents also have been used to activate carriers containing amino or hydrazide groups to bind enzymes through their carboxyl groups. The main difference from the former process is that the carrier and the condensing reagent are added simultaneously to the enzyme solution and stirred. Amide bonds are formed between amino groups of the carrier and carbonyl groups of the enzyme, thereby effecting an immobilization.

Alkylation and Arylation. The alkylation and arylation methods based on the alkylation of amino, phenolic, or thiol groups of an enzyme with reactive

Table 7.12. Principal Examples of Immobilized Enzymes Prepared by Alkylation and Arylation

Carrier	Enzyme	Reference
Halogeno Acetyl Derivatives		
Chloroacetyl cellulose	Aminoacylase (EC 3.5.1.14)	239
Bromoacetyl cellulose	Aminoacylase (EC 3.5.1.14)	239
	Glucoamylase (EC 3.2.1.3)	240
Iodoacetyl cellulose	Aminoacylase (EC 3.5.1.14)	239
	Glucoamylase (EC 3.2.1.3)	240
Triazinyl Derivatives		
Cellulose	Phospho-D-glucomutase (EC 2.7.5.1)	241, 242
	D-Glucose-6-phosphate dehydrogenase (EC 1.1.1.49)	241, 242
Filter paper	Dextransucrase (EC 2.4.1.5)	127
Agarose	Chymotrypsin (EC 3.4.21.1)	243
	Trypsin (EC 3.4.21.4)	243
	Lactate dehydrogenase (EC 1.1.1.27)	243
Polystyrene	α-Amylase (EC 3.2.1.1)	244
Silochrome	Amylosubtilin	215
Oxirane Derivatives		
Glycidylmethacrylate	Glucoamylase (EC 3.2.1.3)	245
Polyacrylamide	β-D-Galactosidase (EC 3.2.1.23)	246
Vinylketo Derivatives		
Polyhydroxyalkylmethacrylate	Chymotrypsin (EC 3.4.21.1)	247
	Trypsin (EC 3.4.21.4)	247

supports containing active halides, oxirane, vinylsulfonyl or vinylketo groups. Enzymes that have been immobilized by this method are listed in Table 7.12.

Supports with halogen-substituted aromatic rings containing nitro groups, such as copolymers of methacrylate and methacrylate-5-fluoro-2,4-dinitroanilide, have been used to couple the amino groups of enzyme via arylation, as indicated in Scheme 19. This support is synthetized from monomeric units that contain reaction groups, and no activation is required (Scheme 20).

However, this method can be used with less reactive supports, in which a halogen-substituted heterocyclic ring is incorporated by reaction with cyanuric chloride or its derivatives (Scheme 21).

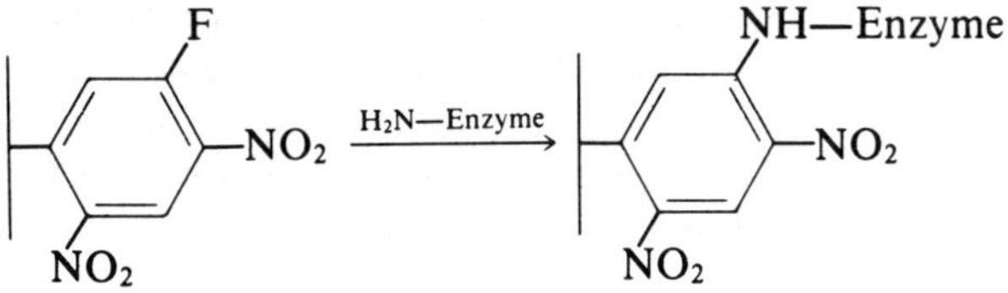

$$\text{Polymer–}C_6H_2(F)(NO_2)_2 \xrightarrow{H_2N\text{—Enzyme}} \text{Polymer–}C_6H_2(NH\text{—Enzyme})(NO_2)_2$$

Scheme 19

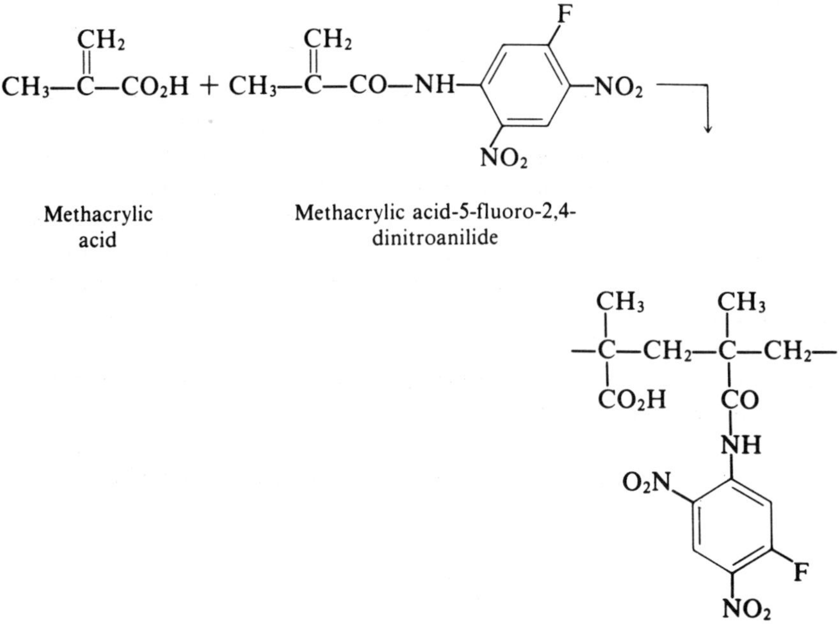

$$CH_3\text{—}C(=CH_2)\text{—}CO_2H + CH_3\text{—}C(=CH_2)\text{—}CO\text{—}NH\text{—}C_6H_2(F)(NO_2)_2 \longrightarrow$$

Methacrylic acid

Methacrylic acid-5-fluoro-2,4-dinitroanilide

$$\text{—}C(CH_3)(CO_2H)\text{—}CH_2\text{—}C(CH_3)(CO\text{—}NH\text{—}C_6H_2(O_2N)(F)(NO_2))\text{—}CH_2\text{—}$$

Scheme 20

$$\text{—OH} + \text{Cl—}C_3N_3(Cl)(X) \longrightarrow \text{—O—}C_3N_3(Cl)(X) \xrightarrow{H_2N\text{—Enzyme}} \text{—O—}C_3N_3(NH\text{—Enzyme})(X)$$

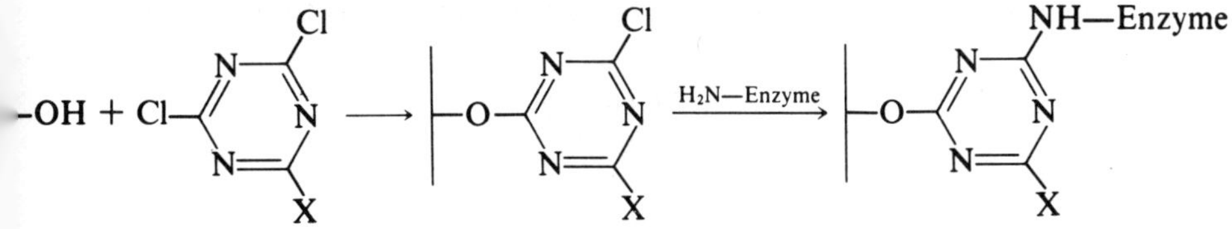

X = —Cl,—NH_2, etc.

Scheme 21

Jagendorf et al. (248) introduced a method using carriers with monohalogenoacetyl functional groups which for immobilization of enzymes involved alkylation of the amino groups on the enzyme protein (Scheme 22). Although other nucleophiles on the enzyme can be used, such as thiol, this method has been applied in a limited number of cases.

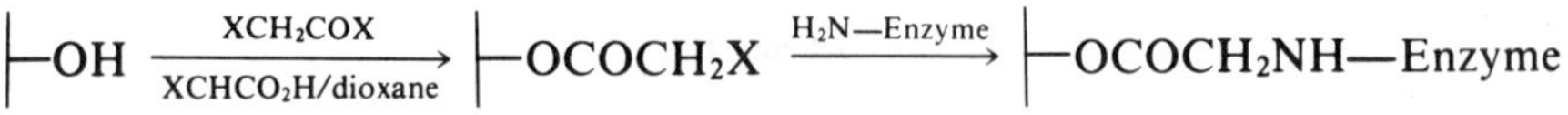

$$\vdash\!\text{OH} \xrightarrow[\text{XCHCO}_2\text{H/dioxane}]{\text{XCH}_2\text{COX}} \vdash\!\text{OCOCH}_2\text{X} \xrightarrow{\text{H}_2\text{N—Enzyme}} \vdash\!\text{OCOCH}_2\text{NH—Enzyme}$$

$$\text{X} = \text{—Cl,—Br,—I}$$

Scheme 22

Among these three halides, iodide confers the best reactivity stability properties to the carrier. The use of X = chloride is disadvantageous; although the activation is performed readily in organic solvents, the transfer into aqueous solution when the enzyme immobilization proceeds is accompanied by a competitive hydrolysis reaction of the chloride, which decreases the activity of carrier and introduces a large negatively charged carboxyl group. A substantial disadvantage of this technique overall is the formation between the carrier and the halogenoacetyl group of ester bonds that are sometimes insufficiently stable.

A more stable bond can be introduced by the reaction of a support matrix with dihalogenoketone or epichlorohydrin (Scheme 23).

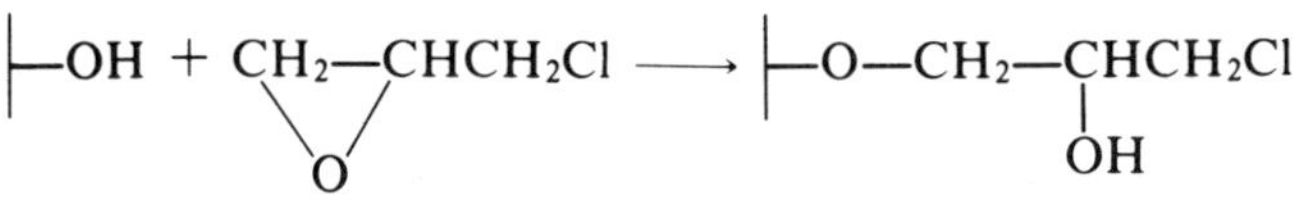

$$\vdash\!\text{OH} + \underset{\text{(epoxide, O)}}{\text{CH}_2\text{—CHCH}_2\text{Cl}} \longrightarrow \vdash\!\text{O—CH}_2\text{—}\underset{\text{OH}}{\underset{|}{\text{C}}}\text{HCH}_2\text{Cl}$$

Scheme 23

Another possibility for activation of carriers containing hydroxy groups involves the introduction of oxiranes, highly strained three-membered electrophilic cycles (Scheme 24), into the matrix. However, epithio groups are relatively stable and not directly suited for bonding enzymes.

The most frequently used are epoxide rings, particularly for their easy ac-

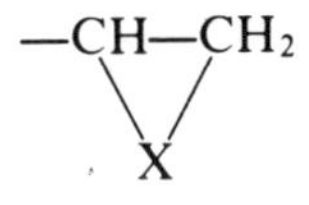

$$\text{—CH—CH}_2 \text{ (three-membered ring with X)}$$

$$\text{X} = \text{O, S, NH}$$

Scheme 24

cessibility. The activation of the carrier with epoxide compounds is shown in Scheme 25.

The reactivity of this epoxide derivative towards nucleophilic groups of the proteins has the known order of thiol > amino > hydroxyl. Nucleophilic attack on aliphatic hydroxyl groups requires a strongly basic medium (pH

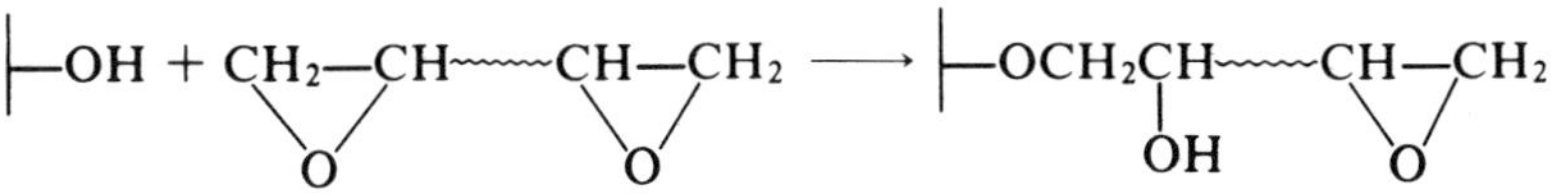

$$\vdash\!\text{OH} + \underset{\diagdown\text{O}\diagup}{\text{CH}_2\text{—CH}}\sim\!\sim\underset{\diagdown\text{O}\diagup}{\text{CH—CH}_2} \longrightarrow \vdash\!\text{OCH}_2\underset{\text{OH}}{\underset{|}{\text{CH}}}\sim\!\sim\underset{\diagdown\text{O}\diagup}{\text{CH—CH}_2}$$

Scheme 25

~11), whereas the amino groups react at a lower pH and thiol groups at a still lower pH value (pH 7). The reaction is depicted in Scheme 26. Aromatic hydroxy (L-tyrosine), guanidino (L-arginine), and imidazole (L-histidine) may also react.

The linkages of type C—N, C—O and C—S that are formed between the support and protein are extremely stable, and no charged groups arise dur-

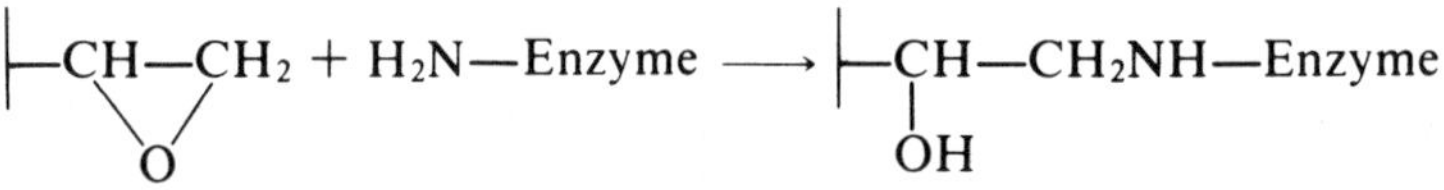

$$\vdash\!\underset{\diagdown\text{O}\diagup}{\text{CH—CH}_2} + \text{H}_2\text{N—Enzyme} \longrightarrow \vdash\!\underset{\text{OH}}{\underset{|}{\text{CH}}}\text{—CH}_2\text{NH—Enzyme}$$

Scheme 26

ing their realization, thus reducing or avoiding the nonspecific adsorptions (249). The reaction with proteins is slow, however, and the reaction time may have to be extended to several days or even weeks.

Vinylsulfonyl groups can be introduced into hydroxyl-containing polymers by treatment of the latter with divinylsulfone in alkali (250) (Scheme 27).

$$\vdash\!\text{OH} + \text{CH}_2\text{=CHSO}_2\text{CH=CH}_2 \longrightarrow \vdash\!\text{OCH}_2\text{CH}_2\text{SO}_2\text{CH=CH}_2$$

Scheme 27

The unavoidable simultaneous crosslinking improves the properties of original carrier (251). The nucleophilic attack of functional groups of enzyme follows the same order as for oxirane: thiol > amino > hydroxyl. They proceed at somewhat lower pH values, but also at somewhat higher reaction velocity. The main disadvantage is a slow release of the low molecular weight ligand at pH > 8. The interaction of the activated support with amino groups of enzyme can be seen in Scheme 28.

The hydroxyl groups of the carrier can also be more readily attacked by

$$\vdash OCH_2CH_2SO_2CH{=}CH_2 \xrightarrow{H_2N\text{—Enzyme}} \vdash OCH_2CH_2SO_2CH_2CH_2NH\text{—Enzyme}$$

Scheme 28

quinones (252). (Scheme 29). Enzymes are bound analogously to Scheme 28, that is, through their nucleophilic groups with high coupling yields. Actually, the coupling reaction can occur in the broad region of pH (3–10). However, undesirable side reactions are obtained, indicated by the color of the products.

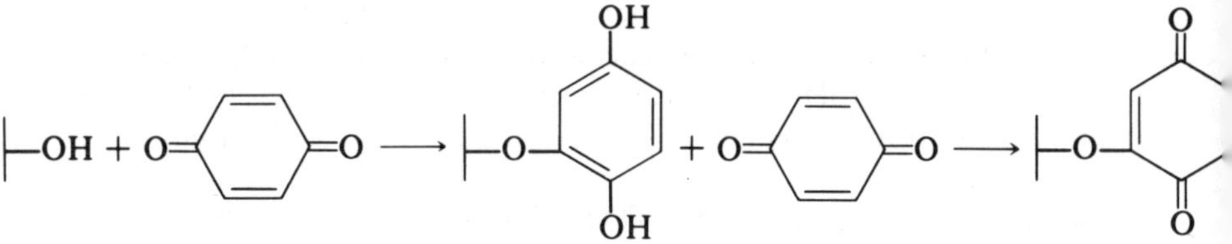

Scheme 29

Schiff's Base Formation. This method is based on the formation of a Schiff base (aldimine) linkage between carbonyl groups of the activated support and free amino groups on the enzyme. Functional aldehyde carriers are prepared from synthetic polymers (polyacryloylamino acetaldehyde (290) or copolymers of allyl alcohol and vanillin methacrylate (291) or polymeric dialdehyde obtained by oxidation of polysaccharides (292) with periodate or dimethylsulfoxide (Scheme 30).

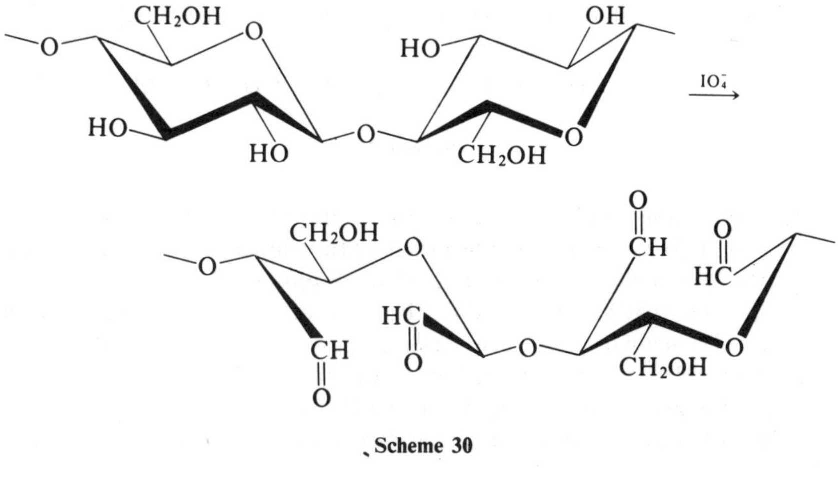

Scheme 30

The oxidation is rapid and proceeds in an aqueous solution at normal temperature; however, the close proximity of the pair of carbonyl groups can cause deformation of the enzyme molecule by two-point binding.

Supports containing primary amino groups have been activated with glutaraldehyde, initially introduced as a crosslinking reagent, to yield an active carbonyl derivative by formation of a Schiff base (Scheme 31).

$$\vdash NH_2 + OHC(CH_2)_3CHO \longrightarrow \vdash N{=}CH(CH_2)_3CHO$$

Scheme 31

Aldehyde derivatives react with amino groups on the protein to form a Schiff base linkage (Scheme 32). Sulfydryl and imidazole groups may undergo similar reactions. One disadvantage of this method is the reversibility of Schiff base formation in aqueous media, particularly at low pH values (293). This can be avoided, however, by hydrogenation of the imine bonds with $NaBH_4$ to stable alkylamino groups.

$$\vdash CHO + H_2N\text{—Enzyme} \longrightarrow \vdash CH{=}N\text{—Enzyme}$$

Scheme 32

Although the activation of amino carriers with glutaraldehyde is based on the idea that one aldehyde group reacts with the solid matrix forming a Schiff base and the other remains for the further reaction with the enzyme, the nature of the reaction with enzymes and amino carriers is not fully understood, as mentioned in Section 7.3.1. Enzyme proteins are bound irreversibly to the glutaraldehyde-treated carrier by a reaction presumably analogous to that occurring during intermolecular crosslinking with the same reagents (see Table 7.13).

Ugi Reaction. In 1962, Ugi (294) described a reaction involving four functional groups simultaneously, carboxylate, amine, aldehyde, and isocyanide, resulting in the formation of an *N*-substituted amide (Scheme 33). In the first step, a protonated Schiff base (immonium ion) (I) is produced by the reaction between the amino and the carbonyl compounds. This ion complexes with isocyanide compound and with a carboxylate ion; the three-component complex then undergoes an intramolecular rearrangement to give the *N*-substituted amide (II). The reaction proceeds in acidic medium and is mild enough to be used for enzyme immobilization.

This method was optimized for protein by Axén et al. (295, 296). It is versatile with respect to the location of the participating functional groups. Any

Table 7.13. Principal Examples of Immobilized Enzymes Prepared by Schiff Base Formation

Carrier	Enzyme	Reference
Cellulose	Penicillinase (EC 3.5.2.6)	253
	β-D-Galactosidase (EC 3.2.1.23)	173
	D-Glucose oxidase (EC 1.1.3.4)	173
	Trypsin (EC 3.4.21.4)	173
	Papain (EC 3.4.22.2)	173
	Pepsin A (EC 3.4.23.1)	173
	L-Glutamate dehydrogenase (EC 1.4.1.2)	254
	Urease (EC 3.5.1.5)	254
Agarose	Acid phosphatase (EC 3.1.3.2)	104
Nylon	Urate oxidase (EC 1.7.3.3)	255
	D-Glucose dehydrogenase (EC 1.1.1.47)	256
	L-Arginase (EC 3.5.3.1)	257, 258
	Lactate dehydrogenase (EC 1.1.1.27)	227
Nylon/polyethylene	D-Glucose dehydrogenase (EC 1.1.1.47)	259
Nylon/polyacrylonitrile	β-D-Fructofuranosidase (EC 3.2.1.26)	228
	Pepsin A (EC 3.4.23.1)	228
	Acid phosphatase (EC 3.1.3.2)	228
	Alkaline phosphatase (EC 3.1.3.1)	228
Polyacrylonitrile	Choline oxidase (EC 1.1.3.17)	260
	Glucoamylase (EC 3.2.1.3)	269
Chitosan	Malate dehydrogenase (EC 1.1.1.37)	261
	β-D-Galactosidase (EC 3.2.1.23)	262
	Pepsin A (EC 3.4.23.1)	263
	Alkaline phosphatase (EC 3.1.3.1)	263
	Trypsin (EC 3.4.21.4)	265
Chitin	Glucoamylase (EC 3.2.1.3)	266
	Urease (EC 3.5.1.5)	267
Polyacrylamide	β-D-Xylosidase (EC 3.2.1.37)	97
	Glucoamylase (EC 3.2.1.3)	274
Amberlite® XAD-7	Penicillin amidase (EC 3.5.1.11)	78, 268, 269
	Cholesterol oxidase (EC 1.1.3.6)	135
Glycidylmethacrylate	Glucoamylase (EC 3.2.1.3)	245
	Penicillin amidase (EC 3.5.1.11)	178
Styrene/maleic anhydride	Chymotrypin (EC 3.4.21.1)	177
Polyaminostyrene	Pyruvate decarboxylase (EC 4.1.1.1)	271
Controlled pore glass	Formate dehydrogenase (EC 1.2.1.2)	272
	Urease (EC 3.5.1.5)	273
	Glucoamylase (EC 3.2.1.3)	274, 275

Table 7.13. *(continued)*

Carrier	Enzyme	Reference
	Alcohol dehydrogenase (EC 1.1.1.1)	276
	Xanthine oxidase (EC 1.2.3.2)	200
	Acetylesterase (EC 3.1.1.6)	281
Controlled pore silica	Trypsin (EC 3.4.21.4)	277, 278
	Glucoamylase (EC 3.2.1.3)	274, 282, 283
	β-Amylase (EC 3.2.1.2)	279, 282
	Hydrogenase (EC 1.18.3.1)	183
	Dextransucrase (EC 2.4.1.5)	127, 280
	β-D-Xylosidase (EC 3.2.1.37)	97
	Catalase (EC 1.11.1.6)	283
	D-Glucose oxidase (EC 1.1.3.4)	283
	D-Glucose isomerase (EC 5.3.1.5)	283
Silica gel	Pepsin A (EC 3.4.23.1)	284
Sand	Trypsin (EC 3.4.21.4)	285
Brick	Acetylesterase (EC 3.1.1.6)	281
Alumina	Glucoamylase (EC 3.2.1.3)	147
Hornblende	Alcohol dehydrogenase (EC 1.1.1.1)	276
	Xanthine oxidase (EC 1.2.3.2)	200
	Urate oxidase (EC 1.7.3.3)	200
	Glucoamylase (EC 3.2.1.3)	158
Titania	Ribonuclease (EC 3.1.27.2)	286
Ferrite	Chymotrypsin (EC 3.4.21.1)	287
	Trypsin (EC 3.4.21.4)	287
	Ribonuclease (EC 3.2.1.17)	287
	Lysozyme (EC 3.2.1.17)	287
Attapulgite	D-Aminoacid oxidase (EC 1.4.3.3)	288
Glass beads	D-Glucose oxidase (EC 1.1.3.4)	289
	Catalase (EC 1.11.1.6)	289
	Hydrogenase (EC 1.18.3.1)	290

of the reacting groups (—CO_2H, —NH_2, —CHO, —NC) may be present in the original matrix. Direct coupling of enzymes can be achieved toward either amino or carboxyl groups on the enzyme using carboxyl and amino functional group carriers respectively. Also, enzymes can be insolubilized in an isocyanide containing carrier through their amino or carbonyl group in the presence of a water-soluble aldehyde.

The major disadvantage of this process is that the final product in the reaction contains four different functional groups. However, it is precisely these reactive groups that provide versatility to the reaction and allow one to attach enzymes by different ways. This process, therefore, can be used ad-

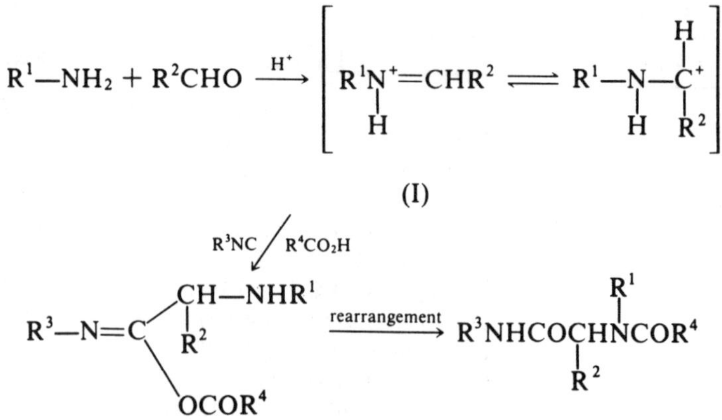

$$R^1{-}NH_2 + R^2CHO \xrightarrow{H^+} \left[R^1\underset{\large H}{\underset{|}{N^+}}{=}CHR^2 \rightleftharpoons R^1{-}\underset{\large H}{\underset{|}{N}}{-}\overset{\large H}{\overset{|}{\underset{\large R^2}{\underset{|}{C^+}}}} \right]$$

(I)

$$\downarrow R^3NC \,/\, R^4CO_2H$$

$$R^3{-}N{=}C\begin{cases} CH{-}NHR^1 \\ \;|\; R^2 \\ OCOR^4 \end{cases} \xrightarrow{\text{rearrangement}} R^3NHCO\underset{\large R^2}{\underset{|}{C}}H\overset{\large R^1}{\overset{|}{N}}COR^4$$

(II)

Scheme 33

vantageously when an enzyme cannot be immobilized to an active aldehyde carrier.

Amidination Reactions. Supports containing imidoester functional groups have been employed for immobilization of enzymes (297). Such carriers can be prepared by treating polyacrylonitrite with absolute ethanol and bubbling hydrogen chloride through the mixture to produce the corresponding imidoester derivative (Scheme 34).

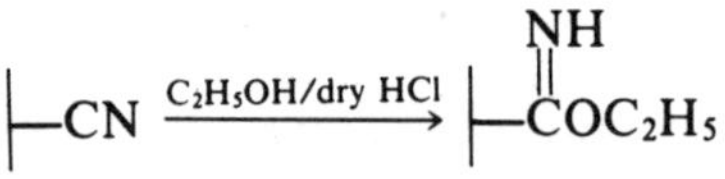

$$\vdash CN \xrightarrow{C_2H_5OH/\text{dry } HCl} \vdash \overset{\large NH}{\overset{\|}{C}}OC_2H_5$$

Scheme 34

Imidoesters are readily attacked by nucleophiles and react selectively (Scheme 35) with amino groups of enzyme protein at basic pH to yield amidines (1), which are stable in neutral or acidic solution, but are hydrolyzed slowly at high pH values (172).

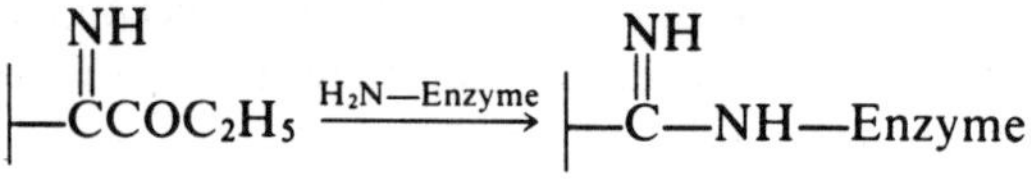

$$\vdash \overset{\large NH}{\overset{\|}{C}}COC_2H_5 \xrightarrow{H_2N\text{—Enzyme}} \vdash \overset{\large NH}{\overset{\|}{C}}\text{—NH—Enzyme}$$

(I)

Scheme 35

In another procedure, carriers containing amino groups can be activated with CNBr and reacted with enzyme amino groups, obtaining guanidine linkages between the amino groups of the enzymes and carriers (298) (Scheme 36).

$$\vdash NH_2 \xrightarrow{CNBr} \vdash NH{-}CN \xrightarrow{H_2N{-}Enzyme} \vdash NH{-}\overset{\overset{\displaystyle NH}{\|}}{C}{-}NH{-}Enzyme$$

Scheme 36

Thiol-Disulfide Interchange Reactions. This method is used for enzyme bonding via thiol groups of both carrier and enzyme, which works in an acidic medium (299–301). A first step involves the preparation of a mixed disulfide derivative of a thiol-containing carrier by treating the thiolated carrier with 2,2′-dipyridyldisulfide according to Scheme 37.

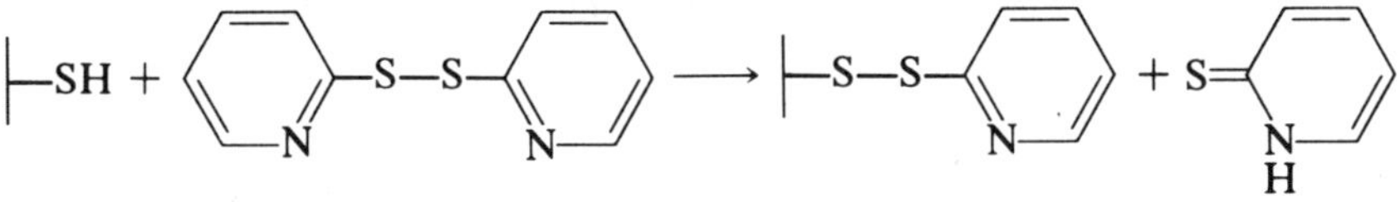

$$\vdash SH + Py{-}S{-}S{-}Py \longrightarrow \vdash S{-}S{-}Py + S{=}C_5H_4NH$$

(I)

Scheme 37

The coupling of enzyme through its thiol groups is accomplished with the liberation of 2-thiopyridone (I) (Scheme 38). However, the linkage between the enzyme and the support via —S—S— bonds is only stable under nonreducing conditions, and the coupling can be reversed with low molecular weight reagents. An alternative procedure with probably more versatility is

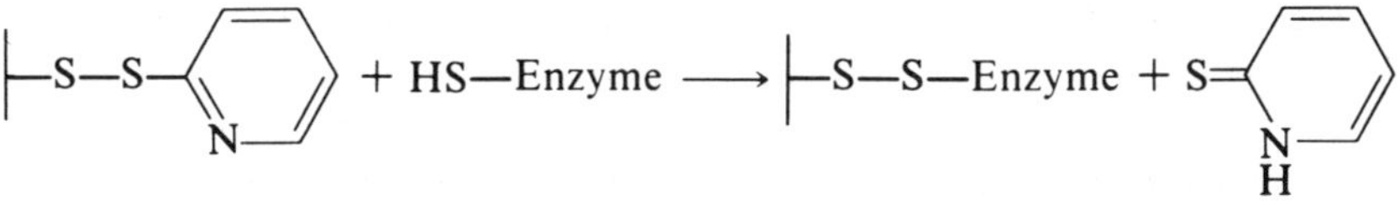

$$\vdash S{-}S{-}Py + HS{-}Enzyme \longrightarrow \vdash S{-}S{-}Enzyme + S{=}C_5H_4NH$$

(I)

Scheme 38

the reaction of the enzyme protein with *N*-acetylhomocysteinethiolactone (Scheme 39), which provides a spacer arm (302) and utilizes the more readily available amino group(s) of the enzyme. Again, reversibility is possible via oxidation–reduction to a third type of matrix, which may be as above or cellulose xanthate (302) (Scheme 39).

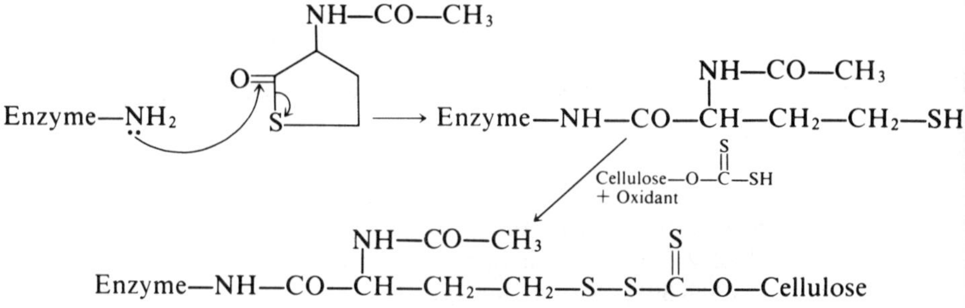

$$\text{Enzyme—}\ddot{\text{N}}\text{H}_2 + \text{(N-acetylhomocysteine thiolactone)} \longrightarrow \text{Enzyme—NH—CO—}\underset{}{\overset{\text{NH—CO—CH}_3}{\text{CH}}}\text{—CH}_2\text{—CH}_2\text{—SH}$$

$$\xrightarrow[\text{+ Oxidant}]{\text{Cellulose—O—C(=S)—SH}} \text{Enzyme—NH—CO—}\overset{\text{NH—CO—CH}_3}{\text{CH}}\text{—CH}_2\text{—CH}_2\text{—S—S—}\overset{\text{S}}{\overset{\|}{\text{C}}}\text{—O—Cellulose}$$

Scheme 39

Mercury-Enzyme Interactions. This method is based on the interaction of mercury derivative carriers and the thiol groups of the enzyme (303) (Scheme 40).

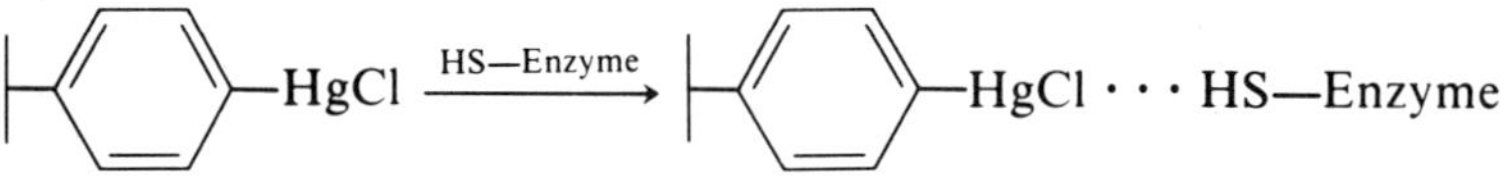

$$\text{|—C}_6\text{H}_4\text{—HgCl} \xrightarrow{\text{HS—Enzyme}} \text{|—C}_6\text{H}_4\text{—HgCl} \cdots \text{HS—Enzyme}$$

Scheme 40

The linkage between carrier and enzyme is formed at pH 4–8, but it is reversed by low molecular weight thiol reagents. This method can also be classified as physical adsorption, since the complex bond formed does not have a purely covalent character.

γ-Irradiation-Induced Coupling. This method has been used by Brandt and Anderson (304) to couple enzymes to agarose and dextran by γ-irradiation of the enzyme in the presence of the carrier. The radicals formed on both components then combine, giving rise to the covalent bond (Scheme 41).

This method is nonspecific, however, the immobilization yields are low and there is a high loss of activity arising from radiation damage. The advantage of this method is its independence of temperature and pH.

Matrices for Covalent Binding. This section surveys the structure and properties of various classes of materials commonly used as matrices for the immobilization of enzymes (Figure 7.2). As discussed in the introduction to the carrier binding method, the potential support for immobilization of enzymes

$$MH \xrightarrow{\gamma\text{-ray}} M\cdot + H\cdot$$

$$EH \xrightarrow{\gamma\text{-ray}} E\cdot + H\cdot$$

$$M\cdot + E\cdot \longrightarrow M - E \qquad \text{where } M = \text{matrix},\ E = \text{enzyme}$$

Scheme 41

ORGANIC CARRIERS: NATURAL POLYMERS—*Polysaccharides:* Cellulose
Starch
Dextran and agarose
Chitin and chitosan
Natural acidic polysaccharides
—*Proteins:* Collagen
Silk
—*Carbon materials*
SYNTHETIC POLYMERS—*Polystyrenes*
—*Polyacrylates:* Polyacrylate and polymethacrylate
Polymethacrylic acid anhydride
Polyacrylamides
Polyhydroxylalkyl methacrylate
Polyglycidyl methacrylate
Polyacrylonitrite
—*Maleic acid anhydride polymers*
—*Polypeptides*
—*Vinyl and allyl polymers*
—*Polyamides*
INORGANIC CARRIERS: MINERALS Attapulgite clays
Bentonite
Kieselguhr
Pumice stone
FABRICATED MATERIALS Controlled pore glass
(Hydrous) metal oxides
Metals
Nonporous glass

Fig. 7.2. Matrices for covalent immobilization of enzymes.

can be classified according to various criteria, for example, morphology and chemical composition. Carriers can be classified according to their chemical composition as organic and inorganic supports, and the former can be further classified into natural and synthetic matrices. The next paragraphs describe these different types of carriers and the type of reactions involved with enzymes.

Organic Carriers. Although some authors have suggested that in industrial processing inorganic carriers have many advantages over their organic counterparts (due to undesirable physical properties of the organic supports that lead to poor stability of enzyme preparations against physical, chemical, thermal, and microbial degradation), practically all commercially available immobilized enzymes are obtained with organic matrices. The reason for this is that there are a wide variety of functional reactive groups on organic polymers. In contrast, inorganic supports are practically always activated with substituted alkylsilane reagents and transition metals and thus it may be that there are limitations on them for covalent attachment.

Organic carriers are classified as either natural macromolecules, polymers (e.g., polysaccharides, proteins, and carbon) or synthetic polymers, and as having hydrophilic or hydrophobic characteristics. Hydrophilic synthetic polymers are usually preferred because of their physical and chemical characteristics and inertness to microbial degradation, although natural polymers based on agar possess some superior characteristics for use in immobilized enzyme systems.

NATURAL POLYMERS. *Polysaccharides.* Among polysaccharides, the most important supports to be used as matrices for immobilization of enzymes are cellulose, dextran, agarose, and starch. The first three are the most utilized and commercialized, while starch (amylose plus amylopectin), one of the most attractive matrices from a purely economic point of view, is considered the least suitable because of the ease with which it can be degradated by microbial attack. Recently several researchers (262, 265) have explored the applicability of polysaccharides that contain amino sugars, namely chitin and chitosan. Some acid polysaccharides, such as pectic acid and alginic acid, have also been used for immobilization of enzymes.

The hydroxyl groups of polysaccharides can be activated directly by introduction of an electrophilic group, reactive towards enzyme, into the matrix. However, the nucleophilic characters of dextran, agarose, cellulose, and similar potential enzyme carriers is so weak that pendant functional groups such as aliphatic or aromatic amino groups, carboxyl, or thiol groups have to be introduced as the activation for coupling or before activation for coupling (indirect coupling).

The different active derivatives that have been used for coupling of enzymes are listed in Table 7.14. The major advantage of the polysaccharide derivatives for immobilization of enzymes is the existence of residual hydroxyl groups, which provide a hydrophilic character, protecting the attached enzyme.

A. Cellulose. The polysaccharide cellulose was one of the first materials to be used as a matrix for the covalent binding of enzymes (305). Chemically, it is a vegetable fiber composed of β-D-glucopyranosyl units linked by $(1 \rightarrow 4)$ bonds and with additional intrachain interaction through hydrogen bonds. Although the hydroxy groups of polysaccharides are not reactive enough to form covalent bonds between the enzyme and the support without previous activation, cellulose undergoes all the reactions associated with polyhydric alcohols so that a wide range of active celluloses can be prepared.

Several types of chemically modified celluloses are commercially available (Table 7.15) and originally were used as ion-exchangers. These modified celluloses have permitted a wide range of covalent binding methods, on which the enzyme binds mainly through amino groups. Some of the most common cellulose derivatives used for immobilization of enzymes are described below.

Triazinyl-cellulose has been prepared (241, 242) by treatment of cellulose with 2,4,6-trichloro-*sym*-triazine or dichloro-*sym*-triazine derivatives. The halogenotriazinyl functional group is attached to the hydroxyl groups of cellulose through stable ether bonds and binds enzyme through enzyme amino groups.

Bromoacetyl-cellulose is formed (239, 240) by reacting the hydroxyl groups of cellulose with bromoacetylbromide, which is subsequently reacted with enzyme.

Cellulose trans-2,3-carbonate is prepared (220–222) by reacting cellulose or an *O*-substituted cellulose (e.g., methylcellulose, hydroxyethylcellulose, carboxymethylcellulose, or diethylaminoethylcellulose) with alkyl or arylchloroformate in anhydrous organic solvents. The enzyme is attached by a stable covalent bond via free amino groups, resulting in the opening of the carbonate ring and the formation of *N*-substituted carbonate.

Cellulose imidocarbonate is prepared (210) by reaction of cellulose with cyanogen bromide, although this reagent has been used mainly for covalent binding of enzymes to dextran and agarose. Reaction with enzyme is similar to the case of cellulose *trans*-2,3-carbonate.

Cellulose azide is one of the first polysaccharide derivatives used as support for enzyme immobilization. This derivative is obtained by activation of the carboxyl groups of modified cellulose (carboxymethylcellulose) successively with methanol, hydrazine, and nitrous acid. The binding of enzymes is mainly through their amino groups.

Table 7.14. Active Polysaccharides for Enzyme Immobilization

Nucleophilic Group of Matrix	Activating Reagent	Active Derivative
Direct Coupling		
$\vdash OH$ $\vdash OH$	CNBr	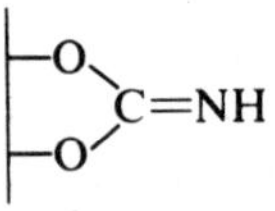 $\vdash O \diagdown C{=}NH$ $\vdash O \diagup$
$\vdash OH$ $\vdash OH$	$ClCO_2C_2H_5$	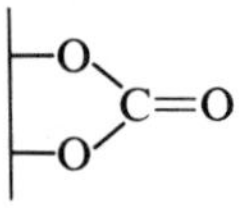 $\vdash O \diagdown C{=}O$ $\vdash O \diagup$
$\vdash OH$	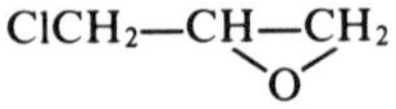 $ClCH_2{-}CH{-}CH_2$ (epoxide, O bridging CH–CH₂)	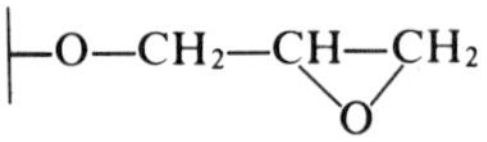 $\vdash O{-}CH_2{-}CH{-}CH_2$ (epoxide, O bridging CH–CH₂)
$-OH$	$CH_2{=}CH{-}SO_2{-}CH{=}CH_2$	$\vdash O{-}CH_2CH_2SO_2CH{=}CH_2$
$\vdash OH$	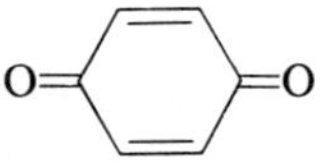 $O{=}C_6H_4{=}O$ (p-benzoquinone)	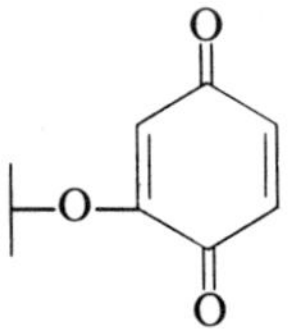 $\vdash O{-}C_6H_3O_2$ (2-substituted p-benzoquinone)
$\vdash OH$	$BrCOCH_2Br$	$\vdash OCOCH_2Br$
$\vdash OH$	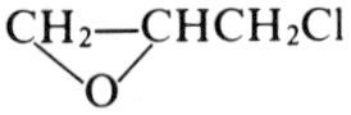 $CH_2{-}CHCH_2Cl$ (epoxide, O bridging CH₂–CH)	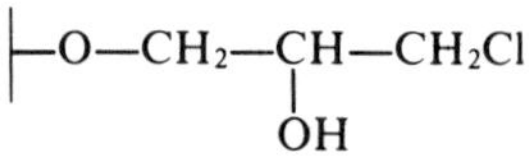 $\vdash O{-}CH_2{-}CH(OH){-}CH_2Cl$
$\vdash OH$	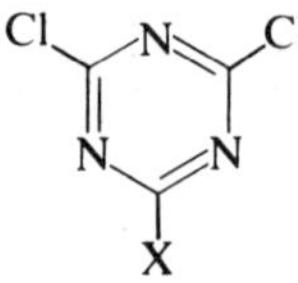 2,4-dichloro-6-X-1,3,5-triazine (Cl, Cl, X on triazine ring N)	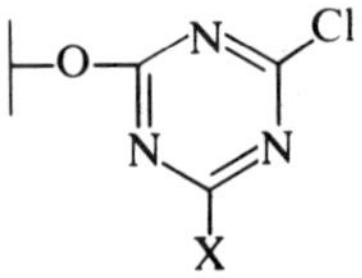 $\vdash O$–triazine(Cl)(X)
$\vdash CH_2OH$	IO_4^-	$\vdash CHO$
Indirect Coupling		
$\vdash NH_2$ $\vdash NH_2$	Cl_2CS $OHC{-}(CH_2)_3{-}CHO$	$\vdash NCS$ $\vdash N{=}CH{-}(CH_2)_3{-}CHO$
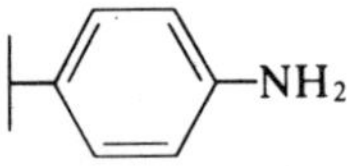 $\vdash C_6H_4{-}NH_2$	Cl_2CS	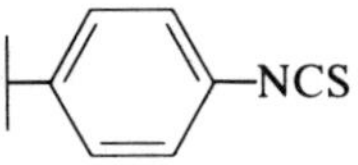 $\vdash C_6H_4{-}NCS$
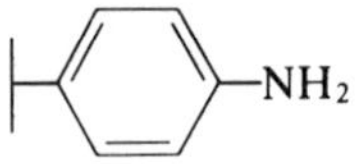 $\vdash C_6H_4{-}NH_2$	Cl_2CO	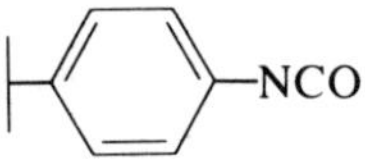 $\vdash C_6H_4{-}NCO$

Table 7.14. (*continued*)

Nucleophilic Group of Matrix	Activating Reagent	Active Derivative
$\vdash C_6H_4-NH_2$	$NaNO_2$,HCl	$\vdash C_6H_4-N_2^+Cl^-$
$\vdash CONHNH_2$	$NaNO_2$	$\vdash CON_3$
$\vdash COOH$	$R-N=C=N-R^1$, H^+	$\vdash COO-C(NHR)=\overset{+}{N}HR'$
$\vdash COOH$	CH_3OH, NH_2NH_2, HNO_2	$\vdash CON_3$
$\vdash SH$	$C_5H_4N-S-S-C_5H_4N$	$\vdash S-S-C_5H_4N$

Cellulose carbonyl is obtained from an aliphatic amino derivative of cellulose (e.g., aminoethylcellulose), by reaction with glutaraldehyde (158).

Diazo-cellulose is prepared (173) from various arylamino derivatives of cellulose by reaction with nitrous acid and binds enzyme through L-tyrosyl or L-histidyl residues.

Isocyanato-cellulose can be obtained from the azide derivative by treatment with acid chloride, or more commonly, from alkyl or arylamino derivatives of cellulose by treatment with phosgene (210).

Cellulose is currently employed as a carrier to a lesser extent than dextran and agarose. The reasons are its biodegradability by microorganisms and the forms in which it is commercially available. It is available in the form of fibers of various sizes that are relatively nonporous. However, macroporous cellulose in the form of beads was more recently obtained (306, 307), with good permeation, mechanical properties, and higher capacities of binding, and it is competing well with agarose, although agarose is already well established in the market. (see Table 7.15).

B. Starch. Starch is the least suitable polysaccharide carrier for enzyme immobilization owing to its ease of degradability by microorganisms; it is rarely employed as a carrier compared with other polysaccharides. Starch has been mainly used in the derivatized form of dialdehyde, which is obtained by oxidation of its chain with periodic acid or its salts (308). Although the dialdehyde starch is active for immobilization, the presence of the pair of carbonyl groups in a close proximity can cause, as already stated, deformation of the enzyme molecule. Other active conjugates have been prepared by first

Table 7.15. Commercially Available Modified Celluloses

4-Aminobenzyl-
Aminoethyl- (AE-)
Diethylaminoethyl- (DEAE-)
Carboxymethyl- (CM-)
Epichlorohydrin triethanolamine- (ECTEOLA-)
Oxy-
Phospho-
Sulfoethyl-
Triethylaminoethyl- (TEAE-)

condensing the dialdehyde starch with alkyl or aryldiamines to produce amino derivatives, and then activating the derivatives by treatment with glutaraldehyde or nitrous acid.

C. Dextran and Agarose. Porous polysaccharides based on dextran and agar possessing molecular sieving properties were originally developed as supports for gel filtration chromatography and are available commercially, in grades characterized by their water regain and molecular exclusion limits (309). Because of these properties they have gained wide acceptance in enzyme immobilization.

Dextran is a linear, water-soluble polysaccharide composed of (1 → 6)-linked α-D-glucopyranonyl units. The commercially available dextran gels (Sephadex) are prepared by crosslinking the water-soluble polymer with epichlorohydrin (310). By control of linear dextran and the degree of crosslinking, a range of gels of well-defined water regain and molecular sieving properties has been obtained (311).

Agar and agarose (fractionated agar of low charge density) are D-galactans that form rigid gels when their solutions are cooled to temperatures below about 45°C. Agar and agarose in bead form have been used by Polsan (312) and Hjertein (313) as support media in zone electrophoresis and as molecular sieves. Somewhat later, agarose began to be used as support for immobilized enzymes.

Agarose is obtained from natural agar by dispergation of a 6–10% hot-water solution of agar in an organic solvent in the presence of a suitable emulsifier; the particles that result from cooling below 45°C are separated.

Agarose gels are mechanically more stable and have greater pore size than other gels. Unlike other polysaccharides, agar and agarose are resistant to microbial degradation, as agar-degrading enzymes (agarases) have been found only in certain microorganisms living on seaweed. Despite their superior properties as macroporous hydrophilic and nonadsorbing supports, however, they suffer from several disadvantages: Agarose gels cannot be heat sterilized; they disintegrate in strong alkaline solutions and in organic solvents; even at neutral pH values there is a possibility of solubilization; and they must be stored in wet form since they shrink irreversibly on drying. But improvement of the mechanical properties and chemical resistance may be achieved by crosslinking with bifunctional reagents (e.g., epichlorohydrin), which eliminates most of these deficiencies.

Despite its superior properties, maeroporous crosslinked agarose is only used for analytical purposes because it is a very expensive matrix and unsuitable economically for industrialization. Inexpensive crude agar, which contains negatively charged sulfate and carboxy groups that may undesirably affect its applicability, can be treated with alkali to remove these groups, and thus the material may be rendered suitable for industrial application as a matrix for immobilized enzymes.

Crosslinked dextran and agarose have to be activated for use as supports of immobilized enzymes. One of the most commonly and extensively used procedures is the cyanogen bromide (CNBr) activation method in which a cyclic *trans*-2,3-imidocarbonate derivative is obtained (314, 315). The method involving the formation of cyclic *trans*-2,3-carbonate derivative (205–207), analogous to the cyclic *trans*-2,3-imidocarbonate structure successfully applied to cellulose and synthetic polymers, is unsuitable for agarose gels, owing to structured limitations and to the harsh treatment of the support with ethyl chloroformate in anhydrous organic solvents.

Some other covalent coupling methods of immobilization of enzymes that are used with these polysaccharides involve the use of cyanuric chloride for activation of hydroxyl groups (243), carbodiimides for carboxyl derivatives (80, 97), benzoquinone for hydroxyl groups (252), epichlorohydrin for hydroxyl groups (316), and 2-pyridine disulfide (299) for thiol derivatives. (See Chapter 1 for a description of the new sulfonyl chloride or "Mosbach" activation procedure).

D. Chitin and Chitosan. In an attempt to develop inexpensive supports for enzyme immobilization, and in conjunction with the increasing need to treat beverages, food processing wastes, and agricultural products or byproducts, chitin and its derivatives have been recently used as support for immobilization of enzymes.

Chitin is an abundant by-product of the fishing (crab, shrimps, and prawn) and fermentation (citric acid and pharmaceuticals) industries. Chitin is a polysaccharide composed of (1 $\rightarrow$ 4)-linked 2-acetamido-2-deoxy-β-D-glucopyranosyl residues. About one of every six residues is not acetylated, whereas in chitosan essentially all the residues are not acetylated (2-amino-2-deoxy-D-glucose). Water-soluble chitosan can be obtained from chitin by deacetylation in concentrated sodium hydroxide solutions.

Enzymes can be bound to chitin by adsorption but this is usually followed by crosslinking with glutaraldehyde (265). Covalent linkage onto a carbonyl derivative, obtained by a previous treatment with glutaraldehyde, is also possible (266).

Chitosan in a soluble form has been mixed with an enzyme solution; by adding a multifunctional crosslinking agent, usually glutaraldehyde, a gel is formed, which is then reacted with a reducing agent to form immobilized enzymes in granular form (246–248). In other techniques (262–264), reprecipitated chitosan obtained from chitosan acetate, or epichlorohydrin-crosslinked chitosan, is first treated with glutaraldehyde to obtain a carbonyl derivative, and this links covalently the enzyme.

All the authors who used chitin and chitosan have reported favorable characteristics of their derivatives and qualified them as attractive supports for enzyme immobilization. Most research workers have only used glutaraldehyde to bind enzymes (261–267), and the use of some other reagents and techniques on these polysaccharides are as yet unexplored.

E. Natural Acidic Polysaccharides. Natural acidic polysaccharides can be used as supports for immobilization on enzymes. Examples of suitable acidic polysaccharides are pectin, pectic acid, and alginic acid. Pectin is a polymer predominantly of D-galacturonic acid in which some of the carboxyl groups are esterified. Pectic acid is the free acid obtained by saponification of pectin. Alginic acid is isolated from algae and consists of a polymer predominantly of D-mannuronic acid and L-guluronic acid.

Enzymes can be covalently linked to these acidic polysaccharides either by reaction with the polysaccharide carboxyl groups or by reaction with the polysaccharides after activation of their carboxyl groups with carbodiimides to form an *O*-acylurea derivative (317), or with hydrazine followed by a treatment with nitrous acid to form an azide derivative (296).

Proteins. Proteins that are not enzymes have been used in several methods of immobilization, such as complexation of enzymes with collagen (318), entrapment of enzymes within a gelatin matrix (319), and crosslinking of enzymes together with one or more inactive proteins with or without a preexisting support. The use of some inactive proteins as supports, such as collagen or silk *per se,* has also recently been reported (193, 320).

Collagen is the most abundant protein constituent of higher vertebrates. It is easily isolated from many biological sources and can be reconstituted into various forms without losing its native structure. This, in conjunction with its ready availability from a large number of biological species, from fish to cattle, makes it a useful and inexpensive carrier.

Collagen has several advantages as an enzyme carrier. Its hydrophilicity facilitates the accessibility of enzyme to the binding sites. Its proteinaceous nature makes possible a strong physical adsorption of enzymes. Its open internal structure provides a high concentration of binding sites. Its fibrous structure and high swellability in aqueous solutions also contribute to its popularity as an enzyme support matrix.

Recently, Coulet and Gautheron and their coworkers (191–194) described a covalent binding technique in which an azide derivative of collagen was used for the preparation of several different collagen–enzyme complexes. Because of its biological origin, collagen may be particularly useful in biomedical applications.

Natural or raw silk is composed of two proteins, the water-insoluble fibroin and the soluble sericin, and it is used as an enzyme support in the form of industrial woven silk. Grasset et al. (320) reported some physical and chemical properties of silk that make it useful as a support, namely, thermal stability, resistance to acid and alkaline attacks from pH 3 to 8, microbial resistance, availability in a useful woven form that is easy to manipulate, has low compaction, and is resistant to abrasion.

The use of silk as support requires activation which, in turn, requires knowledge of the theoretically available functional groups. Although several processes of activation based on the presence of the functional groups are available, these authors (299) made a choice, based on the amino acid composition of the fibroin protein of silk, of enzyme fixation on the carboxyl groups after activation through the azide derivative. This derivative was obtained by acidic methylation, in a first step, producing methylated silk (which itself serves as support for adsorption of enzymes). The methylated silk is successively reacted with hydrazine and then nitrous acid to form an active azide silk.

Carbon Materials. Carbon materials are attractive as immobilized enzyme supports because of their reasonable price and mechanical strength, and because they are obtainable in several forms, including porous structures. Carbon materials are already used in numerous medical, food, and fine-chemical processing operations, which are also the potential fields of application of immobilized enzymes.

One of the most-used carbon materials for immobilization of enzymes is activated carbon, which is a highly porous carbonaceous material prepared by carbonizing and activating organic substances of mainly vegetable origins. Its pore structure has been reported (321) as tridispersed—containing micropores (0–0.2 mm radii), transitional pores (2.5–50 nm), and macropores (50–2000 nm). Activation of carbon is performed with O_2, chemicals, CO_2, or steam at high temperatures; carboxyl groups, phenolic hydroxyl groups, and other oxidized carbon atoms can be detected by its surfaces.

Almost all of the research on enzyme immobilization on activated carbon has dealt with adsorption, but recently methods for immobilization of enzymes by covalent coupling have been reported by Cho and Bailey (110, 230). They show that enzymes may be immobilized in a stable form on activated carbon, using carbodiimides for activation of its carboxyl groups. Other derivatives were also proposed, such as amino carbon, obtained by vigorous oxidation and nitration followed by reduction with a sodium dithionite solution. The amino carbon is subsequently activated with thiophosgene to give an isothiocyanate derivative. Reactive, aminosilanized carbon can be prepared by treatment of the isothiocyanate with a solution of 3-aminopropyltriethoxysilane. Diazotized derivatives can also be prepared from the amino and isothiocyanate derivatives.

SYNTHETIC POLYMERS Among the large number of supports available for the immobilization of enzymes, the synthetic carriers possess a great field of application. This is due to their physical and chemical characteristics and the ease of preparing various polymers for a particular enzyme and application, and with regard to the degree of porosity and chemical composition that can be achieved by either copolymerization of very different available monomers or by chemical modification of performed polymers. One of the major advantages of this type of carrier is inertness to microbial attack.

The main synthetic polymers that can be used in immobilization of enzymes are polystyrene, polyacrylates, polyvinyls, and copolymers based on maleic anhydride and ethylene, or styrene, polyamide, polyaldehyde, and polypeptide structures. The most important of these synthetic polymers are described in the following sections.

Polystyrene. Historically, polystyrene was the first synthetic polymer to be used for the immobilization of proteins (11–17). The main intermediate derivative is the polyaminostyrene, which is obtained from polystyrene by nitration and reduction. This derivative was used in the early research with synthetic carriers. It had been activated by diazotization (11) or by reaction with thiophosgene (13–15) or glutaraldehyde (271).

Despite the rather high concentration of reactive groups on these polymers, the bound protein and coupling yields for these polymers are generally poor because of the inherent hydrophobicity of the polymer. This shortcoming is bypassed by copolymerization with hydrophilic monomers, such as acrylic acid, methacrylic acid and others (322, 323). Recently there has been renewed interest in polystyrene as a support, mainly because of its low cost and ready availability.

Manecke and coworkers (322, 323) reported satisfactory results in terms of bound protein and retention of activity with nitrated terpolymers of 3-fluorostyrene, methacrylic acid, and divinylbenzene (I); of 3-isothiocyanato styrene, acrylic acid, and divinylbenzene (II); and of methacrylic acid-3-fluoroanilide, methacrylic acid, and divinylbenzene (III) (Scheme 42), in which divinylbenzene is used in lesser amounts as the crosslinking agent. These carriers are immediately able to react with proteins with a retention of enzymic activity of 30–50%.

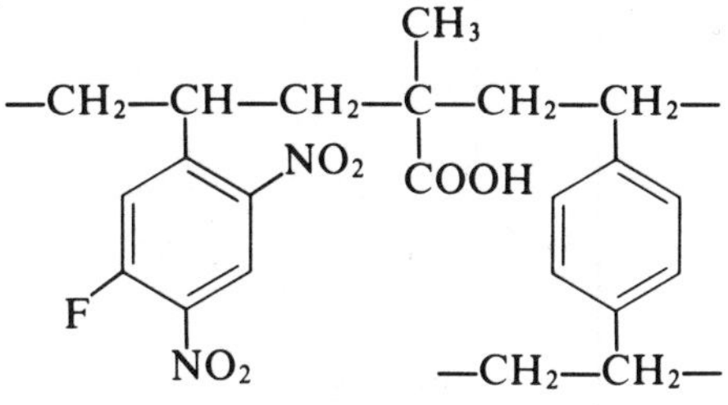

(I)

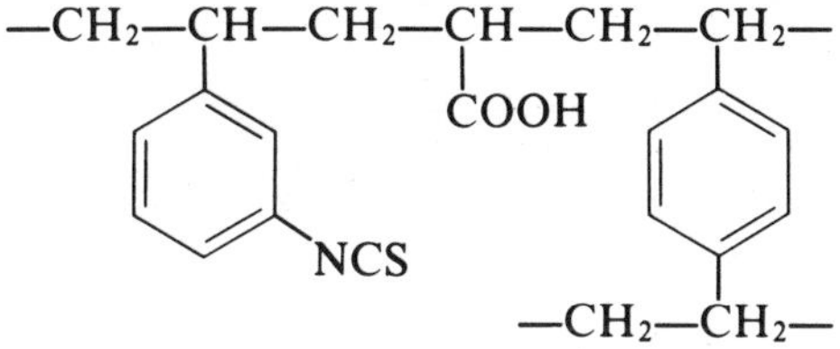

(II)

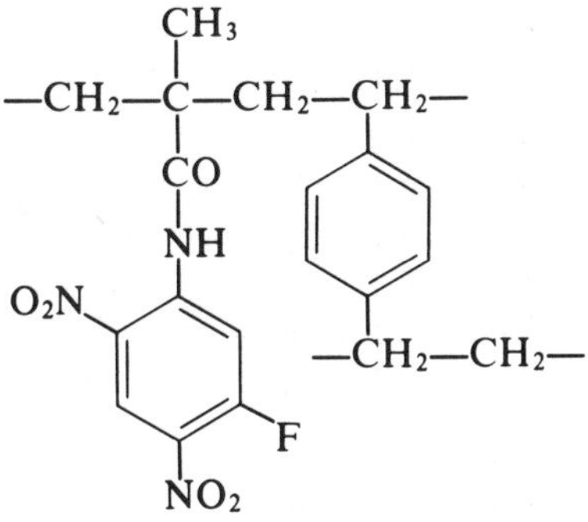

(III)

Scheme 42

Polyacrylate Types. Acrylic polymers are among the most used synthetic polymers in the field of immobilization. They have been used in other immobilization methods, particularly entrapment.

These carriers generally exhibit a fair chemical and mechanical stability and are microbially inert; their major advantage is the ease of preparation of various derivatives, some of them allowing direct coupling of enzymes. Many of them, such as polyacrylamides, poly(hydroxyalkylmethacrylate) and derivatives, are available from commercial sources. A number of polyacrylates that have been applied in the covalent coupling of enzymes are listed in Table 7.16.

Table 7.16. Polyacrylate Derivatives for Enzyme Immobilization

Functional Group of Original Matrix	Activating Reagent	Active Matrix
$\vdash CO_2H$ (Acrylic acid, Methacrylic acid)	$R—N{=}C{=}N—R^1$, H^+	$\vdash COO—C(NHR){=}\overset{+}{N}HR^1$
$\vdash CO\text{—}O\text{—}CO \dashv$ (cyclic anhydride) (Methacrylic acid anhydride)	—	—
$\vdash CONH_2$ (Acrylamide)	$OHC(CH_2)_3CHO$	$\vdash CON{=}CH(CH_2)_3CHO$
$\vdash CONH_2$ (Acrylamide)	H_2NNH_2, HNO_2	$\vdash CON_3$
$\vdash CONH_2$ (Acrylamide)	$H_2N(CH_2)_2NH_2$, $OHC(CH_2)_3CHO$	$\vdash CONH(CH_2)_2N{=}CH(CH_2)_3Cl$

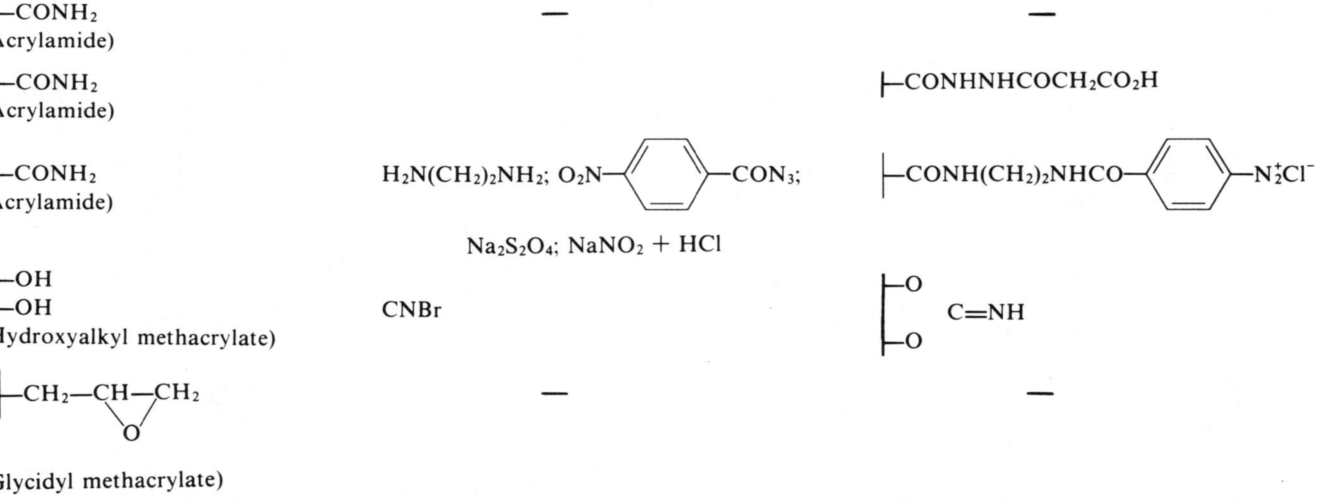

⊢$CONH_2$ (Acrylamide)	—	—
⊢$CONH_2$ (Acrylamide)		⊢$CONHNHCOCH_2CO_2H$
⊢$CONH_2$ (Acrylamide)	$H_2N(CH_2)_2NH_2$; $O_2N-C_6H_4-CON_3$; $Na_2S_2O_4$; $NaNO_2 + HCl$	⊢$CONH(CH_2)_2NHCO-C_6H_4-N_2^+Cl^-$
⊢OH ⊢OH (Hydroxyalkyl methacrylate)	CNBr	⊢O, ⊢O $C{=}NH$
⊢$CH_2-CH-CH_2$ (epoxide, O) (Glycidyl methacrylate)	—	—

A. Polyacrylates and Polymethacrylates. These polymers are obtained by polymerization of acrylic acid and methacrylic acid. They have been used mainly in copolymers with other organic compounds, namely with acrylamide, to form a more hydrophobic matrix or to prepare matrices with negative charges. These types of carriers are usually activated with a soluble carbodiimide (229).

A polymer containing free aldehydo and carboxyl groups, obtained by copolymerization of a mixture of acrylic acid and acrolein, was referred recently by Leemputten (324) for attachment of enzymes via carbonyl groups of the matrix. However this support is easily dissolved at ambient temperature in solutions with a pH value of 8 and is only completely insoluble at a pH below 4.5.

B. Polymethacrylic Acid Anhydride. This polymer of methacrylic acid anhydride was used by Conte and Lehmann (325) to effect direct linkage between enzyme and the carrier. However, as it was already referred, the anhydride ring is opened by the immobilization of enzyme, and the carboxyl groups that are formed do not react with enzyme. This imparts undesirable ionogenic properties to the matrix, which in turn may unfavorably affect the activity of the enzyme.

C. Polyacrylamides. These polymers and their derivatives are among the synthetic matrices most often employed for the immobilization of enzymes. One of the reasons for their popularity is their chemical structure, the primary amide groups affording a definite hydrophilic character to the polymers. The polyacrylamide or copolymers of acrylamides have low physical adsorption towards the enzyme molecules, which means that any immobilized enzyme is chemically bound and not easily removed by subsequent reaction conditions.

Because of the solubility of the linear polymers in water, to use these polymers as matrices they have to be insolubilized by crosslinking with bifunctional compounds. The procedure for the formation of insoluble gel networks is identical to that employed for the preparation of gel commonly used for disk electrophoresis. This method is based on the free-radical polymerization of acrylamide in an aqueous solution containing crosslinking agent, usually, *N,N′*-methylene bisacrylamide (bis). Polyacrylamide can also be obtained by irradiating a frozen monomer solution, using as irradiation sources ^{60}Co or ^{137}Cs. One advantage of this technique is that several shapes of polymer can be obtained, by varying the type of the container in which the polymerization is conducted.

Polyacrylamide can be activated by several of the methods described in the discussion of general coupling reactions for covalent binding. Some of these chemically modified polyacrylamides are listed in Table 7.16. Polyacrylamide can be activated with glutaraldehyde (97, 274) to produce a carbonyl derivative that can couple enzymes directly via their amino groups.

In other procedures (326), polyacrylamide is treated with hydrazine or 1,2-diaminoethane to give the corresponding acylhydrazide or aminoethyl derivatives (Scheme 43). The acyl hydrazide can be activated with nitrous

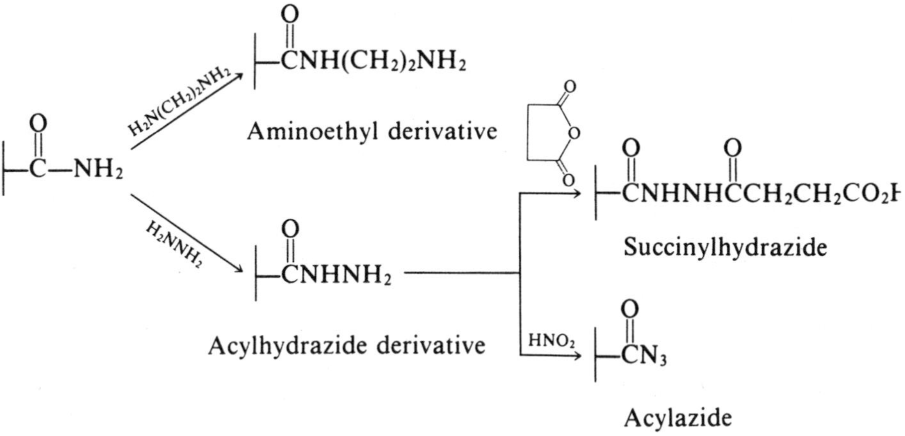

$-C(=O)-NH_2$ —($H_2N(CH_2)_2NH_2$)→ $-C(=O)NH(CH_2)_2NH_2$

Aminoethyl derivative

$-C(=O)-NH_2$ —(H_2NNH_2)→ $-C(=O)NHNH_2$

Acylhydrazide derivative

$-C(=O)NHNH_2$ —(succinic anhydride)→ $-C(=O)NHNHC(=O)CH_2CH_2CO_2H$

Succinylhydrazide

$-C(=O)NHNH_2$ —(HNO_2)→ $-C(=O)N_3$

Acylazide

Scheme 43

acid to form acylazide or may also be converted with succinic anhydride to a derivative containing a carboxyl group, that is, a succinyl hydrazide derivative (Scheme 43). Succinyl hydrazide and aminoethyl derivatives couple to protein amino groups and carboxyl respectively in the presence of water-soluble carbodiimides. The aminoethyl derivative can be treated with glutaraldehyde and used for coupling of enzymes, via amino groups, or further modified with 4-nitrobenzoylchloride and reduced with sodium dithionite to the 4-aminobenzoylaminoethyl derivative, which couples enzymes via diazotization.

Polyacrylamide and some of its chemically modified derivatives are commercially available (Bio-Gel®, Bio-Rad Laboratories Ltd). In addition to direct chemical modifications of preformed polyacrylamides, other chemically modified polyacrylamides are obtained by copolymerization of acrylamide and crosslinked with other monomers. Such copolymers, based on acrylamide have been prepared and are also commercially available (Enzacryls®, Koch Light Laboratories Ltd). (See Table 7.17).

Enzacryl AA contains aromatic amino groups and is obtained by copolymerization of acrylamide with 4-amino acrylamide; it can be activated either

Table 7.17. Polyacrylamide and Derivatives

Name	Functional Groups of Original Matrices
Acrylamide (Bio-Gel® P)	$-CH(CONH_2)-CH_2-CH(CONH_2)-$
Enzacryl® AA	$-CH(CONH_2)-CH_2-CH(CONH-C_6H_4-NH_2)-$
Enzacryl® AH	$-CH(CONH_2)-CH_2-CH(CONHNH_2)-$
Enzacryl® Polythiol	$-CH(CONH_2)-CH_2-CH(CONHCH(COOH)-CH_2SH)-$
Enzacryl® Polythiolactone	$-CH(CONH_2)-CH_2-CH(CONHCH-CH_2)-$, with CH and CH_2 joined through $CO-S$ (ring)
Enzacryl® Polyacetal	$-CH(CO-NH-CH_2-CH-(OCH_3)_2)-CH_2-CH(CO-NH-CH_2-CH-(OCH_3)_2)-$

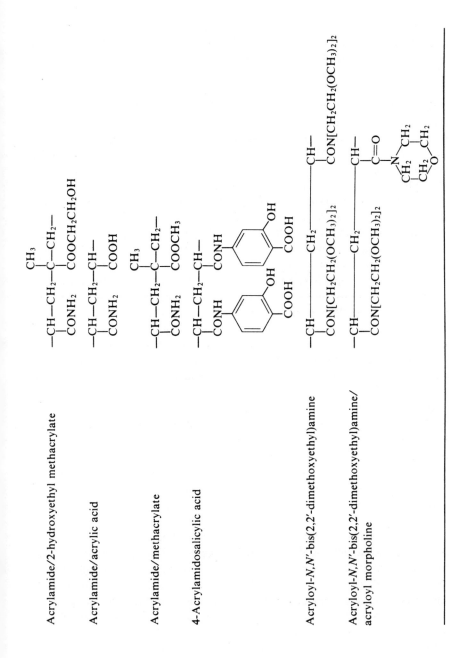

Acrylamide/2-hydroxyethyl methacrylate	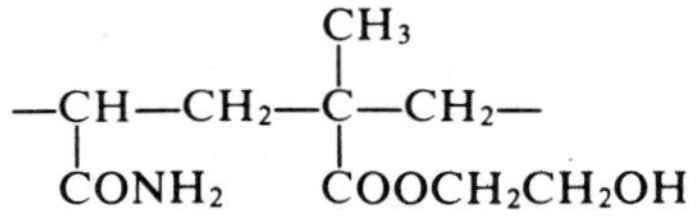 $-CH(CONH_2)-CH_2-C(CH_3)(COOCH_2CH_2OH)-CH_2-$
Acrylamide/acrylic acid	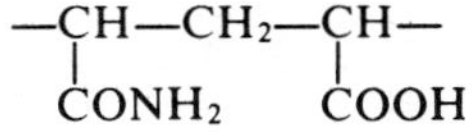 $-CH(CONH_2)-CH_2-CH(COOH)-$
Acrylamide/methacrylate	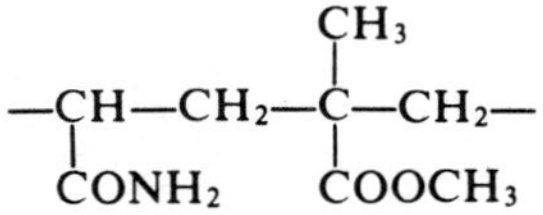 $-CH(CONH_2)-CH_2-C(CH_3)(COOCH_3)-CH_2-$
4-Acrylamidosalicylic acid	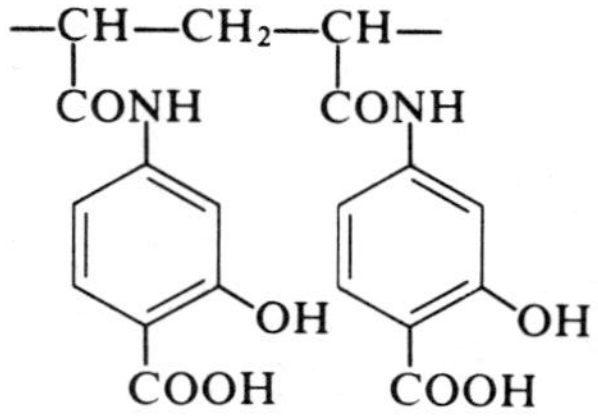 $-CH(CONH-C_6H_3(OH)(COOH))-CH_2-CH(CONH-C_6H_3(OH)(COOH))-$
Acryloyl-*N,N′*-bis(2,2′-dimethoxyethyl)amine	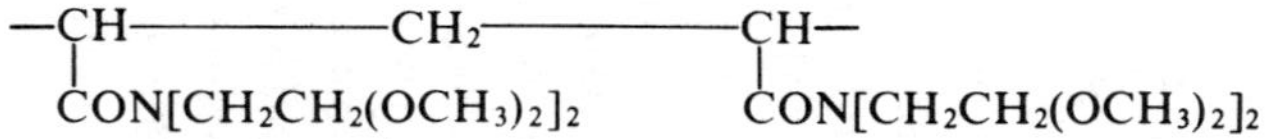 $-CH(CON[CH_2CH_2(OCH_3)_2]_2)-CH_2-CH(CON[CH_2CH_2(OCH_3)_2]_2)-$
Acryloyl-*N,N′*-bis(2,2′-dimethoxyethyl)amine/ acryloyl morpholine	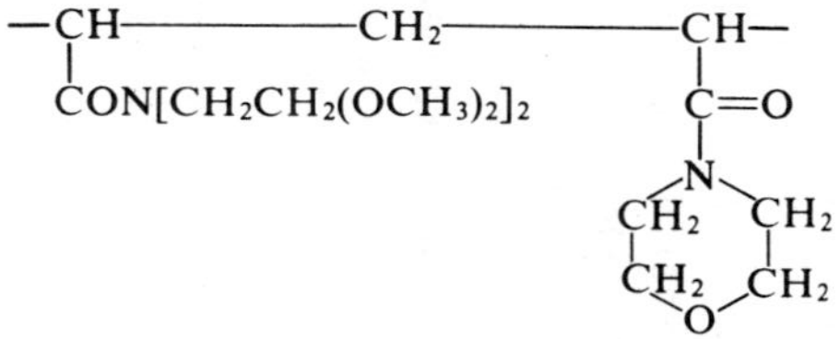 $-CH(CON[CH_2CH_2(OCH_3)_2]_2)-CH_2-CH(C(=O)-N(CH_2CH_2)_2O)-$

by nitrous acid or thiophosgene (97, 127, 174). Enzacryl AH is obtained with acrylolhydrazides and can be activated via the azide method (97, 127, 174). By using *N*-acryloyl-4-carboxy-2,2-dimethyl thiazolidine as comonomer, Enzacryl Polythiol is obtained after hydrolysis of the thiazolidine; it can be used for the reversible attachment of enzymes via their thiol groups. Enzacryl Polythiol can be modified by action of carbodiimide to a reactive thiolactone derivative (Enzacryl Polyactone). This latter derivative reacts with free amino groups and hydroxy groups of L-serine and L-threonine through the thiolactone ring. Also Enzacryls Polythiol and Polythiolactone can be prepared by copolymerization using as comonomers *N*-acryloylcysteine and *N*-acryloylcysteine thiolactone respectively.

Other acrylamide based copolymers have been prepared for immobilization. Some of these are (i) copolymers of acrylamide and 2-hydroxyethylmethacrylate, containing amide groups and hydroxyl groups that can be activated with cyanogen bromide (by a process analogous to that used for polysaccharides), (ii) "substrate attracting" carriers such as copolymers of acrylamide and acrylic acid for immobilization of enzymes via carbodiimide activation, which are useful for reactions with positively charged substrates, or copolymers of acrylamide and methylacrylate, which are useful for more hydrophobic substrates. Other useful *N*-substituted amides have been used for enzyme immobilization. For example, poly(4- and 5-acrylamidosalicyclic acids) are effective in binding enzymes via metal ion chelation (327). Another example is the crosslinked poly(acryloylaminoacetaldehyde dimethylacetal) (Enzacryl Polyacetal). This carrier is activated for enzyme coupling by acid hydrolysis, giving an aldehydo group (328).

Closely related products are obtained by polymerization of acrylo-*N,N'*-bis(2,2-dimethoxyethyl) amine (329) and by copolymerization of this monomer with *N*-acrylomorpholine (330). The applications of these matrices for the immobilization of enzymes are possible via the azide method, using acid hydrolysis in the presence of tartaric dihydrazide, followed by treatment with nitrous acid.

D. Hydroxyalkyl Methacrylate. Hydroxyalkylmethacrylates are hydrophilic organic carriers used in chromatography and in enzyme immobilization. They correspond, on account of their content of hydroxyl groups, to polysaccharide carriers but possess better mechanical properties and biological resistance. Macroporous carriers are prepared in the form of spherical particles by radical suspension copolymerization of hydroxyalkyl methacrylate and a crosslinking agent. Most widely used are the copolymers of 2-hydroxyethyl methacrylate and ethylene dimethacrylate (Spheron®, Lachema).

Like other organic polymeric carriers, chemically modified derivatives may be obtained by (i) direct chemical modification of preformed polymer or (ii) copolymerization with monomers containing reactive groups or precursors of functional groups. By the first method, virtually all formerly described procedures for polysaccharides can be used and conducted under much more drastic reaction conditions than the natural carriers. The main activation procedure involves cyanogen bromide for direct coupling of enzyme, although other derivatives such as amino (331, 332) and carboxyl (331, 332) have been reported.

The ternary copolymerization of hydroxylalkylmethacrylate, crosslinking agent, and a monomer with active group or precursor of these same groups can be conducted. One monomer, 4-acetaminophenylethoxymethacrylate, has been reported (333) as precursor of several reactive intermediates. Other copolymers with an oxirane (epoxy) group for direct coupling of enzymes can be obtained with glycidylmethacrylate.

E. Glycidyl Methacrylate. Carriers containing oxirane (epoxide) groups, enabling both the direct coupling of enzyme and also numerous modifications formerly described, can be obtained by polymerization of a monomer, a suitable example of which is glycidyl methacrylate. The oxirane group is neutral and does not give rise to charge on reactions with enzyme, but the resultant enzyme–carrier bond is very stable.

Copolymerization of glycidyl methacrylate with other monomers has been reported. Krämmer (334) obtained a copolymer with acrylamide as matrix for immobilization, but it was necessary to store this copolymer in the cold when not in use, since there was considerable risk of reaction of amide groups with oxirane groups. Švec and Kalal and colleagues (245, 335) reported with very promising results the preparation of similar copolymers using ethylenedimethacylate as a monomer.

F. Acrylonitrile. Zaborsky (336) has used an insoluble acrylonitrile-based polymer as a carrier for covalent binding of enzymes. Polyacrylonitrile is activated by hydrogen chloride gas in cooled dry methanol, giving an imidoester derivative that covalently binds enzymes by the reaction of imidoester functional groups with enzyme amino groups.

Maleic Anhydride–Based Polymers. One of the more commonly employed supports for the immobilization of enzymes is the copolymer of maleic anhydride and ethylene. The use of this carrier was first reported by Levin et al. (337, 338) in 1964.

Direct coupling of enzymes (337, 338) to this support results in an immobilized enzyme preparation with physically adsorbed enzyme owing to its highly negatively charged groups. Neutralization of the negative charge can be accomplished by the addition of diamines during the immobilization process, which also acts as a crosslinking agent to produce a highly water-insoluble enzyme preparation.

Maleic anhydride copolymers can also serve as starting materials for further chemical modifications. These chemical modifications are listed in Scheme 44.

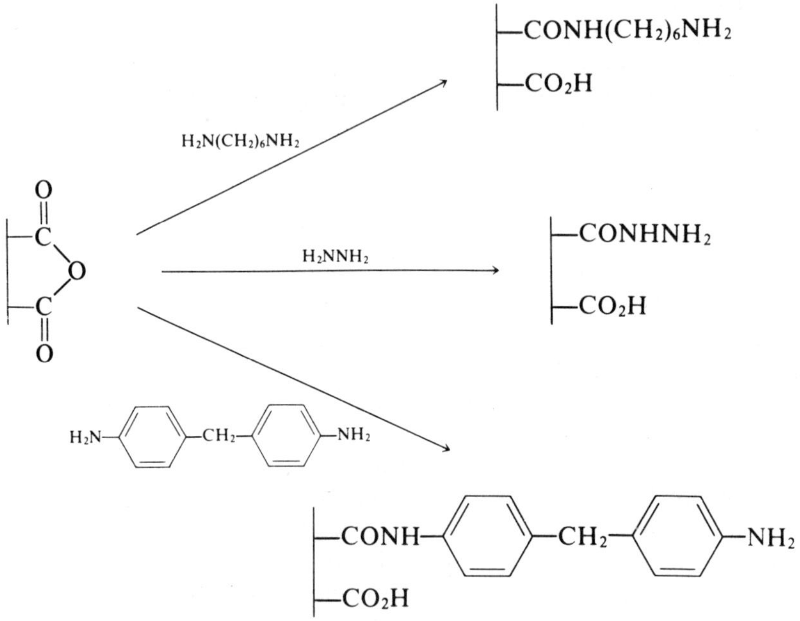

$H_2N(CH_2)_6NH_2$ → —$CONH(CH_2)_6NH_2$, —CO_2H

H_2NNH_2 → —$CONHNH_2$, —CO_2H

H_2N—C_6H_4—CH_2—C_6H_4—NH_2 → —CONH—C_6H_4—CH_2—C_6H_4—NH_2, —CO_2H

Scheme 44

Peptides and Polypeptides. Water-insoluble polypeptides have been used widely for the immobilization of many different enzymes. The polypeptides most commonly used are copolymers of L-leucine and 4-amino-DL-phenylalanine, which is produced by copolymerization of α,*N*-carboxy-4-amino,*N*-benzyloxycarbonyl-DL-phenylalanine anhydride and *N*-carboxy-L-leucine anhydride in dioxane (16); see Scheme 45. This copolymer is activated with nitrous acid to give the diazonium salt.

Using similar techniques of condensation, immobilization of enzymes has been performed by the copolymerization of enzymes with *N*-carboxy amino-acid anhydride. This immobilized enzyme preparation was first described by Stahmann and Becker (339), who showed that the polymerization of reactive

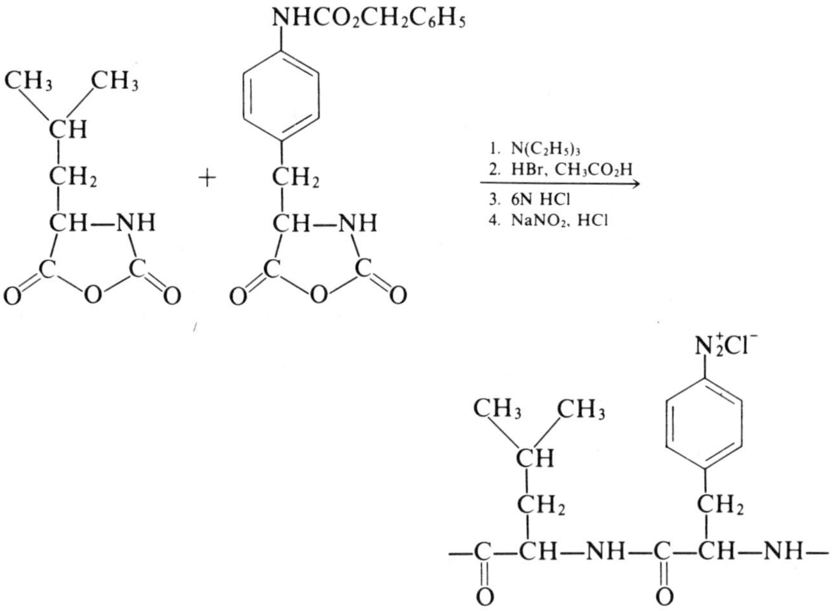

$NHCO_2CH_2C_6H_5$

CH_3 CH_3 CH CH_2 CH—NH C O C O O + CH_2 CH—NH C O C O O

1. $N(C_2H_5)_3$
2. HBr, CH_3CO_2H
3. 6N HCl
4. $NaNO_2$, HCl

$N_2^+Cl^-$

CH_3 CH_3 CH CH_2 CH_2

—C—CH—NH—C—CH—NH—
O O

Scheme 45

anhydride is initiated by condensation with free amino groups of the enzyme to yield the corresponding polypeptidyl derivative. However, these derivatives are only slightly modified proteins and remain water soluble.

Vinyl and Allyl Polymers. Manecke and Schlunsen (340) have recently reported that carriers with neutral and hydrophilic characteristics can be obtained by chemical modification of polyvinylalcohol, polyallylalcohol, or vinyl ether copolymers. Reactive carriers based on polyvinylalcohol are obtained by crosslinking soluble polyvinylalcohol with terephthaldehyde (341) and by reaction of this crosslinked polymer with 2-(3-aminophenyl)-1,3-dioxolane followed by diazotization (342) or with 2,4,6-trichloro-*sym*-triazine (342) (Scheme 46).

Polyvinylethers are prepared by cationic copolymerization, initiated with BF_3, of vinyl ethers with divinyl ether. The most important carrier is obtained using 2-vinyloxyethyl 4-nitro-benzoate as reactive monomer (Scheme 47). Nitro groups of the resulting carrier are first reduced to form an arylamine derivative which is then diazotized.

Poly(allyl alcohol) is obtained by reduction of polyacrolein with sodium

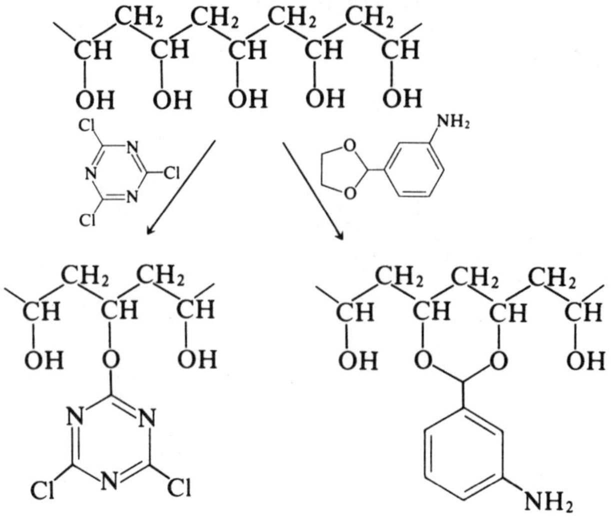

Scheme 46

borohydride (343). It can be chemically modified to yield reactive forms with isocyanato compounds, which serve as crosslinking agent (340). The insoluble polymers can couple directly enzyme to the isocyanato groups. Treatment of poly(allyl alcohol) with ethyl chloroformate produces poly(allyl cyclic carbonate) (Scheme 48), which reacts with enzyme (344, 345) in a way analogous to cellulose *trans*-2,3-carbonate.

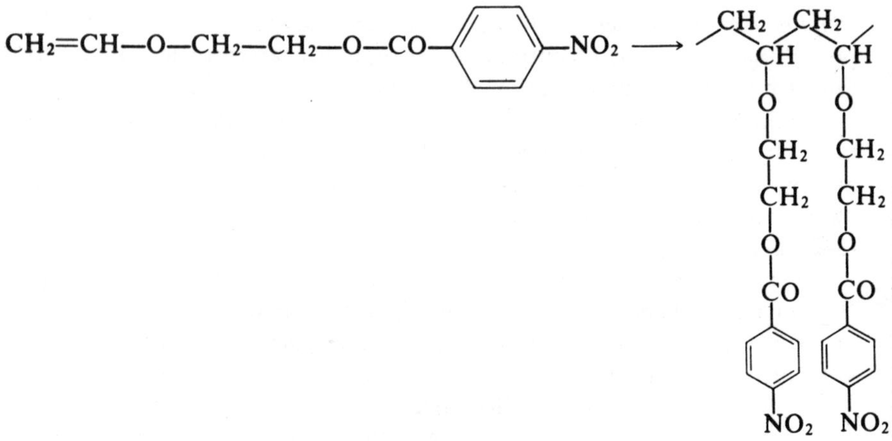

$CH_2{=}CH{-}O{-}CH_2{-}CH_2{-}O{-}CO{-}C_6H_4{-}NO_2 \longrightarrow$

Scheme 47

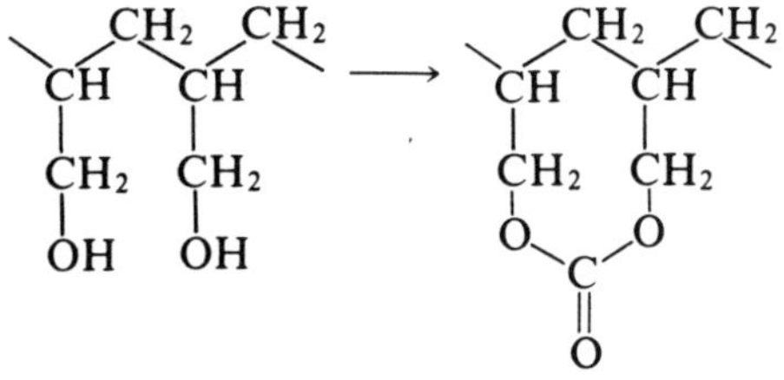

```
  CH2  CH2             CH2  CH2
\/   \/   \     →    \/   \/   \
CH   CH              CH   CH
|    |               |    |
CH2  CH2             CH2  CH2
|    |               |    |
OH   OH              O    O
                      \  /
                       C
                       ||
                       O
```

Scheme 48

Polyamides. Synthetic polyamides, known as nylons, are a family of condensation polymers of α,ω-dicarboxylic acids (or their diacid chlorides) and α,ω-diamines.

Several types of nylon (e.g., nylon 6 and nylon 6,6), differing only in the number of methylene groups in the repeating alkane segments, are available in a variety of physical forms, such as fibers, hollow fibers, foils, membranes, powders, and tubes. These types of supports have several advantages, such as mechanical strength, biological resistance, and relative hydrophilicity. For covalent immobilization of enzymes, however, the chemical inertness of the polyamide backbone leaves only the terminal carboxyl and amino groups as possible reactive groups. Thus, to increase the binding capacity of nylon, it is necessary to treat the support by one of three approaches: (i) partial depolymerization of amide bonds by acid hydrolysis, followed by activation of either resultant amino or carboxyl groups (Scheme 49); (ii) introduction of reactive centers via *O*-alkylation of the support, yielding the reactive imidate salt (I) of the nylon without necessitating depolymerization of the support (Scheme 50); (iii) introduction of reaction site centers via *N*-alkylation of the support. By a mild acid depolymerization, carboxyl and amino pairs are generated on the surface of the nylon structure, and condensing reactions involving an aldehyde and an isocyanide in conjunction with those groups may occur, yielding the polyisonitrile nylon (346) (I) (Scheme 51).

As mentioned above in the discussion of the Ugi reaction, enzymes can be coupled to the cyanide group of polyisonitrite nylon by a Ugi reaction, carried out in the presence of water-soluble aldehyde and the enzyme, which supplies either the amino or carboxyl compound.

When the immobilization of enzymes via Ugi reaction is undesirable, owing to sensitivity of the enzyme to aldehydes, the isocyanide group on nylon can be modified to other functional groups (Scheme 52).

Major disadvantages of this type of carrier in general are that its chemical structure must be modified and that the surface areas of all activated forms are relatively small.

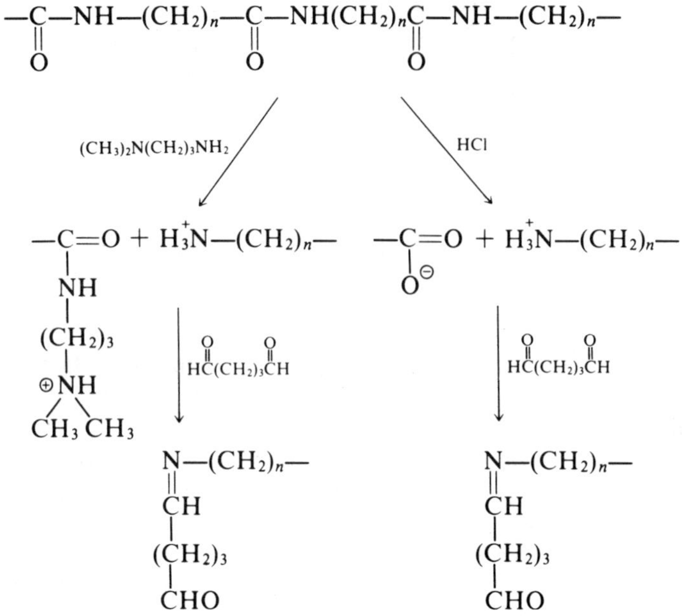

$$-\underset{\underset{O}{\|}}{C}-NH-(CH_2)_n-\underset{\underset{O}{\|}}{C}-NH(CH_2)_n\underset{\underset{O}{\|}}{C}-NH-(CH_2)_n-$$

$(CH_3)_2N(CH_2)_3NH_2$ (left branch); HCl (right branch)

$$-\underset{\underset{\underset{\underset{\underset{CH_3\ CH_3}{\oplus NH}}{(CH_2)_3}}{NH}}{|}}{C}{=}O + H_3\overset{+}{N}-(CH_2)_n- \qquad -\underset{O^{\ominus}}{C}{=}O + H_3\overset{+}{N}-(CH_2)_n-$$

$\downarrow$ $\overset{O}{\overset{\|}{H C}}(CH_2)_3\overset{O}{\overset{\|}{C}}H$ (both branches)

$$\underset{\underset{\underset{CHO}{(CH_2)_3}}{CH}}{\overset{\|}{N}}-(CH_2)_n- \qquad \underset{\underset{\underset{CHO}{(CH_2)_3}}{CH}}{\overset{\|}{N}}-(CH_2)_n-$$

Scheme 49

Inorganic Carriers. Although some of the earliest works on enzyme immobilization were with inorganic matrices (3), organic polymers were used more widely in the subsequent stages of immobilized enzyme development. This situation probably contributed to the interest in enzyme structure, the chemical modification of enzymes, and the ease with which enzymes can be linked to organic materials (11–19).

Research work is not generally limited by economical considerations in terms of materials and techniques chosen, but in industrial application the

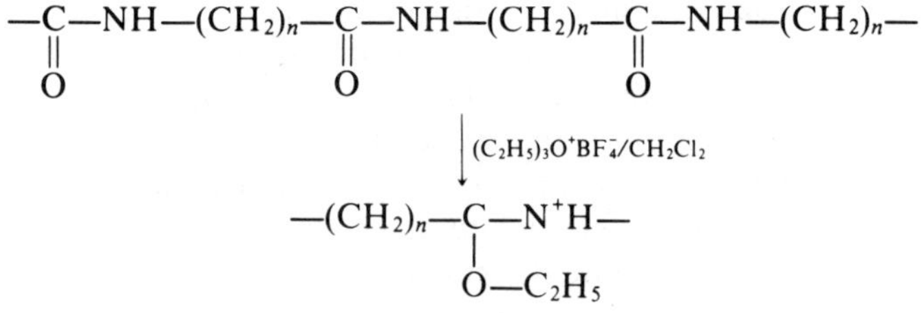

$$-\underset{\underset{O}{\|}}{C}-NH-(CH_2)_n-\underset{\underset{O}{\|}}{C}-NH-(CH_2)_n-\underset{\underset{O}{\|}}{C}-NH-(CH_2)_n-$$

$$\Big\downarrow (C_2H_5)_3O^+BF_4^-/CH_2Cl_2$$

$$-(CH_2)_n-\underset{O-C_2H_5}{\underset{|}{C}}-N^+H-$$

(I)

Scheme 50

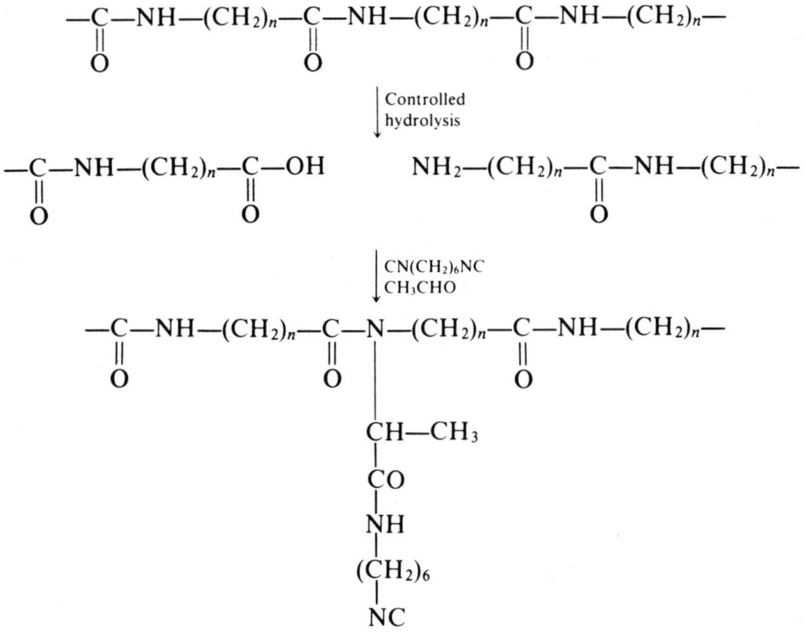

$$-\underset{\overset{\|}{O}}{C}-NH-(CH_2)_n-\underset{\overset{\|}{O}}{C}-NH-(CH_2)_n-\underset{\overset{\|}{O}}{C}-NH-(CH_2)_n-$$

$$\Big\downarrow \text{Controlled hydrolysis}$$

$$-\underset{\overset{\|}{O}}{C}-NH-(CH_2)_n-\underset{\overset{\|}{O}}{C}-OH \qquad NH_2-(CH_2)_n-\underset{\overset{\|}{O}}{C}-NH-(CH_2)_n-$$

$$\Big\downarrow CN(CH_2)_6NC,\ CH_3CHO$$

$$-\underset{\overset{\|}{O}}{C}-NH-(CH_2)_n-\underset{\overset{\|}{O}}{C}-\underset{\substack{|\\ CH-CH_3\\ |\\ CO\\ |\\ NH\\ |\\ (CH_2)_6\\ |\\ NC}}{N}-(CH_2)_n-\underset{\overset{\|}{O}}{C}-NH-(CH_2)_n-$$

(I)

Scheme 51

$$\vdash NC \xrightarrow[CH_3CHO,\ CH_3CO_2H]{H_2N-C_6H_4-CH_2-C_6H_4-NH_2}$$

Polyisonitrile nylon

$$\vdash NHCO\underset{}{\overset{\overset{CH_3}{|}}{C}H}-\overset{\overset{\overset{CH_3}{|}}{CO}}{|}{N}-C_6H_4-CH_2-C_6H_4-NH_2$$

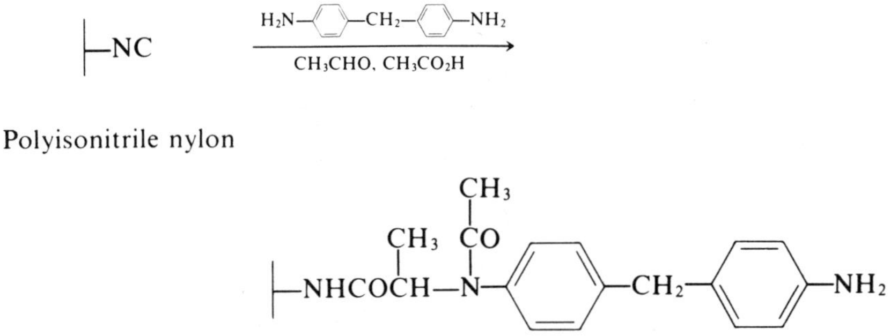

Polyaminoaryl nylon

Scheme 52

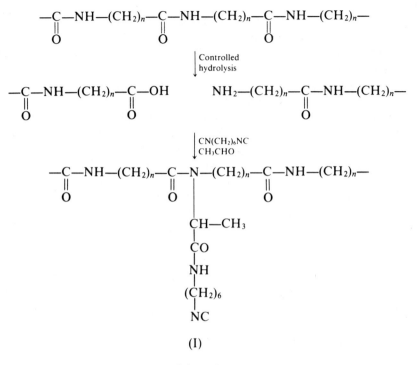

cost of process is of vital importance. Hence, reuse of materials and reagents may have great importance, as may compactness of plant, flexibility and ease of control, nontoxicity of the matrix, microbial resistance, and other characteristics.

Many inorganic materials, because of their physical properties, are very suitable for industrial use and offer several advantages over their organic counterparts: high mechanical strength, thermal stability, resistance to organic solvents and microbial attack, easy handling, excellent shelf life, and easy regenerability by a simple pyrolysis process. Moreover, inorganic materials do not change in structure over wide ranges of pH, pressure, and temperature.

The inorganic support materials may be classified as porous and nonporous. Porous carriers have a high surface area per unit weight, which allows a high enzyme loading and thus makes them ideal for large industrial reactors. A major disadvantage of porous carriers, however, is that most of the surface available for enzyme coupling is internal surface. Porous supports must have an internal morphology that allows not only the bonding of enzyme but also an easy accessibility to substrate molecule in order to minimize diffusional constraints. On the other hand, enzyme, bound on the internal surface, is protected from the turbulent and harsh external environment.

Porous carriers can have a broad or controlled pore distribution. With a broad pore carrier, only a limited number of the pores will be large enough to accommodate both enzyme and substrate and thus only a small portion of the total surface is effectively usable. To overcome this problem, controlled pore supports with a wide range of pore sizes are available. One major factor in choosing a porous carrier is the inverse relationship between the pore size and the surface area (Table 7.18), which makes it possible to optimize the pore diameter for a particular enzyme system.

Table 7.18. Relationships of Pore Size to Surface Area for Porous Particles[a]

Pore diameter (nm)	Surface Area (Pore Volume = 1.0 mL/g)
7.5	249
12.5	149
17.5	107
24.0	78
37.0	50
70.0	27
125.0	15
200.0	9

[a] From H. H. Weetall, in *Methods of Enzymology*, Vol. 44 (K. Mosbach, Ed.), Academic Press, New York (1976), p. 134.

CARRIER TYPES. Controlled pore glass (CPG) (see Chapter 11) is a macroporous high-silica glass prepared from borosilicate glass. This type of glass undergoes a phase separation by heat treatment (500–700°C), and subsequent acid leaching of the borate component leaves a porous structure of very high silica content. However, CPG has proved to be unstable during prolonged dynamic operation because of the continuous leaching of silica from the particles, especially in alkaline environments. To increase the durability of this support, Tomb and Weetall (347) applied a coating of zirconia to the glass surface. This coating is produced by vacuum-impregnating the porous glass with a solution of zirconium salt and subsequent calcination to give an oxide layer of about 4% by weight of the particles.

In spite of this, CPG is unsuitable for industrial purposes owing to its high price (348). Other carriers, controlled pore ceramics, have been developed much more cheaply (349). The main ceramics are: silica, alumina, and titania. One factor demanding attention in the choice of one of those carriers is their chemical durability at several pH solutions. Thus titania and alumina are suitable for use in an immobilized enzyme system at alkaline solutions and silica is better suited to acidic solutions.

Naturally occuring porous minerals, such as kieselguhr (351), attapulgite clays (350, 351), pumice stone (352), and bentonite (353) also can be used as carriers, but they have a broad pore distribution.

Nonporous carriers have the major disadvantage of a low surface area and therefore a low available surface for enzyme coupling. Although this problem may be partly overcome by using fine particles, the particles are hard to remove from the solution and in continuous reactors lead to high pressure drops. One advantage of employing nonporous carriers is the reduction or elimination of diffusional constraints with respect to the substrate, although this advantage does not effectively compete with the advantages of the porous carriers.

Recently there has been an increasing interest in pure metals or metal oxides that exhibit ferromagnetic properties, such as magnetic iron oxide (354), nickel (355), and stainless steel (356), which can be easily separated from reaction mixture and in conjunction with their density could make them useful in fluidized bed enzyme reactors. Another nonporous inorganic support of interest for economic reasons is sand (357).

Inorganic materials that have been used as carriers for enzyme immobilization are listed in Table 7.19.

SPECIALIZED COUPLING REACTIONS. As already stated, a major advantage of using inorganic supports is their relative inertness; conversely, this makes difficult the immobilization of enzymes by covalent coupling. Practically all of the work on covalent binding of enzymes to inorganic supports involves the use of trialkoxysilane derivatives containing an organic func-

Table 7.19. Inorganic Supports for Enzyme Immobilization

Minerals	Fabricated Materials
Bentonite	Controlled pore glass
Attapulgite	Porous titania
Pumice stone	Porous alumina
Horneblende	Porous silica
Sand	Alumina catalyst
Diatomaceous earth	Silochrome
	Aluminosilicate
	Iron oxide
	Nickel/nickel oxide
	Stainless steel

tional group (silanization method) (358). These reagents have proved to be the most successful compounds for activation of inorganic carriers for generating a covalent bond between the enzyme and the carrier. Coupling of these reagents to the carrier takes place presumably by displacement of the alkoxy residues on the silane, by hydroxyl groups on the oxidized surface of the inorganic support to form a metal-O-Si linkage (358) (Scheme 53).

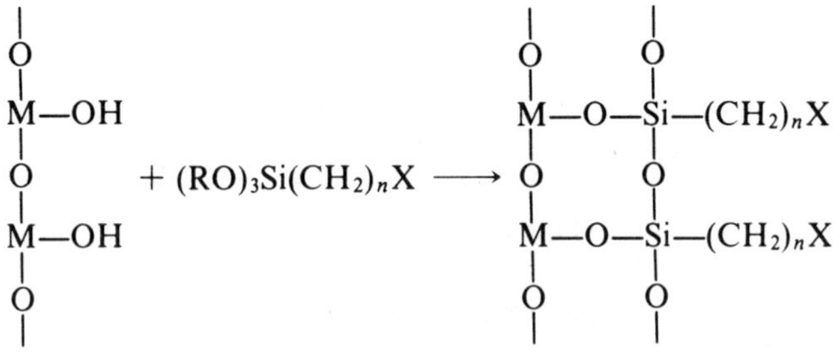

```
 |                                      |     |
 O                                      O     O
 |                                      |     |
 M—OH                                   M—O—Si—(CH2)nX
 |                                      |     |
 O       + (RO)3Si(CH2)nX  ——→          O     O
 |                                      |     |
 M—OH                                   M—O—Si—(CH2)nX
 |                                      |     |
 O                                      O     O
 |                                      |     |
```

Scheme 53

A variety of organic functional groups is available in silane coupling agents (Table 7.20). In some cases the silanized supports can be reacted directly with enzymes, but in most cases the organic groups are modified to produce activated intermediates, which in turn are reacted with enzymes. The most popular and versatile silane compound is the 3-aminopropyltriethoxysilane that produces an alkylamine derivative. The preparation of its reactive intermediates has been discussed extensively in the literature. The pioneering work of silanization was done by Weetall and his collaborators at Corning Glass Works; they introduced aminoalkyl controlled pore glass as

Table 7.20. Organic Functional Groups of Silane Coupling Agents for Enzyme Immobilization

Amine (aliphatic)	$—NH_2$
Amine (aromatic)	$—C_6H_4—NH_2$
Halide	—I; —Br; —Cl
Aldehyde	—CHO
Acetal	$—CH(OC_2H_5)_2$

an enzyme support and described methods of preparing reactive intermediates for enzyme immobilization (358) (see page 334).

Activation of glass can be achieved by either organic or aqueous silanization. With the organic silanization, the support is activated by refluxing a 10% (w/v) solution of 3-aminopropyltriethoxysilane in toluene (358) or xylene (277) or by an evaporative technique using acetone as solvent (358). Aqueous silanization at pH 3–4 gives lower amine loading than organic solvent techniques, but experience has shown that greater carrier durability with slightly lower enzyme loading is achieved by the latter technique. After either activation method there is usually a heat treatment (115°C) stage.

Once the "covalent" link between the inorganic support and the amino group is formed, the amino glass can be reacted with enzymes by several methods, already mentioned in the discussion on the activation of amino supports. Some enzyme derivatives obtained from the alkylamine glass are listed in Table 7.21.

One of earliest but complicated activation methods involves the previous preparation of an arylamine derivative (337) by reaction in chloroform of 4-nitrobenzoyl chloride with the alkylamine glass and subsequent reduction of the nitroaryl group with an aqueous solution of sodium dithionite. The resultant aminoaryl group can be activated by several procedures used for organic arylamines, but the most commonly used is that of diazotization with nitrous acid leading to a diazo salt. The enzyme is coupled to this reactive derivative at an alkaline pH through its L-tyrosyl groups as described previously in the discussion of diazotization.

The most widely used reactive derivative is obtained by activation of alkylamino glass with glutaraldehyde (358); the binding of enzyme to the carbonyl activated glass then occurs at neutral pH values. The relative simplicity of this method is the reason for its being preferred to other processes, even when a slightly less active preparation is obtained.

The isothiocyanate derivative is obtained by treating the alkylamine or

Table 7.21. **Silane Derivatives of Inorganic Carriers for Enzyme Immobilization**

Original Derivatives	Activating Reagents	Reactive Derivative	Enzyme Coupling Method
$\vdash OSi(CH_2)_3NH_2$	$NO_2-C_6H_4-COCl$, $Na_2S_2O_4$, HNO_2	$\vdash OSi(CH_2)_3NHC(=O)-C_6H_4-N_2^+Cl^-$	Diazotization
	$OHC(CH_2)_3CHO$	$\vdash OSi(CH_2)_3N{=}CH(CH_2)_3CHO$	Schiff base formation
	ClCSCl	$\vdash OSi(CH_2)_3NCS$	Peptide bond formation (isothiocyanato derivative)
	H_2NNH_2, HNO	$\vdash OSi(CH_2)_3N_3^+$	Peptide bond formation (azide derivative)
	$F-C_6H_3(NO_2)-N_3^+$	$\vdash OSi(CH_2)_3-C_6H_3(NO_2)-N_3^+$	Peptide bond formation (azide derivative)
	Terephthaloyl chloride, H_2NNH_2, HNO_2	$\vdash OSi(CH_2)_3NHCO-C_6H_4-CON_3^+$	Peptide bond formation (azide derivative)

	$\text{Cl}_3\text{C}_3\text{N}_3$ (2,4,6-trichloro-1,3,5-triazine)	$\vdash\text{OSi(CH}_2)_3\text{NH—C}_3\text{N}_3\text{Cl}_2$	Arylation
	$\text{(CH}_2\text{CO)}_2\text{N—O—C(=O)—CH}_2\text{Br}$	$\vdash\text{OSi(CH}_2)_3\text{NHCOCH}_2\text{Br}$	Alkylation
$\vdash\text{OSi(CH}_2)_3\text{NH}_2$	$\text{(CH}_2\text{CO)}_2\text{O}$, R	$\vdash\text{OSi(CH}_2)_3\text{NHCO(CH}_2)_4\text{CO}_2\text{H}$	—
$\vdash\text{OSi(CH}_2)_3\text{NHCO(CH}_2)_4\text{CO}_2\text{H}$	R—N=C=N—R^1	$\vdash\text{OSi(CH}_2)_3\text{NHCO(CH}_2)_4\text{COOC(—NH—R)(=}^{+}\text{NH—R}^1)$	Peptide bond formation (condensing agents)
	SOCl_2	$\vdash\text{OSi(CH}_2)_3\text{NHCO(CH}_2)_4\text{COCl}$	Peptide bond formation (acyl chloride derivative)
	$\text{HN(CH}_2)_3\text{C=O}$, R—N=C=N—R^1	$\vdash\text{O—Si(CH}_2)_3\text{NHCO(CH}_2)_4\text{COO—N(CH}_2)_3\text{C=O}$	Peptide bond formation

arylamine carrier with thiophosgene in chloroform (358). Several azide derivatives have also been obtained; the most important are the alkylazides (358). Another intermediate is the carboxyl derivative, usually formed by the reaction between the alkylamine glass and succinic anhydride. This latter derivative has been further activated with carbodiimide (337) or thionyl chloride (358). Also, the alkylamine amino derivative can react directly with enzyme in the presence of a carbodiimide on Woodward's reagent K (359), but carbodiimide coupling to a carboxyl derivative is generally preferable because with alkylamine carrier, crosslinking can occur.

In an attempt to increase the operational stability of immobilized enzyme prepared by the metal linking method, Cabral et al. (361, 362) have developed a method that permits binding covalently of enzyme to a titanium coating on porous glass. In this method, carrier and titanium(IV) chloride are dried together and the titanium-activated carrier is subsequently reacted with a diamine in an organic hydrophobic solvent. This yields an alkylamine derivative similar to that obtained with the silanization process. The operational stability of this material can be compared favorably with that obtained when using the silane method with a titanium(IV) oxide coated CPG (362).

By analogy with the organic carriers, Weetall and Detar (338) have suggested that porous glass can be activated with cyanogen bromide with the formation of a cyclic iminocarbonate, which reacts with enzyme (Scheme 54).

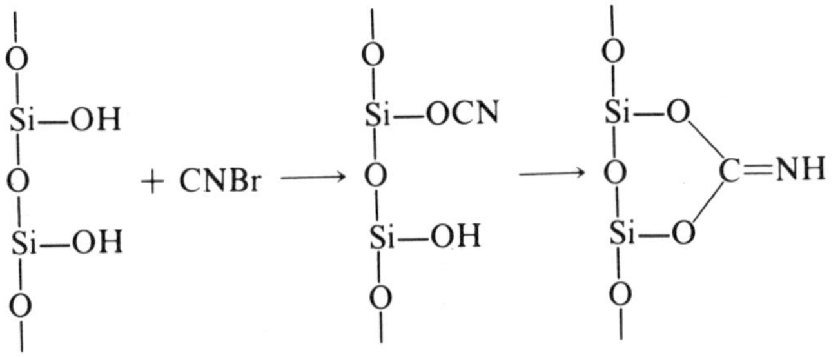

```
 |                           |                    |
 O                           O                    O
 |                           |                    |
Si—OH                       Si—OCN               Si—O
 |                           |                    |    \
 O        + CNBr  ——→        O          ——→       O      C=NH
 |                           |                    |    /
Si—OH                       Si—OH                Si—O
 |                           |                    |
 O                           O                    O
 |                           |                    |
```

Scheme 54

In an attempt to substitute for the expensive CPG, the wide range of natural or fabricated inorganic carriers such as nonporous glass, sand, metal oxides, and clays have been silanated in the same way as CPG. Besides the silanation method, inorganic carriers have been activated by reaction of the hydroxy and oxide groups on the carrier surfaces with several organic activating reagents.

Brick, bentonite, and glass have been activated (360) by refluxing the dry support with thionyl or sulfuryl chloride in dioxan and obtaining a chloride derivative that can link protein. Cyanuric halides have also been used to activate bentonite in organic media (353).

7.3.3. Entrapment Method of Immobilization

The entrapment method is based on the occlusion of an enzyme within a constraining structure (lattice of a polymer matrix or membrane) tight enough to prevent the release of protein while allowing penetration of substrate. Consequently, only reactions involving relatively small reactants may be carried out successfully by using entrapped enzyme preparations.

This method differs from the chemical coupling methods (covalent binding and crosslinking) in that the enzyme itself does not bind to the gel matrix or membrane. Thus it can be most generally applied to entrapping of any kind of enzymes, or of microbial cells and organelles of different sizes and properties, with little destruction of biological activity as compared with the chemical coupling methods. So far, entrapping of microbial cells, especially into polymer matrices, has been the most extensively used method for the immobilization of cells (see Chapter 15). In this chapter, the entrapment method is classified into gel entrapment, fiber entrapment and microencapsulation.

7.3.3a. Gel Entrapment

The gel entrapment method involves entrapping the enzyme within the interstitial spaces of crosslinked water-insoluble polymer gels. Occlusion within crosslinked polyacrylamide gels was the first employed technique of entrapment and was the one with which Bernfeld and Wan (363) in 1963 immobilized trypsin (EC 3.4.21.4), chymotrypsin (EC 3.4.21.1), papain (EC 3.4.22.2), α-amylase (EC 3.2.1.1), and other enzymes. This polymer matrix has subsequently been the most commonly used for enzyme immobilization by entrapment.

The procedure for the formation of polyacrylamide is identical to that employed for the preparation of the gel commonly used for covalent attachment of enzyme. In this method, the free-radical polymerization of acrylamide is conducted in an aqueous solution containing the soluble enzyme and a crosslinking agent, usually *N,N'*-methylene bisacrylamide (BIS). Polymerization is commonly carried out in the absence of oxygen and at lower than usual temperatures (10 to 25 °C) in order to prevent thermal denaturation of enzyme during this operation. The polymerization is initiated by potassium persulfate ($K_2S_2O_8$) or riboflavin and is catalyzed by β-dimethylamino propionitrile (DMAPN) or *N,N,N',N'*-tetramethyl ethylene diamine (TEMED).

The resulting gel block can be mechanically dispersed into particles of defined size. Bead polymerization is performed in a two-phase system similar to a procedure described for the preparation of crosslinked polyacrylamide used in gel choromatography. The aqueous medium, containing the enzyme and the monomer, is dispersed in a hydrophobic phase (*i.e.*, toluene or chloroform). Subsequent polymerization of the emulsion yields well-defined spherical beads (364).

These gels, however, are quite weak in a mechanical sense and have an open network with a broad distribution of pore sizes that often allows the enzyme to leak out. These disadvantages are overcome by optimizing the degree of crosslinking. Chibata et al. (365) optimized the conditions most suitable for achieving a highly active, stable, and mechanically strong immobilized enzyme preparation; they obtained a polyacrylamide gel with a pore size of 10–40 Å, which permits the transport of small substrate and product molecules but prevents leaking out of the large enzyme molecules.

Recently, X- or γ-radiation have been used for the polymerization of acrylamide (366, 367). The gel-entrapped enzyme is produced by freezing a solution of monomer and enzyme at −5°C and polymerizing the resultant composition by irradiating with rays, using as irradiation sources ^{60}Co or ^{137}Cs. One advantage of this technique is that the polymerization can be carried out in a frozen state, where less inactivation of the enzyme is expected to occur. Another advantage is that this method can be used again to prepare various shapes of entrapped enzyme by varying the type of container. However, the activation of the immobilized enzymes prepared by irradiation has not been significantly higher than that obtained with chemical catalysts, and this method also has disadvantages with regard to the cost of processing and the equipment needed. The enzymes to be immobilized must be stable to the radiation required in the immobilization process.

Besides the use of polyacrylamide, gel entrapment of enzymes has been carried out by γ-irradiation (368) with other matrices. This method includes the immobilization of enzymes with resin monomers or polymers (368). In the first technique, of which entrapment in polyacrylamide is an example, hydrophilic monomers can be used as starting materials for the production of hydrophilic polymers. On γ-irradiating the monomer–enzyme solution, both the polymerizing and the crosslinking reactions occur at the same time, therefore a crosslinking agent is not necessary.

Besides acrylamide, monomers used in this technique are dimethylacrylamide (368), 2-hydroxyethylacrylate (368), and *N*-vinylpyrrolidone (368). However, only 2-hydroxyethylacrylate is converted to a very rigid gel, even when small doses of γ-irradiation are employed. To overcome this lack of rigidity, polymers such as poly(vinylalcohol) and poly(vinylpyrrolidone) are used (368). The entrapping is carried out by dissolution of the polymer in hot water (80–95°C) and, after cooling, enzyme solution is added with stirring in an N_2 atmosphere and the mixture is γ-irradiated. Considerable attention has been given to the development of this method for gel entrapment with nontoxic synthetic polymers [poly(vinylalcohol)] or monomers (2-hydroxyethylacrylate). These are preferred by several authors (368) who feel that it is difficult to use polyacrylamide gel in the food industry because of the toxicity of the acrylamide monomer.

Recently, in addition to synthetic polymers, enzymes have been gel-entrapped in naturally occurring gels. Among these, of particular importance are proteins such as collagen and gelatin, and polysaccharides such as agar, calcium alginate, and κ-carregeenan. These materials are usually dissolved in aqueous media at 40–60°C and are mixed with the enzyme solution and cooled to around 10°C. In order to obtain a particulate form in the case of gelatin (319), the aqueous enzyme–gelatin suspension is stirred into an organic liquid that is poorly miscible or immiscible in water and is compatible with the enzyme and gelling protein (usually butan-l-ol) at 50°C. The resultant suspension is rapidly cooled to 10°C, whereby enzyme-gelatin-containing particles are formed. This gel does not possess sufficient physical stability for use at normal operational temperatures; therefore, it is necessary to incorporate a crosslinking agent, glutaraldehyde, to increase its stability.

With κ-carrageenan (369) the gel strength is obtained by soaking the resulting gel with cold potassium chloride solution. The resultant stiff gel is granulated into suitable particle sizes for utilization in immobilized enzyme systems. The stability of the immobilized enzyme can be increased by treating the gel with hardening agents such as tannin, glutaraldehyde, or diamines. This process, using κ-carrageenan, is applicable not only for immobilization of many enzymes but also for many kinds of microbial cells, and it is considered to be better for industrial purposes than the polyacrylamide method since the procedure is very simple, is carried out under mild conditions, and gives immobilized enzymes or cells with excellent stability.

In the case of calcium alginate, the enzyme is mixed with an aqueous solution of sodium alginate, and this mixture is then added, dropwise with agitation, to a solution of calcium chloride (370). Beadlets of calcium alginate are obtained and can be used without further treatment.

Starch is also used for entrapping enzymes (371) as follows: A mixture of enzyme and starch–glycine buffer solutions are dispersed in polymethane foam pads at 47°C and dried.

A silicon polymer (Scheme 55), silastic, has been reported for entrapment of enzymes (372). The immobilized enzyme preparation is obtained by adding the enzyme to an excess of silastic, stirring the mixture for several minutes, and then adding a catalyst, usually stannous octanoate. Under normal circumstances, a gel is formed within half an hour. The enzymes entrapped

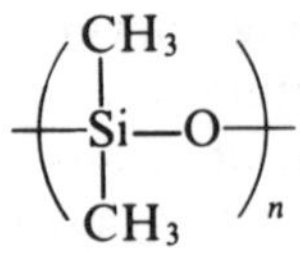

$$-\!\left(\begin{matrix}CH_3\\ |\\ Si-O\\ |\\ CH_3\end{matrix}\right)_{\!n}\!-$$

Scheme 55

by this technique exhibit good thermal stability. Unfortunately, the permeability of substrates into this enzyme preparation seems to be low on the basis of the published data.

Table 7.22 shows matrices used for gel entrapment of enzymes, and Table 7.23 shows the principal examples of gel entrapped enzymes.

7.3.3b. Fiber Entrapment

A method of immobilizing enzymes by entrapment within microcavities of synthetic fiber has been developed by Dinelli et al. (387, 388). Enzymes can be entrapped in fibers and continuously produced by the conventional wet spinning techniques for the manufacture of synthetic fibers, using apparatus very similar to that used in the textile industry.

With this technique, the physical entrapment of an enzyme is achieved by dissolving a fiber-forming polymer in an organic solvent immiscible in water and emulsifying this solution with the aqueous solution of the enzyme. The emulsion is extruded through a spinneret into a liquid coagulant (toluene or petroleum ether) that precipitates the polymer in filamentous form, with microdroplets of enzyme solution entrapped in the fiber.

A high surface area for enzyme binding can be obtained by using very fine fibers. The fibers are resistant to weak acids or alkalis, high ionic strength, and some organic solvents. Depending on the polymer used, the fiber-enzyme conjugate can show good resistance to microbiological attack. It is possible to entrap aqueous solutions containing more than one enzyme for the production of immobilized multienzyme systems. Their use is limited to low

Table 7.22. Matrices for Gel-entrapped Enzymes

Natural Matrices	Synthetic Matrices
Collagen	Polyacrylamide
Gelatin	Poly(vinyl alcohol)
κ-Carrageenan	Poly(vinyl pyrrolidone)
Agar	Poly(2-hydroxyethylmethacrylate)
Calcium alginate	Silastic
Starch	

Table 7.23. Principal Examples of Gel-entrapped Enzymes

Material	Enzyme	Reference
Polyacrylamide	Acetylcholinesterase (EC 3.1.1.7)	373
	Chymotrypsin (EC 3.4.21.1)	202, 374, 375, 376
	Asparginase (EC 3.5.1.1)	377
	D-Glucose dehydrogenase (EC 1.1.1.47)	378
	Alcohol dehydrogenase (EC 1.1.1.1)	378
	β-D-Galactosidase (EC 3.2.1.23)	197
	Phenol-2-monooxygenase (EC 1.14.13.7)	80, 129
	Urease (EC 3.5.1.5)	379
	β-D-Fructofuranosidase (EC 3.2.1.26)	380
	Penicillin amidase (EC 3.5.1.11)	381
	Catalase (EC 1.11.1.6)	382
	D-Glucose oxidase (EC 1.1.3.4)	382
Acrylamide/ glycidylmethacrylate	AMP-deaminase (EC 3.5.4.6)	383
Ethyleneglycol/2-hydroxy ethylmethacrylate	β-D-Fructofuranosidase (EC 3.2.1.2.6)	384
	D-Glucose isomerase (EC 5.3.1.5)	384
2-Hydroxyethylmethacrylate	α-Amylase (EC 3.2.1.1)	385, 386
	Glucoamylase (EC 3.2.1.3)	386
	Cellulase (EC 3.2.1.4)	386
	Glucosidase (EC 3.2.1.20/21))	386
	D-Glucose oxidase (EC 1.1.3.4)	386
κ-Carrageenan	Aminoacylase (EC 3.5.1.14)	369
	Aspartate ammonia-lyase (EC 4.3.1.1)	369
	Fumurate hydratase (EC 4.2.1.2)	369

molecular weight substrates, however, and the necessity of using water-immiscible liquid as polymer solvents and coagulants may in some cases cause inactivation of the enzyme.

Fiber entrapped enzymes are applied industrially in the pharmaceuticals industry [penicillin amidase (EC 3.5.1.11)] and the food industry [β-D-galactosidase (EC 3.2.1.23)]. The polymer most commonly used in this procedure is cellulose acetate, (Table 7.24) which yields preparations with low cost, good biological resistance, and chemical resistance to weak acids and solvents. Other polymers, mainly cellulose based, have been used.

Table 7.24. Principal Examples of Fiber-entrapped Enzyme

Material	Enzyme	Reference
Cellulose acetate	Aminoacylase (EC 3.5.1.14)	389
	Fumarate hydratase (EC 4.2.1.2)	390
	Glucoamylase (EC 3.2.1.3)	391
	D-Glucose isomerase (EC 5.3.1.5)	392
	Dihydropyrimidinase (EC 3.5.2.2)	393
	β-D-Fructofuranosidase (EC 3.2.1.26)	394
	β-D-Galactosidase (EC 3.2.1.23)	395
	Penicillin amidase (EC 3.5.1.11)	396
	Tryptophan synthetase	397, 398
	Dipeptidyl peptidase (EC 3.4.14.½)	399

7.3.3c. Microencapsulation

Microencapsulated enzymes are formed by enclosing enzymes within semipermeable polymer membranes (the usual diameter range is $1 \sim 100\ \mu m$). The enveloped enzymes are physically contained within the semipermeable membrane and cannot leak out, while external substrates can diffuse across the membrane to be processed by the immobilized enzymes.

The first report of the immobilization of enzymes by entrapping the molecules within microcapsules came from Chang in the mid-1960s (400); prior to that report microencapsulation had been used for entrapping drugs, chemicals, dyes (401). Various enzymes have now been immobilized in microcapsules of different chemical composition, as can be seen in Table 7.25.

Advantages of this method of immobilization include the extremely large surface area for contact of substrate and enzyme within a relatively small volume, and the real possibility of simultaneous immobilization of many enzymes in a single step, whether the enzymes are soluble or were previously immobilized by another method. The main disadvantages are: the method can only be applied to low molecular weight substrates, occasionally inactivation of enzyme occurs during the immobilization procedure, a high enzyme concentration is required for microcapsule formation, and the enzyme may become incorporation into the membrane wall. With some of the immobilization techniques used in this method, leakage of enzyme from the microcapsule may also take place.

The techniques used to obtain microencapsulated enzymes can be separated into phase separation methods, interfacial polymerization methods, liquid-surfactant membrane methods, and liquid drying methods.

Phase Separation Method. Microencapsulated enzymes are generally prepared by the phase separation method (402–407). It is based on coacerva-

Table 7.25. Principal Examples of Microencapsulated Enzymes

Material	Enzyme	Reference
Phase Separation Method		
Nitro-cellulose	β-D-Galactosidase (EC 3.2.1.23)	402
	Asparaginase (EC 3.5.1.1)	403
Collodion	Alcohol dehydrogenase (EC 1.1.9.1)	404
	Malate dehydrogenase (EC 1.1.1.37)	404
	Catalase (EC 1.11.1.6)	405
	Pyruvate kinase (EC 2.7.1.40)	406
	Hexokinase (EC 2.7.1.1)	406
	β-D-Galactosidase (EC 3.2.1.23)	407
	Urease (EC 3.5.1.5)	405
Interfacial Polymerization Method		
Nylon	β-D-Galactosidase (EC 3.2.1.23)	408
	Asparaginase (EC 3.5.1.1)	409
Polyurea	Asparaginase (EC 3.5.1.1)	409
Liquid Membrane Method		
	Glucoamylase (EC 3.2.1.3)	410
	Nitrate reductase (EC 1.7.99.4)	411
Liquid Drying Method		
Ethyl-cellulose	Triacylglycerol lipase (EC 3.1.1.3)	412
Polystyrene	Catalase (EC 1.11.1.6)	412
	Triacylglycerol lipase (EC 3.1.1.3)	412
	Urease (EC 3.5.1.5)	412

tion, a physical phenomenon used for the purification of polymers that involves dissolution of the polymer in an organic solvent followed by its reprecipitation by adding another solvent which is miscible with the first but which does not dissolve the polymer.

To immobilize enzymes by this technique, an aqueous solution of the enzyme is first emulsified in a water-immiscible organic solvent containing the polymer, and then to the vigorously stirred emulsion is slowly added another water-immiscible organic solvent in which the polymer is insoluble. The polymer is concentrated, and membranes are formed around the microdroplets of aqueous enzyme solution. This technique is carried out under relatively mild conditions, but it has the disadvantage that it is difficult to completely remove organic solvent remaining on the polymer membrane.

Interfacial Polymerization Method. The interfacial polymerization method is based on a chemical process, the synthesis of a water-insoluble copolymer at the interface of a microdroplet. One monomer, a hydrophilic monomer, is partly soluble in both the aqueous and organic phases; the other, a hydrophobic monomer, is only soluble in the organic phase.

The technique used is very similar to the phase separation method. In the first step the aqueous solution of the enzyme and the hydrophilic monomer are emulsified in a water-immiscible organic solvent. To the emulsion, the hydrophobic monomer is added in the same solvent with stirring. The monomers react by condensation or addition polymerization at the interface between the aqueous and organic solvent phases in the emulsion. This results in an immobilized enzyme by enclosing of the enzyme in the aqueous solution within a membrane of a polymer.

A major factor is the partition coefficient of the hydrophilic monomer, which determines the properties of the membrane that is produced. One disadvantage of this procedure is the possible inactivation of the enzyme by some monomers.

The usual copolymers used for this type of microencapsulation of enzymes include the well-known nylon 6,10 obtained by polymerization of the hydrophilic 1,6 diaminohexane and the hydrophobic monomer 1,10 decanoyl chloride (sebacoyl chloride). The solvent used is a chloroform–cyclohexane mixture (1:4 v/v) containing usually 1% w/v of an organic soluble surfactant (Span 85).

Liquid Surfactant Membrane Method. The liquid surfactant membrane method, also known as the immobilization method with nonpermanent microcapsules, is based on the "liquid surfactant membrane" concept develop by Li (413). Unlike the other methods of microencapsulation, where there is a water-insoluble semipermeable membrane, encapsulation of enzymes is obtained by means of a liquid membrane. An aqueous enzyme solution is emulsified with a surfactant to form the liquid-membrane-encapsulated enzymes.

The major advantages of this procedure are its nonchemical nature and its reversibility. Disadvantages are possible leakage of the enzyme and the fact that diffusion of substrates and products through the membrane is solubility-dependent yet independent of the pore size of the membrane (as this is a liquid membrane).

Liquid Drying Method. A process similar to the liquid surfactant membrane method, known as the liquid drying method, has been used for encapsulation of enzymes (412). However, with this technique a solid permanent membrane is formed.

An aqueous enzyme solution is emulsified in an organic solvent with a boiling point lower than that of water (usually benzene, cyclohexane, or chloroform) containing a membrane-forming polymer and using an oil-soluble surfactant. This emulsion, containing aqueous enzyme microdroplets, is then dispersed in aqueous media with protective colloidal substances [gelatin, poly(vinylchloride) and surfactants] and a secondary emulsion is formed. The final immobilized enzyme preparation is obtained by removing the organic solvent, using, for instance, a vacuum rotary evaporator.

Like the liquid surfactant membrane method, the major advantage is that little or no inactivation of enzyme occurs during microcapsule preparation, owing to the fact that preformed polymers are used and no reactive reagents are necessary.

In the process of preparing the second emulsion, however, some problems may arise with its formation and the yield of microcapsules may become low. Considerable time is also necessary to remove organic solvents completely, and total solvent removal is necessary for the formation of the solid membrane.

7.3.4. Immobilized Soluble Enzyme Method

7.3.4a. Restriction of Enzyme Movement Without Enzyme Derivatization

All the methods of immobilization of enzymes described thus far involve the modification of the enzyme or its microenvironment, with subsequent alteration of its kinetics and sometimes of its pH and temperature profiles. Often a reduced activity relative to the corresponding free enzyme results from these alterations. In order to utilize an enzyme in its native state continuously over a long period of time, enzymes have been immobilized by physically confining them within semipermeable membranes or hollow bore films or ultrafiltration membranes. The membrane or film is impermeable to the enzymes but permeable to products and in some cases to substrates. The use of this method of immobilization was first reported by Rony (414).

This method of immobilization offers several advantages relative to other immobilization methods. Chemical modification of the enzyme is not necessary (although it can be used to modify enzymes in or onto hollow fiber devices and ultrafiltration cells). Thus this method allows the study of soluble enzymes, including their operational stability and their application in continuous enzyme reactors.

The method is especially suited for conversion of high molecular weight water-soluble or insoluble substrates, as it allows the intimate contact of the soluble enzyme with substrate, achieving an efficient conversion of these types of substrate, unlike the immobilized insoluble enzymes which usually have lower catalytic efficiencies towards the same substrates.

Other advantages of this method are: the simplicity of method required to immobilize the enzymes, since this immobilization consists of placing the enzyme in solution on one side of a semipermeable membrane; the simultaneous immobilization of many enzymes; the selectivity control of substrates and products through membrane selectivity; large ratio of surface area to volume (hollow fibers); the protection of enzyme from access by microorganisms; the absence of enzyme leakage when properly constructed membranes are chosen; and the ease with which the membrane reactors can be loaded with enzymes, operated, cleaned, sterilized, and regenerated compared with other methods of immobilization.

However, some disadvantages are inherent to the method: the possible reduction of reaction velocity as a result of the permeability resistance of the membrane; the difficulty of working with very low substrate concentrations owing to substrate adsorption by membranes; the possibility of enzyme inactivation because of high shear forces or vigorous agitation (ultrafiltration membrane cells); and, among others, the need for a careful control of the residence time of low molecular substrates in order to achieve high conversions.

Table 7.26 lists principal examples of enzymes immobilized by this method. However, this method can of course be combined with other methods of immobilization, for example adsorption in which the resultant enzyme is physically adsorbed within the membrane.

7.3.4b. Restriction of Enzyme Movement with Enzyme Derivatization

Recently several reports (423–432) have described the chemical modifications of enzymes without insolubilization, using low or high molecular weight modifiers. Although the modification of enzymes with low molecular weight compounds is often of limited utility, in some situations it can serve a specific and useful purpose; for example, the acylation of enzymes with low molecular weight reagents can have a stabilizing effect (423).

Table 7.26. Principal Examples of Immobilized Soluble Enzymes

Procedure	Enzyme	Reference
Ultrafiltration	β-D-Fructofuranosidase (EC 3.2.1.26)	415, 416
Membrane method	Amidase (EC 3.5.1.4)	417
	Acid phosphatase (EC 3.1.3.2)	418, 419
	β-D-Galactosidase (EC 3.2.1.23)	422
Hollow fiber devices	α-D-Galactosidase (EC 3.2.1.22)	420
	β-D-Fructofuranosidase (EC 3.2.1.26)	420
	D-Glucose oxidase (EC 1.1.3.4)	421

In the preparation of water-soluble enzyme–polymer conjugates, reactions similar to those employed in the chemical coupling of enzymes onto insoluble polymers are used. The bonding of enzymes to soluble polymers may be achieved by one of the following procedures: reaction of the enzyme with an activated soluble polymer, reaction of the enzyme with an activated insoluble polymer followed by solubilization of the enzyme–polymer conjugate, or copolymerization of monomers with enzyme.

The first reports on water-soluble derivatized enzymes for synthetic purposes were made by Katchalski and co-workers (424, 425), who prepared them in order to elucidate the interrelationship between the electrostatic potential of a polymer chain and the displacement of the optimal pH of the enzyme immobilized on it. In other reports (427–429), the water-soluble enzyme derivatives have been prepared in order to increase the effective molecular size of the enzyme to prevent its release from membrane dependent devices, and to improve the mechanical properties and operational stability of the enzyme.

In fact, when using a native soluble enzyme in a membrane device, the instability of the enzyme over long periods and the (associated) need to limit the porosity of the membrane to prevent loss of enzyme in some applications can pose problems. These difficulties can be overcome by using a water-soluble enzyme–polymer conjugate that allows the use of ultrafiltration membranes of higher porosity with consequently faster diffusion of products and therefore reduction of end product inhibition. At the same time a stabilization of the enzyme can be achieved by attachment of the enzyme to a polysaccharide (427–429), or by forming a stable environment of definite electrostate nature around the enzyme (430).

Another advantage of using soluble derivatized enzymes is in the hydrolysis of macromolecular or insoluble particulate substrates, for example, cellulose. The treatment of such substrates with conventional immobilized enzyme catalysts can be limited by severe diffusional and steric resistances between the large substrate and insoluble enzyme.

The derivatized enzymes, such as glycosylated enzymes, may also be chemically bound to an insoluble support through their carbohydrate constituents (429), and highly charged conjugates of enzymes with polymers of ethylene-maleic anhydride may be absorbed on ion-exchange resins to yield immobilized enzymes with better characteristics than the native enzyme absorbed at the same conditions (430). In other words, derivatization of enzyme with retention of solubility may be a useful precursor to insolubilization.

The main limitation to preparing these water-soluble conjugates is the laborious purification needed after polymer activation and reaction with the enzyme as excess reagent and unreacted enzyme must be separated by precipitation, gel filtration, ultrafiltration, or dialysis.

Some principal examples of derivatized enzymes are listed in Table 7.27.

Table 7.27. Some Principal Examples of Soluble Derivatized Enzymes

Soluble Polymer	Enzyme	Reference
Dextran	α-Amylase (EC 3.2.1.1)	431
	Chymotrypsin (EC 3.4.21.1)	426, 428
	β-D-Glucosidase (EC 3.2.1.21)	428
	Lysozyme (EC 3.2.1.17)	428
	Trypsin (EC 3.4.21.4)	427
Ethylene-maleic anhydride copolymer	Glucoamylase (EC 3.2.1.3)	429
Styrene-maleic anhydride copolymer	Glucoamylase (EC 3.2.1.3)	429
Alginic acid	Lysosyme (EC 3.2.1.17)	431

7.3.5. Miscellaneous Methods of Enzyme Immobilization

Although enzymes have been immobilized by one of the earlier described methods, sometimes efficient immobilization of enzymes is achieved by combinations of methods. Combinations are largely used for crosslinking of enzymes previously immobilized by adsorption, as adsorption alone yields immobilized enzymes with poor stability. This same hybrid of immobilization methods also increases the mechanical strength of the preparations obtained with only crosslinking with multifunctional reagents. By crosslinking physically adsorbed enzymes, a monolayer of covalently immobilized enzymes can be formed; however, the experimental conditions must ensure good adsorption of the enzyme on the support, and it is also necessary that no aggregation of individual colloidal particles occurs.

As referred to previously (page 310), collagen is a protein that has been used as a carrier in several methods of immobilization. Apart from its use in the "covalent binding method", the major work with this carrier has been performed by Vieth and collaborators (433), who have used collagen for the immobilization of several enzymes by three different procedures: complexation, electrodeposition, and impregnation. Although these authors consider the procedures to be individual processes, they can be integrated within a broader classification. Thus the complexation method can be and several times has been classified as an entrapment process, as the enzyme is added directly to an aqueous collagen dispersion and comixed before casting the membrane. The membrane impregnation method can be classified as physical adsorption, as the major forces responsible for binding the enzyme to the collagen matrix are multiple salt linkage, hydrogen bonds, and van der Waals interaction. The electrodeposition method can also be classified as an adsorption process, as the enzyme and the protein, collagen, can interact under appropriate pH conditions to form macromolecular complexes under the influence of an external electrical field gradient.

Mainly in conjunction with the complexation method, to add to the mechanical strength of the collagen-enzyme membrane and to increase the amount of enzyme attached to the membrane by covalent links, a crosslinking step is performed by addition of suitable bifunctional reagents, usually glutaraldehyde, to the dried complex membrane, collagen-enzyme (433). With this step the immobilized enzyme is obtained by entrapment followed by intermolecular crosslinking between enzyme and matrix. Enzymes entrapped in gelatin usually must be treated with bifunctional reagents to obtain sufficient mechanical stability for withstanding operational conditions. A very similar process of gelatin entrapped enzymes, known as the co-crosslinking method, has been developed by Broun and collaborators (73). In this method, enzymes at low concentrations are mixed with an inert L-lysine-rich protein (bovine serum albumin) and then a bifunctional agent (glutaraldehyde) is added; crosslinking occurs between the enzyme and the inert protein. The resultant immobilized enzyme can be obtained in membrane, particle, or other forms. This method also has been used when, owing to its chemical nature, insolubilization of enzymes with glutaraldehyde cannot be achieved.

Other miscellaneous methods of enzyme immobilization are used on the covalent bonding of enzymes to inorganic supports. As already mentioned, inorganic carriers are inert, and practically all processes of covalent binding involve the silanization of the inorganic carrier. One alternative is to use inorganic carriers with functional organic coatings that bind protein. Several attempts at this have been made. Alkylamine derivative of porous alumina and glass have been coated with maleic anhydride–methylvinylether copolymer (434, 402A); 3-aminobenzene is diazotized in the presence of a solid support, and the resulting coated carrier binds enzymes (435). Polyionic pellicular enzyme resins have been prepared by coreticulating the protein with a copolymer of maleic anhydride and methylvinylether *in situ* on the surface of glass beads (436); poly(ethyleneimine) has been strongly adsorbed to the carrier and the resulting carrier is activated with glutaraldehyde (434).

Principal examples of these miscellaneous methods of immobilization and the immobilized enzymes produced thereby are listed in Table 7.28.

7.3.6. Comparison of Different Immobilization Techniques

Although a number of immobilization techniques have been applied to many enzymes, it is recognized that no one particular process can be classified as an ideal general method for enzyme immobilization, because of the widely different composition and chemical characteristics of the enzymes and their

Table 7.28. Principal Examples of Miscellaneous Methods of Immobilization

Procedure	Enzyme	Reference
Entrapment and crosslinking (Collagen)	Triacylglycerol lipase (EC 3.1.1.3)	437
	Urokinase (EC 3.4.21.31)	438
	Glucoamylase (EC 3.2.1.3)	439
	Asparaginase (EC 3.5.1.1)	440
	D-Glucose oxidase (EC 1.1.3.4)	441
	D-Glucose isomerase (EC 5.3.1.5)	441
	Alcohol dehydrogenase (EC 1.1.1.1)	442
	Lactate dehydrogenase (EC 3.5.1.5)	443
Entrapment and crosslinking (Gelatin)	Urease (EC 3.5.1.5)	443
Co-crosslinking (Albumin and glutaraldehyde)	Urate oxidase (EC 1.7.3.3)	444
	L-Glutamate dehydrogenase (EC 1.4.1.2)	445, 446
	β-D-2-Acetamido-2-deoxy-hexosidase (EC 3.2.1.52)	447
	Catechol-1,2-dioxygenase (EC 1.13.11.1)	79
	α-Steroid dehydrogenase	448
	Urease	449
Adsorption and crosslinking (Collagen + glutaraldehyde)	β-D-Fructofuranosidase (EC 3.2.1.26)	450
Adsorption and crosslinking (NiO, FeO_2, Mn-Zn Ferrite)	Chymotrypsin (EC 3.4.21.1)	451
Adsorption and crosslinking (Kieselguhr + glutaraldehyde)	D-Glucose oxidase (EC 1.1.3.4)	452
Adsorption and crosslinking (Glass + glutaraldehyde)	D-Glucose oxidase (EC 1.1.3.4)	453
	Catalase (EC 1.11.1.6)	453

substrate and product properties. Therefore each method of immobilization has specific limitations, and for any particular application it is necessary to find an immobilization procedure that is simple and inexpensive and that yields an immobilized enzyme with a good retention of activity and a proper operational stability. One can make a general comparison, however, of the different enzyme immobilization processes, based on the main characteristics of these methods and on the support matrix. Table 7.29 summarizes some of the relative advantages and disadvantages of the different processes of enzyme immobilization.

Table 7.29. Comparison of the Attributes of Different Classes of Immobilization Techniques

Characteristic	Crosslinking	Physical Adsorption	Ionic Binding	Metal Binding	Covalent Binding	Entrapping
Preparation	Intermediate	Simple	Simple	Simple	Difficult	Difficult
Binding force	Strong	Weak	Intermediate	Intermediate	Strong	Intermediate
Enzyme activity	Low	Intermediate	High	High	High	Low
Regeneration of carrier	Impossible	Possible	Possible	Possible	Rare	Impossible
Cost of immobilization	Intermediate	Low	Low	Intermediate	High	Intermediate
Stability	High	Low	Intermediate	Intermediate	High	High
General applicability	No	Yes	Yes	Yes	No	Yes
Protection of enzyme from microbial attack	Somewhat	No	No	No	No	Yes

Immobilization of enzyme by chemical methods, that is, by intermolecular crosslinking and covalent binding, involves chemical modification of the enzyme, which may cause conformational changes of its structure and partial inactivation when its active center participates directly in the immobilization of enzyme. To minimize these problems, these processes must be carried out under the gentlest possible conditions. Because the bonds between enzyme–enzyme or enzyme–carrier are not easily destroyed by substrate or salt, the operational stability of the immobilized enzyme preparation is high. On the other hand, crosslinking is generally an unsuitable method owing to the poor mechanical properties of the crosslinked enzyme preparations. Covalently bonded enzyme preparations, when using organic matrices, are difficult or even impossible to regenerate.

Immobilized enzymes produced by the other three methods—physical adsorption, ionic binding, and metal binding—can be prepared easily under mild conditions, and with generally weak forces between the enzyme and the carrier. Leakage of the enzyme from the matrix can easily occur during operation with changes in ionic strength or pH of the solution. Therefore these preparations have low operational stability although relatively good retention of initial activity, since there is little chemical change of the enzyme structure during immobilization. One benefit from these enzyme preparations is that the carrier may be readily regenerated once the enzyme activity has degenerated to the point where the catalyst needs to be replaced.

With the entrapping method, no binding between enzyme and carrier should occur in theory and the preparations should present a high retention of activity. However, the enzyme activity is limited to small molecular weight substrate and product molecules.

7.4. EFFECTS OF IMMOBILIZATION METHODS ON THE BEHAVIOR, KINETICS, AND PROPERTIES OF ENZYMES

Although enzyme immobilization can be very profitable, immobilization may also change the kinetics and other properties of the enzyme, usually with a decrease of enzyme-specific activity. This may be ascribed to several factors (420):

(i) Conformational and steric effects are present—even when the enzyme is bound without loss of activity—when conformational change of the enzyme molecule occurs by binding to a carrier or when the interaction of the substrate with the enzyme is affected by steric hindrance.

(ii) Partitioning effects, related to the chemical nature of the support material, may arise from electrostatic or hydrophobic interactions between the matrix and low molecular weight species present in the solution, leading to a modified microenvironment.

(iii) Mass transfer or diffusional effects arise from diffusional resistance to the translocation of substrate from the bulk solution to the catalytic sites and from the diffusion of products of the reaction back to the bulk solution. These diffusional resistances may be classified as internal or intraparticle mass transfer effects when the enzyme is located in a porous medium and external or interparticle mass transfer effects occur between the bulk solution and the outer surface of the enzyme-matrix particle.

When the kinetic behavior of immobilized enzyme can be controlled by one or more of the above effects, it is useful to distinguish among (i) intrinsic rate parameters, the kinetic parameters inherent in a particular immobilized enzyme and that are different from those of free enzyme (because of conformational change and steric effects); (ii) inherent rate parameters, the kinetic parameters that are observed in the absence of any diffusional effects; (iii) effective rate parameters, the kinetic parameters when mass transfer effects are present and in the presence or the absence of partition effects.

7.4.1. Conformational and Steric Effects

The decrease of specific activity of enzymes, which occurs on their binding either to solid supports or upon intermolecular crosslinking, is usually attributed to conformational changes in the tertiary structure of the enzymes. For instance, covalent bonds between the enzyme and the matrix can stretch the enzyme molecule and thus alter the three-dimensional structure at the active site. The specific activity decrease may also be attributed to steric hindrance resulting in limits on the accessibility of the substrate. In these two cases the decrease in enzyme activity can be reduced or prevented by choosing suitable conditions for immobilization. Thus the active center of the enzyme can be protected with a specific inhibitor, substrate, or product, and the shielding effect of the matrix that causes steric hindrances can be reduced by the introduction of "spacers" that keep the enzyme at a definite and certain distance from the matrix. In addition to these problems, denaturation of the enzyme can arise by the action of reagents used in entrapment methods.

In addition to their influence on the enzyme activity, any physical or chemical matrix–enzyme interactions may additionally modify the selectivity and stability of the bound enzyme from that which it normally possesses in free solution (42).

7.4.2. Partition Effects

In the carrier binding method, when the support matrix is charged the kinetic behavior of the immobilized enzyme may differ from that of the free enzyme even in the absence of mass transfer effects. This difference is commonly attributed to partition effects that cause different concentrations of charged species, substrates, products, hydrogen ions, hydroxyl ions, and so on, in the domain of the immobilized enzyme and in the domain of the bulk solution, owing to electrostatic interactions with fixed charges on the support.

These differences in the equilibrium concentrations of charged soluble compounds may be described by the partition coefficient, P, given by

$$P = C_i/C_0$$

where C_i and C_0 are the local and bulk concentrations, respectively. The main consequences of these partition effects is a shift in the optimum pH, with a displacement of the pH-activity profile of the immobilized enzyme towards more alkaline or acidic pH values for negatively or positively charged carriers, respectively (338).

Goldstein et al. (338) gave mathematical expression to the qualitative considerations on the displacement of pH activity profiles of immobilized enzymes. Assuming the Boltzmann distribution, the partitioning of hydrogen ions between the local activity ($a_i^{H^+}$) and the bulk activity ($a_0^{H^+}$) is given by

$$P_H = a_i^{H^+}/a_0^{H^+} = \exp(-e\psi/kT)$$

or by the definition of pH

$$\Delta pH = pH_i - pH_0 = 0.43\ (e\psi/kT)$$

where e is the electronic charge, ψ is the electrostatic potential, k is the Boltzmann constant, T is the absolute temperature, and pH_i and pH_0 are the local and the bulk pH values. This equation shows that the local pH is higher if the support is negatively charged.

By similar considerations, the partitioning of charged compounds, substrate or product, between a charged enzyme particle and the bulk solution can be represented in the following form:

$$S_i = S_0 \exp(-Ze\psi/kT)$$

where Ze is the substrate charge.

Thus for positively charged substrate, when using a negatively charged enzyme particle, a higher concentration of substrate is obtained in the local environment or microenvironment than in the bulk solution, and a higher value of relative activity is obtained than with a neutrally charged matrix.

However when effects other than partitioning are present, it is possible to have no shift of the enzyme's pH optimum on charged supports.

7.4.3. Internal and External Mass Transfer Effects

When an enzyme is immobilized on or within a solid matrix, mass transfer effects may exist because the substrate must diffuse from the bulk solution to the active site of the immobilized enzyme. If the enzyme is attached to non-porous carriers there are only external mass transfer effects on the catalytically active outer surface; in the reaction solution, being surrounded by a stagnant film, substrate and product are transported across the Nernst layer by diffusion. The driving force for this diffusion is the concentration difference between the surface and the bulk concentration of substrate and product.

For instance, the rate of flow of substrate V_{dif} from the bulk solution to the enzymic surface is given by:

$$V_{dif} = k_L a(S_B - S_S)$$

where k_L is the mass transfer coefficient; a is the particle surface area per unit of volume, and S_B and S_S are the bulk and surface concentrations of the substrate respectively.

In a surface reaction, the flow of substrate to the enzyme surface and the enzyme reaction take place consecutively. At steady state the rate of external mass transfer of substrate, V_{dif}, will be equal to its internal removal by reaction. Hence, for an enzyme reaction, which obeys Michaelis–Menten kinetics, the overall rate of reaction, V_{obs}, will be

$$V_{obs} = k_L a(S_B - S_S) = \frac{V_{max} S_S}{K_m + S_S}$$

This equation may be solved for S_S if $k_L a$ and the kinetic constants are known (42), or S_S may be obtained graphically (42) using the following dimensionless equation:

$$\frac{V_{obs}}{V_{max}} = \frac{\beta_B - \beta_S}{\mu} = \frac{\beta_S}{1 + \beta_S}$$

where β is the dimensionless substrate concentration ($\beta = S/K_m$) and μ is the dimensionless substrate modulus, ($\mu = V_{max}/k_L a K_m$). The dependence of V_{obs}/V_{max} on β for different values of μ is shown in Figure 7.3.

The external mass transfer effects on the activity of an immobilized enzyme can be quantitatively expressed by the effectiveness factor η, defined as the ratio of the observed reaction rate V_{obs} to the kinetic rate V_{kin}:

$$\eta = V_{obs}/V_{kin}$$

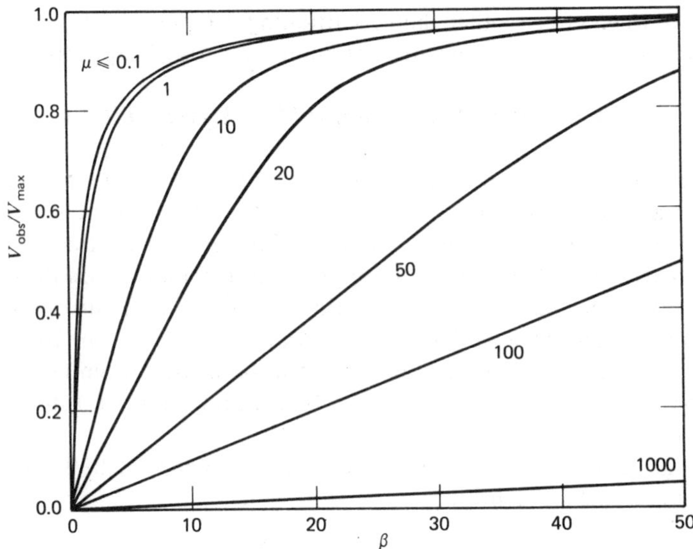

Fig. 7.3. V_{obs}/V_m against the dimensionless bulk concentration β for different values of the substrate modulus μ for external diffusion. From C. Horvath and J. M. Engasser, *Biotechnol. Bioeng.* **16,** 909 (1974).

Figure 7.4 shows the dependence of the effectiveness factor η on β and μ. Similar plots have been obtained by other authors (42).

The rate flow of substrate V_{dif}, or the rate of enzyme reaction V_{kin}, may play predominant roles, depending on their relative magnitudes, as the lower rate step will be the controller step.

As can be seen in Figure 7.5, when the reaction is zero order, V_{obs} will always be equal to V_{max} and the reaction will be kinetically controlled. For first-order reactions, the reaction can be kinetically or diffusionally controlled, depending on $k_La \gg V_{max}/K_m$, that is, mass transport is much faster than the enzyme reaction, or $k_La \ll V_{max}/K_m$, the enzyme reaction is much faster than the diffusion of substrate.

Decreasing of the resistance to external mass transfer is achieved with an increase of linear velocity of the fluid, as this velocity reduces the resistance to a point that S_B and S_S may be considered equals, and the reaction rate, V_{kin}, will be the controlling step.

When an enzyme is immobilized within a porous support, in addition to possible external mass transfer effects there could also exist resistances to internal diffusion of substrate, as this must diffuse through the pores in order to reach the enzyme, and resistance of product, as it must diffuse to the bulk solution. Consequently a substrate concentration gradient is established

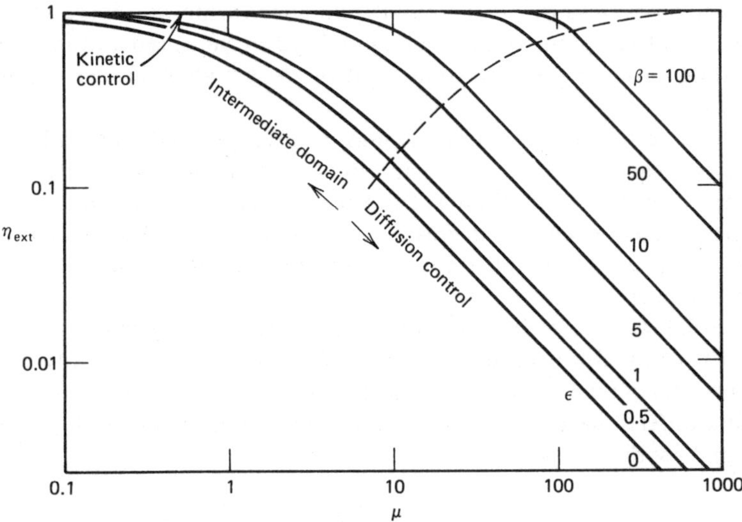

Fig. 7.4. Plots of the external effectiveness factor η_{ext} as a function of the substrate modulus μ for different values of the bulk substrate concentration. β, ϵ is the limiting first order effectiveness factor attained at sufficiently low concentrations. From C. Horvath and J. M. Engasser. *Biotechnol. Bioeng.* **16**, 909 (1974).

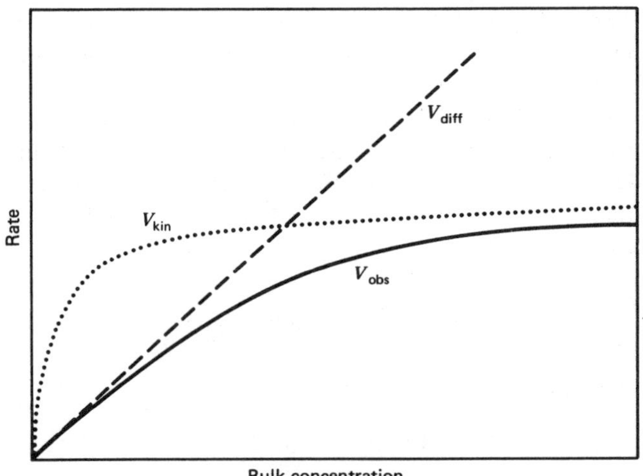

Fig. 7.5. Schematic plot of the overall rate of reaction V, catalyzed by a surface-bound enzyme against the bulk substrate concentration. From C. Horvath and J. M. Engasser, *Biotechnol. Bioeng.* **16**, 909 (1974).

within the pores, resulting in a concentration decreasing with increased distance (in depth) from the surface of immobilized enzyme preparation. A corresponding product concentration gradient is obtained in the opposite direction.

Unlike external diffusion, internal mass transfer proceeds in parallel with the enzyme reaction and takes into account the depletion of substrate within the pores with increasing distance from the surface of the enzyme support. The rate of reaction will also decrease, for the same reason. The overall reaction is dependent on the substrate concentration and the distance from the outside support surface.

The usual way to study this problem is by considering that there is a coupled reaction–diffusion process that can be solved, at the steady state, when the rates of internal diffusion and enzyme reaction are equal, using appropriate differential equations for the various geometries considered and isothermal conditions (42):

$$\frac{d^2S}{dx^2} + \frac{(p+1)}{x} \cdot \frac{dS}{dx} = \frac{V_{\text{kin}}}{D_{\text{eff}}} = \frac{V_{\max}S}{D_{\text{eff}}(K_m + S)}$$

where S is the substrate concentration; x is the distance from the outer surface; p is a geometrical factor with the values of $+1$ for spherical pellet, 0 for cylindrical pellet, and -1 for rectangular membranes; and D_{eff} is the effective diffusivity of the substrate inside the support, given by:

$$D_{\text{eff}} = D \cdot \epsilon / \tau$$

where D is the substrate diffusivity; ϵ is the void fractionation in the porous matrix; and τ is a tortuosity factor that takes into account the pore geometry and by definition is larger than unity.

The analytical solutions of these equations are easily obtained for first order or constant order reactions, but numerical solutions are required for Michaelis–Menten type reactions. The above equations, in these cases, are usually rewritten in terms of dimensionless variables. For a spherical pellet this equation is:

$$\frac{d^2\beta}{dZ^2} + \frac{2}{Z} \cdot \frac{d\beta}{dZ} = 9\phi^2 \frac{\beta}{1+\beta}$$

with the boundary conditions:

$$\beta = \beta_S \qquad \text{for } Z = 1$$

$$\frac{d\beta}{dZ} = 0 \qquad \text{for } Z = 0$$

In this equation β is the dimensionless substrate concentration, Z is the dimensionless position in the porous support given by $Z = x/R$, R is the radius of the spherical pellet, and ϕ is the substrate modulus (a modified Thiele modulus) defined by:

$$\phi = R/3\left(\frac{V_{\max}}{K_m D_{\text{eff}}}\right)^{1/2}$$

Numerical integration yields the effective rate of reaction, V_{obs}, as a function of the concentration β with the modulus ϕ (Figure 7.6). The same results can also be represented in the form of graphics of effectiveness factor η against the modified thiele modulus ϕ (Figure 7.7).

The internal mass transfer effects can be reduced, however, by decreasing the particle dimensions of the porous support containing the enzyme. Particle diameter decrease results in a reduction of the distance from the outer support surface that the substrate must cross, and consequently also results in a decrease of the substrate concentration gradient.

When external and internal diffusion resistances simultaneously affect the rate of the enzymic reaction, the relative contributions of each effect must be estimated separately and quantified by the corresponding effectiveness factors. Hence, the overall reaction rate is given by:

$$V_{\text{obs}} = \eta_{\text{ext}} \cdot \eta_{\text{int}} \cdot V_{\text{kin}}$$

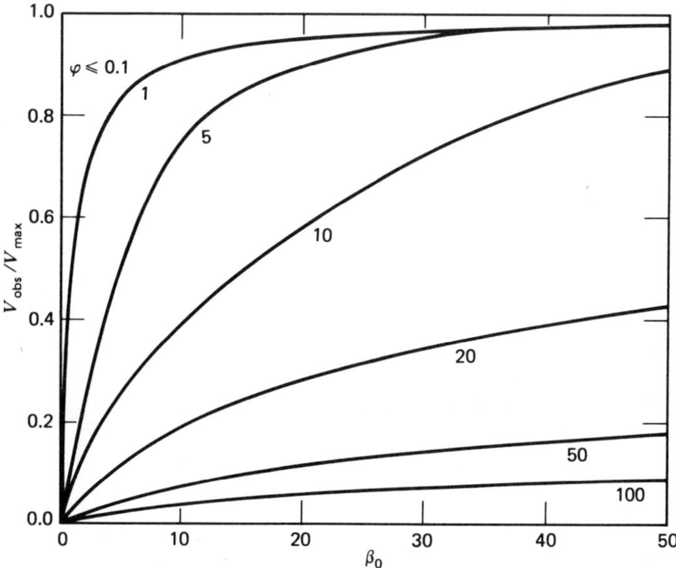

Fig. 7.6. Normalized overall rate as a function of the dimensionless bulk concentration of substrate β for different values of ϕ, the substrate modulus for internal diffusion in an enzyme membrane. From C. Horvath and J. M. Engasser, *Biotechnol. Bioeng.* **16,** 909 (1974).

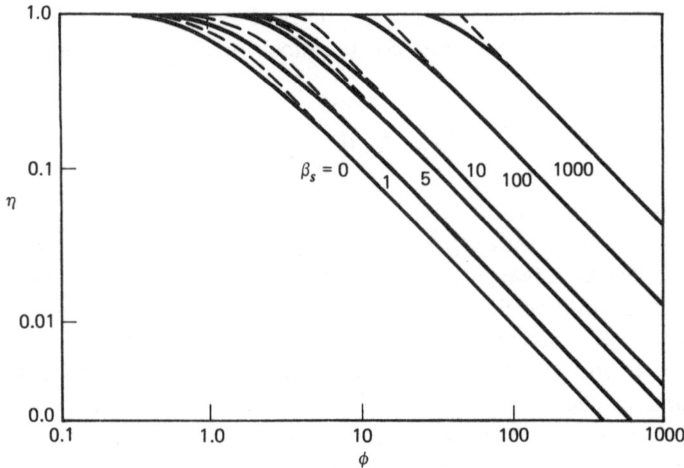

Fig. 7.7. Graph illustrating the effectiveness factor η_{int} as a function of the modulus ϕ with the dimensionless surface concentration as the parameter. The effectiveness factor for spherical particles and membranes is represented by solid and broken lines, respectively. From J. M. Engasser and C. Horvath, *J. Theor. Biol.*, **42**, 437 (1973).

7.4.4. Miscellaneous Effects

Apart from the influence of these factors on the activity of the enzyme, other properties of the enzyme can change. Thus substrate specifity alters, particularly when using high molecules substrate, by effect of steric hindrance and diffusional resistances. The kinetics constants K_m and V_m of the immobilized enzyme are different from the free enzyme as a consequence of conformational changes of the immobilized form, which affect the affinity between enzyme and substrate. The increase of activity energy for some immobilized enzymes may be attributed to diffusional resistances, mainly in porous supports.

One of the major properties of the immobilized enzymes is their stability, especially the operational stability. This enhanced stability is particularly valuable in the continuous use of immobilized enzymes, in industrial operations.

Such increased stability has been observed in many cases, although a decreased stability has been observed in a few systems (42).

7.5. IMMOBILIZED ENZYME REACTORS

Among the applications of immobilized enzymes, their utilization in industry is perhaps the most important and consequently the most frequently dis-

Table 7.30. Classification of Enzyme Reactors

Mode of Operation	Flow Pattern	Type of Reaction
Batch	Well mixed	Batch stirred tank reactor (BSTR)
	Plug flow	Total recycle reactor
Continuous	Well mixed	Continuous stirred tank reactor (CSTR)
		CSTR with continuous ultrafiltration membrane
	Plug flow	Packed bed reactor (PBR)
		Fluidized bed reactor (FBR)
		Tubular reactor (other)
		Hollow fiber reactor

cussed. The use of immobilized enzymes in industrial processes is performed in basic chemical reactors. Several classifications of enzyme reactors have been proposed (26, 31, 40, 46), based on the mode of operation and the flow characteristics of substrate and product. A possible classification is presented in Table 7.30.

7.5.1. Batch Reactors

Batch reactors are the most commonly used type of reactor when soluble enzymes are used as catalysts. The soluble enzymes are not generally separated from the products and consequently are not recovered for reuse.

Since one of the main goals of immobilizing an enzyme is to permit its reuse, the application of immobilized enzymes in batch reactors requires a separation (or an additional separation) to recover the enzyme preparation. During this recovery process, appreciable loss of immobilized enzyme material may occur as well as loss of enzyme activity, therefore the use of immobilized enzymes in a batch operation is generally limited to the production of rather small amounts of fine chemicals. Traditionally, the stirred tank reactor has been used for batchwise work. Composed of a reactor and a stirrer, it is the simplest type of reactor that allows good mixing and relative ease of temperature and pH control. However, some matrices, such as inorganic supports, are broken up by attrition in such vessels, and alternative designs have therefore been attempted. A possible laboratory alternative is the basket reactor, in which the catalyst is retained within a "basket," either forming the impeller "blades" or the baffles of the tank reactor.

Another alternative is to change the flow pattern, using a plug flow type of reactor: the total recycle reactor or batch recirculation reactor, which may be a packed bed or fluidized bed reactor, or even a coated tubular reactor.

This type of reactor may be useful where a single pass gives inadequate conversions. However, it has found greatest application in the laboratory for the acquisition of kinetic data, when the recycle rate is adjusted so that the conversion in the reactor is low and it can be considered as a differential reactor. One advantage of this type of reactor is that the external mass transfer effects can be reduced by the operational high fluid velocities.

7.5.2. Continuous Reactors

Most immobilized enzymes are used in continuous operation. This has some advantages when compared with batch processes, such as ease of automatic control, ease of operation, and quality control of products.

Continuous reactors can be divided into two basic types: the continuous feed stirred tank reactor (CSTR) and the plug flow reactor (PFR).

In the ideal CSTR the conversion degree is independent of the position in the vessel, as a complete mixing is obtained with stirring and the conditions within the CSTR are the same as the outlet stream, that is, low substrate and high product concentrations. With the ideal PFR the conversion degree is dependent on the length of the reactor as no mixing device at all exists and the conditions within the reactor are never uniform.

While a nearly ideal CSTR is readily obtained, since it is only necessary to have good stirring to obtain complete mixing, an ideal PFR is very difficult to obtain. Several factors adverse to obtaining an ideal PFR often occur, such as temperature and velocity gradients normal to the flow direction and axial dispersion of substrate.

Several considerations influence the type of continuous reactor to be chosen for a particular application. One of the most important criteria is based on kinetic considerations. As can be seen from Figure 7.8 for Michaels–Menten kinetics, the PFR is preferable to the CSTR as the CSTR requires more enzyme to obtain the same degree of conversion as a PFR. If product inhibition occurs, this problem is accentuated, as in a CSTR high product concentration is always in direct contact with all of the catalyst. There is only one situation where a CSTR is more favorable kinetically than a PFR, namely, when substrate inhibition occurs.

The form and characteristics of the immobilized enzyme preparations also influence the choice of reactor type, and operational requirements are still another factor to be taken into account. Thus, when pH control is necessary, for instance with penicillin acylase, the CSTR or batch stirred tank reactor are more suitable than PFR reactors. Due to possible disintegration of support through mechanical shearing, only durable preparations of immobilized enzymes should be used in a CSTR. With very small immobilized

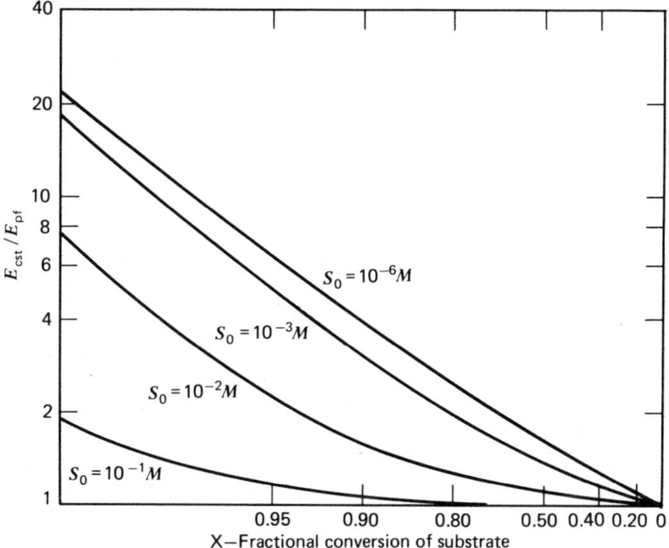

Fig. 7.8. Variation in the ratio (E_{CST}/E_{PF}) with conversion (X) for various values of feed substrate concentration (S_0), at the same K_m ($= 10^{-3}$ M) and constant flow rate. From M. D. Lilly and A. K. Sharp, *Chem. Eng.* (*London*) **215,** CE12 (1968).

enzyme particles, problems such as high pressure drop and plugging arise from the utilization of this catalyst in packed bed reactors (the most used type of PFR). To overcome these problems, a fluidized bed reactor, which provides a degree of mixing intermediate to the CSTR and the ideal PFR, can be used with low pressure drop.

Reactant characteristics can also influence the choice of reactor. Insoluble substrates and products and highly viscous fluids are preferably processed in fluidized bed reactors or CSTR, where no plugging of the reactor is likely to occur, as would be the case in a packed bed reactor.

As can be deduced from this outline, there are no simple rules for choosing the reactor type and the different factors mentioned must be analyzed individually for a specific case.

7.6. APPLICATIONS OF IMMOBILIZED ENZYMES

A number of reviews (26–35) discuss uses for immobilized enzymes in the industrial, analytical, and medical fields. In this section these fields are presented in a series of tables (7.31–7.34) that illustrate the present state of the art in the application of a wide variety of enzymes in these fields.

Table 7.31. Industrial Uses of Immobilized Enzymes

Use	Enzyme	Immobilization method	Reference
Sugar industry			
Hydrolysis of starch (production of D-glucose)	α-Amylase (soluble) (EC 3.2.1.1) + gluco-amylase (EC 3.2.1.3)	Adsorption (copolymer of phenylene diamine and glutaraldehyde)	454
		Covalent binding ($CPSiO_2$ + APTS + glutaraldehyde)	455
		Covalent binding (CPG + APTS + glutaraldehyde)	456
		Covalent binding (CM-cellulose-azide)	457, 458
		Covalent binding (DEAE-cellulose triazinyl derivative)	458, 466
		Ionic binding (DEAE-cellulose)	459, 460
		Physical adsorption (activated carbon)	461
		Ultrafiltration membrane method	462, 463
Hydrolysis of starch (production of maltose)	β-Amylase (EC 3.2.1.2) + amylo-1,6-D-glucosidase (EC 3.2.1.33)	Physical adsorption (active coal)	464
		Entrapment (acrylamide/acryl acid)	465
		Covalent binding (porous cellulose)	466
Hydrolysis of cellulose	Cellulase (EC 3.2.1.4)	Gel entrapping (collagen + glutaraldehyde)	467
		Ultrafiltration membrane method	468, 469
Isomerization of D-glucose (production of D-fructose)	D-Glucose isomerase (EC 5.3.1.5)	Crosslinking (glutaraldehyde)	470
		Physical adsorption (aluminum)	471
		Ionic binding (DEAE-cellulose)	472, 473
		Ionic binding (DEAE-dextran)	474
		Covalent binding (CPG + diazo coupling)	475
		Gel entrapping (polyacrylamide)	472, 476
		Fiber entrapping (cellulose triacetate)	477

Hydrolysis of sucrose (production of invert sugar)	β-D-Fructofuranosidase (EC 3.2.1.26)	Gel entrapping (polyacrylamide)	478
		Fiber entrapping (cellulose triacetate)	387, 394
		Crosslinking (tannic acid)	479
		Covalent binding (CPG + diazo coupling)	480
		Ionic binding	481
Hydrolysis of raffinose	α-D-Galactosidase (EC 3.2.1.22)	Hollow fiber method	482
Production of D-gluconic acid	D-Glucose oxidase (EC 1.1.3.4)	Adsorption (copolymer of phenylene diamine and glutaraldehyde)	454
		Covalent binding (AE cellulose + glutaraldehyde)	173
		Covalent binding (CPSiO_2 + APTS + glutaraldehyde)	283
		Covalent binding (activated carbon + carbodiimides)	110, 230
		Covalent binding (collagen azide derivative)	192
Dairy Industry			
Hydrolysis of lactose	β-D-Galactosidase (EC 3.2.1.23)	Gel entrapping (polyacrylamide)	483, 484
		Fiber entrapping (cellulose triacetate)	395, 485
		Covalent binding (CPG + APTS + glutaraldehyde)	486
		Covalent binding (aluminum + APTS + glutaraldehyde)	487
Cheese manufacturing	Chymosin (EC 3.4.23.4)	Covalent binding (CNBr-activated agarose)	488
		Covalent binding (AE-cellulose + glutaraldehyde)	488
	Pepsin (EC 3.4.23.1)	Adsorption + crosslinking (alumina)	489
Sterilization of milk	Catalase (EC 1.11.1.6)	Ionic binding (DEAE-cellulose)	490
		Crosslinking (glutaraldehyde)	490

Table 7.31. (*continued*)

Use	Enzyme	Immobilization method	Reference
		Gel entrapment (collagen)	491
	Peroxidase (EC 1.11.1.7)	Covalent binding (CNBr-activated agarose)	492
Production of Amino acids			
Resolution of racemic mixtures	Aminocyclase (EC 3.5.1.14)	Ionic binding (DEAE-Sephadex)	493, 494
		Covalent binding (CPG + APTS + glutaraldehyde)	495
		Fiber entrapping (cellulose triacetate)	389
	Leucine aminopeptidase (cytosol) (EC 3.4.11.1)	Covalent binding (CNBr-activated agarose)	496
L-Aspartic acid	Aspartate ammonia-lyase (EC 4.3.1.1)	Gel entrapping (polyacrylamide)	497
		Fiber entrapping (viscose fiber)	498
		Gel entrapping (collagen)	499
L-Tryptophan	Tryptophanase (EC 4.1.99.1)	Fiber entrapping (cellulose triacetate)	500
L-Tyrosine	Tyrosine phenol-lyase (EC 4.1.99.2)	Covalent binding (CNBr-activated agarose)	501
L-Lysine	L-α-Amino-ϵ-caprolactam hydrolase + amino-ϵ-caprolactam racemase	Ionic binding (DEAE-dextran)	502
	Diaminopimelate decarboxylase (EC 4.1.1.20)	Gel entrapping (polyacrylamide)	503
Beer Industry			
Chill proofing	Papain (EC 3.4.22.2)	Gel entrapping (collagen)	504
		Covalent binding (hydroxyalkyl methacrylate)	505

Juice and Wine Industries			
Clarification	Protease	Covalent binding (CNBr-activated agarose)	506
	Pepsin (EC 3.4.23.1)	Covalent binding (bromoacetyl cellulose)	507
	Acid proteinases	Covalent binding (bromoacetyl cellulose)	507
Production of Nucleotides (IMP, GMP)			
Flavor enhancement of meat	Phosphodiestease 1 (EC 3.1.4.1)	Covalent binding (porous ceramic)	508
	AMP deaminase (EC 3.5.4.6)	Covalent binding (porous ceramic)	508
Pharmaceutical Industry			
Penicillins	Penicillin amidase (EC 3.5.1.11)	Fiber entrapping (cellulose acetate)	396, 509
		Covalent binding (CNBr-activated dextran)	510
		Covalent binding (polymethacrylate + glutaraldehyde)	511
		Covalent binding (copolymer of acrylamide and maleic acid anhydride)	512, 513
		Adsorption (bentonite)	514, 515
		Gel entrapping (polyacrylamide)	516, 517
Cephalosporin	Deacylase	Adsorption (celite)	518, 519
		Adsorption (calcium phosphate + dicalite)	520
		Fiber entrapping (cellulose acetate)	396
		Adsorption (brick)	521
		Ultrafiltration membrane method	522
Steroids	3β-Hydroxysteroid dehydrogenase (EC 1.1.1.51 etc.)	Adsorption (copolymer of 1,4-diaminobenzene and glutaraldehyde)	523

Table 7.31. *(continued)*

Use	Enzyme	Immobilization method	Reference
α-Ketoacids	L-Amino-acid oxidase (EC 1.4.3.2) + Catalase (EC 1.11.1.6)	Covalent binding (alkylamine glass + glutaraldehyde)	524
L-Malic acid	Fumarate hydratase (EC 4.2.1.2)	Fiber entrapping (cellulose acetate)	390
Fine Chemicals			
Glutathione	Glutathione synthetase (EC 6.3.2.3)	Covalent binding	525, 526
Porphobilinogen	Porphobilinogen synthase (EC 4.2.1.24)	Covalent binding (CNBr-activated agarose)	527, 528
Waste Treatment			
Detoxification of pesticides	Parathion hydrolase	Covalent binding (CPG + APTS + glutaraldehyde)	195
Ultrafiltration of raw sewage	Proteases + cellulase (EC 3.2.1.4)	Ultrafiltration membrane method	529

Table 7.32. Application of Immobilized Enzymes in Automated Analysis in Combination with a Spectrophotometer

Determination	Enzyme	Immobilization Method	Reference
D-Glucose	D-Glucose oxidase (EC 1.1.3.4)	Gel entrapping (polyacrylamide)	530
		Covalent binding (nylon + glutaraldehyde)	531, 532
	Hexokinase (EC 2.7.1.1) + D-Glucose-6-phosphate dehydrogenase (EC 1.1.1.49)	Covalent binding (nylon + glutaraldehyde)	533
Maltose	Glucoamylase (EC 3.2.1.3) + D-Glucose oxidase (EC 1.1.3.4)	Covalent binding (nylon + glutaraldehyde)	534
Sucrose	β-D-Fructofuranosidase (EC 3.2.1.26) +D-Glucose oxidase (EC 1.1.3.4)	Covalent binding (nylon + glutaraldehyde)	534
Lactose	β-D-Galactosidase (EC 3.2.1.23) + D-Glucose oxidase (EC 1.1.3.4)	Covalent binding (nylon + glutaraldehyde)	534
Urea	Urease (EC 3.5.1.5)	Diazotization (aminoacid copolymer)	535
		Covalent binding (nylon + glutaraldehyde)	531
Nitrate	Nitrate reductase (EC 1.7.99.4)	Diazotization (porous glass)	536
Phosphate	Alkaline phosphatase (EC 3.1.3.1)	Covalent binding (porous glass)	537
Ethanol	Alcohol dehydrogenase (EC 1.1.1.1)	Covalent binding (nylon + glutaraldehyde)	538
		Covalent binding (sand + APTS + glutaraldehyde)	539

Table 7.32. (*continued*)

Determination	Enzyme	Immobilization Method	Reference
Lactic acid	Lactate dehydrogenase (EC 1.1.1.27)	Gel entrapping (polyacrylamide)	530
		Covalent binding (nylon + glutaraldehyde)	538
D-Amino acid	D-Amino acid oxidase	Peptide binding (CNBr-agarose)	540
L-Aspartic acid	Aspartic amino transferase (EC 2.3.1.17)	Peptide binding	541
L-Tryptophan	Tryptophanase (EC 4.1.99.1)	Peptide binding (CNBr-agarose)	542

In the industrial area most of the work has been done with hydrolases, as there are many kinds of hydrolases and they are generally stable and do not require coenzymes. The main areas of application of immobilized enzymes in chemical processes are the nutritional applications (the sugar and food industry) and the pharmaceutical industry, although other fields such as waste treatment and organic synthesis (see Chapter 8) are of importance too. In these fields, however, relatively few applications are on an industrial scale, and only the isomerization of D-glucose to D-fructose is a large-scale operation.

In the analytical field, several recent advances in enzyme technology in conjunction with the potential analytical applications of enzymes (see Volume 56 in this series) make commercially available several immobilized enzymes, particularly as enzyme electrodes (see Chapter 9).

Biomedical applications of immobilized enzymes are still in a state of basic studies, as many necessary prerequisites—absence of toxicity, absence of haemolysis and allergenicity, chemical stability "in vivo" and immunological reactions, must be achieved very precisely before real applications can be profitably realized.

7.7. FUTURE TRENDS IN IMMOBILIZED ENZYME TECHNOLOGY

Although the first artificial immobilized enzyme was presented almost 65 years ago, it was only in the late 1960s that academic institutions and industry began to make great strides toward the widespread use of immobilized

Table 7.33. Immobilized Enzyme–Electrochemical Applications in Analytical Chemistry

Determination	Enzyme	Immobilization Method	Sensor	Reference
D-Glucose	D-Glucose oxidase	Gel entrapping (polyacrylamide)	O_2	543, 544
		Covalent binding (glassy carbon)	O_2	111
		Gel entrapping (collagen)	O_2	545
Sucrose	β-D-Fructofuranosidase (EC 3.2.1.26) +D-Glucose oxidase (EC 1.1.3.4)	Gel entrapping (collagen)	O_2	546
Urea	Urease (EC 3.5.1.5)	Gel entrapping (polyacrylamide)	Cation	547
		Gel entrapping (collagen)	Cation	548
		Ultrafiltration membrane method	CO_2	549
Lactic Acid	Lactate dehydrogenase (EC 1.1.1.27)	Gel entrapping (collagen)	Pt	550, 551
Ethanol	Alcohol dehydrogenase (EC 1.1.1.1)	Gel entrapping (collagen)	Pt	550, 551
D-Aminoacid	D-Amino-acid oxidase (EC 1.4.3.3)	Gel entrapping (polyacrylamide)	Cation	552, 553
Cholesterol	Cholesterol oxidase (EC 1.1.3.6)	Gel entrapping (collagen)	O_2	554
Penicillin	Penicillinase (EC 3.5.2.6)	Ultrafiltration membrane method	pH	555
		Gel entrapping (polyacrylamide)	pH	556

Table 7.33. (*continued*)

Determination	Enzyme	Immobilization Method	Sensor	Reference
Amygdalin	β-D-Glucosidase (EC 3.2.1.21)	Gel entrapping (polyacrylamide)	CN	557, 558
L-Phenylalanine	L-Amino-acid oxidase (EC 1.4.3.2)	Gel entrapping (polyacrylamide)	NH_4^+	559
	L-Amino-acid oxidase (EC 1.4.3.2) + peroxidase (EC 1.11.1.7)	Gel entrapping (polyacrylamide)	I_2	559
Monoamines	Amine oxidase (copper containing) (EC 1.4.3.4)	Gel entrapping (collagen)	O_2	560
Glutamine	Glutaminase (EC 3.5.1.2)	Ultrafiltration membrane method	Cation	552

Table 7.34. Medical Uses of Immobilized Enzymes

Condition	Enzyme	Immobilization Method	Reference
Acatalasemia	Catalase (EC 1.11.1.6)	Microencapsulation (phase separation method)	561, 562
Leukemia	Asparaginase (EC 3.5.1.1)	Microencapsulation (phase separation method)	563
		Gel entrapping (polyacrylamide)	564, 565
		Covalent binding (nylon + glutaraldehyde)	566
		Covalent binding (polymethacrylate + APTS + glutaraldehyde)	567
		Microencapsulation (interfacial polymerization method)	561
Hyperuricemia	Urate oxidase (EC 1.7.3.3)	Covalent binding (glass beads + APTS + glutaraldehyde)	568
Uremic disorders	Urease (EC 3.5.1.5)	Microencapsulation (phase separation method)	569, 570
Artificial lungs	Carbonate dehydratase (EC 4.2.1.1) + catalase (EC 1.11.1.6)	Co-crosslinking (enzyme + albumin + glutaraldehyde)	571
Glycogen storage disease	Glucoamylase (EC 3.2.1.3)	Microencapsulation (liquid membrane method)	572

enzyme technology. To date, however, only three processes based on immobilized enzymes are operative on an industrial scale:

(i) Production of L-amino acids by the optical resolution of acetyl-DL-amino acids using immobilized aminoacylase (EC 3.5.1.14), in Japan.

(ii) Production of fructose syrup using immobilized D-glucose isomerase (D-xylose isomerase) (EC 5.3.1.5), in Europe, Japan, and the U.S.

(iii) Production of 6-aminopenicillanic acid using immobilized penicillin amidase (EC 3.5.1.11), in Europe, Japan, and the U.S.

Besides these few processes, there are two others, based on immobilized microbial cells, for the production of L-aspartic acid and L-malic acid (both in Japan).

This limited number of industrial applications, in contrast with the large number of published reports, is due chiefly to economic factors, for example, the carrier or reagent costs for the immobilization procedure may be too expensive. Other factors may be low efficiency of immobilization, poor operational stability, relatively complicated equipment for continuous operation, and demand for the products being insufficient to justify large-scale production.

The investigated applications are almost exclusively hydrolytic or single-step conversion processes; this puts the immobilized enzyme technology in competition with high-performance chemical or soluble enzyme processes. Further expansion of immobilized enzyme technology must take into account different and more complex processes and at the same time must try to overcome its economic and technical limitations.

The use of microbial cells instead of enzymes in the immobilization procedure can solve several economic problems, such as isolation and extraction of the enzyme from microbial cells and minimization of the enzyme loss (see Chapter 15). Inexpensive and promising matrices such as chitin, κ-carrageenan, inorganic materials (porous ceramics and naturally occuring minerals), and some hydrophilic synthetic polymers (polyacrylamides and hydroalkylmethacrylates) are expected to be used on a more widespread basis to minimize the cost of immobilization procedure. More thermostable enzymes are expected to be obtained from microorganisms; they should confer a better operational stability to immobilized enzyme preparations.

But the most active field of immobilized enzyme application will probably be in the development and evaluation of new reactors using immobilized multienzyme systems requiring multiphase environmental and/or cofactor regeneration, which have not yet been developed for industrial use. These applications include the use of immobilized microbial cells for the production of materials now obtained by fermentation, and the syntheses of new useful products using immobilized forms of enzymes and cofactors. The use of immobilized enzymes or microbial cells (biocatalysts) for waste-water treatment is also expected to be developed. Rising energy costs and the need for new energy sources, another area where much research remains to be done, may stimulate the application of immobilized biocatalysts to energy production from sunlight to chemical energy and from chemical energy to electric energy (biochemical fuel cells).

In the medical area, immobilized enzymes are expected to be developed for the diagnosis and treatment of diseases, in artificial cells and organs, and in the coating of artificial materials to minimize rejecting by the body's immune system.

REFERENCES

1. J. B. Sumner, *Science* **108,** 410 (1948).
2. L. Goldstein and E. Katchalski-Katzir, in *Applied Biochemistry and Bioengineering,* Vol. 1 (L. B. Wingard Jr., L. Goldstein, and E. Katchalski-Katzir, Ed.), Academic Press, New York (1976), p. 1.
3. J. M. Nelson and E. G. Griffin, *J. Amer. Chem. Soc.* **38,** 1109 (1916).
4. H. C. Isliker, *Advances in Protein Chem.* **12,** 387 (1957).
5. K. Landsteiner and J. Van der Scheer, *J. Exp. Med.* **63,** 325 (1936).
6. F. Michael and J. Evers, *Makromol. Chem.* **3,** 200 (1949).
7. D. H. Campbell, F. Luescher and L. S. Lerman, *Proc. Nat. Acad. Sci. U. S.* **37,** 575 (1951).
8. H. C. Isliker, *Ann. N. Y. Acad. Sci.* **57,** 225 (1953).
9. G. Manecke and K. E. Gilbert, *Naturwissenschaften* **42,** 212 (1955).
10. G. Manecke, S. Singer, and K. E. Gilbert, *Naturwissenschaften* **45,** 440 (1958).
11. N. Grubhofer and L. Schleith, *Naturwissenschaften* **40,** 508 (1953).
12. G. Manecke and S. Singer, *Makromol. Chem.* **37,** 119 (1960).
13. H. Brandenberger, *Angew. Chem.* **67,** 661 (1955).
14. H. Brandenberger, *J. Polymer. Sci.* **20,** 215 (1956).
15. H. Brandenberger, *Helv. Chim. Acta* **40,** 61 (1957).
16. A. Bar-Eli and E. Katchalski-Katzir, *Nature* **188,** 856 (1960).
17. A. Bar-Eli and E. Katchalski-Katzir, *J. Biol. Chem.* **238,** 1690 (1963).
18. J. J. Cebra, D. Givol, and E. Katchalski-Katzir, *J. Biol. Chem.* **237,** 751 (1962).
19. E. Katchalski-Katzir, in *Polyamino Acids, Polypeptides and Proteins* (M. A. Stahman, Ed.), Univ. of Wisconsin Press, Madison (1962), p. 283.
20. G. Manecke, *Pure Appl. Chem.* **4,** 507 (1962).
21. T. Tosa, T. Mori, N. Fuse, and I. Chibata, *Enzymologia* **31,** 214 (1966).
22. P. Bernfeld and J. Wan, *Science* **142,** 678 (1963).
23. T. M. S. Chang, *Science* **146,** 524 (1964).
24. M. D. Lilly, W. E. Hornby, and E. M. Crook, *Biochem. J.* **100,** 718 (1966).
25. M. D. Lilly and A. K. Sharp, *Chem. Eng.* (London), CE 12 (1968).
26. L. B. Wingard Jr., Ed., *Enzyme Engineering,* Vol. 1, Wiley-Interscience, New York (1972).
27. E. K. Pye and L. B. Wingard Jr., Eds., *Enzyme Engineering,* Vol. 2, Plenum Press, New York (1974).
28. E. K. Pye and H. H. Weetall, Eds., *Enzyme Engineering,* Vol. 3, Plenum Press, New York (1978).
29. G. B. Broun, G. Manecke, and L. B. Wingard Jr., Eds., *Enzyme Engineering,* Vol. 4, Plenum Press, New York (1978).
30. H. H. Weetall and G. P. Royer, Eds., *Enzyme Engineering,* Vol. 5, Plenum Press, New York (1980).
31. O. R. Zaborsky, *Immobilized Enzymes,* CRC Press, Cleveland (1973).
32. G. R. Stark, Ed., *Biochemical Aspects of Reactions on Solid Supports,* Academic Press, New York (1971).
33. A. C. Olson and C. L. Cooney, Eds., *Immobilized Enzymes in Food and Microbial Processes,* Plenum Press, New York (1974).

34. M. Salmona, C. Saronio, and S. Garattini, Eds., *Insolubilized Enzymes,* Raven Press, New York (1974).
35. B. R. Dunlap, Ed., *Immobilized Biochemicals and Affinity Chromatography,* Plenum Press, New York (1974).
36. S. J. Gutcho, *Immobilized Enzymes,* Noyes Data Corporation (1974).
37. H. H. Weetall, Ed., *Immobilized Enzymes, Antigens, Antibodies, and Peptides,* Marcel Dekker, New York (1975).
38. R. A. Messing, Ed., *Immobilized Enzymes for Industrial Reactors,* Academic Press, New York (1975).
39. H. H. Weetall and S. Suzuki, Eds., *Immobilized Enzyme Technology,* Plenum Press, New York (1975).
40. L. B. Wingard Jr., E. Katchalski-Katzir, and L. Goldstein, Eds., *Applied Biochemistry and Bioengineering,* Vol. 1, Academic Press, New York (1976).
41. L. B. Wingard Jr., E. Katchalski-Katzir, and L. Goldstein, Eds., *Applied Biochemistry and Bioengineering,* Vol. 2, Academic Press, New York (1979).
42. K. Mosbach, Ed., *Methods in Enzymology,* Vol. 44, Academic Press, New York (1976).
43. J. F. Kennedy, *Advances in Carbohydr. Chem. Biochem.* **29,** 497 (1973).
44. A. Wiseman, Ed., *Handbook of Enzyme Biotechnology,* Ellis Horwood, Chichester (1975).
45. A. Wiseman, Ed., *Topics in Enzyme and Fermentation Biotechnology,* Vol. 1, Ellis Horwood, Chichester (1977).
46. A. Wiseman, Ed., *Topics in Enzyme and Fermentation Biotechnology,* Vol. 2, Ellis Horwood, Chichester (1978).
47. A. Wiseman, Ed., *Topics in Enzyme and Fermentation Biotechnology,* Vol. 3, Ellis Horwood, Chichester (1979).
48. A. Wiseman, Ed., *Topics in Enzyme and Fermentation Biotechnology,* Vol. 4, Ellis Horwood, Chichester (1980).
49. A. Wiseman, Ed., *Topics in Enzyme and Fermentation Biotechnology,* Vol. 5, Ellis Horwood, Chichester (1981).
50. D. Perlman, Ed., *Annual Reports on Fermentation Processes,* Vol. 1, Academic Press, New York (1977).
51. C. A. White and J. F. Kennedy, *Polymer* **25,** 1 (1980).
52. T. K. Ghose, A. Fiechter, and N. Blakebrough, Eds., *Advances in Biochemical Engineering,* Vol. 10, Springer-Verlag, Berlin (1978).
53. K. Venkatsubramanian, Ed., *Immobilized Microbial Cells,* ACS Symposium Series No. 106, American Chemical Society, Washington D.C. (1979).
54. T. K. Ghose, A. Fiechter, and N. Blakebrough, Eds., *Advances in Biochemical Engineering,* Vol. 12, Springer-Verlag, Berlin (1979).
55. C. M. Sturgeon and J. F. Kennedy, *Enzyme and Microbial Technology,* tables in every issue from Vol. 1 (1979).
56. J. C. Johnson, *Immobilized Enzymes,* Noyes Data Corporation, New Jersey (1979).
57. I. Chibata, Ed., *Immobilized Enzymes,* Kodansha, Tokyo (1978).
58. J. F. Kennedy in *Specialist Periodical Reports, Carbohydrate Chemistry,* Part II, Vols. 4–13, The Chemical Society, London (1971–1982).

59. C. A. White and J. F. Kennedy, *Enzyme Microb. Technol.* **2,** 82 (1980).
60. W. H. Pitcher, Jr., Ed., *Immobilized Enzymes in Food Processing,* CRC Press, W. Palm Beach, Florida (1979).
61. *Immobilized Enzymes and Cells and Enzymes,* International Technical Information Institute, Tokyo (1979).
62. P. Dunnill, A. Wiseman, and N. Blakebrough, Eds., *Enzymic and Nonenzymic Catalysis,* Ellis Horwood, Chichester (1980).
63. *Dechema Monographs,* Vol. 82, *Biotechnology,* Verlag Chemie, Weinheim (1979).
64. L. B. Wingard, I. V. Berezin, and A. A. Klyosov, Eds., *Enzyme Engineering: Future Directions,* Plenum Press, New York (1980).
65. *Dechema Monographs,* Vol. 84, *Characterization of Immobilized Biocatalysts,* Verlag Chemie, Weinheim (1980).
66. O. Zaborsky and J. F. Kennedy, Eds., *Enzyme and Microbial Technology,* Butterworth Scientific, Guildford (1979 onwards).
67. S. A. Barker and J. F. Kennedy, in *Handbook of Enzyme Biotechnology,* (A. Wiseman, Ed.), Ellis Horwood, Chichester (1975), p. 203.
68. F. A. Quiocho and F. M. Richards, *Proc. Nat. Acad. Sci. U. S.* **52,** 833 (1964).
69. F. M. Richards and J. R. Knowles, *J. Mol. Biol.* **37,** 231 (1968).
70. E. F. Jansen and A. C. Olson, *Arch. Biochem. Biophys.* **129,** 221 (1969).
71. E. F. Jansen, Y. Tomimatsu, and A. C. Olson, *Arch. Biochem. Biophys.* **144,** 394 (1979).
72. Y. Tomimatsu, E. F. Jansen, W. Gaffield and A. C. Olson, *J. Colloid Interface Sci.* **36,** 51 (1971).
73. G. B. Broun, in *Methods in Enzymology,* Vol. 44 (K. Mosbach, Ed.), Academic Press, New York (1976), p. 263.
74. H. Negoro, *J. Ferment. Technol.* **50,** 136 (1972).
75. A. C. Olson and W. L. Stanley, in *Immobilized Enzymes in Food and Microbial Processes* (A. C. Olson and C. L. Cooney, Eds.), Plenum Press, New York (1979), p. 51.
76. G. Sodini, V. Baroncelli, M. Canella, and P. Renzi, *Ital. J. Biochem.* **23,** 121 (1974).
77. B. K. Ahn, S. K. Wolfson Jr., and S. J. Yao, *Bioelectrochem. Bioenergetics* **2,** 142 (1975).
78. S. W. Carleysmith, P. Dunnill, and M. D. Lilly, *Biotechnol. Bioeng.* **22,** 735 (1980).
79. H. Y. Neujahr, *Biotechnol. Bioeng.* **22,** 913 (1980).
80. K. G. Kejellén and H. Y. Neujahr, *Biotechnol. Bioeng.* **21,** 715 (1979).
81. Y. Tashiro and T. Matsuda, Japanese Patent Kokai 78,32191 (1978).
82. F. A. Quiocho and F. M. Richards, *Biochemistry* **5,** 4062 (1966).
83. P. S. Marfey and M. V. King, *Biochim. Biophys. Acta* **105,** 178 (1965).
84. S. Avrameas and T. Ternynck, *Immunochem.* **6,** 53 (1969).
85. R. Ohba, H. Chaen, S. Hayashi, and S. Ueda, *Biotechnol. Bioeng.* **20,** 665 (1978).
86. R. A. Messing, in *Immobilized Enzymes for Industrial Reactors* (R. A. Messing, Ed.), Academic Press, New York (1975), p. 63.

87. J. F. Kennedy and A. Zamir, *Carbohyd. Res.* **41,** 227 (1975).
88. H. H. Weetall, in *Methods of Enzymology,* Vol. 44 (K. Mosbach, Ed.), Academic Press, New York (1976).
89. R. A. Messing, in *Advances in Biochemical Engineering,* Vol. 10 (T. K. Ghose, A. Fietcher, and N. Blakebrough, Eds.), Springer-Verlag, Berlin, (1978).
90. L. K. James and L. Augenstein, *Advances in Enzymol. Relat. Areas Mol. Biol.* **28,** 1 (1966).
91. C. A. Zittle, *Advances in Enzymol. Relat. Areas Mol. Biol.* **14,** 319 (1953).
92. A. D. McLaren, *J. Phys. Chem.* **58,** 129 (1954).
93. A. D. McLaren, G. H. Peterson, and I. Barshad, *Soil Sci. Soc. Amer. Proc.* **22,** 239 (1958).
94. A. D. McLaren and L. Packer, *Advances in Enzymol. Relat. Areas Mol. Biol.* **33,** 245 (1970).
95. J. P. Hummel and B. S. Anderson, *Arch. Biochem. Biophys.* **112,** 443 (1965).
96. J. Dixon, P. Andrew, and L. G. Butler, *Biotechnol. Bioeng.* **21,** 2113 (1979).
97. G. B. Ogumtimein and P. J. Reilly, *Biotechnol. Bioeng.* **22,** 1127 (1980).
98. Y. Horiuti and S. Imamura, *J. Biochem.* (Tokyo) **83,** 1381 (1978).
99. T. Watanabe, T. Mori, T. Tosa, and I. Chibata, *Biotechnol. Bioeng.* **21,** 477 (1979).
100. M. Ono, T. Tosa, and I. Chibata, *Agr. Biol. Chem.* **42,** 1847 (1978).
101. R. Ohba and S. Ueda, *Biotechnol. Bioeng.* **22,** 2137 (1980).
102. H. Schiger, E. H. Teufel, and M. Philipp, *Biotechnol. Bioeng.* **22,** 55 (1980).
103. J. Woodward and A. Wiseman, *Biochim. Biophys. Acta* **527,** 8 (1978).
104. V. P. Torchilin, M. Galka, and W. Ostrowski, *Biochim. Biophys. Acta* **483,** 331 (1977).
105. R. L. Van Etten and M. S. Saini, *Biochim. Biophys. Acta* **484,** 487 (1977).
106. B. Mattiasson and C. Borrebaeck, *FEBS Lett.* **85,** 119 (1978).
107. E. G. Sud'ina, M. A. Samartsov, M. G. Golod, and E. E. Dovbysh, *Ukr. Biokhim Zh.* **51,** 404 (1979).
108. D. L. Garbers, *J. Cyclic Nucleotide Res.* **4,** 271 (1978).
109. J. Tramper, S.A.G.F. Angelino, F. Müller, C. Henk, and H. C. van der Plas, *Biotechnol. Bioeng.* **21,** 1767 (1979).
110. Y. K. Cho and J. E. Bailey, *Biotechnol. Bioeng.* **21,** 461 (1979).
111. C. Bourdillon, J.-P. Bourgeois, and D. Thomas, *Biotechnol. Bioeng.* **21,** 1877 (1979).
112. R. N. Mukherjea, P. Bhattacharya, T. Gangopadhyary, and B. K. Ghosh, *Biotechnol. Bioeng.* **22,** 543 (1980).
113. P. G. Pifferi, C. Pasquali, M. G. Tocco, and I. M. Domini-Pellerano, *Tecnol. Aliment.* **3,** 9 (1980).
114. K. Buchholz, S. K. Dergerd, and A. Borchet, *Dechema Monogr.* **84,** 169 (1979).
115. Z. V. Mikelsone, A. N. Mitrofanova, O. M. Poltorak, and A. K. Arens, *Vestn. Mosk. Univ.* Ser. 2, **20,** 109 (1979).
116. W. Melander and C. Horvath, in *Enzyme Engineering,* Vol. 4 (G. B. Broun, G. Manecke, and L. B. Wingard Jr., Eds.), Plenum Press, New York (1978), p. 355.
117. E. G. Sud'ina and M. G. Golod, *Ukr. Biokhim Zh.* **51,** 400 (1979).

118. P. Grunwald, W. Gunssen, F. R. Heiker, and W. Roy, *Anal. Biochem.* **100,** 54 (1979).
119. M. A. Mitz, *Science* **123,** 1076 (1956).
120. L. Y. Bessermertnaya and V. K. Antonov, *Khim. Proteoliticheskikh Fermentov,* Mater. Vses. Simp. (1973), p. 43.
121. G. A. Yarovaya, T. N. Gulynskaya, V. L. Dotsenko, L. Y. Bessermertnaya, L. V. Kozlov, and V. K. Antonov, *Biorg. Khim.* **1,** 646 (1975).
122. B. Solomon and Y. Levin, *Biotechnol. Bioeng.* **16,** 1161 (1974).
123. H. Maeda, L. F. Chen and, G. T. Tsao, *J. Ferment. Technol.* **57,** 238 (1979).
124. M. F. Aukati, T. I. Kalashnivoka, S. N. Bubenshchikova, V. K. Kagramanova, and L. A. Baratova, *Vestn. Mosk. Univ.,* Ser. 2, **19,** 350 (1978).
125. C. Huitron and J. Limon-Lason, *Biotechnol. Bioeng.* **20,** 1377 (1978).
126. J. Baratti, R. Couderc, C. L. Cooney, and D. I. C. Wang, *Biotechnol. Bioeng.* **20,** 333 (1978).
127. H. Kaboli and P. J. Reilly, *Biotechnol. Bioeng.* **22,** 1055 (1980).
128. E. Keller, J. Eberspächer, and F. Lingens, *Z. Physiol. Chem.* **360,** 19 (1979).
129. K. G. Kjellén and H. Y. Neujahr, *Biotechnol. Bioeng.* **22,** 299 (1980).
130. Y. Miura, K. Miyamoto, H. Urabe, H. Tanaka, and T. Yasuda, *J. Ferment. Technol.* **57,** 440 (1979).
131. H. E. Klei, D. W. Sundstrom, and R. Gargano, *Biotechnol. Bioeng.* **20,** 611 (1978).
132. H. Ooshima, M. Sakimoto, and Y. Harano, *Biotechnol. Bioeng.* **22,** 2155 (1980).
133. S. Adachi, Y. Kawamura, K. Nakanishi, R. Matsuno, and T. Kamikubo, *Agr. Biol. Chem.* **42,** 1707 (1978).
134. S. V. Klinov, N. P. Sugrobova, and B. T. Kurganov, *Molek. Biol.* **13,** 559 (1979).
135. P. S. J. Cheetham, *J. Appl. Biochem.* **1,** 51 (1979).
136. H. Yamada, *J. Biochem.* (Tokyo) **83,** 1577 (1978).
137. Y. Gouges, J. Amen, and S. Sebesi, French Patent 2,420,542 (1979).
138. J. M. Novais, PhD. Thesis, University of Birmingham (1971).
139. A. N. Emery, J. S. Hough, J. M. Novais, and T. P. Lyons, *Chem. Eng.* (London) **258,** 71 (1972).
140. J. F. Kennedy and V. W. Pike, *J. Chem. Soc. Perkin Trans. 1,* 1058 (1978).
141. J. F. Kennedy and C. E. Doyle, *Carbohydr. Res.* **28,** 89 (1973).
142. J. F. Kennedy and P. M. Watts, *Carbohydr. Res.* **32,** 155 (1974).
143. D. B. Johnson and M. Costelloe, *Biotechnol. Bioeng.* **18,** 421 (1976).
144. C. Kent, A. Rosevear, and A. R. Thomson, in *Topics in Enzyme and Fermentation Biotechnology,* Vol. 2 (A. Wiseman, Ed.), Ellis Horwood, Chichester (1978), p. 12.
145. J. P. Cardoso, M. F. Chaplin, A. N. Emery, J. F. Kennedy, and L. P. Revel-Chion, *J. Appl. Chem. Biotechnol.* **28,** 775 (1978).
146. A. Flynn and D. B. Johnson, *Int. J. Biochem.* **8,** 243 (1977).
147. B. R. Allen, M. Charles, and R. W. Coughlin, *Biotechnol. Bioeng.* **21,** 689 (1979).
148. J. F. Kennedy, *Chem. Soc. Rev.* **8,** 221 (1979).

149. J. F. Kennedy and I. M. Kay, *J. Chem. Soc. Perkin Trans. 1,* 329 (1976).
150. J. F. Kennedy, S. A. Barker, and J. D. Humphreys, *J. Chem. Soc. Perkin Trans. 1,* 962 (1976).
151. J. F. Kennedy and J. D. Humphreys, *Antimicrob. Agents Chemother.* **9,** 776 (1976).
152. J. F. Kennedy, J. D. Humphreys, and S. A. Barker, *Enzyme Microbiol. Technol.* **3,** 129 (1981).
153. J. F. Kennedy, S. A. Barker, and C. A. White, *Carbohydr. Res.* **54,** 1 (1977).
154. R. Messing, in *Methods of Enzymology,* Vol. 44 (K. Mosbach, Ed.), Academic Press, 1976, p. 148.
155. J. F. Kennedy and I. M. Kay, *Carbohydr. Res.* **59,** 553 (1977).
156. J. M. S. Cabral, J. P. Cardoso, and J. M. Novais, *Enzyme Microb. Technol.* **3,** 41 (1981).
157. J. M. S. Cabral, J. F. Kennedy, and J. M. Novais, *Enzyme Microb. Technol.* **4,** 337 (1982).
158. A. Flynn and D. B. Johnson, *Biotechnol. Bioeng.* **20,** 1445 (1978).
159. R. W. Coughlin, M. Charles, and B. R. Allen, U. S. Patent 4,115,198 (1978).
160. K. Rokugawa, T. Fujiishima, A. Kuminaka, and H. Yoshino, *J. Ferment. Technol.* **57,** 570 (1979).
161. W. H. Hanish, P. A. D. Rickard, and S. Nyo, *Biotechnol. Bioeng.* **20,** 95 (1978).
162. J. F. Kennedy and V. W. Pike, *Enzyme Microb. Technol.* **1,** 31 (1979).
163. J. F. Kennedy and V. W. Pike, *Enzyme Microb. Technol.* **2,** 288 (1980).
164. T. Kobayashi, I. Kato, K. Ohmiya, and S. Shimizu, *Agr. Biol. Chem.* **44,** 413 (1980).
165. R. R. Bhatt, S. Joshi, and R. M. Kothari, *Enzyme Microb. Technol.* **1,** 113 (1979).
166. A. N. Emery and J. P. Cardoso, *Biotechnol. Bioeng.* **20,** 1903 (1978).
167. J. F. Kennedy and M. F. Chaplin, *Enzyme Microb. Technol.* **1,** 197 (1979).
168. A. N. Volkova, L. V. Ivanova, A. P. Matveenko, V. I. Yakovlev, and S. I. Kol'tsov, *Khim. Khim. Tekhnol.* **22,** 844 (1979).
169. A. V. Bogatskii, T. I. Davidenko, A. V. Chuenko, V. V. Yanishpol'skii, V. A. Tertykh and A. A. Chuiko, *Ukr. Biokhim. Zh.* **51,** 315 (1979).
170. E. Bisse and D. J. Vondershmitt, *FEBS Lett.* **93,** 102 (1978).
171. J. F. Kennedy, V. W. Pike, and S. A. Barker, *Enzyme Microb. Technol.* **2,** 126 (1980).
172. G. Means and R. E. Feeney, *Chemical Modification of Proteins,* Holden-Day, San Francisco (1971).
173. C. G. Beddows, R. A. Mirauer, and J. T. Guthrie, *Biotechnol. Bioeng.* **22,** 311 (1980).
174. J. L. Iborra, A. Manjón, and J. A. Lozano, *J. Solid-Phase Biochem.* **2,** 85 (1977).
175. H. G. Vogt and G. Manecke, in *First European Enzymes on Biotechnology,* Part 1, Dechema, Frankfurt am Main (1978), p. 96.
176. N. B. Kozlova, L. V. Roze, and P. L. Vul'fson, *Biochemistry* (USSR), **43,** 403 (1978).
177. T.-S. Lai and P.-S. Cheng, *Biotechnol. Bioeng.* **20,** 773 (1978).

178. J. Drobnik, V. Saudek, F. Švec, J. Kálal, V. Vojtíšek, and M. Bárta, *Biotechnol. Bioeng.* **21,** 1317 (1979).
179. L. Shemer, R. Granot, A. Freeman, M. Sokolovsky, and L. Goldstein, *Biotechnol. Bioeng.* **21,** 1607 (1979).
180. D. Blassberger, A. Freeman, and L. Goldstein, *Biotechnol. Bioeng.* **20,** 309 (1978).
181. C. Y. Lee, *J. Solid-Phase Biochem.* **3,** 71 (1978).
182. M. Adamich, H. F. Voss, and E. A. Dennis, *Arch. Biochem. Biophys.* **189,** 417 (1978).
183. E. C. Hatchikian and P. Monsan, *Biochem. Biophys. Res. Commun.* **92,** 1091 (1980).
184. R. A. Zingaro and M. Uziel, *Biochim. Biophys. Acta* **213,** 371 (1970).
185. L. Goldstein, A. Lifshitz, and M. Sokolovsky, *Int. J. Biochem.* **2,** 440 (1971).
186. W. Brümmer, N. Hennrich, M. Klockow, H. Lang, and H. D. Orth, *Eur. J. Biochem.* **25,** 129 (1972).
187. R. Epton, M. E. Hobson, and G. Marr, *Enzyme Microb. Technol.* **1,** 37 (1979).
188. G. Manecke, R. Pohl, J. Schluensen, and H. G. Vogt, in *Enzyme Engineering,* Vol. 4 (G. Broun, G. Manecke, and L. B. Wingard Jr., Eds.), Plenum Press, New York (1978), p. 409.
189. S. Miyairi, *Biochim. Biophys. Acta* **571,** 374 (1979).
190. C. G. Beddows, R. A. Mirauer, J. T. Guthrie, F. I. Abdel-Hay, and C. E. J. Morrish, *Polym. Bull.* (Berlin) **1,** 749 (1979).
191. F. Paul, P. R. Coulet, D. G. Gautheron, and J.-M. Engasser, *Biotechnol. Bioeng.* **20,** 1785 (1978).
192. P. R. Coulet, R. Sternberg, and D. R. Thévenot, *Biochim. Biophys. Acta* **612,** 317 (1980).
193. P. R. Coulet, C. Godinot, and D. C. Gautheron, *Biochim. Biophys. Acta* **391,** 272 (1975).
194. N. Arrio-Dupont and P. R. Coulet, *Biochem. Biophys. Res. Commun.* **89,** 345 (1975).
195. D. M. Munnecke, *Biotechnol. Bioeng.* **21,** 2247 (1979).
196. A. Mukherjee and P. A. Srere, *J. Solid-Phase Biochem.* **3,** 85 (1978).
197. B. Danielsson, B. Mattiasson, R. Karlsson, and F. Winqvist, *Biotechnol. Bioeng.* **21,** 1749 (1979).
198. G. Carrea, F. Colombi, G. Mazzola, P. Cremonesi, and E. Antonini, *Biotechnol. Bioeng.* **21,** 39 (1979).
199. J. Tramper, F. Miller, and H. C. van der Plas, *Biotechnol. Bioeng.* **20,** 1507 (1978).
200. D. B. Johnson and M. P. Coughlan, *Biotechnol. Bioeng.* **20,** 1085 (1978).
201. N. Erekin and M. E. Friedmann, *J. Solid-Phase Biochem.* **4,** 123 (1979).
202. K. Martinek, V. V. Mozhaev, and I. V. Berezin, *Biochim. Biophys. Acta* **615,** 426 (1980).
203. J. S. Patton, P.-A. Albertson, C. Erlanson, and B. Borgstrom, *J. Biol. Chem.* **253,** 4195 (1978).
204. K. Rokugawa, T. Fujishima, A. Kuminaka, and H. Yoshino, *J. Ferment. Technol.* **57,** 570 (1979).

205. J. A. Jackson, H. R. Halvorson, J. W. Furlong, K. D. Lucast, and J. D. Shore, *J. Pharmacol. Exp. Ther.* **209,** 271 (1979).
206. V. V. Mozhaev, K. Martinek, and I. V. Berezin, *Mol. Biol.* **13,** 73 (1979).
207. L. Rexova-Benkova, M. Mrackova, and K. Babor, *Collect. Czech. Chem. Commun.* **45,** 163 (1980).
208. F. Pittner, T. Miron, G. Pittner, and M. Wilchek, *J. Amer. Chem. Soc.* **102,** 2451 (1980).
209. V. Janasik, F. Bartha, K. Krettschmer, and J. Lash, *J. Solid-Phase Biochem.* **3,** 291 (1978).
210. H. Maeda, G. T. Tsao, and L. F. Chen, *Biotechnol. Bioeng.* **20,** 383 (1978).
211. J. L. Iborra, A. Manjón, M. Tari, and J. A. Lozano, *Gen. Pharmacol.* **10,** 143 (1979).
212. S. Fukushima, T. Nagai, K. Fujita, A. Tanaka, and S. Fukui, *Biotechnol. Bioeng.* **20,** 1465 (1978).
213. T. E. Lipatova, O. L. Konoplystka, L. N. Chupryna, and D. V. Vasyl'chenko, *Ukr. Biokhim. Zh.* **51,** 319 (1978).
214. L. O. Kolesnyk, I. P. Halych, and T. A. Koval'chuck, *Ukr. Biokhim. Zh.* **51,** 369 (1979).
215. S. V. El'chyts, V. V. Yanyshpol'sky, M. B. Abaimova, and L. A. Zatorska, *Ukr. Biokim. Zh.* **51,** 374 (1979).
216. H. R. Schreiner, U. S. Patent 3,282,702 (1966).
217. H. Brandenberg, *Rev. Ferment. Ind. Aliment.* **11,** 237 (1956).
218. S. Kinoshita, M. Tanaka, and N. Nakamura, Japanese Patent 66-13785 (1966).
219. N. W. H. Cheetham and G. N. Richards, *Carbohydr. Res.* **30,** 99 (1973).
220. J. F. Kennedy and A. Zamir, *Carbohydr. Res.* **29,** 497 (1973).
221. C. J. Gray and T. H. Yeo, *Carbohydr. Res.* **27,** 235 (1973).
222. S. A. Barker, S. H. Doss, C. J. Gray, J. F. Kennedy, M. Stacey, and T. H. Yeo, *Carbohydr. Res.* **20,** 1 (1971).
223. S. A. Barker, J. F. Kennedy, and C. J. Gray, British Patent 1,289,549 (1972).
224. A. H. Merrill, Jr. and D. B. McCormick, *Biotechnol. Bioeng.* **21,** 1629 (1979).
225. J. Kucera and M. Kuminkova, *Collect. Czech. Commun.* **45,** 298 (1980).
226. J. Kucera, *Coll. Czech. Chem. Commun.* **44,** 804 (1979).
227. N. J. Daka and K. J. Laidler, *Biochim. Biophys. Acta* **612,** 305 (1980).
228. F. T. Abdel-Hay, C. G. Beddows, and J. T. Guthrie, *Polym. Bull. (Berlin)* **2,** 607 (1980).
229. V. P. Torchilin, E. G. Tischenko, and V. N. Smirnov, *J. Solid-Phase Biochem.* **2,** 19 (1977).
230. Y. K. Cho and J. E. Bailey, *Biotechnol. Bioeng.* **20,** 1651 (1978).
231. A. Borchet and K. Buchholz, *Biotechnol. Lett.* **1,** 15 (1979).
232. R. Axén, J. Porath, and S. Ernback, *Nature* **214,** 1302 (1967).
233. B. Svensson, *FEBS Lett.* **29,** 167 (1973).
234. R. Jost, T. Miron, and M. Wilchek, *Biochim. Biophys. Acta* **362,** 75 (1974).
235. M. Wilchek, T. Oka, and Y. J. Topper, *Proc. Nat. Acad. Sci. U. S.* **72,** 1055 (1975).
236. J. F. Kennedy, J. A. Barnes and J. B. Matthews, *J. Chromatog.* **196,** 379 (1980).

237. L. A. Cohen, *Ann. Rev. Biochem.* **37,** 695 (1968).
238. G. R. Stark, *Advances in Protein Chem.* **24,** 261 (1970).
239. T. Sato, T. Mori, T. Tosa, and I. Chibata, *Arch. Biochem. Biophys.* **147,** 788 (1972).
240. H. Maeda and H. Suzuki, *Agr. Biol. Chem.* (Tokyo) **36,** 1581 (1972).
241. S. Y. Shimizu and H. M. Lenhoff, *J. Solid-Phase Biochem.* **4,** 75 (1979).
242. S. Y. Shimizu and H. M. Lenhoff, *J. Solid-Phase Biochem.* **4,** 95 (1979).
243. T. H. Finlay, V. Troll, M. Levy, A. J. Johnson, and L. T. Hodgins, *Anal. Biochem.* **87,** 77 (1978).
244. J. Fischer, R. Ulrich, and A. Schellenberg, *Acta Biol. Med. Ger.* **37,** 1413 (1978).
245. F. Švec, J. Kálal, I. I. Menyailova, and L. A. Nakhapteyan, *Biotechnol. Bioeng.* **21,** 1317 (1978).
246. O. H. Friedrich, M. Chun, and M. Sernetz, *Biotechnol. Bioeng.* **22,** 15 (1980).
247. N. Stambolieva and J. Turkova, *Collect. Czech. Chem. Commun.* **45,** 1137 (1980).
248. A. T. Jagendorf, A. Patchornik, and M. Sela, *Biochim. Biophys. Acta* **78,** 516 (1963).
249. R. F. Murphy, J. M. Conlon, A. Iman, and G. J. C. Kelly, *J. Chromatogr.* **135,** 427 (1977).
250. J. Porath, in *Methods in Enzymology,* Vol. 34, Academic Press, New York (1974), p. 13.
251. J. Porath, T. Laas, and J. C. Janson, *J. Chromatogr.* **103,** 49 (1975).
252. J. Brandt, L. O. Anderson, and J. Porath, *Biochim. Biophys. Acta.* **386,** 196 (1975).
253. J. Klemes and N. Citri, *Biotechnol. Bioeng.* **21,** 897 (1979).
254. P. V. Sundaram and K. Joy, *J. Solid-Phase Biochem.* **3,** 223 (1978).
255. P. V. Sundaram, M. P. Igloi, R. Wassermann, and W. Hinsch, *Clin. Chem.* **24,** 1813 (1978).
256. E. Bisse and D. J. Vonderschmidtt, *FEBS Lett.* **81,** 326 (1977).
257. N. Carvajal, J. Martínez, F. M. de Oca, J. Rodríguez, and M. Fernández, *Biochim. Biophys. Acta* **527,** 1 (1978).
258. N. Carvajal, J. Martínez, and M. Fernández, *Biochim. Biophys. Acta* **481,** 177 (1977).
259. P. V. Sundaram, B. Blumenberg, and W. Himsh, *Clin. Chem.* **25,** 1436 (1979).
260. K. Matsumoto, H. Seijo, I. Karube, and S. Suzuki, *Biotechnol. Bioeng.* **22,** 1071 (1980).
261. P. Spettoli, A. Botlacin, and A. Zamorani, *Technol. Aliment.* **3,** 31 (1980).
262. J. L. Leuba and F. Widmer, *J. Solid-Phase Biochem.* **2,** 257 (1979).
263. S. Hirano and O. Miura, *Biotechnol. Bioeng.* **21,** 711 (1979).
264. J. L. Leuba and F. Widmer, *Biotechnol. Lett.* **1,** 109 (1979).
265. W. L. Stanley, G. G. Watters, S. H. Kelly, and A. C. Olson, *Biotechnol. Bioeng.* **20,** 135 (1978).
266. W. H. Liu, S. D. Wang, and Y. C. Su, *Proc. Nat. Sci. Council* (Taiwan) **2,** 275 (1978).
267. L. Iyengar and A. V. S. P. Rao, *Biotechnol. Bioeng.* **21,** 1333 (1979).

268. S. W. Carleysmith, M. B. L. Eames, and M. D. Lilly, *Biotechnol. Bioeng.* **22,** 957 (1980).
269. S. W. Carleysmith and M. D. Lilly, *Biotechnol. Bioeng.* **21,** 1057 (1979).
270. J. A. Berenson and J. R. Benemann, *FEBS Lett.* **76,** 105 (1977).
271. J. Beitz, A. Schellenberger, J. Lasch, and J. Fischer, *Biochim. Biophys. Acta* **612,** 451 (1980).
272. Y. V. Rodinov, T. V. Avilova, and V. O. Popov, *Biochemistry* (USSR) **42,** 1594 (1977).
273. B. Mattiasson, B. Danielsson, C. Hermansson, and K. Mosbach, *FEBS Lett.* **85,** 203 (1978).
274. A. A. Klyosov and V. B. Gerasimas, *Biochim. Biophys. Acta* **571,** 162 (1979).
275. D. D. Lee, G. K. Lee, P. J. Reilly, and Y. Y. Lee, *Biotechnol. Bioeng.* **22,** 1 (1980).
276. D. B. Johnson, *Biotechnol. Bioeng.* **20,** 1117 (1978).
277. P. Monsan, *Eur. J. Appl. Microb. Biotechnol.* **5,** 1 (1978).
278. P. Monsan and G. Durand, *Biochim. Biophys. Acta* **523,** 477 (1978).
279. C. C. Hon and P. J. Reilly, *Biotechnol. Bioeng.* **21,** 505 (1979).
280. A. Lopez and P. Monsan, *Biochimie* **62,** 323 (1980).
281. J. Konecny and M. Sieber, *Biotechnol. Bioeng.* **22,** 2013 (1980).
282. C. G. Bohnenkamp and P. J. Reilly, *Biotechnol. Bioeng.* **22,** 1753 (1980).
283. H. N. Chang and P. J. Reilly, *Biotechnol. Bioeng.* **20,** 243 (1978).
284. K. I. Voivodov, B. R. Galunski, and S. S. Dyankov, *J. Appl. Biochem.* **1,** 442 (1979).
285. P. Puvanakrishnan and S. M. Bose, *Biotechnol. Bioeng.* **22,** 919 (1980).
286. B. E. Dale and D. H. White, *Biotechnol. Bioeng.* **21,** 1639 (1979).
287. P. J. Halling, J. A. Asenjo, and P. Dunnill, *Biotechnol, Bioeng.* **21,** 2359 (1979).
288. K. Parkin and H. O. Hultrin, *Biotechnol. Bioeng.* **21,** 939 (1979).
289. B. P. Wasserman, H. O. Hultrin, and B. S. Jacobson, *Biotechnol. Bioeng.* **22,** 271 (1980).
290. R. Epton, J. V. McLaren, and T. H. Thomas, *Carbohydr. Res.* **22,** 301 (1972).
291. E. Brown and A. Racois, *Tetrahedron Lett.* 5077 (1972).
292. F. B. Weakley and C. L. Mehltretter, *Biotechnol. Bioeng.* **15,** 1189 (1973).
293. W. P. Jencks, *Prog. Phys. Org. Chem.* **2,** 63 (1964).
294. I. Ugi, *Angew. Chem.* **74,** 9 (1962).
295. R. Axén, P. Vretblad, and J. Porath, *Acta Chem. Scand.* **25,** 1129 (1971).
296. P. Vretblad and R. Axén, *FEBS Lett.* **18,** 254 (1971).
297. O. Zaborsky, in *Immobilized Enzymes in Food and Microbial Processes* (A. E. Olson and C. L. Cooney, Eds.), Plenum Press, New York (1974), p. 187.
298. J. Schnapp and Y. Shalitin, *Biochem. Biophys. Res. Commun.* **70,** 8 (1976).
299. J. Carlsson, R. Axén, K. Brocklehurst, and E. M. Crook, *Eur. J. Biochem.* **44,** 189 (1974).
300. R. Axén, H. Drevin, and J. Carlson, *Acta Chem. Scand.* **29,** 471 (1975).
301. V. C. Borlazza, N. W. H. Cheetham, and P. T. Sowthwell-Kelly, *Carbohydr. Res.* **79,** 125 (1980).
302. J. F. Kennedy and A. Zamir, *Carbohydr. Res.* **41,** 227 (1975).
303. J. R. Shainoff, *J. Immunol.* **100,** 187 (1968).

304. J. Brandt and L. O. Anderson, *Acta Chem. Scand.* **B 30** 815 (1976).
305. F. Michael and J. Ewers, *Makromol. Chem.* **3,** 200 (1949).
306. J. Peska, J. Stamberg, and Z. Blace, Czech. Patent 172640 (1976).
307. L. F. Chen and G. T. Tsao, *Biotechnol. Bioeng.* **18,** 1507 (1976).
308. L. Goldstein, M. Pecht, S. Blumberg, D. Atlas, and Y. Levin, *Biochemistry* **9,** 2322 (1970).
309. J. Reiland, in *Methods in Enzymology,* Vol. 22 (W. B. Jakoby, Ed.), Academic Press, New York (1971), p. 287.
310. J. Porath and P. Flodin, *Nature* **183,** 1657 (1959).
311. J. Porath, *Biochim. Biophys. Acta* **39,** 193 (1960).
312. A. Polson, *Biochim. Biophys. Acta* **50,** 565 (1961).
313. S. Hjertein, *Biochim. Biophys. Acta* **79,** 393 (1969).
314. R. Axén, J. Porath, and S. Ernback, *Nature* **214,** 1362 (1967).
315. J. Porath, R. Axén, and S. Ernback, *Nature* **215,** 1491 (1967).
316. J. Porath and R. Axén, in *Methods in Enzymology,* Vol. 44 (K. Mosbach, Ed.), Academic Press, New York (1976), p. 19.
317. P. J. Hill, M. A. Cresswell, and J. G. Feinberg, U. S. Patent 4,003,792 (1977).
318. W. R. Vieth and K. Venkatasubramanian, in *Methods in Enzymology,* Vol. 44 (K. Mosbach, Ed.), Academic Press, New York (1976), p. 223.
319. A. G. van Velzen, U. S. Patent 3,838,007 (1974).
320. L. Grasset, D. Cordier, and A. Ville, *Process Biochem.* **14,** 2 (1979).
321. M. M. Derbinin, *Usp. Khim.* **24,** 3 (1955).
322. G. Manecke and H. J. Förster, *Makromol. Chem.* **91,** 136 (1966).
323. G. Manecke and G. Fünzel, *Naturwissenschaften* **54,** 531 (1967).
324. E. Van Leemputten, U. S. Patent 4,017,364 (1977).
325. A. Conte and K. Lehmann, *Z. Physiol. Chem.* **352,** 533 (1971).
326. J. K. Inman and H. M. Dintzis, *Biochemistry* **8,** 4074 (1969).
327. J. F. Kennedy and J. Epton, *Carbohydr. Res.* **27,** 11 (1973).
328. R. Epton, J. V. McLaren, and T. H. Thomas, British Patent 1,365,886 (1974).
329. R. Epton, B. L. Hibbert, and G. Marr, *Polymer* **16,** 314 (1975).
330. R. Epton, C. Holloway, and J. V. McLaren, British Patent 1,448,364 (1975).
331. P. Cuatrecasas, *J. Biol. Chem.* **245,** 3059 (1970).
332. O. Valentova, J. Turkova, R. Lapka, J. Zima, and J. Coupek, *Biochim. Biophys. Acta* **403,** 192 (1975).
333. J. Turkova, in *Methods in Enzymology,* Vol. 44 (K. Mosbach, Ed.), Academic Press, New York (1976), p. 66.
334. D. M. Krämmer, K. Lehman, H. Pennerviss, and H. Plainer, *23rd Colloq. "Part. Biol. Fluids"* **23,** 505 (1976).
335. F. Švec, J. Hradill, J. Coupek, and J. Kalal, *Angew. Makromol. Chem.* **48,** 135 (1975).
336. O. R. Zaborsky, in *Immobilized Enzymes in Food and Microbial Processes* (A. C. Olson and C. C. Cooney, Eds.), Plenum, New York (1974), p. 187.
337. Y. Levin, M. Pecht, L. Goldstein, and E. Katchalski-Katzir, *Biochemistry* **3,** 1905 (1964).
338. L. Goldstein, Y. Levin, and E. Katchalski-Katzir, *Biochemistry* **3,** 1913 (1964).
339. M. A. Stahmann and R. R. Becker, *J. Amer. Chem. Soc.* **74,** 288 (1952).

340. G. Manecke and J. Schlünsen, in *Methods in Enzymology,* Vol. 44 (K. Mosbach, Ed.), Academic Press, New York (1976), p. 107.
341. W. Kuhn and G. Balmer, *J. Polymer. Sci.* **57,** 311 (1962).
342. G. Manecke and H. G. Vogt, *Makromol. Chem.* **177,** 725 (1976).
343. R. C. Schulz, J. Kovacs, and W. Kern, *Makromol. Chem.* **54,** 146 (1962).
344. S. A. Barker, J. F. Kennedy, and A. Rosevear, *J. Chem. Soc., C,* 2726 (1971).
345. J. F. Kennedy, S. A. Barker, and A. Rosevear, *J. Chem. Soc., Perkin Trans 1,* 2568 (1972).
346. L. Goldstein, A. Freeman, and M. Sokolovsky, *Biochem. J.* **143,** 497 (1974).
347. W. H. Tomb and H. H. Weetall, U. S. Patent 3,783,101 (1979).
348. R. A. Messing, *Process Biochem.* **9,** 26 (1974).
349. R. A. Messing, U. S. Patent 3,910,851 (1975).
350. R. A. Burns, U. S. Patent 3,953,292 (1976).
351. P. F. Greenfield and R. L. Laurence, *J. Food Sci.* **40,** 906 (1975).
352. J. M. S. Cabral, J. M. Novais, and J. P. Cardoso in 3rd International Chemical Engineering Symposium, Chempor '81, Porto, April 1981.
353. E. L. F., British Patent 1,363,526 (1974).
354. P. J. Robinson, P. Dunnill, and M. D. Lilly, *Biotechnol. Bioeng.* **15,** 603 (1975).
355. P. A. Munro, P. Dunnill, and M. D. Lilly, *Biotechnol. Bioeng.* **19,** 101 (1977).
356. M. Charles, R. W. Coughlin, E. K. Parachuri, B. R. Allen, and F. X. Hasselberger, *Biotechnol. Bioeng.* **17,** 302 (1975).
357. J. E. Brotherton, A. Emery and V. W. Rodwell, *Biotechnol. Bioeng.* **18,** 527 (1976).
358. H. H. Weetall, in *Methods in Enzymology,* Vol. 44 (K. Mosbach, Ed.), Academic Press, New York (1976), p. 134.
359. H. H. Weetall and C. C. Detar, *Biotechnol. Bioeng.* **17,** 295 (1975).
360. P. Monsan and G. Durand, *C. R. Acad. Sci., Paris, Ser. C.* **273,** 33 (1971).
361. J. M. S. Cabral, J. M. Novais, and J. P. Cardoso, *Biotechnol. Bioeng.* **23,** 2083 (1981).
362. J. M. S. Cabral, J. M. Novais and J. P. Cardoso, paper presented at the 2nd European Conference of Biotechnology, Eastbourne, April 1981.
363. P. Bernfeld and J. Wan, *Science* **142,** 678 (1963).
364. A. Dahlquist, B. Mattiasson, and K. Mosbach, *Biotechnol. Bioeng.* **15,** 395 (1973).
365. T. Mori, T. Sato, T. Tosa, and I. Chibata, *Enzymology* **43,** 213 (1974).
366. H. Maeda, A. Yamaguchi, and H. Suzuki, *Biochim. Biophys. Acta* **315,** 18 (1973).
367. J. Dobo, *Acta Chim. Acad. Sci. Hung.* **63,** 453 (1970).
368. H. Maeda and H. Suzuki, *Process Biochem.* **12,** 9 (1977).
369. T. Tosa, T. Sato, T. Mori, K. Yamamioto, I. Takata, Y. Nishida, and I. Chibata, *Biotechnol. Bioeng.* **21,** 1697 (1979).
370. J. Schovers and W. E. Sandine, U. S. Patent 3,733,205 (1973).
371. E. K. Bauman, L. H. Goodson, G. G. Guilbault, and D. N. Kramer, *Anal. Chem.* **37,** 1378 (1965).
372. G. G. Guillbault and J. Das, *Anal. Biochem.* **33,** 341 (1970).
373. T. T. Ngo and K. J. Laidler, *Biochem. Biophys. Acta* **525,** 93 (1978).

374. K. Martinek, A. M. Klibanov, V. S. Goldmacher, and I. V. Berezin, *Biochim. Biophys. Acta* **485,** 1 (1977).
375. K. N. Kuan, Y. Y. Lee, and P. Melius, *Biotechnol. Bioeng.* **22,** 1725 (1980).
376. W. Halwachs, C. Wandrey, and K. Schügerl, *Biotechnol. Bioeng.* **20,** 541 (1978).
377. T. Mori, R. Sano, Y. Iwasawa, T. Tosa, and I. Chibata, *J. Solid-Phase Biochem.* **1,** 15 (1976).
378. A. K. Chen, C. C. Liu, and J. G. Schiller, *Biotechnol. Bioeng.* **21,** 1905 (1979).
379. E. Sada, S. Katoh, and M. Terashima, *Biotechnol. Bioeng.* **22,** 243 (1980).
380. S. Adachi, K. Hashimoto, R. Matsuno, K. Nakanishi, and T. Kamikubo, *Biotechnol. Bioeng.* **22,** 779 (1980).
381. A. Szewczuk, E. Ziomek, M. Mordarski, M. Siewínski, and J. Wieczorek, *Biotechnol. Bioeng.* **21,** 1543 (1979).
382. K. Buchholz and B. Gödelmann, *Biotechnol. Bioeng.* **20,** 1201 (1978).
383. I. Karube, K. I. Hirano, and S. Suzuki, *J. Solid-Phase Biochem.* **2,** 41 (1977).
384. S. Fukui, A. Tanaka, and G. Gelff, in *Enzyme Engineering,* Vol. 4 (G. B. Broun, G. Manecke, and L. B. Wingard Jr., Eds.), Plenum Press, New York, 1978, p. 299.
385. M. Kumakura, M. Yoshida, M. Asano, and I. Kaetsu, *J. Solid-Phase Biochem.* **2,** 279 (1977).
386. I. Kaetsu, M. Kumakura, and M. Yoshida, *Biotechnol. Bioeng.* **21,** 847 (1979).
387. D. Dinelli, *Process Biochem.* **7,** 9 (1972).
388. D. Dinelli, W. Marconi, F. Cecere, G. Galli, and F. Morisi, in *Enzyme Engineering,* Vol. 3 (E. K. Pye and H. H. Weetall Eds.), Plenum Press, New York (1978), p. 47.
389. F. Bartoli, G. E. Bianchi, and D. Zaccardelli, in *Enzyme Engineering,* Vol. 4 (G. B. Broun, G. Manecke, and L. B. Wingard, Jr. Eds.), Plenum Press, New York (1978), p. 279.
390. W. Marconi, F. Morisi, and R. Mosti, *Agr. Biol. Chem.* **39,** 1323 (1975).
391. C. Corno, G. Galli, F. Morisi, M. Bettonte, and A. Stopponi, *Stärke* **24,** 420 (1972).
392. S. Giovenco, F. Morisi, and P. Pansolli, *FEBS Lett.* **36,** 57 (1973).
393. S. P. A. Snamprogetti, U. S. Patent 3,969,990 (1976).
394. W. Marconi, S. Gulinelli, and F. Morisi, *Biotechnol. Bioeng.* **16,** 501 (1974).
395. F. Morisi, M. Pastone, and A. Viglia, *J. Dairy Sci.* **56,** 1123 (1973).
396. W. Marconi, F. Cecere, F. Morisi, G. Della Penna, and B. Rapperolli, *J. Antibiot.* **26,** 226 (1973).
397. W. Marconi, F. Bartoli, F. Cecere, and F. Morisi, *Agr. Biol. Chem.* **38,** 1343 (1974).
398. P. Zaffaroni, N. Oddo, R. Olivieri, and L. Formiconi, *Agr. Biol. Chem.* **39,** 1875 (1975).
399. M. Pardin, C. D. Bello, A. Marani, F. Bartoli, and F. Morisi, *J. Solid-Phase Biochem.* **2,** 251 (1977).
400. T. M. S. Chang, *Science* **146,** 524 (1964).
401. J. A. Herbig, *Encyl. Polymer Sci. Technol.* **8,** 719 (1968).
402. D. T. Waliack and R. G. Carbonell, *Biotechnol. Bioeng.* **17,** 1157 (1975).

403. T. M. S. Chang, *Enzyme* **14,** 95 (1973).
404. J. Campbell and T. M. S. Chang, *Biochem. Biophys. Res. Commun.* **69,** 562 (1976).
405. A. D. Mogensen and W. R. Vieth, *Biotechnol. Bioeng.* **15,** 467 (1973).
406. J. Campbell and T. M. S. Chang, *Biochim. Biophys. Acta* **397,** 101 (1975).
407. M. A. Paine and R. G. Carbonell, *Biotechnol. Bioeng.* **17,** 617 (1975).
408. J. C. W. Ostergaard and S. C. Martiny, *Biotechnol. Bioeng.* **15,** 561 (1973).
409. T. Mori, T. Tosa, and I. Chibata, *Biochim. Biophys. Acta* **321,** 653 (1973).
410. G. Gregoriadis, P. D. Leathwood, and B. E. Ryman, *FEBS Lett.* **14,** 95 (1971).
411. R. R. Mohan and N. N. Li, *Biotechnol. Bioeng.* **16,** 513 (1974).
412. M. Kitajema, S. Miyano, and A. Kondo, *Kogyo Kageku Zasshi* **72,** 493 (1969).
413. N. N. Li, *Ind. Eng. Chem., Process Des. Develop.* **10,** 215 (1971).
414. P. R. Rony, *Biotechnol. Bioeng.* **13,** 431 (1971).
415. M. Cantarella, L. Gianfreda, R. Palescondolo, V. Scardi, G. Greco, F. Alfani and G. Iori, *J. Solid-Phase Biochem.* **2,** 167 (1977).
416. G. Greco, F. Alfani, G. Iori, M. Cantarella, A. Formisano, L. Gianfreda, R. Palescondolo, and V. Scardi, *Biotechnol. Bioeng.* **21,** 1921 (1979).
417. C. Wandrey, E. Flaschel, and K. Schùgerl, *Biotechnol. Bioeng.* **21,** 1649 (1979).
418. G. Greco, D. Albanesi, M. Cantarella, and V. Scardi, *Biotechnol. Bioeng.* **22,** 215 (1980).
419. F. Alfani, G. Iori, G. Greco, M. Cantarella, M. H. Renny, and V. Scardi, *Chem. Eng. Sci.* **34,** 1213 (1979).
420. R. W. Silman, L. T. Black, J. E. McGhee, and E. B. Bagley, *Biotechnol. Bioeng.* **22,** 533 (1980).
421. H. Besserdich, D. Kinstein, and E. Kahrig, *Chem. Tech.* (Leipzig) **32,** 243 (1980).
422. L. Roger, J. L. Thapon, J. C. Maubois, and G. Burle, *Le Lait* **551,** 56 (1970).
423. N. N. Ugarova, L. Y. Bronko, G. D. Rozhkora, and I. V. Berezin, *Biokhimiya* **42,** 943 (1977).
424. E. Katchalski and M. Sela, *Advances in Protein Chem.* **13,** 243 (1958).
425. A. N. Glazer, E. Bar-Eli, and E. Katchalski, *J. Biol. Chem.* **237,** 3458 (1962).
426. S. P. O'Neill, J. R. Wykes, P. Dunnill, and M. D. Lilly, *Biotechnol. Bioeng.* **13,** 319 (1971).
427. J. J. Marshall and M. L. Rabinowitz, *J. Biol. Chem.* **251,** 1081 (1976).
428. G. Vegarud and T. B. Christensen, *Biotechnol. Bioeng.* **17,** 1391 (1977).
429. B. Solomon and Y. Levin, *Biotechnol. Bioeng.* **16,** 1161 (1974).
430. I. R. Wykes, P. Dunnill, and M. D. Lilly, *Biochim. Biophys. Acta* **250,** 522, (1971).
431. M. Charles, R. W. Coughlin, and F. X. Hasselberger, *Biotechnol. Bioeng.* **16,** 1553 (1974).
432. I. M. Tereshin and B. V. Moskivchev, in *Enzyme Engineering: Future Directions,* (L. B. Wingard Jr., I. V. Berezin, and A. A. Klyosor, Eds.), Plenum Press, New York (1980), p. 295.
433. W. R. Vieth, S. G. Gilbert, S. S. Wang and K. Venkatusubramanian, U. S. Patent 2,809,613 (1979).
434. G. P. Royer and R. Uy, *J. Biol. Chem.* **248,** 2627 (1973).

435. C. J. Gray, C. M. Livingstone, C. M. Jones, and S. A. Barker, *Biochim. Biophys. Acta* **341,** 457 (1974).
436. C. Horvath, *Biochim. Biophys. Acta* **358,** 164 (1974).
437. I. Sato, I. Karube, and S. Suzuki, *J. Solid-Phase Biochem.* **2,** 1 (1977).
438. I. Karube, S. Suzuki, T. Kusano, and I. Sato, *J. Solid-Phase Biochem.* **2,** 273 (1977).
439. S. Gondo and H. Koya, *Biotechnol. Bioeng.* **20,** 2007 (1978).
440. Y. Morikawa, I. Karube, S. Suzuki, Y. Nakano, and T. Taguchi, *Biotechnol. Bioeng.* **20,** 1143 (1978).
441. S. Gondo, M. Morishita, and T. Osaki, *Biotechnol. Bioeng.* **22,** 1287 (1980).
442. Y. Morikawa, I. Karube, and S. Suzuki, *Biochim. Biophys. Acta* **523,** 263 (1978).
443. J. P. Bollmeier and S. Middleman, *Biotechnol. Bioeng.* **21,** 2303 (1979).
444. M.-H. Remy, A. David, and D. Thomas, *FEBS Lett.* **88,** 332 (1978).
445. J.-N. Barbotin and B. Thomasset, *Biochim. Biophys. Acta* **570,** 11 (1979).
446. J.-N. Barbotin and M. Breuil, *Biochim. Biophys. Acta* **525,** 18 (1978).
447. K.-K. Yeung, A. J. Owen, and J. A. Dain, *Carbohydr. Res.* **75,** 295 (1979).
448. M. D. Legoy, V. L. Garde, J. M. LeMoullec, F. Ergan, and D. Thomas, *Biochemie* **62,** 341 (1980).
449. D. Vallin and C. Tran-Minh, *Biochim. Biophys. Acta* **571,** 321 (1979).
450. R. A. Ludolph, W. R. Vieth, K. Venkatasubramanian, and A. Constantinides, *J. Mol. Cat.* **5,** 1977 (1979).
451. P. J. Halling and P. Dunnill, *Biotechnol. Bioeng.* **21,** 393 (1979).
452. S. Krishnaswamy and J. R. Kittrel, *Biotechnol. Bioeng.* **20,** 821 (1978).
453. B. D. Wassermann, H. O. Hultrin, and B. S. Jacobson, *Biotechnol. Bioeng.* **22,** 271 (1980).
454. V. Krasnobajew and R. Böniger, *Chimia* **29,** 123 (1975).
455. D. D. Lee, Y. Y. Lee, P. J. Reilly, E. V. Collin Jr., and G. T. Tsao, *Biotechnol. Bioeng.* **18,** 253 (1976).
456. S. Swanson, A. Emery, and H. C. Lim, *J. Solid-Phase Biochem.* **1,** 119 (1976).
457. I. Christison, *Chem. Ind.* **4,** 215 (1972).
458. H. Maeda, S. Miyamichi, and H. Suzuki, *Hakko Kyokaishi* **28,** 391 (1970).
459. M. J. Bachler, G. W. Stranberg, and K. L. Smiley, *Biotechnol. Bioeng.* **12,** 85 (1970).
460. K. L. Smiley, *Biotechnol. Bioeng.* **13,** 309 (1971).
461. A. Kimura, H. Shirasaki, and S. Usami, *Kogyo Kageku Zasshi* **72,** 489 (1969).
462. T. A. Butterworth, D. I. C. Wang, and A. J. Sinskey, *Biotechnol. Bioeng.* **12,** 615 (1970).
463. J. J. Marshall and W. J. Whelan, *Chem. Ind.* (London) **25,** 701 (1971).
464. Y. Takasaki and Y. Takahara, Japanese Patent 76-70275 (1976).
465. K. Martenson, *Biotechnol. Bioeng.* **16,** 579 (1979).
466. H. Maeda, *Biotechnol. Bioeng.* **20,** 383 (1978).
467. I. Karube, S. Tanaka, T. Shirai, and S. Suzuki, *Biotechnol. Bioeng.* **19,** 1183 (1977).
468. T. K. Ghose and J. A. Kostick, *Biotechnol. Bioeng.* **12,** 921 (1970).
469. M. Mandels, J. Kostick and R. Panizek, *J. Polym. Sci., Part C* **36,** 445 (1971).

470. L. Zittan, P. B. Poulsen, and S. H. Hemmingsen, *Stärke* **27,** 236 (1975).
471. R. A. Messing, *Biotechnol. Bioeng.* **16,** 897 (1974).
472. J. C. Davis, *Chem. Eng.* **19,** 52 (1970).
473. N. H. Mermelstein, *Food Technol.* **29,** 201 (1975).
474. N. Tsumura and M. Ishikawa, *Shokuklin Kogyo-Gakkaishi* **19,** 539 (1967).
475. G. W. Strandberg and K. L. Smiley, *Biotechnol. Bioeng.* **14,** 509 (1972).
476. T. Kasumi, K. Kawashima, and N. Tsumura, *Hakko Kogaku Zasshi* **51,** 321 (1974).
477. S. Glovenco, F. Morisi, and P. Pansolli, *FEBS Lett.* **36,** 57 (1973).
478. M. Kreen, A. Köstner and K. Kask, *Tr. Tallin Politekh. Inst.* **331,** 131 (1973).
479. H. Negoro, *J. Ferment. Technol.* **48,** 689 (1970).
480. R. D. Mason and H. H. Weetall, *Biotechnol. Bioeng.* **14,** 637 (1972).
481. H. Suzuki, Y. Ozawa, and H. Haeda, *Agr. Biol. Chem.* **30,** 807 (1966).
482. R. A. Korus and A. C. Olson, *Biotechnol. Bioeng.* **19,** 1 (1977).
483. T. Kobayashi, K. Ohmiya, and S. Shimizu, in *Immobilized Enzyme Technology,* (H. H. Weetall and S. Suzuki, Eds.), Plenum, New York (1975) p. 169.
484. K. E. Pappel, E. K. Siimer, A. I. Köstner, E. V. Letunova, and A. S. Tikhominova, *Appl. Biochem. Microb.* **12,** 173 (1976).
485. M. Pastore, F. Marisi, and L. Leali, *Milchwissenchaft* **31,** 362 (1976).
486. W. H. Pitcher Jr., J. R. Ford, and H. H. Weetall, in *Methods in Enzymology,* Vol. 44 (K. Mosbach, Ed.), Academic Press, New York (1976), p. 792.
487. R. W. Coughlin and M. Charles, in *Enzyme Engineering,* Vol. 4 (G. B. Broun, G. Manecke, and L. B. Wingard Jr., Eds.), Plenum Press, New York (1978), p. 273.
488. M. L. Green, and G. Crutchfield, *Biochem. J.* **115,** 183 (1969).
489. M. J. Taylor, M. Cheryan, T. Richardson, and N. F. Olson, *Biotechnol. Bioeng.* **19,** 683 (1977).
490. L. K. Ferrier, T. Richardson, N. F. Olson, and C. L. Hicks, *J. Dairy Sci.* **55,** 726 (1972).
491. S. S. Wang, G. E. Gallili, S. G. Gilbert, and J. G. Leeder, *J. Food Sci.* **39,** 338 (1974).
492. P. Cuatrecasas, M. Wilchek, and C. B. Anfinsen, *Proc. Nat. Acad. Sci.* U. S. **61,** 636 (1968).
493. I. Chibata, T. Tosa, T. Sato, and T. Mori, in *Methods in Enzymology,* Vol. 44 (K. Mosbach, Ed.), Academic Press, New York (1976), p. 746.
494. I. Chibata and T. Tosa, in *Applied Biochemistry and Bioengineering,* Vol. 1 (L. B. Wingard, Jr., E. Katchalski-Katzir and L. Goldstein, Eds.), Academic Press, New York (1976), p. 329.
495. Y. Yokote, M. Fujita, G. Shimura, S. Noguchi, K. Kimura, and H. Samejima *J. Solid-Phase Biochem.* **1,** 1 (1976).
496. R. Koelsch, *Enzymologia* **42,** 257 (1972).
497. T. Tosa, T. Sato, T. Mori, Y. Matuo, and I. Chibata, *Biotechnol. Bioeng.* **15,** 69 (1973).
498. M. Matsui, T. Yoneya, and N. Nagura, Japanese Patent 75-100285 (1975).
499. W. R. Vieth and K. Venkatasubramanian, in *Methods in Enzymology,* Vol. 44 (K. Mosbach, Ed.), Academic Press, New York (1976), pp. 76 and 243.

500. P. Zaffaroni, V. Vitobello, F. Cecere, E. Giacomozzi, and F. Morisi, *Agr. Biol. Chem.* **32,** 1335 (1974).
501. S. Fukui, S. Ikeda, M. Fujimura, H. Yamada, and H. Kumagai, *Eur. J. Appl. Microbiol.* **1,** 25 (1975).
502. T. Fukumura, Japanese Patent 74-15795 (1974).
503. O. Kanemitsu, Japanese Patent 75-132179 (1975).
504. K. Venkatasubramanian, R. Saini, and W. R. Vieth, *J. Food Sci.* **40,** 109 (1975).
505. G. Basarova and J. Turkova, *Brausviss.* **30,** 204 (1977).
506. Y. Nunokawa and J. Saito, *J. Soc. Brew. Japan* **71,** 286 (1976).
507. B. S. Gaina, N. M. Pavlenko, E. N. Datunashvili, Y. I. Krylova, L. V. Kozlov, and V. K. Antonov, *Appl. Biochem. Microbiol.* **12,** 167 (1976).
508. S. Noguchi, G. Shimura, K. Kimura, and H. Samejima, *J. Solid-Phase Biochem.* **1,** 105 (1976).
509. E. Brandl and F. Knansider, German Patent. 2503584 (1975).
510. B. Ekström, E. Lagerlöf, L. Nathorst-Westfelt, and B. Sjöberg, *Svensk Farm. Tidskr.* **78,** 531 (1974).
511. T. Savidge, L. W. Powell, and K. B. Warren, German Patent 2,336,829 (1979).
512. F. Hueper, German Patent 2,157,970 (1973).
513. F. Hueper, German Patent 2,157,972 (1973).
514. L. J. Hensen, C. Chiang, and C. F. Anderson, U. S. Patent 3,446,705 (1969).
515. D. Y. Ryu, C. F. Brunno, B. K. Lee, and K. Venkatasubramanian, in *Fermentation Technology Today* (G. Ierui, Ed.), Society of Fermentation Technology, Osaka (1972), p. 307.
516. M. O. Mandel, A. I. Köstner, E. Kh. Siimer, G. I. Kleiner, L. M. Elizarovskaya, and V. Y. Shtaner, *Appl. Biochem.* **11,** 197 (1975).
517. G. Kleiner, L. Elizarovskaya, M. Mandel, and A. I. Köstner, *Tr. Tallin, Politekh. Inst.* **383,** 31 (1975).
518. T. Fujii, K. Matsumoto, and T. Watanabe, *Process Biochem.* **11,** 21 (1976).
519. T. Yamaguchi and H. Ishii, German Patent 2,331,295 (1974).
520. H. Takeda, I. Matsumoto and Y. Maisuda, Japanese Patent 75-3588 (1975).
521. J. Konecny, in *Enzyme Engineering,* Vol. 4 (G. B. Broun, G. Manecke, and L. B. Wingard Jr., Eds.), Plenum Press, New York (1978) p. 253.
522. B. J. Abbott, B. Cerimele, and D. S. Fukuda, *Biotechnol. Bioeng.* **18,** 1033 (1976).
523. A. Szentirmai, G. Kerenyi, and M. Natonek, in *Enzyme Engineering.* Vol. 4 (G. B. Broun, G. Manecke and L. B. Wingard Jr., Eds.), Plenum Press, New York (1978), p. 155.
524. D. J. Fink, R. D. Falb, and M. K. Bean, in *Advances in Enzyme Engineering,* Vol. 2 (G. T. Tsao, Ed.), Purdue University, West Lafayette, Indiana (1976), p. 79.
525. N. Miwa, Japanese Patent 77-144,789 (1977).
526. I. Miyamoto and N. Miwa, Japanese Patent 77-51089 (1977).
527. D. Gurner and D. Shemin, in *Methods in Enzymology,* Vol. 44 (K. Mosbach, Ed.), Academic Press, New York (1976), p. 844.
528. A. M. Stella, E. Wider de Xifra, and A. M. del. C. Batlie, *Mol. Cell. Biochem.* **16,** 97 (1977).

529. C. Y. Jenq, S. S. Wang, and B. Davidson, *Enzyme Microb. Technol.* **2,** 145 (1980).
530. G. P. Hicks and P. J. Updike, *Anal. Chem.* **38,** 726 (1966).
531. D. J. Inman and W. E. Hornby, *Biochem. J.* **129,** 255 (1972).
532. J. Campbell, W. E. Hornby, and D. L. Morris, *Biochim. Biophys. Acta* **384,** 307 (1975).
533. M. H. Keyes, F. E. Semersky, and D. N. Gray, *Enzyme Microb. Technol.* **1,** 91 (1979).
534. D. J. Inman and W. E. Hornby, *Biochem. J.* **137,** 25 (1974).
535. E. Riesel and E. Katchalski, *J. Biol. Chem.* **239,** 1521 (1964).
536. D. R. Senn, P. W. Carr, and L. N. Klatt, *Anal. Chem.* **48,** 954 (1976).
537. H. H. Weetall, in *Immobilized Biochemicals and Affinity Chromatography* (R. Bruce Dunlap, Ed.), Plenum Press, New York, 1974, p. 191.
538. W. E. Hornby, D. J. Inman, and A. McDonald, *FEBS Lett.* **23,** 114 (1972).
539. R. W. Coughlin, M. Aizawa, and B. E. Alexander, U.S.G.R.A., No 22 (1975).
540. T. Tosa, R. Sano, and I. Chibata, *Agr. Biol. Chem.* **38,** 1529 (1974).
541. S. Ikeda, Y. Sumi, and S. Fukui, *FEBS Lett.* **47,** 295 (1974).
542. S. Ikeda and S. Fukui, *FEBS Lett.* **41,** 216 (1974).
543. S. J. Updike and G. P. Hicks, *Nature* **214** 986 (1967).
544. L. B. Wingard, C. C. Liu, and N. L. Nagda, *Biotechnol. Bioeng.* **13,** 629 (1971).
545. M. Aizawa and S. Suzuki, *Denki Kagaku* **44,** 279 (1976).
546. I. Satoh, I. Karube, and S. Suzuki, *Biotechnol. Bioeng.* **18,** 269 (1976).
547. G. G. Guilbault and J. G. Montalvo, *J. Amer. Chem. Soc.* **92,** 2533 (1970).
548. Y. Nakamoto, I. Karube, and S. Suzuki, *Biotechnol. Bioeng.* **17,** 1387 (1975).
549. G. G. Guilbault and F. R. Shu, *Anal. Chem.* **44,** 2161 (1972).
550. M. Aizawa, I. Karube, and S. Suzuki, *Anal. Chem. Acta* **69,** 431 (1974).
551. S. Suzuki, F. Takahashi, I. Satoh, and N. Sonobe, *Bull. Chem. Soc. Jpn.* **48,** 3246 (1975).
552. G. G. Guilbault and F. R. Shu, *Anal. Chem. Acta* **56,** 333 (1971).
553. G. G. Guilbault and E. Hrabankova, *Anal. Chem. Acta* **56,** 285 (1971).
554. I. Satoh, I. Karube, and S. Suzuki, *Biotechnol. Bioeng.* **19,** 1095 (1977).
555. H. Nilsson, K. Mosbach, S. O. Enfors, and N. Molin, *Biotechnol. Bioeng.* **20,** 527 (1979).
556. G. J. Papaniello, A. K. Mukherji, and C. M. Sheare, *Anal. Chem.* **45,** 790 (1973).
557. G. A. Rechnitz and R. Llenado, *Anal. Chem.* **43,** 283 (1971).
558. R. A. Llenado and G. A. Rechnitz, *Anal. Chem.* **43,** 1457 (1971).
559. G. G. Guilbault and E. Hrabankova, *Anal. Lett.* **3,** 53 (1970).
560. I. Karube, I. Satoh, Y. Araki, S. Suzuki, and H. Yamada, *Enzyme Microb. Technol.* **2,** 117 (1980).
561. T. M. S. Chang, *Science* **146,** 524 (1964).
562. T. M. S. Chang and M. J. Poznansky, *Nature* **218,** 243 (1968).
563. T. M. S. Chang, *Nature* 229, 117 (1971).
564. T. Mori, T. Tosa, and I. Chibata, *Biochim. Biophys. Acta* **321,** 653 (1973).
565. T. Mori, T. Tosa, and I. Chibata, *Cancer Res.* **34,** 3066 (1974).

566. J. P. Allison, L. Davidson, A. G. Hartman, and G. B. Kitto, *Biochem. Biophys., Res. Commun.* **47,** 66 (1972).
567. D. Sampson, L. S. Hersh, D. Cooney, and G. P. Murphy, *Trans. Amer. Soc. Artific. Intern. Organs* **18,** 54 (1972).
568. J. C. Venter, B. R. Venter, J. E. Dixon, and N. O. Kaplan, *Biochem. Med.* **12,** 79 (1975).
569. T. M. S. Chang, *Trans. Amer. Soc. Artific. Intern. Organs* **12,** 13 (1966).
570. D. L. Gardner, R. D. Falb, B. C. Kim, and D. C. Emnerling, *Trans. Amer. Soc. Artific. Intern. Organs* **17,** 239 (1971).
571. G. Broun, C. Tran-Minh, D. Thomas, D. Domurado, and E. Selegny, *Trans. Amer. Soc. Artific. Intern. Organs* **17,** 341 (1971).
572. G. Gregoriadis, P. D. Leathwood, and B. E. Ryman, *FEBS Lett.* **14,** 95 (1971).

CHAPTER

8

ORGANIC SYNTHESIS USING IMMOBILIZED ENZYMES

J. TRAMPER

Department of Process Engineering
Agricultural University
De Dreyen 12
6703 BC Wageningen
The Netherlands

Enzymes are in many ways much more skillful at synthesizing organic compounds than ordinary chemical catalysts, and numerous potential synthetic applications are documented (1). *Specificity* is in this respect the most characteristic property that makes enzymes superior to the chemical catalysts (2). *Immobilization* can make possible an efficient use of enzymes, and over a hundred have already been immobilized by at least an equal number of different methods (3). Rather than trying to cover all the enzymes that have synthetic potential, the emphasis in this chapter is put on examples of immobilized enzymes that distinctly illustrate the possibility of exploiting the various aspects that enzyme-specificity encompasses. A detailed description of the development of an immobilized *xanthine oxidase* system for application in organic syntheses forms the last section of the chapter.

8.1. BIOCATALYSIS AND APPLICATIONS

8.1.1. Enzymes for in vitro syntheses

A question that might arise when reading the title of this chapter is, why should a chemist want to use an enzyme as a catalyst in an organic synthesis? The answer is simple. Enzymes catalyze a large number of interesting chemical reactions that are otherwise difficult to establish, if possible at all. Furthermore, using enzymes has several important practical advantages (Table 8.1). The first major advantage of enzymes is their high specificity, a concept that is discussed in more detail in Section 8.1.4. Thus a very pure product solution is obtained when a reaction is performed by means of an

Table 8.1. Reasons for the Application of Enzymes in Organic Syntheses

1. Enzymes are highly specific.
2. Very pure products are obtained.
3. Enzymic reaction mixtures are easy to work up.
4. Acceleration factors are extremely high.
5. There are many advantages with respect to economy, execution of reaction, energy requirements, and environment.

enzyme, provided that the substrate solution is free of contaminants. When the latter condition is satisfied, enzymic reaction mixtures are easy to work up, the part in conventional organic syntheses that is usually difficult and laborious.

The fourth advantage of enzymes is that they are highly active. Turnover numbers in the range 10^{-3} to 10^{3} molecules substrate per second per molecule enzyme are typical at 10^{5} Pa (1 atm) and between 0 and 37°C, with some enzymes exhibiting turnover numbers up to 10^{5} under similar conditions (4).

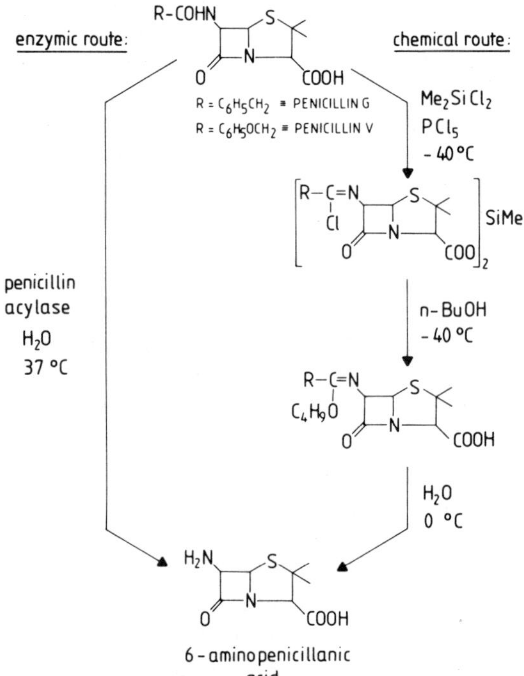

Fig. 8.1. Production of 6-aminopenicillanic acid. (Reproduced from Ref. 5 with permission.)

This means that enzyme-catalyzed reactions are relatively very fast in the presence of only small quantities of the enzymes. Consequently, enzymic reaction times are generally short. Another advantage is the simplicity of execution of enzymic reactions; they are also much less dangerous than most chemical reactions that do not use enzymes. Nearly all enzymic reactions can be performed in aqueous solution at atmospheric pressure and room temperature or close to this temperature, and usually no by-products are formed that are hard to remove. As such, enzymic reactions are potentially energy-saving and less environmentally polluting, currently important considerations even on the smallest scale.

The production of 6-aminopenicillanic acid (6-APA) is a good example of the usually greater complexity of chemical synthesis in comparison with enzymic synthesis. The key compound in the preparation of semisynthetic antibiotics, (6-APA), is synthesized by hydrolysis of penicillin G (benzyl penicillin) or penicillin V (phenoxymethyl penicillin), which are both produced by fermentation. Two processes, one chemical and one enzymic, are available for the hydrolysis of these penicillins (Figure 8.1). Instead of the one reaction step at very mild conditions in the biocatalytic process, three reaction steps at more extreme conditions are necessary for the chemical synthesis. When, in addition, the chemicals necessary to carry out the reaction are examined for the two methods (Table 8.2), the contrast is even more clear. The advantages of using the enzymic route in this example become even more obvious if the utilities are also taken into consideration (Table 8.3).

Despite these advantages, enzymes have been used relatively infrequently as catalysts in organic syntheses. This has both scientific and practical causes, but the unfamiliarity of chemists with enzymology should not be underestimated too, and a multidisciplinary cooperation is probably the best way to solve the problem. Among other causes the limited availability of enzymes must take first place. Isolation from their natural environment is usually de-

Table 8.2. Chemicals Needed for the Production of 6-Amino-penicillanic Acid (5)

Enzymic Synthesis	Chemical Synthesis
Penicillin acylase	Dimethyl aniline
Phosphate buffer	Dimethyl dichlorosilane
Sodium hydroxide	Ammonia
Hydrochloric acid	Methylene chloride
Methanol	Butanol
	Acetone
	Methanol

Table 8.3. Utilities Needed for the Production of 1 kg of 6-Aminopenicillanic Acid (5)

Utility	Enzymic route	Chemical route
Electric power	6.1 kWh	13 kWh
Liquid nitrogen	—	17.5 kg
Low pressure steam	73.7 kg	23.1 kg
Process water	0.11 m^3	0.19 m^3
Cooling water	4.3 m^3	1.25 m^3

sirable and in many cases even indispensable. This isolation is often difficult and laborious, making the enzymes scarce and expensive, so that efficient utilization is a requisite. The solubility of enzymes in water and the necessity to use this solvent as reaction medium hinders efficiency. At the end of the reaction, the enzyme is difficult to recover from the reaction mixture with retention of activity. Consequently, only single-batch use is usually possible. Also, their stability many times leaves a lot to be desired, especially when they are removed from their natural environment. These problems confined the use of enzymes for a long time, and still do. In the last two decades, however, much effort has been spent to solve these problems and several new techniques have been developed that strongly stimulate the application of enzymes. In the first place is affinity chromatography, a one-step purification procedure based on the specific interaction of a specific enzyme with a specific ligand that is bound to an otherwise inert solid support. This technique can improve and simplify the isolation of enzymes considerably, making enzymes available in larger quantities and in purer form. Detailed accounts of the various aspects of affinity chromatography are found in other chapters of this volume.

8.1.2. Application of Immobilized Enzymes

Another new technique, at least as important as affinity chromatography, is the immobilization of enzymes. Its relative importance is apparent from a computer literature search over 1980 on affinity chromatography and immobilized enzymes; it yielded about 800 references (40%) for the former and about 1200 references (60%) for the latter. Enzyme immobilization may be defined (6) as the imprisonment of an enzyme in a distinct phase that allows exchange with, but separation from, the bulk phase in which substrate, effector, or inhibitor molecules are dispersed and monitored. The enzyme phase is usually insoluble in water and is often a high molecular weight, hydrophilic polymer. This so-called support, and the various means by which the imprisonment of the enzyme may be achieved, is discussed in more detail in Section 8.2 and 8.3.

Table 8.4. Advantages of Immobilized Enzymes in Comparison with Free Enzymes

1. Repeated use is possible.
2. Continuous application is feasible.
3. They are often more stable.
4. Less-contaminated products are obtained.
5. They can be tailor-made for specific purpose.
6. They are less labor intensive.

8.1.2a. Multiple and Continuous Application

Immobilization yields an enzyme preparation with a number of distinct advantages over the free, water-soluble enzyme (Table 8.4). Reuse is simply realizable, since at every desired moment the immobilized enzyme can easily and rapidly be recovered from the reaction mixture, for example by filtration or centrifugation, and added to the next batch or stored until later usage. The solid-supported enzymes can also be used in a continuous fashion, that is, in a packed bed reactor or in a stirred tank reactor.

8.1.2b. Stability: Free vs. Immobilized Enzymes

Immobilization often improves the stability of the enzyme, or at least apparently does. If the immobilization concerns proteases, or if these are present as impurities, the fixation onto the support prevents autolysis and proteolysis, respectively (7). This is simply because most of the immobilized enzyme molecules, although usually having high local concentrations, cannot get together any more in the right manner for lysis, when the fixation onto the support occurs randomly. The local "togetherness" of the immobilized enzyme molecules brings about an aspect that makes comparison between the stabilities of free and immobilized enzymes rather trivial. The stability of an enzyme in solution is not just dependent on the physical conditions in which it is placed, but may as well depend upon the enzyme or even total protein concentration. Thus, while an enzyme may be relatively unstable in the dilute solution in which it is usually studied, raising the enzyme's concentration can enhance its stability—and raising the enzyme's local concentration is what usually happens as result of immobilization.

Few studies have actually been performed taking such "safety in number" (6) considerations into account and, of those studies, it would seem that stabilization due to immobilization per se is an exception. This sheds considerable doubt on the validity of a suggestion heard many times for explaining the increased enzyme stability after immobilization, that binding of the enzyme to the support by more than one linkage (multipoint attachment), or confining it in a limited space, fixes the enzyme in its active conformation and provides it with a protective shell, so that it better stands extremes in pH, temperature, and denaturing agents like urea, guanidine-HCl, and organic solvents. It has indeed been observed that trypsin immobilized on a copolymer of ethylene and maleic anhydride remains active in 8 molar urea, when the soluble enzyme is completely denatured (7, 8). Similarly, the thermostability of chymotrypsin immobilized in a 50 w/w% polymethacrylate gel dramatically increases (9). The first-order constant for thermoinactivation of this enzyme at 60°C is 10^{-5} of that in water, which is ascribed to multipoint noncovalent interaction with the polymeric support. This kind of stabilization, however, does seem to be the exception rather than the rule (6).

Another factor that may apparently increase the enzyme's stability and obscure comparison of free and immobilized enzyme stability is the existence of substrate diffusion limitation in the heterogeneous immobilized enzyme system. When diffusion of substrate to (through stagnant layer) and in an immobilized enzyme particle is slower than the conversion of substrate by the immobilized enzyme, a substrate concentration gradient will develop in the particle and/or outside it (Figure 8.2, situation 1). This gradient gets steeper as the activity per unit volume of support increases, and a situation may arise (Figure 8.2, situation 2) in which only part of the immobilized enzyme molecules "sees" substrate, because all the substrate molecules penetrating the particle are converted before they reach the center of the particle. In this situation, the activity initially appears to remain about constant, as inactivated enzyme is replaced by enzyme bound more to the center of the particle. Eventually, if the experiment is continued long enough, the enzyme activity drops below the point at which diffusion of substrate is rate-limiting and the activity rather suddenly starts to decrease. It is obvious that care must be taken in drawing any conclusions about increased stability of an enzyme as result of immobilization.

8.1.2c. Cleaner Processes

Compounds that are present as impurities in the enzyme preparation can, if they are not also immobilized, easily be washed from the immobilized-enzyme system before the actual catalysis and thus cannot contaminate the desired product. In contrast with reactions executed with free enzymes, immobilized enzymes are not present in the reaction mixture in a soluble form, making the work-up even simpler. This is illustrated by the preparation of (radioactive) CMP-sialic acids using immobilized CMP-acylneuraminate synthase from frog liver by Corfield et al. (10). Pure CMP-glycosides of sialic acids, either radioactively labeled in the sialic acid moiety or unlabeled, are required in glycoconjugate biosynthesis studies. Corfield et al. partially purified CMP-acylneuraminate synthase from frog livers and coupled it to

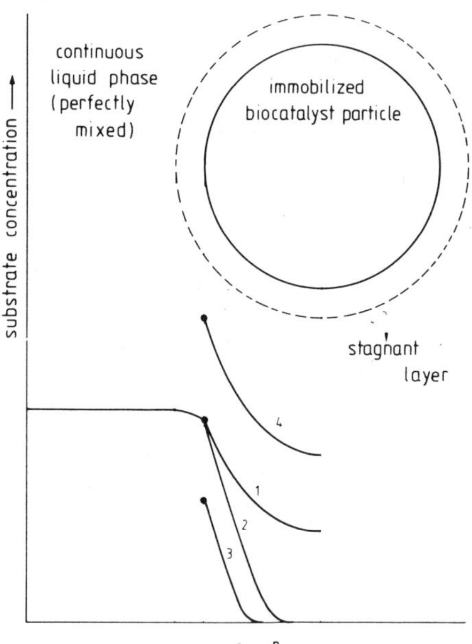

Fig. 8.2. Schematic concentration profiles.

CNBr-activated Sepharose 4B (for the general procedure see Section 8.3, "Immobilization Techniques"). The actual synthetic reaction (Figure 8.3) was executed by incubation of the washed immobilized enzyme preparation with CTP and (radioactive) *N*-acetylneuraminate at 37°C, pH 9 for 2 hr. After filtering off the immobilized enzyme, the product could be isolated from the supernatant in 80% yield. The authors conclude:

> The advantages of such an immobilized CMP-acylneuraminate synthase preparation are evident in the rapidity and ease of removal of enzyme activity from the reactants, eliminating the problems of contamination of the CMP-glycosides with low and high-molecular-weight substances other than the reactants, as has been observed with the use of soluble-enzyme preparations, e.g., tissue extracts.

Another example in this respect, from Cooper and Gelbard (11), is the synthesis of ^{13}N-amino acids by immobilized glutamate dehydrogenase:

$$\alpha\text{-Ketoglutarate} + \text{NADH} + {}^{13}\text{NH}_4^+ \xrightleftharpoons{\text{immobilized glutamate dehydrogenase}} \text{L-}{}^{13}N\text{-glutamate} + \text{NAD}^+ + \text{H}_2\text{O}$$

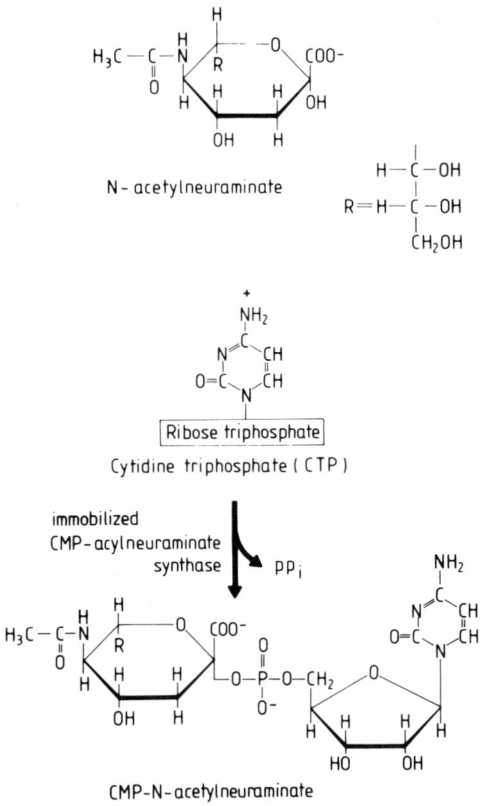

N-acetylneuraminate

Ribose triphosphate

Cytidine triphosphate (CTP)

immobilized CMP-acylneuraminate synthase

PP_i

CMP-N-acetylneuraminate

Fig. 8.3. Synthesis of (radioactive) CMP-sialic acids by immobilized CMP-acylneuraminate synthase (10).

Such pure positron-emitting radionuclides with short half-lives (10 min) are uniquely suited for quantitative in vivo biochemical and physiological studies. The short half-lives necessitate the use of rapid synthesis techniques, which makes enzymes very appropriate in this case. Because many of the labeled compounds are being synthesized as potential organ-imaging agents for clinical studies, immobilization of the biocatalyst is advisable in order to prevent possible pyrogenic or antigenic reactions resulting from the presence of protein in the final product. Cooper and Gelbard (11) bound glutamate dehydrogenase to CNBr-activated Sepharose and used the resulting preparation for the facile synthesis of six ^{13}N-amino acids (Table 8.5) with a radiochemical purity greater than 99%. According to these authors the ^{13}N-amino acid solutions obtained are potentially suitable and in sufficient quantity for whole-body and organ imaging in large animals and humans.

Table 8.5. **^{13}N-Amino Acids Synthesized from (^{13}N) Ammonia using Immobilized Glutamate Dehydrogenase**[a]

α-Keto acid Substrate used (concentration m*M*)	Corresponding L-Amino Acid	Apparent K_m(m*M*)	V (μmol/min/mg)	Radiochemical Yield (%)
α-Ketoglutarate (1.7)	L-Glutamate	1.2	15.9	90
α-Ketobutyrate (27)	L-Amino butyrate	50	0.13	58
α-Ketoisovalerate (24)	L-Valine	83	0.11	40–60
α-Keto-γ-methiolbutyrate (20)	L-Methionine	17	0.12	41
α-Ketoisocaproate (22)	L-Leucine	52	0.036	23
Pyruvate (27)	L-Alanine	26	0.015	16

[a] Adapted from Ref. 11.

8.1.2d. "Tailor-made" Biocatalysts

Every application with immobilized enzymes has its specific demands, which can be taken into account when the support and coupling procedures are chosen. For example, if the experimenter wishes to use the immobilized enzyme in a packed bed reactor, he or she requires preferably a rigid bead. In situations where the pH optimum of the enzyme activity is far from those pH values that are thermodynamically conducive to the synthesis of the desired product, utilization of an appropriately charged (polyelectrolyte) support can apparently shift the pH optimum of the immobilized enzyme in a favorable direction. Immobilization on a charged support makes the local concentration of hydrogen ions in the microenvironment of the enzyme molecules different from the concentration in the continuous liquid phase. It follows from the general theory (12) that this causes a pH optimum shift in the catalytic activity of the immobilized enzyme (compared with the enzyme in solution) toward acidic or alkaline pH values depending on whether a negatively or positively charged support, respectively, is employed. The practical applicability of this partition effect is limited, however, as the possible pH shift is not great (not over 1–2 pH units). Furthermore, since this partition is caused by electrostatic effects, it can occur solely in the case when the ionic strength of the solution is low (not over 10^{-2} *M*).

A principally different approach to the solution of this problem has been proposed by Martinek et al. (13–17), who employ the biphasic system "water/water-immiscible organic solvent." This method does not affect the enzyme but the equilibrium of the catalyzed reaction. Such reactions are, for example, synthesis of esters, polymerization of amino acids and sugars, and dehydration reactions. Formation of the final product in such processes is accompanied by formation of water, and so in aqueous solution the equilibrium is strongly shifted toward the starting reagents. An obvious approach to overcome this difficulty is to use nonaqueous solvents as reaction medium instead of water. There have been reported (references cited in Ref. 13) many attempts to perform enzymic reactions in organic solvents or water/organic solvent mixtures with a high concentration of a nonaqueous component, but dramatic decrease in enzyme activity and substrate specificity limits this approach. Immobilization can partially solve this problem. An example, which shows that immobilization can produce an enzyme derivative that is catalytically active in ester synthesis in low water concentrations where the free enzyme is inactive, concerns α-chymotrypsin. Ingalls et al. immobilized α-chymotrypsin on polyacrylamide gel via the succinyl hydrazide derivative (18). This preparation catalyzes esterification of *N*-acetyltyrosine in a medium containing high concentrations of alcohols. According to the authors, the hydrophilic support and inclusion of glycerol protect the enzyme activity and allow catalysis to proceed in the presence of only 10% (v/v) water. However, in all studies with immobilized enzymes in nonaqueous media (references cited in Ref. 13) the enzyme becomes inactive if the organic compound exceeds 90%.

The method proposed by Martinek and colleagues is based on the fact that enzymes can entirely retain their catalytic activity in a water/water-immiscible organic solvent system, even at a quite low water content (14). These biphasic systems can be an emulsion of an aqueous solution of the enzyme in an organic medium or—and this is much more advantageous (13) both methodologically and technologically—as a suspension of porous particles saturated with the aqueous solution of the enzyme; the latter may be immobilized onto the support, but encapsulation techniques or entrapment in hollow fibers (see Section 8.3 on immobilization techniques) can be used as well. Martinek et al. support their novel approach to preparative enzymic synthesis in nonaqueous solvents with several examples. They impregnated porous glass with aqueous buffer solution of chymotrypsin and suspended it in chloroform containing *N*-acetyl-L-tryptophan and ethanol. In water the yield of *N*-acetyl-L-tryptophan ethyl ester is about 0.01%; in the biphasic system, they found it to be practically 100% (13). α-Chymotrypsin, trypsin, pyrophosphatase, peroxidase, lactate dehydrogenase, and pyruvate kinase were also used to demonstrate their approach; they were immobilized by entrapment into reversed micelles formed by surfactants in an organic solvent. Enzymes immobilized in this fashion retain their catalytic activity and substrate specificity as well (17).

Another factor restricting the role of both free and immobilized enzymes as catalysts of many important chemical reactions is the poor solubility of many organic compounds in aqueous solutions. When this is the case for either the substrate (or substrates) or the product, an efficient synthesis by means of an immobilized enzyme is hindered. Large quantities of water are needed, making large reaction vessels necessary and work-up laborious. The approach of Martinek et al. is also applicable in this situation, and the transformation of steroids in biphasic water/organic solvent systems by immobilized hydroxysteroid dehydrogenases illustrates this (19). Carrea et al. showed that immobilization of hydroxysteroid dehydrogenases (β-HSDH, 20β-HSDH, and 3α-HSDH) by coupling to CNBr-activated Sepharose 4B yields immobilized enzyme preparations suitable for preparative transformation of steroids in water/ethyl acetate systems. The presence of cofactor during the immobilization reaction increased the activity recovery (40–60% of the total) and also led to immobilized enzyme preparations highly stable in the presence of organic solvents. For example, β-HSDH maintained 60% of its original activity two months after continuous use in water/ethyl acetate mixtures. Inclusion of substrates, inhibitors, effectors, cofactors, or products in the immobilization mixture may indeed stabilize enzymes and prevent severe losses in activity during the immobilization reaction.

Partition effects imposed by the support can be used to advantage in a similar manner to the apparent pH shifts when substrate or product activation or inhibition, or severe diffusion limitation, is involved. If, for example, substrate activation occurs, one should choose a support with a partition coefficient for substrate greater than one—that is, substrate solubility in the support is higher than in the bulk (Figure 8.2, situation 3), for example, as result of opposite charges of support and substrate. If, on the other hand, substrate inhibition is involved, one should choose a support with a partition coefficient for substrate smaller than one (Figure 8.2, situation 4).

8.1.2e. Less Labor-intensive Production

The last advantage of immobilized enzymes mentioned in Table 8.4 results from (i) the fact that immobilized enzymes can be used repeatedly or continuously and for a longer time than free enzymes, and (ii) the absence of enzyme in the product mixture. Consequently, processes with immobilized enzymes require less labor and a minimal product isolation.

8.1.3. Requirements for Routine Use of Immobilized Enzymes

Although a large number of enzymes have been immobilized, relatively few have reached the actual application stage; most remain unused either as a routine catalyst in the organic laboratory or as an industrial catalyst. Many problems, for example, efficient cofactor regeneration, remain to be solved before immobilized enzymes will gain their permanent place as a shelf reagent in the laboratory. Nevertheless, the future of immobilized enzymes as catalysts in organic syntheses seems to be a bright one and the organic chemist is becoming increasingly aware of the great potential of immobilized enzymes. For eventual acceptance as a routine catalyst, an immobilized enzyme will have to meet the requirements listed in Table 8.6.

Table 8.6. Requirements for Acceptance of an Immobilized Enzyme as Routine Catalyst in Organic Chemistry[a]

1. Must catalyze a reaction of preparative interest that is otherwise difficult to establish.
2. At least the free enzyme should be commercially available, or easy to obtain.
3. Should be sufficiently stable.
4. Enough specificity data should be available to enable reliable predictions to be made.

[a] From Ref. 2 (p. 832), with modifications.

8.1.4. Enzyme Specificity

The most characteristic properties of enzymes that distinguish them from ordinary chemical catalysts are those associated with their specificity. The concept of specificity in this context requires further explanation. In fact, it encompasses four distinct aspects of enzyme-catalyzed reactions, as listed in Table 8.7.

8.1.4a. Reaction and Constitutional Specificity

Reaction specificity implies that in general an enzyme only catalyzes one type of reaction. Constitutional specificity of enzymes is the result of the dimension, shape, and structure of their active site. The synergistic interactions between the chemistry of the active sites and their elegant geometrical precision endow many enzymes with great specificity for particular substrates and with high catalytic activity at moderate temperatures and pressures. Only when the important structural features of a substrate complement those of the corresponding binding loci at the active site will formation of a catalytically productive enzyme–substrate complex be favored. Such reaction and constitutional specificity make it possible that in the living cell, which contains many enzymes, enzymes do not disturb each other's functions. Consequently, many series and parallel reactions can occur concurrently in living cells. Although this property of enzyme catalysis has been exploited to a very limited extent in synthetic organic chemistry, it needs little imagination to visualize the practical implication. Series and parallel reactions in a multistep reaction scheme can in principle be performed in a single reaction vessel by adding all the enzymes to the necessary substrates at the same time.

A distinct example of the use of this property of enzymes is the preparation of the antibiotic gramicidin S in a reactor containing immobilized enzymes, a system studied by Wang and colleagues at the Massachusetts Institute of Technology (6, 20, 21). Gramicidin S is a cyclic decapeptide (Figure 8.4), composed of two identical pentapeptide chains that consist of four L-amino acids and D-phenylalanine. The enzyme complex responsible for the synthesis of the drug gramicidin S synthetase can be isolated from the bacterium *Bacillus brevis*. For this purpose the MIT group uses a new protein separa-

Table 8.7. The Four "Specificities" of Enzymes (1)

1. Reaction specificity
2. Constitutional (or structural) specificity
3. Stereospecificity
4. Kinetic or rate-acceleration specificity

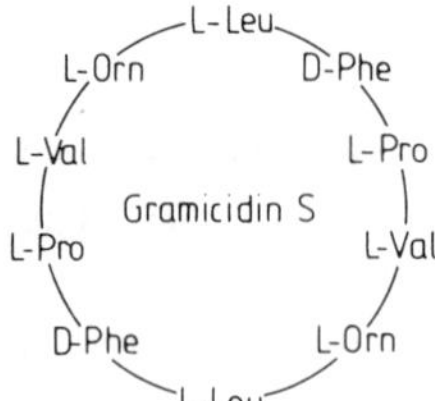

Fig. 8.4. Structure of gramicidin S.

tion technique based on two-phase partitioning in high molecular weight polymer systems. When two polymers—for example, polyethyleneglycol and dextran—are dissolved in water at appropriate concentrations, they are mutually immiscible. Certain enzymes will preferentially dissolve in one or the other of the polymer phases and thus may be separated and purified. By using 1.6% polyethyleneglycol 4,500 and 6.5% dextran 82,000, Wang and his colleagues achieved in one step an 8.5-fold purification with 70% recovery of gramicidin S synthetase complex. This multienzyme complex consists of at least five different proteins, which racemize L-phenylalanine and activate the formed D-isomer and the other L-amino acids involved. For activation to occur, ATP and magnesium ions are required. Adsorption of the enzyme complex onto ion exchange beads yields a preparation that can in principle be used for synthesis of gramicidin S, according to the following reaction equation:

$$\begin{matrix} 2\ \text{L-Phe} \\ 2\ \text{L-Pro} \\ 2\ \text{L-Val} \\ 2\ \text{L-Orn} \\ 2\ \text{L-Leu} \end{matrix} + 10\ \text{ATP} \xrightarrow[\text{Mg}^{++}]{\text{immobilized gramicidine S synthetase}} \begin{matrix} 10\ \text{AMP} \\ \text{gramicidine S} \\ 10\ \text{PP}_i \end{matrix}$$

To make this synthesizing system economically more feasible, the MIT scientists, couple it to an ATP regenerating system consisting of immobilized acetate kinase and immobilized adenylate kinase. The first enzyme converts AMP to ADP, while the second transfers a phosphate group from acetyl phosphate to ADP to produce ATP:

$$\text{AMP} + \text{ATP} \underset{\text{kinase}}{\overset{\text{immobilized adenylate}}{\rightleftharpoons}} 2\ \text{ADP}$$

$$\text{AcOP}_3 + \text{ADP} \underset{\text{kinase}}{\overset{\text{immobilized acetate}}{\rightleftharpoons}} \text{CH}_3\text{CO}_2^- + \text{ATP}$$

Adenylate kinase is isolated from baker's yeast by a purification scheme developed by the MIT group. Acetate kinase is obtained from a cell-free homogenate of *E. Coli* via a one-step affinity chromatography procedure that achieves a 40-fold purification. For that purpose, AMP, an inhibitor of acetate kinase, is bound to agarose, a polysaccharide support, to form the affinity matrix.

Immobilization of these enzymes has been accomplished by several methods. One involves a magnetic support made from polyacrylamide beads containing iron particles. Using this support, acetate and adenylate kinase have been immobilized in yields of 65 and 90%, respectively. Another immobilization technique that has been studied is entrapment in hollow fibers. A third method, entrapment in polyacrylamide, is described in general terms in the following section. With the systems coupled, the MIT group has been able to produce sufficient quantities of gramicidine S to perform all the necessary chemical and biological analyses to prove authenticity. They have estimated that such a continuous process based on immobilized enzymes should be economically feasible. The importance of their work lies especially in demonstrating the viability of immobilized multienzyme systems for organic syntheses. Furthermore, this is an excellent example of how biotechnological research should be performed: chemical engineers integrating their know-how with that of researchers from various biodisciplines and with that of chemists. The chemists at the same time have developed an efficient synthesis for acetyl phosphate, the substrate necessary in the cofactor regeneration. From ketene and phosphoric acid, this compound can be prepared cheaply in 85 to 90% yields. The overall synthesis thus consists of three components, as depicted in Figure 8.5.

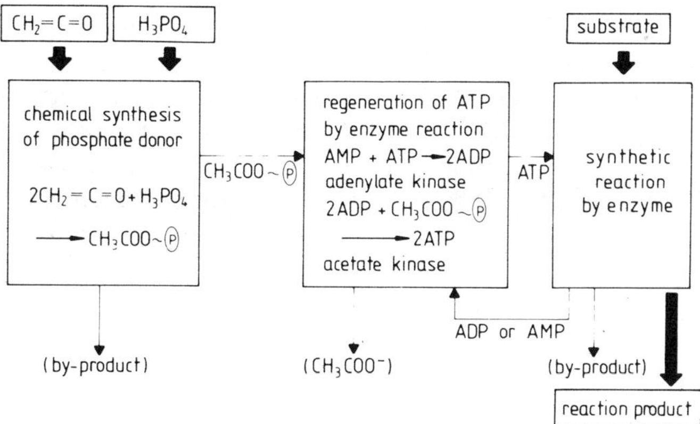

Fig. 8.5. Schematic representation of biotechnological synthesis of gramicidin S. (Reproduced from Ref. 3 with permission.)

8.1.4b. Stereospecificity

The third specificity in Table 8.7, stereospecificity, concerns the property of many enzymes to be able to discriminate between enantiomers, a feature not often found in ordinary chemical catalysis. In free, nonimmobilized form, numerous enzymes are used by chemists in studies of asymmetrical syntheses (1). Such applications with immobilized enzymes, on the contrary, are relatively scarce. The most prominent and best described example in this respect is the production of L-amino acids using immobilized aminoacylase. Chemical synthesis yields a racemic mixture of D- and L-isomers, and optical resolution is necessary to obtain the desired L-amino acid from it. One such separation procedure makes use of the enzyme aminoacylase, a hydrolase isolated from molds. Aminoacylase catalyzes the hydrolysis of acyl-L-amino acids (Figure 8.6). In a chemically synthesized acyl-D, L-amino acid mixture the L-enantiomer is selectively hydrolyzed by aminoacylase. When the reaction is completed, the mixture is concentrated and the precipitated L-amino acid is collected by filtration. The acyl-D-amino acid can be once more racemized and recirculated.

As a result of the solubility of the aminoacylase preparation and its contamination with impurities, a labor-intensive batch process and several difficult separation and purification steps must be performed. In order to solve these problems, Tanabe Seiyaku Company in Japan started in 1969 a continuous process (Figure 8.7) in which the optical resolution is established by means of immobilized aminoacylase (in fact, it was the first industrial process based on immobilized enzymes). After testing many immobilization procedures, aminoacylase immobilized by electrostatic binding to the anion-exchange support DEAE-Sephadex was chosen for this purpose. In this manner, the L-amino acids mentioned in Table 8.8 are produced with a reduction in costs of about 40% compared with the conventional batch process using the soluble enzyme.

Several other hydrolases have been immobilized for application in the synthesis of L-amino acids by means of optical resolution of racemic mixtures of corresponding amino acid derivatives, for example, carboxypeptidase has been immobilized by covalent binding to the diazonium derivative

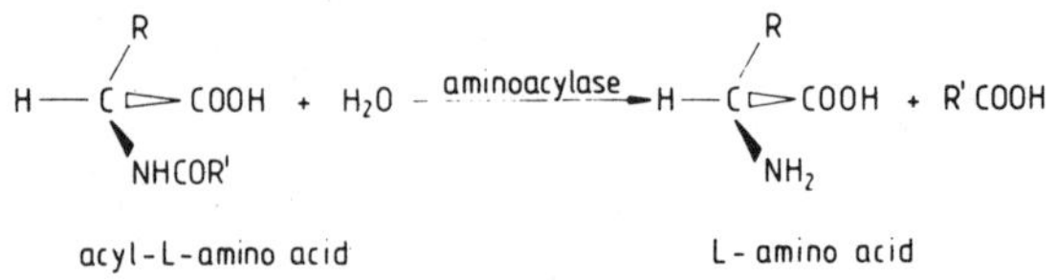

$$\text{H—C(R)(NHCOR')—COOH} + H_2O \xrightarrow{\text{aminoacylase}} \text{H—C(R)(NH}_2\text{)—COOH} + \text{R'COOH}$$

acyl-L-amino acid   L-amino acid

Fig. 8.6. Asymmetric hydrolysis of acyl-L-amino acids by aminoacylase.

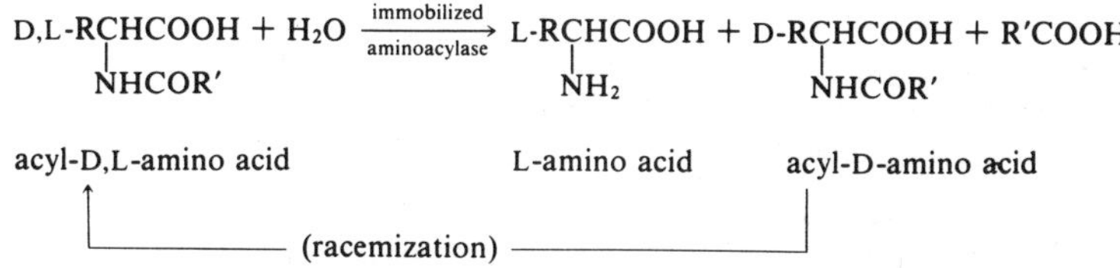

Fig. 8.7. Enzymic synthesis of L-amino acids. (Reproduced from Ref. 3 by permission.)

of polyaminostyrene, and leucine aminopeptidase has been covalently bound to CNBr-activated Sepharose (3).

In a similar way, immobilized enzymes are used for the production of D-amino acids, as can be illustrated by the synthesis of D-(−)-phenylglycine and several of its derivatives substituted in the para-position of the phenyl group, e.g. D-(−)-4-hydroxyphenylglycine. These compounds are raw materials for semisynthetic penicillins and cephalosporins. Olivieri et al. (22) have developed a method starting from racemic hydantoins (Figure 8.8), which are easily prepared from aldehydes by the Bucherer synthesis. Hydantoins are stereospecifically hydrolyzed by dihydropyrimidinase to the corresponding carbamoyl derivatives of D-amino acids. The advantage of using hydantoins, instead of phenylglycine derivatives for which enzymic resolutions also have been patented, is that they very easily undergo spontaneous racemization (see Figure 8.8), even under the mild conditions at which the enzymic catalysis is performed. Consequently, enzymic cleavage of the ring and chemical racemization occur simultaneously, so that D, L-hydantoins are completely converted to the D-carbamoyl derivatives. The latter can be further hydrolyzed to the corresponding amino acids by chemical methods, under conditions where complete retention of configuration is maintained. These methods, however, have several disadvantages, and therefore Olivieri and his colleagues also studied an enzyme capable of hydrolyzing *N*-carbamoyl de-

Table 8.8. Yields of L-Amino Acids Produced by Means of Aminoacylase Immobilized onto DEAE-Sephadex (21)

	Yield (kg)	
L-Amino Acid	per 24 hr	per 20 days
L-Alanine	214	6,420
L-Methionine	715	21,450
L-Phenylalanine	594	17,820
L-Tryptophan	441	13,320
L-Valine	505	15,150

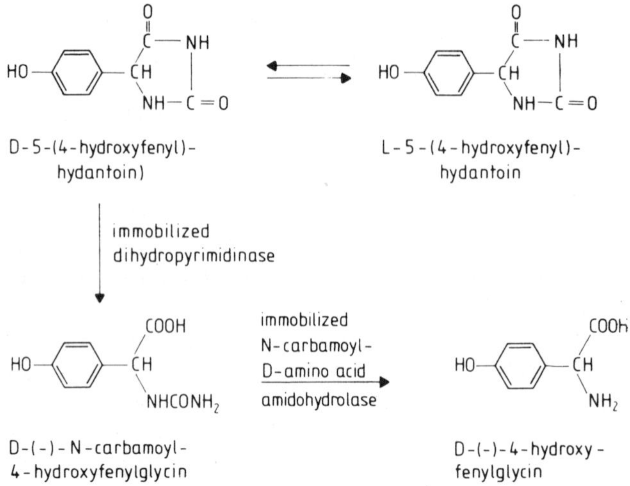

D-5-(4-hydroxyfenyl)-hydantoin) ⇌ L-5-(4-hydroxyfenyl)-hydantoin

↓ immobilized dihydropyrimidinase

D-(-)-N-carbamoyl-4-hydroxyfenylglycin —(immobilized N-carbamoyl-D-amino acid amidohydrolase)→ D-(-)-4-hydroxy-fenylglycin

Fig. 8.8. Enzymic synthesis of D-(−)-4-hydroxyfenylglycin (22).

rivatives to the corresponding D-amino acids (Figure 8.8). This enzyme, *N*-carbamoyl-D-amino acid amidohydrolase, is stereospecific as well, but because of the strong inhibition exerted by the L-antimer, it is not practical to use this enzyme for the separation of D, L-carbamoyl derivatives, especially as racemization of the unhydrolyzed L-antimer is difficult to accomplish by chemical means. For these reasons, this enzyme was studied in combination with dihydropyrimidinase, taking racemic hydantoins as starting materials. A mixture of these enzymes has been immobilized in cellulose triacetate fibers, and the results indicate that it is feasible to develop a one-step process for the preparation of D-amino acids. This example nicely illustrates the possibility of performing multistep stereospecific syntheses by means of enzymes in "one pot."

Several other papers on applications of immobilized enzymes, and especially immobilized microbial cells for asymmetric synthesis, have been published (3) and are described below. L-aspartic acid can be prepared from fumaric acid and ammonia, employing the catalytic action of aspartase:

$$\underset{\text{Fumaric acid}}{HOOCCH{=}CHCOOH} + NH_3 \xrightarrow[\quad]{\text{immobilized aspartase}} \underset{\text{L-Aspartic acid}}{HOOCCH_2CH(NH_2)COOH}$$

Immobilization in polyacrylamide gel has been found particularly useful for immobilization of both the isolated enzyme and cells with a high aspartase activity. Similarly, cells having high fumarase activity were immobilized by the polyacrylamide gel method. Treatment of the immobilized cells with bile extract makes the cell membrane more permeable, resulting in a marked increase in L-malic acid formation, the desired product:

$$\underset{\text{Fumaric acid}}{\mathrm{HOOCCH{=}CHCOOH}} + H_2O \xrightleftharpoons[]{\text{immobilized fumarase}} \underset{\text{L-Malic acid}}{\mathrm{HOOCCH_2\underset{\displaystyle OH}{\underset{|}{C}}HCOOH}}$$

This treatment at the same time effectively suppresses formation of the by-product, succinic acid. Recently, Takata and his colleagues reported (23) an improved immobilization procedure with carrageenan, using a microorganism with a higher fumarase activity. Both fumarase and aspartase in the form of immobilized cells have been used for about eight years now by Tanabe Seiyaku Ltd for the continuous production of L-malic acid and L-aspartic acid, respectively. The same company uses, for the synthesis of urocanic acid, microbial cells having high L-histidine ammonia lyase activity immobilized by the polyacrylamide gel method:

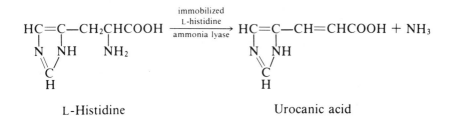

$$\underset{\text{L-Histidine}}{\text{(imidazol-4-yl)}\mathrm{-CH_2CH(NH_2)COOH}} \xrightarrow[\text{ammonia lyase}]{\text{immobilized L-histidine}} \underset{\text{Urocanic acid}}{\text{(imidazol-4-yl)}\mathrm{-CH{=}CHCOOH}} + NH_3$$

Another example of asymmetric synthesis by means of immobilized enzymes was described as long ago as 1965 (24). D-Oxynitrilase immobilized by ionic binding to ECTEOLA-cellulose was packed into a column, and a solution of benzaldehyde and hydrogen cyanide dissolved in 50% methanol was passed through it. Continuous conversion of benzaldehyde to D-(+)-mandelonitrile proceeded according to the reaction:

$$\underset{\text{Benzaldehyde}}{C_6H_5\mathrm{-CHO}} + HCN \xrightarrow{\text{immobilized D-oxynitrilase}} \underset{\text{D-(+)-Mandelonitrile}}{C_6H_5\mathrm{-C^*H(OH)-CN}}$$

The product was obtained in high yield (95%) and was of high optical purity (97% D-form and 3% L-form). In addition to stereospecificity the enzyme possesses a very broad substrate specificity, a property that makes the application of this enzyme in organic synthesis attractive. Conversion of aliphatic, aromatic, and heterocyclic aldehydes to the corresponding nitriles is accomplished in the presence of hydrogen cyanide and this enzyme, almost all in high yield.

The preparation of L-lysine using two kinds of immobilized enzymes has been reported too (3). L-α-Amino-ϵ-caprolactam hydrolase and α-amino-ϵ-caprolactam racemase were immobilized by ionic binding to anion exchange polysaccharides such as DEAE-Sephadex, and both immobilized enzymes were allowed to react with D, L-α-amino-ϵ-caprolactam at the same time. This substrate is easily produced by chemical synthesis from cyclohexane, a by-product of nylon production. The L-antimer is hydrolyzed to L-lysine by the first enzyme, while at the same time the D-form is racemized by the second enzyme:

$$
\begin{array}{c}
\text{L-}\left[\begin{array}{l}CH_2-CH_2-NH \\ CH_2 \quad\quad\quad\;\; | \\ CH_2-CH-CO \\ \quad\quad\;\; NH_2\end{array}\right]
\underset{\text{lactam racemase}}{\overset{\text{immobilized } \alpha\text{-amino-}\epsilon\text{-capro-}}{\rightleftarrows}}
\text{D-}\left[\begin{array}{l}CH_2-CH_2-NH \\ CH_2 \quad\quad\quad\;\; | \\ CH_2-CH-CO \\ \quad\quad\;\; NH_2\end{array}\right] \\
\downarrow \text{L-}\alpha\text{-Amino-}\epsilon\text{-caprolactam hydrolase} \\
\text{L-}H_2NCH_2CH_2CH_2CH_2\underset{\displaystyle NH_2}{\underset{|}{C}}HCOOH \\
\text{L-Lysine}
\end{array}
$$

In this way essentially all the D, L-α-amino-ϵ-caprolactam can be converted to L-lysine.

In order to hydrogenate enoates stereospecifically, enoate reductase and formate dehydrogenase have been immobilized on controlled pore glass by Tischer et al. (25). Enoate reductase is very versatile for the stereospecific reduction of many different 1-enoates (Figure 8.9). This enzyme is NADH-dependent, and efficient regeneration of this cofactor is desired. In contrast to the often used alcohol and lactate dehydrogenases, with formate dehydrogenase the NADH formation is strongly favored and the product CO_2 or HCO_3^- may hardly interfere with other components of a coupled system; further-

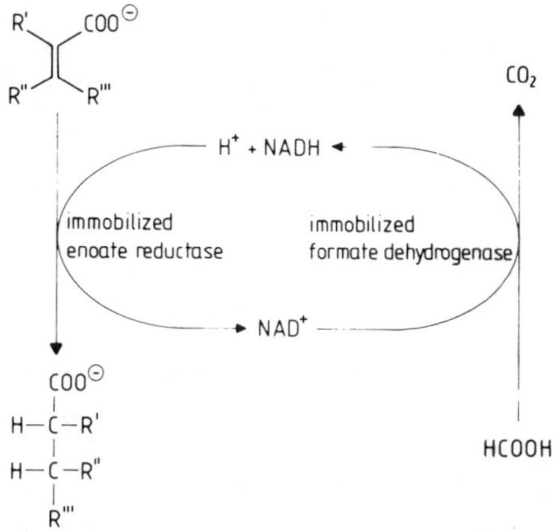

$$\text{R}'\text{R}''\text{C}{=}\text{C}(\text{COO}^{\ominus})\text{R}''' \longrightarrow \text{H}{-}\text{C}(\text{COO}^{\ominus})(\text{R}'){-}\text{C}(\text{H})(\text{R}'')\text{R}'''$$

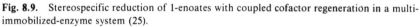

Fig. 8.9. Stereospecific reduction of 1-enoates with coupled cofactor regeneration in a multi-immobilized-enzyme system (25).

more, it leads to no difficulties during the product isolation. Tischer et al. are studying this coupled immobilized enzyme system because it is of interest from both a preparative and mechanistic point of view.

8.1.4c. Kinetic or Rate Acceleration Specificity

The last-mentioned specificity in Table 8.7 is kinetic, or rate acceleration, specificity. Although it cannot yet be stated that the mechanism of action of any enzyme is fully unraveled, the most important factors determining the rate of acceleration are known (Table 8.9). A dramatic example of rate acceleration is, for instance, the conversion of hydrogen peroxide into water and oxygen by the action of catalase:

$$H_2O_2 \xrightleftharpoons{K^*} H_2O_2^* \xrightarrow{k} H_2O + O$$

Table 8.9. Factors Determining the Dramatically Lower Free Energies of Activation of Enzymic Processes Compared with Their Uncatalyzed Analogs (1)

1. Approximation of reactants and their correct orientation for reaction at the active site.
2. Entropy effects.
3. Formation of covalent intermediates.
4. General acid- and base-catalysis by active-site functional groups.
5. Strain or distortion in the substrate or the enzyme.

Table 8.10. Fraction Activated Molecules K^* (= $[H_2O_2^*]/[H_2O_2]$), Activation Energy ΔG^* (= $-RT\ln K^*$), First-Order Reaction Constant k (= kTK^*/hr) and half time $t_{1/2}$ of the Hydrogen-Peroxide-Dissociation Reaction (20°C)

Catalyst	G^*(kJ/mol)	K^*	$k(s^{-1})$	$t_{1/2}$(s)
—	76	3.7×10^{-14}	0.23	8.7
Pt	55	2.0×10^{-10}	1200	1.6×10^{-3}
Catalase	29	6.0×10^{-6}	3.7×10^{7}	5.5×10^{-8}

This reaction is indeed spontaneous at room temperature but relatively slow in aqueous solutions. The addition of catalase increases the rate of reaction by many orders of magnitude, much more than can be effected by a chemical catalyst (Table 8.10). This property of catalase has been extensively exploited, especially in immobilized enzyme systems with oxidases such as glucose oxidase or xanthine oxidase as the biocatalyst. These enzymes need oxygen as a hydrogen acceptor in dehydrogenation reactions and produce H_2O_2 as a by-product, which in turn inactivates enzymes because of its reactivity with proteins, among other substances. From the stoichiometry of the reactions it follows (26) that with progressing conversion more H_2O_2 is produced, and in continuous reactions the stationary H_2O_2 concentration will increase with oxidase concentration. Since the inactivation is first order with respect to H_2O_2 concentration, the product formation and inactivation rates are coupled. Especially in immobilized oxidase systems, mass-transfer limitations can increase the microenvironmental by-product concentration, resulting in a rapid deterioration of the catalytic activity. Stabilization of such systems can be afforded by catalytic decomposition of H_2O_2 by means of catalase, which can be coimmobilized with the oxidase. The stability depends on the relative activities of both enzymes.

The immobilized glucose oxidase/catalase system has been studied extensively both for analytical (27) and preparative (28) purposes, for example, in gluconic acid production according to the reaction scheme:

$$\alpha\text{-D-Glucose} \rightleftharpoons \text{D-glucose} \rightleftharpoons \beta\text{-D-glucose}$$

$$\beta\text{-D-Glucose} + O_2 \xrightarrow[\text{oxidase}]{\text{immobilized glucose}} \text{D-gluconolactone} + H_2O_2$$

$$H_2O_2 \xrightarrow{\text{immobilized catalase}} H_2O + \tfrac{1}{2}O_2$$

$$\text{D-Gluconolactone} + H_2O \longrightarrow \text{D-gluconic acid}$$

The straight chain D-glucose is unstable in solution, where it can exist only in minute concentrations. In solution, equilibrium is set up between the α

and β forms of D-glucose, the latter only being oxidized to yield D-glucono-lactone, which spontaneously hydrolyses to gluconic acid. This is another example of a stereospecific multiple immobilized-enzyme reaction in which the most prominent characteristic of enzyme reactions, specificity in all respects, is exploited.

8.1.5. Section Summary

In this section an overview has been given of the application of immobilized enzymes in organic synthesis. By no means has complete coverage been pursued; rather, the vast potential of immobilized enzymes has been suggested on the basis of the most characteristic property of enzymes, specificity. Of especially great potential is the employment of immobilized enzymes in fine organic syntheses, in the preparation of drugs, and in the synthesis of important biochemical compounds. As shown in this section, a number of processes based on immobilized enzymes have already been developed on a laboratory, semitechnological, and even industrial scale, and more exist. Doubtless in the future immobilized enzymes will find a still wider use in preparative organic synthesis.

8.2. THE SUPPORT

The choice of the support is one of the most important factors to consider in developing an immobilized-enzyme system as it strongly influences the eventual characteristics of the immobilized enzyme. An ideal, preformed, solid support should possess the following properties:

1. Sufficient Permeability and a Large Surface Area for Enzyme Attachment. In order to be able to immobilize significant amounts of enzyme, the support must have a large surface area ($>10\ m^2 \cdot g^{-1}$). Since surface area is inversely related to the pore size, the ideal support will have a high porosity and small pores. Obviously, to be effective, the pores must have a minimum diameter to allow easy access of enzymes and substrate.

2. Presence of Functional Groups Allowing Attachment of the Enzyme Under Mild Conditions. The support must possess a sufficient number of chemical groups that can be activated or modified such that they are able to bind enzymes under nondenaturing conditions.

3. Hydrophilic character. A hydrophilic character of the support is generally desirable because it permits relatively unhindered substrate diffusibility of the type normally met in aqueous systems (29). Hydrophilic features, often enhance the stability of the attached enzyme. A hydrophobic support tends to decrease the stability and activity of immobilized enzymes by a mechanism similar to that of the denaturation of enzymes by organic solvents.

4. Insolubility. Insolubility of the support is essential, not only for prevention of enzyme loss but also to prevent contamination of the product by dissolved support and enzyme.

5. Chemical, Mechanical, and Thermal Stability. The support must be chemically resistant under the conditions of its activation (if necessary), during the immobilization process of the enzyme and during catalysis. The latter two processes must generally be performed under mild conditions, since enzymes themselves have a limited stability and operate optimally under moderate circumstances. The mechanical stability should be sufficiently high to withstand treatments such as filtration, centrifugation, and stirring, since the immobilization process and the repeated or continuous use of the immobilized enzyme usually require such manipulations. The thermal stability of a support is also important. The enzyme-active site may be distorted or destroyed when a support has a large expansion coefficient, leading to contraction or expansion during temperature changes. Such changes can occur during a multistep immobilization at varying temperatures, or when the immobilized enzyme is brought from storage to operational temperature and vice versa. When bound at several places (multipoint attachment), these changes are particularly likely to "pull" the enzyme out of its active conformation.

6. High Rigidity and a Suitable Form of Particles. The need for rigid particles is connected with the problem of fluid flow. Rigid, spherical, and uniformly sized particles (beads) are generally most suitable for continuous packed bed reactors, since they cause only a small pressure drop and have good flow properties. A rigid pore structure also protects the enzyme against a turbulent external environment. In addition, once an enzyme has been immobilized in its active form on a rigid congruent surface via multipoint attachment, the tertiary structure of the enzyme is stabilized by the lack of deformation of the support (7).

7. Resistance to Microbial Attack. Another consideration in selecting a support is the inertness to microbial degradation. Obviously, if the support is attacked and metabolized by microbes, the enzyme is lost either by release into solution or by direct microbial consumption.

8. *Regenerability.* Tightening budgets and increased public awareness with regard to pollution and finite resources have made regeneration and recycling of prime importance. The possibility of regeneration and reuse must be considered in the total economics of the immobilized-enzyme system, especially when considering a relatively expensive support for large-scale application.

At present an ideal and universal enzyme support does not exist, and it is doubtful whether one will ever be found. It is necessary, therefore, to evaluate all aspects and determine which support best meets the demands set. Eaton (30) has presented a "Support Study Decision Tree," enabling the researcher to evaluate whether or not the pertinent support is suitable for the immobilization of the enzyme of interest. The key role of the support in the immobilization of enzymes is apparent from the extensive research efforts and the rapid progress in this area. Growing numbers of various types of supports are being developed and commercialized. In the following sections the support materials agarose, alginate, carrageenan, cellulose, gelatin, acrylic copolymers, activated carbon, and controlled-pore glass are treated in some detail, taking into account the factors discussed above.

8.2.1. Polysaccharide Supports

The most widely used supports are those having a polysaccharide backbone, especially those extracted from algae (Figure 8.10), namely agarose, alginate, and lately carrageenan. Cellulose and its derivatives have also proved to be suitable supports.

8.2.1a. Agarose

Agarose is a purified linear-galactan hydrocolloid isolated from agar, which is a mixture of polysaccharides extracted from certain red seaweeds (Rhodophyceae). In 1956 the formula for the repeating subunit, agarobiose, was presented by Araki (31), showing an alternating 1,3-linked β-D-galactopyranose and 1,4-linked 3,6-anhydro-α-L-galactopyranose structure (Figure 8.11).

Sulfate, methoxyl, ketal pyruvate, and/or carboxyl groups are also present as substituents in the agar molecule in an almost infinite number of combinations. These substituents influence the properties of the various agar molecules considerably.

Duckworth and Yaphe (32), as a result of their comprehensive chromatographic and enzymic studies, recommend as a practical definition of agarose: ". . . that mixture of agar molecules with the lowest charge content

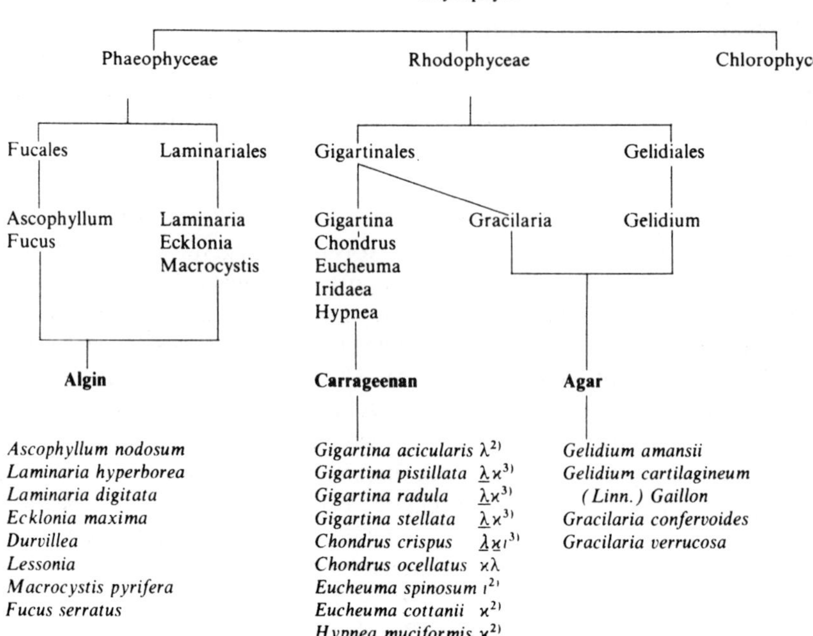

Fig. 8.10. Sources of algae extracts—general classification. (Reproduced from Ref. 41 with permission.)

and therefore the greatest gelling ability, fractionated from a whole complex of molecules, called agar, all differing in the extent of masking with charged groups." Most agarose solutions gel at temperatures around 40°C. The precise gelling temperature, measured during cooling at a fixed rate, has been found to be directly related to the methoxyl content: the higher the degree of substitution, the higher the temperature of gelling (33). Once formed, a gel remains "stable" up to its melting point of about 90°C (34). These and other properties have been explained by a model of parallel double helices (35). The double helices are extensively aggregated and form cavities that extend

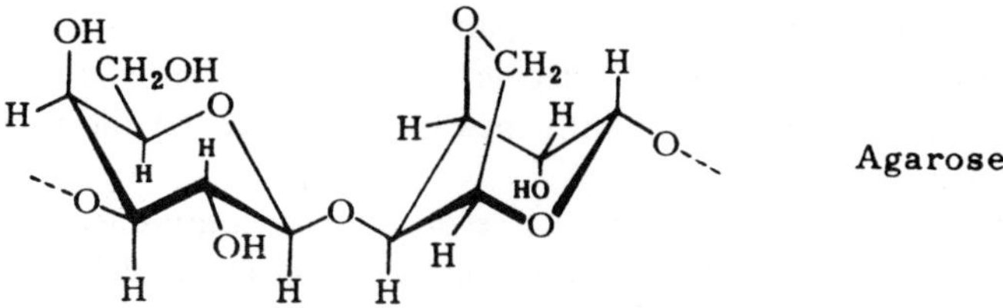

Fig. 8.11. Structure of agarobiose, the repeating subunit of agarose. Reproduced from "Affinity Chromatography: Principles and Methods," with permission of Pharmacia Fine Chemicals.

along the length of the helix axis (Figure 8.12). The relatively large voids are occupied with water molecules, which contribute to the stability of the structure through hydrogen bonding; no covalent crosslinks are present. Many procedures for the isolation of agarose exist, and some have been developed into commercial processes for preparing beaded agarose (36). The ready availability of beaded agarose and the fact that it meets many of the requirements of an ideal support have made agarose the support most widely used.

The effective pore size of beaded agarose is inversely related to the weight concentration of agarose used during gelation. Agarose forms useful gels in the range of 2 to 6% (4% is commonly used; e.g., Sepharose 4B), and these gels allow free access to even the largest enzymes. Each agarobiose unit has 4 hydroxyl groups, which in principle could be used for activation and coupling. The average number of hydroxyl groups available for reaction is somewhat lower, however, owing to the presence of the above-mentioned naturally occurring substituents. It has further been found that derivatization in aqueous solution never approaches the theoretical limit, at least not when the reactions are carried out in media of alkalinity lower than 2 *M* sodium or potassium hydroxide. Supports in which one of the four hydroxyl groups has been substituted can occasionally be obtained (37). This means that agarose gels have a moderately high capacity for substitution (38). Several procedures exist for introducing reactive groups into agarose, making the support ap-

Fig. 8.12. Schematic representation of an agarose gel. Reproduced from "Affinity Chromatography: Principles and Methods," with permission of Pharmacia Fine Chemicals.

propriate for immobilization of enzymes under mild conditions (see Section 8.3.1). Since the hydroxyl groups in agarobiose are never fully substituted, the derivatized support remains very hydrophilic in nature. Agarose is chemically stable in aqueous media in the pH range 4–9 and at room-temperature for a short time (2–3 hr) in 0.1 *M* NaOH and 1 *M* HCl solutions.

For reaction with substances weakly soluble in water, 50% dimethylformamide in water or 50% ethylene glycol in water can be employed. Temperatures below 0°C and above 40°C should be avoided. Lyophilization can be carried out only after the addition of protective substances such as dextran, glucose, or serum albumin. The mechanical stability of agarose (4%) is adequate for most purposes, although magnetic stirring is not recommended. The rigidity of the beads is only moderate and high pressures cause compaction, which results in poor fluid flow. Agar polysaccharides are very resistant to microbial attack, since agar-degrading enzymes (agarases) are only found in certain microbes living in a marine environment. Regeneration of agarose supports is usually impossible.

Beaded agarose has thus many attractive features as a support, but there are a number of factors that limit its use: (i) lack of thermal and mechanical stability, (ii) poor rigidity, (iii) shrinkage or swelling due to changes in ionic strength or dielectric constant of the medium (iv) inability to be frozen or dried easily, (v) ready solubility in the presence of denaturing or chaotropic ions, (vi) drastic and irreversible changes in structure that occur in many organic solvents, (vii) fairly high cost coupled with impossibility of regeneration. However, the stability of agarose can be considerably increased in all respects by crosslinking with epichlorohydrin, 2,3-dibromopropanol or divinylsulfone (39). The crosslinked agarose beads are mechanically much stronger, more rigid, have better flow properties, and can be employed with a far wider range of buffers and solvents. As a consequence of the crosslinking, the porosity is somewhat reduced and the hydroxyl groups available for coupling decrease by about 50%. This can be compensated for by the addition of sorbitol or phloroglucinol during the crosslinking reaction (40).

In summary, the chemistry of agarose is developed sufficiently well to make agarose suitable as a starting support for most purposes. The relatively high costs of pure agarose limit large-scale applications, but in many cases it may be that less pure and therefore cheaper agarose could be used.

8.2.1b. Alginate gels

Alginate is a glycuronan extracted from brown algae according to the scheme given in Figure 8.13 (41). It is an unbranched copolymer consisting of residues of D-mannuronic acid and L-guluronic acid. In principle three structural elements occur (Figure 8.14), α-1,4-L-guluronan (G) blocks, β-

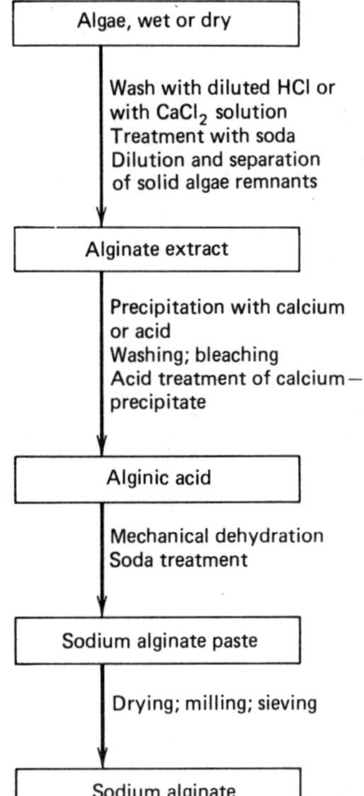

Fig. 8.13. Process scheme of alginate production. (Reproduced from Ref. 41 with permission.)

1,4-D-mannuronan (M) blocks, and a polyuronid consisting of alternating guluronic and mannuronic acid residues arranged in a blockwise fashion along the polymer chain:

$$\text{M—G—M—(M—M)}_n\text{—M—G—(M—G)}_q\text{—M—G—(G—G)}_p\text{—G—M—G}$$

The formation of gels in the presence of calcium ions or other multivalent counterions mainly depends upon autocooperatively formed junctions between G-blocks (42). The axial–axial bond between guluronic acid residues (Figure 8.14) force the G-blocks in a zigzag chain and several of these chains can form strands, which are "glued" together by calcium ions (Figure 8.15). The mechanical properties of calcium alginate gels thus formed are related to the distribution of M and G residues and to the molecular weight and degree of polydispersity within the sample. As the concentration of poly-G-blocks increases, the gel becomes harder and more friable.

When evaluating the properties of calcium alginate gels using the criteria outlined above, we note that only substrate permeability is relevant in this

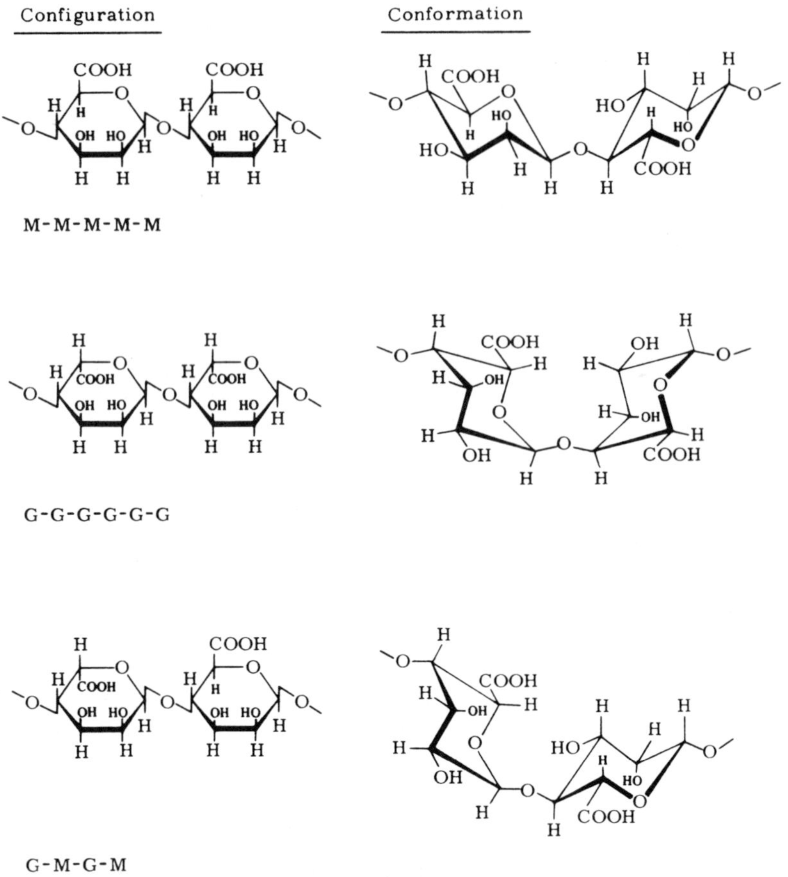

Fig. 8.14. Structural elements in alginate. (Reproduced from Ref. 41 with permission.)

case, since this support is prepared from a homogeneous solution of enzyme and sodium alginate (see Section 8.3.3a). Substrates with a low molecular weight will generally have good access to the entrapped enzyme. For macromolecular substrates the access will largely depend on the density of the gel and on the dimension of the pertinent substrate molecules. It has been found (43) that moderately dense calcium alginate gels totally retain inulase (molecular weight > 100,000) but do not exclude the substrate inulin (molecular weight 3000–5000). The amount of enzyme that can be immobilized in this kind of gels is high. In the above example, 100 mg enzyme protein were used in 10 mL of sodium alginate solution (1–2% w/v), and in the case of yeast cells 10 g of sodium alginate were required to immobilize 1 kg wet weight of

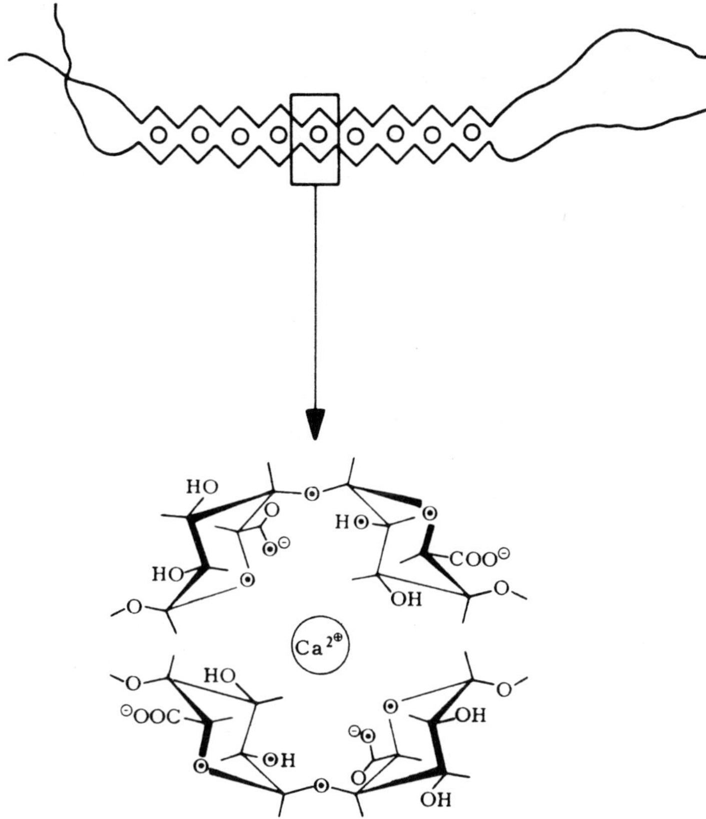

Fig. 8.15. Chain association by complex formation with calcium ions; the oxygen atoms that form chelate bonds are designated by ⊙. (Reproduced from Ref. 41 with permission.)

yeast cells (equivalent to about 200 g of dry weight of cells; Ref. 43). The immobilization procedure can be performed under very mild conditions (see Section 8.3.3a), and alginate has sufficient functional groups in case an experimenter wants to modify it. Also the hydrophilic character works in favor of this support. Solubility, however, can be a problem as the gel goes into solution in phosphate-containing media or in the presence of other strong calcium-ion-binding compounds. Introduction of an intermediate gel drying step (44), polycation treatment (45), replacement of calcium with barium ions (46), or hardening with glutaraldehyde or glutaraldehyde and gelatin or persimmon tannin (47) yields, however, a more stable gel.

Between pH 5 and 10, alginates are chemically very stable. High acid concentrations as well as high temperatures cause decarboxylation of alginates.

Direct-oxidation agents such as halogens or periodate and redox systems such as polyphenols, ascorbic acid or thiols degrade alginates (45). Alginate gels (48, 49) are mechanically strong except when high concentrations of phosphate, Mg^{2+}, K^{+}, or chelating agent are present. Columns of alginate pellets are resistant to compression and abrasion even at high flow rates. Resistance to compression, channeling, abrasion, and the generation of large pressure drops is conferrred by the even size and smooth spherical shape in which they can be manufactured and by the resulting uniform close-packing of the pellets. Columns of these pellets are biologically inert.

A disadvantage of the alginate gels is that regeneration is not usually realizable. Also, leakage of biocatalyst and substrate diffusion limitation limit the applicability of this type of support.

To sum up (48), immobilization in calcium alginate is a safe, fast, mild, simple, cheap and versatile technique, which may be applied to a wide range of biocatalysts.

8.2.1c. Carrageenan

Carrageenan is the collective name for the galactans, extracted from red algae, which consist of galactose units that are partly in the anhydro form and partly esterified with sulfuric acid (Figure 8.16; Ref. 41). Generally speaking, carrageenan is a relatively new product, but the application of support for enzyme immobilization is one of the youngest. Because it has properties very

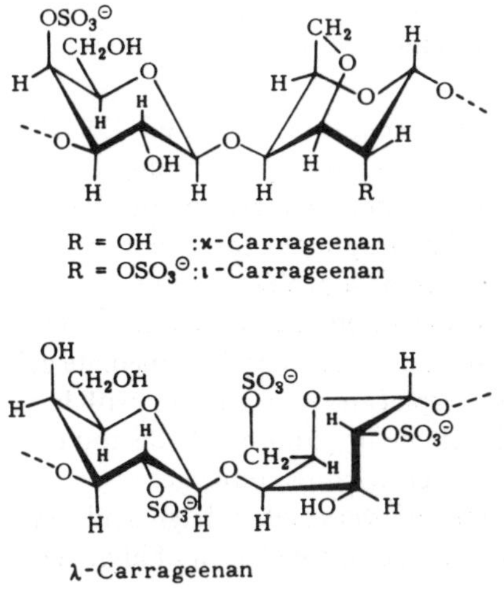

R = OH :κ-Carrageenan
R = $OSO_3^{\ominus}$:ι-Carrageenan

λ-Carrageenan

Fig. 8.16. Structure of the three main forms of carrageenan, kappa-, iota-, and lambda-carrageenan. (Reproduced from Ref. 41 with permission.)

similar to agarose and alginate, the potential for this particular application is high and the first reports are promising (47, 50). Of the three known types of carrageenan, gamma, kappa, and iota, only the latter two are suitable as a support for enzyme immobilization. Furcellaran, also a carrageenan fraction is very similar to the kappa form (it has fewer sulfate groups), and as such is potentially useful as a support.

The sodium salts are soluble in water, but the potassium and calcium salts form gels at ambient temperatures as a result of the formation of double helices (Figure 8.17). Increasing the potassium ion concentration and adding sugar yields a stronger gel. Mixing κ-carrageenan with carubin results in a much stronger and more elastic gel (41).

Carrageenan is very unstable at low pH because of the weak 1,4-glycosidic bond connecting the galactose residues. On the other hand, above pH 4.5 it is stable even under sterilization conditions.

Table 8.11 gives an impression of the strength of carrageenan gels in comparison with other materials used as enzyme support and Takata et al. (47) conclude that especially the κ-form is an excellent support for the immobilization of biocatalysts.

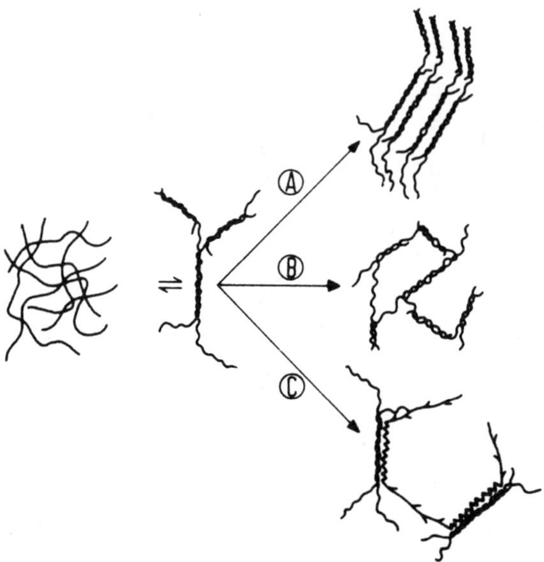

Fig. 8.17. Gelation of carrageenan as result of helix formation: (A) κ-carrageenan; (B) ι-carrageenan; (C) Interaction of λ-carrageenan with carubin. (Reproduced from Ref. 41 with permission.)

Table 8.11. Activity, Operational Stability, and Gel Strength of Immobilized Glucose Isomerase Before and After Hardening Treatments (47)

Polymer	Hardening Reagent[a]	Gel Strength[b]	Enzyme Activity (μmol/h/g of cells) and Yield (%)[c]	Operational Stability[d] at 60°C (half-life, days)
Kappa-carrageenan	None	+++	4440[e] (57.3)	150
	Gelatin + GA	+++	3280 (42.4)	532
	HMDA + GA	++++	4530 (58.6)	260
Iota-carrageenan	GA	++++	1110 (14.4)	—
Sodium alginate	None	+++	300[e] (3.9)	—
	GA	++++	2140 (27.7)	241
	Gelatin + GA	++++	1800 (23.3)	—
	Persimmon tannin	+++	1530 (19.2)	137
Furcellaran	None	++	3090[e] (40.0)	—
	GA	++++	2640 (34.1)	—
ESC[f]	None	++	2490[e] (32.2)	—
	Gelatin + GA	+++	1840 (23.8)	—

SZ[g]	None	++	1800[e] (23.3)	—
	Persimmon tannin	+++	2840 (36.7)	—
MPM-47[h]	None	++	2360 (30.5)	—
	Gallotannin	++++	3110 (40.2)	121
	Persimmon tannin	++++	3670 (47.4)	89
	Gelatin + GA	++++	2180 (28.1)	89
	Gelatin + persimmon tannin	++++	3690 (47.1)	96
Polyacrylamide	None	++++	4220 (57.2)	150

[a] GA = Glutaraldehyde; HMDA = Hexamethylenediamine.

[b] Symbol for gel strength corresponds to respective load for gel crush as follows: ++ 200–500 g/cm^2; +++ 500–1000 g/cm^2; ++++ 1000–1500 g/cm^2.

[c] Parentheses contain activity yield (%).

[d] Stability is shown with apparent half-life.

[e] During enzyme reaction, a stabilizer for gel was added to the substrate solution at a concentration of 0.3 *M* as follows: KCl for kappa-carrageenan and furcellaran; $CaCl_2$ for sodium alginate; $MgCl_2$ for ESC and SZ.

[f] Ethyl succinylated cellulose.

[g] Succinylated zein.

[h] 2-Methyl-5-vinylpyridine-methylacrylate-methacrylic acid copolymer.

8.2.1d. Cellulose

Cellulose is a vegetable fiber and one of the most abundant natural organic compounds. It consists of linear polymers of varying length. Each molecule is composed of 1,4-linked β-D-glucose units (Figure 8.18).

The polymers strongly interact through hydrogen bonds, and about 40 of them form so-called elementary fibers, cellulose strands with a diameter of about 3.5 nm. Elementary fibers contain highly ordered crystalline regions and more accessible amorphous regions of a low degree of order. Swelling of cellulose can occur when it is suspended in a solvent. Two types of swelling can be distinguished: intercrystalline and intracrystalline. In the latter type the crystalline zones are also penetrated and a definite change in structure occurs. Cellulose goes into solution when treated with strong alkali. When native cellulose is regenerated from solution or passed through a highly swollen state following alkali treatment, the linear polymers shift to a different, less ordered, and more accessible configuration, and the cellulose is referred to as mercerized. The largest intercrystalline swelling is established in water (51).

The permeability, the surface area available for enzyme attachment, and the reactivity of cellulose largely depend on the degree of crystallinity, the size of the compound to be bound, and the swelling-inducing capacity of the reaction medium. In principle, cellulose can undergo all the usual reactions associated with polyhydric alcohols. Thus a wide range of modification and activation reactions is theoretically possible (52). Obviously, for reaction with the hydroxyl groups the reagent must have access to these groups. The first requirement is that cellulose is swellable in the reaction medium. Although a high degree of substitution of the three available hydroxyl groups in each subunit is possible, the ultimate capacity for enzyme binding is determined by the size of the enzyme itself. And although cellulose may be able to couple 500 times more small molecules than agarose, the capacity of cellulose to bind macromolecules (such as serum albumin) is actually lower than that of agarose (53). If the hydroxyl groups are not substituted to a large extent, cellulose is very hydrophilic, easily wettable, but insoluble in aqueous medium. Substitution to a high degree with ionic functions may lead to

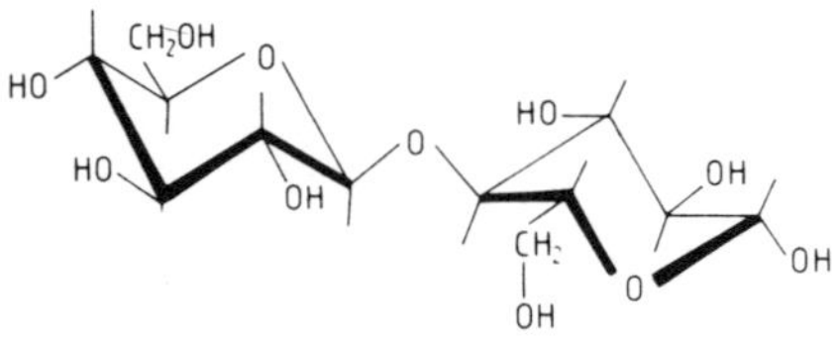

Fig. 8.18. Structure of cellobiose, the repeating subunit of cellulose.

water solubility, for example, when about one of every three hydroxyl groups is substituted with sodium-carboxymethyl groups, this cellulose derivative is water soluble (52). Extremes in pH must be avoided: strong acids cause hydrolysis of the glucosidic bonds; strong alkali causes dissolution. The chemical, mechanical, and thermal stability is sufficient to withstand most manipulations and conditions in the preparation and operation of enzyme/cellulose systems. Cellulose is commercially available in fibrous and granular form. In fibrous form it is easily clogged by particulate material and compressed by the application of even moderate pressure. The granular form is composed of microscopic cigar-shaped, fairly rigid particles of uniform size (about 35 μm) (54), which can be packed into columns having reasonable flow properties. Cellulose in bead form with good mechanical strength and considerable porosity has also been described by various authors (40). Cellulose is not completely inert to microbial attack, as it can be hydrolyzed by extracellular microbial cellulases. Regeneration of cellulose supports is in most instances impossible, but this is not very relevant since its natural abundance makes it a relatively cheap material.

To recapitulate, cellulose is an acceptable support provided that binding capacity, homogeneous enzyme distribution, and rapid high-pressure application are not essential. Cellulose can easily be derivatized in a variety of organic solvents, is hydrophilic, water insoluble, sufficiently stable and rigid for most applications, and commercially available in various inexpensive forms. It is therefore not surprising that cellulose and modified celluloses were some of the first supports to be used for enzyme immobilization and that over 40 different enzymes have now been immobilized successfully on cellulose supports (52).

8.2.2. Gelatin

The protein gelatin has several properties that make it attractive as a support for the immobilization of enzymes. In the first place, it is cheap and available in large quantities. It is easily obtained from collagen, the principal protein of skin and connective tissue of mammals, by boiling with water (Figure 8.19). Glycine, proline, and hydroxyproline are the main amino acid building blocks of the collagen molecule. The regular repetition of proline and hydroxyproline residues force the peptide chain into a peculiar spiral, and in native collagen three of these chains are intertwined to form a triple helix (Figure 8.19), which gives collagen its strength. Actually, collagen itself has been employed successfully as a support for the immobilization of several enzymes as well (55). An advantage of gelatin over collagen is that one starts with a solution of the enzyme and gelatin and thus a gel results in which the enzyme is homogeneously distributed. Crosslinking of this gel is necessary in

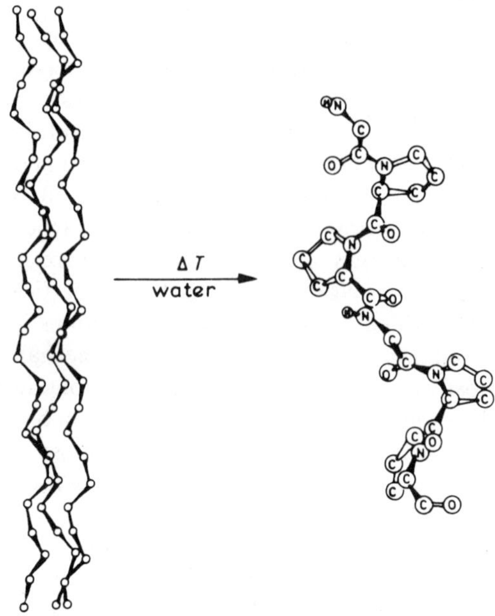

Fig. 8.19. When boiled in water, the three strands of the collagen "rope" come apart and gelatin is produced.

order to prepare a water-insoluble and mechanically stable enzyme support. Broun et al. (56) showed in their studies on immobilization procedures that the most promising method consists of crosslinking the enzyme molecules together with inactive protein by means of a bifunctional reagent.

The formation of a structure in which the enzyme is bound to other proteins more closely resemble the in vivo conditions: Most enzymes in vivo function while embedded in membranes, adsorbed to interfaces, or entrapped in other solid-state assemblages. When enzymes are isolated from their natural environment they usually are less stable. The presence of many surface charges and a large number of other potential binding sites for enzyme attachment and stabilization favor the use of gelatin as a support for enzyme immobilization.

When evaluating the properties of gelatin using the criteria outlined in Section 8.2, the same can be said with respect to permeability as for alginate. The amount of enzyme immobilized in gelatin can be very high (up to 20%) (57) and in the case of the immobilization of whole cells, even much higher (dry weight cells : weight gelatin = 1 : 1) (58). Gelatin has many functional groups allowing crosslinking to various degrees. Gelatin is also very hydrophilic in nature, and when sufficiently crosslinked it becomes insoluble in water. Chemical, mechanical, and thermal stability largely depend on the degree of crosslinking. Crosslinked gelatin is, however, more stable than enzymes, such that under operational conditions the gelatin support is chemically inert and thermally stable. At least two immobilization procedures exist at present that produce mechanically stable gelatin preparations that can withstand manipulations such as stirring, filtration, and centrifugation (57, 58). The rigidity of the gelatin support again depends on the degree of crosslinking, but both immobilization techniques produce particles that cause low pressure drops, even in packed beds with large dimensions. As a proteinaceous support, gelatin is subject to microbial attack. However, the pores of the crosslinked gelatin matrix do not allow entrance of microbes or proteases, so that microbial breakdown can only start at the outer surface and is thus likely to be slow. Van Velzen (57) found, when operating an invertase/gelatin reactor with a saccharose solution as substrate, that during the ninth week of operation the conversion rate dropped as a result of microorganisms growing on the enzyme particles. The original conversion rate could be restored, however, by thoroughly washing and sterilizing the column with a 2% glutaraldehyde solution. This most probably indicates that the gelatin support did not act as a substrate but merely as a support for the growth of the microbes. Regeneration of gelatin is impossible.

In conclusion, gelatin appears to be quite generally applicable as a support for the immobilization of enzymes and whole cells, since the existing immobilization procedures using it are mild and yield an immobilized-enzyme preparation with favorable properties.

8.2.3. Acrylic Copolymers

Solid supports derived from acrylic monomers such as acrylamide, acrylic acid, methacrylic acid, *N*-acryloylmorpholine, and hydroxyethyl methacrylate comprise the largest class of wholly synthetic polymers used for the immobilization of enzymes (59). In one respect the acrylic copolymers are outstanding: that is, in the possibility of preparing various polymers "tailor-made" for a specific application. Parameters that can easily be controlled are: (i) the degree of porosity and (ii) the chemical composition, which can be achieved by either copolymerization of different monomers among the large number available or by chemical modification of preformed polymers. Acrylic copolymers are available from commercial sources and are also relatively easily prepared in the laboratory. By choosing the ratio of the participating monomers it is possible to prepare a polymer with the desired amount of a specific functional group. The structure of the gel (and thus the pore size and pore-size distribution within the gel) is dependent not only on polymerization kinetics but also on other factors, mainly on the total concentration of

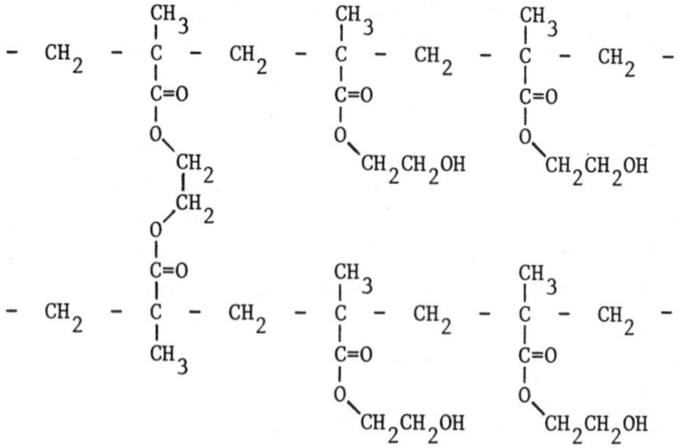

```
           CH3               CH3               CH3
           |                 |                 |
- CH2  -   C   -  CH2  -     C   -  CH2  -     C   -  CH2  -
           |                 |                 |
           C=O               C=O               C=O
           |                 |                 |
           O                 O                 O
            \                 \                 \
             CH2               CH2CH2OH          CH2CH2OH
             |
            CH2
           /
           O
           |
           C=O               CH3               CH3
           |                 |                 |
- CH2  -   C   -  CH2  -     C   -  CH2  -     C   -  CH2  -
           |                 |                 |
           CH3               C=O               C=O
                             |                 |
                             O                 O
                              \                 \
                               CH2CH2OH          CH2CH2OH
```

Fig. 8.20. The chemical structure of Spheron.

monomer, such as acrylamide, and on the relative concentration of the cross-linking agent, such as *N,N'*-methylene bisacrylamide (Bis). A minimal average pore size is obtained when Bis comprises about 5% by weight of the monomers. Above and below 5%, the pore size will be larger. Preparations above 5% become turbid, indicating microprecipitation, which leads to larger pores. However, gel structure can still be found in the microparticles of this coherent disperse system. These so-called macroreticular gels are characterized by having "interparticular" pores, which allow relatively unhindered diffusion into this type of gel because they are large in comparison with normal gel pores.

Another method of preparing bodies of varying porosity involves polymerization in different solvent systems. Regardless of how they are made, polyacrylic gels possess attractive features and are commercially available in beaded form, pregraded in sizes and porosities (60). The permeability and surface area of acrylic copolymers depend on the monomeric mixture used and on modifications following the polymerization. For example, the commercially available polyacrylamide gel with the highest porosity is permeable for proteins with a molecular weight up to 300,000, but during derivatization reactions the pores shrink so drastically that the largest protein that can be bound has a molecular weight of only about 30,000 (53). This problem can be eliminated by changing the composition of the monomeric mixture. Spheron for instance, a copolymer of hydroxyethyl methacrylate and ethylene bismethacrylate, has large and stable pores with exclusion limits up to 5 million (Figure 8.20). The principal advantage of polyacrylic gels is that they possess an abundant supply of easily modifiable groups which, together with a versatility in derivatization techniques, allow the covalent attachment of enzymes in many different ways. Reactive groups can even be introduced immediately by using monomers having active groups that do not participate in the polymerization reaction.

Enzacryl was the first of these modified acrylamide supports to be marketed commercially, and it is available with various functional groups preattached to it (Figure 8.21). Spheron is available with hydroxyl, carboxyl, sulfonyl, and aminoaryl functions. Anhydride and oxirane-acrylic beads are also available commercially (Röhm). Polyacrylic supports are usually very hydrophilic in nature and insoluble in water. Their chemical stability is excellent, which is mainly the result of the polyethylene backbone, since carbon–carbon bonds are very stable. The mechanical stability of polyacrylamide is poor, but this property is greatly improved in Enzacryl, Spheron, and Röhm, and manipulations such as filtration, centrifugation, and stirring are possible. The thermal stability is sufficient to withstand all temperatures likely to

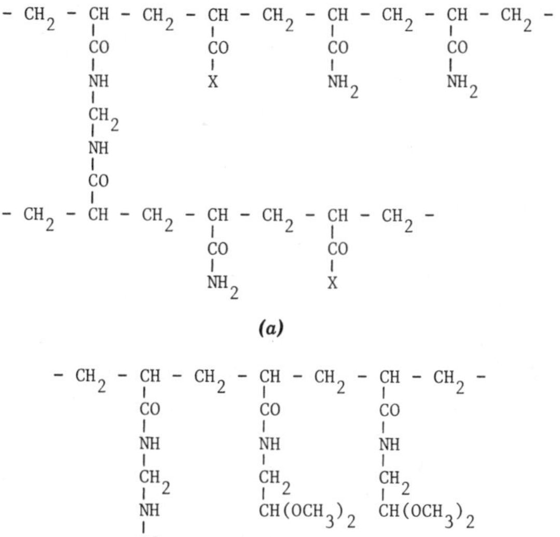

```
- CH2 - CH - CH2 - CH - CH2 - CH - CH2 - CH - CH2 -
        |          |          |          |
        CO         CO         CO         CO
        |          |          |          |
        NH         X          NH2        NH2
        |
        CH2
        |
        NH
        |
        CO
        |
- CH2 - CH - CH2 - CH - CH2 - CH - CH2 -
                   |          |
                   CO         CO
                   |          |
                   NH2        X
```

(a)

```
- CH2 - CH - CH2 - CH - CH2 - CH - CH2 -
        |          |          |
        CO         CO         CO
        |          |          |
        NH         NH         NH
        |          |          |
        CH2        CH2        CH2
        |          |          |
        NH         CH(OCH3)2  CH(OCH3)2
        |
        CO
        |
- CH2 - CH - CH2 - CH - CH2 -
                   |
                   CO
                   |
                   NH
                   |
                   CH2
                   |
                   CH(OCH3)2
```

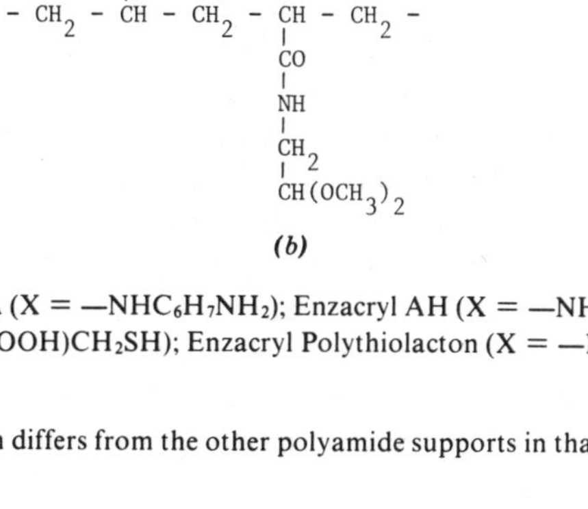

(b)

Fig. 8.21. Enzacryl AA (X = $—NHC_6H_7NH_2$); Enzacryl AH (X = $—NHNH_2$); Enzacryl Polythiol (X = $—NHCH(COOH)CH_2SH$); Enzacryl Polythiolacton (X = —NHCH—CO, with CH and CO joined through CH_2—S). (*b*) Enzacryl polyacetal, which differs from the other polyamide supports in that all the amide groups are substituted.

occur in immobilized-enzyme systems (e.g., Spheron is stable up to 250°C). Enzacryl is commercially available as rough particles and Spheron and the Röhm products as beads. The rigidity of these supports is such that these materials produce columns with good flow properties. Polyacrylic copolymers are resistant to microbial breakdown. Regeneration of these types of supports is generally impossible.

To sum up, polyacrylic copolymers form an interesting group of solid supports for the immobilization of enzymes, and the range of commercially available products is wide. The particular products are well characterized so that a sensible choice can be made. Many more variations are possible when the support is prepared in the laboratory. For a specific purpose, "tailor-made" supports can then be made, although this demands a thorough knowledge of polyacrylic chemistry and much technical skill and experience. Making one's own polyacrylic support also offers the possibility of immobilization by entrapment. In this case, careful control of the temperature is desirable during the exothermic polymerization reaction in order to prevent thermal denaturation of the enzyme.

8.2.4. Activated Carbon

Activated carbon is a highly porous carbonaceous material with a large internal pore surface (61). It is prepared by dehydration and carbonization, followed by activation, of organic substances such as coconut shells, wood, coal and petroleum coke, bone, molasses, peat, and paper-mill waste (lignin) (62, 63). The source of carbon and the precise conditions of the various preparation steps determine the ultimate properties of the activated carbon. The permeability of activated carbon is largely determined by its pore structure, which has been reported as "tridispersed", that is, containing micropores (0–0.2 nm radii), transitional pores (0.2–50 nm), and macropores (50–200 nm) (64). Activated carbon may be manufactured with high surface areas (600–1000 $m^2 \cdot g^{-1}$) and a significant fraction of its pore volume in the 30–100 nm range of pore diameter, suitable for enzyme immobilization. Thus, the morphology of activated carbon should be conducive to large loadings of immobilized enzyme (65). The existence of vinyl, carboxylic, phenolic groups and other oxides on the carbon surface has been shown, and many details on the surface chemistry of activated carbon are known (62). The type of oxidation process used in the activation step of the carbon determines to a large extent the nature of these functional groups and the possible modifications of a specific activated-carbon preparation in order to make it suitable for covalent binding of enzymes. The manner in which the carbon is activated also influences the degree of hydrophilicity. When the carbon is activated at 1000°C either in pure CO_2 or under vacuum, followed by exposure to oxygen at room temperature, a hydrophobic surface results. In contrast, the oxidation of carbon by exposure to gaseous oxygen at temperatures between 200 and 400°C, or by adding it to an aqueous oxidizing solution, produces a hydrophilic surface. Activated carbon is insoluble and chemically very resistant except for the functional groups on the surface, which can be modified under mild conditions. Cho and Bailey (65) claimed that activated carbon possesses a mechanical strength comparable to porous-glass materials, although we found that a minimal amount of material easily abrases from the carbon, causing the reaction medium to become slightly turbid. The thermal stability is more than sufficient for all applications with immobilized enzymes. The rigidity strongly depends on the carbon source and manufacturing procedure. The granular form usually allows operation in large packed beds. Activated carbon is completely inert to microbial breakdown. Finally, regeneration can easily be established in various ways (62).

In summary, activated carbon can be produced in such a way that it has many advantageous properties for use as a support in the immobilization of enzymes. So far it has not been specifically designed for this purpose and not all the necessary specifications are given for the various preparations, making a sensible choice difficult. Nevertheless, several enzymes have been successfully immobilized on activated carbon. In fact, activated carbon was the first material to be used as enzyme support (64).

8.2.5. Controlled-Pore Glass (CPG)

Many different kinds of inorganic materials have been used as a support for the immobilization of enzymes. Materials ranging from fabricated particles with specifically tailored properties to inexpensive minerals have been examined as inorganic supports. The variety in these inorganic substances is indeed wide, including aluminas, clays, sand, glass, charcoal, celite, kieselguhr, hornblende, nickel oxide, silicas, titanias, and zirconias (66). The distinct advantages of these supports are: high mechanical strength, resistance to solvent or microbial attack, reusability, and easy handling. The scarcity of natural macroporous materials has made it desirable to prefabricate porous inorganic supports for the immobilization of enzymes.

The most important prefabricated inorganic solid support is undoubtedly controlled-pore glass (CPG). CPG is prepared by heating certain borosilicate glasses to 500–700°C for prolonged periods of time, followed by various physical and chemical treatments (67). Careful control of these treatments allows the production of glass beads of various diameters with narrow pore-size distributions in the range from 4.5 to 400 nm. Accurate determination of the pertinent nominal pore size is possible.

The commercially available CPG is well characterized with regard to mesh size, pore size, and surface area. Consequently, if the dimensions of the enzyme and substrate to be used are known, an optimal choice with respect to enzyme or substrate permeability and binding capacity can be made beforehand. The surface of CPG is mainly composed of silanol functions:

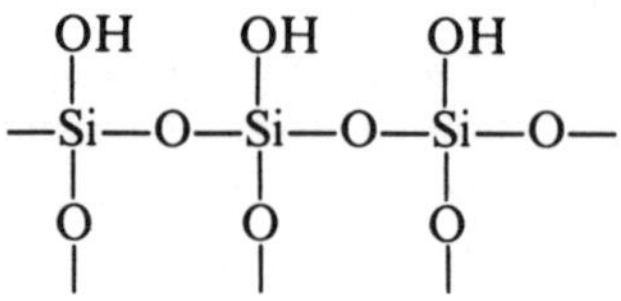

```
  OH     OH     OH
  |      |      |
—Si—O—Si—O—Si—O—
  |      |      |
  O      O      O
  |      |      |
```

These groups provide a mildly reactive surface for activation and enzyme binding. Several activation and coupling procedures have been developed (see Section 8.3.1c).

The binding capacity is generally less than that of agarose (68), but the best properties of CPG and agarose can be combined by covalently covering the CPG surface with a hydrophilic carbohydrate monolayer (Glycophase, Pierce). In addition to the SiOH functions, the surface of CPG contains tightly bonded water molecules, making the glass both hydrophilic in nature and well wettable:

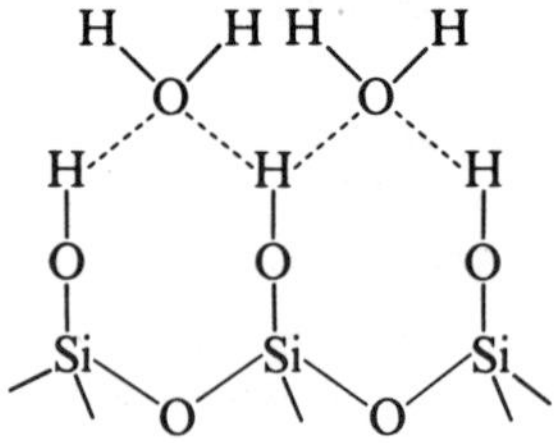

```
  H   H H   H
   \ /   \ /
    O     O
 .·´ `·..´ `·.
H      H      H
|      |      |
O      O      O
|      |      |
Si     Si     Si
/ \  /  \  /   \
    O      O
```

CPG also has a wide solvent compatibility (no swelling or shrinkage in changing environments), a property that may be important for activation of the support. Glass is slightly soluble in alkali ($pH > 8$), but because of the large surface area, CPG is appreciably more soluble, actually going into complete solution during prolonged operation under alkaline conditions. Since the enzyme is attached to the surface it goes into solution when the first silica surface layer dissolves and is thus rapidly lost. This solubility of CPG in alkali severely limits its application as enzyme support. Derivatization of CPG with a crosslinked organic coating such as a polysaccharide minimizes this problem of alkali solubility. Coating with metal oxides, less soluble in alkali, can also be employed. Zirconium-clad CPG, CPG with a ZrO_2 surface, is available (Zirclad, Pierce) and can be derivatized in the same fashion as CPG, with the resulting support more stable at alkaline pH. Except for the effect of alkali and hydrofluoric acid, the bulk of the CPG is chemically resistant and can withstand harsh conditions, so that various procedures for activation of the moderately reactive surface silanols can be applied (see Section 8.3.1c). The thermal stability is excellent, and contraction or expansion is minimal upon temperature changes. The mechanical stability of CPG strongly depends on pore size: CPG with very large pores is very friable and easily breaks under pressure. Magnetic stirring also breaks CPG as a result of grinding of the particles. Furthermore, small glass particles, inhomogeneous in size, can easily become clogged with particulate matter. CPG is very rigid and can be prepared as almost spherical, uniformly sized particles. Therefore, CPG with moderate pore (<250 nm) and mesh size (40–80 mesh) (53) has very good flow properties even under high-pressure operation. Other advantages of glass are its complete inertness with respect to microbial attack and the possibility of regeneration of the CPG by pyrolysis of the organic matter. For laboratory purposes most CPG products are reasonably priced. They are too expensive for applications on a larger scale; for these purposes porous ceramics, although less uniform in pore size, are an alternative, since they can be fabricated more cheaply.

As with agarose, the advantages of using CPG often outweigh its disadvantages. The choice will eventually be determined by the specific purpose.

8.2.6. Other Supports

The support materials discussed above comprise the largest (>90%) and most successful part of applications with immobilized enzymes and are therefore representative for this field. The number of materials used one or more times as a support in immobilized-enzyme studies is, however, very large. Among these are gluten, starch, stainless steel, polystyrene copolymers, dacron, nylon, chitin, and many more (3). All have specific advantages, but no one is ideal in all respects and suits all applications best. Therefore, for a particular application a thorough evaluation of various supports, using the "Support Study Decision Tree," is advisable before a definite choice is made.

8.3. IMMOBILIZATION TECHNIQUES*

The properties of immobilized enzymes, especially the specific activity, are influenced not only by the characteristics of the support to which the enzyme is bound but also by the way in which an enzyme is coupled to the support.

* (See also Chapter 7).

The most important concern is to avoid loss of enzymic activity during the immobilization process. Consequently, it is necessary that both during and after immobilization the amino acid residues in the active center are not altered and that the tertiary structure is maintained. The three-dimensional structure is based on the presence of relatively weak binding forces such as hydrogen, hydrophobic and ionic bonds, and sometimes a few disulfide bridges. Therefore it is essential to perform the coupling reaction under very mild and precise conditions. High temperatures, strong acid, or strong alkali must be avoided if the structural integrity of enzymes is to be preserved. Organic solvents or high salt concentrations may also cause denaturation and loss of activity.

It is thus logical that a large number of studies on immobilized enzymes have been devoted to immobilization techniques as such; the literature has now proliferated with a variety of methods for immobilizing enzymes. The various methods can be classified into three basically different approaches to immobilization:

1. *Support-binding methods.* The binding of enzymes to solid supports.
2. *Crosslinking method.* Intermolecular crosslinking of enzymes by means of multifunctional reagents.
3. *Entrapping method.* Incorporating enzymes into the lattice of a semipermeable polymer gel or enclosing the enzymes in a semipermeable polymer membrane.

These three categories can be further subdivided, as shown in Figure 8.22. None of the applied immobilization procedures depends on a single mechanism. In fact, each one is probably a combination of two or more of the binding modes. For example, in the case of covalent coupling to supports, it is likely that part of the enzyme molecules is adsorbed to the support surface. In the description given below, it is merely implied that the proposed attachment is predominantly of the type discussed. Excellent reviews (3, 6, 26, 27, 69–74) discuss the various techniques extensively and are the main literature source for this chapter.

8.3.1. The Support-Binding Method

The support-binding method is the oldest immobilization technique, and hundreds of papers have been published on this type of immobilization. The support-binding method can be further divided into three categories according to the binding mode of the enzyme; van der Waals binding, ionic binding, or covalent binding (Figure 8.22).

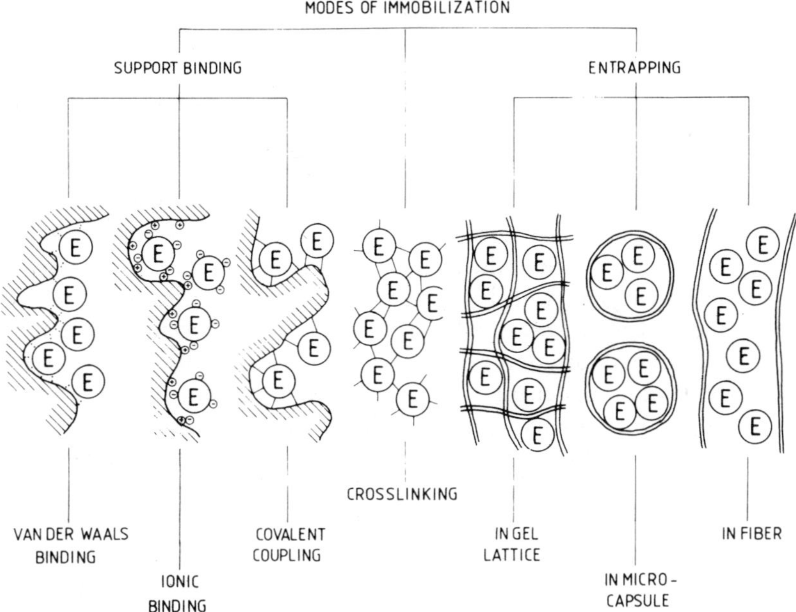

Fig. 8.22. Division and schematic representation of immobilized enzymes.

8.3.1a. van der Waals Binding

Adsorption based on van der Waals attractive forces of enzymes and the surface of solid supports is undoubtedly one of the easiest immobilization procedures. Adsorption is established by simply contacting the enzyme solution with the surface of the support material and washing the resulting immobilized-enzyme preparation to remove any nonadsorbed enzyme. The binding forces between enzyme and support are generally weak, so that adsorbed enzymes usually are liable to desorption during the utilization. To minimize this problem it is advisable to wash the adsorbed-enzyme preparation thoroughly with the substrate solution under the conditions to be applied. The adsorption process is nonspecific and sometimes results in partial or total inactivation of the enzyme. Only supports that have a high affinity for the enzyme and also cause minimal denaturation are suitable. The adsorption of an enzyme is dependent on such experimental variables as pH, the nature of the solvent, ionic strength, concentration of enzyme and adsorbent, and temperature. It is important to recognize these factors and to control them for optimal adsorption and retention of activity. The more commonly employed adsorbents are alumina, activated carbon, clays, collagen, glass, diatomaceous earth, and hydroxylapatite.

8.3.1b. Ionic binding

Like van der Waals binding, ionic binding is an old and simple way to prepare immobilized enzymes. This method is based on the ionic binding of enzymes to water-insoluble materials containing ion exchange residues. The binding of enzyms to this type of support is easily performed, and the conditions are mild in comparison with those necessary in most methods of forming covalent bonds. The ionic binding method usually causes little or no change in the active site of the enzyme or tertiary structure and often yields immobilized enzymes with high activity. The binding forces are relatively weak, and leakage of enzyme from the support may occur in solutions of high ionic strength or upon variations in pH. Thoroughly washing with the substrate solution under the operational conditions may minimize leakage. Increased binding strength can be achieved by increasing the charge of the enzyme by chemical modification. For example, the slightly anionic enzyme glucoamylase was modified to a polyanionic derivative by covalent binding with a water-soluble copolymer of ethylene and maleic anhydride (75). This polyanionic enzyme derivative was strongly adsorbed to cationic supports such as DEAE-cellulose and DEAE-Sephadex. Cellulose derivatives are the ion exchange materials most widely used as enzyme support.

In practice, both ionic binding and van der Waals binding usually occur concurrently. This is illustrated by the method of immobilization we have used most extensively in our own studies, in which xanthine oxidase is adsorbed by both hydrophobic and ionic forces to an agarose derivative (see Section 8.4.6a). Agarose is activated with cyanogen bromide, yielding mainly a reactive imidocarbonate (76) (Figure 8.23). This activated agarose is reacted with *n*-octylamine, and the support thus formed strongly adsorbs, in

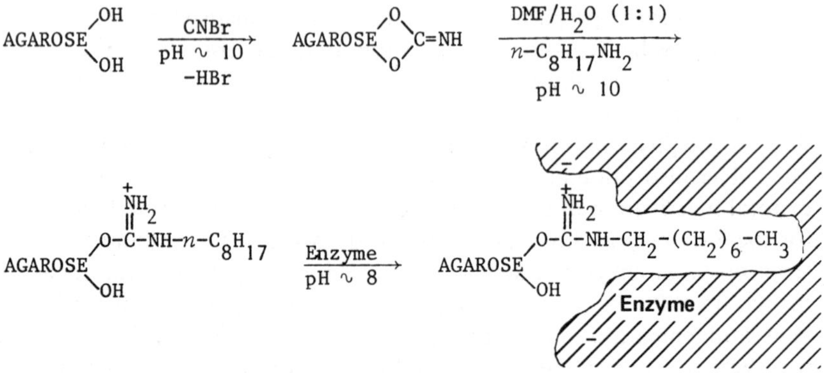

AGAROSE(OH)(OH) —CNBr, pH ∿ 10, −HBr→ AGAROSE(O)(O)C=NH —DMF/H_2O (1:1), n-$C_8H_{17}NH_2$, pH ∿ 10→

AGAROSE(OH)–O–C(=$\overset{+}{N}H_2$)–NH–n-C_8H_{17} —Enzyme, pH ∿ 8→ AGAROSE(OH)–O–C(=$\overset{+}{N}H_2$)–NH–CH_2–$(CH_2)_6$–CH_3 [bound in Enzyme]

Fig. 8.23. CNBr activation of agarose, reaction with *n*-octylamine and immobilization of an enzyme.

weak alkaline solutions, enzymes having an isoelectric point in the acid region (77). The adsorption is based on both ionic and hydrophobic attractive forces. That is, in weak alkaline solution (pH ~ 8) the isourea derivative ($pK_a = 10.4$) is protonated and the enzyme has an overall negative charge resulting in an ionic attraction between them. At the same time the freely mobile *n*-alkyl tail can reach the hydrophobic areas of the enzyme, and thus hydrophobic attraction occurs in addition to the ionic interactions (Figure 8.23). This dual binding character has the advantage that in solutions of high ionic strength the hydrophobic forces are largest, whereas at low ionic strength the ionic binding is strongest. An enzyme immobilized in this fashion is thus less liable to leakage from the support in a wider range of aqueous media than enzymes bound only by ionic forces.

8.3.1c. Covalent Binding

Covalent binding of an enzyme to a solid support is achieved by a chemical reaction between groups on the support and one or more functional groups of the enzyme. Usually ε-amino groups of lysine residues take part in the binding, but other nitrogen-containing groups such as the *N*-terminal amino group, the imidazole side chain of histidine, and the guanidine group of arginine residues can also take part in the binding. The carboxyl group of glutamate and aspartate, the C-terminal carboxyl group, the hydroxyl group of serine and threonine, the hydroxyphenyl group of tyrosine, and the sulfhydryl group in cysteine residues can also serve as coupling sites. It is important that minimal reaction occurs with the amino acid residues in the active center. The probability that reaction in the active center occurs is small, since the pertinent amino acids are practically always present more than once in a nonessential form in the protein chain. By random chemical reaction it is thus likely that only part of the enzyme molecules will be inactivated, unless a large excess of reagent is present. It is nevertheless useful, before trying a specific covalent binding method, to have knowledge of the effect of several reagents on enzyme activity and to know whether the enzyme can withstand the coupling conditions of that particular procedure, since covalent couplings cannot always be performed under mild conditions. The advantage of immobilization by covalent binding is that the binding is usually irreversible, so that the probability of enzyme loss by leakage is very small.

According to the mode of linkage, the method of covalent binding can be classified into diazo, peptide and alkylation methods, coupling by means of multifunctional reagents, and some miscellaneous procedures.

The Diazo Method. This coupling procedure is often used for immobilizing enzymes. The method is based on the reaction of the enzyme with diazonium derivatives of solid supports, that is, supports containing aromatic amino groups diazotized with nitrous acid, according to the overall reaction:

$$R{-}NH_2 \xrightarrow[HCl]{NaNO_2} (R{-}\overset{+}{N}{\equiv}N)Cl^- \xrightarrow{Enzyme} R{-}N{=}N{-}Enzyme$$

Functional groups of proteins likely to participate in diazo coupling include free amino groups, the imidazole side chain of histidine, and, especially, the phenolic group of tyrosine residues. Commonly employed as supports for this method are aromatic amino derivatives of polysaccharides, of copolymers of amino acids, of polyacrylamide, of polystyrene, of copolymers of ethylene and maleic acid, and of porous glass. See as an example the scheme in Figure 8.24, showing the preparation of the diazo derivative of CPG and its coupling to the enzyme.

Peptide binding Method. This method is an application of the peptide synthesis technique and is based on the formation of peptide bonds between enzyme and solid support. It can be categorized as follows.

1. Supports containing carboxyl groups can be converted to reactive derivatives such as acyl azide, acid chloride, or isocyanate. These derivatives form with free amino groups of the enzyme peptide bonds.
2. Peptide bonds are formed between free carboxyl or amino groups of the enzyme and, respectively, carboxyl or amino groups of the support, using condensing agents such as carbodiimides and Woodward's reagent K.

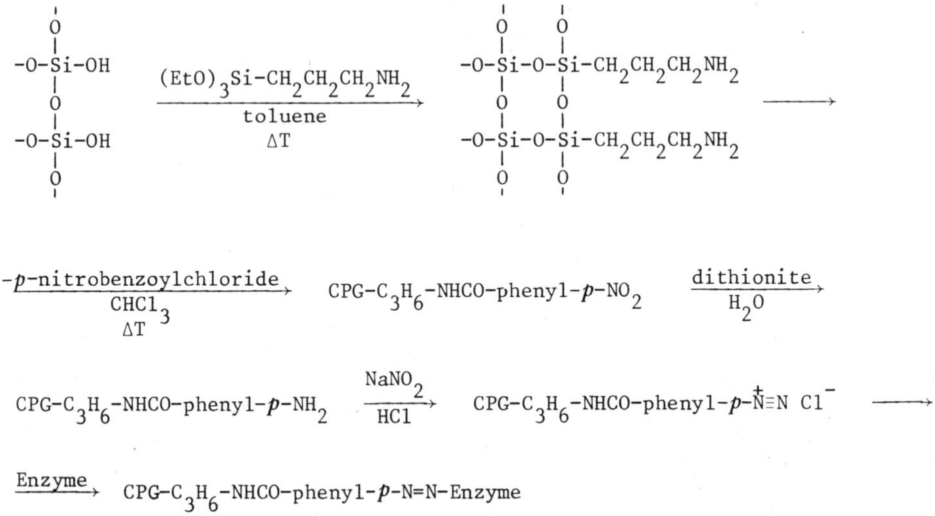

```
  |                                                    |     |
  O                                                    O     O
  |                                                    |     |
-O-Si-OH                                            -O-Si-O-Si-CH2CH2CH2NH2
  |        (EtO)3Si-CH2CH2CH2NH2                       |     |
  O       ------------------------>                    O     O            ------->
  |              toluene                               |     |
-O-Si-OH           ΔT                               -O-Si-O-Si-CH2CH2CH2NH2
  |                                                    |     |
  O                                                    O     O
  |                                                    |     |
```

$\xrightarrow[\substack{CHCl_3 \\ \Delta T}]{p\text{-nitrobenzoylchloride}}$ CPG-C_3H_6-NHCO-phenyl-*p*-NO_2 $\xrightarrow[H_2O]{\text{dithionite}}$

CPG-C_3H_6-NHCO-phenyl-*p*-NH_2 $\xrightarrow[HCl]{NaNO_2}$ CPG-C_3H_6-NHCO-phenyl-*p*-$\overset{+}{N}{\equiv}N\ Cl^-$ $\longrightarrow$

$\xrightarrow{\text{Enzyme}}$ CPG-C_3H_6-NHCO-phenyl-*p*-N=N-Enzyme

Fig. 8.24. Diazo coupling of enzymes to controlled-pore glass (CPG).

$$\text{CELLULOSE-OCH}_2\text{-COOH} \xrightarrow[\text{HCl}]{\text{CH}_3\text{OH}} \text{CELLULOSE-OCH}_2\text{-COOCH}_3 \xrightarrow{\text{NH}_2\text{-NH}_2}$$

$$\text{CELLULOSE-OCH}_2\text{-CONHNH}_2 \xrightarrow[\text{HCl}]{\text{NaNO}_2} \text{CELLULOSE-OCH}_2\text{-CON}_3 \xrightarrow{\text{Enzyme-NH}_2}$$

$$\text{CELLULOSE-OCH}_2\text{-CO-NH-Enzyme}$$

Fig. 8.25. Azide coupling of an enzyme to carboxymethylcellulose.

An example of the first category is the acyl azide derivative of carboxy-methyl-cellulose; the preparation is shown in Figure 8.25. Activation of polysaccharides with CNBr (Figure 8.23) also belongs to the first category. Instead of reacting the reactive imidocarbonate with *n*-alkylamine, the reaction can immediately be performed with the enzyme, yielding mainly the isourea derivative. In the second category, activated carbon and controlled-pore glass are examples (Figure 8.26).

Amines form amide linkages with carbodiimide-activated carboxyl groups. The reaction can be carried out with both water-soluble and insoluble carbodiimides. They are performed under mildly acidic conditions, unlike most other amide-forming reactions.

Alkylation Method. The alkylation method involves the alkylation of amino, phenolic, or sulfhydryl groups of an enzyme with a reactive group of the support, such as a halide. The alkylation method used to illustrate this principle involves a reaction with *s*-trichlorotriazine followed by an alkylation of a primary amine in the protein chain, primarily the ϵ-amino group of

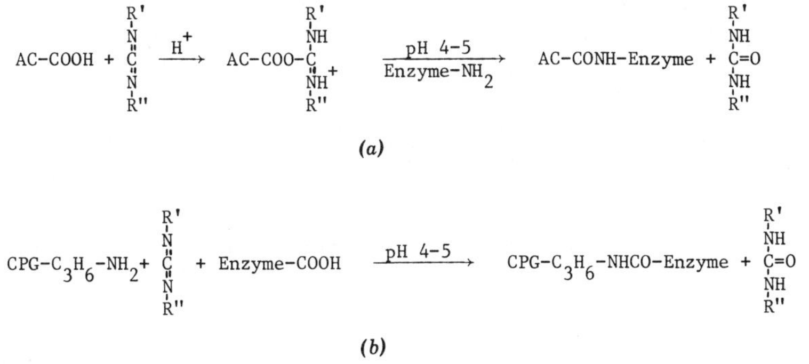

$$\text{AC-COOH} + \text{R'-N=C=N-R''} \xrightarrow{\text{H}^+} \text{AC-COO-C(=NH}^+\text{-R'')(-NH-R')} \xrightarrow[\text{Enzyme-NH}_2]{\text{pH 4-5}} \text{AC-CONH-Enzyme} + \text{R'-NH-C(=O)-NH-R''}$$

(*a*)

$$\text{CPG-C}_3\text{H}_6\text{-NH}_2 + \text{R'-N=C=N-R''} + \text{Enzyme-COOH} \xrightarrow{\text{pH 4-5}} \text{CPG-C}_3\text{H}_6\text{-NHCO-Enzyme} + \text{R'-NH-C(=O)-NH-R''}$$

(*b*)

Fig. 8.26. (*a*) Covalent immobilization of an enzyme onto activated carbon (AC) using a carbodiimide derivative. (*b*) Covalent immobilization of an enzyme onto the γ-aminopropyl derivative of controlled-pore glass (CPG) using a carbodiimide derivative.

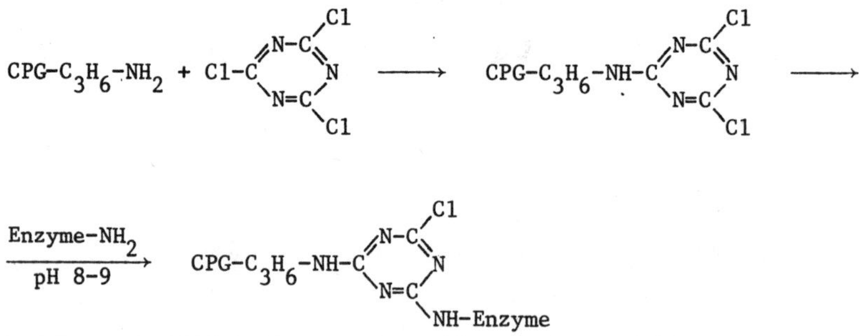

$$\mathrm{CPG{-}C_3H_6{-}NH_2} + \text{Cl-C}_3\text{N}_3\text{Cl}_2 \longrightarrow \mathrm{CPG{-}C_3H_6{-}NH{-}C_3N_3Cl_2} \xrightarrow[\text{pH 8–9}]{\text{Enzyme–NH}_2} \mathrm{CPG{-}C_3H_6{-}NH{-}C_3N_3(Cl)(NH{-}Enzyme)}$$

Fig. 8.27. Covalent immobilization of an enzyme onto the γ-aminopropyl derivative of controlled-pore glass (CPG) by means of *S*-trichlorotriazine.

a lysine residue. The reaction is relatively simple and straightforward and appears to be a potential coupling method of choice (Figure 8.27). For large-scale applications, cost may be a limiting factor.

Another example of the alkylation method is binding via an epoxy group, as shown in Figure 8.28.

Derivatives of cellulose, agarose, porous glass, bentonite, and acrylic copolymers have mainly been used as support.

Support Binding with Multifunctional Reagents. This method is based on the formation of crosslinks between functional groups of support and enzyme by means of multifunctional reagents. Glutaraldehyde is by far the bifunctional reagent most commonly employed, although its mechanism of binding is the least well understood. The reaction is usually presented as a Schiff's base formation, but according to Richard and Knowles (78) the actual reaction is much more complicated. They found that commercial glutaraldehyde solutions are invariably impure; they can be contaminated with acids, significant amounts of α,β-unsaturated aldehydes, or polymeric mate-

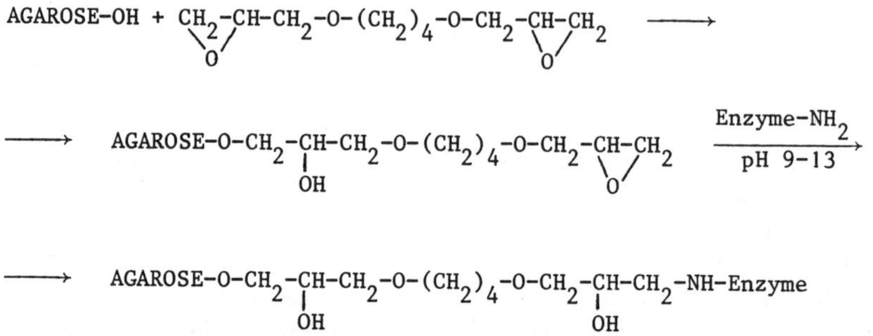

$$\text{AGAROSE-OH} + \underset{\text{(epoxide)}}{\mathrm{CH_2{-}CH}}\mathrm{{-}CH_2{-}O{-}(CH_2)_4{-}O{-}CH_2{-}}\underset{\text{(epoxide)}}{\mathrm{CH{-}CH_2}} \longrightarrow$$

$$\longrightarrow \text{AGAROSE-O-}\mathrm{CH_2{-}\underset{OH}{CH}{-}CH_2{-}O{-}(CH_2)_4{-}O{-}CH_2{-}}\underset{\text{(epoxide)}}{\mathrm{CH{-}CH_2}} \xrightarrow[\text{pH 9–13}]{\text{Enzyme–NH}_2}$$

$$\longrightarrow \text{AGAROSE-O-}\mathrm{CH_2{-}\underset{OH}{CH}{-}CH_2{-}O{-}(CH_2)_4{-}O{-}CH_2{-}\underset{OH}{CH}{-}CH_2{-}NH{-}Enzyme}$$

Fig. 8.28. Covalent immobilization of an enzyme onto epoxy-activated agarose.

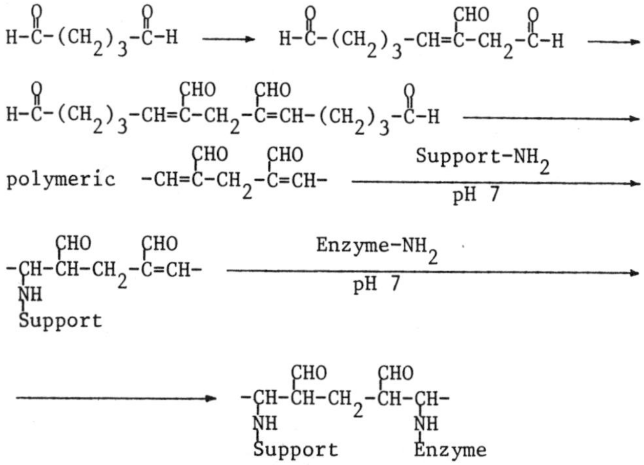

$$\mathrm{H{-}\overset{O}{\overset{\|}{C}}{-}(CH_2)_3{-}\overset{O}{\overset{\|}{C}}{-}H} \longrightarrow \mathrm{H{-}\overset{O}{\overset{\|}{C}}{-}(CH_2)_3{-}CH{=}\overset{CHO}{\overset{|}{C}}{-}CH_2{-}\overset{O}{\overset{\|}{C}}{-}H} \longrightarrow$$

$$\mathrm{H{-}\overset{O}{\overset{\|}{C}}{-}(CH_2)_3{-}CH{=}\overset{CHO}{\overset{|}{C}}{-}CH_2{-}\overset{CHO}{\overset{|}{C}}{=}CH{-}(CH_2)_3{-}\overset{O}{\overset{\|}{C}}{-}H} \longrightarrow$$

$$\text{polymeric}\quad \mathrm{{-}CH{=}\overset{CHO}{\overset{|}{C}}{-}CH_2{-}\overset{CHO}{\overset{|}{C}}{=}CH{-}} \xrightarrow[\text{pH 7}]{\text{Support-NH}_2}$$

$$\mathrm{{-}\underset{\underset{Support}{|}}{\underset{NH}{|}}{CH}{-}\overset{CHO}{\overset{|}{C}H}{-}CH_2{-}\overset{CHO}{\overset{|}{C}}{=}CH{-}} \xrightarrow[\text{pH 7}]{\text{Enzyme-NH}_2}$$

$$\longrightarrow \mathrm{{-}\underset{\underset{Support}{|}}{\underset{NH}{|}}{CH}{-}\overset{CHO}{\overset{|}{C}H}{-}CH_2{-}\overset{CHO}{\overset{|}{C}H}{-}\underset{\underset{Enzyme}{|}}{\underset{NH}{|}}{CH}{-}}$$

Fig. 8.29. Proposed mechanism of enzyme immobilization with glutaraldehyde (78).

rials originating from glutaraldehyde itself. As a mechanism of the reaction of this bifunctional reagent with proteins, they propose a Michael-type addition of amino groups of the protein and support to the α,β-unsaturated aldehydic polymer, as shown in Figure 8.29.

The reaction is gentle and can be carried out at a pH of about 7. By performing the reaction sequentially with an intermediate washing step, intermolecular enzyme crosslinking can be prevented. The reaction is very simple, extremely gentle, and relatively inexpensive to scale up.

Miscellaneous Covalent Binding Methods. Among covalent binding methods, the *Ugi reaction* is unusual because it involves four different functional groups—carbonyl, amino, isocyano, and carboxyl groups—leading to the formation of an *N*-substituted amide according to the scheme presented in Figure 8.30. If one of the R_1–R_5 groups is a solid support, immobilization of enzymes is possible in principle. The enzymes can be coupled either through their amino groups or carboxyl groups. The reaction also works with a hydroxyl group replacing the carboxyl group. It has been successfully applied with polysaccharide, polyacrylic, and nylon supports, mainly using α-chymotrypsin as the enzyme to be immobilized.

Reversible immobilization by thiol–disulfide interchange is a relatively new immobilization technique based on thiol–disulfide interchange between thiol groups of enzymes and mixed disulfide residues of supports. The preparation of a mixed disulfide derivative of a support and its use for enzyme immobilization are depicted in Figure 8.31. The special advantage of this technique is that the enzyme can be removed with, for instance, dithiothreitol or 2-mer-

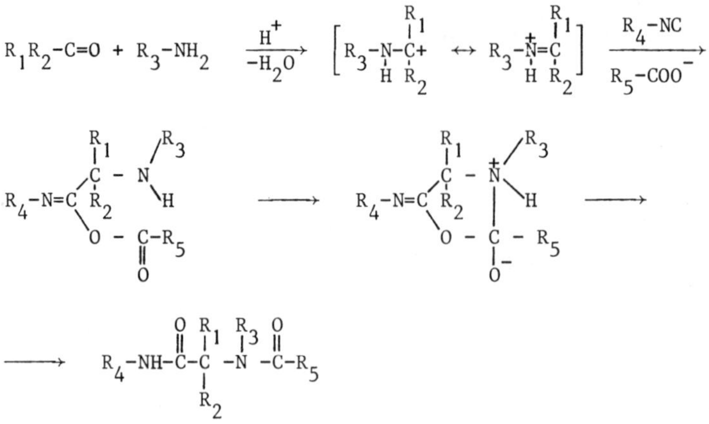

$$R_1R_2\text{-C=O} + R_3\text{-NH}_2 \xrightarrow[-H_2O]{H^+} \left[R_3\text{-}\underset{H}{N}\text{-}\overset{R_1}{\underset{R_2}{C}}{+} \leftrightarrow R_3\text{-}\underset{H}{\overset{+}{N}}\text{=}\overset{R_1}{\underset{R_2}{C}}\right] \xrightarrow[R_5\text{-COO}^-]{R_4\text{-NC}}$$

$$\longrightarrow R_4\text{-NH-}\overset{O}{\overset{\|}{C}}\text{-}\overset{R_1}{\underset{R_2}{C}}\text{-}\overset{R_3}{N}\text{-}\overset{O}{\overset{\|}{C}}\text{-}R_5$$

Fig. 8.30. The Ugi reaction.

captoethanol and the support subsequently reactivated. Thiolation of enzymes not containing thiol groups and subsequent immobilization via disulfide interchange has been reported (79). Thiolation can be performed with methyl-3-mercaptopropioimidate, as shown in Figure 8.32. As activated thiol support, agarose-glutathione-2-pyridyldisulfide is commercially available.

Several other covalent binding methods have been used for the immobilization of enzymes. These methods have mostly found limited use and are therefore not treated here. They can be found in the more extensive reviews (3, 6, 26, 27, 69–74).

8.3.2. The Crosslinking Method

A water-insoluble enzyme preparation can be produced by intermolecular crosslinking of the enzyme molecules in the absence of a solid support. If possible, a crosslinking agent is chosen that specifically binds functional groups on the enzyme not involved in the catalysis at a concentration suitable for complete insolubilization with retention of activity. Although the

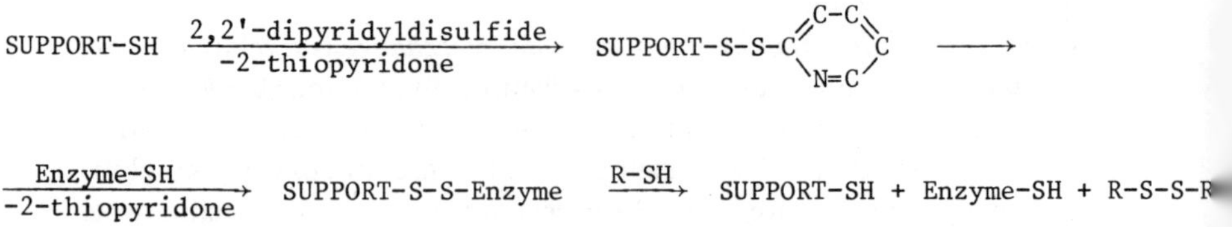

$$\text{SUPPORT-SH} \xrightarrow[\text{-2-thiopyridone}]{\text{2,2'-dipyridyldisulfide}} \text{SUPPORT-S-S-(2-pyridyl)} \longrightarrow$$

$$\xrightarrow[\text{-2-thiopyridone}]{\text{Enzyme-SH}} \text{SUPPORT-S-S-Enzyme} \xrightarrow{\text{R-SH}} \text{SUPPORT-SH} + \text{Enzyme-SH} + \text{R-S-S-R}$$

Fig. 8.31. Reversible immobilization of enzyme molecules by thiol–disulfide interchange.

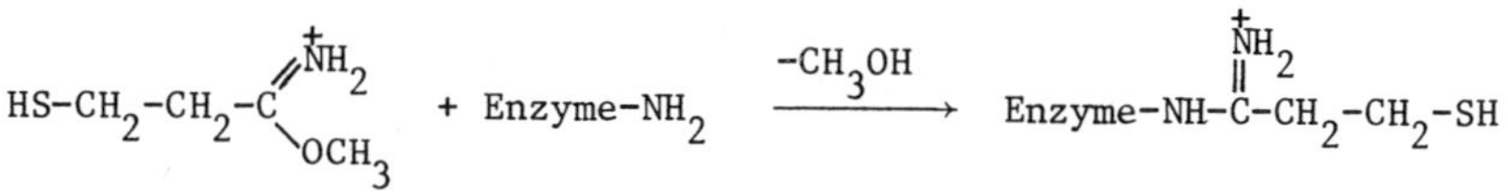

$$HS{-}CH_2{-}CH_2{-}C(=\overset{+}{N}H_2)(OCH_3) + \text{Enzyme-}NH_2 \xrightarrow{-CH_3OH} \text{Enzyme-}NH{-}C(=\overset{+}{N}H_2){-}CH_2{-}CH_2{-}SH$$

Fig. 8.32. Thiolation of enzyme molecules with methyl-3-mercaptopropioimidate.

procedure is simple to perform, establishing the suitable conditions to achieve this goal is difficult and must be determined empirically. Only a few thorough studies have been reported, but these show that insolubilization of enzymes with maximum retention of activity depends on a delicate balance of factors such as concentration of enzyme and crosslinking reagent, pH, ionic strength, temperature, and reaction time. The flow properties of the resulting preparations are generally poor, as they are gelatinous in nature. This method is therefore usually combined with another immobilization technique. Adsorption to a solid support followed by crosslinking of the adsorbed enzyme molecules has worked for a variety of enzymes. Glutaraldehyde has been the most extensively applied crosslinking agent, but many other multifunctional reagents have also been used (71).

8.3.3. The Entrapping Method

The entrapping method is based on confining enzymes to the lattice of a polymer matrix or enclosing them in semipermeable membranes. This method is subdivided into entrapping in a lattice, a microcapsule, or a fiber (Figure 8.22).

8.3.3a. Lattice Type

This type of immobilization involves entrapping enzymes within the interstitial spaces of a crosslinked water-insoluble polymer. The existing methods can be further subdivided on the basis of the manner in which the polymers are crosslinked (44).

1. Aggregation of polymers with concomitant gelation or precipitation of the enzyme–polymer solution as a result of the formation of nonspecific secondary bindings. This can be established by changing the solubility variables, for example, changing temperature, ionic strength, pH, and solvent. Examples of gels formed in this manner are agar, carrageenan, collagen, and cellulose. Since the network formation is based on secondary valence forces, ranging from dispersion to hydrogen bonding, it is a reversible process. A more permanent character can be given to the network by further reaction steps, especially

those involving covalent crosslinking. The general procedure given in Scheme 1 shows the simplicity of this technique.

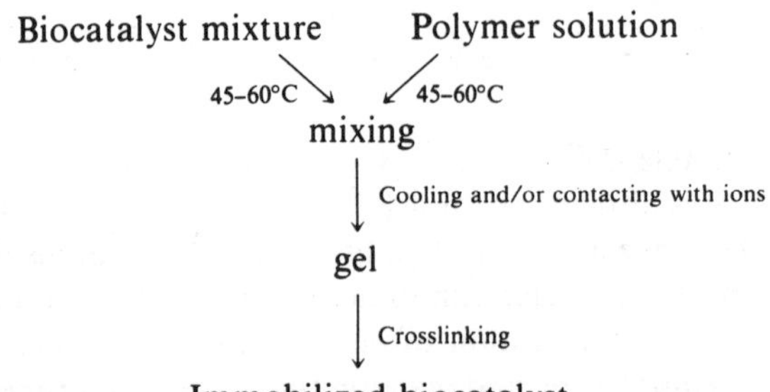

Scheme 1. Procedure for immobilization by precipitation (Adapted from Ref. 44).

2. Crosslinking of polymers by means of multivalent ions such as Ca^{2+} and Al^{3+}, resulting in gelation of the enzyme–polymer solution. The most prominent example of this type is immobilization in calcium-alginate gel. This immobilization procedure is so easy, mild, and versatile, and has been applied successfully in so many cases, that it is advisable to try out this technique first. Except when dissolving ions (the network formation is of course reversible) must be present in the substrate solution, this procedure is likely to yield an immobilized enzyme preparation with a high retention of activity and in a suitable form and strength. A general preparation is given in Figure 8.33. In Section 8.2.1b, procedures for strengthening of the gel are given for the exceptional cases.
3. Entrapment of enzymes in a gel by polymerization or polycondensation of an aqueous solution of the monomers or oligomers, respectively, and the enzyme in the presence of a crosslinking agent. Several procedures exist, for example, immobilization in gelatin crosslinked with glutaraldehyde, which can be performed under mild conditions (see Section 8.4.6d). The technique of this type most widely used so far, entrapment in a polyacrylamide gel, is also carried out under mild circumstances, but when polymerized in bulk the heat of reaction may denature the enzyme. A particular disadvantage of these techniques is that the polymerization and polycondensation must be performed in the presence of the enzyme, which may lead to chemical modification with concomitant inactivation of the enzyme by the reactive monomers or crosslinking agent.

Advantages of the lattice type of immobilization are the general applicability and the resulting homogeneous distribution of the enzyme in the support.

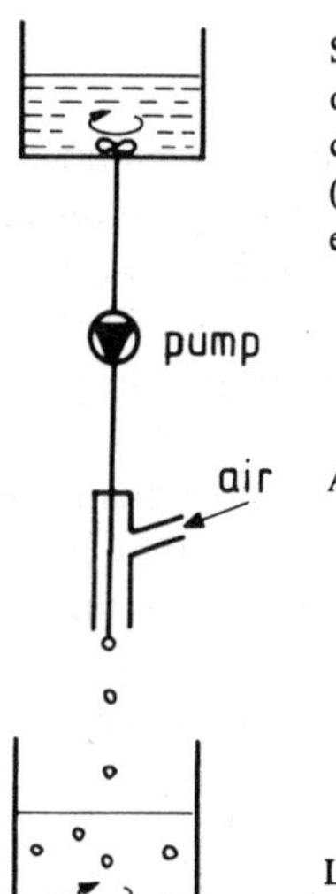

Sodium alginate/biocatalyst mixture (0–40°C): 1% solution (w/v) of viscous type of alginate for immobilization of large enzyme complexes, cell organelles, and whole cells, up to an 8% solution (w/v) of a less viscous type of alginate for immobilization of small enzymes.

Air flow rate and needle diameter determine particle size.

Large excess of calcium chloride solution (0.05–0.2 *M*; 0–25°C). Hardening of gel is practically completed within 5 hr.

Fig. 8.33. Schematic representation of immobilization in calcium alginate gel.

Disadvantages are that a small amount of enzyme usually leaks from the support and that the diffusion limitation of substrate and product in the gel pores generally is high. This can lead to retardation in reaction velocity, especially for high molecular weight substrates. Even complete inactivity has been observed for amylases as a result of the inability of the substrate (starch) to penetrate into the gel.

8.3.3b. Microcapsule Type

Enzymes can be immobilized within microcapsules prepared from organic polymers. Microencapsulation is usually achieved by dispersing an aqueous enzyme solution containing 1,6-diaminohexane into a solution of adipoylchloride in an organic solvent immiscible with water (e.g., chloroform, tetra, toluene). They polymerize upon contact at the water/organic-solvent interface, forming a thin polyamide (nylon 6,6) membrane around aqueous droplets of enzyme solution:

$$n\ NH_2(CH_2)_6NH_2 + n\ ClCO(CH_2)_6COCl \xrightarrow{-HCl} —(NH(CH_2)_6NHCO(CH_2)_6CO)_n—$$

A solution of cellulose acetate in an organic solvent gives a precipitate and forms a membrane when contacted with water. This is also a common method used for the microencapsulation of enzymes.

8.3.3c. Fiber Type

A method strongly related to microencapsulation is fiber entrapment, a procedure developed by Dinelli in Italy (80). In this process an emulsion is formed of a synthetic polymer in an organic solvent (e.g. cellulose acetate or polyvinylchloride in methylene chloride) with an aqueous enzyme solution. The emulsion is extruded through a spinneret into a precipitate, droplets of enzyme solution being trapped in the fiber.

The applicability of both microencapsulation and fiber entrapment is limited by the necessity of using water-immiscible solvents, which may in some cases cause inactivation of the enzyme. Both methods also result in immobilized enzyme systems generally exhibiting diffusion-limited kinetics, hence they are best suited for enzyme systems that involve substrates and products with a low molecular weight.

8.3.4. Overview

None of the above discussed immobilization procedures (the examples only covering part of the large number described in the literature) has pronounced advantages for all enzymes. The immobilization of a particular enzyme aimed at a specific application still requires an empirical, essentially trial-and-error approach. Of course, the choice of methods to be investigated should be made on the basis of need and on the available knowledge. In many cases immobilization in calcium alginate gel may be a good start because of the mildness and simplicity of the procedure.

8.4. IMMOBILIZED XANTHINE OXIDASE FOR ORGANIC SYNTHESIS

The study of "the application of immobilized enzymes in organic synthesis" was initiated in 1975 in the Departments of Organic Chemistry and Biochemistry of the Agricultural Unviersity, Wageningen in the Netherlands.* The objective was to show that immobilized enzymes can indeed be conveniently and profitably used in organic synthesis. To limit the number of potentially useful enzymes, the following restrictions were set:

1. A close relation with the research already going on in these laboratories was desired.
2. The enzyme to be chosen must have a broad substrate specificity and be available commercially.
3. Research leading to new immobilization procedures was not intended, at least not in the initial stage.

*I worked in these laboratories on this project from 1975 through 1979.

These three restrictions appeared to be sufficient to allow the choice of the enzyme xanthine oxidase.

In the laboratory of the Organic Chemistry department, there is strong interest in the chemistry of azaheterocyclic compounds like pyrimidines, purines, pteridines (Figure 8.34), and related compounds (81, 82), while flavoproteins are a main subject of study in the laboratory of Biochemistry (83). The flavoprotein xanthine oxidase shows an interestingly wide substrate specificity, as appears from the fact that it efficiently catalyzes the oxidation of many azaheterocycles (84), which by other means are sometimes very reluctant to oxidation. Xanthine oxidase from bovine milk is available commercially and can easily be immobilized (77, 85, 86). Immobilized xanthine oxidase was therefore developed and evaluated for application in organic synthesis.

8.4.1. Properties and Kinetics of Xanthine Oxidase

Xanthine oxidase (EC 1.2.3.2) and xanthine dehydrogenase (EC 1.2.1.37) form a very closely related group of enzymes. Their wide specificities overlap to a large extent and all of them, from all sources so far as is known, contain FAD, together with molybdenum and iron-sulfur centers. The reactions they catalyze are generally of the form

$$RH + H_2O + E \longrightarrow ROH + E'$$

$$E' + \text{oxidizing substrate} \longrightarrow E + \text{reduced substrate}$$

where RH is the reducing substrate. The oxygen introduced into RH is derived from water. When molecular oxygen is preferentially used as the terminal electron acceptor the enzyme is referred to as "oxidase," with other oxidizing substrates (e.g. NAD^+) called "dehydrogenase."

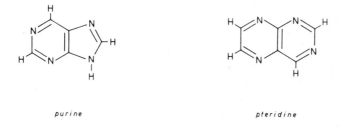

purine

pteridine

Fig. 8.34. Examples of azaheterocyclic compounds.

Xanthine oxidase from cow's milk has been studied most extensively, and several reviews (e.g., Refs. 87–89) were used as sources for this section. The enzyme can easily be isolated from cow's milk, which usually contains around 50 mg of enzyme per liter of milk.

Bovine xanthine oxidase is able to oxidize a wide variety of compounds. A large number of purines containing hydroxyl, amino, methyl, mercapto, and halogeno groups and purine *N*-oxides are oxidized, although at greatly differing rates. Also effective as substrates are 2- or 8-azapurines. Replacement of the imidazole ring of the purines by a pyrazole ring gives a series of compounds that can be good substrates, although they are perhaps more remarkable with respect to the dramatic inhibitions they show in the oxidation of xanthine to uric acid by xanthine oxidase. A wide variety of pyrimidines, pteridines, and other heterocyclic compounds are also oxidized, some of them quite rapidly. A comparison of the rates of oxidation is difficult, since the measurements of the activities towards reducing substrates were performed under widely different conditions using a variety of terminal electron acceptors. Furthermore, apart from the normal complications of two-substrate enzyme-catalyzed reactions, xanthine oxidase reactions are often particularly sensitive to inhibition by excess substrate. An overview was written by Massey (88), and Massey and coworkers (90) have proposed a model of the catalysis, which is considered a fair approximation of the true mechanism (87). The kinetic and chemical scheme representing the oxidation of xanthine by xanthine oxidase as they propose is given in Figure 8.35.

In the Michaelis type of complex (I), a nucleophilic attack by the disulfide occurs at C-8 of the purine ring. In complex (II) a two-electron donation to Mo^{VI} takes place leading to Mo^{IV}, and a proton is transferred to the nitrogen ligand, yielding complex (III). In this complex the N═C—S—S moiety is highly susceptible to a nucleophilic attack. Reaction with water produces uric acid and a reduced enzyme, which by reaction with molecular oxygen is oxidized to E. It is assumed that water and not the hydroxide ion is involved in the formation of uric acid in view of the fact that the oxidation rate is independent of the pH over a wide range. It is further assumed that the formation of the Michaelis complex (I) (k_1) and the reoxidation of the enzyme (k_5) are relatively fast.

Xanthine dehydrogenase from chicken liver is very similar to milk xanthine oxidase, except that it preferentially uses NAD^+ as the terminal electron acceptor. It also has a broad substrate specificity, but substrate inhibition has been observed only at very high concentrations of substrate.

Xanthine oxidase has also recently been isolated from Arthrobacter and studied (91). The properties of this bacterial enzyme seem to resemble more the chicken-liver enzyme than milk xanthine oxidase, except that it efficiently uses O_2 as terminal electron acceptor but not NAD^+. According to

Step 1. *Xanthine* (S) + *Xanthine Oxidase* (E) $\underset{k_{-1}}{\overset{k_1}{\rightleftharpoons}}$
Complex **I**

Step 2. *Complex* **I** $\xrightarrow{k_2}$ *Complex* **II**

Step 3. *Complex* **II** $\xrightarrow{k_3}$ *Complex* **III**

Step 4. *Complex* **III** + H_2O $\xrightarrow{k_4}$ *Uric Acid (UA)* +
Reduced Enzyme (E′)

Step 5. $E' + O_2 \xrightarrow{k_5} E + H_2O_2$

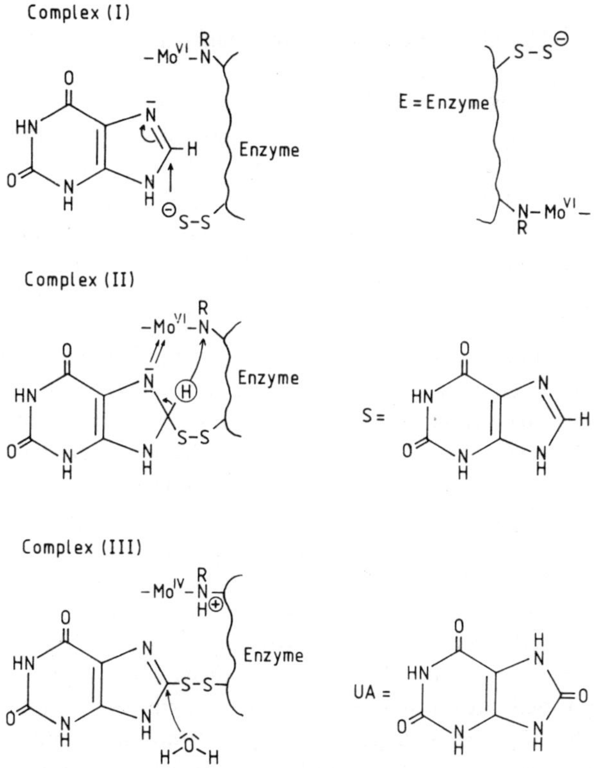

Complex (I)

Complex (II)

Complex (III)

Fig. 8.35. Representation of the oxidation of xanthine by xanthine oxidase as proposed by Olsen et al. (90). Complex II is introduced as additional intermediate complex to make the model chemically more feasible.

Woolfolk and Downard (91) the bacterial enzyme is relatively specific, but only a few substrates have been investigated. Substrate activation instead of inhibition has been observed for xanthine (the opposite is true for hypoxanthine). The specific activity is about 50 times that of the milk enzyme. Arthrobacter xanthine oxidase is thus also potentially useful for synthesis, and its application was therefore studied. The high specificity of the subsequent enzyme in the purine-oxidative-pathway sequence, uricase (92), and the development of several procedures for the immobilization of whole cells (3), induced us to immobilize the Arthrobacter cells without any prior enzyme isolation.

8.4.2. Specificity of Free and Immobilized Xanthine Oxidase

Organic chemists have become accustomed to being able to forecast the course and rate of reaction of a given compound with a chemical reagent, and also to be provided with information on the limitation of the method. Realistically or not, they will expect to be able to do the same with biocatalysts. Therefore, sufficient specificity data should be available to enable reliable prediction to be made (Table 8.6) for any previously unevaluated potential substrates, before general acceptance as a routine catalyst for organic syntheses is likely to occur (2).

Bergmann and his colleagues (references cited in 87) have extensively studied the substrate specificity of bovine milk xanthine oxidase, but the complexity of this enzyme has allowed few definite regularities thus far. Such studies did not exist at all for immobilized xanthine oxidase at the time we choose this system as a model to develop. We therefore studied various series of new substrates in order to determine systematically the substrate limits of both free and immobilized xanthine oxidase.

In these studies, agarose was used as support material for the immobilization of xanthine oxidase. It has the particular advantage for this type of work that it can be used in a cuvette for spectrophotometric enzyme activity assays in a similar manner as the soluble enzyme, provided that the spectrophotometer is equipped with a magnetic device to stir in the cuvette in order to maintain a homogeneous suspension of the immobilized enzyme (93). The enzyme was either covalently bound to CNBr-activated Sepharose 4B according to the general procedure (Section 8.3.1c) or adsorbed to *n*-octyl-substituted Sepharose 4B (Section 8.3.1b). The retention of activity was at the most 30% for the covalently bound enzyme and up to 90% for the adsorbed enzyme (93). The latter was therefore mainly used for the determination of the substrate specificity.

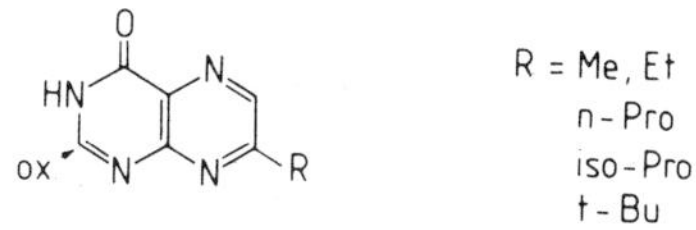

O
HN N
ox N N R

R = Me, Et
n-Pro
iso-Pro
t-Bu

Fig. 8.36. Oxidation of 7-R-pteridin-4-ones by xanthine oxidase.

8.4.2a. Effects of Substrate–Substituent Size on Activity

The first series of substrates was investigated (94) with the aim of establishing the influence of the size of the substituent in 7-R-pteridin-4-ones (R = Me, Et, *n*-Pro, *i*-Pro, *t*-Bu) on the rate of oxidation and on the affinity of the biocatalyst for the substrate as reflected in V_m and K_m, respectively (Figure 8.36). With the exception of the *t*-butyl derivative, which is very slowly oxidized and apparently too bulky to fit easily into the active site of the enzyme, the 7-alkylpteridin-4-ones are good substrates with maximum oxidation rates comparable to that of the "natural" substrate xanthine. The V'_m-values with immobilized xanthine oxidase are almost all smaller than the V_m values obtained with the free enzyme in the pH-range tested (7.6–8.9), but they show the same pattern.

The difference in size and electron donating properties of the substituents are not reflected in the kinetic constants, neither in the V_m and V'_m values nor in the K_m and K'_m values. The latter are an order of magnitude larger than that of xanthine and in that sense poorer substrates. The behavior of K'_m for immobilized xanthine oxidase around the pK_a (8.2) of the 7-R-pteridin-4-ones is slightly different from that of K_m for the free enzyme (Figure 8.37). This difference can be explained by the electrostatic attraction of the negatively charged substrate (above pH 8.2) by the positively charged support, which has a pK_a of about 10.4. The substrate concentration will thus be higher in the support than in the liquid continuous phase, provided that dif-

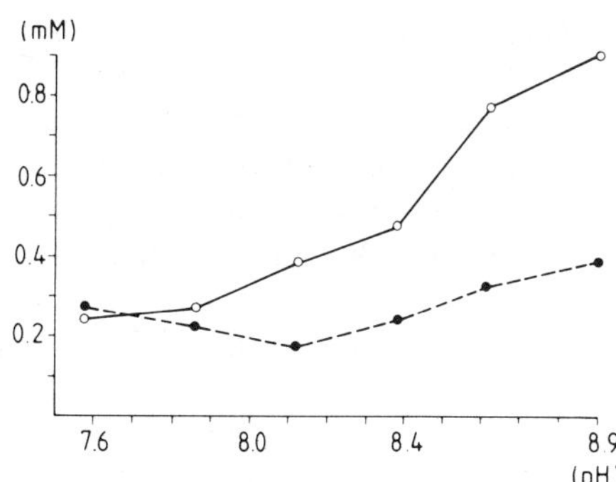

Fig. 8.37. K_m (○) and K'_m (●) of 7-isopropylpteridin-4-one with free and immobilized xanthine oxidase, respectively.

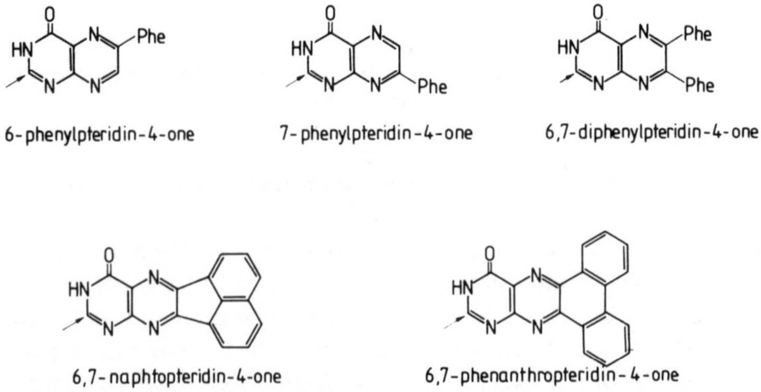

Fig. 8.38. Place of oxidation (/) in series of "two-dimensionally" increasing substrates of xanthine oxidase.

fusion of substrate does not limit the rate of reaction. This result in an apparent decrease in Michaelis constant in comparison to the free enzyme. No further significant differences were found between free and immobilized xanthine oxidase.

In the following series of substrates tested, the substituents were larger "two-dimensionally" (Figure 8.38). All these compounds appeared to be very poor substrates with the exception of 7-phenylpteridin-4-one, which is rapidly oxidized. Therefore, this type of substrate was studied (95).

8.4.2b. Oxidation of 7-(p-X-Phenyl)pteridin-4-ones

In addition to 7-phenylpteridin-4-one (X = H), four parasubstituted (X = OMe, Br, CN, NO_2) derivatives were investigated in order to establish if substituents with electron donating or withdrawing properties affect the biocatalysis.

It seems reasonable to assume that the same mechanism is involved in the oxidation of 7-(p-X-phenyl)pteridin-4-ones as with xanthine (Section 8.4.1). The disulfide group is then attached to position 2 of the pteridine system and in that position the nucleophilic attack of water can take place. Analogous to the model for xanthine, the two-electron donation to Mo^{VI} can take place from N-1 or N-3 but may in this case occur from N-8, which can also accommodate a negative charge (Figure 8.39).

In order to obtain more detailed insight into the mechanism of this enzymic oxidation, we tried to establish the influence of substituent X on the maximum oxidation rate (V) and on the affinity of the enzyme for the substrate, as reflected in the K_m value. Detailed studies were carried out at pH 9.1. Since the pK_a value of these compounds vary between 8.0 and 8.2, they

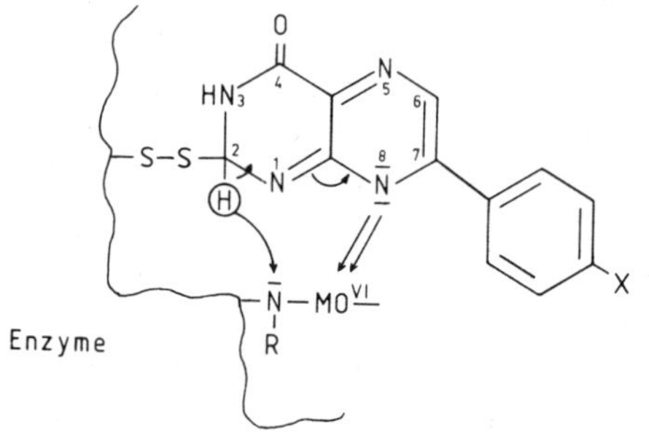

Fig. 8.39. Proposed mechanism for oxidation of 7-(p-X-phenyl)pteridin-4-ones by xanthine oxidase (95).

are largely present in the anionic forms at this pH. The negative charge is delocalized over both rings, oxygen, N-3, N-1, and N-8 having the greatest electron density (Figure 8.40). If the nucleophilic attack of either —S—S⁻ or H_2O (step 2 or step 4 in Figure 8.35) is rate limiting, it can be expected that the influence of X on the oxidation rate is relatively small, as there is no important contribution of a mesomeric effect of X on C-2 and the influence of the inductive effect of X on the electron density of C-2 can be considered to be insignificant. On the other hand, if the electron transfer to the metal ion is rate limiting, one can expect a relatively large effect of X on the rate: If X has a +M effect it will promote the electron flow from one of the substrate nitrogens to Mo^{VI} and rate acceleration will be observed; when X is electron-withdrawing, the oxidation rate will be decreased.

According to Bergmann et al. (96) the groupings $(\overset{1}{N}—C—\overset{8}{N})$ and $(O=\overset{4}{C}—C—\overset{5}{N})$ are important in the association of the substrate with the enzyme. Baker et al. (97) showed that there must be a hydrophobic group in or near

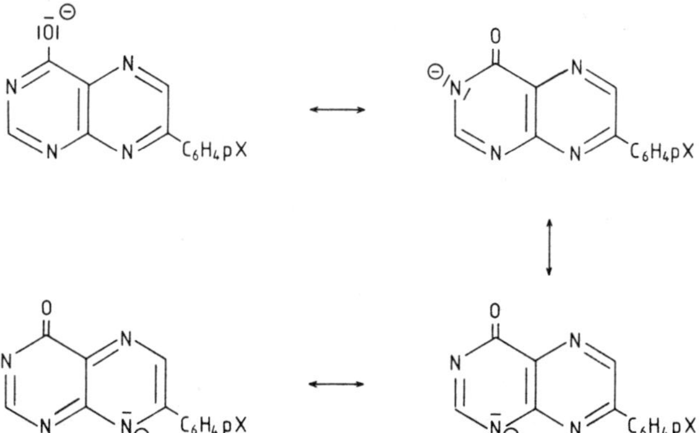

Fig. 8.40. Delocalization of negative charge of 7-(p-X-phenyl)pteridin-4-ones (95).

the active site of the enzyme. Therefore, in the 7-(*p*-X-phenyl)pteridin-4-ones the hydrophobicity of the phenyl group possibly determines the formation of the enzyme–substrate complex to a certain extent. An influence of X on K_m can be expected as X has both an effect on the groupings involved in the binding and on the hydrophobicity of the phenyl group.

The Influence of Substituent X on the Rate of Oxidation. The results of the kinetic experiments show that the rate of oxidation with free and immobilized xanthine oxidase are nearly the same (see Table 8.12). A plot of the logarithm of the ratio of the maximal oxidation rate of the 7-(*p*-X-phenyl)pteridin-4-ones and 7-phenylpteridin-4-one vs. the substituent constant σ of X is given in Figure 8.41. The lines were obtained by linear regression; a reaction constant ρ of about −0.5 for free and immobilized enzyme is calculated from the slopes. This means that the rate-limiting step in both the free and immobilized enzyme systems is facilitated by a high electron density at the reaction site and is moderately sensitive to substituent effects. This indicates that neither the nucleophilic attacks by the active-site disulfide and water or hydroxide ion (the more likely nucleophile in alkaline solution) nor the C—H bond breaking is the rate-limiting step. That the C—H bond breaking is not

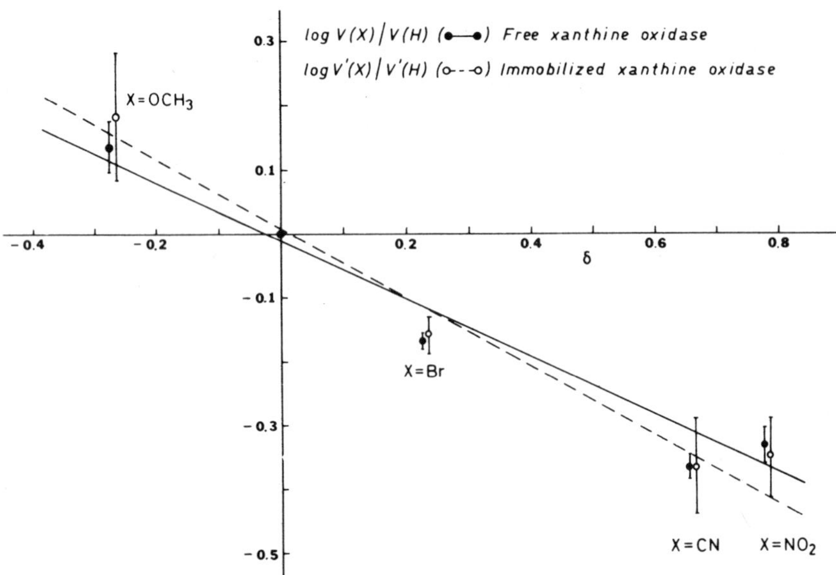

Fig. 8.41. Plot of the logarithm of the ratio of the maximal oxidation rate of the 7-(p-X-phenyl)pteridin-4-ones and 7-phenylpteridin-4-one versus the substituent constant σ of X. Oxidation by free and immobilized xanthine oxidase at pH 9.1 and 20°C.

rate limiting is confirmed by our observation that there is no significant difference in the maximum rate of oxidation of 2-D-7-phenylpteridin-4-one and the parent compound 7-phenylpteridin-4-one. A similar observation has been made for xanthine and 1-methylxanthine (87). Apparently the electron transfer from the substrate nitrogen to the Mo^{VI} in the active center is rate limiting. Accordingly, the rate of oxidation of the anion is higher than that of the neutral substrate.

The Effect of Substituent X on K_m and K_m'. The K_m and K_m' values for free and immobilized xanthine oxidase, respectively, are given in Table 8.12. As we showed previously for xanthine (93), the K_m' values of the immobilized enzyme are larger than the K_m values of the free enzyme as a result of diffusion limitation in the pores of the support material. All the K_m and K_m' values for the 7-(*p*-X-phenyl)pteridin-4-ones are much lower than the corresponding values found for xanthine and 7-*n*-propylpteridin-4-one (Table 8.12). This increase in affinity is probably to a large extent due to the presence of the phenyl group at position 7. This is in agreement with the proposal of Baker et al. (97) that there is a hydrophobic group in the vicinity of the active center that strongly increases the binding of inhibitors substituted with a phenyl group. Apparently the 7-(*p*-X-phenyl)pteridin-4-ones fit with the right orientation in the active center of the enzyme, since they are good substrates. That a hydrophobic interaction is indeed responsible for the high affinity of the enzyme for the 7-(*p*-X-phenyl)pteridin-4-ones is supported by the data in Table 8.12, showing that if the polarity of X increases (NO_2, CN), K_m tends to increase and thus the affinity decreases. Accordingly, the affinity also decreases upon anion formation.

As an interesting contrast, Bergmann et al. (99) reported that the corresponding 8-(*p*-X-phenyl)purin-6-ones (X = OCH_3, H, NO_2) are strongly bound but clearly in the wrong orientation, since they act as very poor substrates but strong inhibitors. That the position of the phenyl groups indeed plays an essential role is underlined by our observation that the isomeric 6-(*p*-X-phenyl)pteridin-4-ones are also poor substrates of xanthine oxidase and inhibit the oxidation of xanthine to uric acid too (100).

The observation that the K_m value of the deuterated substrate is higher than that for ordinary 7-phenylpteridin-4-one is in agreement with a similar observation for 1-methylxanthine (87).

8.4.2c. Specificity of Bacterial Xanthine Oxidase

Based on a small number of substrates investigated, Woolfolk and Downard (91) conclude that xanthine oxidase from Arthrobacter is more specific than the milk enzyme. However, our results—see Table 8.13—show the opposite.

Table 8.12. Kinetic Constants of Free and Immobilized Xanthine Oxidase for Various Substrates in Tris-HC1 Buffer pH 9.1 (I = 0.01; 0.1 m*M* EDTA) at 20°C (95)

Substrate	pK_a	λ ($\Delta\epsilon$) Used in Assay nm($mM^{-1}cm^{-1}$)	Free Xanthine Oxidase		Immobilized Xanthine Oxidase	
			V(units mg^{-1})	K_m(μM)	V'(units mg^{-1})	K'_m(μM)
7-Phenylpteridin-4-one	8.17 ± 0.05	315 (3.7)	0.20	1.3	0.19	4
7-(*p*-Methoxyphenyl)pteridin-4-one	8.2 ± 0.1	335 (1.9)	0.27	1.3	0.30	7
7-(*p*-Broomphenyl)pteridin-4-one	8.12 ± 0.05	320 (4.7)	0.14	0.7	0.14	3
7-(*p*-Nitrophenyl)pteridin-4-one	8.0 ± 0.1	320 (5.3)	0.09	3.6	0.09	6
7-(*p*-Cyanophenyl)pteridin-4-one	8.0 ± 0.1	320 (4.2)	0.09	3.3	0.09	6
2-D-7-phenylpteridin-4-one[a]		315 (3.7)	0.20	3.5	0.18	7
7-*n*-Propylpteridin-4-one[b]	8.11 ± 0.03	269 (6.2)	0.9	3×10^2	0.8	3×10^2
Xanthine[b]	7.5[c]	296 (8.9)	0.64	27	0.46	31

[a] ^2H-2 content: 65%.
[b] Unpublished results.
[c] Bergmann and Levene (98).

Table 8.13. Specific Activities of Milk Xanthine Oxidase and Immobilized Arthrobacter Cells with Various Substrates at 20°C (58)

	Immobilized Arthrobacter		Milk Xanthine Oxidase	
Substrate	Units/100 mg of Column Material	%	Units × 10^3/mg Protein	%
1-Methylxanthine	10.7	100	240	100
8-Phenylpurin-6-one	0.8	7.5	1.9	0.8
7-Phenylpteridin-4-one	2.9	27	65	27
6-Phenylpteridin-4-one	7.0	66	5.8	2.4

Evidently the specificity of these enzymes is different. Therefore, a wider spectrum of substrates can be conveniently and efficiently oxidized when both the immobilized milk xanthine oxidase and immobilized Arthrobacter xanthine oxidase are available.

8.4.3. Stability of Immobilized Xanthine Oxidase

If it is to be useful in organic synthesis, the immobilized enzyme should remain active during long periods of catalytic turnover and during storage. The immobilized xanthine oxidase (both covalently bound and adsorbed) was found to be more stable than the soluble enzyme when incubated at 4° and 30°C in Tris-HCl or borate buffer at pH 8.2. The immobilized enzyme lost no activity during 6 months of storage at 4°C, whereas the activity of the soluble enzyme (0.2 mg protein · mL^{-1}) decreased by about 30% in 4 months. The half-lives ($t_{1/2}$) for the free and immobilized enzyme at 30°C were 55 hr and up to 300 hr, respectively. Note that, although the overall concentration of the immobilized enzyme was lower, the local concentration, that is, per mL support (about 2 mg protein/mL gel) was higher than that of the soluble enzyme. Effects as described in Section 8.1.2b are therefore not excluded. The stability of the immobilized enzyme was not affected by storage in the presence of EDTA or the enzyme product (0.2 m*M* uric acid).

The immobilized enzyme rapidly lost activity during continuous turnover, either in suspensions with substrate or in reactor columns in which a solution of substrate was continuously passed over the enzyme. The half-lives of the adsorbed enzyme (Table 8.14) and the covalently bound enzyme at 25°C during continuous turnover in a reactor column were about 1 hr. The soluble enzyme (0.14 mg protein · mL^{-1}) lost all its activity in less than 3 hr incubation with substrate (0.7 m*M* xanthine) at 25°C. The rapid inactivation was unexpected since Hofstee and Otillio (77) found, in an experiment where the

Table 8.14. Half-lives of Various Xanthine-Oxidase Preparations During Catalysis at 25°C (93)

Enzyme system	Half life (h)
Free xanthine oxidase	<1
Sepharose preparations	
Covalently bound xanthine oxidase	1.3
Adsorbed xanthine oxidase	1.3
Xanthine oxidase and superoxide dismutase coadsorbed	2
Xanthine oxidase and catalase coadsorbed	3.3
Xanthine oxidase, superoxide dismutase, and catalase, coadsorbed	5.5
Covalently coimmobilized xanthine oxidase, superoxide dismutase, and catalase	7.5
Covalently coimmobilized xanthine oxidase and albumin	2
Gelatin preparations	
Xanthine oxidase entrapped in gelatin	26
Xanthine oxidase, superoxide dismutase and catalase entrapped in gelatin	26
Xanthine oxidase, superoxide dismutase, and catalase crosslinked and then entrapped in gelatin	50

strength of binding between support and enzyme was tested, at least 90% conversion of substrate during several days at 5°C. When we repeated Hofstee's experiment a similar high conversion was obtained during one week, but this sustained high conversion was due to the fact that the column initially contained much more enzyme than was needed for complete conversion (see also Section 8.1.2b) and, at the constant pressure applied, the column became more tightly packed and consequently the flow rate continuously decreased. A constant flow rate was obtained by flowing the substrate from the bottom to the top at lower speed or by using a pump. When, in addition, less enzyme was used, it was possible to determine the operational half-life at 4°C. It was found to be 4 hr. A similar value was obtained for the covalently bound enzyme under identical conditions.

Several attempts were made to determine the cause of the rapid inactivation of soluble and immobilized enzyme that occurs in the presence of substrate (93). The effluents from the columns contained no activity or protein, and therefore no significant leaching of enzyme occurred as is sometimes observed with other systems (101). Accordingly, the preparations had the same protein content before and after continuous use. Figure 8.42 clearly shows

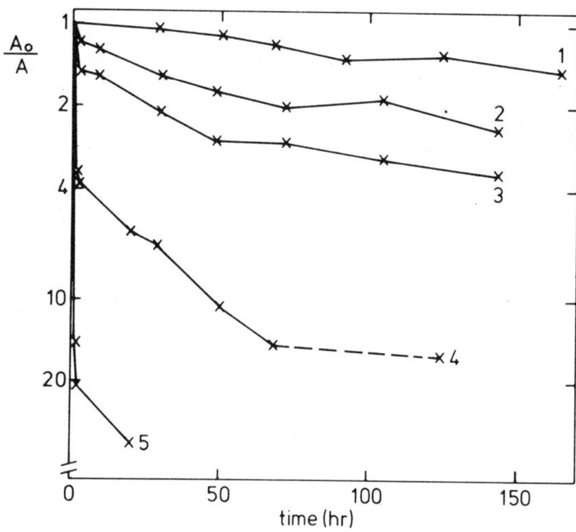

Fig. 8.42. Dependence of the stability of immobilized xanthine oxidase on the concentration of xanthine at 30°C. Assay consisted of aliquots containing 0.4 mg protein in 10 mL Tris-HCl buffer, pH 8.2 (I = 0.01; 0.1 m*M* EDTA) and various xanthine concentrations: 1, 0 μmol; 2, 0.6 μmol; 3, 1.2 μmol; 4, 4 μmol; and 5, 6 μmol. Ratio of the activity at $t = 0$ (A_0) and $t(A)$ is plotted as a function of time (93).

that the rapid drop in activity is proportional to the absolute amount of substrate converted. When all the substrate is converted, the decay is comparable with that in buffer alone. The rapid inactivation could be decreased by adding superoxide dismutase and/or catalase to the incubation mixture. This observation suggested that $O_2^{\overline{\cdot}}$ and H_2O_2 produced in the enzyme reaction, or OH· produced by the chemical reaction of $O_2^{\overline{\cdot}}$ with H_2O_2 (102), might be involved in the inactivation phenomenon. It has been observed that $O_2^{\overline{\cdot}}$ has a damaging effect in biological systems (103) and that in dilute xanthine oxidase solutions the enzyme is inactivated by relatively high H_2O_2 concentrations (104). In the microenvironment of the immobilized enzyme, because of the slight positive charge of the matrix and as a result of the diffusion limitation, accumulation of $O_2^{\overline{\cdot}}$ and H_2O_2 is likely to occur. Therefore, we expected that optimal protection of this system could be achieved by coimmobilization of superoxide dismutase and catalase with xanthine oxidase.

8.4.4. Stabilization of Xanthine Oxidase by Coimmobilization with Superoxide Dismutase and/or Catalase

Superoxide dismutase and catalase have an isoelectric point in the acid region at pH 4.95 (105) and 5.4 (106), respectively, which makes them, like xanthine oxidase, suitable for immobilization by the procedure described by

Hofstee and Otillio (77). Both enzymes could indeed be coimmobilized with xanthine oxidase by adsorption on Sepharose 4B substituted with *n*-octylamine.

As the data of Figure 8.43 show, either dismutase or catalase improved the stability of xanthine oxidase, and coimmobilization of both dismutase and catalase gave a further improvement. The protection given by catalase

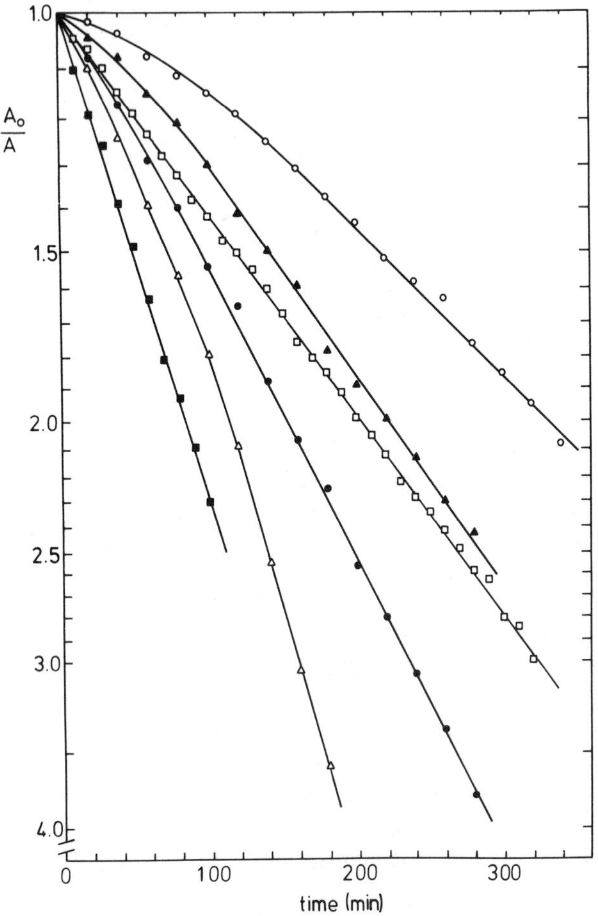

Fig. 8.43. Stability of immobilized xanthine oxidase (XO) under continuous operation at 25°C (constant flow rate: 46 mL h^{-1}). Amount of XO was in all cases 0.4 mg (0.7 units mg^{-1}) per column (1 cm high × 0.5 i.d.). Feed: 40 μM xanthine in Tris-HCl buffer (pH 8.2; I = 0.01; 0.1 m*M* EDTA. Ratio of the conversion at $t = 0$ (A_0) and $t(A)$ is plotted. (■) Immobilized XO; (Δ) XO coimmobilized with SOD (0.4 mg); (●) XO coimmobilized with SOD (0.4 mg) and CAT (0.5 mg); (□) XO coimmobilized with CAT (1.0 mg); (▲) XO coimmobilized with SOD (0.2 mg) and CAT (1.0 mg); (O) XO coimmobilized with SOD (0.4 mg) and CAT (1.0 mg). SOD = superoxide dismutase (3800 units mg^{-1}); CAT = catalase (3100 units mg^{-1})

was constant during the time of operation, and although a detectable amount of catalase activity was found in the effluent, the catalase activity of the immobilized-enzyme preparation was still very high after 8 hr of operation. The protection afforded by superoxide dismutase activity decreased with time, and the dismutase activity of the column material was zero after 8 hr of operation. Low levels of superoxide–dismutase activity were detected in the first fractions of the effluent, and the loss of protection can thus be explained by leakage of dismutase from the column.

To prevent the leakage of catalase and superoxide dismutase from the column material, the enzymes were reacted with CNBr-activated Sepharose 4B to covalently bind them. When all three enzymes were reacted with Sepharose at the same time, xanthine oxidase and catalase but not superoxide dismutase were bound. Accordingly, this preparation showed a half-life comparable with that of xanthine oxidase coadsorbed with only catalase. When superoxide dismutase was first reacted with Sepharose (xanthine oxidase and catalase were added after one hr), about 10% of the activity applied was bound. The half-life of xanthine oxidase in this system improved to $7\frac{1}{2}$ hr, and no superoxide dismutase, catalase or xanthine oxidase activity was found in the effluent of the continuous reactor. Accordingly, the protein content of this preparation was the same before and after 24 hr of operation.

8.4.5. Stabilization of Xanthine Oxidase by Entrapment in Gelatin

In order to check if the protection afforded by superoxide dismutase and catalase was indeed due to the catalytic action of these enzymes on O_2^- and H_2O_2 rather than a stabilizing effect due to protein–protein interactions, xanthine oxidase was coimmobilized with albumin. The operational stability of this albumin–xanthine oxidase preparation was slightly increased ($t_{1/2} = 2$ hr) compared with the xanthine oxidase preparation, but it was less than that of the three-enzyme system. To test further a possible effect of protein on the stability of xanthine oxidase, this enzyme was completely surrounded by protein by immobilization in glutaraldehyde crosslinked gelatin. This improved the operational stability considerably ($t_{1/2} = 26$ hr). No further improvement was achieved by coimmobilization of a mixture of xanthine oxidase, superoxide dismutase, and catalase in gelatin. When, on the other hand, the three enzymes were first incubated with glutaraldehyde to accomplish interenzyme crosslinking, the stability was further increased ($t_{1/2} = 50$ hr). This observation again suggests that O_2^- and H_2O_2 are involved in the inactivation, and that superoxide dismutase and catalase protect xanthine oxidase by catalyzing the destruction of these species. Operational half-lives of some preparations are collected in Table 8.14.

The gelatin system has, however, several disadvantages. To obtain a mechanically stable support, the amount of glutaraldehyde needed causes a severe inactivation of xanthine oxidase (retention of activity less than 10%). This observed low activity may also be the result of the poor flow properties of this material in a packed bed reactor. Another disadvantage is that a small amount of gelatin continuously leaks from the column.

The properties of the gelatin preparations can be greatly improved, however, by a slight modification in immobilization procedure (107). The modification consists of lyophilization of the enzyme–gelatin solution prior to the crosslinking with glutaraldehyde (see Section 8.4.6c). The resulting preparation has good mechanical stability and shows no protein leakage, and the yield of active immobilized xanthine oxidase is high and stable (Figure 8.44). The same simple procedure can be used for direct immobilization of unpurified xanthine oxidase in milk protein, with or without the addition of gelatin. Whole milk without any preceding treatment or skimmed milk can be used as starting solution. Consequently, the financial advantage is dramatic: the cost is reduced about 2000 times.

8.4.6. Application of Immobilized Xanthine Oxidase in Syntheses

To test their suitability for organic synthesis, several of the above-described immobilized xanthine oxidase preparations were employed for oxidation of various substrates on a preparative scale.

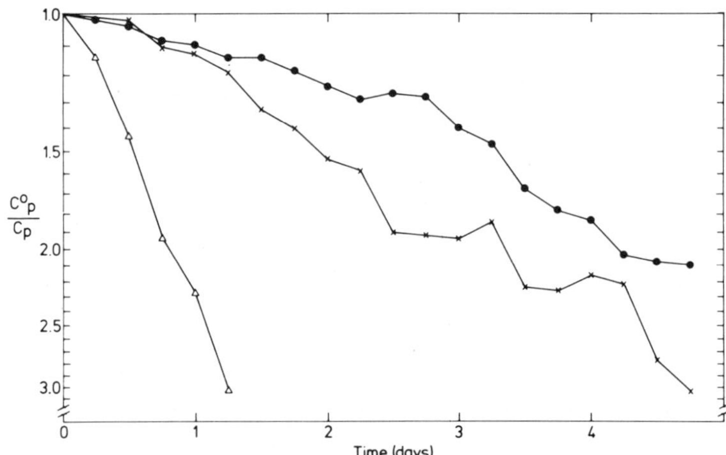

Fig. 8.44. Stability of immobilized milk xanthine oxidase under continuous operation at 5°C. Substrate: 200 μM xanthine in Tris-HCl buffer pH 8.2 (I = 0.02; 0.1 mM EDTA). Flow rate: 14 ml/hr. (●) Gelatin/fresh whole milk preparation; (X) Skimmed milk preparation; (Δ) Commercial xanthine oxidase adsorbed to *n*-octyl substituted sepharose 4B.

8.4.6a. Oxidation of 7-p-X-Phenylpteridin-4-one (X = H and OMe)

Immobilized xanthine oxidase was prepared by adsorption of xanthine oxidase to *n*-octyl-substituted Sepharose 4B, analogous to the procedure described by Hofstee and Otillio (77). For that, one gram of freeze-dried CNBr-activated Sepharose 4B (Pharmacia) is thoroughly washed with 200 mL 10^{-3} *M* HCl and then with distilled water. The washed material—or an equivalent amount of Sepharose 4B activated with CNBr as described Chapter 12, Section 12.3—is reacted with *n*-octylamine by suspending it in 60 mL of a solution consisting of 30 mL DMF, 5 mL distilled water, and 2mL *n*-octylamine, which after adjustment of the pH to 9 is brought to 60 mL with distilled water. The suspension is shaken or rotated at room temperature for 2–3 hr, and the resulting *n*-octyl-substituted Sepharose 4B is washed at least three times with DMF/distilled water (1 : 1), distilled water, ethylene glycol/distilled water (1 : 1)/NaCl (1 *M*), and finally, washed thoroughly with distilled water.

The *n*-octyl-Sepharose 4B is shaken or rotated with 30 mL of xanthine oxidase solution (0.2 mg protein · mL^{-1}; 0.03 *M* borate buffer pH 8.2) at room temperature for 2–3 hr, and then thoroughly washed with Tris/HCl buffer pH 8.1 (I = 0.01; 0.1 m*M* EDTA). Oxidation of the title compounds was performed (95) by using this adsorbed xanthine oxidase preparation either as a suspension in a stirred batch reactor that is continuously fed with concentrated substrate or by packing the enzyme system in a column through which a solution of the substrate is recirculated:

1. One-tenth mmol of the 7-(*p*-X-phenyl)pteridin-4-one is dissolved in a small excess of alkali and diluted to 50 mL with distilled water. This concentrated substrate solution is slowly added (in ~8 hr) to 100 mL of a stirred suspension of immobilized xanthine oxidase (corresponding to 10 mg of protein). The suspension is buffered at pH 9.1 (and adjusted as needed) and contains 0.1 m*M* EDTA. After the addition of all the substrate, the reaction is allowed to continue overnight. The suspension is filtered and the filtrate acidified with concentrated acetic acid. The product is filtered, washed with water, and dried.
2. Immobilized enzyme containing 10 mg of protein is packed in a small column. One-tenth mmol of the 7-(*p*-X-phenyl)pteridin-4-one dissolved in a small excess alkali and diluted with 1 L buffer pH 9.1, containing 0.1 m*M* EDTA, is slowly (2 mL · min^{-1}) passed through the column and recirculated overnight. The effluent is acidified (pH 3), evaporated to about 100 mL, and the precipitated lumazine is then filtered, washed with water, and dried.

The structures of the resulting 7-(*p*-X-phenyl)lumazines were proved by UV, IR, and MS-comparison with compounds prepared by an independent chemical synthesis. Method (1) has the advantage that, after the immobilized enzyme is removed by filtration, the reaction mixture can be immediately acidified without prior concentration. During the reaction the pH must be controlled. The recirculation reactor is easier to operate, but the solution of the product must be concentrated in order to obtain maximum yield. In both procedures the yields (>90%) are high (95).

8.4.6b. Preparative-scale Conversion of Xanthine

Freeze-dried CNBr-activated Sepharose 4B (1.2 g) was substituted with *n*-octylamine as described above and reacted with 18 mL of xanthine oxidase (0.2 mg protein $\cdot$ mL^{-1}) and 18 mL of catalase (0.2 mg protein $\cdot$ mL^{-1}) solution. The washed enzyme preparation was suspended in 50 mL buffer, pH 8.2, and 5 mL of the suspension was packed into a column and the half-life determined (4 hr). The rest of the suspension was rotated in a round-bottomed flask and continuously fed with concentrated xanthine solution (50 mL 2 m*M*; i.e., 15 mg xanthine) at such speed that no substrate accumulated (based on the half-life). This was regularly checked by taking small samples and measuring the extinction at 290 nm. After 8 hr of operation the addition was stopped and the reaction continued for another hr. The immobilized enzyme was filtered off, the filtrate (100 mL) acidified and concentrated to 10 mL. The precipitate formed upon cooling was filtered, washed with ethanol and ether, dried, weighed, and analyzed. The uric acid isolated after work-up was 14.2 mg, which represented an 84% yield (UV and mass spectra were identical with those of authentic material).

8.4.6c. Synthesis of 7-R-Lumazines (R = Me, Et, n*-Pro,* i*-Pro)*

The immobilized xanthine oxidase preparation used for the oxidation of 7-R-pteridin-4-ones into the corresponding lumazines was made by lyophilization of a mixture of gelatin and fresh whole milk as follows. Fifty grams of gelatin were dissolved in 100 mL of water of 100°C. After cooling to about 80°C, the solution was slowly mixed with 1 L of fresh whole milk under continuous stirring and the mixture lyophilized. The freeze-dried material was granulated to a suitable size (about 1 mm) and crosslinked in a glutaraldehyde/acetone/distilled-water mixture. For that, the granules were gradually added to a vigorously stirred solution consisting of 0.5 mL of 25% glutaraldehyde in water, 12 mL of distilled water, and 12.5 mL of acetone per g of dry material added. The stirring was continued for 30 min, all at room temperature, and the suspension then poured into a tubular reactor. The drained particles were at least 24 hr at 4 °C exhaustively washed by passing a large quantity of Tris/HCl

Table 8.15. Synthesis of 7-R-Lumazines from 7-R-Pteridin-4-ones by Means of Immobilized Xanthine Oxidase

R	Amount of Biocatalyst (L)	Elution Rate (mL/hr)	Substrate Converted After 1st, 2nd and 3rd Cycle (%)	Amount of Substrate Used (mg)
Me	0.5	100	75	134
			93	
			98	
Et	0.5	120	75	120
			97	
n-Pro	0.3	50	78	129
			91	
			100	
i-Pro	0.3	55	70	116
			90	
			95	

buffer pH 8 (I = 0.01; 0.1 m*M* EDTA) through the reactor. Finally, the reactor was drained and the remains of buffer washed away with demiwater of pH 8.

The substrate solution was prepared by dissolving 120–150 mg (Table 8.15) of the pertinent 7-R-pteridin-4-one in 1 L of distilled water, adjusting the pH to 8.2 with 1 *M* NaOH, and making up to an end-volume of 3 L with distilled water. After cooling to 4°C, this solution was fed to the reactor and recycled as many times (Table 8.15) as necessary to accomplish at least 95% conversion, which was checked spectrophotometrically. The effluent was then evaporated to 200 mL, acidified to pH 3 with 1 *M* HCl, and further evaporated to dryness. To the residue 200 mL of methanol and 5 g of SiO_2 were added, and the methanol subsequently evaporated. The mixture was suspended in ethyl acetate and applied to an SiO_2-column. The 7-R-lumazines were eluted with ethyl acetate/methanol (2:1) and the effluent evaporated to dryness. According to NMR, UV, IR and mass-spectroscopic analysis, the products were pure. The yields varied between 60 and 100 mg.

8.4.6d. Synthesis of 1-Methyluric Acid by Immobilized Bacterial Xanthine Oxidase

Arthrobacter cells having high xanthine oxidase activity were directly immobilized in gelatin as follows. Washed cells were suspended in a 10% gelatin solution (dry-weight ratio of cells and gelatin of 1) and the mixture lyophilized, granulated, and crosslinked as described above for milk xanthine oxidase.

Immobilized biocatalyst prepared in this fashion was packed in columns, thoroughly washed with Tris/HCl buffer of pH 7.9 (I = 0.01; 0.1 m*M* EDTA) at 4°C for about 24 hr, and used for activity, specificity, and stability assays, as well as for synthetic purposes. The results of the assays are shown in Tables 8.13 and 8.16 (58).

Activity and Stability. When xanthine was used as the substrate, uric acid was formed as intermediate product but eventually disappeared completely. As expected, if 1-methylxanthine was used as the substrate, the specificity of uricase prevented further oxidations and 1-methyluric acid was the end product. The activity at various temperatures is given in Table 8.16. From these data an activation energy of about 22 kJ mol^{-1} is calculated as being close to the one of immobilized xanthine dehydrogenase (108). The half-life ($t_{1/2}$) and the amount of substrate converted in the first half-time at these temperatures are also given in Table 8.16. At a temperature of about 20°C, immobilized Arthrobacter xanthine oxidase performs optimally.

Oxidation of 1-Methylxanthine on a Preparative Scale. Fifty mg of 1-methylxanthine were dissolved in a minimal amount of NaOH solution (4 *M*) and diluted to 300 mL with Tris-HCl buffer pH 7.9 (I = 0.01; 0.1 m*M* EDTA). This solution was recycled through a column containing immobilized Arthrobacter cells (270 mg of cell material) until complete conversion was reached (after about a week) and then acidified to pH about 5 with HCl (2 *M*). The solution was evaporated *in vacuo* to a small volume, cooled (0°C), filtered, and the residue extensively washed with distilled water. The yield of analytically pure 1-methyluric acid was 33 mg (60%). UV and mass spectra were identical with those of authentic material.

Table 8.16. Temperature Dependency of Xanthine Oxidase Activity and Stability of Immobilized Arthrobacter Cells, and Amount of 1-Methylxanthine Converted in the First Half-time per 100 mg of Column Material (dry weight)

T(°C)	$t_{1/2}$(days)	Total Conversion in $t_{1/2}$ in mg/100 mg of Column Material	Initial activity μmol/min per 100 mg of Column Material
4	9.5	14.3	6.4
17	8.4	26.1	9.9
28	4.2	16.5	16.3
36	2.4	10.4	20.4

8.4.7. Conclusion

The operational stability of immobilized xanthine oxidase is rather low, but the example in Section 8.4.6a shows that it can be used for synthetic purposes. It means, however, that a relatively large amount of enzyme is needed per mole of substrate oxidized (in the synthetic experiments 1 mole of enzyme converts about 3000 moles of substrate, assuming that all the protein applied is xanthine oxidase). This is not necessarily a disadvantage, since milk xanthine oxidase is commercially available at a relatively cheap price or can easily be isolated from milk. Also, stabilization by means of special immobilization techniques is possible, as shown in Section 8.4.4 and illustrated by means of the example in Section 8.4.6b. The gain in enzyme, however, is not much. Isolation and immobilization of chicken liver xanthine dehydrogenase also did not yield a significantly more stable preparation, despite the fact that no superoxide radicals and hydrogen peroxide are formed in this case (108). Because of the more inconvenient source of this enzyme and the need for the cofactor NAD^+, this system was not further studied. Direct immobilization of milk xanthine oxidase in whole-milk protein enriched with gelatin as described in Sections 8.4.5 and 8.4.6c yields a highly active and stable preparation with good mechanical stability and flow properties. The method is very simple and can be performed in any laboratory equipped with a freeze-drying apparatus. Moreover, the costs are very low. Consequently, as shown in Section 8.4.6c, this immobilized xanthine oxidase preparation is very suitable for application in organic synthesis. Since the described immobilization procedure is very mild it would appear to be applicable to many enzymes. Section 8.4.6c discussed the extension to the immobilization of whole cells, that is, Arthrobacter cells, containing xanthine oxidase activity. These cells were chosen because of their high specific activity, their substrate activation instead of inhibition, and their expected higher stability. Immobilized Arthrobacter xanthine oxidase is indeed highly active, more stable, and has a different substrate specificity than milk xanthine oxidase. Therefore, a wider spectrum of substrates can be conveniently and efficiently oxidized.

All the collected products were of high purity, showing that reactions executed with immobilized enzymes are not only clean in theory. This is nicely illustrated in Figure 8.45, which shows UV spectra of reaction mixtures at various stages of completion. The occurrence of several sharp isosbestic points strongly indicates the absence of by-product formation in the reaction mixtures.

The conclusion is that xanthine oxidase, either from whole milk or Arthrobacter, when immobilized in glutaraldehyde-crosslinked gelatin, can

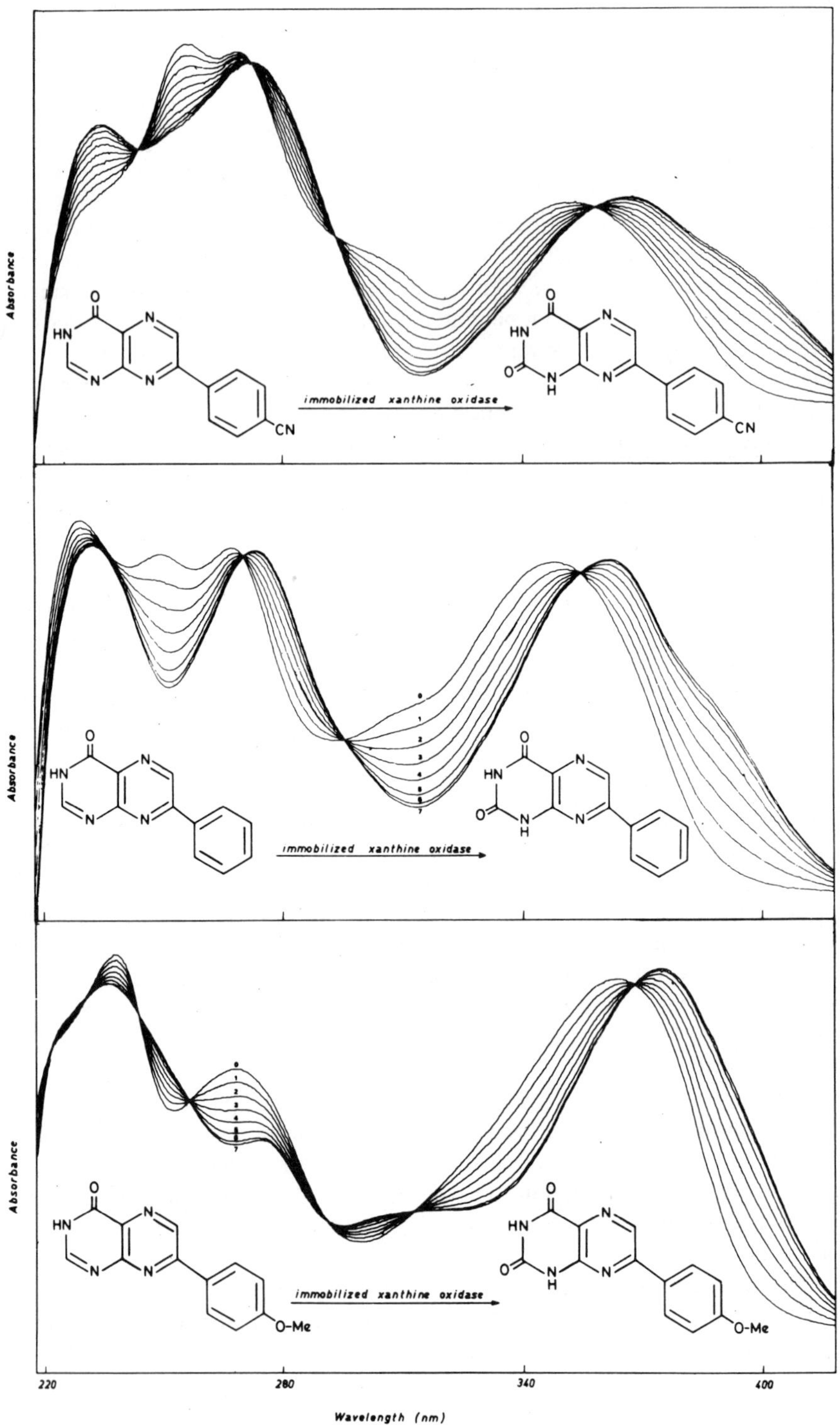

Fig. 8.45. Oxidation of 7-(*p*-X-phenyl)pteridin-4-ones by immobilized xanthine oxidase. Spectra were continuously scanned during the reaction (cycle time: 1 min).

be conveniently used for the oxidation of a wide range of substrates in a continuous fashion.

REFERENCES

1. J. B. Jones, C. J. Sih, and D. Perlman, *Applications of Biochemical Systems in Organic Chemistry,* Parts I and II, Wiley, New York (1979).
2. J. B. Jones in *Immobilized Enzymes,* Vol. 44, *Methods in Enzymology* (K. Mosbach, ed.), Academic Press, New York (1976), pp. 831–844.
3. I. Chibata, *Immobilized Enzymes: Research and Development,* Wiley, New York (1978).
4. J. Bailey, *Chem. Eng. Sci.* **35,** 1854–1886 (1980).
5. W. Marconi, in *Biotechnology,* Dechema Monographs Nos. 1693–1703, Vol. 82, Verlag Chemie, Weinheim and New York (1978), pp. 88–141.
6. M. D. Trevan, *Immobilized Enzymes: An Introduction and Applications in Biotechnology,* Wiley, Chichester and New York (1980).
7. A. M. Klibanov, *Anal. Biochem.* **93,** 1–25 (1979).
8. Y. Levin, M. Pecht, L. Goldstein, and E. Katchalski, *Biochemistry* **3,** 1905–1913 (1964).
9. K. Martinek, A. M. Klibanov, V. S. Goldmacher, A. V. Tchernysheva, V. V. Mozhaev, I. V. Berezin, and B. O. Glotov, *Biochim. Biophys. Acta* **485,** 13–28 (1977).
10. A. P. Corfield, R. Schauer, and M. Wember, *Biochem. J.* **177,** 1–7 (1979).
11. A. J. L. Cooper and A. S. Gelbard, *Anal. Biochem.* **111,** 142–48 (1981).
12. L. Goldstein, Y. Levin, and E. Katchalski, *Biochemistry* **3,** 1913–1919 (1964).
13. A. M. Klibanov, G. P. Samokhin, K. Martinek, and J. V. Berezin, *Biotechnol. Bioeng.* **19,** 1351–1361 (1977).
14. K. Martinek, A. N. Semenov, and J. V. Berezin, *Biotechnol. Bioeng.* **23,** 1115–1120 (1981).
15. K. Martinek, A. N. Semenov, and J. V. Berezin, *Biochim. Biophys. Acta* **658,** 76–89 (1981).
16. K. Martinek and A. N. Semenov, *Biochim. Biophys. Acta* **658,** 90–101 (1981).
17. K. Martinek, A. V. Levashov, N. L. Klyachko, V. J. Pantin, and J. V. Berezin, *Biochim. Biophys. Acta* **657,** 277–294 (1981).
18. R. G. Ingalls, R. G. Squires, and L. G. Butler, *Biotechnol. Bioeng.* **17,** 1627–1637 (1975).
19. G. Carrea, F. Colombi, G. Mazzola, P. Cremonesi, and E. Antonini, *Biotechnol. Bioeng.* **21,** 39–48 (1979).
20. K. J. Skinner, *Chem. Eng. News,* Aug. 18, 1975, 22–41.
21. G. M. A. van Beynum, in *Biotechnology: A Hidden Past, a Shining Future,* Proceedings of the 13th International TNO Conference, Rotterdam, 1980, pp. 59–71.

22. R. Olivieri, E. Fascetti, L. Angelini, and L. Degen, *Enzyme Microb. Technol.* **1,** 201–204 (1979).
23. I. Takata, K. Yamamoto, T. Tosa, and I. Chibata, *Enzyme Microb. Technol.* **2,** 30–36 (1980).
24. W. Becker, H. Freund, and E. Pfeil, *Angew. Chem. Int. Ed.* **4,** 1079–1080 (1965).
25. W. Tischer, W. Tiemeyer, and H. Simon, *Biochimie* **62,** 331-339 (1980).
26. K. Buchholz, *Characterization of Immobilized Biocatalysts,* Dechema Monographs Nos. 1724–1731, Vol. 84, Verlag Chemie, Weinheim and New York (1979), pp. 258–262.
27. P. W. Carr and L. D. Bowers, *Immobilized Enzymes in Analytical and Clinical Chemistry: Fundamentals and Applications,* Wiley, New York (1980).
28. M. Nelboeck, in *Biotechnological Applications of Proteins and Enzymes,* (Z. Bohak and N. Sharon, Eds.) Academic Press, New York (1977), pp. 279–301.
29. W. R. Vieth and K. Venkatasubramanian, *Chem. Technol.* Jan. 1974, 47–55.
30. D. L. Eaton, in *Immobilized Biochemicals and Affinity Chromatography,* (R. B. Dunlap, Ed.), Plenum Press, New York (1974), pp. 241–258.
31. C. Araki, *Bull. Chem. Soc. Jpn* **29,** 543–544 (1956).
32. M. Duckworth and W. Yaphe, *Carbohydr. Res.* **16,** 189–199 (1971).
33. K. B. Guiseley, *Carbohydr. Res.* **13,** 247–256 (1970).
34. Pharmacia, *Agarose for Electrophoresis and Immunological Techniques,* April 1977, Rahms 1, Lund, Sweden.
35. S. Arnott, A. Fulmer, W. E. Scott, J. C. M. Dea, R. Moorhouse, and D. A. Rees, *J. Mol. Biol.* **90,** 269 (1974).
36. K. B. Guiseley and D. W. Renn, *Agarose: Purification, Properties, and Biomedical Applications,* Marine Colloids, Rockland (Maine), 1975; and references cited therein.
37. J. Porath and R. Axén, in *Immobilized Enzymes,* Vol. 44, *Methods in Enzymology* (K. Mosbach, Ed.), Academic Press, New York (1976), pp. 19–45.
38. P. Cuatrecasas and C. B. Anfinsen, in *Enzyme Purification and Related Techniques, Methods in Enzymology,* Vol. 2 (W. B. Jakoby, Ed.), Academic Press, New York (1971), pp. 345–378.
39. J. Turková, *Affinity Chromatography,* Vol. 12, *J. Chrom. Library,* Elsevier, Amsterdam (1978).
40. T. C. J. Gribneau, *Coupling of Effector Molecules to Solid Supports* (Ph. D. Thesis), Van Mameren B. V., Nijmegen (1977).
41. W. Pilnik, F. Voragen, H. Neukom, and E. Nittner, in *'Ullmonns Encyklopädie der technischen Chemie,* Vol. 19, Verlag Chemie, Weinheim (1980), pp. 233–263.
42. A. Haug, B. Larsen, and O. Smidsrod, *Carbohydr. Res.* **32,** 217–225 (1974); and references cited therein.
43. M. Kierstan and C. Bucke, *Biotechnol. Bioeng.* **19,** 387–397 (1977).
44. J. Klein and F. Wagner, in *Biotechnology,* Dechema Monographs Nos. 1693–1703, Vol. 82, Verlag Chemie, Weinheim and New York (1978), pp. 142–164.
45. I. A. Veliky and R. E. Williams, *Biotechnol. Lett.* **3,** 275–280 (1981).
46. F. Paul and P. M. Vignais, *Enzyme Microb. Technol.* **2,** 281–287 (1980).

47. I. Takata, T. Tosa, and I. Chibata, *J Solid-Phase Biochem.* **2,** 225–236 (1977).
48. P. S. J. Cheetham, K. V. Blunt, and C. Bucke, *Biotechnol. Bioeng.* **21,** 2155–2168 (1979).
49. P. S. J. Cheetham, *Enzyme Microb. Technol.* **1,** 183–188 (1979).
50. T. Tosa, T. Sato, T. Mori, K. Yamamoto, I. Takata, Y. Nishida, and I. Chibata, *Biotechnol. Bioeng.* **21,** 1697–1709 (1979).
51. A. J. Stamm, *Wood and Cellulose Science,* Ronald Press, New York (1964).
52. M. D. Lilly, in *Immobilized Enzymes,* Vol. 44, *Methods in Enzymology* (K. Mosbach, Ed.), Academic Press, New York (1976), pp. 46–53.
53. W. H. Scouten, *Affinity Chromatography: Bioselective Adsorption on Inert Matrices,* Vol. 59, *Chemical Analysis,* Wiley, New York (1981), p. 36.
54. K. Mosbach, in *Applications of Biochemical Systems in Organic Chemistry,* Part II, (J. B. Jones, C. J. Sih, and D. Perlman, Eds.), Wiley, New York (1976), pp. 969–994.
55. W. R. Vieth and K. Venkatasubramanian, *Chem. Technol.* Nov. 1973, 677–684.
56. G. Broun, D. Thomas, G. Gellf, D. Somurado, A. M. Berjonneau, and C. Guillon, *Biotechnol. Bioeng.* **15,** 359–375 (1973).
57. J. C. Johnson, *Industrial Enzymes,* NDC, Park Ridge, N.J. (1977), p. 47.
58. J. Tramper, H. C. Van der Plas, A. Van der Kaaden, F. Müller, and W. J. Middelhoven, *Biotechnol. Lett.* **1,** 391–396 (1979).
59. R. Mosbach, A. C. Koch-Schmidt, and K. Mosbach, in *Immobilized Enzymes,* Vol. 44, *Methods in Enzymology* (K. Mosbach, Ed.), Academic Press, New York (1976), pp. 53–65.
60. G. R. Lowe and P. D. G. Dean, *Affinity Chromatography,* Wiley, New York (1974).
61. W. A. Beverloo and S. Bruin, in *New Processes of Waste Water Treatment and Recovery* (G. Mattock, Ed.), Ellis Horwood, Chichester (1978), pp. 307–323.
62. J. S. Mattson and H. B. Mark, *Activated Carbon: Surface Chemistry and Adsorption from Solution,* Marcel Dekker, New York (1971).
63. R. L. Culp, G. M. Wesner, and G. L. Culp, *Handbook of Advanced Wastewater Treatment,* Van Nostrand Reinhold, New York (1978), pp. 166–248.
64. Y. K. Cho and J. E. Bailey, *Biotechnol. Bioeng.* **21,** 461–476 (1979).
65. Y. K. Cho and J. E. Bailey, *Biotechnol. Bioeng.* **19,** 769–775 (1977).
66. C. Kent, A. Rosevaer, and A. R. Thomson, in *Topics in Enzyme and Fermentation Technology 2* (A. Wiseman, Ed.) Ellis Horwood, Chichester (1978), pp. 12–119.
67. A. M. Filbert, in *Immobilized Enzymes for Industrial Reactors* (R. A. Messing, Ed.), Academic Press, New York (1975), pp. 39–61.
68. G. P. Royer, *Chem. Technol.,* Nov. 1974, 694–700.
69. L. Goldstein and G. Manecke, in *Immobilized Enzyme Principles,* Vol. 1, *Applied Biochemistry and Bioengineering* (L. B. Wingard, E. Katchalski-Katzir, and L. Goldstein, Eds.), Academic Press, New York (1976), pp. 23–127.
70. K. Mosbach, Ed., *Immobilized Enzymes,* Vol. 44, *Methods in Enzymology* Academic Press, New York (1976).
71. R. A. Messing, *Immobilized Enzymes for Industrial Reactors,* Academic Press, New York (1975).
72. O. R. Zaborsky, *Immobilized Enzymes,* CRC Press, Cleveland (1972).

73. H. H. Weetall, *Immobilized Enzymes, Antigens, Antibodies, and Peptides,* Marcel Dekker, New York (1975).
74. J. Turková, *Affinity Chromatography,* Vol. 12, *J. Chrom. Library,* Elsevier, Amsterdam (1978).
75. B. Solomon and Y. Levin, *Biotechnol. Bioeng.* **16,** 1161–1177 (1974).
76. M. Wilchek, T. Oka, and Y. J. Topper, *Proc. Nat. Acad. Sci. USA,* **72,** 1055–1058 (1975).
77. B. H. J. Hofstee and N. F. Otillio, *Biochem. Biophys. Res. Commun.* **53,** 1137–1144 (1973).
78. F. M. Richards and J. R. Knowles, *J. Mol. Biol.* **37,** 231–233 (1968).
79. J. Carlsson, R. Axén, and T. Unge, *Eur. J. Biochem.* **59,** 569–572 (1975).
80. D. Dinelli, *Process Biochem.* **7,** 9–12 (1972).
81. H. C. Van der Plas, *Acc. Chem. Res.* **11,** 462–468 (1978).
82. H. C. Van der Plas, *Heterocycles* **9,** 33–78 (1978).
83. F. Müller, *Chem. Weekbl.* **71**(8), 25–26 (1975).
84. C. J. Suckling and K. E. Suckling, *Chem. Soc. Rev.* **3,** 387–406 (1974).
85. M. P. Coughlan and D. B. Johnson, *Biochim. Biophys. Acta* **302,** 200–204 (1973).
86. G. Broun, D. Thomas, G. Gellf, D. Domurado, A. M. Berjonneau, and C. Guillon, Biotechnol. Bioeng. **15,** 359–375 (1973).
87. R. C. Bray, in *The Enzymes,* Vol. 12 (P. D. Boyer, Ed.), Academic Press, New York (1975), pp. 299–419.
88. V. Massey, in *Iron–Sulfur Proteins,* Vol. 1 (W. Lovenberg, Ed.), Academic Press, New York (1973), pp. 301–360.
89. R. C. Bray, in *The Enzymes,* Vol. 7 (P. D. Boyer, H. Lardy, and K. Myrbäck, Eds.), Academic Press, New York (1963), pp. 533–556.
90. J. S. Olson, D. P. Ballou, G. Palmer, and V. Massey, *J. Biol. Chem.* **249,** 4363–4382 (1974).
91. C. A. Woolfolk and J. S. Downard, *J. Bacteriol.* **135,** 422–428 (1978).
92. G. D. Vogels and C. Van der Drift, *Bacteriolog. Rev.* **40,** 403–468 (1976).
93. J. Tramper, F. Müller, and H. C. van der Plas, *Biotechnol. Bioeng.* **20,** 1507–1522 (1978).
94. J. Tramper, H. C. van der Plas, and W. E. Hennink, *Abstracts,* 2nd European Congress of Biotechnology, Eastbourne, 1981, p. 133; *J. Applied Biochem.,* in press.
95. J. Tramper, A. Nagel, H. C. van der Plas, and F. Müller, *Rec. Trav. Chim.* (Pays Bas) **98,** 224–231 (1979).
96. F. Bergmann, L. Levene, and J. Tamir, in *Chemistry and Biology of Pleridines* (W. Pfleiderer, Ed.), de Gruyter, Berlin (1975), p. 603.
97. B. R. Baker, W. F. Wood, and J. A. Kozma, *J. Med. Chem.* **11,** 661–666 (1968).
98. F. Bergmann and L. Levene, *Biochim. Biophys. Acta* **481,** 359–363 (1977).
99. F. Bergmann, L. Levene, and H. Govrin, *Biochim. Biophys. Acta* **484,** 275–289 (1977).
100. J. Tramper, H. S. D. Naeff, H. C. van der Plas, and F. Müller, *Advances in Biotechnology,* Vol. III, Fermentation Products, Proceedings of the Sixth

International Fermentation Symposium, London, Canada 1980 (M. Moo-Young, C. Vezina, and K. Singh, Eds.) Pergamon Press, Toronto (1981), pp. 383–387.

101. G. I. Tesser, H.-U. Fisch and R. Schwitzer, *Helv. Chim. Acta* **57,** 1718–1730 (1974).
102. J. W. Peters and C. S. Foole, *J. Am. Chem. Soc.* **98,** 873–875 (1976).
103. I. Fridovich, *Acc. Chem. Res.* **5,** 321–326 (1972).
104. F. Bergel and R. C. Bray, *Biochem. J.* **73,** 182–192 (1959).
105. J. Bannister, W. Bannister, and E. Wood, *Eur. J. Biochem.* **18,** 178–186 (1971).
106. D. G. Priest and J. R. Fisher, *Eur. J. Biochem.* **10,** 439–444 (1969).
107. J. Tramper, H. C. van der Plas, and F. Müller, *Biotechnol. Lett.* **1,** 133–138 (1979).
108. J. Tramper, S. A. G. F. Angelino, F. Müller, and H. C. van der Plas, *Biotechnol. Bioeng.* **21,** 1767–1786 (1979).

CHAPTER

9

IMMOBILIZED ENZYME ELECTRODE PROBES

GEORGE G. GUILBAULT

Department of Chemistry
University of New Orleans
New Orleans, Louisiana 70148

The role of enzymes in biological systems is one of the most interesting and promising areas of science at present under investigation. Enzymes are the regulators of many biological systems upon which depend the very existence of life as we know it. It is not surprising, then, that there is a need for methods to determine the activity of these extremely important agents. Enzymes are proteins that catalyze a given chemical conversion. As catalysts, they are effective at very low concentrations, they enhance the rate of a reversible reaction without affecting the equilibrium of that reaction, and are unchanged in the reaction. The term "unchanged" is naive. Since enzymes bind the substrate that is undergoing a chemical change they are altered. The enzymes, however, subsequently release the modified substrate and return to their original form. As proteins, the activity of enzymes is influenced by those factors which are critical in biological systems: temperature, pH, the presence or absence of inhibitors, ionic strength, and the concentration of substrate are factors that govern the activity of an enzyme.

9.1. USE OF ELECTRODES TO MEASURE ENZYME REACTIONS

The determination of enzyme activity is of great importance in clinical studies and in the area of biochemistry. Electrochemical methods have been applied for these assays and probably the most common electrochemical method for the assay of enzymes that produce or consume an acid is to follow the pH change of the reaction mixture as a measure of the activity of the enzyme. This method is not usually employed directly, since the activity of a given enzyme is affected by changes in pH. Instead, a "pH stat" method is used in which the pH of the assay mixture is maintained by the addition of an acid or base. The rate of the addition of reagent gives the reaction velocity.

The activity of an enzyme in a system in which oxygen is consumed can be determined using an oxygen electrode. The electrode is a gold cathode separated by an epoxy casting from a silver anode. The inner sensor body is housed in a plastic casing and comes into contact with the assay solution only through an intervening membrane. When oxygen diffuses through the membrane it is electrically reduced at the cathode by an applied potential of 0.8 V. This reaction causes a current to flow between the anode and cathode that is proportional to the partial pressure of oxygen in the sample. The rate of uptake of oxygen can be related to the activity of the enzyme or the concentration of substrate in the assay mixture. Good correlation between glucose values determined in blood by a measurement of oxygen uptake compared with those obtained by standard chemical tests was found by Kadish and Hall (1) and by Makino and Koono (2).

Ion-selective electrodes have been used to determine the activity of rhodanase and cholinesterase. In the rhodanase system, a cyanide-selective electrode was used to follow the decrease of cyanide ion during the reaction

$$CN^- + S_2O_3^{2-} \xrightarrow{\text{rhodanase}} SCN^- + SO_3^{2-} \tag{1}$$

which is catalyzed by rhodanase (3). Results obtained by this method are comparable to those obtained by spectrophotometric procedures. The method is easily adapted to automated systems. The cholinesterase assay was performed using a sulfide-selective electrode to monitor the amount of thiocholine released under the influence of cholinesterase:

$$\text{Acetylthiocholine} + H_2O \xrightarrow{\text{cholinesterase}} \text{thiocholine} + CH_3COOH \tag{2}$$

The amount of thiocholine released is proportional to the activity of cholinesterase (4). Llenado and Rechnitz (5) used systems very similar to those described above for the assay of β-glucosidase, rhodanase, and glucose oxidase. An ion-selective electrode for cyanide was used to follow the production of cyanide ion in the assay of β-glucosidase (3).

$$\text{Amygdalin} + H_2O \xrightarrow{\beta\text{-glucosidase}} \text{benzaldehyde} + \text{glucoside} + HCN \tag{3}$$

An iodide-selective electrode was used with the glucose oxidase assay:

$$\beta\text{-D-glucose} + H_2O + O_2 \xrightarrow[\text{oxidase}]{\text{glucose}} \text{D-gluconic acid} + H_2O_2 \tag{4}$$

$$H_2O_2 + 2H^+ + 2I^- \xrightarrow{\text{Mo(VI)}} 2H_2O + I_2 \tag{5}$$

to measure the decrease in iodide concentration resulting from oxidation of iodide to iodine by hydrogen peroxide.

9.2. GENERAL CHARACTERISTICS OF IMMOBILIZED ENZYMES

In addition to their application in the determination of enzyme activity, electrochemical methods have been combined with enzymatic systems to provide highly selective and sensitive procedures for the determination of the concentration of a given substrate. This is possible because under controlled conditions, the rate of an enzyme catalyzed reaction is proportional to the concentration of substrate. The concept of using an enzyme as a reagent in conjunction with an electrode was introduced by Clark and Lyons (6). The first working enzyme electrode was reported by Updike and Hicks (7); they used glucose oxidase immobilized in a gel that was layered over a polarographic oxygen electrode to measure the concentration of glucose in biological solutions and in tissues. The use of immobilized enzymes is a more recent development, and it overcomes several of the objections that were associated with the use of enzymes as reagents. The high cost of enzymes makes their routine use impractical and, moreover, the activity of a particular enzyme preparation often varies from manufacturer to manufacturer and from batch to batch. However, by immobilizing the enzyme, the amount of the material required to perform routine analysis is greatly reduced and frequent assay of the enzyme preparation becomes unnecessary. Furthermore, the stability of the enzyme is often improved when it is incorporated in a suitable gel matrix. An electrode for the determination of glucose prepared by covering a platinum electrode with chemically bound glucose oxidase has been used for more than 300 days (8).

Two general methods are used to immobilized an enzyme. First, the enzyme may be modified by the introduction of insolubilizing groups (see Chapter 7). This technique, which results in a chemically bonded enzyme, is in practice sometimes difficult to achieve because the insolubilizing groups often attach to the active site, thus destroying the activity of the enzyme. Second, the enzyme may be physically entrapped in an inert matrix such as starch or polyacrylamide gels. The latter procedure is often faster and simpler than chemical immobilization methods. The technique of chemical bonding with the bifunctional reagent glutaraldehyde is very simple—the enzyme is simply treated with the aldehyde and an inert support (albumin glass beads, etc.); a rigid layer of bound enzyme results, which is quite stable for several months and thousands of assays. The immobilized enzyme is then placed over the sensor of an electrode that is sensitive to the product of the enzyme–substrate reaction. When the enzyme electrode is placed in a solution that contains the substrate for which the electrode is designed, the substrate diffuses into the enzyme layer where the enzyme-catalyzed reaction takes place, producing an ion that is detected by the electrode. Excellent chemical analysis can be performed with enzymes, and there are many advantages of immobilized enzymes in analyses using electrochemical probes or other methods of analysis. One advantage of the immobilized enzyme over free enzymes in solution is a pH shift: by choosing the right support for immobilization, the pH optimum can be shifted to that region at which one wants to make a measurement. For example an enzyme may have a narrow pH range of say, 6–8; this can be shifted upon immobilization down toward the acidic pH values or, conversely, up toward basic pH's. Such an immobilized enzyme may also be much more stable than the free enzyme. In work at Edgewood Arsenal, Maryland, we actually heated our immobilized enzymes to 150°F and brought them back down to room temperature with very little loss of activity. No soluble enzyme could be treated in this fashion.

One advantage of enzyme immobilization often overlooked is that better selectivity can be realized with the enzyme immobilized; this insolubilized reagent becomes much more selective for an inhibitor, and only the most powerful inhibitor can actually bind to the enzyme. We demonstrated this several years ago in an immobilized cholinesterase alarm for the assay of organophosphorus compounds in air and water. No other common cholinesterase inhibitors disturbed the alarm; it responded only to organophosphorus compounds. This increased selectivity was no doubt due to a shielding of the active site of the enzyme by the polymeric support such that only the strongest inhibitors had access to the active site of the immobilized enzyme.

In 1961 at Edgewood Arsenal, I first experimented with some soluble enzymes, such as glucose oxidase, and developed an electrochemical assay for glucose. This led to the use of an immobilized enzyme coupled with a commercially available ion-selective electrode sensor to form a single self-contained sensor that could be used to measure either organic or inorganic compounds that are primary or secondary substrates for the immobilized enzyme. The base sensor can be glass, that is, the cation response can be measured (the ammonium ion, for example) or the pH change in a penicillin electrode can be measured, as done by Mosbach and Papariello and others. Or a gas membrane can be used as the base sensor, such as the ammonia or the CO_2 membrane. Also possible are the polarographic sensors that measure peroxide or oxygen, or any of the solid membrane electrodes, such as the cyanide electrode. For example, the enzyme can be placed on top of a flat glass electrode sensor and a membrane put over the outside of this sensor to hold the enzyme in and keep out undesired materials such as catalase and bacteria; this protects the enzyme from bacterial spoilage, which is one of the main reasons for loss of enzyme activity.

With potentiometric and amperometric devices, we can measure the response either by a steady-state (i.e., equilibrium) method, which measures millivolts or microamperes, or by a rate method, which senses the change in millivolts or microamperes per minute. Measurements of substrate concentration can be performed by either a steady-state or a rate method, but measurements of enzyme activity must be done by a rate method. This is a point often hazy in the literature and one can find many reports of measurement of enzyme activity by steady-state methods. Such measurements are contrary to the basic definition of enzyme activity. Enzymes are catalysts and must be measured by a rate method, although one may use either an interrupted or a continuous measurement of rate.

Table 9.1 is a compilation of some examples of enzyme electrodes (this list is not by any means complete). It is an expansion of a table published in a recent book of mine, *Handbook of Enzymatic Analysis* (Dekker, New York, 1977). In this table are listed the enzymes that act on these various materials, and some of the base sensors that might be useful. A typical example is glucose, which can be assayed with glucose oxidase:

$$O_2 + \text{glucose} \xrightarrow{\text{oxidase}} H_2O_2 + \text{gluconic acid} \quad (6)$$

One can measure the uptake of oxygen with a gas membrane electrode, a technique pioneered by Clark (6) and perfected by Hicks and Updike (7), or one can record the peroxide or oxygen concentrations polarographically. There are other ways to follow this reaction: one can measure the gluconic acid by a pH change, as Mosbach showed very nicely (at low buffer capacity), or use an iodide membrane (the latter is much less recommended). The point is that there are many ways to measure a particular substrate, and one should choose the one best for the application. For example, one would not choose to measure urea in biological fluid with an ammonium cation electrode because of the interference of potassium and sodium. One would prefer an ammonia gas membrane electrode, in which there is no interference from sodium and potassium.

An enzyme electrode operates via a five-step process: (i) the substrate must be transported to the surface of the electrode. (ii) The substrate must diffuse through the membrane to the active site. (iii) Reaction occurs at the active site. (iv) Product formed in the enzymatic reaction is transported through the membrane to the surface of the electrode. (v) Product is measured at the electrode surface. The first step, transport of the substrate is most critically dependent on the stirring rate of the solution, so that rapid stirring will bring the substrate very rapidly to the electrode surface. If the membrane is kept very thin, using highly active enzyme, then steps 2 and 4 are eliminated or minimized; since step 3 is very fast, the response of an enzyme electrode should theoretically approach the response time of the base sensor. Many researchers have shown with experimental data that one can approach this behavior by using a thin membrane and rapid stirring. For example, Figure 9.1 shows a comparison of the characteristics of the amygdalin electrode using β-glucosidase; on the left-hand side are data obtained by Mascini

Table 9.1. Various Electrodes and Their Characteristics

Type	Enzyme	Sensor	Immobilization[a]	Stability	Response Time	Amount of Enzyme (U)	Range (mol/1)[b]
1. Urea	Urease (EC 3.5.1.5)	Cation	Physical	3 weeks	30 sec–1 min	25	10^{-2}–5×10^{-5}
		Cation	Physical	2 weeks	1–2 min	75	10^{-2}–10^{-4}
		Cation	Chemical	>4 months	1–2 min	10	10^{-2}–10^{-4}
		pH	Physical	3 weeks	5–10 min	~100	5×10^{-3}–5×10^{-5}
		Gas(NH_3)	Chemical	>4 months	2–4 min	10	5×10^{-2}–5×10^{-5}
		Gas(NH_3)	Chemical	20 days	1–4 min	0.5	10^{-2}–10^{-4}
		Gas(CO_2)	Physical	3 weeks	1–2 min	25	10^{-2}–10^{-4}
2. Glucose	Glucose oxidase (EC 1.1.3.4)	pH	Soluble	1 week	5–10 min	~100	10^{-1}–10^{-3}
		Pt(H_2O_2)	Physical	6 months	12 sec kinetic[c]	10	2×10^{-2}–10^{-4}
		Pt(H_2O_2)	Chemical	>14 months	1 min, steady state	10	2×10^{-2}–10^{-4}
		Pt(H_2O_2)	Soluble		1–2 min	~10	10^{-2}–10^{-4}
		Pt(quinone)	Soluble	<1 week[d]	3–10 min	10	2×10^{-2}–10^{-3}
		Pt(O_2)	Chemical	>4 months	1 min	10	10^{-1}–10^{-5}
		I^-	Chemical	>1 month	2–8 min	10	10^{-3}–10^{-4}
		Gas(O_2)	Physical	3 weeks	2–5 min	20	10^{-2}–10^{-4}
	Glucose oxidase (EC 1.1.3.4) and Catalase (EC 1.11.1.6)	Gas(O_2)	Chemical	>3 weeks	2–5 min	10	2×10^{-2}–10^{-4}
3. L-Amino acids (general)[e]	L-AA oxidase (EC 1.4.3.2)	Pt(H_2O_2)	Chemical	4–6 months	12 sec kinetic[c]	10	10^{-3}–10^{-5}
		Gas(O_2)	Chemical		2 min	10	10^{-2}–10^{-4}
		Pt(O_2)	Chemical	>4 months	1 min	10	10^{-2}–10^{-4}
		Cation	Physical	2 weeks	1–2 min	10	10^{-2}–10^{-4}
		NH_4^+	Chemical	>1 month	1–3 min	10	10^{-2}–10^{-4}
		I^-	Chemical	>1 month	1–3 min	10	10^{-3}–10^{-4}
L-Tyrosine	L-Tyrosine decarboxylase (EC 1.1.25)	Gas(CO_2)	Physical	3 weeks	1–2 min	25	10^{-1}–10^{-4}

L-Glutamine	Glutaminase (EC 3.5.1.2)	Cation	Soluble	2 days[d]	1 min	50	10^{-1}–10^{-4}
L-Glutamic acid	Glutamate dehydrogenase (EC 1.4.1.3)	Cation	Soluble	2 days[d]	1 min	50	10^{-1}–10^{-4}
L-Asparagine	Asparaginase (EC 3.5.1.1)	Cation	Physical	1 month	1 min	50	10^{-2}–5×10^{-5}
4. D-Amino acids (general)[f]	D-AA oxidase (EC 1.4.3.3)	Cation	Physical	1 month	1 min	50	10^{-2}–5×10^{-5}
5. Lactic acid	Lactate dehydrogenase (EC 1.1.1.27)	$Pt[Fe(CN)_6^{4-}]$	Soluble	<1 week	3–10 min	2	2×10^{-3}–10^{-4}
6. Succinic acid	Succinate dehydrogenase (EC 1.3.99.1)	$Pt(O_2)$	Physical	<1 week	1 min	10	10^{-2}–10^{-4}
7. Acetic, formic acids	Alcohol oxidase (EC 1.1.3.13)	$Pt(O_2)$	Chemical	>4 months	30 sec	10	10^{-1}–10^{-4}
8. Alcohols[g]	Alcohol oxidase (EC 1.1.3.13)	$Pt(H_2O_2)$	Soluble	1 week	12 sec kinetic[c]	10	0.5–100 mg%
		$Pt(H_2O_2)$	Soluble	1 day[d]	1 min	~1	0.5–50 mg%
		$Pt(O_2)$	Chemical	>4 months	30 sec	10	0.5–100 mg%
9. Penicillin	Penicillinase (EC 3.5.2.6)	pH	Physical	1–2 weeks	0.5–2 min	400	10^{-2}–10^{-4}
			Soluble	3 weeks	2 min	~1000	10^{-2}–10^{-4}
10. Uric acid	Uricase (EC 1.7.3.3)	$Pt(O_2)$	Chemical	4 months	30 sec	~10	10^{-2}–10^{-4}
11. Amygdalin	β-Glucosidase (EC 3.2.1.21)	CN^-	Physical	3 days[g]	10–20 min	100	10^{-2}–10^{-5}
12. Cholesterol	Cholesterol oxidase (EC 1.1.3.7)	$Pt(H_2O_2)$	Soluble		2 min		10^{-2}–10^{-4}
13. Phosphate	Phosphatase/ glucose oxidase (EC 3.1.3.1/ 1.1.3.4)	$Pt(O_2)$	Chemical	4 months	1 min	10 each	10^{-2}–10^{-4}
14. Nitrate	Nitrate reductase/ nitrite reductase (EC 1.9.6.1/ 1.6.6.4)	NH_4^+	Soluble		2–3 min	10	10^{-2}–10^{-4}

Table 9.1. *(continued)*

Type	Enzyme	Sensor	Immobilization[a]	Stability	Response Time	Amount of Enzyme (U)	Range (mol/1)[b]
15. Nitrite	Nitrite reductase (EC 1.6.6.4)	NH_3(gas)	Chemical	3–4 months	2–3 min	10	5×10^{-2}–10^{-4}
16. Sulfate	Aryl sulphatase (EC 3.1.6.1)	Pt	Chemical	1 month	1 min	10	10^{-1}–10^{-4}

[a] "Physical" refers to polyacrylamide gel entrapment in all cases; "chemical" is attachment chemically to glutaraldehyde with albumin, to polyacrylic acid, or to acrylamide, followed by physical entrapment.
[b] Analytically useful range, either linear or with reasonable change if curvature is observed.
[c] "Kinetic," rate of change in current measured after 12 s; "steady state," current reaches a maximum in 1 min.
[d] Preparation lacks stability as evidence by constant decrease in signal each day.
[e] Electrode responds to L-cysteine, L-leucine, L-tyrosine, L-tryptophan, L-phenylalanine, and L-methionine.
[f] Electrode responds to D-phenylalanine, D-alanine, D-valine, D-methionine, D-leucine, D-norleucine, and D-isoleucine.
[g] Time required for signal to return to baseline before reuse.

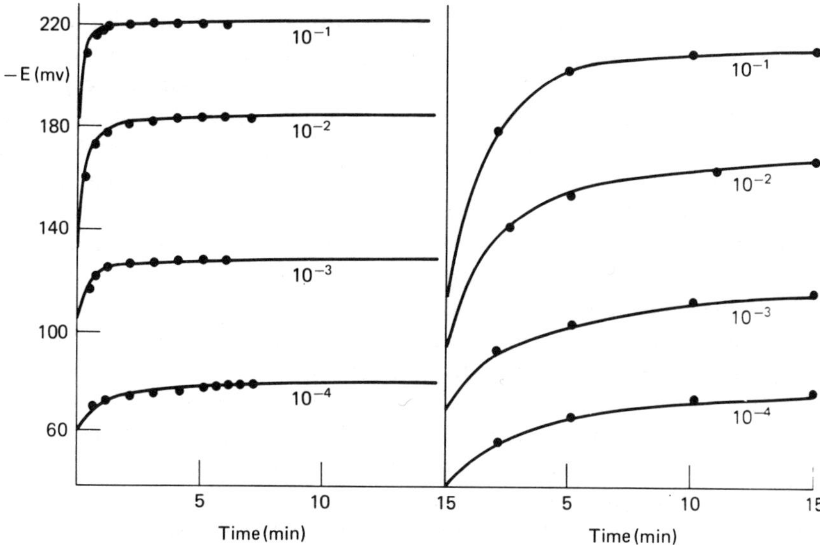

Fig. 9.1. Amygdalin response-time curves for an electrode containing 1 mg of β-glucosidase immobilized by a dialysis paper. (A) At pH 7; (B) at pH 10. From M. Mascini and A. Liberti, *Anal. Chim. Acta* **68,** 177 (1974); reprinted with permission of the authors.

and Liberti at the optimum pH for the cyanide sensor, pH 9. Realizing that there is no free cyanide except at very high pH, say 9 or 10, Rechnitz reasoned that this optimum pH should be used in order to get the best sensitivity. But at this high pH the enzyme reaction is killed and the rate of conversion becomes very slow. If the optimum pH of the enzyme reaction is used, even though there is very little cyanide at this pH the response is almost instantaneous, since the enzyme layer is in such intimate contact with the base probe. This has been shown in many cases; for example, Anfalt et al. in Sweden showed with the urea–urease reaction that the ammonia liberated from this reaction could be measured more effectively at pH 7 or 7.5 that at pH 9 or 10, because the enzyme reaction was functioning much better at this low pH. Another factor often overlooked in the use of ion-selective electrodes is that the stirring rate not only will promote a faster response at the enzyme electrode or at any probe but also will affect the equilibrium potential or the equilibrium pH that is measured. This becomes very critical: If one is going to stir, one has to stir at a constant rate; otherwise, a different potential value will be obtained every time the assay is performed.

The stability of the electrode depends on the type of entrapment. Here again there is much ambiguous reporting of immobilization data in the literature. Some researchers use dry storage for a long period of time and then report a fantastically long lifetime. Realistically, the immobilization characteristics and the stability of the enzyme should be defined in terms of dry storage and use storage. The lifetime of most soluble enzymes, except perhaps some types of glucose oxidase which are quite stable in the crude form, is generally about 1 week or 25–50 assays. (However, one must realize that there are potential problems in the use of soluble enzymes which are not found in the use of an entrapped enzyme.) The physically entrapped enzyme lasts about 3–4 weeks or 50–200 assays. For the chemically bound enzyme, 200–1000 assays is a good range. In many cases, we and others have achieved at least this. Furthermore there are many immobilized enzymes available, bound onto nylon tubes (such as those Technicon is producing for use on SMAC or the Auto Analyzer, those Boehringer has been experimenting with, and those Miles is selling under the trade name Catalink), which are very stable. These tubes have been used for more than 10,000 assays.

Stability is also dependent on the content of enzyme in the gel, on the optimum conditions, as was mentioned, and on the stability of the base sensor itself. Figure 9.2 shows the stability of some electrodes using immobilized glucose oxidase prepared by various methods; entrapment of the solubilized enzyme on an electrode surface; and two types of immobilization, physical entrapment in a gel and covalent bonding. The type of chemical bonding serves two purposes: (i) it selects the pH range, (ii) it provides the best im-

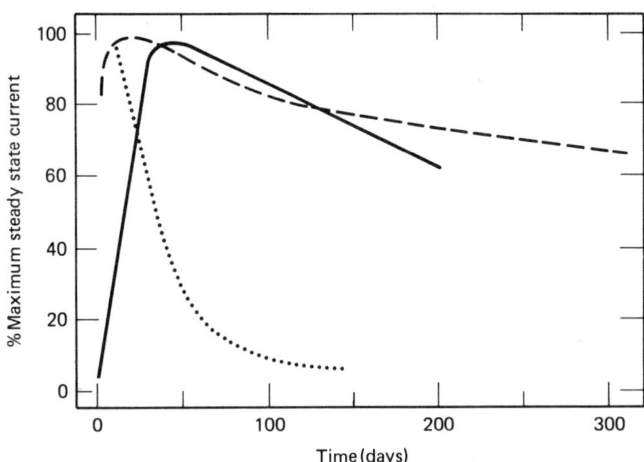

Fig. 9.2 Long-term stability of glucose electrodes by the steady state method. ---, Type 1 electrode; ——, Type 2 electrode; · · · ·, Type 3 electrode. Type 1, chemically bound—glutaraldehyde; Type 2, chemically bound—polyamide; Type 3, physically bound. From G. G. Guilbault and G. J. Lubrano, *Anal. Chim. Acta* **64,** 439 (1973); reprinted with permission of the authors.

mobilization method for each enzyme. This is shown below for some studies of cholesterol.

Interferences can be in the sensor itself or from other substrates for the enzyme. For example, alcohol oxidase can be used for an excellent acetic acid electrode; an interference would be the native substrate, ethanol. Inhibitors of the enzyme are also interferences, and here immobilizing enzyme makes it much less susceptible to environmental factors.

9.3. PREPARATION OF ENZYME ELECTRODE

This section is devoted to examples of preparation of an enzyme electrode. Various types of chemically bound enzyme and physically entrapped and soluble enzyme electrodes are discussed.

9.3.1. Construction of Glucose Enzyme Electrodes

Figure 9.3 contains cross-sectional views of two enzyme electrode embodiments that have been developed. The main body of each is constructed of a platinum inlay electrode (Beckman model 39273) consisting of a platinum disk, 2, of approximately 5 mm in diameter, housed in one end of a cylindri-

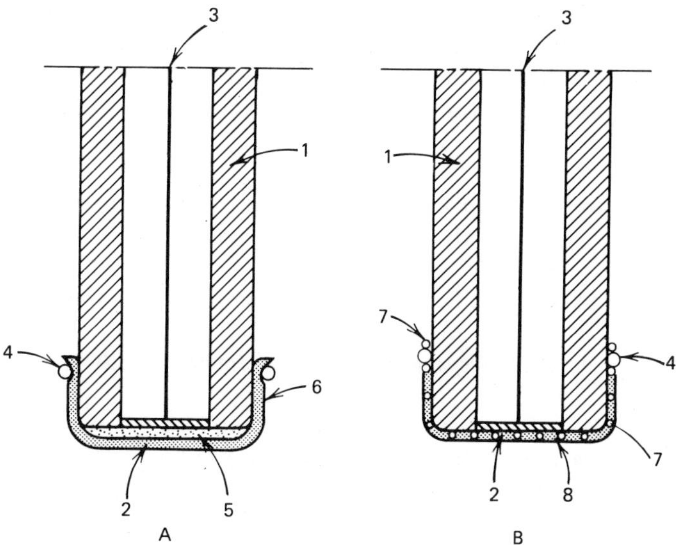

Fig. 9.3 Cross-sectional views of two enzyme electrode embodiments. 1, Insulating support body; 2, Platinum disk; 3, Connecting wire; 4, "O" ring; 5, Enzyme layer; 6, Cellophane membrane; 7, Nylon net; 8, Enzyme layer.

cal electrically insulating support body, 1, of plastic or glass. Positioned within the support body is a conducting wire, 3, which is attached to the platinum disk to allow electrical contact with detection instrumentation. Configuration *A* represents an enzyme electrode whereby an enzyme layer, 5, consisting of either solubilized enzyme in buffer or insolubilized enzyme on a gel matrix, is held over the electrode surface by means of a cellophane membrane, 6 (Will Scientific), approximately 25 μm thick. The cellophane membrane is in turn attached to the electrode body by means of a rubber "O"-ring, 4. Configuration *B* represents an enzyme electrode whereby a nylon net, 7, approximately 90 μm thick, is used as a support to anchor the enzyme layer, 8, over the platinum electrode and to control the thickness of the layer. The enzyme layer may be either enzyme occluded in a gel matrix or enzyme crosslinked with a bifunctional reagent. The nylon net is supported over the electrode by means of the rubber O-ring, 4.

Chemically Insolubilized Enzyme Electrodes

Type 1A and 1B electrodes were prepared by using glucose oxidase insolubilized by means of azide coupling to polyacrylamide beads and diazonium coupling to polyacrylic acid, respectively. The gel layer was placed in position between the cellophane membrane and platinum surface (configuration *A*) and made uniformly thin by pressing the layer lightly on a glass plate. Type 1C electrodes were constructed using configuration *B*. These electrodes were prepared by dipping netted bodies into the glucose oxidase–glutaraldehyde solution, removing them, and allowing them to dry. They are then washed in distilled water for 10 min and in glycine buffer (0.13 *M*, pH 7.0) for 10 min.

9.3.2. Preparation of Covalently Bound Glucose Oxidase via Acyl Azide Derivative of Polyacrylamide

This preparation is a modification of one given by Inman and Dintzis.

9.3.2a. Preparation of Polyacrylamide Beads

Acrylamide (21 g), *N,N'*-methylenebisacrylamide (0.55 g), tris(hydroxymethyl) aminomethane (Sigma Chemical Co.) (13.6 g), *N,N,N',N'*-tetramethylethylenediamine (Eastman Chemical Co.) (0.086 mL), and HCl (1 *M*, 18 mL) were dissolved in 300 mL of distilled water. Ammonium persulfate was then added as a catalyst. Free radical polymerization was brought about by stirring magnetically under a nitrogen atmosphere and irradiating with a 150 watt projector spotlight (Westinghouse). After polymerization was complete, the stirring bar was removed and the polyacrylamide was broken into large pieces. The polymer was then made into small spherical beads by blending at high speed with 300 mL water for 5 min. The resulting bead suspension was stored under refrigeration.

9.3.2b. Preparation of Hydrazide Derivative

Three-hundred mL of the polyacrylamide bead suspension were placed into a siliconized three-necked round bottom flask. The flask and a beaker containing anhydrous hydrazine (Matheson, Coleman and Bell) (30 mL) were immersed in a 50°C constant-temperature oil bath. After about 45 min the hydrazine was added to the gel, the flask stoppered, and the mixture stirred magnetically for 12 hr at 50°C. The gel was then centrifuged and the supernate discarded. The gel was washed with 0.1 *M* NaCl by stirring magnetically for several minutes, centrifuging, and discarding the supernatant. The washing procedure was repeated until the supernatants were essentially free of hydrazine as indicated by a pale violet color after 5 min when tested by mixing 5 mL with a few drops of a 3% solution of sodium 2,4,6-trinitrobenzenesulfonate (Eastman Chemical Co.) and 1 mL of saturated sodium tetraborate (Mallinckrodt). The hydrazide derivative was then stored under refrigeration.

9.3.2c. Coupling of Glucose Oxidase

The hydrazide derivative (100 mg) was placed in a plastic centrifuge tube and washed with 0.3 *M* HCl. It was then suspended in HCl (0.3 *M*, 15 mL), cooled to 0°C, and sodium nitrite (1 *M*, 1 mL), also at 0°C, was added. After stirring magnetically in an ice bath for 2 min, the azide derivative formed was rapidly washed with phosphate buffer (0.1 *M*, pH 6.8) at 0°C by centrifugation and decantation until the pH of the supernatant was close to 6.8. The azide gel was then suspended in phosphate buffer (0.1 *M*, pH 6.8, 10 mL) containing 100 mg glucose oxidase (Sigma Chemical Co., Type II, from *Aspergillus niger*). The mixture was stirred magnetically for 60 min at 0°C, after which time glycine (10 mL, 0.5 *M*) was added to couple with unreacted azide groups. After stirring for an additional 60 min at 0°C, the enzyme gel was washed several times with phosphate buffer (0.1 *M*, pH 6.8) and stored under refrigeration.

The general reaction scheme for this preparation is given by the equations below:

$$CH_2{=}CHC\begin{matrix}{\scriptstyle /\!/}O \\ \diagdown NH_2\end{matrix} + (CH_2{=}CHCONH)_2CH_2 \xrightarrow{(NH_4)_2S_2O_8} \text{⫼}\!-\!C\begin{matrix}{\scriptstyle /\!/}O \\ \diagdown NH_2\end{matrix} \qquad (7)$$

$$\text{⫼}\!-\!C(=O)NH_2 + H_2NNH_2 \xrightarrow[12\ \text{hr}]{50^\circ\text{C}} \text{⫼}\!-\!C(=O)NHNH_2 \tag{8}$$

$$\text{⫼}\!-\!C(=O)NHNH_2 + HNO_2 \xrightarrow[2\ \text{min}]{0^\circ\text{C}} \text{⫼}\!-\!C(=O)N_3 \tag{9}$$

$$\text{⫼}\!-\!C(=O)N_3 + \boxed{\text{Enzyme}}\!-\!NH_2 \xrightarrow{0^\circ\text{C}} \text{⫼}\!-\!C(=O)NH\!-\!\boxed{\text{Enzyme}} \tag{10}$$

9.3.3. Preparation via Diazonium Derivative of Polyacrylic Acid

9.3.3a. Polymerization of Acrylic Acid

Approximately 50 mL of reagent-grade acrylic acid (Aldrich Chemical Co.) were dissolved in 20 mL hexane and placed in a round-bottomed flask. A few milligrams of ammonium persulfate were added as a free radical initiator, and the system was kept in a dry nitrogen atmosphere. The flask was heated with a heating mantle until rapid polymerization was observed. The mantle was then quickly removed and the flask allowed to cool to room temperature.

9.3.3b. Preparation of Copolymer

The polymer was broken into small particles and neutralized with sodium hydroxide. Sodium salt was evaporated to dryness in a rotary evaporator and ground to a fine powder. The powder (approx. 3.6 g) was suspended to 6 mL hexane and cooled to approximately 4°C. The acid was converted to the acyl chloride by the addition of $SOCl_2$ (2.8 mL) with stirring in an ice bath for 1 hr (removing generated gases by suction). The acylchloride polymer was washed with ether and dried under vacuum. Then nitroaniline (0.5 g) and ether (6 mL) were added, and the mixture was allowed to stir overnight. The product formed was filtered, washed with ether, and air dried.

9.3.3c. Coupling of Glucose Oxidase

The *p*-nitroaniline derivative (150 mg) was dissolved in 10 mL of distilled water and the solution adjusted to pH 5 with dilute acetic acid. Ethylenediamine was added slowly with stirring until a fine white precipitate was observed. The precipitate was then washed three times with distilled water and suspended in 5 mL distilled water. The polymer was then reduced by the addition of $TiCl_3$ and washed several times with distilled water by centrifugation and decantation. The reduced derivative was converted to a diazonium salt by addition of nitrous acid (0.5 *M*, 10 mL) at approximately 4°C with stirring for 2 min. The diazonium salt intermediate was flushed with cold distilled water and rapidly washed several times with phosphate buffer (0.1 *M*, pH 6.8) by centrifugation and decantation. It was then mixed with a phosphate buffer solution (0.1 *M*, pH 6.8) containing 100 mg glucose oxidase (Sigma Chemical Co., Type II) at approximately 4°C for 1 hr. The resulting gel was washed several times with buffer and stored under refrigeration.

The general reaction scheme for this preparation is given by the equations below:

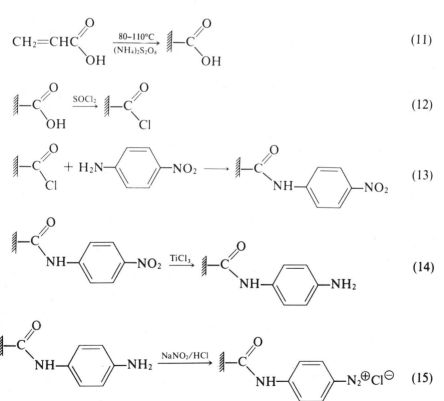

$$CH_2{=}CHC(=O)OH \xrightarrow[(NH_4)_2S_2O_8]{80\text{–}110°C} \text{[polymer]}{-}C(=O)OH \qquad (11)$$

$$\text{[polymer]}{-}C(=O)OH \xrightarrow{SOCl_2} \text{[polymer]}{-}C(=O)Cl \qquad (12)$$

$$\text{[polymer]}{-}C(=O)Cl + H_2N{-}C_6H_4{-}NO_2 \longrightarrow \text{[polymer]}{-}C(=O)NH{-}C_6H_4{-}NO_2 \qquad (13)$$

$$\text{[polymer]}{-}C(=O)NH{-}C_6H_4{-}NO_2 \xrightarrow{TiCl_3} \text{[polymer]}{-}C(=O)NH{-}C_6H_4{-}NH_2 \qquad (14)$$

$$\text{[polymer]}{-}C(=O)NH{-}C_6H_4{-}NH_2 \xrightarrow{NaNO_2/HCl} \text{[polymer]}{-}C(=O)NH{-}C_6H_4{-}N_2^{\oplus}Cl^{\ominus} \qquad (15)$$

$$\text{—C(=O)—NH—}C_6H_4\text{—}N_2^{\oplus}Cl^{\ominus} \xrightarrow{\text{HO—}C_6H_4\text{—Enzyme}} \text{—C(=O)—NH—}C_6H_4\text{—N=N—}C_6H_3(OH)\text{—Enzyme} \tag{16}$$

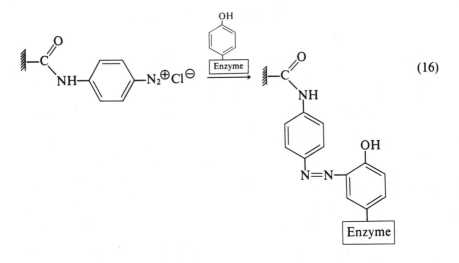

9.3.4. Preparation via Glutaraldehyde Crosslinking

To a solution containing phosphate buffer (0.1 *M*, pH 6.8, 2.7 mL) and bovine serum albumin (Nutritional Biochemicals Corp.) (17.5%, 1.5 mL) was added 50 mg glucose oxidase (Sigma Chemical Co., Type II). The resulting solution was rapidly mixed with glutaraldehyde (Sigma Chemical Co.) (2.5%, 1.8 mL) for several seconds and then immediately applied to an electrode as described in Section 9.3.4a. The general reaction for this preparation is given by the equation below:

$$\begin{array}{llllll}
CHO & & OH & & OH & \\
| & & | & & | & \\
CH_2 & & HC\text{———} & NH & HC\text{———} & NH \\
| & & | & | & | & | \\
CH_2 + & NH_2 & CH_2 & \boxed{\text{Enzyme}} & CH_2 & \text{Enzyme} \\
| & | & | & & | & \\
CH_2 & \boxed{\text{Enzyme}} \longrightarrow & CH_2 & \longrightarrow & CH_2 & \\
| & & | & & | & \\
CHO & & CH_2 & NH_2 & CH_2 & \\
 & & | & | & | & \\
 & & CHO & \text{Albumin} & HC\text{—NH—Albumin} & \\
 & & & & | & \\
 & & & & OH &
\end{array} \tag{17}$$

9.3.4a. Physically Insolubilized Enzyme Electrodes

1. Type 2 electrodes were prepared using configuration *B*. Netted bodies were covered with a thin film of polyacrylamide gel solution. They were then

placed in a water-jacketed cell at 0 to 5°C. Oxygen inhibits the polymerization and was removed by purging with nitrogen before and during the polymerization. Polymerization was then completed by irradiating with a 150 watt projector spotlight (Westinghouse) for 1 hr. After polymerization, the electrodes were soaked in phosphate buffer (0.1 *M*, pH 6.0) for several days.

2. *N,N′*-methylene-bis-acrylamide (Eastman Chemical Co.) (1.15 g) was dissolved in phosphate buffer (0.1 *M*, pH 6.8, 40 mL) by heating to 60°C. The solution was then cooled to 35°C and acrylamide monomer (Eastman Chemical Co.) (6.06 g) was added. After mixing, the solution was filtered into a 50 mL volumetric flask containing riboflavin (Eastman Chemical Co.) (5.5 mg) and potassium persulfate (5.5 mg) and diluted to the mark. The resulting gel solution is stable for months if stored in the dark.

9.3.4b. Solubilized Enzyme Electrodes

Type 3 electrodes were prepared using configuration *A*. A cellophane cup was constructed by placing wetted cellophane around the end of a bare main body and allowing the cellophane to dry. The cup was then removed, and powdered glucose oxidase (Sigma Chemical Co., Type II) (2.5–10 mg) was uniformly spread on the bottom. The cup was again placed on the main body and secured with an O-ring, thus trapping the enzyme. The electrode was then placed in phosphate buffer ($\mu = 0.1$, pH 6.0) for a few days to allow the formation of a solubilized enzyme-buffer layer and the loss of trapped air.

9.3.5. Collagen Membrane Electrodes

Coulet et al. (48–53) have reported the use of a highly polymerized collagen, prepared under industrial conditions, as the binding site of various enzymes, and have discussed the advantages of using a collagen membrane as matrix. Because of its protein nature, collagen has free amino groups available for covalent coupling with the enzymes. The membrane also possesses high polyol content (33%), which enhances film hydrophilicity, making it very supple and mechanically strong.

This same immobilization technique reported by Coulet is followed with minor modification. Collagen membranes (obtained from Centre Technique du Cuir, Lyon, France, or Rutgers University, New Brunswick, N.J.) with a diameter of 2.5 cm and 0.1 mm thick in the dry state (0.3 to 0.5 mm thick when swollen), are immersed in 60 mL of methanol containing 0.2 *N* HCl for 3 days at room temperature. The membranes are rinsed thoroughly after this period with distilled water, and transferred to 100 mL of 1% hydrazine. They are kept immersed for 12 hr at room temperature and washed again in distilled water at 0°C, then immersed into a mixture of 0.5 *M* KNO_2 and 0.3 *N* HCl for 15 min at 0°C. Then the membranes are washed with buffer solution (phosphate buffer, pH 7.5, 0.05 *M*). At this step the five membranes, activated by acyl azide formation, are immersed in a solution containing 10 units of enzyme (e.g. glucose oxidase) in 2 mL phosphate buffer (pH 7.5, 0.05 *M*) and stored in a refrigerator overnight at 4°C. Finally, the membranes are washed with phosphate buffer, and store in the same buffer until used.

A general scheme for enzymes immobilized on collagen is shown in Figure 9.4.

Upon use, the collagen enzyme layer is mounted onto a base electrode to be used (eg., a CO_2 or NH_3 gas membrane electrode or a Pt electrode) and the enzyme layer is secured in place with a rubber O-ring of appropriate diameter. The electrode is placed into 2 mL of buffer containing the sample to be assayed (eg., glucose) and the substrate concentration is measured from an appropriate calibration curve. After each analysis, the colagen membrane is washed several times with phosphate buffer, pH 7.5, 0.05 *M*.

Part 1. Activation: Acyl Azide Formation

$$\text{—COOH} \xrightarrow[23°\text{C, 3 days}]{CH_3OH/0.2N\ HCl} \text{—COOCH}_3$$

Washing in doubly distilled water, 23°C

$$\text{—COOCH}_3 \xrightarrow[23°\text{C, 12 hr}]{1\%\ NH_2\text{—}NH_2} \text{—CONH—NH}_2$$

Washing in doubly distilled water, 0°C

$$\text{—CONH—NH}_2 \xrightarrow[0°\text{C, 15 min}]{0.5\ M\ NaNO_2/0.3\ HCl} \text{—CON}_3$$

Washing in phosphate buffer, pH 7.5, 0.05 *M*

Part 2. Coupling of the Enzymes

$$\text{—CON}_3 \xrightarrow[\text{and diaphorase) 4°C, 12 hr}]{H_2N\text{—Enzymes (Glycerol dehydrogenase}} \text{—CO—NH—Enzymes}$$

Washing in phosphate buffer, pH 7.5 and 0.05 *M*. Storage of enzymically active film in the same buffer at 4°C.

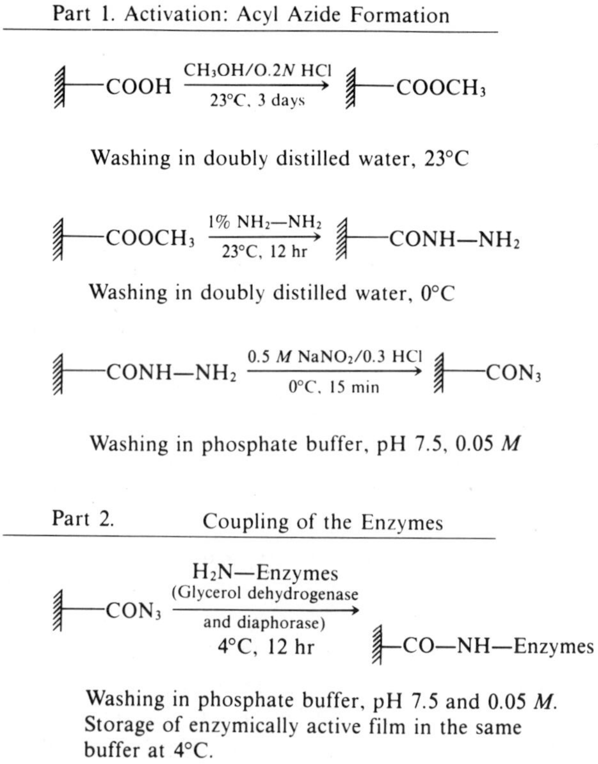

Fig. 9.4. Schematic diagram of the enzyme immobilization procedure.

9.4. VARIOUS TYPES OF ENZYME ELECTRODES

To gain an overall view of what has been done and how broad the field really is, we shall first discuss inorganic ion determination. Nonenzymatic electrodes have been designed for phosphate, sulfate, and nitrate, but these have undesirable characteristics. The phosphate and sulfate electrodes are totally useless except for titrations of sulfate or phosphate, and other commercial electrodes such as the nitrate electrode suffer serious interferences. An enzyme-based phosphate electrode has been formulated using alkaline phosphatase and glucose oxidase immobilized in a membrane, with measurement of the oxygen uptake by the glucose reaction:

$$\text{Glucose phosphate} \underset{\text{phosphatase}}{\overset{\text{alkaline}}{\rightleftharpoons}} \text{phosphate} + \text{glucose} \tag{18}$$

$$\text{Glucose} + O_2 \xrightarrow[\text{oxidase}]{\text{glucose}} \text{gluconic acid} + H_2O_2 \tag{19}$$

The amount of glucose present at any given time is controlled by the phosphate concentration, which reacts with glucose in this reversible reaction, forming glucose phosphate. So, by a measurement of the rate of oxygen uptake, one can measure the phosphate concentration, since the rate is proportional to the concentration of this anion. The curve is similar to that of an inhibitor of an enzyme reaction and is extremely reproducible, but what is remarkable is its specificity. We analyzed about 50 anions and cations, and very few interfered. Tungstate and arsenate are the only materials that gave an appreciable interference, but fortunately these are almost never found in blood or estuaries. Molybdate responds somewhat; the selectivity is about 10:1. Borate and EDTA give a slight interference. The selectivity of the electrode for phosphate over sulfate ion is about 500:1. Other ions—chloride, acetate, bromide, and so on—did not respond at all. So indeed the phosphate electrode is very selective. Another electrode we built was a sulfate electrode based on the hydrolysis of 4-nitrocatechol sulfate, catalyzed by arylsulfatase, to produce 4-nitrocatechol and sulfate. We looked at the polarographic wave for oxidation of the product of the enzymatic reaction, 4-nitrocatechol, to the corresponding quinone. Hence the rate of oxidation to the quinone is proportional to the sulfate concentration. Such an enzymatic sulfate electrode is extremely difficult to fabricate. Many people have tried but had no success for one primary reason: the measuring technique is critically dependent on both the type of substrate (some substrates do not work at all) and the concentration of substrate. All these factors have to be considered in formulating the enzyme electrode. However, the system does work very nicely. Figure 9.5 shows some curves demonstrating the specificity of the electrode. The response to sulfate is quite good. Note the line for

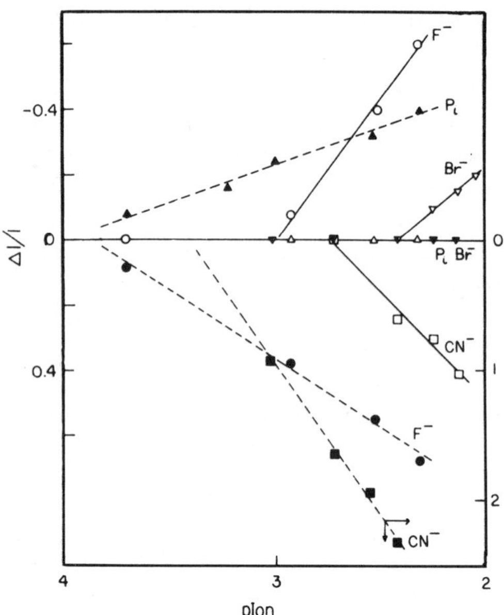

Fig. 9.5. Anion effects on the sulfatase electrode. $\Delta I/I$ is plotted against p ion. Electrode type IV; 10% (w/w) enzyme in the active layer; pH 4.10; $E = +0.8$ V vs. SCE. Open marks with the solid lines are at a 2×10^{-4} M NCS concentration; closed (solid) marks with dashed lines are at 3.8×10^{-5} M NCS concentration. Δ Phosphate; □ cyanide; ○ fluoride; ∇ bromide. From G. G. Guilbault and T. Cserfalvi, *Anal. Chim. Acta* **84,** 259 (1976); reprinted with permission of the authors.

phosphate, which gives a slight response at 10^{-4} M to a very moderate response at 10^{-2} M (3.8×10^{-5} M NCS). Normally, the response to sulfate is at least 10 times as great. The response to cyanide depends on the concentration of nitrocatechol sulfate. By using a 2×10^{-4} M concentration of nitrocatechol sulfate, we could eliminate much of the interference of ions like cyanide or phosphate.

We shall now discuss one more inorganic system before going to the organics: an assay of nitrate–nitrite ions in foodstuff. Both of these ions are important in food technology. Recent studies have shown that besides being the cause of methemoglobinemia, nitrate and nitrite can react with secondary and tertiary amines to form a carcinogenic reagent, *N*-nitrosamine. Thus the analysis of nitrate and nitrite in the environment has become increasingly important. A specific and sensitive method has been developed by using an enzymatic approach and newly developed electrochemical and fluorometric approaches. MVH(methyl viologen, reduced form)-nitrate reductase (EC 1.9.6.1), induced from *Escherichia coli*, and MVH-nitrite reductase

(EC 1.6.6.4), isolated from spinach leaves, were purified and immobilized. A newly developed air gap electrode was adapted to monitor the ammonia formed by reduction of nitrate and nitrite with MVH-nitrate and MVH-nitrite reductase. The analytical characteristics of both methods have been thoroughly studied. The detection limit using the air gap electrode is 5×10^{-5} *M* for nitrite or nitrate, and using the fluorometric monitoring method the limit is 1×10^{-6} *M* nitrate. Meat samples and water samples have been analyzed. The results obtained are very satisfactory, especially those obtained by the fluorometric method.

Most recently an NADH(nicotinamide adenine dinucleotide, reduced form)-nitrate reductase (EC 1.6.6.1) induced from *Chlorella vulgaris* was highly purified by affinity chromatography and was used for a highly specific assay of NO_3^-:

$$NO_3^- + NADH \xrightarrow{E} NO_2^- + NAD + H_2O \tag{20}$$

The decrease in the absorbance, fluorescence, or diffusion current of NADH is measured and related to the concentration of nitrate.

Another enzyme was isolated from *Azotobacter crococcum* and used for the assay of NO_2^-:

$$NO_2^- + NADH_2 \longrightarrow NO + NAD + H_2O \tag{21}$$

No other anions respond in this enzyme sequence. At least 10^{-6} *M* concentrations are assayable.

An enzyme electrode was prepared for the substrate urea by immobilizing urease in a polyacrylamide matrix on 100μm dacron and nylon nets. These nets were placed over the Beckman 39137 cation-selective electrode (which responds to NH_4^+ ion) (9). The resulting "enzyme" electrode responds only to urea. The urea diffuses to the urease membrane where it is hydrolyzed to NH_4^+ ion. This NH_4^+ ion is monitored by the ammonium ion-selective electrode and the potential observed is proportional to the urea content of the sample in the range 1.0–30 mg of urea/100 mL of solution. This enzyme electrode appears to possess stability (the same electrode has been used for weeks with little change in potential readings or drift), sensitivity (as little as 10^{-4} mol/L urea is determinable), and specificity. Results are available to the analyst in less than 100 sec after initiation of the test, and the electrode can be used for individual samples or in continuous operation.

In a later publication, Guilbault and Montalvo (10) described an improved urea-specific enzyme electrode that was prepared by placing a thin film of cellophane around the enzyme gel layer to prevent leaching of urease into the surrounding solution. The electrode could be used continuously for

21 days with no loss of activity. A full discussion of the parameters that affect the polymerization of urease as well as of the stability of 4 types of urease electrode was published by Guilbault and Montalvo (11).

Guilbault and Montalvo (12) described the preparation of a sensitized cation-selective electrode. By placing a film of urease over the outside of an ordinary cation-selective glass electrode, these workers obtained an electrode with increased sensitivity. Enzyme electrodes for the assay of urine uric acid (13) and serum urea (14) have also been reported. Adams and Carr have developed a flow-immobilized enzyme analyzer for urea (15). An improved urea electrode was reported by Guilbault and Nagy (16), using an ammonium ion-selective electrode in place of the cation electrode.

Enzyme electrodes for the determination of L-amino acids were developed by Guilbault and Hrabankova (17) who placed an immobilized layer of L-amino acid oxidase over a monovalent cation electrode to detect the ammonium ion formed in the enzyme-catalyzed oxidation of the amino acid. Two different kinds of enzyme electrodes were prepared by Guilbault and Nagy for the determination of L-phenylalanine (18). One of the electrodes used a dual enzyme reaction layer, L-amino acid oxidase with horseradish peroxidase, in a polyacrylamide gel over an iodide-selective electrode. The electrode responds to a decrease in the activity of iodide at the electrode surface due to the enzymatic reaction and subsequent oxidation of iodide.

$$\text{L-Phenylalanine} \xrightarrow{\text{L-amino acid oxidase}} H_2O_2 \tag{22}$$

$$H_2O_2 + 2H^+ + I^- \xrightarrow[\text{peroxidase}]{\text{horseradish}} I_2 + H_2O \tag{23}$$

The other electrode was prepared using a silicone rubber base nonactin type ammonium ion-selective electrode covered with L-amino oxidase in a polyacrylic gel. The same principle of diffusion of substrate into the gel layer, enzymatic reaction, and detection of the released ammonium ion applies to the system.

Another type of enzyme electrode makes use of a platinum electrode to detect the hydrogen peroxide produced in an enzymatic reaction. The enzyme electrodes for glucose, one of which was mentioned previously (8), used preparations of glucose oxidase that were nonbonded, physically bonded, or chemically immobilized as chemical transducers over a platinum electrode. The electrode was poised at a potential of +0.6V vs. S.C.E. Determinations of the glucose concentration of serum samples made using these electrodes required less than 12 sec, and the only reagent required was a phosphate buffer. Of the enzyme preparations studied, the chemically immobilized gel was the most stable.

Alcohol oxidase catalyzes the oxidation of lower primary aliphatic alcohols.

$$RCH_2OH + O_2 \xrightarrow[\text{oxidase}]{\text{alcohol}} RCHO + H_2O_2$$

The hydrogen peroxide produced in these reactions may be determined amperometrically with a platinum electrode, as in the determination of glucose above. Guilbault and Lubrano (19) used the alcohol oxidase obtained from *Basidiomycete* to determine the ethanol concentration of 1-mL samples over the range 0–10 mg/100 mL with an average relative error of 3.2% in the 0.5–7.5 mg/100 mL range. This procedure should be adequate for clinical determinations of blood ethanol, since normal blood from individuals who have not ingested ethanol ranges from 10–50 mg/100 mL. Methanol is a serious interference in the procedure since the alcohol oxidase is more active for methanol than ethanol. The concentration of methanol in blood is negligible, however, compared with that of ethanol.

Papariello et al. (20) used an immobilized preparation of penicillinase and a pH electrode to determine penicillin concentrations over the range 10^{-1}–10^{-4} *M*. The reproducibility of this work is poor in comparison with the procedure described above. A better electrode was described by Mosbach et al. (21).

Williams et al. (22) prepared electrodes for analyzing glucose and lactate by entrapping the enzymes glucose oxidase and lactate dehydrogenase, respectively, between an electrochemical sensor (a platinum electrode) and a dialysis membrane. Today, one of the most common applications of enzyme electrodes is for the clinical assay of glucose (23–26). An immobilized enzyme stirrer (27–28), a rotating ring-disk enzyme electrode (29), and a glucose oxidase amperometric (30) electrode have also been developed for the analysis of plasma and blood glucose. Enzyme electrodes have also been used for the assay of sucrose (31) and for the amperometric determination of lactate (32). Riechel and Rechnitz (33) used a nucleotide-selective enzyme electrode to record D-fructose-1,6-diphosphatase-AMP binding measurements.

Several recent publications have described the use of specific amino acid enzyme electrodes for the determination of L-glutamic acid (34), L-asparagine (35), L-phenylalanine (36), L-lysine (37), and tyrosine (38). Highly selective enzyme electrodes for the assay of 5′-adenosine monophosphate (39), blood ethanol (40), and creatine (41) have also been developed. Enzyme electrodes have been used for the measurement of cholesterol and its esters in blood using soluble enzyme (42, 43), and for total cholesterol in serum using chemically bound enzyme (44). An electrochemical system with immobilized tyrosinase has been used to determine phenol concentrations in industrial effluents and surface waters (45).

A novel application of enzyme electrodes has been for the selective assay of several anions and heavy metal ions. Ogren and Johansson (46) have reported the determination of traces of mercury(II) by inhibition of an enzyme reactor electrode loaded with immobilized urease.

Several recent attempts have been devoted to the design of stable, self-contained enzyme electrode probes that can be easily fabricated on a large scale. Guilbault and Lubrano (47) described the production of such electrodes for glucose and L-amino acids, using various membrane films. A generally useful mild coupling method for enzyme or collagen membranes, using acyl azide activation, has been reported by Coulet et al. (48–52). Stable, very sensitive glucose sensors have been described that can measure concentrations as little as 1×10^{-8} mol/L (53).

9.5. ANTIGEN-ANTIBODY PROBES

Another possible application of biological probes is the construction of sensor probes utilizing bound antibodies or antigens.

We have, for example, successfully immobilized creatine kinase M antibody as a pretreatment for the detection of cardiospecific CK-MB isoenzyme. Goat antihuman CK-M IgG was immobilized on a carrier (glass beads) through glutaraldehyde coupling, and the immobilized carrier was packed into a magnetic stirring device that is a rotating porous cell with a removable lid (54). The bound antibody could be used for several hundred assays and is regenerable; excellent results in use of this "immuno-stirrer probe" were obtained in the assay of CPK-MB specifically.

An alternative approach was presented by Suzuki (55), who bound an antigen and developed an assay for syphilis in blood. The contact potential was measured, with very low ΔmV changes (e.g., 1–3 mV).

9.6. PROBES USING IMMOBILIZED WHOLE CELLS

Still another possibility is to use the whole immobilized cell or tissues for electrodes. This field has been reviewed recently by Mattiasson (56).

This technique has the advantages (i) there is no need to purify an enzyme, (ii) the system can catalyze a series of reactions, and (iii) the whole cells can be even kept alive for long periods of time. Disadvantages are (i) generally the response is slow unless the whole cells contain very high enzyme activity, which is not common, and (ii) the response of the probe is not as selective as that obtained with a highly purified enzyme. Some examples of analytical uses are: sliced procine kidney (glutaminase) for assay of glutamine (57), whole living cells of *sarcina flava* (glutaminase) for glutamine (58), intact living cells (*streptococcus faecelis*:-L-arginase) for arginine (59), *Azotobacter vinelandii* (nitrate reductase) for nitrate (60), and immobilized nitrifying bacteria for ammonia (61).

9.7. FUTURE APPLICATIONS

The immobilized enzymes are likely to bring a new face to the future of enzymic analysis and to biochemistry and medicine in general. Enzyme electrodes (transducers with immobilized enzymes) would allow direct, simple, continuous in vivo analysis of important body chemicals. A glucose electrode, for example, would permit a continuous analysis of blood glucose levels in patients, or the analysis of glucose in urine or blood samples in a hospital or clinical laboratory in as simple a manner as a pH measurement. Similarly, implanted transducers using immobilized enzymes could be used for patient therapy. The use of immobilized enzymes in synthesis and therapy and their application in automated analysis, using the enzyme tubes of Technicon, Miles, and Carla, will open new horizons for this very exciting area.

Finally, it should be mentioned that immobilized enzymes, together with electrochemical sensors, are used in several instruments available commercially. Owens-Illinois (Kimble) has designed a urea instrument using immobilized urease and an ammonia electrode probe, and a glucose instrument using insolubilized glucose oxidase and a Pt electrode (62, 63). Patent rights to this system have been purchased by Technicon. Yellowsprings Instrument Company (Ohio) markets a glucose instrument with an immobilized glucose oxidase pad placed on a Pt electrode and has other instruments available for triglycerides, lipase, and amylase. Leeds and Northrup (North Wales, Pa.) has instruments for glucose, galactose, sucrose, maltose, and lactose using glucose oxidase, galactose oxidase, invertase-mutarotase-glucose oxidase, maltase-glucose, and lactase-glucose oxidase respectively.

Self-contained electrode probes for glucose, urea, creatinine and amino acids are available from Universal Sensors (P.O. Box 736, New Orleans, La. 70148).

REFERENCES

1. A. H. Kadish and D. A. Hall, *Clin. Chem.* **9,** 869 (1965).
2. Y. Makino and K. Koono, *Rinsho Byori* (*Japan*) **15,** 391 (1967).

3. W. R. Hussein, L. H. von Storp, and G. G. Guilbault, *Anal. Chim. Acta* **61,** 89 (1972).
4. L. H. van Storp and G. G. Guilbault, *Anal. Chim. Acta* **62,** 425 (1972).
5. R. A. Llenado and G. A. Rechnitz, *Anal. Chem.* **45,** 826 (1973).
6. L. Clark and C. Lyons, *Ann. N. Y. Acad. Sci.* **102,** 29 (1962).
7. S. J. Updike and G. P. Hicks, *Nature* (*London*) **214,** 986 (1971).
8. G. G. Guilbault and G. J. Lubrano, *Anal. Chim. Acta* **64,** 439 (1973).
9. G. G. Guilbault and J. G. Montalvo, *J. Amer. Chem. Soc.* **91,** 2164 (1969).
10. G. G. Guilbault and J. G. Montalvo, *J. Anal. Lett.* **2,** 283 (1969).
11. G. G. Guilbault and J. G. Montalvo, *J. Amer. Chem. Soc.* **92,** 2533 (1970).
12. G. G. Guilbault and J. G. Montalvo, *J. Anal. Chem.* **41,** 1897 (1969).
13. G. G. Guilbault and M. Nanjo, *Anal. Chem.* **46,** 1769 (1974).
14. G. G. Guilbault and W. Stokbro, *Anal. Chim. Acta* **76,** 237 (1975).
15. R. E. Adams and P. W. Carr, *Anal. Chem.* **50,** 944 (1978).
16. G. G. Guilbault and G. Nagy, *Anal. Chem.* **45,** 417 (1973).
17. G. G. Guilbault and E. Hrabankova, *Anal. Chem.* **42,** 1779 (1970).
18. G. G. Guilbault and G. Nagy, *Anal. Lett.* **6,** 301 (1973).
19. G. G. Guilbault and G. J. Lubrano, *Anal. Chim. Acta* **69,** 189 (1974).
20. G. J. Papariello, A. K. Mukherji, and C. M. Shearer, *Anal. Chem.* **45,** 790 (1973).
21. K. Mosbach, H. Nilsson, and A. Akerlund, *Biochim. Biophys. Acta* **320,** 529 (1973).
22. D. L. Williams, A. R. Doig, and A. Korosi, *Anal. Chem.* **42,** 118 (1970).
23. J. Kulis and B. S. Panavo, *Metody Biokhim.* Mater. S'ezdu Biokhim, Lit. 2nd Akad. Nauk Lit. SSR Inst. Biokhim. Vilnius, USSR, 1975.
24. S. Nakane, S. Murai, and A. Takasaka, *Rinsho Kensa* **20,** 1416 (1976).
25. F. Scheller, D. Pfeiffer, I. Seyer, H. J. Pruemke, and M. Jaencher, East German Patent 123,125 (Classification GOIN 27/46 dated November 20, 1976).
26. G. G. Guilbault and G. J. Lubrano, *Anal. Chim. Acta* **97,** 229 (1978).
27. S. W. Kiang, J. W. Kuan, S. S. Kuan, and G. G. Guilbault, *Clin. Chim. Acta* **78,** 495 (1977).
28. J. W. Kuan and G. G. Guilbault, *Clin. Chem.* **23,** 1058 (1977).
29. F. R. Shu and G. S. Wilson, *Anal. Chem.* **48,** 1679 (1976).
30. L. D. Mell and J. T. Maloy, *Anal. Chem.* **48,** 1597 (1976).
31. I. Satoh, I. Karube, and S. Suzuki, *Biotech. Bioeng.* **18,** 269 (1976).
32. H. Durliat, M. Comtat, J. Mahenc, and A. Baudras, *J. Electroanal. Chem. Interfacial. Electrochem.* **66,** 73 (1975).
33. T. L. Riechel and G. A. Rechnitz, *Biochem. Biophys. Res. Commun.* **74,** 1377 (1977).
34. B. K. Ahn, S. K. Wolfson, and S. J. Yao, *Bioelectrochem. Bioenerg.* **2,** 142 (1975).
35. R. Wawro and G. A. Rechnitz, *J. Membrane Sci.* **1,** 143 (1976).
36. C. P. Hsiung, S. S. Kuan, and G. G. Guilbault, *Anal. Chim. Acta* **90,** 45 (1977).
37. G. G. Guilbault and C. White, *Anal. Chem.* **50,** 1481 (1978).
38. C. Calvot, A. M. Berjonneau, G. Gellf, and D. Thomas, *FEBS Lett.* **59,** 258 (1975).

39. D. S. Papastathopoulos and G. A. Rechnitz, *Anal. Chem.* **48,** 862 (1976).
40. M. Nanjo and G. G. Guilbault, *Anal. Chim. Acta* **75,** 169 (1975).
41. M. Meyerhoff and G. A. Rechnitz, *Anal. Chim. Acta* **85,** 277 (1976).
42. L. C. Clark and C. R. Emory, in *Ion Enzyme Electrodes in Biology and Medicine* (M. Kessler and L. C. Clark, Eds.), International Workshop Proceedings, Univ. Park Press, Baltimore (1976).
43. I. Satoh, I. Karube, and S. Suzuki, *Biotechnol. Bioeng.* **19,** 1095 (1977).
44. H. S. Huang, S. S. Kuan, and G. G. Guilbault, *Clin. Chem.* **23,** 671 (1977).
45. J. G. Schiller, A. K. Chen, and C. C. Liu, *Anal. Biochem.* **85,** 25 (1978).
46. L. Ogren and G. Johansson, *Anal. Chim. Acta* **96,** 1 (1978).
47. G. G. Guilbault and G. J. Lubrano, *Anal. Chim. Acta* **60,** 254 (1972).
48. P. R. Coulet, J. Julliard and D. Gautheron, *Biotechnol. Bioeng.* **16,** 1055 (1974).
49. D. R. Thevenot, P. R. Coulet, R. Sternberg, and D. C. Gautheron, in *Enzyme Engineering,* Vol. 4, Plenum Press, New York (1978), p. 221.
50. P. R. Coulet, C. Godinot, and D. Gautheron, *Biochim. Biophys. Acta* **391,** 272 (1977).
51. J. M. Engasser, P. R. Coulet, and D. C. Gautheron, *J. Biol. Chem.* **252,** 7919 (1977).
52. J. M. Brillouet, P. R. Coulet, and D. C. Gautheron, *Biotechnol. Bioeng.* **18,** 1821 (1976); **19,** 125 (1977).
53. D. R. Thevenot, P. R. Coulet, R. Sternberg, J. Laurent, and D. C. Gautheron, *Anal. Chem.* **51,** 96 (1979).
54. C. Yuan, G. Guilbault, and S. S. Kuan, *Anal. Chim. Acta* **124,** 169 (1981).
55. S. Suzuki, *J. Solid Phase Biochem.* **4,** 25 (1979).
56. B. Mattiasson, in *Immobilized Microbial Cells* (1979), Chap. 14.
57. G. Rechnitz, *Nature* **278,** 466 (1979).
58. G. Rechnitz, *Science* **199,** 440 (1978).
59. G. Rechnitz, *Anal. Chim. Acta* **94,** 357 (1977).
60. R. Kobos, D. Rice, and D. S. Flournoy, *Anal. Chem.* **51,** 1122 (1979).
61. S. Suzuki, M. Hikuma, T. Kubo, and T. Yasuda, *Anal. Chem.* **52,** 1020 (1980).
62. D. N. Gray, M. H. Keyes, and B. Watson, *Anal. Chem.* **49,** 1067A (1977).
63. D. N. Gray and M. H. Keyes, *Chemtech.* **7,** 642 (1977).

CHAPTER

10

SOLID PHASE PEPTIDE AND PROTEIN SYNTHESIS

JOHN M. STEWART

Department of Biochemistry
University of Colorado Medical School
Denver, Colorado 80262

The idea for solid phase peptide synthesis (SPPS)* was conceived by Bruce Merrifield in 1957. He envisioned that procedures for synthesis and purification could be greatly simplified if the amino acid residue forming one terminus of the chain of a desired peptide could be anchored to an insoluble, solid support and the peptide chain "grown" on that support. At that time Merrifield was experiencing a great deal of difficulty with the synthesis of an insulin fragment in solution by standard methods available at that time. If a suitable system of blocking groups for all of the functional groups of amino acids could be devised, and if a means could be found to make all of the reactions involved in peptide synthesis go absolutely to completion, one might hope to obtain a homogeneous product from SPPS. The chemical reactivity of the blocking groups would also be designed so that a final application of a reagent to cleave the peptide–support link would also remove all the blocking groups from side-chain functional groups of the various amino acids in the peptide. In an ideal system where all of the reactions did indeed go to completion, there would be no intermediate purification, and the growing peptide chain would remain attached to the insoluble support throughout the synthesis while all excess reagents and by-products could be removed simply by washing. All of the laborious steps of purification of the many intermediates in solution phase peptide synthesis could then be avoided. Moreover, such a system was ideally suited for automatic operation.

---

* An Abbreviations section appears at the end of this chapter. Note that abbreviations for the amino acids and for representation of peptide structure are generally those recommended by the IUPAC–IUB Commission [*J. Biol. Chem.* **250,** 3215 (1975)]. In brief, peptides are always written with the amino end at the left, the carboxyl at the right; for example, Arg-Pro represents arginyl-proline and Arg-Pro-NH_2 represents arginyl-prolyl-amide. Blocking groups on the α-amino group of an amino acid are to the left of the amino acid symbol; those for side-chain functions follow the amino acid symbol, and are in parentheses. Boc-Glu(Bzl) is α-Boc-glutamic acid γ-benzyl ester.

After approximately five years' work, Merrifield announced (1) the synthesis of a tetrapeptide by the SPPS. After approximately one year of additional work, he announced the synthesis of the peptide hormone bradykinin (2). When it became apparent that the new system of SPPS was practical, I, working with Merrifield, undertook to develop an automatic instrument to carry out the operations of the synthesis. The announcement of the fully automatic synthesis of bradykinin (3) appeared in 1965. Application of SPPS to the synthesis of a large series of bradykinin analogs (4) helped call attention to the new method and convinced critics that it was practical. Indeed, the initial enthusiasm for SPPS was almost unbounded, and was quite uncritical. Many investigators, assuming that "going through the motions" of SPPS would inevitably give them their desired peptide in high yield and purity, were shocked and disappointed when they found this not to be the case. In a few years the initial wave of naive enthusiasm was replaced by discouragement and outspoken criticism on the part of many peptide chemists. This emotional reaction has persisted for years, and has unfortunately slowed the application of serious chemical investigation to SPPS. Reasoned evaluation of the initial reports of successes with SPPS should have warned experienced peptide chemists that all of the problems could not possibly have been evaluated and overcome at that time. Fortunately, several investigators systematically examined the problems encountered in SPPS and through diligent work have overcome them, one by one. At the present time, nearly two decades after the initial announcement of SPPS, materials and procedures are available that allow the facile synthesis of a very large number of peptides without serious difficulty. Nonetheless, problems do remain. Certain amino acids, such as histidine and aspartic acid, still cause problems, and improved methods are needed for using them. As many more peptides are synthesized and studied, chemists become ever more aware that peptides, especially larger ones, frequently have properties that could not be predicted solely on the basis of the amino acid composition. These unexpected properties can still cause problems in the synthesis of certain specific peptides, and they continue to afford challenges to peptide chemists.

Peptide chemists using SPPS since its introduction, have reported many outstanding accomplishments. Probably the most audacious project, particularly considering the early stage of development of SPPS at the time, was the synthesis of a protein, ribonuclease A, by Gutte and Merrifield (5). The ribonuclease molecule, having 124 amino acids, was synthesized in a yield and purity that allowed critical studies of the product to be made. A later attempted synthesis of human growth hormone, with 191 amino acid residues, was marred by use of the wrong amino acid sequence but did yield material with some biological activity. A report, later shown to be erroneous, prompted the synthesis by Peña and Stewart of a 128-residue fragment of growth hormone. This synthesis yielded chemically acceptable material, but unfortunately without biological activity. More recently, the 91-amino acid sequence of β-lipotropin has been synthesized in good yield and purity (6). In addition to such attempts to synthesize very long peptides, and indeed even proteins, SPPS has had its major usefulness in allowing the rapid and convenient synthesis of large numbers of smaller peptides for physical, chemical, and biological studies.

It can be safely said that SPPS was a signal accomplishment of synthetic chemistry. Not only has it revolutionized the field of peptide chemistry but it has caused other chemists to consider the use of polymer-supported techniques in organic synthesis. The greatest excitement at the present time is in solid phase synthesis of nucleic acids. Such synthesis of short DNA "probes" or "primers," in conjunction with cloning techniques, is revolutionizing our understanding of biochemical genetics and protein biosynthesis (see Chapter 13).

The general plan of SPPS is indicated in Scheme 1. The insoluble polymeric carrier is functionalized to permit attachment of the first amino acid.

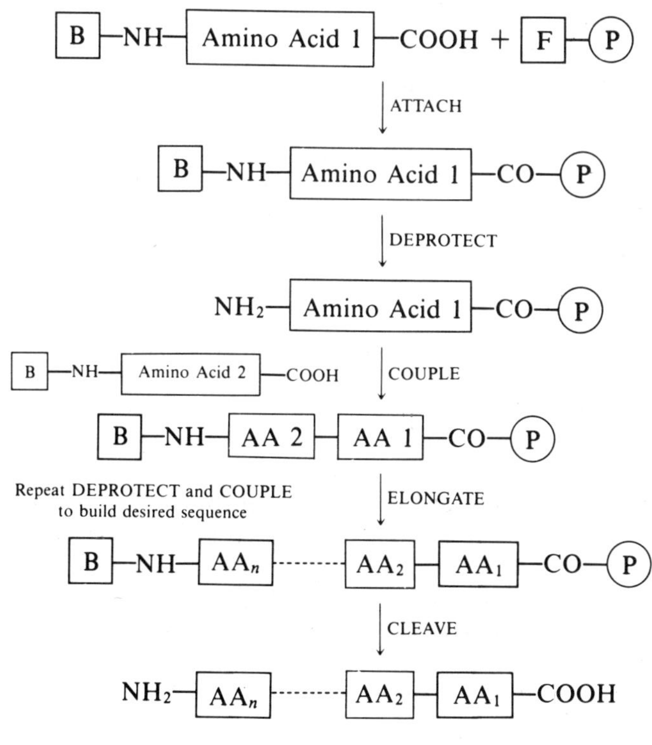

Scheme 1

During this process, the amino group of that amino acid must be blocked to prevent its interference with the desired reaction. Following the attachment reaction, the blocking group is removed ("deprotection"), and a new blocked amino acid is coupled onto the polymer bearing the first amino acid. This sequence of deprotection and coupling is then repeated as many times as is necessary, using the desired amino acid derivatives, to build up the desired sequence of amino acids on the polymer. At the end of the assembly process, a different reagent is used to cleave the peptide from the polymeric carrier. The first amino acid attached to the polymer then becomes the carboxy-terminal residue in the synthetic peptide. While the process of SPPS could in principle be carried out in the opposite direction, starting with attachment of the amino-terminal residue to the polymer, this has so far not been practical, principally due to problems with racemization of the optically active amino acids during the coupling reaction.

10.1. THE MERRIFIELD SYSTEM OF SOLID PHASE PEPTIDE SYNTHESIS

At the time of the ribonuclease synthesis in 1969 (5), the initial problems in designing a workable system of SPPS had been overcome, and that system (7) has served as the basis for most of the SPPS work done until the present time. That system, outlined in Scheme 2, was based on the use of amino acids blocked with the *t*-butyloxycarbonyl (Boc) group as the labile, easily removable blocking group of the α-amino group and blocking groups derived from benzyl (Bzl) for the side-chain functional groups of the amino acids as well as for attachment of the C-terminal residue to the polymer. The labile Boc group was removed by a solution of trifluoroacetic acid (TFA) in dichloromethane (DCM) or by anhydrous HCl in an organic solvent. Anhydrous HF or HBr in TFA was then used as the final cleavage reagent to remove the peptide from the resin and simultaneously remove side-chain blocking groups. Selectivity in removal of the labile and stable groups was thus achieved by use of acidic reagents of different strengths. Although this system has proved to be very satisfactory for the synthesis of small and medium-sized peptides, it does not provide for an adequate differential stability for ideal synthesis of large peptides and proteins. As a matter of fact, in the ribonuclease synthesis, inadequate stability of the ester bond linking the peptide to the resin caused loss of approximately 85% of the peptide from the resin during the synthesis. We now know (8) that this loss of peptide from the resin is not merely an inconvenience that diminishes the yield but it can also contribute to chain termination reactions.

Anhydrous HF proved to be an almost ideal reagent for cleavage and

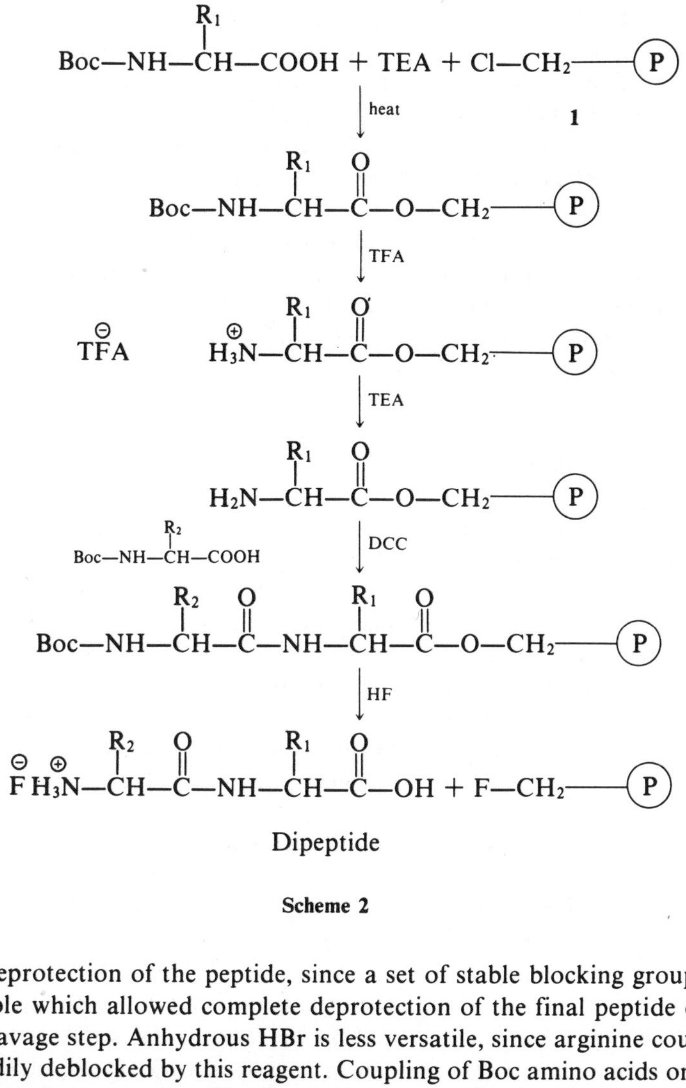

$$\text{Boc—NH—}\underset{}{\overset{R_1}{\overset{|}{\text{CH}}}}\text{—COOH} + \text{TEA} + \text{Cl—CH}_2\text{—(P)} \quad \mathbf{1}$$

$$\downarrow \text{heat}$$

$$\text{Boc—NH—}\overset{R_1}{\overset{|}{\text{CH}}}\text{—}\overset{O}{\overset{\|}{\text{C}}}\text{—O—CH}_2\text{—(P)}$$

$$\downarrow \text{TFA}$$

$$\overset{\ominus}{\text{TFA}} \quad \overset{\oplus}{\text{H}_3\text{N}}\text{—}\overset{R_1}{\overset{|}{\text{CH}}}\text{—}\overset{O}{\overset{\|}{\text{C}}}\text{—O—CH}_2\text{—(P)}$$

$$\downarrow \text{TEA}$$

$$\text{H}_2\text{N—}\overset{R_1}{\overset{|}{\text{CH}}}\text{—}\overset{O}{\overset{\|}{\text{C}}}\text{—O—CH}_2\text{—(P)}$$

$$\text{Boc—NH—}\overset{R_2}{\overset{|}{\text{CH}}}\text{—COOH} \quad \downarrow \text{DCC}$$

$$\text{Boc—NH—}\overset{R_2}{\overset{|}{\text{CH}}}\text{—}\overset{O}{\overset{\|}{\text{C}}}\text{—NH—}\overset{R_1}{\overset{|}{\text{CH}}}\text{—}\overset{O}{\overset{\|}{\text{C}}}\text{—O—CH}_2\text{—(P)}$$

$$\downarrow \text{HF}$$

$$\overset{\ominus}{\text{F}}\,\overset{\oplus}{\text{H}_3\text{N}}\text{—}\overset{R_2}{\overset{|}{\text{CH}}}\text{—}\overset{O}{\overset{\|}{\text{C}}}\text{—NH—}\overset{R_1}{\overset{|}{\text{CH}}}\text{—}\overset{O}{\overset{\|}{\text{C}}}\text{—OH} + \text{F—CH}_2\text{—(P)}$$

Dipeptide

Scheme 2

final deprotection of the peptide, since a set of stable blocking groups was available which allowed complete deprotection of the final peptide during the cleavage step. Anhydrous HBr is less versatile, since arginine could not be readily deblocked by this reagent. Coupling of Boc amino acids onto the peptide-resin was mediated by dicyclohexylcarbodiimide (DCC) except in the cases of asparagine and glutamine, which were coupled as the active 4-nitrophenyl (ONp) esters to avoid the possibility of dehydration of the amide function and incorporation of nitriles into the peptide chain. At the time of the ribonuclease synthesis histidine and tryptophan were major synthetic problem amino acids, but with the introduction of the *p*-toluenesulfonyl (Tosyl, or Tos) group for protection of the imidazole group of histidine and of the use of indole in the TFA deprotection reagent for protection of tryptophan, the situation was much improved. With these modifications, the Merrifield system of SPPS has been used by a large number of investigators for the synthesis of many hundreds of peptides during the last decade. While many improvements have been made during this time, the basic system is still a very practical method for the synthesis of many small and medium-sized peptides.

Solid phase synthesis has been discussed in several reviews, the most recent and most comprehensive being that of Barany and Merrifield (9). The laboratory manual of Stewart and Young (7) provided practical guidance for over a decade; it is now, unfortunately, out of print, but second edition of this work (10) is in progress.

10.2. IMPROVEMENTS ON THE ORIGINAL SYSTEM

Research in many laboratories during the past decade has shed light on many of the problems that existed in the classic Merrifield system and has suggested means for overcoming these difficulties. These improvements have been in areas of (i) the nature of the polymer support itself, (ii) the various means of attachment of the first amino acid to the polymer, (iii) the nature of protecting groups for various amino acid functions and their removal, (iv) improvement in the coupling reaction and monitoring of coupling, and (v) methods for cleaving the peptide from the polymeric support at the end of the synthesis. These areas are considered in the following sections.

10.2.1. The Polymer Support

The original polymer support used by Merrifield was prepared by copolymerizing a mixture of styrene with 2% divinylbenzene. This matrix was functionalized by chloromethylation to yield a chloromethyl polystyrene (**1**) (see structure **1** in Scheme 2, upper right). Variability in different batches of this chloromethyl polymer caused problems in many of the early syntheses. It soon became apparent that the degree of crosslinking of the polymer was critical for swelling and consequent penetration of solvents and reagents into the polymer matrix. Moreover, it was realized that additional crosslinking could be introduced during the chloromethylation reaction unless conditions were carefully controlled. Problems were sometimes encountered in syntheses where the degree of loading of the first amino acid on the resin was unusually high. It also became clear that at all stages of the synthetic process it was important to use reagents and solvents that swelled and expanded the matrix. The practical result of these investigations is that at the present time polystyrene resins crosslinked with 1% divinylbenzene are generally used, and the degree of chloromethylation is held low, usually around 0.75 mmol per g resin. The extent of swelling of a sample of dry resin in dichloromethane (DCM) can be used as a rough estimate of the degree of crosslinking (11). The low limit of crosslinking of polystyrene for practical use has probably been reached by Birr (12), who has used a 0.5% crosslinked resin. This resin is so soft, however, that it cannot be used in the usual synthesis apparatus but requires a special centrifugal reactor.

Some years ago, Sheppard (13) proposed a paradox in solid-phase synthesis. He suggested that the growing peptide chain and the polystyrene matrix were mutually incompatible in any given solvent, and he suggested that better synthesis might be achieved by use of a more polar matrix as the support. Following his own suggestion, he has developed a series of polyamide resins based upon crosslinked dimethylacrylamide, in which the pendant groups on the polyhydrocarbon chain are carboxamides rather than phenyl rings as is the case with polystyrene. This resin has the general form indicated in structure **2,** and is used in dimethylformamide (DMF) or dimethylacetamide as

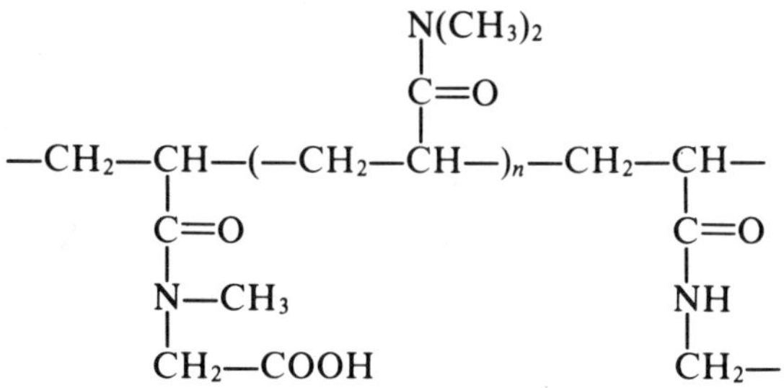

```
                      N(CH3)2
                      |
                      C=O
                      |
—CH2—CH—(—CH2—CH—)n—CH2—CH—
     |                     |
     C=O                   C=O
     |                     |
     N—CH3                 NH
     |                     |
     CH2—COOH              CH2—
```

2

solvent. Sheppard and his colleagues have reported very good success with these resins, although the syntheses required the use of extremely high excesses of amino acid derivatives in the coupling reactions. At the same time improvements were made in polystyrene-based resins, particularly the phenylacetamidomethyl (PAM) resins (see Section 10.2.2). Most recently, Kent and Merrifield (14) have described an experiment in which the PAM-polystyrene and Sheppard polyamide resin were compared for the synthesis of a sequence from the acyl carrier protein which had been reported to be "impossible" to make by SPPS. When the best of contemporary techniques were used, both resins were capable of yielding the desired peptide in comparable yield and purity. Doubtless many other changes and new polymer supports will be proposed in the coming years, but polystyrene still seems to be quite useful for SPPS.

10.2.2. Attachment of the Peptide to the Polymer

The original method for attaching the triethylammonium salt of the first amino acid to the chloromethyl polymer had several limitations. In the first place, some of the TEA used to form the salt of the amino acid reacts directly with chloromethyl groups on the polymer, introducing quaternary triethylammonium groups on the polymer, thus giving it some ion exchange properties (see structure **3**). These ion exchange groups proved to be particularly

$$(CH_3CH_2)_3N + Cl{-}CH_2{-}\text{Ⓟ} \longrightarrow (CH_3CH_2)_3\overset{\oplus}{N}{-}CH_2{-}\text{Ⓟ}$$

3

troublesome after the introduction of TFA for removal of Boc blocking groups, since the TFA could be activated by DCC in subsequent coupling reactions, leading to termination of the peptide chain. Other techniques were proposed to overcome this difficulty, such as the use of potassium salts of Boc-amino acids at this step.

Most chloromethyl polystyrenes available even today have a degree of chloromethylation of at least 0.7 mmol per g polymer, while the most desirable degree of substitution of peptide on the polymer appears to be much less for most syntheses; 0.4 mmol/g appears to be optimal for many syntheses. The residual chloromethyl groups, those left on the resin in the original attachment procedure, can react slowly with TEA at every neutralization step during the synthesis (see Scheme 2), thus leading to gradual introduction of additional ion exchange groups on the polymer as the synthesis proceeds. Moreover, these chloromethyl groups can also react with the free amino group of the peptide chain if the peptide-resin is left for an extended period of time after the neutralization step. In at least one early synthesis, complete termination of a peptide occurred in this way.

Although residual chloromethyl groups have been present on resins during synthesis of many small peptides, their presence is undesirable for attempts to synthesize large peptides. Many investigators feel that yields of peptides can probably be improved by eliminating all such residual chloromethyl groups. This can be accomplished either by complete replacement of chloromethyl groups with Boc-amino acids, as can be done by the Gisin procedure (15), by reaction of residual chloromethyl groups with potassium acetate, as was done by Merrifield in the original SPPS work, or by use of a different technique for introduction of the first amino acid. Chloromethyl resin may be hydrolyzed completely to hydroxymethyl resin and the first amino acid attached to these hydroxyl groups. Both carbonyldiimidazole and DCC-dimethylaminopyridine coupling reactions have been used (9), as well as symmetrical anhydrides. A recent communication (16) suggests that racemization may be a problem in certain methods of attaching amino acids to hydroxymethyl resins; further investigation seems warranted. In any case, once the amino acid derivative has been introduced to the desired degree of substitution, all remaining free hydroxyl groups must be covered by acylation. Benzoylation with benzoyl chloride and pyridine in DCM appears to be the most practical procedure for this purpose (17). If this is not done, subsequent amino acid coupling steps can lead to acylation of these hydroxyl groups by aminoacyl residues with the formation of new, shorter peptide chains on the polymer support.

As mentioned above, the classic Merrifield resin link, a benzyl ester substituted in the para position by the electron-donating alkyl chain, does not provide completely adequate stability in the system where Boc groups are removed from the peptide at every stage of synthesis by acidolysis with HCl or TFA. In addition to causing loss of peptide from the resin, with consequent reduction in yield of final product, this undesired acidolytic cleavage of these peptide-resin esters exposes hydroxyl groups on the resin, which can then be acylated by TFA during the deprotection steps. These resin-bound trifluoroacetyl groups can subsequently be transferred to the peptide amino groups, terminating the chain and reexposing the hydroxyl group on the resins for repetition of this harmful process (8). This serious problem can be avoided either by use of much more labile α-blocking groups, by use of a more stable peptide-resin link, or by use of a system of orthogonal groups that are not susceptible to cleavage by the same type of reagent (18). The first of these approaches, the use of more labile α-blocking groups, is accessible through the use of Bpoc, Ppoc, Ddz, or Tmz amino acids (see Abbreviations section) during the synthesis. This approach is of limited practicality at the present time because of the chemical instability (Bpoc), high cost (Ddz), or lack of commercial availability (Ppoc and Tmz) of the appropriate derivatives. The alternative approach of strengthening the peptide-resin link appears more practical at this time. To achieve this goal, the PAM resin (structure **4**) was introduced by the Merrifield laboratory (19). The Boc-amino

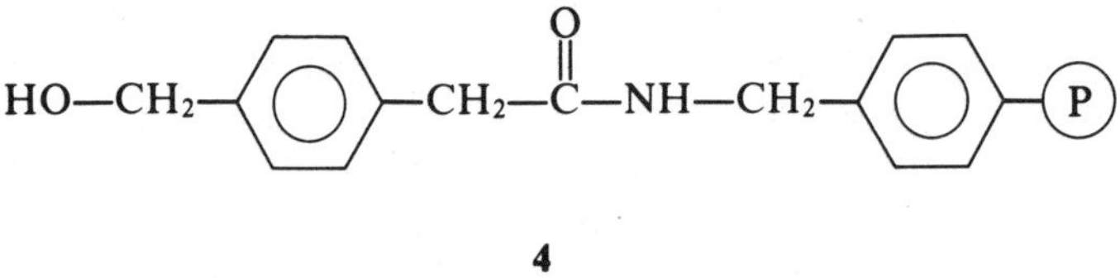

$$HO{-}CH_2{-}C_6H_4{-}CH_2{-}\overset{O}{\overset{\|}{C}}{-}NH{-}CH_2{-}C_6H_4{-}\text{Ⓟ}$$

4

acid is attached to the free hydroxyl group of this polymer as indicated above, or the Boc-aminoacyl-PAM link can be synthesized as a unit and attached to the aminomethyl polymer. The PAM resin provides about a

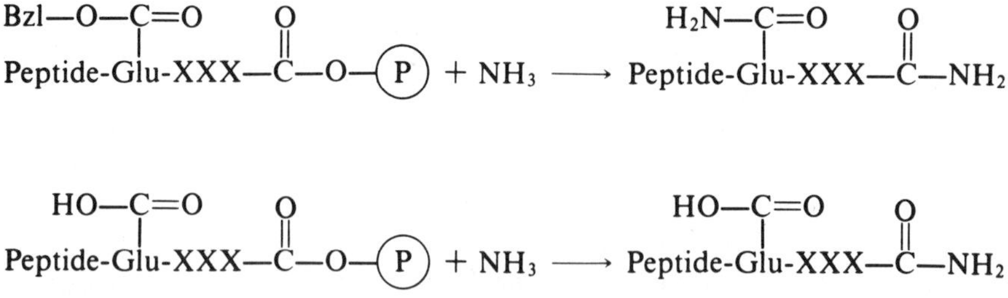

$$\text{Peptide-Glu}(\text{O-Bzl})\text{-XXX—C(=O)—O—}\text{(P)} + NH_3 \longrightarrow \text{Peptide-Glu}(\text{C(=O)—NH}_2)\text{-XXX—C(=O)—NH}_2$$

$$\text{Peptide-Glu}(\text{C(=O)—OH})\text{-XXX—C(=O)—O—}\text{(P)} + NH_3 \longrightarrow \text{Peptide-Glu}(\text{C(=O)—OH})\text{-XXX—C(=O)—NH}_2$$

Scheme 3

hundredfold increased stability of the peptide-resin link toward acidolysis, and this resin has been used recently for some impressive synthetic work. Bromination of the polystyrene was originally proposed by Merrifield as a way of stabilizing the peptide-resin link, and that approach has been reported (6) to be useful, although a recent comparison of the classic Merrifield, brominated, and PAM resins (20) showed that the PAM resin is clearly superior. A resin similar to PAM was proposed by Sparrow (21) and appears to have similar chemical properties.

Since many peptide hormones are blocked at their C-terminal ends by amide groups, the synthesis of peptide amides is of major importance. Many peptide amides have been synthesized from Merrifield peptide-resins by ammonolysis of the peptide-resin bond (see Scheme 3). This procedure may be unsatisfactory if the C-terminal residue is sterically hindered, as is the case with isoleucine or valine, and it also presents problems if the peptide contains aspartic or glutamic acid groups protected as esters. In the latter case, these will be converted to the corresponding asparagine and glutamine residues in the peptide. These problems have led to the development of amine resins suitable for direct synthesis of peptide amides. If the Merrifield chloromethyl resin is converted by ammonolysis to the aminomethyl resin and that used for peptide synthesis, none of the resulting peptide-aminomethyl resin can be cleaved by HBr or HF. For satisfactory acidolysis, the aminomethyl polymer must be labilized by the introduction of a different type of linkage. One way of labilizing this group is to introduce an additional phenyl residue on the amine-bearing carbon. This was originally done by Marshall, who introduced the benzhydrylamine (BHA) resin (structure **5**). This resin has been used for synthesis of a wide variety of peptide amides, using HF for cleavage of the peptide-resin link. The strength of the peptide-resin bond depends in a large measure upon the nature of the C-terminal amino acid. In the case of BHA resins, peptides having C-terminal phenylalanine or valine residues may not be cleaved from BHA resins with HF under the usual conditions. To overcome this problem the MBHA resin (structure **6**) was re-

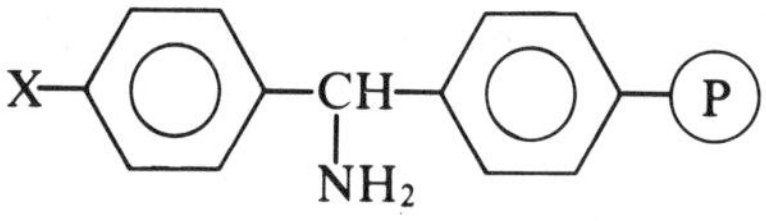

5: X = —H **6:** X = —CH_3

cently introduced (22) and it is now commercially available (23). This resin appears to offer the best balance of properties for the synthesis of peptide amides using α-Boc protection.

Several other peptide-resin links have been designed for different purposes. More labile links are useful if the α-blocking group is more labile than Boc. Other types of resins have been designed for the synthesis of protected peptide fragments for subsequent segment coupling in the synthesis of larger peptides (see ref. 9).

10.2.3. The Coupling Reaction

The coupling agent originally used in SPPS was DCC, and it continues to be the most commonly used reagent for this purpose. It provides a high degree of activation of the Boc-amino acid, and coupling reactions usually go rapidly to completion with a relatively small excess of the activated amino acid. Although this high degree of activation produced by DCC will cause racemization when it is used to activate the carboxyl group of a peptide, such as would be the case in segment coupling, it generally does not racemize urethane-protected single amino acids. The only Boc-amino acid found to be racemized significantly in DCC-mediated coupling reactions in SPPS is Boc-His (Bzl). In this case, racemization is evidently due to the basicity of the imidazole ring, which is not diminished by the presence of the benzyl group. If electron-withdrawing blocking groups are used for the imidazole function, such as Tosyl or Dnp, then racemization is not a problem.

Active esters have also been used for SPPS coupling reactions. In the original SPPS system, *p*-nitrophenyl active esters were used for asparagine and glutamine, since dehydration of the side-chain amide of glutamine to the corresponding nitrile during DCC activation had been described. There have been several investigations of the use of various active esters for all SPPS coupling reactions, but in general the results confirmed that active ester-mediated coupling reactions are slower than DCC-mediated reactions and require longer reaction times and larger excesses of active ester to yield complete coupling.

If certain hydroxyl compounds are added to a DCC-mediated coupling reaction, the corresponding active ester may be formed as a temporary intermediate. Compounds used for this purpose are 1-hydroxybenzotriazole (HBT) (structure **7**) and *N*-hydroxysuccinimide (structure **8**). These additives

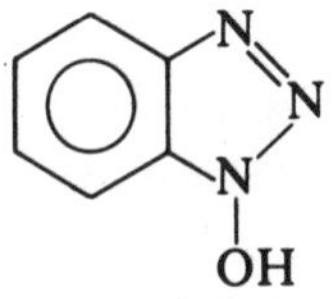

7 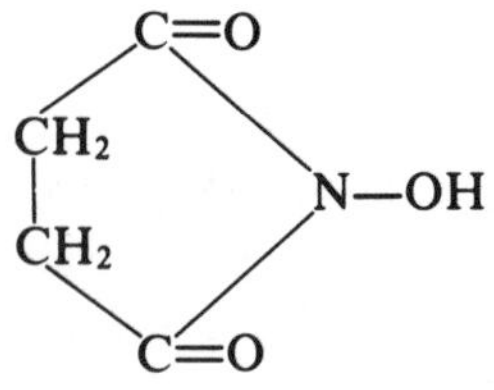 **8**

presumably react immediately with the DCC-activated amino acid, and the coupling reaction to form the peptide bond is then actually carried out by this active ester. Of these two additives, HBT appears to give better results. In certain cases DCC-HBT coupling reactions appear to go faster than those mediated by DCC alone. This has been true in the case of coupling reactions with sterically hindered amino acids, such as isoleucine and threonine. Since the reactive intermediate formed from DCC and a Boc-amino acid is very bulky around the reactive site (see structure **9** in Scheme 4), the HBT-active ester may accelerate the reaction because it is smaller and less hindered even though the degree of activation is lower. It is not yet completely clear whether addition of HBT can be recommended as a universal procedure since there have been several cases in which syntheses were not improved by such addition.

In recent years much mention has been made of the use of symmetrical Boc-amino acid anhydrides in SPPS (9). These anhydrides (structure **10** in Scheme 4) can be either prepared separately or prepared *in situ* by the use of one equivalent of DCC with two equivalents of Boc-amino acid (see Scheme 4). In their recent very careful study of the solid-phase synthesis of the "impossible" decapeptide sequence from acyl carrier protein (14), Kent and Merrifield showed that in DCM the most difficult coupling reaction went somewhat faster with either the symmetrical anhydride or DCC-HBT than with DCC alone. The most rapid coupling was obtained in DMF, where the symmetrical anhydride procedure showed a somewhat faster coupling rate than any other method. It is not yet clear, however, that the use of symmetrical anhydrides in DMF should be recommended as the universally applicable procedure for SPPS. Racemization of activated amino acids clearly proceeds more rapidly in polar solvents such as DMF than in DCM. While the particular coupling reactions studied by Kent and Merrifield proceeded ex-

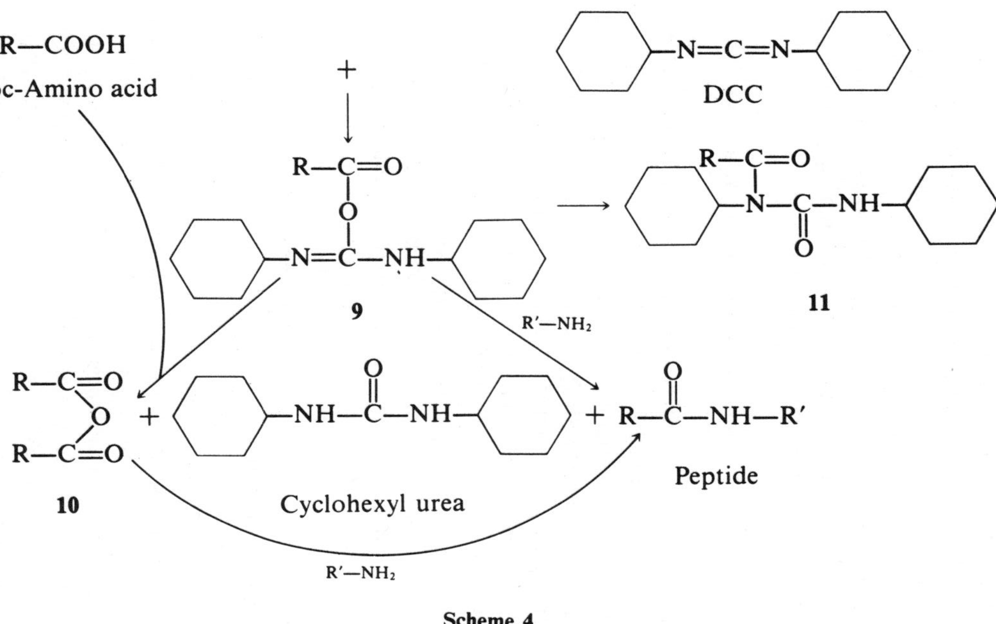
R—COOH
Boc-Amino acid
DCC
9
10
11
R'—NH₂
Cyclohexyl urea
Peptide
R'—NH₂

Scheme 4

tremely rapidly, there may be others in which the coupling is not rapid and the activated intermediate may have time to racemize to a significant degree. Moreover, the symmetrical anhydrides are not stable in DMF and must be used immediately after dissolution in the solvent. This is not a major problem for manual SPPS, since the operator must be present in any case. It causes problems, however, for automatic operation, and will require provisions for storage of the anhydride solutions at low temperature. It is clear that many coupling reactions go virtually to completion in SPPS in the usual DCC-mediated coupling reactions in DCM. In these cases, the additional work necessary to carry out the symmetrical anhydride procedure in DMF is probably not warranted. A modified double coupling procedure (see Section 10.3) will probably be adequate in many cases.

10.2.4. Monitoring Solid Phase Reactions

When the perfect system of SPPS is developed and all deprotection and coupling reactions go absolutely to completion, the need for monitoring procedures will have been eliminated. Even at the present time the synthesis of a small or medium-sized peptide (perhaps up to 15 amino acids) can reasonably be attempted by a straightforward automatic synthesis, using DCC in DCM as solvent. If such an automatic synthesis is done on a relatively small scale (0.4 mmol peptide) the cost of reagents is not large, and little time is invested. If examination of the product after cleavage from the resin reveals a significant degree of failure at one or more points during the synthesis, the results of this initial synthesis may be used to guide a repeat synthesis in which the various coupling reactions thought to give rise to problems are monitored.

Several different procedures are available for detecting unreacted peptide amino groups on the resin at the end of a coupling reaction. The most widely used of these is the Kaiser test (24), which uses the reaction of ninhydrin on a small sample of resin beads drawn from the reaction mixture. If the beads are blue after heating with ninhydrin, some free amino groups remain on the resin, indicating that recoupling is needed. It is possible to make a rough estimate of the amount of amino groups remaining from the intensity of the blue color. Fluorescamine was introduced as a monitoring reagent in SPPS by Felix (25), who showed that it has a much higher sensitivity than does the Kaiser test. In practical terms, the fluorescamine reaction may be so sensitive that it is almost impossible to obtain a negative test. In addition, if any fluorescent groups are present in the peptide-resin, this test can obviously not be used. This is the case if such blocking groups as xanthydryl are used for glutamine and asparagine side-chain protection. Christensen (26) proposed a chloranil monitoring test for SPPS coupling reactions, but the additional problems with this procedure would not appear to allow unequivocal recommendation of it at the present time; it does not appear to be significantly more sensitive than the Kaiser test.

Whereas the previously discussed monitoring procedures involve removal of a small sample of resin beads, two procedures have been used that involve titration of the entire batch of resin used in the synthesis. Hodges and Merrifield (27) used picric acid treatment of the peptide-resin, washing, and elution of the bound picric acid with an amine. Photometric estimation of the amount of picric acid thus removed gave a measure of the free amino groups remaining on the resin. The Beckman 990 automatic synthesizer was modified to carry out this picric acid monitoring automatically and stop the synthesis if the amount of picric acid bound was greater than a prescribed limit. Villemoes, Christensen, and Brunfeldt (28) titrated the peptide-resin with perchloric acid to measure the remaining uncoupled amino groups and designed a modification to their automatic synthesizer that could carry out the monitoring automatically and make a judgment whether to recouple or proceed with addition of the next amino acid. These titration methods can give a quantitative measure of the amount of unreacted amino groups, but it is not clear that such procedures—particularly the perchloric acid titration—can be applied to a synthesis in progress without some damage to the peptide or blocking groups. Perhaps such procedures should be applied to a synthesis in which all of the steps are studied carefully and then the synthesis could be repeated employing the previous results to direct such operations as recoupling and chain termination.

Monitoring of coupling reactions is practical because the desired end point is a negative or null reaction. Monitoring of deprotection methods is inherently more difficult, since the reaction curve for deprotection approaches the desired end point asymptotically. Any monitoring reaction based on the presence of free amino groups will thus be required to detect very small differences in rather large numbers. This has been done by titration, using the Brunfeldt perchloric acid method, and by Birr (12) using photometric detection of the Ddz blocking group released in the deprotection reaction. In both these cases, the problem of accurate estimation of the end point was apparent.

Although most cases of failure of peptide chains to continue growing in SPPS are due to failure of coupling reactions, several cases of failure were apparently due to failure of the removal of Boc-blocking groups. In these cases, apparently the blocked end of the peptide chain interacts with the polymer matrix in such a way that the blocking group is not accessible to the deprotection reagent. In one such case (29), the synthesis of a large growth hormone fragment, chain continuation was much improved by the use of two successive deprotection reactions in solvents that swelled the matrix to a different degree. A common practice in synthesis of larger peptides by SPPS, in which all coupling reactions are monitored, is to recouple once (using DMF if the initial coupling was done in DCM). If the coupling is still not complete, it is often more practical to terminate unreacted chains by acetylation or with a terminating reagent having useful and unique properties (30). Such chain termination will prevent later acylation of the remaining amino groups by a less hindered amino acid residue with consequent formation of deletion sequences in the finished product. Intentionally terminated chains should be much easier to remove from the desired product during purification.

10.2.5. Blocking Groups for Amino Acid Functions

10.2.5a. Temporary α-Protection

Although most SPPS up to the present time has been done with the use of α-Boc-amino acids, it is clear that this is not the ideal protecting group for synthesis of large peptides and proteins. If acidolysis is to be used for deprotection at every step of the synthesis, much more acid-labile groups are needed. The Bpoc protecting group (structure **12**) was proposed many years ago, and

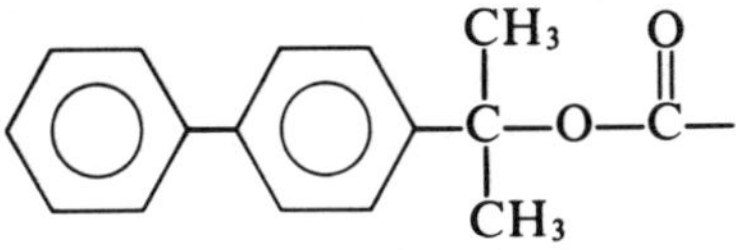

$$C_6H_5-C_6H_4-C(CH_3)_2-O-C(=O)-$$

12

it is extremely labile to acid. It is in fact so labile that the acidity of the Bpoc-amino acid carboxyl group is sufficient to cause slow removal of the blocking group upon standing in the dry solid state. For this reason, Bpoc-amino acids must be stored as the salt and the free acid must be regenerated immediately before use. This added complication has been a great deterrent to the use of these derivatives, especially with automatic synthesizers. The Bpoc group is approximately 100,000 times as labile to acid as is the Boc group. Several α-blocking groups have been developed that have reactivities intermediate between that of the Boc and Bpoc. These groups offer a much greater margin of safety in SPPS procedures based upon selective acidolysis and they are still convenient to use. The Ddz (structure **13**), Ppoc (structure **14**), and Tmz (structure **15**) groups are all of a reactivity that will allow convenient storage of the protected amino acid yet give ready acidolytic cleavage of the blocking group in dilute TFA solutions in DCM. A comparison of these groups (31) showed that they are about 4000 times as labile as Boc.

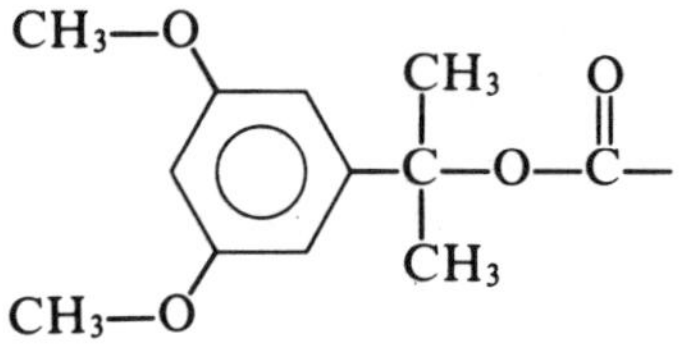

13

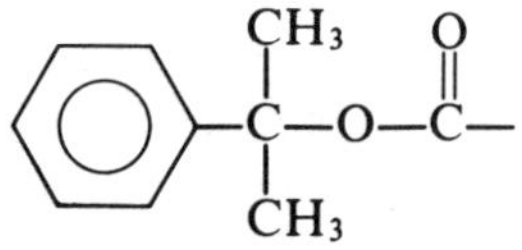

14

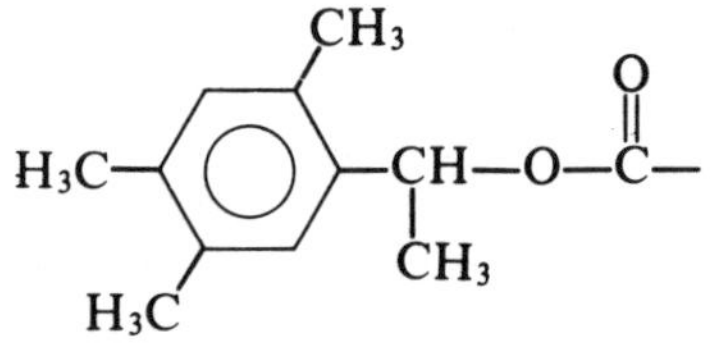

15

While one or more of these groups may offer very significant advantages for the synthesis of large peptides, problems such as lack of commercial availability and high cost have prevented their adoption on a wide scale.

The Fmoc protecting group (structure **16**) allows an "orthogonal" approach to SPPS. This blocking group is stable to acid but is removed by treatment with an amine in a polar solvent such as DMF. A few syntheses

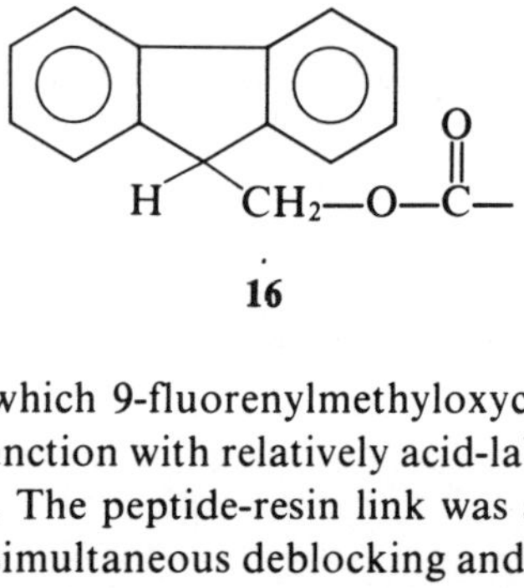

16

have been reported in which 9-fluorenylmethyloxycarbonyl (Fmoc) amino acids were used in conjunction with relatively acid-labile side-chain blocking groups, such as *t*-butyl. The peptide-resin link was also acid labile, so that TFA could be used for simultaneous deblocking and cleavage of the peptide from the resin. Fmoc is probably not compatable with the normal Merrifield SPPS resin, since the peptide-resin link is probably labile to the amine used for cleavage of the Fmoc groups.

10.2.5b. Side-Chain Blocking Groups

In SPPS all reactive side-chain functions of the amino acids must be blocked with groups that will remain intact throughout the entire process of assembly of the desired sequence on the resin and yet be removable at the end of the synthesis by a reagent that will not harm the peptide chain. In the standard system, benzyl-related blocking groups have been used. Of the originally proposed blocking groups, the benzyl esters of aspartic and glutamic acids and the benzyl ethers of serine and threonine are still considered satisfactory for use with Boc α-protection. The sulfhydryl function of cysteine can best be blocked with a *p*-methylbenzyl group if one wishes a blocking group that is removable by the HF at the cleavage step. For synthesis of complex cysteine-containing peptides, there remains a real need for a satisfactory blocking group that will be stable to HF but can be removed later, after purification of the peptide, in order to allow formation of the desired disulfide bridges. The acetamidomethyl group has been proposed for this purpose, but it is not completely stable to the usual HF treatment, and the procedures available for its removal leave much to be desired. Histidine clearly requires blocking of the imidazole group for satisfactory SPPS, and both the Tosyl and 2,4-dinitrophenyl (Dnp) groups are in common use. Neither of these is ideal, since the Tosyl is not quite stable enough to allow some procedures (for example, use of HCl for Boc removal or attachment to chloromethyl resin by the classic procedure) and the Dnp group requires an additional step of thiolysis for removal. Probably imidazole-benzenesulfonyl histidine should be investigated.

The benzyloxycarbonyl group originally used for protection of the lysine side-chain is clearly too labile, and at the present time the 2-chlorobenzyloxycarbonyl group is in general use; it appears to have the desired increased stability to acid. Many investigators are now using DCC-HBT mediated coupling reactions for Boc-asparagine and Boc-glutamine, since this procedure has been reported to avoid any nitrile formation. More cautious investigators prefer to use *N*-blocked amide functions for these amino acids. Of the several groups that have been proposed for this purpose (9), the xanthenyl (Xan) group offers many advantages although it is not ideal. The Xan derivatives (structure **17**) are very easily formed and allow DCC coupling reac-

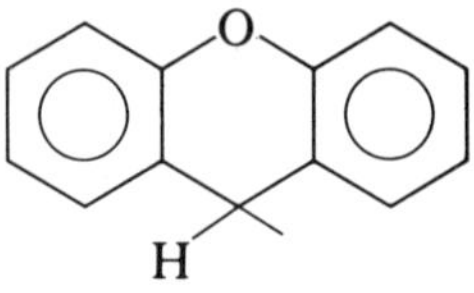

O

H

17

tions to be used in many cases with much improved yields of Asn and Gln peptides. The Xan group does lack some of the desired acid stability and may lead to some steric hindrance in coupling reactions in some sequences.

Arginine is now routinely used in SPPS with the guanidine function blocked as the Tosyl derivative, which is readily removed by HF. Since this group is not cleaved by HBr, it is not the ideal blocking group for use by investigators who do not have facilities for handling HF. For additional comments on arginine, see Section 10.2.7.

Tryptophan is now used routinely in SPPS without any protection of the indole group. Crucial to the success of its use is the avoidance of any oxidizing or electrophilic groups during acidolysis of Boc groups. A major improvement was incorporation of scavengers in the TFA solutions used for deprotection; indole (1 mg/mL), allowed to stand with the reagent overnight before use, appears to be satisfactory. Alkylation of tryptophan residues by *t*-butyl groups during TFA removal of Boc groups has been reported (32) but does not appear to be a significant problem when a scavenger is used. To avoid alkylation of the indole, *N*-formyl tryptophan may be used; the formyl group is removed easily by base treatment after the synthesis. Such base treatment may be harmful in some cases, however.

The use of *O*-benzyl tyrosine leads to the formation of 3-benzyl tyrosine residues in the peptide during HF cleavage, since HF is a good catalyst for Friedel–Crafts reactions. This side reaction has in fact been the principal course of reaction in certain cases. It can be largely eliminated by the use of either dichlorobenzyl or 2-bromobenzyloxycarbonyl (BrZ) groups to block the phenolic hydroxyl. Both of these are removed readily by HF; the BrZ group appears to be better in some cases.

10.2.6. Deprotection of α-Blocking Groups

TFA in DCM (25–50% solution, by volume) remains the most popular deprotection reagent in SPPS, and it is generally satisfactory when used with a scavenger such as indole. The more labile groups such as Ddz, Ppoc, and Tmz can be removed by 15-min treatment with 3% TFA in DCM (31). Erickson (32) has introduced the use of 0.1 *M* TFA containing 0.01 *M* methanesulfonic acid in DCM in a very rapid deprotection procedure. This reagent was developed for use in a continuous flow high-speed synthesizer, and it is not yet clear how satisfactory it may be for synthesis using usual procedures.

10.2.7. Cleavage of the Peptide from the Resin

Anhydrous HF continues to be the most widely used cleavage reagent for removal of the peptide from the resin at the end of SPPS procedures. Its use requires heavy plastic vacuum line equipment for safety and avoidance of side reactions. Moreover, some large peptides, and certainly many proteins, will not be stable to HF. Use of HBr may be preferable in some cases. Several other reagents have been proposed for cleavage of peptide-resins; see Barany and Merrifield (9) for a discussion of these. Of all these non-HF procedures, methanesulfonic acid may prove to be the best (33).

Recently, several interesting biologically active peptides have been prepared in which the C-terminal residue is reduced to an alcohol. Such peptides are not accessible by normal SPPS procedures. In a recently developed procedure (34), peptide-resins were treated with lithium borohydride in dioxane. This procedure caused reductive cleavage of the peptide-resin bond and yielded the peptide alcohol in good yield. If the peptide contains glutamic and aspartic acid residues blocked as esters, they will also be reduced to alcohols by this procedure. If one wishes to synthesize C-terminal alcohols of such peptides, it will be necessary to introduce these residues as, for example, *t*-butyl esters, using α-Fmoc protection. Removal of the *t*-butyl esters with TFA prior to reductive cleavage of the peptide-resin ester would yield a product with these carboxyl groups intact.

10.3. PROBLEMS AND RECOMMENDATIONS

Problems associated with specific amino acid residues have been discussed in the preceding sections. An additional serious problem associated with aspartic acid residues in SPPS is rearrangement of these residues to β-aspartyl residues during cleavage and deprotection (see Scheme 5). This has been a serious problem in cases where the next residue in the carboxyl direction after the aspartic residue was glycine or serine, which do not provide steric hindrance to cyclization to the succinimide intermediate. When this intermediate forms, it is opened preferentially when the peptide is exposed to water to yield the β-aspartyl residue in the peptide. At the present time, no completely satisfactory method is available for preventing this side reaction in standard SPPS. The rearrangement is much less severe if the aspartic residue has a free side-chain carboxyl group during the cleavage step. This could be arranged in cases where the PAM or standard Merrifield resins are used in synthesis by use of α-Bpoc-aspartic-β-*t*-butyl ester. Very dilute TFA could then be used for α-deprotection, and 50% TFA treatment would remove the aspartic ester prior to cleavage of the peptide-resin. A Friedel–Crafts type of acylation of anisole by glutamic acid residues during HF cleavage has also been described (35), and this can be a problem if the temperature is elevated during HF cleavage. Some investigators insist that HF should be used at −20°C instead of the customary 0°C.

Other problems encountered in SPPS are sequence-dependent and cannot be predicted a priori from the amino acid composition. Some histidine-containing sequences have been notorious, particularly a small peptide containing two histidines adjacent to each other. Some tyrosine-containing se-

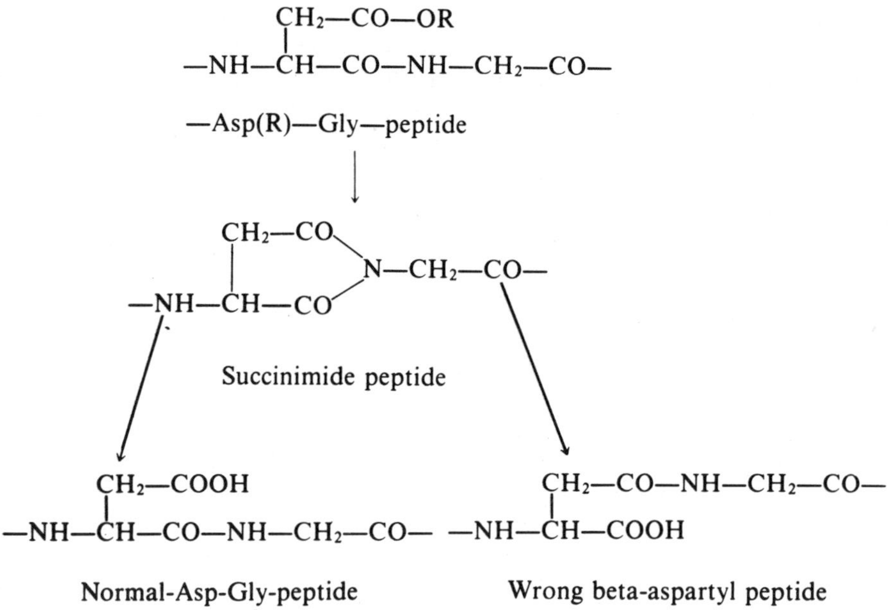

```
          CH2—CO—OR
          |
—NH—CH—CO—NH—CH2—CO—

—Asp(R)—Gly—peptide
          |
          ↓
     CH2—CO
     |       \
     |        N—CH2—CO—
     |       /
—NH—CH—CO

Succinimide peptide

     CH2—COOH                              CH2—CO—NH—CH2—CO—
     |                                     |
—NH—CH—CO—NH—CH2—CO—                  —NH—CH—COOH

Normal-Asp-Gly-peptide              Wrong beta-aspartyl peptide
```

Scheme 5

quences have given problems with failure to couple as well as failure to deprotect. Steric hindrance may be a problem when coupling branched residues, such as isoleucine and threonine, particularly to other hindered residues.

General procedures can be suggested for standard SPPS and approaches to solution of problems when they are encountered. Most small and some medium-sized peptides should be synthesized readily on Merrifield polystyrene resins. Residual chloromethyl groups should not remain on the resin during the synthesis. This can be accomplished either by complete replacement of a low-substituted chloromethyl resin by amino acid or by the use of hydroxymethyl resin. Standard DCC-mediated coupling reactions in DCM should probably be tried first. If problems in complete coupling are encountered, recoupling with DCC in DMF should be done. It appears to be important that a neutralization step be included in every recoupling procedure. If incomplete coupling is a problem at any point, DCC with the addition of HBT can be used. Symmetrical anhydrides may offer help in some truly difficult sequences, but the anhydrides should preferably be prepared in a separate reaction rather than *in situ* in the coupling reaction vessel. The PAM resin, or one related to it, clearly is superior for SPPS (36). The greatly increased work and cost of using this resin suggest that it should probably be reserved for synthesis of large peptides or those which have been found to cause problems by simpler procedures. Commercial availability of components for the Merrifield procedure of PAM synthesis (Pierce) and of the hydroxymethyl PAM (Peninsula) should lead to increased use of PAM resin. The Sheppard polyamide resins are available also (Chemical Dynamics and UCB).

Experience gained in nearly two decades of solid phase synthesis has taught many valuable lessons for avoiding trouble and improving the chances of success. One of the most important lessons is the value of standardization of all SPPS procedures used in the laboratory. This is critically important where several people are using common equipment and supplies, but standardization will improve the efficiency of even an individual chemist. In this laboratory, all SPPS is done on one of two standard scales, 0.4 or 1.0 mmol of peptide. The only exceptions to this rule are cases of very large peptides or extremely rare or expensive materials, when a synthesis will be begun with less than 0.4 mmol of amino acid–resin. Coupling reactions are generally run with 2.5-fold excess of Boc-amino acid and DCC. This means that the only amounts of Boc-amino acid weighed are 1.0 or 2.5 mmol. A small beaker is chosen and weighed, and its tare weight is inscribed on the beaker with a diamond. A chart is constructed containing a list of all the amino acid derivatives commonly used and their molecular weights, and columns containing the gross weight of the tare of the beaker plus 1.0 mmol or 2.5 mmol of Boc-amino acid are entered. This reference chart is kept by the balance where amino acid derivatives are weighed. In those cases where it appears advantageous to do a synthesis on a scale smaller than 0.4 mmol, one will often want to use a larger excess of acylating Boc-amino acids (for example in the synthesis of long peptides) to improve the chances of obtaining quantitative coupling at each step. Such syntheses can still be done so that the amount of Boc-amino acids weighed at each step is 1.0 mmol (or 2.5 mmol). All DCC solutions are 1.0 *M* for manual synthesis or 0.25 *M* for automatic synthesis. Weights of DCC needed to make up convenient volumes of these standard solutions are also recorded on the chart, as well as commonly used amounts of chain-terminating agents such as acetylimidazole and volumes of acetic anhydride and pyridine. With the use of these standardization procedures, the only calculations that need to be made—and these only once for each batch of aminoacyl resin—are those for the amounts of resin to give 0.4 or 1.0 mmol of amino acid. In addition to eliminating calculations, these standardization procedures are particularly valuable with automatic instruments, since level sensors never need to be changed and programs do not need to be rewritten for each synthesis. Standard record forms are used to record each operation in manual synthesis and each residue in automatic synthesis, along with comments on progress of the synthesis—monitoring results, for example.

It is also extremely important to use adequate quantities of wash solvents throughout the synthesis. Standard practice specifies the use of at least 20 mL of wash solvents per gram of 1% crosslinked polystyrene for each of six washes between different synthetic operations (that is, between deprotection and neutralization, or between neutralization and the addition of Boc-amino acid).

Another important lesson learned through experience is the importance of checking all starting materials before using them. Although the quality of amino acid derivatives currently obtainable from commercial sources is generally satisfactory, this cannot be assumed, and each new batch of derivative should be checked for purity by melting point and thin layer chromatography (tlc) (7).

10.4. SOLID PHASE SEGMENT SYNTHESIS

While nearly all the SPPS done up to the present time has used coupling of single amino acid derivatives to the resin, the use of protected small peptides in SPPS offers many advantages, the principal one being easier purification of the product. Peptides differing by a block of several amino acids should be much easier to separate than those differing by a single amino-acid residue. One must weigh the advantages against such problems as difficulties in synthesis of the required blocked peptide segments, difficulty in obtaining complete reaction on the polymer of the larger acylating groups, and racemization of the activated peptide segment. Several interesting SPPS segment syntheses have been reported, the most recent a synthesis of the insulin C-peptide by Niu (37). One can expect to see much additional work in segment condensation for synthesis of large peptides and proteins.

10.5. APPARATUS FOR SOLID PHASE SYNTHESIS

A wide variety of vessels have been described for manual SPPS; these are summarized in the review of Barany and Merrifield (9). The original Merrifield vessel is quite satisfactory if care is taken to assure that all resin particles come completely in contact with the solvents and reagents. Washing down of resin particles from the glass walls is facilitated by previous silanization of the vessel with dichlorodimethylsilane in toluene. The problem of complete wetting of the resin by reagents can be avoided if a vessel is used which is rotated through 180°, similar to that used on the first automatic instrument (38). With this type of vessel one must be sure that the liquid level is at least above the midpoint of the vessel so that the entire inner surface is wetted with each inversion. With these simple manual vessels, all solvents and reagents are poured manually into the vessel. This can be avoided by a suitable system of reservoirs, valves, and tubing. Several such systems have been described, and one is commercially available (Peninsula). These avoid open transfer of noxious reagents and exposure of materials to atmospheric moisture.

The first automatic instrument for SPPS was built by Stewart and described (38) in 1966. This simple instrument, improved by several modifications, is very reliable, and several are still in daily use in different laboratories. It has the advantage of using a rotating selector valve for solvents and reagents. This arrangement avoids cross-contamination of reagents and provides for complete wash-out of liquid lines at every step. Commercial SPPS instruments have been introduced by Schwarz, Beckman, and Vega. The Schwarz instrument has been discontinued, although some are still in use. All of these commercial instruments suffered from overengineering, which led to needless complexities and impaired reliability of operation. There is a great need today for a simple, reliable, inexpensive SPPS instrument. Such an instrument would not need the degree of sophistication and versatility provided by the present commercial instruments, but if it were reliable, it would save chemists a tremendous amount of time. A recent interesting development (39) is the adaptation of a Schwarz synthesizer to control by a laboratory microcomputer. A good automatic synthesizer, when used properly, can generally improve the quality of syntheses in addition to saving much time. The automatic instrument carries out each operation in a reproducible, standard manner, and is not distracted by such things as conversations and telephone calls.

In his original conception of SPPS, Merrifield visualized that a practical procedure for synthesis would be the use of the resins packed in a column, with all solvents and reagents being pumped through the column in a manner similar to chromatography. This principle has recently been applied by the Erickson (32) and Goodman (Chaturvedi et al., 40) groups. These investigators have used high pressure and stainless steel equipment, and they have demonstrated that SPPS can be carried out under these conditions with a dramatic decrease in the amount of time needed for synthesis. Additional research will be needed to determine the practicality of such systems and their merit relative to standard procedures.

10.6. PURIFICATION OF PEPTIDES

If all the reactions of SPPS proceeded to their theoretical limit, a homogeneous product would be obtained from the synthesis and purification would not be a problem. Unfortunately, this is rarely the case, although the best procedures recently developed have yielded nearly homogeneous peptides in some cases (14, 30).

The product from a typical solid phase synthesis will contain, in addition to the desired product, some shorter peptide chains. Some of the first amino acid attached to the resin usually fails to enter all synthetic reactions and contaminates the final product. In addition, a certain fraction of the peptide chains may stop growing during the synthesis, due to heterogeneity of the reaction sites within the polymer matrix. These "truncation sequence" peptides can frequently be removed from large molecular weight products by chromatographic methods based upon size discrimination, such as gel chromatography. Short chains present from deliberate termination of unreacted amino groups will differ in the presence of the terminating residue, and frequently this can be chosen (30) so as to facilitate purification. So-called "deletion sequence" peptides arise when a reaction fails to go to completion but is later completed by coupling of subsequent residues, yielding peptides lacking one or more amino acids from within the synthetic sequence. Such deletion peptides, particularly if they differ from the desired product by a single, simple amino acid, provide a great challenge in purification. Specialized techniques of purification may need to be developed for the individual peptide in question.

Countercurrent distribution (ccd) is the most generally useful primary tool for purification of peptides synthesized by solid phase methods. Although it does not offer extremely high resolving power unless large numbers of transfers are used, it offers the capacity to handle several hundred mg of crude peptide and frequently yields small peptides of adequate purity in a single overnight run of 100 or 200 transfers. Since the purification is based on liquid partition, hydrophobic impurities such as those derived from anisole in the HF cleavage are easily removed. Other commonly encountered side products, such as those derived by alkylation of methionine or tyrosine residues or acylation of anisole by glutamic residues, are readily removed. Partition chromatography in columns can substitute for ccd if the equipment is not available, and it offers the advantage of being able to handle more polar materials. Preparative high performance liquid chromatography (hplc) offers much promise, although many details of this relatively new method still need to be worked out. Hplc is unquestionably extremely valuable for demonstrating purity and homogeneity of synthetic peptides, although it should not be relied upon exclusively but used in conjunction with other procedures, such as electrophoresis and tlc (7).

One alarming feature of hplc reversed-phase supports is their tendency to retain significant amounts of the peptides run through them. A single cycling through extremely polar and nonpolar solvents will not reliably remove this material, and it often appears in subsequent runs as "ghosts." Such traces of contaminants from previously run peptides have already caused serious embarrassments to peptide chemists, and great pains should be taken to guard against these problems.

For large peptides and proteins, methods based on specific affinity may be required to achieve the necessary purification. Several affinity-based methods have been reviewed (9). If the synthesis is of a naturally occurring substance, antibodies raised against the natural material may be attached to gel supports and used in affinity chromatography for purification. Other methods unique to individual peptide and protein sequences will need to be developed by the investigator to solve his or her particular problems. In any case, it is critical that assessment of purity of a synthetic peptide be done by methods based on criteria different from those used for purification. For example, if the peptide was purified by ion exchange methods, assessment of homogeneity should include methods based, for example, on partition (such as tlc or hplc) or size.

Solid phase peptide synthesis has provided a vast host of materials since its introduction. At the same time, the chemistry of SPPS has been greatly improved. We can anticipate further progress in both these areas in coming years.

ABBREVIATIONS

BHA	benzhydrylamine (resin)
Boc	*t*-butyloxycarbonyl
Bpoc	2-(4-biphenylyl)-2-propyloxycarbonyl
BrZ	2-bromobenzyloxycarbonyl
Bzl	benzyl
ccd	countercurrent distribution
DCC	dicyclohexylcarbodiimide
Ddz	2-(3,5-dimethoxyphenyl)-2-propyloxycarbonyl
DCM	dichloromethane
DMF	dimethylformamide
Dnp	2,4-dinitrophenyl
Fmoc	9-fluorenylmethyloxycarbonyl
HBT	1-hydroxybenzotriazole
hplc	high performance liquid chromatography
MBHA	*p*-methylbenzhydrylamine (resin)
ONp	4-nitrophenyl ester
PAM	phenylacetamidomethyl (resin)
Ppoc	2-phenyl-2-propyloxycarbonyl
—Ⓟ	polymeric insoluble support

SPPS	solid phase peptide synthesis
TEA	triethyl amine
TFA	trifluoroacetic acid
tlc	thin layer chromatography
Tmz	α,2,4,5-tetramethylbenzyloxycarbonyl
Tos	*p*-toluenesulfonyl (Tosyl)
Xan	9-xanthenyl

REFERENCES

1. R. B. Merrifield, *J. Amer. Chem. Soc.* **85,** 2149 (1963).
2. R. B. Merrifield, *Biochemistry* **3,** 1385 (1964).
3. R. B. Merrifield and J. M. Stewart, *Nature* **207,** 522 (1965).
4. J. M. Stewart, *Fed. Proc.* **27,** 63 (1968).
5. B. Gutte and R. B. Merrifield, *J. Amer. Chem. Soc.* **91,** 501 (1969).
6. D. Yamashiro and C. H. Li, *J. Amer. Chem. Soc.* **100,** 5174 (1978).
7. J. M. Stewart and J. D. Young, *Solid Phase Peptide Synthesis,* Freeman, San Francisco (1969).
8. S. B. H. Kent, A. R. Mitchell, M. Engelhard, and R. B. Merrifield, *Proc. Nat. Acad. Sci. U. S.* **76,** 2180 (1979).
9. G. Barany and R. B. Merrifield, in *The Peptides* (E. Gross and J. Meienhofer, Eds.), Academic Press, New York (1980), p. 1.
10. J. D. Young and J. M. Stewart, *Solid Phase Peptide Synthesis,* Pierce Chemical Co., Rockford, Ill. (1983).
11. P. Fankhauser and M. Brenner, in *Chemistry of Polypeptides* (P. G. Katsoyannis, Ed.), Plenum, New York (1973), p. 389.
12. C. Birr, in *Reactivity and Structure Concepts in Organic Chemistry,* Vol. 8 (K. Hafner, C. W. Rees, B. M. Trost, J. M. Lehn, P. von Rague Schleyer, and R. Zahradnic, Eds.), Springer-Verlag, New York, 1978.
13. R. C. Sheppard, in *Peptides 1971* (H. Nesvadba, Ed.), North-Holland, Amsterdam (1973), p. 111.
14. S. B. H. Kent and R. B. Merrifield, in *Peptides 1980* (K. Brunfeldt, Ed.), Scriptor, Copenhagen (1981), p. 328.
15. B. F. Gisin, *Helv. Chim. Acta* **56,** 1476 (1973).
16. E. Atherton, N. L. Benoiton, E. Brown, R. C. Sheppard and B. J. Williams, *J. Chem. Soc. Chem. Commun.,* 336 (1981).
17. S. S. Wang, *J. Amer. Chem. Soc.* **95,** 1328 (1973).
18. C. D. Chang and J. Meienhofer, *Int. J. Pept. Protein Res.* **11,** 246 (1978).
19. A. R. Mitchell, S. B. Kent, M. Engelhard, and R. B. Merrifield, *J. Org. Chem.* **43,** 2845 (1978).
20. J. M. Stewart, P. DeArmey, and S. Berga, in *Chemical Synthesis and Sequencing of Peptides and Proteins* (T.-Y. Liu and A. N. Schlechter, Eds.), Elsevier, New York (1981), p. 179.
21. J. T. Sparrow, *J. Org. Chem.* **41,** 1350 (1976).

22. G. R. Matsueda and J. M. Stewart, *Peptides* **2,** 45 (1981).
23. U. S. Biochemical Corp.
24. E. Kaiser, R. L. Colescott, C. D. Bossinger, and P. I. Cook, *Anal. Biochem.* **34,** 595 (1970).
25. A. M. Felix, M. H. Jimenez, R. Vergona, and M. R. Cohen, *Int. J. Peptide Protein Res.* **5,** 201 (1973).
26. T. Christensen, in *Peptides, Structure and Biological Function* (E. Gross and J. Meienhofer, Eds.), Pierce Chemical Co., Rockford, Ill., 1979, p. 385.
27. R. S. Hodges and R. B. Merrifield, *Anal. Biochem.* **65,** 241 (1975).
28. P. Villemoes, T. Christensen, and K. Brunfeldt, *Hoppe-Seyler's Z. Physiol. Chem.* **357,** 713 (1976).
29. C. Peña, J. M. Stewart, A. C. Paladini, J. M. Dellacha, and J. A. Santome, in *Peptides* (R. Walter and J. Meienhofer, Eds.), Ann Arbor Science Pub., Ann Arbor, Mich. (1975), p. 523.
30. R. B. Merrifield, in *Peptides: Structure and Biological Function* (E. Gross and J. Meienhofer, Eds.), Pierce Chemical Co., Rockford, Ill. (1979), p. 27.
31. G. R. Matsueda and J. M. Stewart, in *Peptides: Chemistry, Structure and Biology* (R. Walter and J. Meienhofer, Eds.), Ann Arbor Science Pub., Ann Arbor, Mich. (1975), p. 333.
32. B. W. Erickson, T. J. Lukas, and M. B. Prystowsky, in *Polypeptide Hormones* (R. F. Beers Jr. and E. G. Bassett, Eds.), Raven Press, New York (1980), p. 121.
33. H. Yajima, in *Chemical Synthesis and Sequencing of Peptides and Proteins* (T.-Y. Liu and A. N. Schechter, Eds.), Elsevier, New York (1981), p. 21.
34. J. M. Stewart and D. H. Morris, U. S. Patent 4,254,023 (1981).
35. R. S. Feinberg and R. B. Merrifield, *Tetrahedron* **30,** 3209 (1974).
36. V. K. Sarin, S. B. H. Kent, and R. B. Merrifield, *J. Am. Chem. Soc.* **102,** 5463 (1980).
37. C.-I. Niu, in *Chemical Synthesis and Sequencing of Peptides and Proteins,* (T.-Y. Liu and A. N. Schechter, Eds.), Elsevier, New York (1981), p. 11.
38. R. B. Merrifield, J. M. Stewart, and N. Jernberg, *Anal. Chem.* **38,** 1905 (1966).
39. J. T. Sparrow, personal communication.
40. N. Chaturvedi, G. Sigler, W. Fuller, M. Verlander, and M. Goodman, in *Chemical Synthesis and Sequencing of Peptides and Proteins* (T.-Y. Liu and A. N. Schechter, Eds.), Elsevier, New York (1981).

CHAPTER

11

APPLICATION OF CONTROLLED PORE GLASS IN SOLID PHASE BIOCHEMISTRY

WOLFGANG HALLER

National Bureau of Standards
Washington, DC 20234

Biochemistry is mostly carried out in liquids or on solid–liquid interfaces. In the latter, the interaction of the liquid or its constituents with the solid can span the whole spectrum from being solely physical (electrostatic repulsion or attraction) to being a complicated highly specific chemical reaction. Traditionally, one likes to increase the available surface area, and therefore the solids employed are either in the form of fine powders or fibers, or they achieve the high surface area by being porous. There are exceptions to this: certain detection systems, for instance radioactive or biospecific methods, are so sensitive that they respond to the small amounts of reactants provided by the area of a body with flat surfaces.

The choice of porous solid phases is not always just a way to obtain high surface area in a small volume. The actual geometry of the solid can achieve desired functions. For instance, the size of the pores affects the kinetics of diffusion of a species to or from the interior of the solid phase substrate. In many cases, it is therefore highly desirable to have optimal control of pore diameters, pore-diameter distribution, and pore volume. Preferably these geometric parameters should also remain the same throughout the experiment under all possible chemical and physical insults to the substrate and also, for economy, through cleaning procedures and long-range storage.

Meeting the important requirements of surface condition and geometry, the solid phase substrate is otherwise a passive participant in its uses, or at least should be so. Ideally, it should not change surface area and pore size during use, and it should be chemically inert, but not to the extent that its surface cannot be chemically manipulated to carry the widest possible variety of surface functionalities and ligands.

During the past years a large number of supports that meet the above requirements reasonably well have been developed, and many excellent surveys on their use have been published. In writing this chapter, I do not intend to duplicate these writings. Instead, I will report on a single type of support material that I helped to develop in the late 1950s and early 1960s and which is now widely used in a variety of applications. This support has become known as controlled pore glass, or CPG. Any attempt to survey a field consisting of a large number of publications within an allotted space inevitably requires the arbitrary omission of a fraction of the existing body of work, and I hope that this does not raise too much criticism from the educated readers.

My interest in inert solid phase supports was aroused through a fortuitous meeting with Leonard Lerman in the summer of 1954. Lerman at that time, together with Campbell and Luescher, did some pioneering work (1) on a technique which much later was named affinity chromatography. Lerman used cellulose as the solid substrate for his reactions and complained about its lack of chemical and physical stability. Sharing the favorable opinion of every chemist that glass represents the ultimate in inertness and shaping ability, Lerman asked if glass could not be produced with high and controlled surface area and also be covalently surface-derivatized to accept surface functionalities such as antigens and antibodies. This accidental meeting eventually led me into several research areas: The foremost questions to be solved were how to attach organic groups covalently and stably to a glass surface and how to produce a glass with high surface area, having convenient shapes, and with a maximum of control of its pore morphology.

I approached first the task of covalent linkage of biospecific groups to glass in analogy to Lerman's cellulose system, using available high-surface-area siliceous materials, such as glass powder, sintered glass, fiberglass, "Thirsty Glass," and silica gel (2). These materials accepted covalent ligands very well, but they either had too small a surface area or they had pores that were too large or too small or of an unfavorably heterogeneous nature. To overcome this shortcoming, I turned my interest to developing a suitable porous glass. The completion of this work coincided with two originally unrelated technical developments. One was the emerging commercial synthesis of bifunctional silane coupling agents for the glass fiber reinforced plastic industry (3). The other was the immense success of permeation chromatography (gel filtration) in the biological sciences (4). The porous glass eventually developed (5) proved to have very sharp pore distribution and therefore suggested itself as a medium for permeation chromatography, particularly since the crosslinked dextran gel and the agarose used at this time were not practical for the chromatography of large molecules and viruses. Consequently, I devoted myself to the chromatography of viruses and to mechanistic investigations into permeation chromatography. At the same time, I continued with my earlier interest in glass-derivatization procedures, using the new controlled pore glass and the now readily available bifunctional silane-coupling agents. I found aminopropyl controlled pore glass a particularly stable and versatile material, which I recommended and distributed to workers in the field (6). Amino propyl CPG was subsequently used by many workers to prepare CPG analogous to solid phase materials previously based on cellulose or organic gels.

11.1. WHAT IS GLASS?

Most treatises on glass start with several definitions of the term glass which defines it by its molecular structure, its technical process of manufacture, by a kinetic description, or by a combination of the above. For our purposes, we may do well with Morey's (7) definition as "an inorganic substance in a condition which is continuous with, and analogous to, the liquid state of that substance, but which, as the result of a reversible change in viscosity during cooling, has attained so high a degree of viscosity as to be for all practical purposes rigid." This definition implies that glass is a metastable supercooled liquid, that is, a liquid that has been cooled below its crystallization point without crystallization having occurred. This is correct, except that in order to become a glass the undercooled liquid has to be further cooled through the transformation region until its viscosity is so high that it has become a solid. Contrary to some legends, technical glasses do not flow or crystallize over long periods of time. As a matter of fact and despite its reputation of fragility, glass under compressive loading is stronger than most metals, which are crystalline and, having easily yielding slip planes, deform plastically.

By the above definition, certain high polymers are also glasses in the physical sense, but we will limit our further discussion to glasses composed of inorganic oxides such as the silicate glasses used to make optical lenses, windows, containers, insulating fibers, lamp bulbs, and so on. Glass has four much-used outstanding characteristics: Transparency, chemical inertness, rigidity, and the ability to be shaped in the hot plastic state. To this we may add its high degree of homogeneity, which it derives from its being a solidified liquid. It can be stirred in the liquid state, and its attainable high degree of homogeneity is a prerequisite for its use for optical components and is also essential to the preparation of high performance controlled pore glass.

Fused silica is one of the simplest inorganic glasses. It is also widely used for scientific instruments because of its extremely high chemical durability, its low thermal expansion, good ultraviolet transmittance, and high deformation temperature. Traditionally, it is made by high temperature (~1900°C) melting of quartz, by flame hydrolysis of silicon halide vapors, or by sintering of phase-separated, leached borosilicate glass. These processes are relatively costly, which explains the limited use of fused silica.

The introduction of other inorganic oxides into silica lowers the temperature required for fusion but at the same time lowers its chemical durability and increases the tendency for unwanted crystallization of the glass melt. To balance these opposing effects, it is not uncommon for technical glasses to be composed of at least five different major constituents. Common "soda lime silica" glass as used worldwide for windows, containers, and envelopes for incandescent lights has an approximate composition (weight percent) of 72% SiO_2, 16% Na_2O, 7% CaO, and the residual 5% mainly the oxides of magnesium and aluminum. The glass may have many more minor constituents, which it acquires from contaminants in the raw material or from dissolution of the walls of the containers in which it is molten. Most glasses are molten in ceramic pots or in furnaces with ceramic walls. For more demanding applications, such as special optical components, glass is molten in platinum crucibles or platinum-lined furnaces. The use of platinum contributes to cost, but at the necessary melting temperatures of 1400 to 1500°C, other container material dissolves in the glass and cause serious interferences.

Numerous authors have attempted to develop a molecular model for the structure of glass. Most widely used is the model of Zachariasen (8). Fundamental for this model is the observation that glasses and crystals of the same composition have similar energy. Therefore glass and crystal are composed of identical structural entities held together by the same type of bonds. In the case of crystalline and noncrystalline silica (silica glass), the units are SiO_4-tetrahedra linked by oxygen bridges. In the crystal, the tetrahedra are regularly arranged, whereas in the glass they form an irregular random network. Figure 11.1 shows a two-dimensional rendering of the regular network of crystalline silica, and Figure 11.2 shows the irregular network of silica glass. In both cases, silica is shown with three valences only, the fourth valency extending above or below the paper plane. Zachariasen's concept was supported by X-ray diffraction studies by Warren (9) and others. Glasses have very diffuse X-ray diffraction patterns, similar to those obtained from liquids. Since controlled pore glass is almost pure silica, we will at this point not deal with structural models for more complicate glasses that contain network modifying cations such as sodium.

11.2. SOLUBILITY OF GLASS

In many applications, chemical durability is the most important material property for which glass is chosen. Conventional glasses are resistant to water and to acids of any strength, except hydrofluoric and strong phosphoric acid. They are less resistant to strong alkali. The latter is, for instance, the reason that glass-fiber-reinforced concrete has limited service-life, even

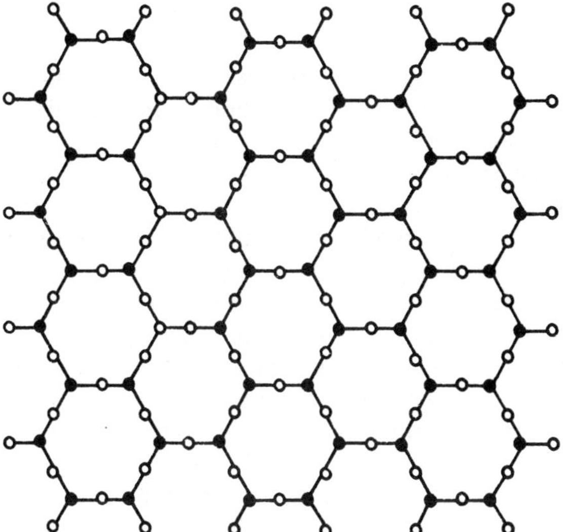

Fig. 11.1. Warren–Zachariasen network model for crystalline silica. ● Silicon; ○ oxygen.

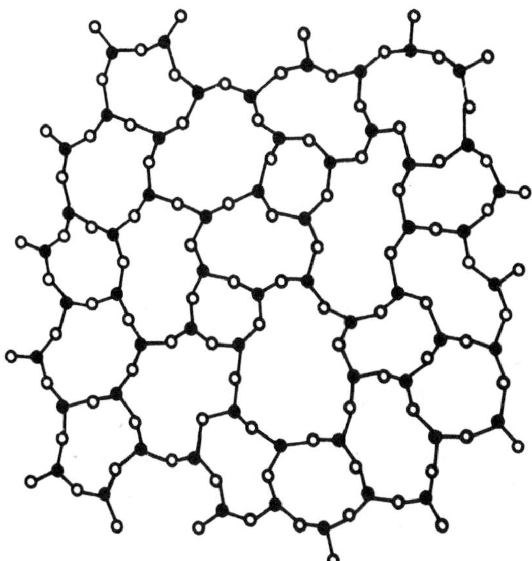

Fig. 11.2. Warren–Zachariasen network model for silica glass. ● Silicon; ○ oxygen.

though the incorporation of zirconia into the glass has led to a modest improvement in alkali resistance.

Not all glasses have high corrosion resistance. "Waterglass" is a binary silicate glass with a high content of sodium oxides. It dissolves easily in water, and such solutions ("waterglass") are used in cements and for impregnating masonary, textiles, and paper. The incorporation of B_2O_3 and P_2O_5 into glass also decreases durability against aqueous reagents. These substances are, however, sometimes added to lower fusion temperatures or to design glasses with certain specific properties such as low or high expansion coefficient, low electric conduction or high ultraviolet transmission.

The mechanism by which glasses dissolve is very complex. It can range from simple dissolution of all components to a selective dissolution of some of the components, leaving behind a more or less coherent porous layer. To complicate matters, the solubility of some glasses is profoundly affected by the thermal history they had experienced. Turner and Winks (10), in a paper published in 1926, reported that glasses in the Na_2O-B_2O_3-SiO_2 system became very soluble in hydrochloric acid when its B_2O_3 content exceeded 20 weight percent. This solubility was selective, that is, the acid left behind a very fragile skeleton composed of almost pure SiO_2. The discovery of Turner and Winks was the starting point of the development of a process by Nordberg and Hood (11) for making the high silica glass "Vycor." In the Vycor process, objects are formed from a low-melting alkali borosilicate glass, and after a heat treatment the objects are leached with acid. The acid removes the Na_2O and the B_2O_3. The residual SiO_2 skeleton is dried and sintered into compact objects of the high-silica Vycor glass. Although the Vycor process was successfully practiced over decades, the physical mechanism underlying the heat-treatment-dependent solubility of the borate glass was not well understood and gave rise to many conflicting theories. Eventually, the phenomenon yielded to scientific curiosity and improvements in electron microscopy. The phenomenon was actually found to be rather common. Heat-treatment-dependent loss of chemical durability in a glass composition widely used in glassware for kitchen and laboratory was, for instance, reported by Howell, Simmons, and Haller (12).

11.3. WAYS TO PREPARE HIGH-SURFACE-AREA MATERIALS

Powders produced by grinding, spraying, or precipitation have high surface area but are difficult to handle because one generally prefers a monolithic material. Powders can be sintered, particularly if they consist of a thermoplastic material such as glass, and sintered glass has indeed occasionally been used as a solid support or sorbent in biochemistry. Traditional ceramic bodies (firebrick, porcelain, etc.) can also be rendered porous, but, similar to sintered glass, the specific surface area is low and the pore size not well controllable. On the other end of the spectrum of pore sizes is silica gel and "Thirsty Glass." Thirsty Glass, an intermediate stage in the production of Vycor glass, is sold by Corning Glass Works under Code #7935. It has an average pore size of 30 to 45 Å, with a rather wide pore distribution, an internal surface area of 200–350 m^2/g, and a pore volume of 28%. It had been used as adsorbent and a solid-phase support in biochemistry (13) and found useful in applications where small pore size is acceptable.

Silica gel is a coherent, rigid three-dimensional network of particles of colloidal silica. The particles form originally by the polymerization of a monomer (silicic acid) and the fluid suspension of the particles is called a sol. The sol converts to a gel by collision and cohesion of the particles, and the particles may become cemented together by further polymerization of silicic acid on the points of contact. Pore size, pore volume, and pore morphology of silica gels are determined by a complicated interplay of monomer concentration, particle growth, coagulation, capillary forces at drying, and so on. The art of making silica gel has been practiced over centuries, and because of its low price it is the mainstay of many industries requiring a porous adsorbent or support. An excellent description of its theory and practice can be found in Iler's book (14).

Organic hydrogels such as polyacrylamide or crosslinked dextrans are formed by the polymerization of a monomer in the presence of a solvent. The concentrations of the monomer and of the crosslinking agent determine the density of the gel and also its pore size and pore volume. Dense gels have small pore size and small pore volume, less dense gels have large pore size and high pore volume. This explains the fact that gels of large pore size are generally less rigid.

Controlled pore glass is produced by a process related to the process of making Thirsty Glass but conceptually quite unrelated to the making of gels. It is discussed in more detail in the next sections.

There may be some interest in producing solid objects (container walls, beads) that are nonporous except for an exterior high-surface-area layer. Such "pellicular" bodies are not economical in terms of surface area per total volume, but they offer possible advantages to the diffusion kinetics in solid phase assays, enzyme reactions, and analytical rather than preparation permeation chromatography where only small amounts of substances have to be processed. Obviously, it is possible to mat-etch glass surfaces with hydrofluoric acid or to create a rough surface by abrasive grinding or airblasting. More sophisticated porous layers can be produced on nonporous surfaces by depositing from the liquid or vapor phase coherent small particles as, for instance, when making silica gel or fumed silica.

11.4. METASTABLE LIQUID–LIQUID IMMISCIBILITY

Turner and Winks's (10) observation of the thermal-history-dependent solubility of borate glasses has been discussed in Section 11.2. Also mentioned was Nordberg and Hood's (11) process of making high silica (Vycor) glass. It is only very recently that satisfactory physical explanations for the phenomena underlying these processes have become available. The gained understanding, aside from being an academically satisfying luxury, has been a prerequisite for the development of controlled pore glass.

It is today understood that the above phenomena are caused by a special and somewhat esoteric case of liquid–liquid phase separation. Liquid–liquid phase separation in binary or multicomponent liquid mixtures has been known for centuries. It is the phenomenon by which two or more liquids are completely immiscible above a certain temperature (the critical immiscibility temperature T_c) but "unmix" below this temperature. The process can be explained by a phase diagram (Figure 11.3). Plotted is the concentration of substances A and B in a mixture against temperature. The parabolic line is the coexistence curve which encloses the miscibility gap. Above T_c, the two liquids are miscible in every ratio. Below T_c, miscible compositions are limited to areas outside of the gap. A liquid M_1 of composition C_1 at temperature T_1 which is being cooled to temperature T_2 will segregate into two liquids M_2 and M_3 of composition C_2 and C_3. In low-viscosity liquids this segregation process is rather fast, and the end result is two separated layers of liquids by

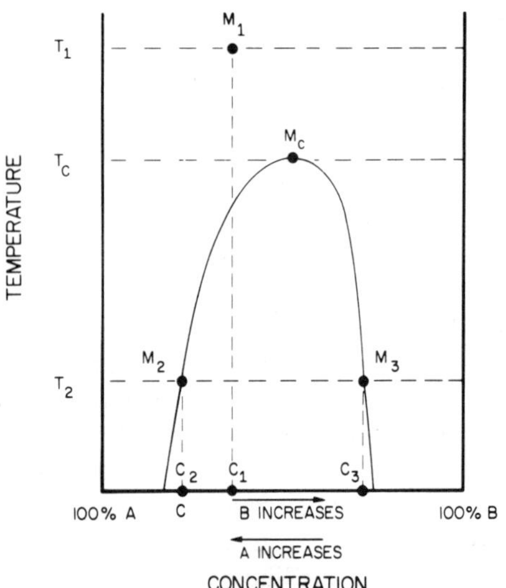

Fig. 11.3. Phase diagram for mixture of two liquids having limited mutual miscibility.

composition C_2 and C_3. The diagram tells only the final result of the segregation, not its kinetics or the morphology of intermediate stages. As a matter of fact, because of the speed with which the unmixing occurs in ordinary liquids, very little is known about these intermediate stages. Liquid–liquid phase separation in glass-forming systems has been known for many years. All of the early work, however, was concerned with immiscibility above the crystallization temperature of the glass melt. At these temperatures the viscosity of the melt is low enough that quenching of the melt, if not leading to crystallization, leads to a two-layered glass or to a turbid glass consisting of a coarse emulsion of two different glasses. In recent years, thanks to the widespread use of the electron microscope, detailed attention has also be given to miscibility gaps that occur in the undercooled-liquid region of glass forming systems. Due to the high viscosity at these temperatures, phase separation proceeds sometimes so slowly that the phase-separated regions are only several angstroms in diameter. Because such gaps occur below the crystallization temperature in the undercooled melt they are termed metastable. Metastable liquid–liquid immiscibility is sometimes very difficult to detect. If the domains are large enough and their compositions are sufficiently different in refractive index, or electron absorption, the phase separation can be detected by light scattering, or transmission electron microscopy. Otherwise one has to resort to the measurement of viscosity anomalies (15) or take advantage of the different solubilities or corrosion resistance of the two phases. Superficial etching, followed by replication and shadowing, shows phase separation. In Figure 11.4 the less soluble phase forms spherical noninterconnected inclusions. In cases where the volumes of both phases are close to equal and the more soluble phase is continuous, the glass can be leached into a porous structure and either the solubility can increase because of heat treatment (12) or the porosity of the residual skeleton can be used as an indication for phase separation.

Binary immiscible systems are relatively easily illustrated and thermodynamically treated. Classic thermodynamics calls for symmetrical gaps, yet most gaps are assymetric (Figure 11.5). Haller, Blackburn, and Simmons (16) proposed that silica is polymeric and that indeed, after allowing for a $(SiO_2)_n$ entity, the gap becomes symmetric (Figure 11.6). It is more difficult to study and display data of systems of more than two components. In a three-component system, the immiscibility curve becomes an immiscibility surface, which can be displayed by curves connecting all points of equal critical immiscibility temperature. Figure 11.7 from Haller et al. (17) shows such a rendering for the B_2O_3-SiO_2-Na_2O system, which is the basis for the Vycor process as well as for the presently most widely used controlled pore glass.

The shape of immiscibility curves or surfaces may be strongly affected by small changes in composition and particularly by certain impurities. In any

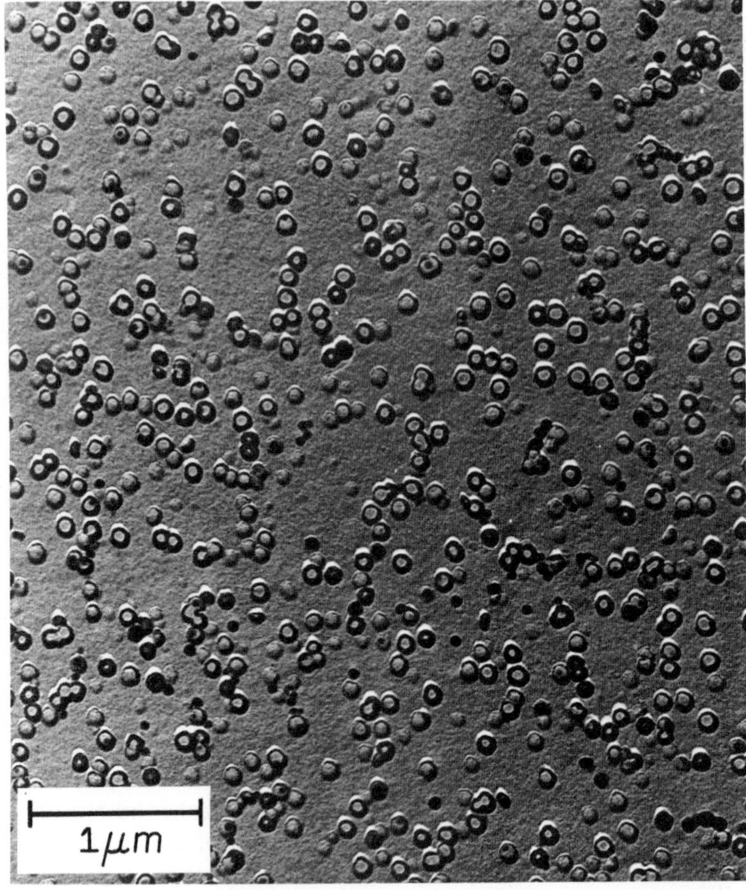

Fig. 11.4. Replica electron micrograph of phase-separated glass with isolated-sphere morphology. Courtesy H. R. Golob, PPG Industries.

such studies, it is therefore of utmost importance to use pure raw materials, the most reliable and clean melting procedures, to hold evaporation losses during melting to a minimum and to homogenize the melt by vigorous stirring.

11.5. PHASE MORPHOLOGY IN LIQUID–LIQUID PHASE SEPARATION

The hypothetical terminal stage of liquid–liquid phase separation are two layers of immiscible liquids. Thermodynamics tells little about how a single

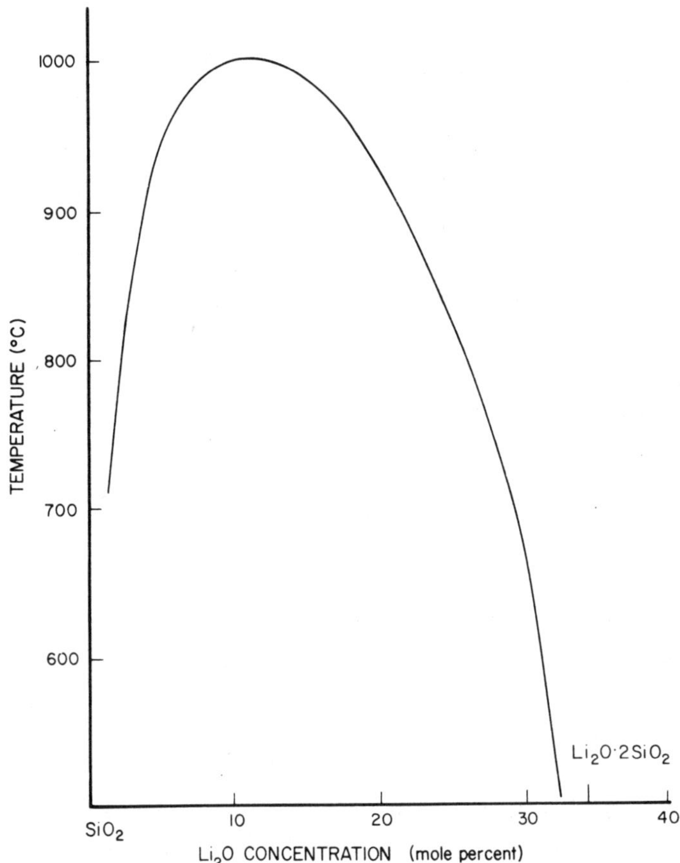

Fig. 11.5. Miscibility gap in Li_2O-SiO_2 system. Courtesy *J. Amer. Ceram. Soc.* (16).

liquid becomes two liquids. The discovery of metastable liquid–liquid phase separation in glass-forming systems created an unique opportunity to study a system from the first appearance of a second liquid to the more advanced and better-known stages. There are two reasons for this. First, the metastable gaps appear at such low melt temperature and therefore high viscosities that diffusion is slow. Second, the undercooled liquid, or liquid dispersion, can further be cooled below the glass transition temperature, where all motion and diffusion freezes, preserving the intermediate stages of unmixing. As the reader may guess, this freezing at various stages is also the process by which pore control is achieved in controlled pore glass (CPG).

Early stages of liquid–liquid phase separation are usually characterized by either a continuous matrix of one glass in which isolated spheres of another

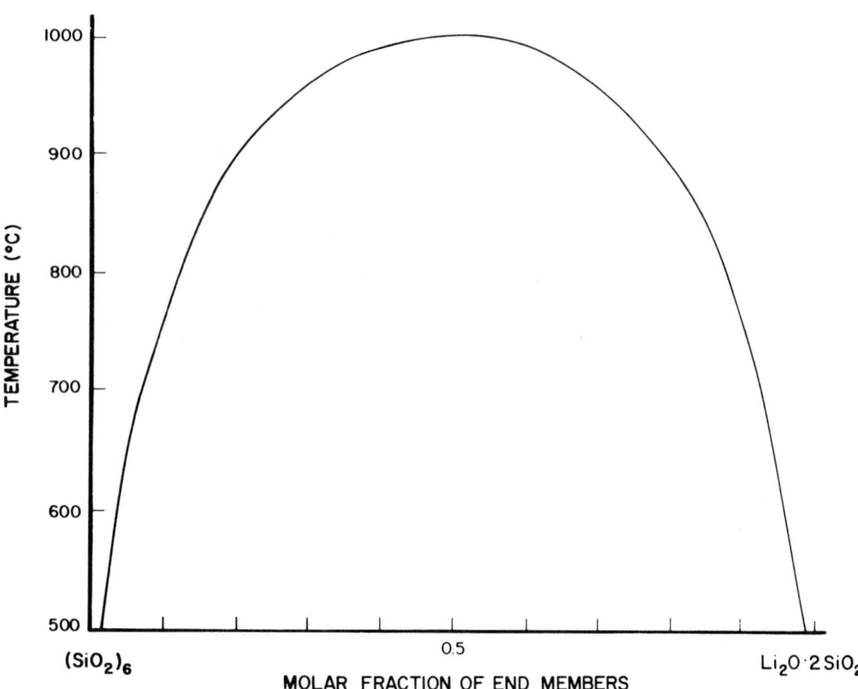

Fig. 11.6. Miscibility gap in Li_2O-SiO_2 system. Curve represents regular mixing thermodynamics, assuming complex end members. Courtesy *J. Amer. Ceram. Soc.* (16).

glass are suspended (Figure 11.4), or by both phases being interconnected (Figure 11.8). Scientists agree that isolated spheres represent homogeneous nucleation. Interconnected phases were believed to be caused by spinodal decomposition. Cahn (18) constructed a mathematical model based on spinodal decomposition and computer-generated cross-sections through two-phase structures that strongly resemble micrographs like Figure 11.7. I claimed (19) that interconnected morphology alone is not a sufficient criterion for spinodal decomposition and suggested that interconnected morphology could also be caused by an "intersecting growth mechanism." The "intersecting growth mechanism" hypothesis has since received support from experimental work by Seward et al. (20). A more extensive discussion of the origin of phase-connecting in microheterogeneous glasses can be found in Haller and Macedo (21). According to the intersecting growth theory, the ratio of volume fractions of the two phases is critical for interconnected morphology, and in designing a porous glass process it is advisable to choose glass compositions that are at or close to the highest point of a miscibility gap or surface.

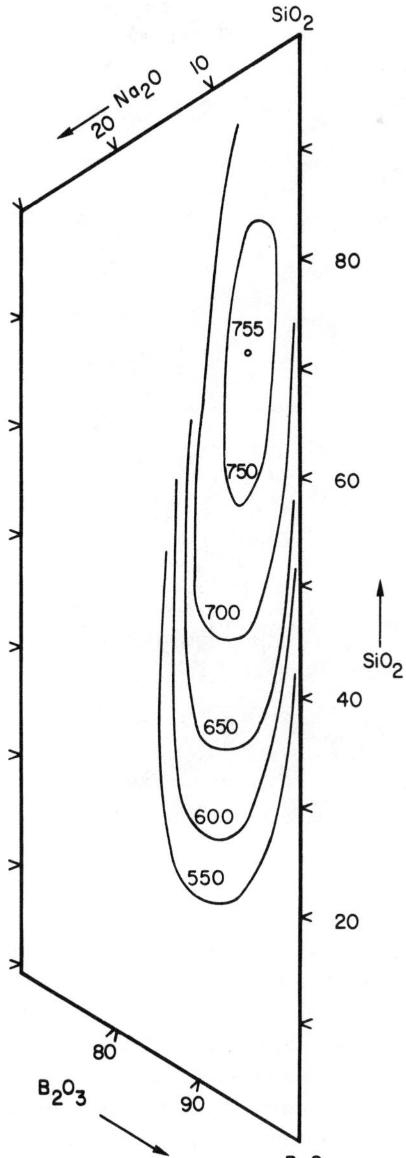

Fig. 11.7. Metastable immiscibility surface for Na_2O-B_2O_3-SiO_2 system. Courtesy *J. Amer. Ceram. Soc.* (17).

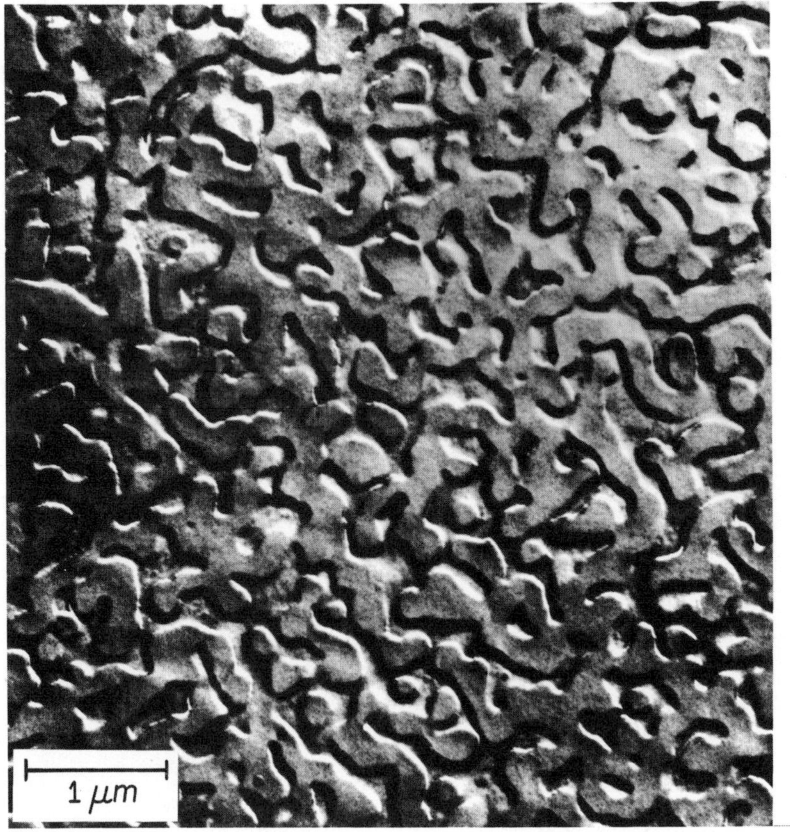

Fig. 11.8. Replica electron micrograph of phase-separated glass with interconnected phase morphology.

After the initial stages of phase separation, an interconnected two-phase system undergoes a coarsening process that is similar to Ostwald ripening of crystals in saturated solutions. While the general morphology does not change, the whole system coarsens. I experimentally studied the kinetics of this coarsening (19) in a glass of the B_2O_3-SiO_2-Na_2O system and derived kinetic coarsening laws for mass transport by volume and bulk diffusion. I came to the conclusion that the dominant process in the studied temperature interval is interfacial-step-controlled bulk diffusion.

11.6. CONTROLLED PORE GLASS PREPARATION

The making of controlled pore glass involves the steps of melting a suitable glass composition, for instance at the top of the immiscibility surface of Figure 11.7. As already mentioned, small deviations in composition and particularly impurities cause shifts in the boundaries of the miscibility gaps. They also cause changes in the phase-separation kinetics. To obtain best results, melting should be done in platinum containers under stirring and observance of all practices to obtain a highly homogeneous glass. The glass is quench-cooled to minimize any phase separation at this stage. Subsequently the glass is reheated for several hours to several days to achieve phase separation and coarsening of the microheterogeneities. This thermal history determines the later size of the pores of the glass. Figure 11.9 shows a typical set of heat-treatment times and temperatures to obtain a number of desired pore diameters. Minor changes in glass composition, and particularly contaminants, cause shifts in the shown curves, and it is advisable to make calibration runs on any new glass batch produced. Next, the glass is crushed and sieved to the desired particle-size fractions and the alkali-borate-rich phase is decomposed in hot acid (e.g., 3*n*-HCl at 50°C for 6 hr). Decomposing the soluble phase removes Na_2O and B_2O_3 but leaves behind a colloidal precipitate of silica from the decomposed phase. In the next step, this colloidal silica plus some of the skeleton is drilled out by either dilute hydrofluoric acid or alkali solu-

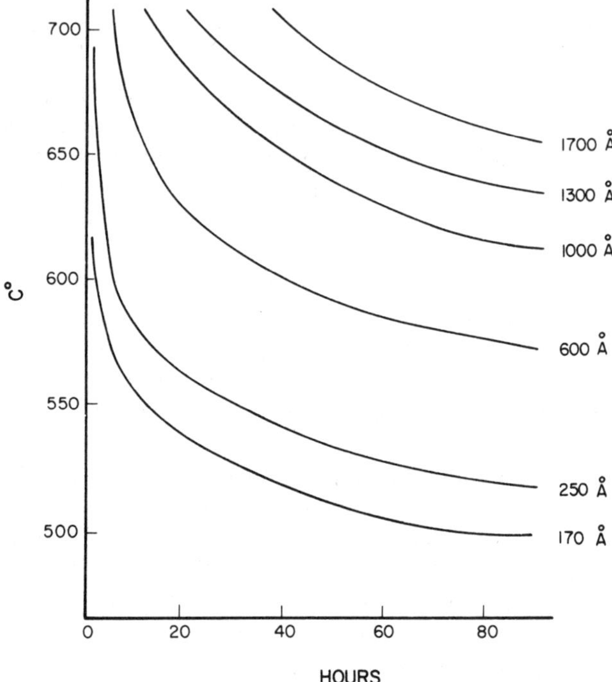

Fig. 11.9. Heat-treatment history vs. resulting pore-diameters for phase-separated and leached Na_2O-B_2O_3-SiO_2 glass.

tions (e.g., 0.5*n*-NaOH at 25°C for 2 hr). The timing of this step is rather critical, otherwise the CPG is either destroyed or colloidal silica is left behind and interferes with its later uses. The fairly high activation energy of silica dissolution makes this process also temperature-sensitive, and its exothermic nature can cause runaways in large-volume operations. The glass is characterized by surface area measurements (BET-nitrogen adsorption) and by high pressure mercury intrusion porosimetry. The latter gives pore volume and pore distribution curves. It has become customary to characterize the average pore size of the glass as the mercury intrusion pore diameter at 50% volume intrusion (5). Compared with electron micrographs, this value corresponds to the narrow parts of the undulating pores (Figure 11.8), which are the determining parameter in the advance of the mercury column through the pores. This diameter is obviously also the determining parameter in permeation chromatography which accounts for the good agreement in the use of mercury-intrusion calibrated CPG for the determination of molecular sizes (22). Glass surfaces adsorb iron, calcium, and magnesium at neutral and alkaline pH, and it is therefore important to use deionized water for all washing steps.

11.7. PROPERTIES AND HANDLING OF CONTROLLED PORE GLASS

Controlled pore glass is a free-flowing powder consisting of rigid porous granules of nearly pure quartz glass, permeated by an interconnected network of pores. The dimensions of the granules are not affected by immersion in organic or aqueous solvents. The rigidity results in constant column parameters, which is particularly useful in repeated or automatized operation. Accidental air bubbles introduced into beds of CPG do not cause bed-deformations or local deswelling, therefore they are easily removed by flushing with degassed solvents, preferably at high flow-rate and pressures and at low temperature. This is achieved, for instance, by flushing the column with degassed (boiled and cooled) water at highest pumping speed while restricting the flow at the column exit by pinching the hose or inserting a fine-gauge hypodermic needle.

Since the grains of CPG do not deform under pressure, flow resistance is low and pressure and flow rates are linearly related to each other. Table 11.1 gives the flow rates through a column of 1 cm^2 cross-section and 100 cm length, at 1 bar working pressure. (1 bar is equivalent to ~1 atm or 14.7 psi, 101 kPa, or the pressure produced by 10 m head of water.) Since flow rate is proportional to cross-section and inversely proportional to length of column, the given figures can easily be used to calculate flow rates, respectively, flow resistance for other column configurations.

Table 11.1. Flow Rate of a Column of 1-cm² Cross-Section and 1-m Length

Grain Diameter (μm)	Mesh Range (ASTM)	Flow Rate[a] cc/min
180–840	20/80	10–12
125–177	80/120	5–7
74–125	120/200	3.5–5.0
37–74	200/400	1.2–1.6

[a] Liquid is water. Pressure is one atmosphere (1 bar, 14.7 psi, 10 m head of water, 101 kPa).

CPG granules made by the described process are crushed bulk product and have the same pore size at the surface of the product as in its center. The granules are chemically inert against water, organic liquids, or exotic eluants and can be subjected to exhaustive derivatization procedures at high temperatures in optimal solvents. They are attacked by strong bases and hydrofluoric acid. They are not attacked by microorganisms. CPG can be heat sterilized by steam or dry heat. Glass that has been contaminated with organic material can be cleaned by heating to 700°C without change in pore size distribution, or by oxidation in hot, concentrated nitric acid. Alkali-contaminated CPG or poorly made CPG may suffer some pore-distribution change by heating to 700°C, and lower temperatures should be used. With good care, it is infinitely reusable. When stirring a slurry of CPG, one has to avoid grinding action between stirrer and bottom of vessel, as with bottom-resting magnetic stirrers. Stirring should be done by hand with a rod, by motion of the vessel (slushing), or by a top-supported stirrer. Because the glass settles on the bottom of containers, slurries should never be heated above the boiling point without adequate stirring. Otherwise bumping occurs, which is particularly dangerous with strong acids or flammable liquids. It is advisable to use a steam bath when cleaning CPG with hot nitric acid. For *in situ* cleaning of CPG columns, one can use acids (unless they attack the components of the system) and mild bases. Slugs of diluted ammonia (1 part conc. NH_4OH, 3 parts H_2O) can be pumped through a protein-contaminated column, but exposure to the base should not exceed several minutes. Such treatment should be followed by extensive flushing with water until the pH of the eluant has returned to neutral. This treatment is not advisable for surface derivatized CPG. Nonoxidizing and nonsilica-dissolving cleaning agents, particularly for derivatized CPG, are denaturing solvents (dodecylsulfate, urea, propylalcohol), chaotropic solvents (KCNS, LiBr), a mixture of the above, or proteolytic enzymes. Fines produced by abrasion during stirring or handling can easily be removed by decanting or flushing in a Buchner funnel or in the column.

Clean CPG is easily wetted by water. To produce an aqueous slurry, as for packing of columns, it is not necessary to evacuate the glass. Instead, one adds degassed water or buffer to the powder, stirs gently, lets settle, and decants the supernatant. This procedure is repeated three or four times until the sizzling noise of displacing air in the pores ceases and the supernatant is completely clear. Degassing of the liquid can be achieved by briefly evacuating it in a safety-taped vacuum flask, or better, by bringing it to a short boil and cooling it. Degassed liquid takes up air quite slowly, providing air bubbles are not entrained by shaking or pouring. Some workers first expose slurries to ultrasonic vibrations and then decant the produced fines. This causes abrasive rounding of the irregular grains. The increased sphericity results in less abrasion during the following column packing under strong vibration, but does not seem to affect column resolution.

Since controlled pore glass is not produced by crosslinking of monomers, its specific pore volume (cc pores per cc CPG) is not dependent on its pore size. Specific volume for all pore sizes is optimally adjusted to around 68%, except that at the low end of the pore size spectrum, specific pore volume is also lower. In addition, there may be some variations from lot to lot. The breadth of the distribution of pore diameters is usually less than ±10%. The customary definition of this distribution is that 80% (and not only 68.27%, as in standard deviations) of the pore volume consists of pores whose size falls within ±10% of average pore size (5). For small pore sizes, this distribution may be somewhat broader. Specific surface area, as measured by nitrogen adsorption (BET), is inversely related to the pore diameter. Table 11.2 gives typical figures of the discussed parameters for 11 different pore sizes.

A well-packed bed of CPG has approximately 40% of void space (V_0) between the granules. The real density of the quartz-skeleton of the CPG is 2.15 g/cc. From this the material parameters of CPG of several pore volumes and of a well-packed bed for CPG have been calculated; they are listed in Table 11.3.

To obtain optimal packing, columns should be slurry packed, subjecting the columns to strong orbital vibration in a horizontal plane while the granules are settling. Vibration should be continued until there is no change in the height of the packed bed. This may take up to 1 hr of vibration. To prevent "bridging," the columns should occasionally be dealt a mechanical shock as, for instance, by giving them each a gentle blow with a piece of wood. Fines produced during packing-vibration are easily removed by flushing of the column. If the porous disk at the end of the column should become plugged up by fines, the column should be turned around and back-flushed. It is good practice to select column closures with porous disks or screens having openings that are only slightly smaller than the size of the substrate particles.

Good column packing is particularly important for permeation chroma-

Table 11.2. Typical Parameters for Controlled Pore Glass of Different Pore Sizes

Mean Pore Diameter (Å)	Typical Specific Pore Volume cc/g	Typical Specific Pore Volume %	Pore Diameter Distribution for 80% of Pore Volume (% of average parameter)	Typical Specific Surface Area (m^2/g)
75	0.4	46	<±20	185
120	0.7	60	<±15	185
170	0.8	63	<±10	140
240	0.9	66	<±10	100
350	1.0	68	<±10	75
500	1.0	68	<±10	50
700	1.0	68	<±10	37
1000	1.0	68	<±10	26
1400	1.0	68	<±10	20
2000	1.0	68	<±10	13
3000	1.0	68	<±10	9

Table 11.3. Typical Parameters of a Column Packed to 40% Void Volume as a Function of the Pore Volume of the Used Controlled Pore Glass

Specific Pore Volume of Controlled Pore Glass cc/g	Specific Pore Volume of Controlled Pore Glass %	Apparent Density of Packed Bed (g/cc)	100 cc Bed Volume Contents: Void Volume (cc)	100 cc Bed Volume Contents: Pore Volume (cc)	100 cc Bed Volume Contents: Total Free Volume (cc)
0.4	46	0.69	40	28	68
0.5	52	0.62	40	31	71
0.6	56	0.56	40	34	74
0.7	60	0.52	40	36	76
0.8	63	0.47	40	38	78
0.9	66	0.44	40	40	80
1.0	68	0.41	40	41	81
1.1	70	0.38	40	42	82
1.2	72	0.36	40	43	83
1.3	74	0.34	40	44	84

tography. It is useful to test the quality of packing by determining chromatographically the exclusion volume (V_0) of a column. It should be around 40% and never larger than 50% of the envelope volume. Columns should be weighed empty and then filled with a liquid of known density before packing, to obtain the precise envelope volume in order to enable calculation of the void volume fraction of the envelope volume.

One further advantage of the rigidity and nonswelling behavior of CPG is the ability to observe and measure accurate pore morphologies with such instruments as the electron microscope and the mercury intrusion porosimeters. Replica microscopy of partially etched samples (Figure 11.8) is suitable for scientific studies of pore shape. Overall micrographs of the porous network are best obtained with the scanning electron microscope (Figure 11.10).

Mercury intrusion is a technique that relies upon the resistance of nonwetting liquids to entering small pores. The pressure necessary to force the liquid into a pore is inversely proportional to pore diameter. In practice, an evacuated porous sample surrounded by mercury is subjected to increasing pressure, and the volume of mercury entering the sample is measured as a

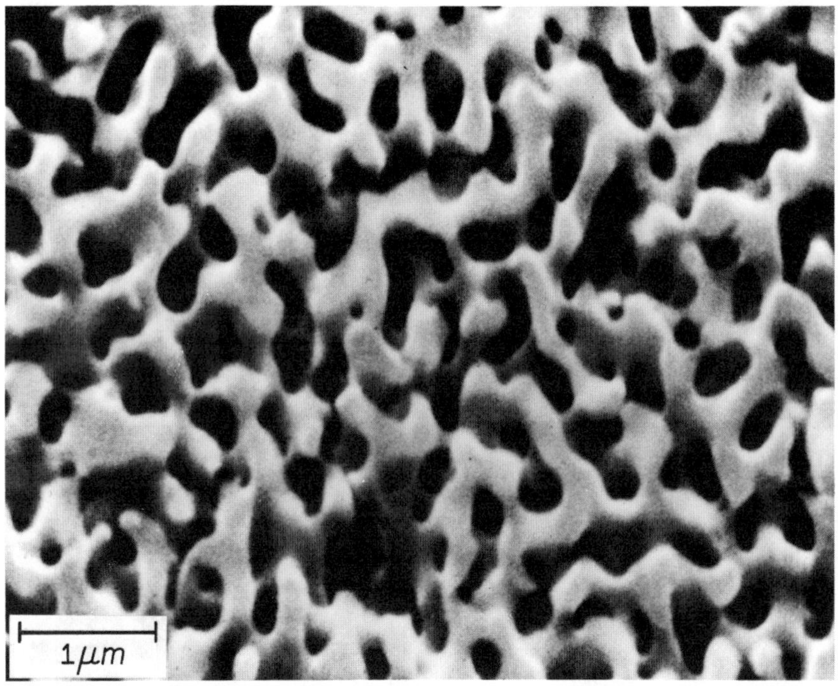

Fig. 11.10. Scanning electron micrograph of controlled pore glass of 3000 Å mean pore diameter.

function of pressure. Because of the relationship of pressure to pore size, the resulting pressure vs. intrusion volume plot is also an integral pore volume histogram. The steeper the rise in the curve, the sharper is the pore distribution. Figure 11.11, from my 1965 paper (5), shows such a plot for controlled pore glass and some other rigid inorganic supports. Such plots are used to determine mean pore diameter, width of pore distribution, and specific pore volume. Actual internal surface area is measured by nitrogen adsorption (BET). Narrow-pore-size distribution is essential for many applications and so is the absence of excess surface area. Excess surface area due to poor CPG fabrication procedure indicates residual silica gel in the pores and is the cause of undesirable adsorption in certain applications.

CPG is nearly pure SiO_2 glass and has very low solubility in aqueous solvents. Liquids exposed to glass surfaces such as the walls of glass bottles become saturated with respect to SiO_2. Because of its high surface area, SiO_2 saturation in CPG columns occurs rapidly, reaching approximately the same levels as in glass bottles. The solubility increases with pH but rarely interferes in normal laboratory operation unless pH exceeds 9.0. Saturated borax solutions (pH ~ 9.5) have been used in routine virus purifications. In some commercial operations in which columns are continuously flushed with large volumes of liquids, loss of CPG at the column head and widening of the pores has been observed. In such applications, the use of a small "sac-

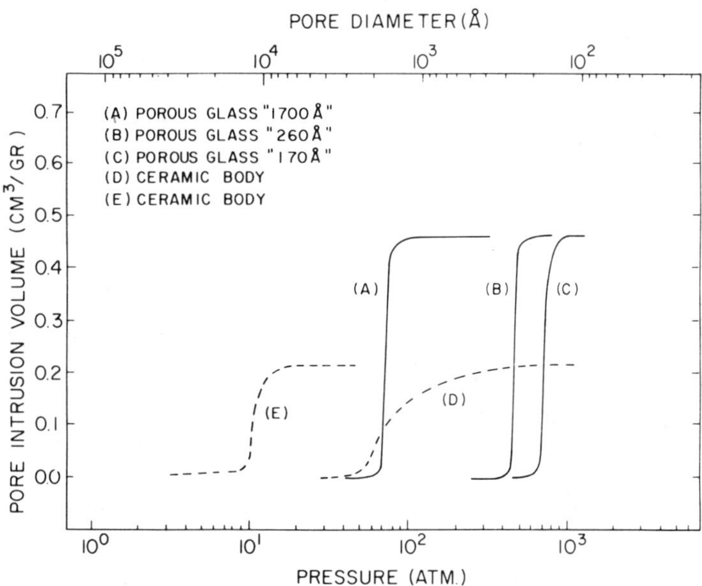

Fig. 11.11. Cumulative pore size distribution curves obtained by mercury intrusion technique. Reprinted by permission from *Nature*. Copyright 1965, Macmillan Journals Ltd. (5).

rificial'' precolumn inserted in the eluant line before the actual column presaturates the eluant with SiO_2 and protects the main column. CPG of small-pore size (~70 Å) and coarse-mesh size is best suited for such a precolumn.

The solubility of the SiO_2 at alkaline pH is of somewhat greater concern when expensive derivatives become lost due to alkaline attack. Derivatization, when exhaustively performed, protects the Si—O—Si bonds at the base of the ligands, and so does SiO_2-presaturation as described in the above paragraph. As mentioned in Section 11.2, zirconia imparts increased alkali resistance to glass, and coating CPG with zirconium oxide has been suggested to improve its alkali resistance. An example of such a procedure is given in Section 11.8.4d.

11.8. APPLICATIONS OF CONTROLLED PORE GLASS

11.8.1. Permeation Chromatography

The rigidity of controlled pore glass, which also extends to large pore sizes, as well as its sharp pore-size distribution, suggested its use for permeation chromatography. This technique, pioneered by Porath and Flodin (4), also became known under the names gel filtration and exclusion chromatography. It is most widely practiced with organic gels of crosslinked dextran, agarose, and polyacrylamid. Inherently, such gels, when having large pore size, are very soft and deform under pressure of bed-height or pressure created by fast flow-rates. This feature limited the use of gels to small and medium molecular weight species and made gel filtrations of high molecular weight substances and viruses impractical. For general treatises on permeation chromatography, see Ackers and Steere (23), Fasold et al. (24), Philipson (25), Altgelt (26), Determann (27), Oss (28), Ackers (29), Harmon (30), and Ackers (31).

I introduced controlled pore glass (5) by demonstrating its ability to separate the components of artificial mixtures of plant viruses and plant viruses with albumin. When performed on organic gels, such separations took many hours; with CPG, they could be completed within fractions of an hour. This first application was soon followed by separation of other viruses, high polymers, proteins, cell components, and so on. The tables in Section 11.8.1a list such uses.

The emergence of gel filtration was accompanied by many theoretical papers, postulating for it a variety of detailed mechanisms. A disturbing (and sometimes charitable) element in these theories was that the pore size and pore distribution of the organic gels could not be well determined and remained therefore an unknown or adjustable parameter. When controlled pore glass, with its readily measurable pore diameter became known, there appeared a series of papers using CPG experiments intended to clarify the mechanism of separation by gel filtration. Although not suggesting any new mechanism, I raised doubts (32) about some previously postulated ones. Later papers relied on empirical or semiempirical equations to describe the behavior of species on CPG columns (33–37). A recent paper by Basedow, Ebert et al. (38) gives a good description of the present status of the field.

An improved quantitative understanding of the behavior of molecular species in microporous materials, and the ability to measure the pore size of porous glass by electron microscopy and mercury intrusion pressure, presented an opportunity to devise some novel means of measuring the size of molecular species by permeation chromatography on controlled pore glass. I described (22) some mathematical and graphical procedures and tested them on dextran fractions of high monodispersity. Dextran was chosen because of its closely ideal behavior according to polymer conformation theory and because it shows a minimum of adsorptive or repulsive behaviors against glass surfaces. In an unrelated later study, similar ideal behavior of dextran in controlled pore glass was found by Day, Alince, and Robertson (39), who performed bulk-penetration experiments. A recent study of an aggregating spherical virus (Haller, Gschwender, and Peters, 40) found that it observed laws similar to those of the dextran molecules. The chromatographic behavior of a species is not only determined by its size (which can be solvent-dependent) but also by its electrostatic interaction with the slightly negatively charged silanol surface of the glass or any other charged porous matrix. Figure 11.12 shows approximate normalized peak positions of particulate species (e.g., virus particles) and polymers of different composition and molecular weight, when chromatographed on controlled pore glass columns. It is evident that elution behavior is not only molecular-weight dependent, but also structure dependent. The curves in Figure 11.12 are plots of molecular weight vs. normalized peak positions. Figure 11.13 shows how such curves are generated from individual chromatographic runs.

The curves in Figure 11.12 qualitatively resemble those obtained with nonrigid gels as chromatographic substrate. They differ quantitatively. A flat curve indicates good resolution for a given column length. The trade-off for high resolution is a narrow separation range. To increase range, columns of several pore sizes can be operated in series. This is widely practiced when CPG is used for high polymer separation and analysis. We have treated the case of serial or mixed-pore-size columns (Haller, Basedow, and Konig, 35).

The upward turn close to exclusion volume of the curves in Figure 11.12 is far less pronounced in CPG than in gels with wide-pore-size distribution. In fact, it relates to sharpness of pore-size distribution and it has been shown (34) that in CPG of outstanding sharpness, the curve makes an abrupt sharp bend at V_0.

Ideally in permeation chromatography, one desires a minimum of adsorp-

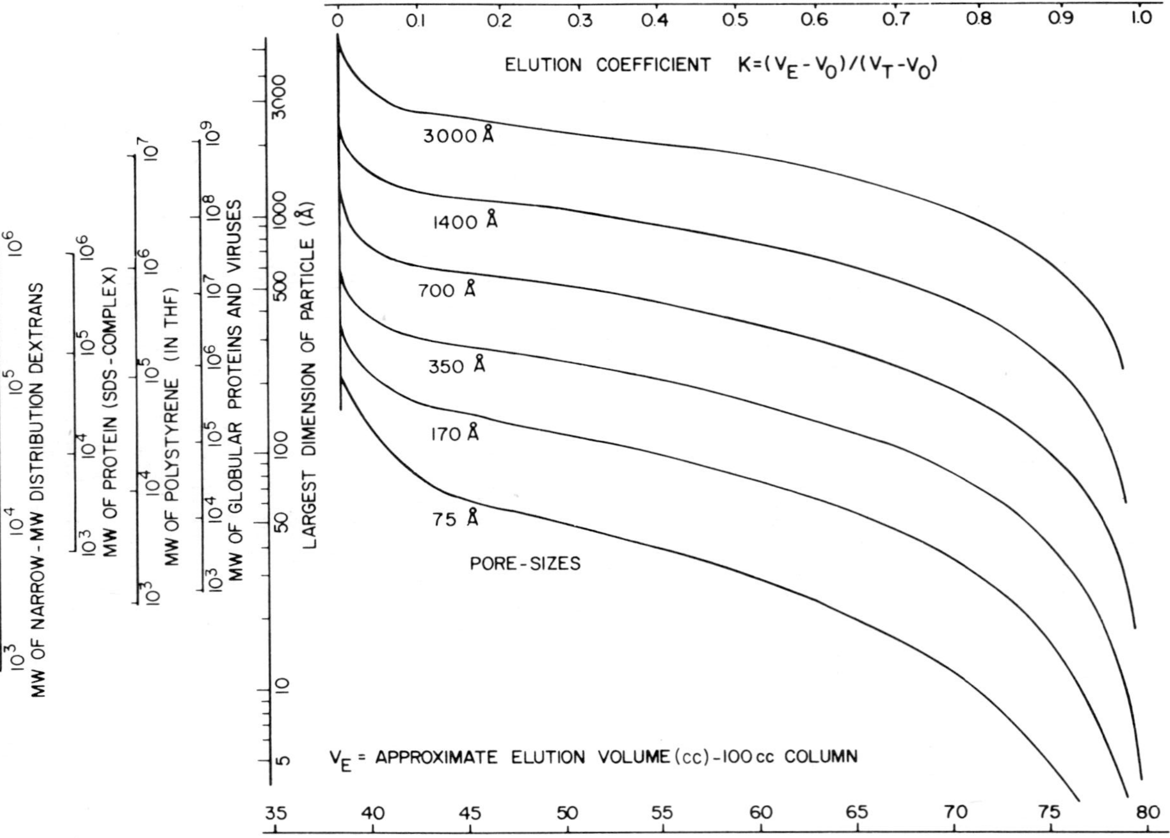

Fig. 11.12. Approximate elution positions of several classes of substances in controlled pore glass chromatography. Molecular weights vs. normalized elution position (elution coefficient) and vs. elution volume on a column of 100 cc envelope volume.

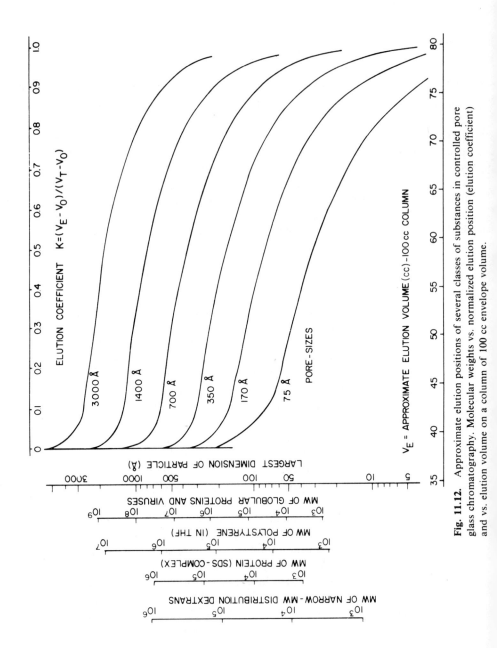

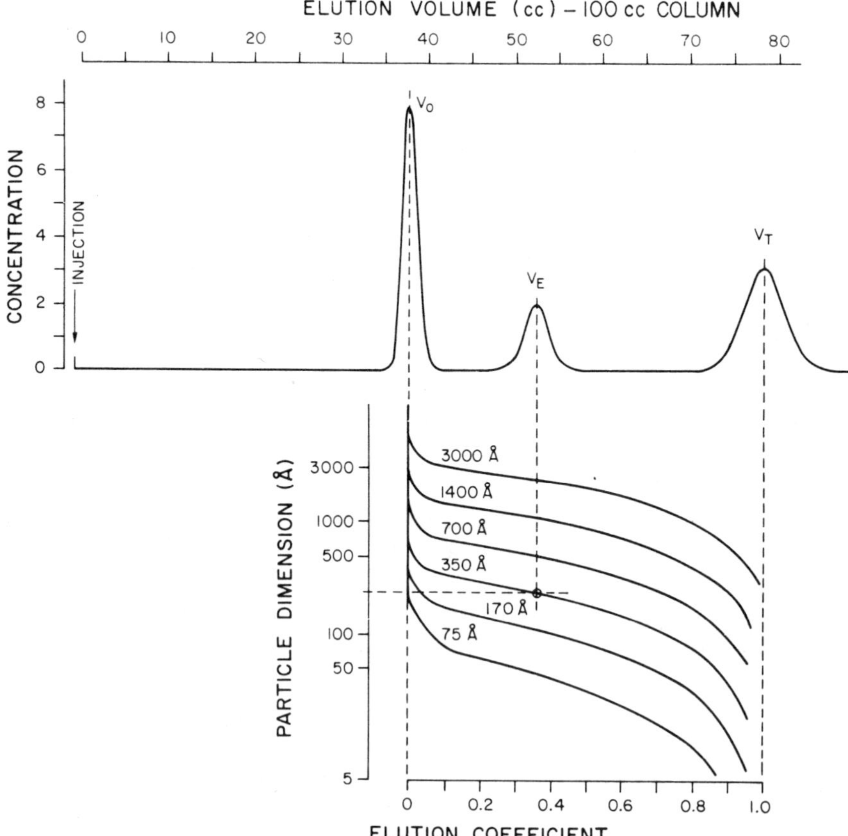

Fig. 11.13. Curve on top: Typical permeation chromatogram for three substances with large excluded size (V_0), very small size (V_t) and medium size (V_e). Curves below are from Figure 11.12 and demonstrate their relation to actual chromatogram.

tive or repulsive interactions between species and pore wall. In practice, this goal is almost never achieved. Interaction cause peak shifts in both directions. Molecular repulsion forces advance peaks, attraction forces delay peaks, or, in severe cases, can cause complete adsorption of some species. Generally one can devise some conditions to overcome adsorption. With proteins and viruses, high pH and high ionic buffer strength generally minimize adsorption. Operation in denaturing or chaotropic solvents (e.g., KCNS-containing buffers) also minimizes adsorption. Although there are exceptions to this rule (e.g., interferon; see Section 11.8.3), it usually pays to experiment and find conditions that prevent adsorption without having to resort to precoating or derivatization. The latter particularly increases cost and requires more gentle column regeneration to preserve the covalently attached groups. Repeatedly used ("aged") columns usually show less adsorption, presumably because of coverage of active sites. When everything fails, one can modify the CPG surface by preadsorption of high-molecular-weight polyethylene glycol (41). Such pretreatments last for hundreds of runs and have the advantage that they can be easily repeated *in situ* after columns have been washed with cleaning agents. To prevent adsorption, covalent surface modification of aminopropyl was used by Haller (6) and by Eltek and Kiselev (42); glyceryl groups by Chang et al. (43, 44) and Regnier (45); bonded gelatin by Gschwender and Traub (46), and bonded albumin and carboxyl-groups by Krasilnikov et al. (47). Glyceryl CPG to this date seems to be the most generally usable of the bonded surface derivatizations for work with aqueous solvents.

When working with organic solvents, it is advisable to use polar solvents to prevent adsorption of polymers on the polar glass surface. Otherwise one can use CPG that has been hydrophobically surface-derivatized (silanized). Silanization procedures were described by Aue and Hastings (48), Cooper and John (49), Iwama and Homma (50), and Derenbach et al. (51).

The chemical resistance of CPG has invited its use with exotic solvent systems, particularly those which ordinarily attack or change the structure of organic gel supports. Collins and Haller (34) determined the size of a large number of protein sodium dodecyl sulfate complexes by controlled pore glass chromatography. An extension of the molecular weight range to smaller sizes was achieved by Blagrove and Frenkel (52) by adding urea to the denaturing solvent. Denatured protein complexes lose structure-dependent size features. A universal equation for their behavior on CPG columns was published (Haller, 33).

Other uses of CPG which challenge the resistance of other solid-phase matrixes are permeation chromatography in strong organic solvents such as methylsulfoxide (53) or even in hot solvents (54).

The low flow-resistance of CPG allows fast flow-rates, which is of benefit in column cleaning and also for speed of operations. As in any permeation chromatography, excess flow-rate causes peak-broadening and loss of resolution for species with similar elution coefficients. Extremely high flow-rates have been used in industrial or clinical separations involving species that have widely different molecular sizes, such as ethanol and human albumin (processing of fraction 5 in column fractionation of plasma), purification of viruses from host proteins and precipitants (NH_4Cl, polyethyleneglycol), and isolation of immunoglobulin M from serum (vaccines, clinical tests). The ability to cover a wide range of flow rates is also useful in the study of slowly shifting equilibria of self-association or solvation (vacancy chromatography).

The rules for operating CPG columns are not very much different from those for operating permeation chromatography columns with other substrates. Columns should have a large length-to-diameter ratio. Since a well-packed bed of CPG does not change its dimensions, it is not necessary to have adjustable volume columns (plunger-type), and cartridge-type columns with fixed end-closures are very satisfactory. One should make the rule to determine the exact column envelope volume before packing and to calibrate packed columns for exclusion volume (V_0) and total free volume (V_t) before any other runs are being made. Expressing peak positions in normalized form as elution coefficients (K) is particularly useful with CPG because of the good reproducibility between runs and also for column scaling purposes.

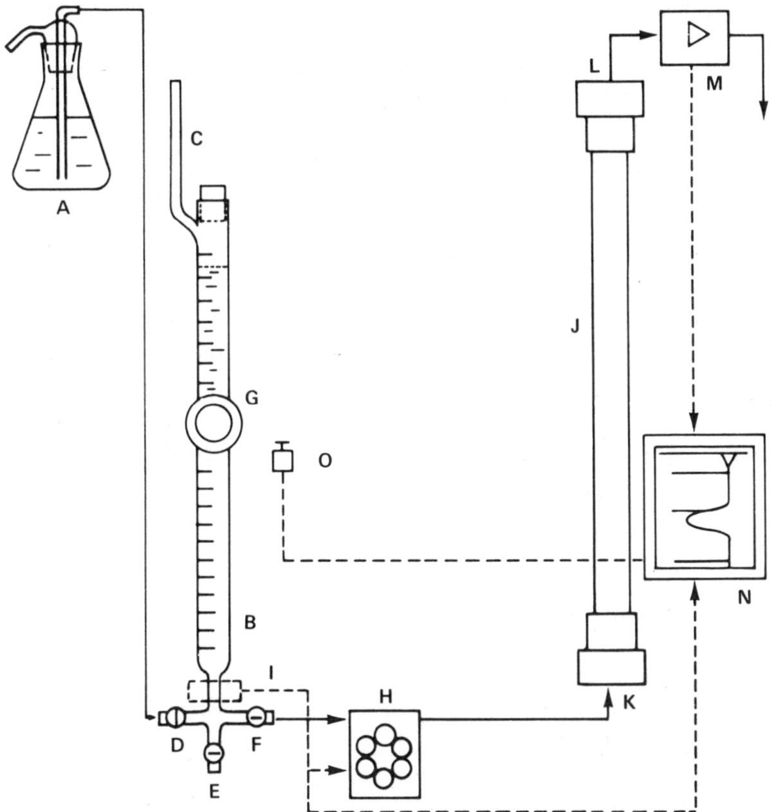

Fig. 11.14. Schematic of typical CPG permeation-chromatography setup. *A*, Eluant reservoir; *B*, delivery burette; *C*, tube to prevent overflow; *D,E,F*, Stopcocks (large samples injected through additional three-way stopcock located at *F*); *G*, magnifier to read burette; *H*, peristaltic pump with adjustable speed control; *I*, liquid level monitor for automatic pump shut-off; *J*, CPG-column; *K,L*, Column closures with septa (small samples injected at *K*); *M*, effluent monitor; *N*, recorder; *O*, event marker.

Calibration for V_0 in aqueous systems is frequently performed with Blue Dextran or tobacco mosaic virus (TMV). Blue Dextran has a tendency to tail with larger pore sizes, whereas TMV is universally usable. Latex suspensions have also been used, but some lots tend to coagulate when the latex spheres separate from their normal suspension liquid while passing through the column. V_t is frequently calibrated with tryptophane, benzylalcohol, phenol red, or trypan blue.

CPG particle size affects flow resistance, and most work has been done with particles in the 50 to 150 μm diameter range. Peak resolution does not change drastically over this size range. Resolution improves when particles are below 40 μm and narrow-size cuts are used. Such columns require high pressure instrumentation and are mostly used for small-volume analytical purposes only.

It is of operational advantage to use eluant pumps that can be operated over a wide range of flow rates. Fast survey runs can be made at very high flow rates, and eluant changes and column flushing can be accomplished in a very short time. Similarly useful is a graduated burette for eluant delivery that allows precise correlation between eluant volume and elution peaks. To prevent aspiration of air into the column, an eluant level guard with automatic pump shut-off is useful. Furthermore, one should be able to flush columns directly from a back-up reservoir. A schematic of such a permeation chromatography system is shown in Figure 11.14.

The following tables show CPG-isolated substances and literature references. To conserve space, only a fraction of published papers could be listed.

11.8.1a. Substances Purified, Isolated, or Characterized by CPG Permeation Chromatography

Plant Viruses

Alfalfa mosaic: 55
Carnation mottle: 56
Cymbidium ringspot: 56
Hibiscus latent ringspot: 57
Honeysuckle latent: 56
Hypochoeris mosaic: 58
Lilac chlorotic leafspot: 56
Narcissus latent: 59
Papaya mosaic: 60
Pepper mottle mosaic: 60
Pepper veinal mottle mosaic: 56
Potato virus and X-virus: 60

Red clover mottle: 55
Red clover necrotic mosaic: 56
Southern bean mosaic: 5
Sweet potato mild mottle: 56
Tobacco mosaic: 5, 34, 55, 56, 61, 62, 63
Tobacco rattle: 56
Tobacco ring spot: 5
Tomato aspermy: 56
Tomato black ring: 56
Turnip crinkle: 56
Turnip yellow mosaic: 56
White clover mosaic: 55, 56
Yellow clover mosaic: 60

Bacterial Viruses (Bacteriophages)

Coliphage
F2: C. W. Hiatt (unpublished)
F2: C. R. Merril (unpublished), 63
λ: 64
M-12: 61
M-13: 61
MS-2: 33, 62
Qβ: 61
T2 and T3: C. W. Hiatt (unpublished)
T4: 61
T6: 65
φX-174: 61, 66

Human and Animal Viruses

Adeno: C. W. Hiatt, unpublished
Australian antigen: see Hepatitis B
Avian myeloblastosis: 67
Eastern equine encephalitis: D. S. Spicer (unpublished)
Foot and mouth disease: H. L. Bachrach (unpublished)
Hamster melanoma: 67
Hepatitis B: 68
Influenza: 69
Kilham rat: 62
Lymphocytic choriomeningitis (LCM): 40, 70
Mengovirus: 46

Moloney sarcoma: 71
Newcastle disease: 72
Pleural-pneumonia-like organism (PPLO): V. H. Zeve (unpublished)
Polio: 41
Rabies: 41
Scrapie agent: 73
Sindbis: 74
Tick-borne encephalitis: 75
Toga: 74
Vaccinia: 76
Vesicular stomatitis (VSV): J. F. Obijesky (unpublished)
West nile: 75

Nucleic Acids and Viroids

DNA (Hamster myeloblastosis virus): 78
DNA (M-13 coliphage replicative form): H. H. Gschwender (unpublished)
DNA (Shark tapeworm): R. D. Tanaka (unpublished)
DNA (ϕX-174 coliphage replicative form): 77
Nucleotides: 78
RNA (5-s): 80
RNA (plant virus): 60
RNA (ribosomal): 80
RNA (transfer): 79, 80
Viroid RNA (Exocortis disease of citrus): 80

Proteins, Peptides, Enzymes, Coenzymes

Acetylcholine: 81
Acetylcholinesterase: 81, 82, 83
Adenosine triphosphate: 81
Albumin dimer: 86
Albumin: 5, 44, 52, 63, 67, 82, 84, 85
Aldolase: 84
Alkaline phosphatase: 67
Apoferritin: 86
Apomyoglobin: 52, 84
Blood plasma and blood serum: 63, 87, 88, 89, 90, 91
Brain protein (human): 92
Carbonic anhydrase: 33, 34
Casein micelles: 93, 94, 95, 96 (and many others)
Catalase: 33, 34, 67, 97

Chymotrypsinogen: 33, 34, 52
Cytochrome c: 33, 34, 44, 84, 52, 86
DNA-polymerase (myeloblastosis virus associated, virus RNA directed): 67
Factor VIII (Von Willerbrand factor, antihemophilic factor): 98, 99, 100, 101
Fatty acid synthetase: D. Osterhelt (unpublished)
Ferredoxin (spinach): S. Jackel (unpublished)
Fibrinogens and polymers: A. P. Fletcher (unpublished), 63, 91
Galactosidase: 33, 34
Glucagon: 84
Glutamate dehydrogenase: 34
Glycylphenylalanine: 44
Hemocyanine (mammalian): 33, 34
Hexokinase: 67
High tyrosine, component 0.62: 52, 84
Immunoglobulins and antibodies
  IgG (gamma globulin): 63, 88, 89, 90, 91, 102
  IgM: 63, 88, 89, 90, 102, 103, 104
  antibodies against:
    E. coli: 63
    rubella: 88, 89, 105
    tick-borne encephalitis: 102
    toxoplasmosis: 90
Insulin and chains: 52, 67, 84
β-Lactoglobulin: 33, 34, 86
Lens-crystallines: 86, 106
Lysozyme: 85
α_2-Macroglobulin: 63
Myoglobin: 33, 34, 67, 86
Ovalbumin: 33, 34, 52, 67, 84, 85
Paramyosin: 52
Pepsin: 33, 34
Phosphorylase: 33, 34
Plasma: see Blood plasma
Plastocyanine (spinach): S. Jackel (unpublished)
Protein (glue): 107
Rhodopsin: see Visual pigments
Serum: see Blood plasma
Sialoglycopeptides (from melanoma): 108
Thyroglobulin: 33, 34, 86
Transferrin: 63

Tropomyosin: 52, 84
Trypsin: 33, 34
Tubulin-like protein from human synaptic membrane: 109
Urease: 81
Visual pigment: 110

Cell Components

Microsomes: 81
Neurohypophysial granules: 111
Vesicles
  brain (aplysia): 116
  intestinal brush border membrane: 117
  synaptic (*Narcine brasiliensis*): 113, 115
  synaptic (torpedo): 81, 112, 113, 114, 115
  synthetic lipid: 118

Polysaccharides and Other Natural High Polymers

Amylose: 53, 119, 120
Chitosan: 121
Dextrans: 22, 32, 35, 38, 122, 123, 124, 125, 126
Glycogen (clam): C. W. Hiatt (unpublished)
Glycosaminoglycans: 127
Halitoxin (marine sponges): 128
Humic acids: 129
Hyaluronic acid (synovial fluid, eye, umbilical cord): 130, 131
Lemon-gum polysaccharides: 132
Lignins and hemicellulose: 133, 134
Lipopolysaccharides: 135
Mannan (*Ceratocystis fagelearum*): 136
Mouse melanoma polysaccharides: 137
Protein glue: 85
Pullulan: 120
Starch (potato): 5
Xanthan gum: J. R. Beckett (unpublished)

Synthetic High Polymers, Lattices, Silica Sols A large number of the above substances have been characterized, using CPG permeation chromatography. They were omitted from this publication since they were not of biological or biochemical use. Silica spheres (sols) with covalently attached virus-specific antibodies have been used as immunospecific markers in electron microscopy. To obtain a sharp size distribution of the markers and to exclude aggregates formed in the derivatization steps, the silica sphere suspension was fractionated by CPG chromatography (138).

Antibiotics, Drugs, Food Components Telepchak reported in several papers (91, 139, 140) on the analysis of liquids by high-speed permeation chromatography on columns of modified controlled pore glass. Analyzed substances were: barbiturates, beer, caffeine, chloramphenicol, hydantoins, neomycine, phenylbutazone, soybean oil, streptomycine, theobromine, theophylline, xanthines.

Bacitracin has been used as a column marker (67).

11.8.2. CPG as Liquid Phase Carrier in Liquid–Liquid Partition Chromatography and Biphasic Reactions

The convenience and multistage nature of column operation has occasionally been applied to separation schemes involving partitioning of substances between immiscible liquids. By immobilizing one liquid phase in particles of a porous solid such as CPG and exposing the particles to a stream of the second liquid phase, one can obtain countercurrent effects without the expense of a large number of discrete mixing and decanting stages. The immobilized liquid is held by capillary forces in the pores of the controlled pore glass. In a liquid–liquid system consisting of one aqueous or polar phase and one hydrophobic less-polar phase, one has either to immobilize the aqueous polar phase in the nonderivatized CPG or one renders the CPG hydrophobic by silanation and uses it to hold the hydrophobic liquid phase.

In materials with very small pores and therefore large surface-area-to-volume ratio, the forces at the solid interface can enter into the equilibrium of the two immiscible bulk phases. Or, more drastically, an apparently homogeneous mixture of two liquids with a composition close to mutual saturation may release one phase to adsorb on the internal surface to which it has affinity and actually partially fill the pores. To understand this phenomenon, one has to remember that mutual solubility relations of bulk liquids bounded by a flat interface are different from those at a sharply curved meniscus in a pore. It is therefore possible that apparent adsorptive chromatographic separations observed with surface-modified or nonmodified porous substrates in complex solvent mixtures are really liquid–liquid partition chromatography. One has to suspect such a mechanism whenever an apparently adsorptive separation in a complex solvent system proceeds equally well with a variety of chemical surface derivatizations.

Liquid–liquid partition chromatography in general is not widely practiced intentionally, and I know of only two cases where CPG was used. Chlorophyll was purified by Glynne-Jones et al. (141), and Williams et al. (142) separated thyroidal iodoamino acids by this technique. It is partly a matter of conjecture if and when phases that are covalently attached to a solid surface are to be considered a solid or a liquid. Certain surface-modified chromatographic substrates are sometimes referred to as bonded stationary liquids. Such material will be discussed in Sections 11.8.3 and 11.8.4b.

CPG has also been used to hold a liquid reagent in biphasic catalytic and enzymatic reactions of industrial interest. As of this writing, no published literature is available.

11.8.3. Adsorption-Permeation Chromatography

In this section, we deal with applications where adsorptive properties of CPG are exploited. Generally, one first creates special conditions which favor the adsorption of one or more substances and elutes this substance by changing the eluant. With some important exceptions, adsorption of protein is favored by low pH and low ionic strength. Columns are loaded under these conditions and the substances are subsequently sequentially eluted in gradients of pH, ionic strength, additions of hydrophobic or chaotropic substances, detergents, and so on. The technology is similar to the one practiced in adsorption chromatography with other substrates, except that each substrate has its own surface properties. Frequently, however, the sharp pore-size distribution of CPG allows one to introduce an additional separation parameter by combining size exclusion with adsorption. When loading the adsorption column, one can exclude substances that have larger molecular size than the desired range and thus improve separation efficiency and column capacity.

Some substances are very unusual adsorbates in that they do not obey the general rule expected from the class of substances to which they belong. This makes them outstanding candidates for adsorptive purification, except that the type of optimal substrate and purification procedure do not follow general rules and are frequently discovered by accident and optimized by tedious trial-and-error procedures. Some industrially important enzymes and a growing number of interferon types belong to these groups of substances. The interferons in particular make a good case history because much of the purification work has been published in the open literature.

Following observation of losses of interferon activity by adsorption on the walls of glass beakers (143), the first attempts to use nonporous glass beads for interferon purification were made by Davies (144) in 1965. Working with human fibroblast (β) interferon and CPG, an elaborate purification scheme was developed by Edy et al. (145), which involved adsorbing interferon from solutions of near-neutral pH and recovering the interferon from the CPG by elution at low pH. This behavior of interferon runs counter to other proteins and accounts for the high purification factors obtained by a single passage through a CPG column. The adsorption conditions are as critical as the desorption conditions, and the nature of the substrate seems just as specific. Edy reports, for instance, that with silica gel he could adsorb but not desorb interferon (146). Edy has since improved and upscaled the procedures reported in the cited papers, and using fibroblasts grown on microcarriers he routinely obtains concentration factors of 44, purification factors of 100, and yields of over 60%. He purifies lots of 44 liters cell-supernatant on a CPG column of 700 cc bed volume, but having never experienced saturation breakthrough, assumes that considerably smaller columns would suffice. It also appears that factors of concentration, purification, and yield can be increased still further by such refinements as closer control of elution profiles, improvements in column design, and so on (146).

Following CPG purification of human fibroblast interferon, other interferon types were purified by adsorption chromatography on CPG. Some interferons lose activity by exposure to low pH, in which case diluted ethylene glycol was used as an eluant. Other interferons required elution with eluants containing chaotropic ions.

In the following section, interferons purified by CPG adsorption chromatography are listed. Also listed are other substances of biological importance.

Substances Isolated by Adsorptive Chromatography on Nonderivatized CPG

Interferons
- Human
  - α: lymphoblastoid (namalva): 160
  - β: fibroblast: 147, 148, 149, 150, 151
    - fibroblastoid: 152, 153, 154
  - γ: lymphocyte (immune): 155, 156, 157
- Mouse
  - α and β: 161, 162, 163, 164
  - γ: lymphocyte (immune): 158
- Amino acids: 165
- Biogenic amines and metabolites: 159
- Bovine serum albumin components: 167
- Chlorophyll-protein complex (light harvesting): 168
- Cholesterol: 169
- Chymotrypsinogen-A: 167
- Echo-virus-12: 170
- Hepatitis B antigen, and antibodies to it: 171, 172, 173
- D-Hydroxybutyrate apodehydrogenase: 167, 174

Influenza virus: 175
Lysozyme: 167
Myoglobin: 167
Polio virus: 170
Poly(L-arginine) and Poly(L-lysine): 162
Procollagen: 176
Staphylococcal α-toxin: 167

11.8.4. Surface-Derivatized CPG

As I mentioned at the beginning of this chapter, my interest in glass surface derivatization and the possible development of a controlled pore glass was aroused in 1954 through a meeting with L. Lerman, who at that time used cellulose supports for the covalent attachment of antibodies and antigens (1). The state of the art of covalent organic derivatization of siliceous surfaces at that time consisted mainly of attaching organic groups via Si—O—C linkage. In his pioneering work on estersils, Iler had produced silica gels having hydrophobic surfaces by reacting the gels at elevated temperatures with alcohols (177). The silanol group of the silica surface reacts with the alcohol under formation of an Si—O—R group, while the water formed in the reaction is removed by azeotropic distillation. The aminoaryl functional glass reported in 1956 by Haller and Duecker (2), in analogy to the cellulose affinity chromatography substrate of Campbell and coworkers (1), also relies on a Si—O—C bond between the glass surface and the organic ligand. The Si—O—C bond is sensitive to hydrolysis. However, with large hydrophobic ligands and exhaustive derivatization leaving a minimum of unreacted silanol groups, the Si—O—C bonded derivatives can be made to be satisfactorily stable. Technologies existing in the mid 1950s for attaching organic ligand to siliceous surfaces via the more stable Si—C bond had been developed by Deuel. The siliceous surface was halogenated with thionylchloride and subsequently reacted with benzene to attach a phenyl group via a Friedel–Crafts type of reaction or was reacted with metal organic reagents such as Grignard compounds or lithium alkyls (178). By far the most promising reagents for derivatizations were silanes of the type R_3SiX, R_2SiX_2, and $RSiX_3$, where X represents halogene, alkoxy, or acyloxy. Such compounds react with the SiOH surface of glass or other siliceous materials giving stable Si—C bonding. Such compounds had been used to "silanize" siliceous surfaces to render them hydrophobic. Generally, R was methyl, ethyl, or phenyl. Extensive studies of reactions of these silanes with SiO_2 were made as early as 1956 (179), but only since the advent of bifunctional silanes (3) with a second reactive group in place of R was it possible to use the silane route as an almost universal first step in the attachment of the widest kind of ligands. The first covalently bound organic derivatives of controlled pore glass, which I distributed in 1965, used aminopropyl triethoxy silane (6) as reagent. The great synthetic versatility of the alkylamino group, its fast reaction with the glass surface (apparently due to electrostatic preadsorption), and the fact that this silane is soluble in organic solvents as well as in water are the reasons that it is still one of the main pathways of inorganic support derivatization chemistry. The second most widely used versatile pathway, via an epoxy-functional silane (43), has been used since around 1975. Ligands can be attached under an opening of the epoxy ring, or the ring can first be opened hydrolytically and the resulting glycyl-CPG can be further derivatized. Figures 11.15 and 11.16 show some of these chemical pathways.

The following discussions deal mainly with covalent surface derivatization based upon Si—O—Si and Si—C bonds as provided by the bifunctional silane technology, which is used in the largest part of existing publications. In a few places, adsorptive bonding and bonding by inorganic bridges is mentioned.

Due to the importance of and rapid developments in affinity chromatography and immobilized-enzyme technology, a considerable number of excellent treatises and surveys on these subjects have been published. Many of these discuss in depth the surface derivatization of glass and other inorganic

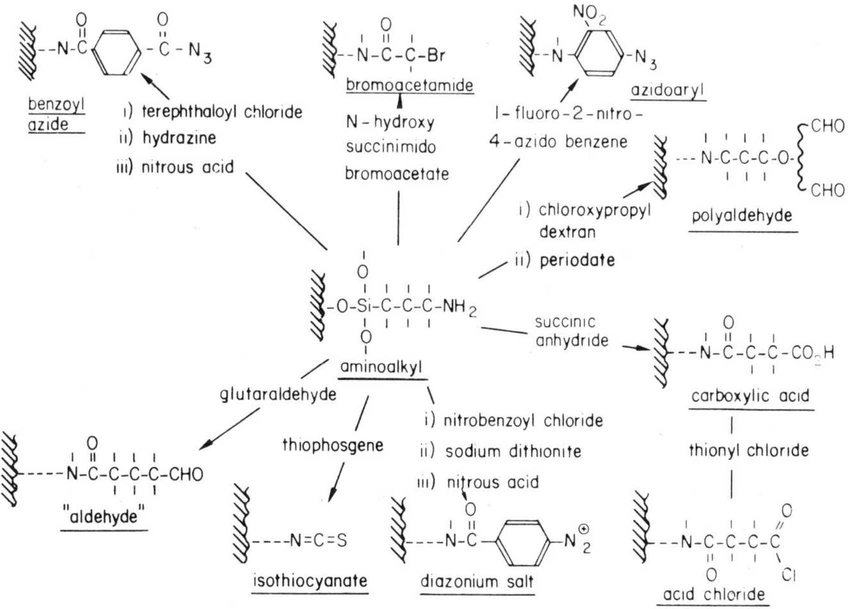

Fig. 11.15. Derivatization pathways starting from aminoalkyl CPG. Courtesy John Wiley and Sons (184).

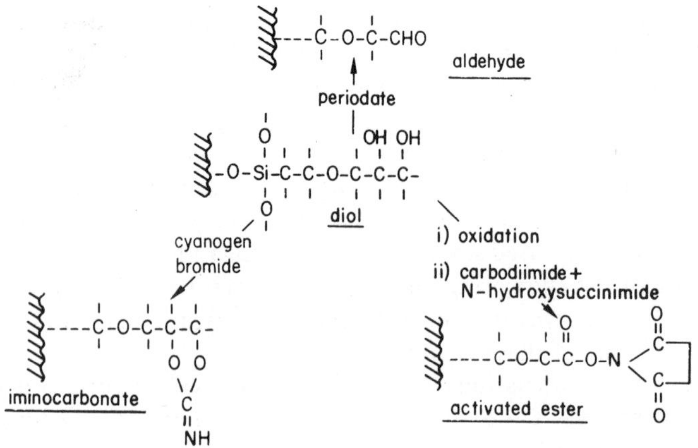

Fig. 11.16. Derivatization pathway starting from glycyl–CPG. Courtesy John Wiley and Sons (184).

supports. Such treatises are listed in the following sections. At this point, however, the publications of Weetall (180, 181, 182, 183), Kent, Rosevear, and Thomson (184), and Carr and Bowers (185) should be specially mentioned.

11.8.4a. Surface Functionalities

Simple Organic Groups: Polymers

Alcohols
- glyceryl: 43, 45, 186
- polyethylene glycol: 43
- unsaturated (for subsequent ozone cleavage in solid phase oligosaccharide synthesis): 187

Aldehydes
- via amine and glutaraldehyde: 182, 188, 189
- via oxidation of glyceryl: 190, 191, 192, 193

Amines
- aminoalkyl-aminopropyl: 200, 201, 202
- aminoaryl (via reduction of nitro benzoyl CPG): 182, 194, and many more
- aminoaryl (via epoxy-CPG and nitrophenol): 195
- aminopropyl: 6, 182, 194, 197, 198, 199, 200, and many more
- diethylaminoalkyl: 43
- diethylaminoethylether: 44

Polyproline: 224
Sulfophenyl: 43
Triazine: 182

Drugs, Hormones

Catecholamines (epinephrin, isoproterenol, norepinephrin): 196, 225, 226, 227, 228, 229, 230, 231, 232, 233, 234, 235
Corticotropin: 226
Heparin: 199
Insulin (β-chain): 236
Propanolol: 227, 234
Thyroxine: 237, 238

Nucleic Acids

DNA: 208, 239, 240, 241
RNA: 208, 240, 241

Enzyme Inhibitors, Substrates, Pseudosubstrates, Coenzymes

Adenosine monophosphate (AMP): 207
p-Aminophenyl-β-D-thiogalactopyranoside: 242, 243
Glycyl-D-phenylalanine: 188
Guanosine monophosphoric acid: 208
Lipoylamide: 219
Mellibiose: 244
Nicotinamide-adenine dinucleotide: 245, 246

Enzymes

Reviews: 184, 180, 182, 185, 247, 248, 249, 250, 251, 252, 253 (adsorption and inorganic bridges), 254
Special techniques
  Peptide stems: 255
  Photochemical bonding: 221, 222, 223
Substances
  Acetylcholinesterase: 256
  β-*N*-Acetylhexosaminidase: 257
  Alcohol dehydrogenase: 203, 257, 258, 259
  Alkaline phosphatase: 194, 215, 260
  L-Amino acid oxidase: 261

Aminoacylase: 263
Aminopeptidase: 262, 264
Amyloglucosidase: 265 (inorganic bridge)
Aryl sulfatase: 266
L-Asparaginase: 222 (photochemical)
Carboxypeptidase: 190, 191, 193
Chymotrypsin: 188, 267 (aging, tertiary structure), 268
Chymotrypsinogen: 267 (aging, tertiary structure), 269
β-Cyanoalanin synthetase: 270
Deoxyribonucease: 271
Esterase (*Bacillus subtilis*): 272
Exonuclease: 273
Ficin: 274
Fructose 1,6-diphosphatase and fructosediphosphate aldolase: 275
Fumarase: 276
β-Galactosidase: 188, 277, 278, 279, 280, 281, 282
Glucan phosphorylase: 283
Glucoamylase: 215, 284, 285, 286, 287, 288, 289
Glucoisomerase: 287, 290
Glucose oxidase: 214, 291, 292, 293, 294, 295
Glutamate dehydrogenase: 296
Glycerylaldehyde-3-phosphate dehydrogenase: 275
Hexokinase: 192, 297
Invertase: 298
Lactase: 299, 300, 301, 302
Lactase dehydrogenase: 267, 303, 304, 305, 306, 307
Lactoperoxidase: 308
Leucine aminopeptidase: 262, 264, 309
Lipoamide dehydrogenase: 310
Luciferase and FMN-reductase (bacterial): 311
Neuraminidase: 312
Nitrate reductase: 313
Papain: 214, 314
Pektin esterase: 315
Penicillinase: 292, 316
Pepsin: 317, 318
Peroxidase: 215
Phosphoglycerate kinase: 275
Phosphoglycerate kinase: 275
Phospholipase A_2 (cobra venom): 319
Phosphoramidate-hexose-transphorylase: 275
Poly(metoxygalacturonide)lyase: 320 (inorganic bridge)

Pronase: 264, 321, 322
Protease: 189, 323, 324, 325
Renin: 326, 327
Rhodanese: 270
Ribulose-diphosphate carboxylase: 328
Steroid esterase: 329
Sulfhydryl oxidase: 330, 331, 332
Triose phosphate isomerase: 275
Trypsin: 190, 191, 193, 206, 215, 255, 285, 292, 305, 312, 314, 333, 334, 335, 336, 337, 338
Urate oxidase: 339, 340
Urease: 197, 292, 342, 343, 344
Uricase: 291, 307, 341
Xanthine oxidase: 340

Antigens and Antibodies

Review: 345
Substances
Albumin: 198, 199, 347
Asparaginase: 346
Carbohydrates: 211
Digoxin: 349
β-D-Galactopyranoside: 211
Gamma globulin (IgG, interspecies): 198, 199, 346, 347
Gamma globulin (IgM): 198, 199
GAT (poly-glu-ala-tyr): 350
α_{2HS}-Globulin: 199
G_C-Globulin: 199
Hepatitis B: 171, 172, 173, 352
α_2-Macroglobulin: 199
Nicotine: 348
Progesterone: 351
Thyroxine-binding protein: 237, 353
Transferrin: 199

11.8.4b. Adsorptive Chromatography on Derivatized CPG

Originally, chromatography was the name given to a separation process that relied on adsorption and desorption on the surface of finely dispersed or porous solids. It was carried out by containing the solids in tubes, called columns, where they were exposed to streams of fluids. The name chromatography was eventually applied to all separation processes involving similar manipulations regardless of the operative separation mechanism.

Since adsorption chromatography is the oldest type of chromatography, the number of publications on the subject are countless and so are the number of substrates that have been employed. Depending upon the type of interaction between the surface of the solid and the substances separated, one frequently classifies adsorptive chromatography further and speaks of ion exchange, hydrophobic, chelating, and so on. Traditionally, substrates for these types of chromatography were substances such as cellulose, alumina powders, silica gels, and ion exchange resins. More recently, the explosive growth of biochemistry has intensified the interest in the isolation and characterization of high molecular weight substances. As a result, adsorptive substrates had to have larger pores, preferably without loss of mechanical properties. For preparative work, the pore volume should also be high to maintain as high a column capacity as possible. While the pore size has to be sufficiently large to give the high molecular weight substances access to the internal surface area, pores that are too large unnecessarily reduce surface area and therefore interfere with column capacity and frequently with the economics of the separation process. Hence, it is understandable that the need for a material with highly uniform pore size, mechanical rigidity, and chemical inertness, and which could also be easily surface derivatized, had originally led to the development of controlled pore glass.

Once CPG was available, it was also used in its derivatized form for various inorganic low molecular weight applications such as the recovery of trace metals by chelation or ion exchange. It performs satisfactorily, but the advantages of CPG over other substrates in these applications are not striking enough to have created a large number of uses and publications.

An interesting biochemical application of a solid phase substrate consisting of a copper-functional CPG column was reported by Masters and Leyden (202). They first prepared ethylenediamine functional CPG by reacting CPG with *N*-β-aminoethyl-α-aminopropyltrimethoxy-silane, then loaded this column with copper ions. These columns were used for the chromatographic separation of amino acids and amino hexoses. It appears that many other biochemical uses of inorganic-ion-modified CPG are possible, particularly since metal complexes of biological substances are well known and have also been exploited for separation purposes. In the cited publication (202), the columns showed some bleeding of copper, which was compensated for by incorporating copper into the eluant. However, the particular diamine does not have a very high affinity for copper and it would not be difficult to synthesize CPG with more specific chelating groups.

In analogy to organic supports, carboxyfunctional CPG was used (Gschwender and Haller, 354) to purify immunoglobulin (IgG), and excellent separations of serum proteins, hemoglobins, nucleotides, and the three isoenzymes of creative phosphokinase on diethlaminoethyl (DEAE) were reported by Chang, Gooding, and Regnier (44). Alkyl-functional (silanized) CPG is used to collect and concentrate organic contaminants in the analysis of water (51).

By far the most important use of derivatized CPG in adsorptive chromatography has been made with biospecific ligands attached to the glass surface. This type of chromatography has more recently become known as affinity chromatography. Traditional substrates had been such substances as cellulose (1) and polymeric gels, and the properties of controlled pore glass have been of particular advantage to this type of chromatography. The ex-

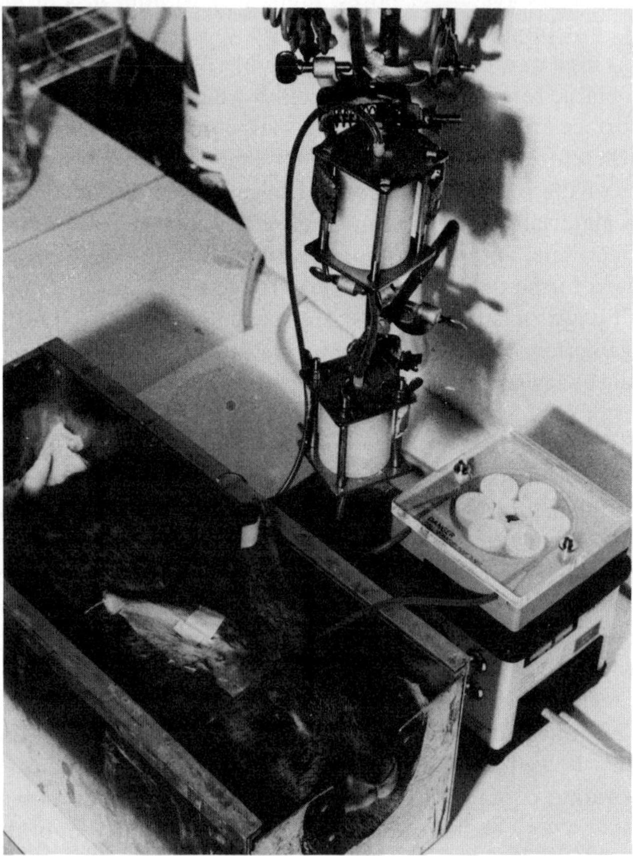

Fig. 11.17. Extracorporal perfusion harvesting of antibodies on controlled pore glass columns. Columns for two different antibodies are series-connected. Courtesy J. Jungfer (199).

tent of its use can be guessed by studying the ligands listed in Section 11.8.4a. In order to make this literature more accessible, and also because some of the ligands listed have been used for other purposes, substances recovered by bioselective adsorption on derivatized CPG are listed in the following section.

The possible high flow-rates in CPG-columns and the inertness and easy sterilization of glass led Jungfer to the use of CPG in the extracorporal *in vivo* harvesting of antibodies from rabbits and sheep (198, 199, 347). Full blood from mildly heparinized, previously inoculated animals is pumped through CPG columns and returned to the animals. The CPG is derivatized with heparin and the respective antigens. Column walls and CPG-retaining nylon screens are also heparin-derivatized. No interference by blood-clotting or ill effects on the animals could be detected. Figure 11.17 shows a view of the experimental set-up. Animals were inoculated against several antigens and the respective antibodies were harvested on separate series-connected columns. Jungfer followed the antibody levels in the perfused animals. Typically, the specific antibody titer fell to 3–5% of the original titer after three times the total blood volume of the animal had passed through the column. The antibody level in the animals recovered in the days after the perfusion harvesting. It took approximately 5 to 14 days to full recovery. With certain antigens, the antibody level after recovery became 150% of the level before the perfusion harvesting. Details of the procedure are reported in Jungfer's Habilitation Report from the University of Heidelberg (199). This technique, which removes only specific antibodies without large blood loss or serum loss, may have good potential for commercial antibody and vaccine production.

Bioselective Adsorption (Affinity Chromatography)

Reviews: 181, 242 (includes discussion on spacer length), 244, 253, 355–361
Substances recovered

- β-Adrenergic receptor adenylate cyclase complex: 227
- Antibodies: 171, 172, 173, 198, 199, 211, 346, 347
- L-Asparaginase: 346
- Carboxypeptidase A: 188
- Digoxin: 349
- DNA-Polymerase: 239
- β-D-Galactosidase: 242, 243
- GAT (Poly-glu-ala-tyr): 350
- Hepatitis B antigen: 352
- Lactate dehydrogenase: 207
- Lipoamide dehydrogenase: 219, 220

Nicotine: 320
Progesterone: 351
Ribonuclease T1: 208
RNA-polymerase: 239
Thyroid hormone receptors: 238
Thyroxine binding globulin: 237, 353
Trypsin inhibitors: 206
Tumor cells: 226

11.8.4c. Derivatized CPG As Reagent and Catalyst

The basic concept of immobilization of reagents on surfaces of insoluble matrices is as old as industrial chemistry itself. Heterogeneous catalysis practiced with catalysts bound to silica-gel, kieselguhr, and similar substances is the backbone of many large-scale industries. Only through the recent explosive growth of enzyme chemistry, however, has this concept been seriously expanded towards catalysis by biospecific substances. The term catalysis implies that the reagent is not consumed in the process, and if there is a decline of reagent activity it is by poisoning, structural aging, dissolution, contamination, or other destructive agents. On the other end of the spectrum of insolubilized reagents are those that undergo some well-defined change. They become exhausted after a certain throughput, and, under optimal circumstance, can be regenerated for reuse. Well-known examples in this group are ion exchange resins and substrates carrying ox–redox functionalities.

The primary advantage of insoluble matrix immobilized reagents is the easy removal of the reagent from the reaction mixture. This saves one step in the further processing of the mixture and also opens up the possibility of using the same reagent over and over. More subtle advantages lie in simplified handling, better possibility of applying automation, and, in some cases, improved stability of the reagent.

In most cases the immobilized reagent is applied to convert some given substance *A* into a new substance *B*, the latter being of actual commercial or scientific interest. Sometimes *A* is a harmful substance and *B* is harmless, and the removal of *A* is the prime objective. Examples of such reactions are the removal of substances interfering with the storage stability of certain products (for instance: milk), or in a more futuristic setting, the removal of a harmful substance from the blood of diseased humans by extracorporal perfusion.

The availability of highly purified and therefore highly active enzymes has furthermore opened exciting possibilities for analytical instrumental analysis of complex mixtures. Utilizing the high chemical specificity of the immobilized enzymes, an enzyme probe is exposed to the mixture to be analyzed. The enzyme facilitates conversion of the substance of interest, *A*, into another substance, *B*. While *B* may not be of particular interest itself, its appearance in the reaction mixture may be easily followed, for instance by a change in pH. The change in pH is quantitatively related to the concentration of *A*, thus allowing quick and highly specific analysis of the mixture. In another type of probe, the heat of reaction of the conversion process itself is being measured and quantitatively related to the concentration of the substance to be analyzed. Such probes are called calorimetric or enthalpic sensors.

A wide variety of solid support matrices, including controlled pore glass, has been investigated and utilized in the above-mentioned applications. Pertinent publications are listed in the following tables. Publications dealing with CPG-immobilized enzymes have been listed in the subsection on enzymes in Section 11.8.4a. To the reader interested in enzyme chemistry, the use of the listed enzymes is rather obvious, and therefore the listing of substances processed on CPG-reactors (see below) has been limited to some selected examples and less obvious uses. Since this section deals with analytical uses, we have also incorporated a brief listing of publications dealing with the use of derivatized CPG for solid phase support assays. Strictly speaking, one practices affinity chromatography in these assays, but because of their analytical nature, we preferred to list them in this section.

We also wish to call attention to extensive recent reviews by Carr and Bowers on the subject "Immobilized Enzymes in Analytical and Clinical Chemistry" (185, 362).

Simple Reagents and Indicators

Fluorescent sulfhydryl reagent: 204
pH-Indicators: 210
Oxidation and reduction reagents: 204, 209, 221

Substances Processed with CPG Bioreactors (See also Section 11.8.4a, subsection on enzymes)

N-Acetyl-L-tyrosine: 188
Beer: 325
Caseine: 214, 314, 321, 325
Cephalosporin: 272 (study of carrier morphology)
Dextrin: 287
Dextrose: 214

DNA: 271
Ethylacetate: 272
Gelatin: 221
Glucose: 192, 287, 291, 297
Lactose: 299, 308
Milk (whey; see also lactose): 318, 330, 335
Nitrates: 313
Penicillin: 316
Phospholipids: 319
Soya protein: 189
Starch: 287
Sucrose: 298
Urea: 197, 342, 343
Uric acid: 291, 339

Derivatized CPG As Extracorporal Drug (See also Section 11.8.4a, subsection on drugs and hormones)

Catecholamines: 225, 227, 228, 231, 232, 235
Hyperuricemia treatment: 341
Removal of cholesterol: 169 (nonderivatized CPG)
Removal of pathogenic antibodies: 189

CPG Bioreactor As Analytical Tool

Reviews: 185, 336, 362, 363, 364
Analyzed substances
  Benzoyl-L-arginine ethyl ester (thermal): 292
  Glucose (chemoluminescence): 295
  Glucose (colorimetric): 291
  Glucose (thermal): 292, 297
  NADH (light emission): 311
  Nitrates (potentiometric): 313
  Penicillin (potentiometric): 316
  Penicillin (thermal): 292
  Phosphate ions (photometric): 260
  Pyruvate (thermal): 307
  Urea (pH, potentiometrically restored): 343
  Urea (thermal): 292, 336, 342, 344, 363
  Uric acid (colorimetric): 291
  Uric acid (thermal): 307

Solid Phase Support Assays on CPG (RIA, etc.)

Antibodies against:
carbohydrates: 211
hepatitis B: 173, 352
Digoxin: 349
GAT (Poly-glu-ala-tyr): 350
Hepatitis B antigen: 352
Human thyroxine-binding globulin: 237, 353
Nicotine: 348
Progesterone: 351

11.8.4d. CPG As Solid Support in Sequential Synthesis of Peptides, Oligosaccharides, and Polynucleotides

A major advance in the synthetic chemistry of peptides and other biologically important molecules was provided in the early 1960s through the methods developed by Merrifield (365). The basic concept of the Merrifield synthesis is the attachment of a starter group to a solid phase matrix, followed by successive attachments of amino acids in a desired and well-controlled sequence. The major advantage of the solid support is the ease with which the intermediates can be isolated by simple filtering and washing of the support. Originally, matrices for these syntheses were high polymer beads. The disadvantages of polymer beads were the heterogeneous location of the anchor groups, which resulted in an unfavorable kinetics, and the fact that the organic polymers suffer under the organic solvents used in the synthetic procedure.

Inorganic supports such as silica gel and controlled pore glass were subsequently used as supports to overcome these difficulties. Parr and coworkers describe the use of CPG as solid phase matrix for polypeptide synthesis (205, 366).

Polypeptides were also attached to CPG to produce surface modifications for other than synthetic purposes. Peptide chains were attached to CPG surfaces (Haller, 224) by initiating the polymerization of *N*-carboxy-α-amino acid anhydrides (Leuch's anhydrides) with amino-propyl-functional CPG. Good yields were obtained with L-proline as the amino acid. In order to create different microenvironments for CPG-bound enzymes, polypeptide chains were attached sequentially to CPG by Taylor and Swaisgood (255).

Sequential syntheses of oligosaccharides on CPG-supports were performed by Holick (367) and Eby and Schuerch (187). The oligosaccharide synthesis necessitates a step that produces high alkalinity, apparently attacking the surface of the glass. Eby and Schuerch (187) overcame this problem by treating the CPG at 350° with zirconium chloride, hydrolyzing the residual zirconium chloride bonds with water and performing the subsequent silane coupling reactions on the zirconia-modified CPG. The zirconia-modified glass exhibited improved resistance to the alkali.

According to a recent article in *Chemical and Engineering News* (368), resins, silica gel, and controlled pore glass supports are being used in several new commercial automatic polynucleotide synthesizers. More detailed information will be available in the future.

11.8.4e. CPG in Solid Phase Sequence Analysis

In solid phase sequence analysis, the peptide or protein to be analyzed is immobilized on an insoluble matrix. The attached ligand is thereafter sequentially cleaved and the dislocated fragments are analyzed. Since the introduction of this technique by Laursen (369), it has become one of the most useful tools of protein analysis. Originally, high polymer resins were used as the supporting matrix. In 1973, Wachter, Machleidt, and Hofner et al. (216) introduced aminopropyl-controlled pore glass as solid phase support and subsequently published several papers on specific uses (370), binding procedures (371), and efficiency studies of the glass support in automated analyzers (372). Aminoethyl-aminopropyl-controlled pore glass was used by Bridgen (218).

In the previously mentioned papers, the amino acids are cleaved from the *N*-terminal. Parham and Loudon describe techniques for degrading carboxyl-terminal peptides, using a carboxyl-functional controlled pore glass as the solid matrix (373, 374).

Pertinent reviews of the field and comparisons of techniques were published by Laursen (375) and Powers (376).

REFERENCES

1. D. H. Campbell, E. Luescher, and L. S. Lerman, *Proc. Nat. Acad. Sci. USA* **37,** 575 (1951).
2. W. Haller and H. C. Duecker, *Nature* (*London*) **178,** 376 (1956).
3. S. Sterman and H. B. Bradley, *SPE-Transactions,* October 1961, p. 224.
4. J. Porath and P. Flodin, *Nature* (*London*) **183,** 1657 (1957).
5. W. Haller, *Nature* (*London*) **206,** 693 (1965).
6. Haller, W. (See dated reference to this 1965 work in footnote on page 362 of Ref. 41).
7. G. W. Morey, *The Properties of Glass,* 2nd ed. Reinhold, New York (1954), p. 28.
8. W. H. Zachariasen, *J. Amer. Chem. Soc.* **54,** 3841 (1932).

9. B. E. Warren, *J. Appl. Phys.* **13,** 602 (1942).
10. W. E. S. Turner and F. Winks, *J. Soc. Glass Technol.* **10,** 102 (1926).
11. M. E. Nordberg and H. P. Hood, U. S. Patent 2,106,744 (1934).
12. B. R. Howell, J. H. Simmons, and W. Haller, *Amer. Ceram. Soc. Bull.* **54,** 707 (1971).
13. R. A. Messing, *Enzymol.* **38,** 39 (1970).
14. R. K. Iler, *Chemistry of Silica,* Wiley, New York (1979).
15. W. Haller, J. H. Simmons, and A. Napolitano, *J. Amer. Ceram. Soc.* **54,** 299 (1971).
16. W. Haller, D. H. Blackburn, and J. H. Simmons, *J. Amer. Ceram. Soc.* **57,** 120 (1974).
17. W. Haller, D. H. Blackburn, F. E. Wagstaff, and R. J. Charles, *J. Amer. Ceram. Soc.* **53,** 34 (1970).
18. J. W. Cahn, *J. Chem. Phys.* **42,** 93 (1968).
19. W. Haller *J. Chem. Phys.* **42,** 686 (1965).
20. T. P. Seward III, D. R. Uhlmann, and D. Turnbull, *J. Amer. Ceram. Soc.* **51,** 634 (1968).
21. W. Haller and P. B. Macedo, *J. Phys. Chem. Glasses* **9,** 153 (1968).
22. W. Haller, *Macromolecules* **10,** 83 (1977).
23. G. K. Ackers and R. L. Steere, in *Methods in Virology* (K. Maramorosch and H. Koprowski, Eds.), Academic Press, New York (1967), p. 325.
24. H. Fasold, G. Gundlach, and F. Turba, in *Chromatography* (E. Heftman, Ed.), Reinhold, New York (1967), p. 405.
25. L. Philipson, in *Methods in Virology* (K. Maramorosch and H. Koprowski, Eds.), Academic Press, New York (1967), p. 179.
26. K. H. Altgelt, in *Chromatography,* Vol. 7 (J. C. Giddings and R. A. Keller, Eds.), Marcel Dekker, New York (1968), p. 3.
27. H. Determann, *Gel Chromatography,* Springer, New York (1968).
28. V. J. Oss, in *Progress in Separation and Purification,* Vol. 1 (E. S. Perry, Ed.), Wiley, New York (1968), p. 187.
29. G. K. Ackers, in *Advances in Protein Chemistry,* Vol. 24 (C. B. Anfinsen Jr., J. T. Edsall, and F. M. Richards, Eds.), Academic Press, New York (1970), p. 343.
30. D. J. Harmon, *Gel Permeation Chromatography,* (K. H. Altgelt and L. Segal, Eds.), American Chemical Society Symposium, 1970, Houston, Texas, Marcel Dekker, New York (1971), p. 13.
31. G. K. Ackers, in *The Proteins,* Vol. 1 (H. Neurath, R. L. Hill, and C. L. Boeder, Eds.), Academic Press, New York (1975), p. 1.
32. W. Haller, *J. Chromatogr.* **32,** 676 (1968).
33. W. Haller, *J. Chromatogr.* **85,** 129 (1973).
34. R. C. Collins and W. Haller, *Anal. Biochem.* **54,** 47 (1973).
35. W. Haller, A. M. Basedow, and B. Konig, *J. Chromatogr.* **132,** 387 (1977).
36. R. D. Hester and P. H. Mitchell, *J. Polym. Sci., Polym. Chem. Ed.* **18,** 1727 (1980).
37. R. A. Messing, L. F. Bialousz, and R. E. Lindner, *J. Solid Phase Biochem.* **1,** 151 (1976).

38. A. M. Basedow, K. H. Ebert, H. J. Ederer, and E. Fosshag, *J. Chromatogr.* **192,** 259 (1980).
39. J. C. Day, B. Alince, and A. A. Robertson, *Can. J. Chem.* **56,** 2951 (1978).
40. W. Haller, H. H. Gschwender, and K. R. Peters *J. Chromatogr.* **211,** 53 (1981).
41. C. W. Hiatt, A. Shelokov, E. J. Rosenthal, and J. M. Galimore, *J. Chromatogr.* **56,** 362 (1971).
42. Y. A. Eltek and A. V. Kiselev, *Chromatographia* **6,** 187 (1973).
43. S. H. Chang, K. M. Gooding, and F. E. Regnier, *J. Chromatogr.* **120,** 321 (1976).
44. S. H. Chang, K. M. Gooding, and F. E. Regnier, *J. Chromatogr.* **125,** 103 (1976).
45. F. E. Regnier and R. E. Noel, *J. Chromatogr. Sci.* **14,** 316 (1976).
46. H. H. Gschwender and P. Traub, *Arch. Virol.* **56,** 327 (1978).
47. I. V. Krasilnikov, L. B. Elbert, L. L. Mamonenko, V. V. Pogodina, G. L. Krutyanskaya, S. E. Bresler, V. M. Kolikov, B. V. Mchedlishvili, V. N. Borisova, N. S. Golovina, and L. A. Nakhapetyan, *Vop. Virusol.* **6,** 685 (1977).
48. W. A. Aue and C. R. Hastings, *J. Chromatogr.* **42,** 319 (1969).
49. A. R. Cooper and J. F. John, *J. Appl. Polym. Sci.* **13,** 1487 (1969).
50. M. Iwama and T. Homma, *Kogyo Kagaku Zasshi* (*J. Chem. Soc. Jpn., Ind. Chem. Sect.*) **74,** 277 (1971).
51. J. B. Derenbach, M. Ehrhardt, C. Osterroth, and G. Petrick, *Marine Chem.* **6,** 351 (1978).
52. R. J. Blagrove and M. J. Frenkel, *J. Chromatogr.* **132,** 399 (1977).
53. F. R. Dintzis and R. Tobin, *J. Chromatogr.* **88,** 77 (1974).
54. J. H. Ross and M. E. Casto, *J. Polym. Sci., Part C* **21,** 143 (1968).
55. K. Marcinka, *Acta Virol.* **16,** 53 (1972).
56. R. J. Barton, *J. Gen. Virol.* **35,** 77 (1977).
57. A. A. Brunt, R. J. Barton, S. Phillips, and A. O. Lana, *Ann. Appl. Biol.,* **96,** 37 (1980).
58. A. A. Brunt and R. Stace-Smith, *Ann. Appl. Biol.* **90,** 205 (1978).
59. A. A. Brunt, *Ann. Appl. Biol.* **87,** 355 (1977).
60. D. Batchelor, Ph.D. Thesis, Univ. of Florida, Gainesville, 1974 (Faculty Supervisor, D. E. Puriciful).
61. H. H. Gschwender, W. Haller, and P. H. Hofschneider, *Biochim. Biophys. Acta* **190,** 460 (1969).
62. W. Haller, *Virology* **33,** 740 (1967).
63. W. Haller, K. D. Tympner, and K. Hannig, *Anal. Biochem.* **35,** 23 (1970).
64. G. I. Bespalova, S. E. Bresler, N. V. Katushkina, and V. M. Kolikov, *Vop. Virus,* **16,** 112 (1971).
65. P. W. Kuhl, *Zbl. Bakt. Paras. Infk. Hyg., I. Orig.* **212,** 358 (1970), in German.
66. A. S. Lee and R. L. Sinsheimer, *J. Virol.* **14,** 872 (1974).
67. T. Darling, J. Albert, P. Russell, D. M. Albert, and T. W. Reid, *J. Chromatogr.* **131,** 383 (1977).
68. A. R. Neurath, L. Cosio, A. M. Prince, A. Lippin, and H. Ikram, *Proc. Soc. Exp. Biol Med.* **143,** 440 (1973).
69. J. T. Heyward, R. A. Klimas, M. D. Stapp, and J. F. Obijeski, *Arch. Virol.* **55,** 107 (1977).

70. H. H. Gschwender, M. Brummund, and F. Lehmann-Grube, *J. Virol.* **15,** 1317 (1975).
71. K. Leung, J. M. Jones, and J. D. Feldman, *J. Immunol.* **121,** 1836 (1978).
72. H. H. Gschwender, G. Rutter, and M. Popescu, *Arch. Virol.* **49,** 359 (1975).
73. S. B. Prusiner, D. F. Groth, S. P. Cochran, M. P. McKinley, and F. R. Masiarz, *Biochemistry* **19,** 4892 (1980).
74. W. Frisch-Niggemeyer, *Acta Virol.* **19,** 381 (1975).
75. W. Frisch-Niggemeyer, *Zbl. Bakt. Parask. Infk. Hyg., I. Ref.* **230,** 306 (1972), in German.
76. K. H. Richter, M. Schwenen, and D. G. Padval, *Zbl. Bakt. Hyg. I. Abt. Orig.* **A223,** 15 (1973), in German.
77. H. H. Gschwender, Doctoral Dissertation, Medical Faculty, University of Munich, 1969.
78. D. S. Gregerson, J. Albert, and T. W. Reid, *Biochemistry* **19,** 301 (1980).
79. B. A. Roe, *Nucleic Acids Res.* **2,** 21 (1975).
80. C. Weitbrecht, Diplomarbeit, University of Giessen, Germany, 1974 (Faculty Supervisor, H. L. Sanger).
81. S. J. Morris, *J. Neurochem.* **21,** 713 (1973).
82. H. D. Crone, R. M. Dawson, and E. M. Smith, *J. Chromatogr.* **103,** 71 (1975).
83. H. D. Crone, and R. M. Dawson, *J. Chromatogr.* **129,** 91 (1976).
84. M. J. Frenkel and R. J. Blagrove, *J. Chromatogr.* **111,** 397 (1975).
85. C. Persiani, P. Cukor, and K. French, *J. Chromatogr. Sci.* **14,** 417 (1976).
86. D. E. S. Truman and A. G. Brown, *Exp. Eye Res.* **12,** 304 (1971).
87. D. H. Boehme, M. W. Fordice, and J. S. Bykowski, *Virchow's Arch., Abt. B. Zellpath.* **15,** 323 (1974).
88. W. Frisch-Niggemeyer, F. Heinz, and H. Stemberger, *Immunitat und Infektion* **2,** 231 (1974), in German.
89. W. Frisch-Niggemeyer, *J. Clin. Microbiol.* **2,** 377 (1975).
90. W. Frisch-Niggemeyer and O. Picher, *Wiener Klinische Wochenschrift* **88,** Suppl. 12 (1976), in German.
91. M. J. Telepchak, *J. Chromatogr.* **83,** 125 (1973).
92. G. D. Miner, J. McSwigan, and L. L. Heston, *Prep. Biochem.* **6,** 1 (1976).
93. R. D. Kearney and T. C. A. McGann, in *Proceedings of the 4th Annual Research Conference on Food Science and Technology,* Cork, Ireland, September 1974.
94. R. D. Kearney and T. C. A. McGann, in *Chromatography of Synthetic and Biological Polymers* Vol. 1 (Roger Epton, Ed.), Ellis Horwood, Chichester, **269** (1978).
95. E. Almlof, M. Larsson-Raznikiewicz, I. Lindquist, and J. Munyua, *Prep. Biochem.* **7,** 1 (1977).
96. M. Larsson-Raznikiewicz, E. Almlof, and B. Ekstrand, *J. Dairy Res.* **46,** 313 (1979).
97. H. Gruft, R. Ruck, and J. Traynor, *Can. J. Biochem.* **56,** 916 (1978).
98. J. Margolis and P. Rhoades, *Vox Sanguinis* **36,** 369 (1979).
99. J. Margolis and P. Rhoades, Lancet *2* (8244), 446 (1981).
100. J. L. Pearson, B. L. Evatt, and R. B. Ramsey, *Thrombosis Haemostasis,* **42,** 223 (1979).

101. N. R. Shulman and K. M. Tack, *Clin. Res.* **25,** 513A (1977).
102. H. Hofmann, W. Frisch-Niggemeyer, and C. Kunz, *Infection* **6,** 154 (1978).
103. F. Mehnert and K. Hummel, *Z. Immun.-Forsch. Immunobiol.* **151,** 316 (1976), in German.
104. K. Zeiller and L. Dolan, *Eur. J. Immunol.* **2,** 439 (1972).
105. P. W. Robertson, J. Giannikos, and G. A. Bishop, *Med. J. Australia,* **2,** 293 (1978).
106. D. C. Beebe and J. Piatigorsky, *Exp. Eye Res.* **23,** 83 (1976).
107. C. Persiani, P. Cukor, and K. French, *J. Chromatogr. Sci.* **14,** 417 (1976).
108. V. P. Bhavanandan, J. Umemoto, J. R. Banks, and E. A. Davidson, *Biochemistry* **16,** 4426 (1977).
109. S. E. Korngut and E. Sunderland, *Biochim. Biophys. Acta* **393,** 100 (1975).
110. L. Y. Fager and R. S. Fager, *Anal. Biochem.* **85,** 98 (1978).
111. J. J. Nordmann, F. Louis, and S. J. Morris, *Neuroscience* **4,** 1367 (1979).
112. A. Nagy, R. R. Baker, S. J. Morris, and V. P. Whittaker, *Brain Res.* **109,** 285 (1976).
113. S. S. Carlson, J. A. Wagner, and R. B. Kelly, *Biochemistry* **17,** 1188 (1978).
114. K. Ohsawa, G. H. C. Dowe, S. J. Morris, and V. P. Whittaker, *Exp. Brain Res.* **24,** 19 (1976).
115. J. A. Wagner, S. S. Carlson, and R. B. Kelly, *Biochemistry* **17,** 1199 (1978).
116. R. T. Ambron, A. A. Sherbany, L. J. Shkolnik, and J. H. Schwartz, *Brain Res.* **207,** 17 (1981).
117. K. Ohsawa, A. Kano, and T. Hoshi, *Life Sci.* **24,** 669 (1979).
118. V. K. Miyamoto and W. Stoeckenius, *J. Membrane Biol.* **4,** 252 (1971).
119. R. C. Jordan and D. A. Brant, *Macromolecules* **13,** 491 (1980).
120. P. R. Straub and D. A. Brant, *Biopolymers* **19,** 639 (1980).
121. A. C. Wu and W. A. Bough, *J. Chromatogr.* **128,** 87 (1976).
122. A. M. Basedow and K. H. Ebert, *Infusionstherapie* **2,** 261 (1975), in German.
123. A. M. Basedow, K. H. Ebert, H. Ederer, and H. Hunger, *Makromol. Chemie.* **177** 1501 (1976), in German.
124. A. M. Basedow and K. H. Ebert, *J. Polym. Sci, Part C* **66,** 101 (1979).
125. A. R. Cooper and D. P. Matzinger, in *Column Packings, GPC, GF, and Gradient Elution,* Vol. 1 (Roger Epton, Ed.), Chem. Soc. Macrom. Group, Engl. Ellis Horwood, Chichester (1978), p. 350.
126. A. L. Spatorico and G. L. Beyer, *J. Appl. Polym. Sci.* **19,** 2933 (1975).
127. E. V. Chandrasekaran and E. A. Davidson, *Cancer Res.* **39,** 870 (1979).
128. F. J. Schmitz, K. H. Hillenbeak, and D. C. Campbell, *J. Org. Chem.* **43,** 3916 (1978).
129. O. H. Danneberg and J. Schmidt, *Bodenkultur* **29,** 1 (1978).
130. S. A. Barker, S. J. Crews, and J. B. Marsters, in *Symposium on the Biochemistry of the Eye,* Tutzing, August 1966, Karger, Basel and New York (1968), p. 481.
131. E. A. Balazs, S. Briller, and J. L. Denlinger, in *Seminars in Arthritis and Rheumatism,* Vol 11 (J. M. Talbott, Ed.), Grune and Stratton, N.Y. (1981), p. 141.
132. J. F. Stoddart and J. K. N. Jones, *Carbohydr. Res.* **8,** 29 (1968).

133. R. Simonson, *Svensk Papperstidn.* **74,** 691 (1971).
134. A. Hutterman, *Holzforschung* **20,** 108 (1978).
135. K. J. Mayberry-Carson, T. A. Langworthy, W. R. Mayberry, and P. F. Smith, *Biochim. Biophys. Acta* **360,** 217 (1974).
136. P. McWain and G. F. Gregory, *Phytochem.* **11,** 2609 (1972).
137. C. Satoh, J. Banks, P. Horst, J. W. Kreider, and E. A. Davidson, *Biochemistry* **13,** 1233 (1974).
138. K.-R. Peters, G. Rutter, H. H. Gschwender, and W. Haller, *J. Cell Biol.* **78,** 309 (1978).
139. M. J. Telepchak, High Speed Molecular Size Exclusion Chromatography of Minute Components in Biological Fluids. FASEB Meeting, April 15–20, 1973.
140. M. J. Telepchak, *Chromatogr. Newsl.* **1,** 10 (1972).
141. E. Glynne-Jones, R. Marshall, R. A. Hann, and G. Read, *J. Chromatogr.* **114,** 232 (1975).
142. A. D. Williams, D. D. Freeman, and W. M. Florsheim, *J. Chromatogr. Science* **9,** 619 (1971).
143. G. P. Lampson, A. A. Tytell, M. M. Nemes, and M. R. Hilleman, *Proc. Soc. Exp. Biol. Med.* **112,** 468 (1963).
144. A. Davies, *Proc. Biochem. Soc.* **95,** 20P (1965).
145. V. G. Edy, I. A. Braude, E. De Clercq, A. Billiau, and P. De Somer, *J. Gen. Virol.* **33,** 517 (1976).
146. V. G. Edy and J. L. Olpin, personal communication.
147. A. Billiau, J. Van Damme, F. Van Leuven, V. G. Edy, M. DeLey, J.-J Cassiman, H. Van den Berghe, and P. De Somer, *Antimicrob. Agents Chemother.* **16,** 49 (1979).
148. A. Billiau, P. De Somer, V. G. Edy, E. De Clercq, and H. Heremans, *Antimicrob. Agents Chemother.* **16,** 56 (1979).
149. J. E. Whitman Jr., G. M. Crowley, and C. L. Hung, in *Interferon Properties and Clinical Uses* (N. O. Hill and G. L. Dorn, Eds.), Wadley Institute Dallas, Leland-Fikes, (1980), p. 561.
150. V. G. Edy, I. A. Braude, E. De Clercq, A. Billiau, and P. De Somer, *J. Gen. Virol.* **33,** 517 (1976).
151. J. W. Heine, M. DeLey, J. Van Damme, A. Billiau, and P. De Somer, *Ann. N.Y. Acad. Sci.* **350,** 364 (1980).
152. Y. H. Tan, H. Smith-Johannsen, and H. Okamura, *Ann. N.Y. Acad. Sci.* **350,** 376 (1980).
153. Y. H. Tan, F. Barakat, W. Berthold, H. Smith-Johannsen, and C. Tan, *J. Biol. Chem.* **254,** 8067 (1979).
154. H. Okamura, E. Berthold, L. Hood, M. Hunkapiller, and M. Inoue, *Biochemistry* **19,** 3831 (1980).
155. J. A. Georgiades, M. L. Langford, G. J. Stanton, and H. M. Johnson, *IRCS Medical Sciences: Biochemistry, Biomedical Technology, Microbiology, Parasitology, and Infectious Disease,* **7,** 559 (1979).
156. J. A. Georgiades, M. P. Langford, L. D. Goldstein, J. E. Blalock, and H. M. Johnston, in *Interferon: Properties and Clinical Uses* (A. Kahn, N. L. Hill, and G. L. Dorn, Eds.), Wadley Institute, Dallas (1979), p. 97.

157. M. P. Langford, J. A. Georgiades, G. J. Stanton, F. Dianzani, and H. M. Johnson, *Infect. Immun.* **26,** 36 (1979).
158. M. Wiranowska-Stewart, L. S. Lin, I. A. Braude, and W. E. Stewart II, *Mol. Immunol.* **17,** 625 (1980).
159. H. G. Bock, P. Skene, and S. Fleischer, *Science* **191,** 380 (1976).
160. J. E. Whitman Jr., C. L. Hung, J. V. Tredway, J. Tou, and G. M. Crowley, in *Proceedings of the 1st International Congress for Interferon Research,* Washington, D.C., Nov. 9–12, 1980.
161. I. A. Braude and E. De Clercq, *J. Chromatogr.* **172,** 207 (1979).
162. I. A. Braude, V. G. Edy, and E. De Clercq, *Biochim. Biophys. Acta* **580,** 15 (1979).
163. H. Taira, R. J. Broeze, B. M. Jayaram, and P. Lengyel, *Science* **207,** 528 (1980).
164. B. Cabrer, H. Taira, R. J. Broeze, T. D. Kempe, K. Williams, E. Slattery, W. H. Konigsberg, and P. Lengyel, *J. Biol. Chem.* **254,** 3681 (1979).
165. H. D. Crone, *J. Chromatogr.* **107,** 25 (1975).
166. Y. Hashimoto and H. Miyazaki, *J. Chromatogr.* **168,** 59 (1979).
167. H. G. Bock, P. Skene, and S. Fleischer, *Science* **191,** 380 (1976).
168. P. R. Dunkley and J. M. Anderson, *Arch. Biochem. Biophys.* **193,** 463 (1979).
169. M. R. Malinow, P. McLaughlin, and L. Papworth, *Atherosclerosis* **22,** 293 (1975).
170. S. Bresler, N. Katushkina, V. Kolikov, M. Chumakov, J. Pervikov, and S. P. Zhdanov, *Virology* **59,** 36 (1974).
171. R. Y. Dodd and J. A. Kobita, *J. Immunol. Methods* **20,** 117 (1978).
172. R. Y. Dodd, *Transfusion* **20,** 212 (1980).
173. C. T. Fang, N. Nath, H. Berberian, and R. Y. Dodd, *J. Immunol. Meth.* **24,** 371 (1978).
174. H. G. Bock and S. Fleischer, in *Methods in Enzymology,* Vol. 32 (S. Fleischer and L. Packer, Eds.), Academic Press, New York (1974), p. 374.
175. S. E. Bresler, N. V. Katushkina, V. M. Kolikov, J. L. Potokin, and G. N. Vinogradskaya, *J. Chromatogr.* **130,** 275 (1977).
176. S. Gerard and W. M. Mitchell, *Anal. Biochem.,* **96,** 433 (1979).
177. R. K. Iler, U. S. Patent 2,657,149 (1953).
178. J. Wartman and H. Deuel, **12,** 82 (1958).
179. W. Stober, *Beitr. Silikose-Forsch. Sonderb.* **2,** 333 (1956), in German.
180. H. H. Weetall, in *Immobilized Biochemicals and Affinity Chromatography* (R. B. Dunlap, Ed.), Plenum Press, New York (1974), p. 191.
181. H. H. Weetall, and A. M. Filbert, in *Methods in Enzymology,* Vol. 34 (W. B. Jakoby and M. Wilcheck, Eds.), Academic Press, New York (1974), p. 59.
182. H. H. Weetall, in *Methods in Enzymology,* Vol. 44 (K. Moosbach, Ed.), Academic Press, New York (1976), p. 134.
183. H. H. Weetall, Ed., *Immobilized Enzymes, Antigens, Antibodies and Peptides Preparation and Chromatography,* Vol. 1, *Enzymology* (H. H. Weetall, Ed.), Marcel Dekker, New York (1975).
184. C. Kent, A. Rosevear, and A. R. Thomson, in *Topics in Enzyme and Fermenta-*

tion Biotechnology, Vol. 2 (Alan Wiseman, Ed.), Ellis Horwood, Chichester, (1978), p. 12.

185. P. W. Carr, and L. D. Bowers, in *Chemical Analysis,* Vol. 56 (P. J. Elving, J. D. Wineforder, and I. M. Volthoff, Eds.), John Wiley, New York (1980).
186. S. H. Chang, R. Noel, and F. E. Regnier, *Anal. Chem.* **48,** 1833 (1976).
187. R. Eby, and C. Schuerch, *Carbohydr. Res.* **39,** 151 (1975).
188. P. J. Robinson, P. Dunnill, and M. D. Lilly, *Biochim. Biophys. Acta* **242,** 659 (1971).
189. R. D. Mason, C. C. Detar, and H. H. Weetall, *Biotech. Bioeng.* **17,** 1019 (1975).
190. P. G. Royer, F. A. Liberatore, and G. M. Green, *Biochem. Biophys. Res. Commun.* **64,** 478 (1975).
191. G. P. Royer, F. A. Liberatore, G. M. Green, W. E. Schwartz, and W. E. Meyers, *Polym. Prepr.* **16,** 76 (1975).
192. L. D. Bowers and P. W. Carr, *Biotechnol. Bioeng.* **18,** 1331 (1976).
193. G. P. Royer and F. A. Liberatore, *Enzyme Eng.* **3,** 43 (1978).
194. H. H. Weetall, *Nature* **223,** 959 (1969).
195. C. Lewis and W. H. Scouten, *J. Chem. Educ.* **53,** 395 (1976).
196. J. C. Venter and J. E. Dixon, in *Methods in Enzymology,* Vol. 38 (B. W. O'Malley and J. G. Hardman, Eds.), Academic Press, New York, (1974), p. 180.
197. H. H. Weetall and L. S. Hersh, *Biochim. Biophys. Acta* **185,** 464 (1969).
198. H. Jungfer and H. P. Geisen, *10th International Congress of Biochemistry,* Hamburg, Germany (1976).
199. H. Jungfer, *Habilitation,* Medical Faculty, University of Heidelberg (Dept. Head: K. Rother) (in German) (1976).
200. D. E. Leyden, G. H. Luttrell, W. K. Nonidez, and D. B. Wekho, *Anal. Chem.,* **48,** 67 (1976).
201. B. B. Jablonski, and D. E. Leyden, in *Advances in X-ray Analysis,* Vol. 21 (C. S. Barrett and D. E. Leyden, Eds.), Plenum Press (1978), p. 59.
202. R. G. Masters and D. E. Leyden, *Anal. Chim. Acta,* **98,** 9 (1978).
203. N. Kelly and A. Flynn, *Biotechnol. Bioeng.,* **19,** 1211 (1977).
204. W. H. Scouten, *Enzyme Eng.* **4,** 391 (1978).
205. W. Parr and K. Grohmann, *Tetrahedron Lett.* **28,** 2633 (1971).
206. L. J. Loeffler and J. V. Pierce, *Biochim. Biophys. Acta* **317,** 20 (1973).
207. L. Jervis, in *Chrom. Synth. Biol. Polym.* Vol. 78 (Roger Epton, Ed.), Ellis Horwood, Chichester (1978), p. 231.
208. L. Jervis and N. M. Pettit, *J. Chromatogr.* **97,** 33 (1974).
209. C. Lewis and W. H. Scouten, *Biochim. Biophys. Acta,* **444,** 326 (1976).
210. G. B. Harper, *Anal. Chem.* **47,** 348 (1975).
211. P. H. Boullanger, A. Nagpurkar, A. Noujaim, and R. U. Lemieux, *Can. J. Biochem.* **56,** 1102 (1978).
212. E. D. Moorehead and P. H. Davis, *Anal. Chem.* **46,** 1879 (1974).
213. K. F. Sugawara, H. H. Weetall, and G. D. Schucker, *Anal. Chem.* **46,** 489 (1974).
214. H. H. Weetall, *Biochim. Biophys. Acta* **212,** 1 (1970).
215. C. Flemming, A. Gabert, and H. Wand, *Acta Biol. Med. Germ.* **32,** 135 (1974).

216. E. Wachter, W. Machleidt, H. Hofner, and J. Otto, *FEBS Lett.* **35,** 97 (1973).
217. W. Machleidt and E. Wachter, *Enzymology* **47,** 263 (1977).
218. J. Bridgen, *FEBS Lett.* **50,** 159 (1975).
219. W. H. Scouten, F. Torok, and W. Gitomer, *Biochim. Biophys. Acta* **309,** 521 (1973).
220. W. H. Scouten, in *Methods in Enzymology,* Vol. 34. (W. B. Jakoby and M. Wilcheck, Eds.), Academic Press, New York (1975), p. 288.
221. W. H. Scouten and G. L. Firestone, *Biochim. Biophys. Acta* **453,** 277 (1976).
222. M. Yaqub and P. Guire, *J. Biomed. Mater. Res.* **8,** 291 (1974).
223. P. Guire, in *Methods in Enzymology,* Vol. 44 (K. Moosbach, Ed.), Academic Press, New York (1976), p. 280.
224. W. Haller, Polyproline Derivatives of Controlled Pore Glass, Max Planck Inst., Munich, unpublished (1968).
225. N. O. Kaplan and J. C. Venter, *Proceedings of the 6th International Congress of Pharmacology* (M. Mattila, Ed.), Pergamon Press, New York (1975), p. 73.
226. B. R. Venter, J. C. Venter, and N. O. Kaplan, *Proc. Nat. Acad. Sci. USA,* **73,** 2013 (1976).
227. J. C. Venter, J. E. Dixon, P. R. Maroko, and N. O. Kaplan, *Proc. Nat. Acad. Sci. USA,* **69,** 1141 (1972).
228. J. C. Venter, J. Ross Jr., J. E. Dixon, S. E. Mayer, and N. O. Kaplan, *Proc. Nat. Acad. Sci. USA,* **70,** 1214 (1973).
229. J. C. Venter and N. O. Kaplan, in *Methods in Enzymology,* Vol. 38 (B. W. O'Malley and J. G. Hardman, Eds.), Academic Press, New York (1974), p. 187.
230. J. C. Venter, N. O. Kaplan, M. S. Yong, and J. B. Richardson, *Science* **185,** 459 (1974).
231. J. C. Venter, Ph.D. Dissertation, Univ. of Calif., San Diego, 1975 (Faculty Supervisor: N. O. Kaplan).
232. J. C. Venter, J. Ross Jr., and N. O. Kaplan, *Proc. Nat. Acad. Sci.* **72,** 824 (1975).
233. J. C. Venter, L. J. Arnold, and N. O. Kaplan, *Mol. Pharm.* **11,** 1 (1975).
234. M. S. Yong, *Science* **182,** 157 (1973).
235. D. C. Woodman, J. C. Venter, and N. O. Kaplan, *Midl. Macromol. Monogr.* **5** (Polym. Delivery Systems), 251 (1978).
236. R. Wade, University of Washington, Seattle, unpublished.
237. A. Castro and E. Ugarte, *Res. Commun. Chem. Pathol. Pharmacol.* **7,** 453 (1974).
238. K. R. Latham, J. W. Apriletti, N. L. Eberhardt, and J. D. Baxter, *J. Biol. Chem.,* **256,** 12088 (1981).
239. W. H. Scouten, *Amer. Lab.* 23 (August 1974).
240. K. G. Skryabin, T. V. Venkstern, V. M. Zaharyev, V. A. Varlamov, S. V. Rogozhin, and A. A. Bayev, *Hoppe-Seyler Z. Physiol. Chem.* **357,** 337 (1976).
241. K. G. Skryabin, V. B. Varlamov, V. M. Zakhariev, S. V. Rogozhin, and Bayev, A. A. *Biorg. Khim.* **2,** 1416 (1976).
242. G. Baum, in *Enzyme Engineering,* Vol. 2 (E. K. Pye and L. B. Wingard Jr., Eds.), Plenum Press, New York (1974), p. 63.
243. ———, *J. Chromatogr.* **104,** 105 (1975).

244. W. H. Scouten, in *Encyclopedia of Polymer Science and Technology*, Supplement, Vol. 2 (N. M. Bikales, Ed.), Wiley, New York (1978), p. 19.
245. M. K. Weibel, H. H. Weetall, and H. J. Bright, *Biophys. Biochem. Res. Commun.* **44,** 347 (1971).
246. M. K. Weibel, E. E. Doyle, A. E. Humphrey, and H. J. Bright, *Biotechnol. Bioeng., Symp. No. 3,* (1972), p. 167.
247. J. Everese, C. L. Ginsburgh, and N. O. Kaplan, in *Methods in Biochemical Analysis,* Vol. 25 (D. Glick, Ed.), Interscience, New York (1979), p. 135.
248. G. P. Royer, J. P. Andrews, and R. Uly, *Enzyme Technol. Digest* **1,** 99 (1973).
249. D. L. Eaton, in *Immobilized Biochemicals and Affinity Chromatography* (R. B. Dunlap, Ed.), Plenum Press, New York (1974), p. 241.
250. K. Mosbach, Ed., *Methods in Enzymology,* Vol. 44, Academic Press, New York, 1976.
251. R. A. Messing, Ed., *Immobilized Enzymes for Industrial Reactors,* Academic Press, New York (1975).
252. H. H. Weetall and R. A. Messing, in *The Chemistry of Biosurfaces* (M. L. Hair, Ed.), Marcel Dekker, New York (1972), p. 563.
253. R. A. Messing, in *Methods in Enzymology,* Vol. 44 (K. Moosbach, Ed.), Academic Press, New York (1976), p. 148.
254. M. Lynn, in *Immobilized Enzymes, Antigens, Antibodies, and Peptides: Preparation and Chromatography,* Vol. 1 (H. H. Weetall, Ed.), *Enzymology,* Marcel Dekker, New York (1975), p. 1.
255. J. B. Taylor and H. E. Swaisgood, in *Immobilized Biochemicals and Affinity Chromatography* (R. B. Dunlap, Ed.), Plenum Press, New York (1974), p. 283.
256. G. F. B. Baum and H. H. Weetall, *Biochim. Biophys. Acta* **268,** 411 (1972).
257. J. R. Ford, A. H. Lambert, W. Cohen, and R. P. Chambers, *Biotechnol. Bioeng. Symp. No. 3,* Wiley, New York (1972), p. 267.
258. M. J. Brougham and D. B. Johnson, *Int. J. Biochem.* **9,** 283 (1978).
259. D. B. Johnson, *Biotechnol. Bioeng.* **20,** 1117 (1978).
260. H. H. Weetall and M. A. Jacobson, *Proc. IV, IFS, Ferment. Technol. Today* (1972), p. 361.
261. H. H. Weetall and G. Baum, *Biotechnol. Bioeng.* **12,** 399 (1970).
262. G. P. Royer and J. P. Andrews, *J. Biol. Chem.* **248,** 1807 (1973).
263. C. Flemming, A. Gabert, and H. Wand, *Acta Biol. Med. Germ.* **32,** 135 (1974).
264. G. P. Royer and J. P. Andrews, *J. Macromol. Sci. Chem.* **A7** 1167 (1973).
265. J. P. Cardoso, M. F. Chaplin, A. N. Emery, J. F. Kennedy, and L. P. Revel-Chion, *J. Appl. Chem. Biotechnol.* **28,** 775 (1978).
266. H. H. Weetall, *Nature* **232,** 473 (1971).
267. H. R. Morton and H. E. Swaisgood, in *Immobilized Biochemicals and Affinity Chromatography* (R. B. Dunlap, Ed.), Plenum Press, New York (1974), p. 345.
268. K. Tanizawa and M. L. Bender, *J. Biol. Chem.* **249,** 2130 (1974).
269. H. R. Horton and H. E. Swaisgood, in *Methods in Enzymology,* Vol. 44 (K. Moosbach, Ed.), Academic Press, New York (1976), p. 516.
270. A. Svenson and B. Anderson, *Anal. Biochem.* **83,** 739 (1977).
271. A. R. Neurath and H. H. Weetall, *FEBS Lett.* **8,** 253 (1970).
272. J. Konecny and W. Voser, *Biochim. Biophys. Acta* **485,** 367 (1977).

273. V. P. Varlamov, T. N. L'vova, D. G. Val'kovskii, V. Y. Mokeev, R. I. Tartarskaya, and S. V. Rogozhin, *Bioorg. Khim.* **1,** 816 (1975).
274. H. H. Weetall, *Biochim. Biophys. Acta* **212,** 1 (1970).
275. D. L. Marshall, in *Immobilized Biochemicals and Affinity Chromatography* (R. B. Dunlap, Ed.), Plenum Press, New York (1974), p. 345.
276. J. V. Jackson, Ph.D. Thesis, Cornell Univ., Ithaca, N.Y. 1974.
277. E. S. Okos and W. J. Harper, *J. Food Sci.* **39,** 88 (1974).
278. L. E. Wierzbicki, V. H. Edwards, and F. V. Kosikowski, *Biotechnol. Bioeng.* **16,** 397 (1974).
279. M. V. Wondolowski and J. H. Woychik, *Biotechnol. Bioeng.* **16,** 1633 (1974).
280. J. H. Woychik and M. V. Wondolowski, *Biochim. Biophys. Acta* **289,** 347 (1972).
281. J. H. Woychik and M. V. Wondolowski, *J. Milk Food Technol.* **36,** 31 (1973).
282. M. J. Byrne and D. B. Johnson, *Biochem. Soc. Trans.* **2,** 496 (1974).
283. F. Wangenmayer, D. Linder, and K. Wallenfels, *Biotechnol. Bioeng.* **19,** 1387 (1977).
284. H. H. Weetall and N. B. Havewala, in *Biotechnol. Bioeng. Symp.* (L. B. Wingard Jr., Ed.), Interscience, (1972), p. 241.
285. J. Fischer, W. Heyer, F. Janowski, F. Wolf, and A. Schellenberger, *Acta Biol. Med. Ger.* **36,** 999 (1977).
286. T. Z. Lai and F. F. Hsu, *J. Chin. Biochem. Soc.* **4,** 36 (1975).
287. G. K. Lee, in *Proc. Ann. Biochem. Eng. Symp.* Kansas State Univ. (1975), p. 32.
288. D. R. Marsh, Y. Y. Lee, and G. T. Tsao, *Biotechnol. Bioeng.* **10,** 483 (1973).
289. A. P. Sinitsyn, A. M. Klibanov, A. A. Klesov, and K. Martinek, *Prikl. Biokhim. i Mikrobiol.* **14,** 236 (1978).
290. Y. Y. Lee, A. R. Fratzke, K. Wunk, and G. T. Tsao, *Biotechnol. Bioeng.* **18,** 389 (1976).
291. J. Endo, M. Tabata, S. Okada, and T. Murachi, *Clin. Chim. Acta* **95,** 411 (1979).
292. K. Mosbach, B. Danielsson, A. Borgerun, and M. Scott, *Biochim. Biophys. Acta* **403,** 256 (1975).
293. M. K. Weibel and H. J. Bright, *Biochem. J.* **124,** 801 (1971).
294. M. K. Weibel, W. Dritschilo, H. J. Bright, and A. E. Humphrey, *Anal. Biochem.* **52,** 402 (1973).
295. D. C. Williams, G. F. Huff, and W. R. Seitz, *Clin. Chem.* **22,** 372 (1976).
296. H. E. Swaisgood, H. R. Horton, and K. Mosbach, in *Methods in Enzymology,* Vol. 44 (V. Mosbach, Ed.), Academic Press, New York (1976), p. 504.
297. L. D. Bowers and P. W. Carr, *Clin. Chem.* **22,** 1427 (1976).
298. R. D. Mason and H. H. Weetall, *Biotech. Bioeng.* **14,** 637 (1972).
299. H. H. Weetall, N. B. Havewala, W. H. Pitcher Jr., C. C. Detar, W. P. Vann, and S. Yaverbaum, *Biotechnol. Bioeng.* **16,** 295 (1974).
300. K. Sukegawa and S. Takahashi, *Eiyo To Shokuryo* **28,** 33 (1975).
301. L. E. Wierzbicki, V. H. Edwards, and F. V. Kosikowski, *J. Food Sci.,* **38,** 1070 (1973).
302. L. E. Wierzbicki, V. H. Edwards, and F. V. Kosikowski, *J. Food Sci.* **39,** 374 (1974).

303. I. C. Cho and H. E. Swaisgood, *Biochim. Biophys. Acta* **258,** 675 (1972).
304. J. E. Dixon, F. E. Stolzenbach, J. A. Berenson, and N. O. Kaplan, *Biochem. Biophys. Res. Commun.,* **52,** 905 (1973).
305. F. Widmer, J. E. Dixon, and N. O. Kaplan, *Anal. Biochem.,* **55,** 282 (1973).
306. J. E. Dixon, F. E. Stolzenbach, C. T. Lee, and N. O. Kaplan, *Israel J. Chem.* **12,** 529 (1974).
307. N. N. Rehak, J. Everese, N. O. Kaplan, and R. L. Berger, *Anal. Biochem.,* **70,** 381 (1976).
308. D. B. Johnson, D. Thorton, and P. D. Ryan, *Biochem. Soc. Trans.,* **2,** 494 (1974).
309. S. W. Sae, *Biotechnol. Bioeng.* **16,** 275 (1974).
310. W. H. Scouten, H. Knowles Jr., L. C. Freitag, and W. Iobst, *Biochim. Biophys. Acta,* **482,** 11 (1977).
311. E. Jablonski and M. Deluca, *Proc. Natl. Acad. Sci. USA* **73,** 3848 (1976).
312. E. C. Lee, G. F. Senyk, and W. F. Shipe, *J. Food Science,* **39,** 927 (1974).
313. D. R. Senn, P. W. Carr, and L. N. Klatt, *Anal. Chem.,* **48,** 954 (1976).
314. H. H. Weetall, *Science,* **166,** 615 (1969).
315. M. K. Weibel, R. Barrios, R. Delotto, and A. E. Humphrey, *Biotechnol. Bioeng.,* **17,** 85 (1975).
316. J. F. Rusling, G. H. Luttrell, L. F. Cullen, and G. J. Papariello, *Anal. Chem.,* **48,** 1211 (1976).
317. W. F. Line, A. Kwong, and H. H. Weetall, *Biochim. Biophys. Acta,* **242,** 194 (1971).
318. L. K. Ferrier, T. Richardson, and N. F. Olson, *J. Dairy Sci.,* **54,** 765 (1971).
319. M. Adamich, H. F. Voss, and E. A. Dennis, *Arch. Biochem. Biophys.,* **189,** 417 (1978).
320. W. H. Hanisch, P. A. Rickard, and S. Nyo, *Biotechnol. Bioeng.,* **20,** 95 (1978).
321. C. C. Detar, R. D. Mason, and H. H. Weetall, *Biotechnol. Bioeng.,* **17,** 451 (1975).
322. G. P. Royer, and G. M. Green, *Chem. Papers* **31,** 357 (1971).
323. J. D. Chapman and H. O. Hultin, *Biotechnol. Bioeng.* **17,** 1783 (1975).
324. A. M. Kalashnikova, I. I. Menyailova, L. A. Nakhapetyan, I. M. Gracheva, and A. A. Konurkina, *Prikl. Biokhim. Mikrobiol.* **11,** 842 (1975).
325. M. Kreen, E. Erin, K. Kivisilla, N. Lukaseviciene, and A. Kestner, *Ref. Dokl. Soobsch;* Mendeleevsk. S'ezd Obshch. *Prikl. Khim. 11th* **6,** 65 (1975).
326. R. J. Brown and H. W. Swaisgood, *J. Dairy Sci.,* **58,** 796 (1975).
327. M. Paquot, P. Thonart, and C. Deroanne, *Lait* **56,** 154 (1976).
328. J. Shapira, C. L. Hanson, J. M. Lyding, and P. J. Reilly, *Biotechnol. Bioeng.* **16,** 1507 (1974).
329. M. J. Grove, G. W. Strandberg, and K. L. Smiley, *Biotechnol. Bioeng.* **13,** 703 (1971).
330. V. G. Janolino and H. E. Swaisgood, *J. Dairy Sci.,* **61,** 393 (1978).
331. H. E. Swaisgood, V. G. Janolino, and P. S. Patrick, *J. Dairy Res.,* **58,** 796 (1975).
332. H. E. Swaisgood, V. G. Janolino, and H. R. Horton, *Am. Inst. Chem. Eng., Symp. Ser.* **74,** 25 (1978).

333. L. J. Berliner, S. T. Miller, R. Uy, and G. P. Royer, *Biochim. Biophys. Acta* **315,** 195 (1973).
334. G. P. Royer and R. Uy, *J. Biol. Chem.* **248,** 2627 (1973).
335. W. F. Shipe, G. Senyk, and H. H. Weetall, *J. Dairy Sci.* **55,** 647 (1972).
336. R. S. Schifreen, D. A. Hanna, L. D. Bowers, and P. W. Carr, *Anal. Chem.* **49,** 1929 (1977).
337. H. H. Weetall and C. C. Detar, *Biotechnol. Bioeng.* **17,** 235 (1975).
338. E. C. Lee, G. F. Senyk, and W. F. Shipe, *J. Food Sci.* **39,** 1124 (1974).
339. D. B. Johnson and M. P. Coughlan, *Biochem. Soc. Trans.* **2,** 1362 (1974).
340. D. B. Johnson and M. P. Coughlan, *Biotechnol. Bioeng.* **20,** 1085 (1978).
341. J. C. Venter, B. R. Venter, J. E. Dixon, and N. O. Kaplan, *Biochem. Med.* **12,** 79 (1975).
342. L. D. Bowers, L. M. Canning Jr., C. N. Sayers, and P. W. Carr, *Clin. Chem.* **22,** 1314 (1976).
343. R. E. Adams and P. W. Carr, *Anal. Chem.* **50,** 944 (1978).
344. L. M. Canning Jr., and P. W. Carr, *Anal. Lett.* **8,** 359 (1975).
345. H. H. Weetall, in *Chemistry of Biosurfaces* (M. L. Hair, Ed.), Marcel Dekker, New York (1972), p. 597.
346. H. H. Weetall, *Biochem. J.* **117,** 257 (1970).
347. H. Jungfer, *Arztl. Lab.* **21,** 80 (1975), in German.
348. A. Castro, and I. Prieto, *Biochem. Biophys. Res. Commun.* **67,** 583 (1975).
349. R. B. Rombauer, C. Frank, M. A. Doran, and R. Piasio, *Clin, Chem.* **20,** 870 (1974).
350. A. A. Luderer, D. M. Hess, and G. Odstrchel, *Mol. Immunol.* **16,** 777 (1979).
351. F. H. Bodley, A. Chapdelaine, G. Flickinger, G. Mikchail, S. Yaverbaum, and K. D. Roberts, *Steroids* **21,** 1 (1973).
352. S. P. O'Neill, T. E. Robb, and P. S. Shenvi, *Fed. Amer. Soc. Exper. Biol.* **36,** 1211 (1977).
353. G. Odstrchel, W. Hertl, F. B. Ward, K. Travis, R. E. Lindner, and R. D. Mason, in *Radioimmunoassays and Related Procedures in Medicine,* Vol. 2, Vienna, 1977, International Atomic Energy Agency (1978), p. 369.
354. H. H. Gschwender and W. Haller, Max Planck Inst. fur Zellbiology, Wilhelmsmaven, Germany, unpublished results (1969).
355. C. R. Lowe and P. D. G. Dean, *Affinity Chromatography,* Wiley, New York (1974).
356. W. B. Jakoby and M. Wilchek, in *Methods in Enzymology,* Vol. 34, (W. B. Jakoby and M. Wilchek, Eds.), Academic Press, New York (1974).
357. R. B. Dunlap, Ed., *Immobilized Biochemicals and Affinity Chromatography,* Plenum Press, New York (1974).
358. J. Turkova, *Affinity Chromatography,* Elsevier, Amsterdam, and New York (1978).
359. W. H. Scouten, *Affinity Chromatography,* Wiley, New York (1981).
360. C. R. Lowe, *An Introduction to Affinity Chromatography.* Elsevier/North-Holland, Amsterdam (1979).
361. G. Baum and S. J. Wrobel, in *Immobilized Enzymes, Antigens, Antibodies and*

Peptides. Preparation and Chromatography, Vol. 1, (H. H. Weetall, Ed.), *Enzymology,* Marcel Dekker, New York (1975), p. 419.
362. L. D. Bowers and P. W. Carr, in *Advances in Biochemical Engineering* (A. Fiechter, Ed.), Springer-Verlag, Berlin (1980), p. 89.
363. P. W. Carr, W. D. Bostick, L. M. Canning Jr., and R. H. Callicott, *Amer. Lab.* 45 (January 1976).
364. B. Danielsson and K. Mosbach, in *Methods in Enzymology,* Vol. 44 (K. Moosbach, Ed.), Academic Press, New York (1976), p. 667.
365. R. B. Merrifield, *J. Amer. Chem. Soc.,* **85,** 2149 (1963).
366. W. Parr, and M. Novotny, in *Bonded Stationary Phases in Chromatography* (E. Grushka, Ed.), Ann Arbor Press (1974), p. 173.
367. S. A. Holick, Dissertation, Univ. Wisconsin, Madison, 1974 (Faculty Supervisor: L. Anderson).
368. S. C. Stinson, *Chem. Eng. News* **17,** Feb. 2 (1981).
369. R. A. Laursen, *Eur. J. Biochem.* **20,** 89 (1971).
370. W. Machleidt, E. Wachter, M. Scheulen, and J. Otto, *FEBS Lett.* **37,** 217 (1973).
371. E. Wachter, H. Hofner, and W. Machleidt, in *Solid Phase Methods in Protein Sequence Analysis* (R. A. Laursen, Ed.), Pierce Chemical Co., Rockford, Ill. (1975), p. 31.
372. W. Machleidt, H. Hofner, and E. Wachter, in *Solid Phase Methods in Protein Sequence Analysis* (R. A. Laursen, Ed.), Pierce Chemical Co., Rockford, Ill. (1975), p. 17.
373. M. E. Parham, and G. M. Loudon, *Biochem. Biophys. Res. Commun.* **80,** 1 (1978).
374. M. E. Parham, and G. M. Loudon, *Biochem. Biophys. Res. Comm.* **80,** 7 (1978).
375. R. A. Laursen, in *Solid Phase Methods in Protein Sequence Analysis,* (R. A. Laursen, Ed.), Pierce Chemical Co., Rockford, Ill. (1975), p. 3.
376. D. A. Powers, in *Solid Phase Methods in Protein Sequence Analysis* (R. A. Laursen, Ed.), Pierce Chemical Co., Rockford, Ill. (1975), p. 99.

CHAPTER

12

ACTIVATION OF POLYSACCHARIDE RESINS BY CNBr

J. KOHN and M. WILCHEK*

Department of Biophysics
Weizmann Institute of Science
Rehovot, Israel

The preparation of an affinity chromatography column or the immobilization of enzymes is a two-step process, as illustrated in Figure 12.1. First, reactive groups have to be introduced into the chemically inert polymeric resin such as Sephadex or Sepharose. This step is generally called "activation." Only then is the desired ligand covalently attached to the resin via the active groups by the "coupling" reaction.

In order to gain a full understanding of the properties of the resulting resin–ligand complex, it is necessary to know the nature of the active group or groups formed upon activation, the nature of the covalent linkages formed during coupling, and the nature of all possible side products. In addition, the kinetic rate constants of formation and decay for the various products of activation and coupling should be known.

Many different methods for the activation of polysaccharide resins have been developed; however, the procedure employing CNBr (1) is the one most widely used. Despite its extensive employment, several important details of its mechanism have been elucidated only recently. The main obstacle to detailed mechanistic studies was the unavailability of suitable analytical methods. Since only trace amounts of active groups are incorporated into the resin (usually about 350 μmol/1 g dry Sepharose 4B, corresponding to about 1–2% by weight), very sensitive methods are required. Unfortunately, the most powerful analytical tools like nmr, UV, or mass spectroscopy cannot be used due to the insolubility of the polymeric resin. Since solid-state ^{13}C NMR was not available at the time, all early investigations (1, 2, 3) had to rely on solid-state IR spectroscopy as the sole means for the direct analysis of

---

* Fogarty Scholar-in-Residence, Fogarty International Center, National Institutes of Health, Bethesda, Maryland.

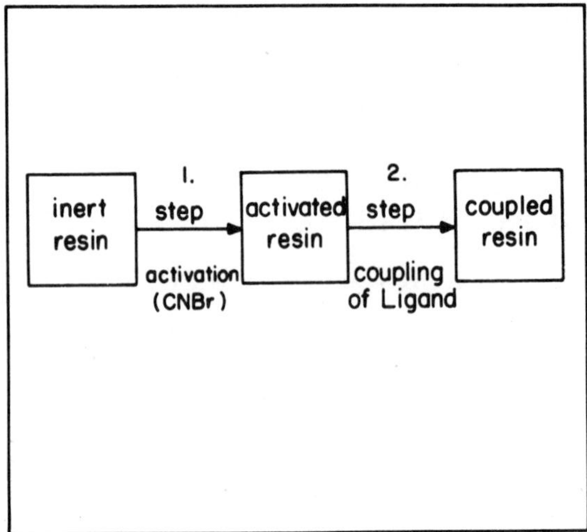

Fig. 12.1. Schematic representation of the process of enzyme immobilization or the preparation of affinity chromatography resins.

activated resins. However, IR spectroscopy cannot be used for quantitative investigations, and due to the partial overlap of several important bands, the strong background absorption of Sepharose itself, and the necessity to dry the activated resin prior to analysis, even qualitative investigations were severely restricted. Consequently, most of our knowledge about the mechanism of CNBr activation and the subsequent coupling reaction is derived from the study of soluble monomeric model compounds like trans-1,2-cyclohexandiol (3) or methyl 4,6-*O*-benzylidine α-D-glucopyranoside (4), which all attempt to imitate the particular structure of Sephadex (Figure 12.2). Based on the assumption that all polysaccharide resins behave in a similar fashion, all the early mechanistic studies were performed on Sephadex, and no substantial investigations were undertaken using model compounds for Sepharose.

The use of soluble model compounds has the advantage that the analytical tools mentioned above can be employed. Unfortunately, one can never be sure whether a soluble model compound indeed behaves like a molecule trapped within the three-dimensional framework of a polymeric resin. A good illustration for the effect of the resin is the change in the activity profile of immobilized enzymes as compared with the free enzyme (5, 6). These changes have been regarded as evidence that the microenvironment within the resin differs from the environment of the bulk solution. In addition to such "microenvironmental" effects, steric factors within the resin may deci-

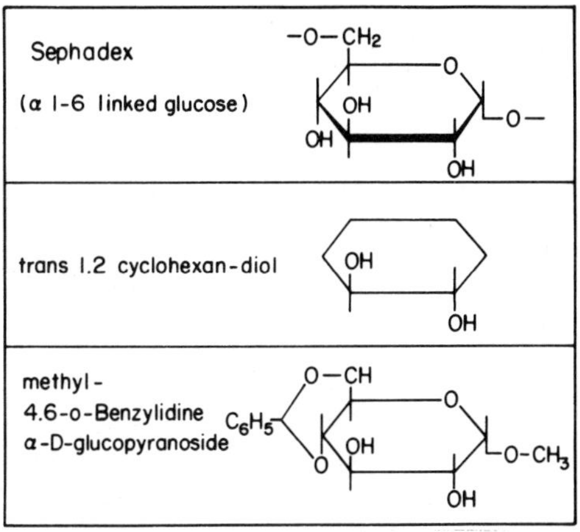

Fig. 12.2. Molecular structures of Sephadex and two of the model compounds employed for mechanistic studies (3, 4).

sively alter reaction pathways, and hence model compounds have only limited value as tools for detailed mechanistic studies. For such studies, suitable analytical techniques must be available, making it feasible to quantitatively determine the products of activation and coupling. Although this prerequisite has not yet been fully satisfied, the approach of direct chemical analysis of the activated resin by selective and sensitive quantitative reactions has lately been successful in elucidating several details of the mechanism of activation (7).

12.1. ACTIVATION BY CNBr

12.1.1. The Mechanism of Activation

The reaction of CNBr with the hydroxyl groups of the polysaccharide resins will result in basic medium in the incorporation of cyanate ester groups into the resin:

$$\boxed{\text{resin}}\!-\!OH + CNBr \longrightarrow \boxed{\text{resin}}\!-\!O{-}C{\equiv}N + HBr \qquad (1)$$

The cyanate ester groups are generally believed to be unstable, short-lived intermediates; however, recent results (7, 8, 9, 10) indicate that the importance of cyanate esters for coupling of the ligand has probably been underestimated. Cyanate esters are very sensitive to hydrolysis by base (11). Most of them can be expected to be hydrolyzed to the inert carbamate moiety:

$$\boxed{\text{resin}}\!-\!O\!-\!C\!\equiv\!N + H_2O \xrightarrow{OH^-} \boxed{\text{resin}}\!-\!O\!-\!\overset{\overset{\displaystyle O}{\|}}{C}\!-\!NH_2 \qquad (2)$$

cyanate ester carbamate

Equation (2) represents the most important side reaction during the activation process. Since carbamates do not contribute to the coupling capacity of the resin, the formation of carbamates not only reduces the overall reaction yield, it also blocks reactive hydroxyl groups and usually limits the coupling capacity of the activated resins to less than 500 μmol/g dry resin. The extent of the undesired hydrolysis of cyanate esters to carbamates can be illustrated by the fact that carbamates usually represent 60–80% of the total amount of nitrogen incorporated into the resin during activation.

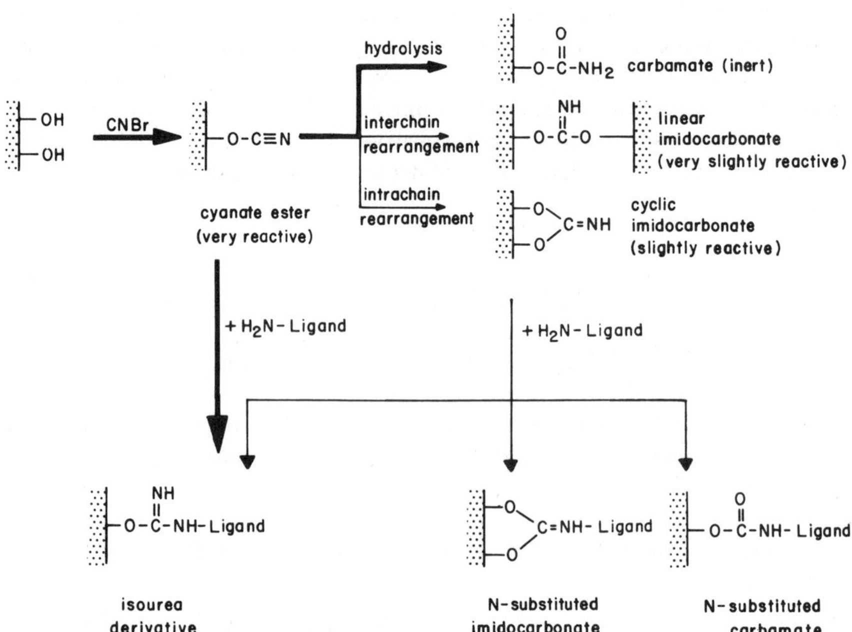

Fig. 12.3. Mechanism of activation and coupling for agarose resins as modified in accordance with recent experimental data (7). Previously cyanate esters were regarded as short-lived intermediates and coupling was assumed to proceed via imidocarbonates only. Heavy lines indicate main reaction pathways.

Cyanate esters are electrophilic enough to react with alcohols (11, 12), forming imidocarbonates (esters of imidocarbonic acid), as shown by:

$$\underset{\text{cyanate ester}}{R{-}O{-}C{\equiv}N} + HO{-}R' \longrightarrow \underset{\text{imidocarbonate}}{R{-}O{-}\overset{\overset{\displaystyle NH}{\|}}{C}{-}O{-}R'} \qquad (3)$$

On the resin, the reaction between cyanate esters and neighboring hydroxyl groups proceeds in an analogeous way. If the reacting hydroxyl group is situated on the same sugar residue, an intramolecular, cyclic imidocarbonate will result. If the reacting hydroxyl group is situated on a different sugar residue, an intermolecular, linear imidocarbonate will be formed, as illustrated by Figure 12.3. When the mechanism of activation was investigated, employing soluble model compounds, the freely movable model compounds produced nearly exclusively imidocarbonates (3, 4). However, on the resin *steric factors* come into play; therefore the extent of imidocarbonate formation depends not only on the experimental conditions but mainly on the molecular structure of the polymeric resin. Most of the differences in the behavior of Sepharose and Sephadex during activation can be explained by the differences in rate and extent of imidocarbonate formation, as discussed in detail below.

Similar to the hydrolysis of cyanate esters, imidocarbonates are also hydrolyzed by base to carbamates:

$$\underset{\text{imidocarbonate}}{\boxed{\text{resin}}\begin{matrix}{-}O\\{-}O\end{matrix}\!\!>\!C{=}NH} + H_2O \xrightarrow{OH^-} \underset{\text{carbamate}}{\boxed{\text{resin}}\begin{matrix}{-}O{-}\overset{\overset{\displaystyle O}{\|}}{C}{-}NH_2\\{-}OH\end{matrix}} \qquad (4)$$

However, imidocarbonates are much more stable toward hydrolysis by base, and equation (4) represents only a minor reaction pathway.

With the exception of one publication (1), it is usually reported that activated resins contain trace amounts of bromine (2, 13). The incorporation of Br can be explained by the reaction of triazines (cyanuric acid derivatives), which are formed by polymerization of CNBr according to:

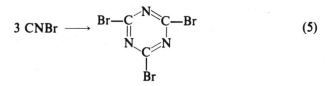

$$3\ CNBr \longrightarrow \underset{\text{tribromotriazine}}{(BrC{=}N)_3\ \text{ring}} \qquad (5)$$

tribromotriazine

The trimerization of CNBr is base catalyzed, and the formation of small amounts of triazines during the activation reaction cannot be avoided, resulting in the incorporation of Br into the resin:

$$\boxed{\text{resin}}\!-\!\text{OH} + \text{C}_3\text{N}_3\text{Br}_3 \text{ (2,4,6-tribromo-1,3,5-triazine)} \longrightarrow \boxed{\text{resin}}\!-\!\text{O}\!-\!\text{C}_3\text{N}_3\text{Br}_2 + \text{HBr} \qquad (6)$$

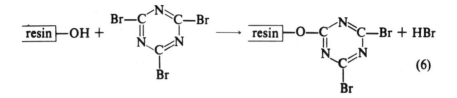

This reaction has actually been used for the activation of polysaccharide resins employing triazines instead of CNBr (14, 15, 16).

Under normal conditions of CNBr activation, however, the incorporation of Br into the resin is a very minor side reaction, which can be neglected for all practical purposes.

Carbamates, imidocarbonates, and cyanate esters are the three major products of activation and are found in various proportions on all freshly activated resins. Because of the lack of suitable analytical techniques, the early mechanistic investigations did not reveal the existence of cyanate esters on the activated resin. Therefore, imidocarbonates were assumed to be the active moiety, responsible for coupling of ligand. In the light of several recent publications (7, 8, 9, 10), it seems that, at least in the case of Sepharose, coupling of ligand proceeds predominantly via cyanate esters (17), contrary to the presently accepted mechanism of activation and coupling (18). The modified mechanism of CNBr activation, based on recent experimental results, is shown in Figure 12.3. The fact that cyanate esters participate in the coupling reaction has important consequences, as discussed in detail below.

12.1.2. Effect of the Different Molecular Structures of Sephadex and Sepharose on the Mechanism of Activation

Sepharose and Sephadex differ not only in the three-dimensional organization of the resin but also in their molecular structure. The differences between the resin configurations of Sephadex and Sepharose are well known and have been reviewed in detail (17, 18). The following discussion is limited, therefore, to the differences between the molecular structures and their effect on the mechanism of activation.

Sephadex (epichlorohydrin-crosslinked dextran) is a resin based on α-1,6-linked polyglucose. The glucopyranose ring is its basic repeating unit, containing three potentially reactive hydroxyl groups at positions 2, 3, 4, as shown by Figure 12.4. Sepharose (agarose) is an alternating copolymer of 3-

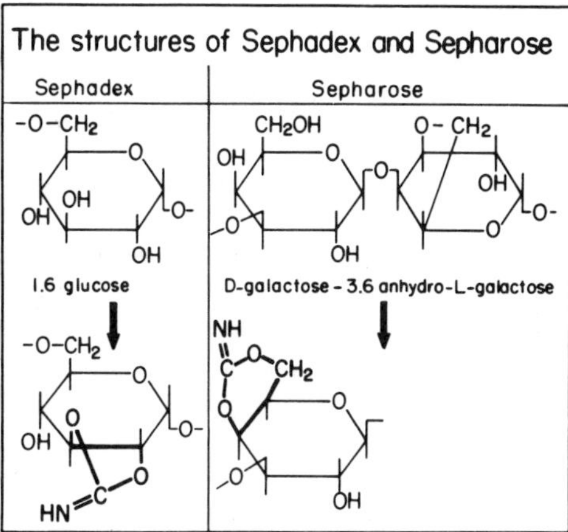

Fig. 12.4. Molecular structures of Sephadex and Sepharose. The glucopyranose ring of Sephadex facilitates the formation of 5-membered cyclic imidocarbonates, whereas the structure of Sepharose allows the formation of 6-membered rings only.

linked β-D-galactopyranose and 4-linked 3,6-anhydro-α-L-galactopyranose. Its basic repeating unit is a galactose–anhydrogalactose disaccharide, as shown in Figure 12.4. The repeating unit contains four potentially reactive hydroxyl groups but no vicinal diol structure, since the hydroxyl groups are positioned at a distance of at least two carbon atoms from each other.

After incorporation of a cyanate ester group at one of the three potentially reactive hydroxyl groups of Sephadex, the molecular structure of the glucopyranose ring facilitates the rapid formation of trans-5-membered cyclic imidocarbonates (Figure 12.4). Sephadex contains about 18 mmol of OH/g dry resin. This relatively high density of reactive OH groups on the resin increases the chances for the formation of intermolecular linear imidocarbonates. Consequently, due to the easy formation of both cyclic and linear imidocarbonates on Sephadex, the amount of cyanate esters on the resin is rapidly reduced to trace quantities.

On Sepharose, no cyclic intramolecular imidocarbonates are possible for cyanate esters incorporated into the resin at positions 2 of the galactopyranose or anhydrogalactopyranose rings, since these positions have no neighboring hydroxyl groups. The only possibility for the formation of a cyclic imidocarbonate is between positions 4 and 6 of the galactopyranose ring, which results in a 6-membered ring as shown in Figure 12.4. Such 6-membered rings, however, are sterically strained and energetically less favorable than the 5-membered rings formed on Sephadex. Although 6-membered cyclic imidocarbonate rings may exist on Sepharose, the rate and extent of their formation is significantly reduced. In addition, Sepharose, containing only

13 mmol OH/g dry resin, has a lower density of potentially reactive hydroxyl groups. This diminishes the chances for the formation of intermolecular linear imidocarbonates. Consequently, the formation of both cyclic and linear imidocarbonates is slower and less pronounced on Sepharose, and one can expect cyanate esters to show a higher stability on Sepharose than on Sephadex.

This expectation is confirmed by experimental data (7, 9, 19). Table 12.1 shows the average amounts of cyanate esters, imidocarbonates, and carbamates as determined by the analysis of dried samples of activated resins; the decreased formation of imidocarbonates on Sepharose is in fact concomitant with an increased relative abundance of both cyanate esters and carbamates.

12.1.3. Chemical Considerations

12.1.3a. The Effect of pH

Since CNBr reacts only with ethoxide ions ($R\text{-}O^-$) but not with un-ionized hydroxyl groups (R-OH), no reaction takes place if CNBr is mixed with alcohols, sugars, or polysaccharide resins in acidic or neutral solution. The presence of a base sufficiently strong to cause the ionization of at least some of the resin hydroxyl groups is required. Since the pK_a of the resin hydroxyl groups is about 12.5 (20), only about 0.1% of all available hydroxyl groups are ionized at pH 10. The lack of ethoxide ions on the resin will make the activation reaction proceed rather sluggishly and with low yield. With increasing pH, the amount of ionized hydroxyl groups increases, causing an in-

Table 12.1. Average Composition of Sephadex and Sepharose After CNBr Activation[a] (19)

Activation product	Sepharose: Amount of Activation Product, µmol/g dry resin	Sepharose: Amount of Activation Product, % of total nitrogen content	Sephadex: Amount of Activation Product, µmol/g dry resin	Sephadex: Amount of Activation Product, % of total nitrogen content
Imidocarbonates	400	20	1400	46
Cyanate esters	250	12	50[b]	1–2
Carbamates	1400	68	1600	52

[a] Activation procedure: March et al. (22).

[b] Strongly fluctuating values from 10 to 100 µmol/g dry resin were obtained, depending on slight variations of experimental conditions.

creased incorporation of cyanate esters into the resin. However, CNBr is rapidly hydrolyzed by base to the unreactive cyanate ion:

$$\mathrm{CNBr} + 2\mathrm{OH}^- \longrightarrow \mathrm{Br}^- + (\mathrm{OCN})^- + \mathrm{H_2O} \qquad (7)$$

The initially formed cyanate esters are also very easily hydrolyzed by base. Performing the activation reaction at or above pH 13 results in the rapid hydrolysis of free CNBr and the nearly complete decay of the initially formed cyanate esters, leading to a considerable waste of CNBr and a very poorly activated resin. Clearly, one must strike the balance between the necessity to work in a strongly basic medium in order to form enough ethoxide ions on the resin and the necessity to keep the excess of base to a minimum in order to prevent the rapid hydrolysis of both CNBr and cyanate esters. These considerations explain the experimental observation that activation proceeds best if the pH of the reaction medium is kept at about 11.5–12. Although both Sepharose and Sephadex are stable at pH 14 for at least 1 hr, activation at extremely high pH values causes irreversible changes in the resin structure. This leads to changes in the physical and mechanical properties of the resin and reduces the reswelling capacity (5, 13). Therefore some investigators recommend not raising the pH above 11.0 during activation.

12.1.3b. The Effect of Temperature

Since the rate constants for the reactions of activation and coupling are not yet determined, no exact analysis of the effect of temperature is feasible. There are, however, certain preliminary observations, which make the following speculations possible: both resin ethoxide ions and free hydroxyl ions compete for the reaction with CNBr:

$$\mathrm{CNBr} \begin{cases} \xrightarrow{+\,\mathrm{OH}^-} (\mathrm{OCN})^- & \text{hydrolysis of CNBr} \\ \xrightarrow{+\,\boxed{\text{resin}}\!-\!\mathrm{O}^-} \boxed{\text{resin}}\!-\!\mathrm{O}\!-\!\mathrm{C}\!\equiv\!\mathrm{N} & \text{activation} \end{cases} \qquad (8)$$

It seems that both reaction pathways of equation (8) are influenced to about the same extend by temperature. Therefore the ratio of "hydrolyzed CNBr" to "incorporated cyanate ester" cannot be improved significantly by decreasing the temperature. The rapid hydrolysis of cyanate esters to carbamates, however, seems to be critically dependent on temperature. The ratio of "incorporated cyanate esters" to "hydrolyzed cyanate esters" can therefore be improved by decreasing the temperature (21). The effect of temperature on the activation yield is more pronounced for Sepharose, since the hydrolysis

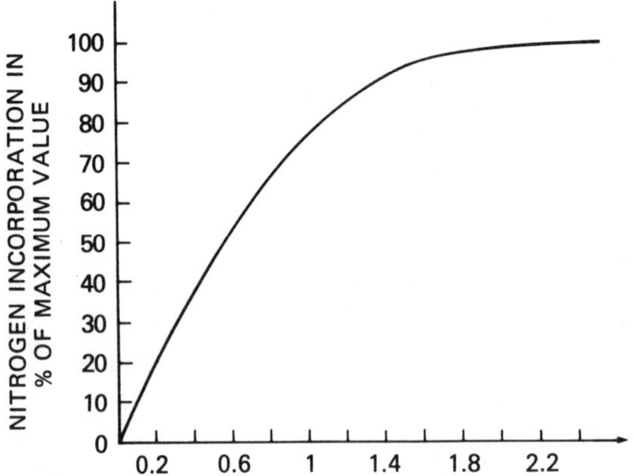

Fig. 12.5. Incorporation of nitrogen into the resin during activation as a function of the employed amount of CNBr. Calculated from experimental data of several sources (19, 21, 23, 24).

of cyanate esters is more important on Sepharose than on Sephadex. This explains why Sephadex can be activated at ambient temperatures without significant loss of coupling capacity, whereas the coupling capacity of Sepharose can be improved by about 30–40% by cooling the reaction medium during activation (22).

12.1.3c. The Effect of Concentration of CNBr

Much of the CNBr employed for activation is actually hydrolyzed by free base to the unreactive cyanate ion [equations (7) and (8)]. Therefore, a large excess of CNBr is required, and the amount of nitrogen incorporated into the resin is strongly dependent on the initial concentration of CNBr used. In several studies (19, 21, 23, 24) the "coupling capacity" of the activated resin was correlated with the amount of CNBr added. Although differing experimental conditions were employed in these studies, quite similar results were obtained. Figure 12.5 is based on those results. It illustrates in a semiquantitative way the effect of the amount of CNBr on the incorporation of nitrogen into Sepharose. At low amounts of CNBr, nitrogen incorporation is a nearly linear function of the amount of CNBr employed. Therefore control of the amount of CNBr is probably the easiest and most effective way to control the degree of activation obtained on the resin. Above a certain saturation point, further increase of the amount of CNBr does not result in an increase in nitrogen incorporation. For Sepharose, saturation occurs at about 2.5 g CNBr/g dry resin. This corresponds roughly to a twofold molar excess of CNBr over the amount of available hydroxyl groups on the resin. In terms of wet resin (all interstitial water removed), the saturation point for Sepharose 2B (2% agarose) is reached at about 0.5 g CNBr/10 g wet resin, for Sepharose 4B (4% agarose) at about 1 g CNBr/10 g wet resin, and for Sepharose 6B (6% agarose) at about 1.5 g CNBr/10 g wet resin (21).

Unfortunately, no comparable investigations of this kind were performed on Sephadex. It seems reasonable to assume, however, that in the case of Sephadex a twofold molar excess of CNBr over the amount of available OH on the resin will lead also to saturation. Thus, 3.5 g CNBr/g dry Sephadex would be required for the highest possible activation.

12.1.3d. Overall Reaction Yield

From the point of view of an organic chemist, activation and coupling are highly unattractive reactions since the overall reaction yields are very low. The following calculations may illustrate this point: Even under optimal conditions, usually only about 300 μmol/g dry resin of ligand can be coupled to Sepharose 4B. Starting with 25 g wet Sepharose 4B (1 g dry resin) and a saturation amount of 2.5 g CNBr (23,500 μmol), 300 μmol of coupled ligand represent in terms of CNBr an overall reaction yield of only about 1–2% (See Table 12.2).

One gram of dry Sepharose resin contains 3.27 mmol of galactose-anhydrogalactose repeating unit, corresponding to 13.1 mmol of available OH groups of which 3.27 mmol are primary ones. Usually 3000 μmol/g resin of nitrogen incorporation are obtained, corresponding to a reaction yield of only about 23% in terms of the total amount of available OH groups. On a statistical basis, only 1 out of 4 OH groups seems to be reactive. No satisfactory explanation is available for this phenomenon. Since it is known that primary alcohols are both more acidic and more reactive than secondary ones, one is tempted to speculate that under the usual conditions of activation only the primary OH groups are reactive. In terms of primary OH groups, the usual levels of nitrogen incorporation would correspond to a reaction yield of over 90%, and this could explain the saturation phenomenon for the amount of CNBr (See Section 12.1.3c) and the difficulty in obtaining Sepharose resins with a degree of nitrogen incorporation higher than 3.27 mmol/g dry resin (4.5% N). Presently no experimental data are available to support this speculation.

12.2. THE ROLE OF CYANATE ESTERS AND IMIDOCARBONATES DURING COUPLING

12.2.1. Brief Survey of the Properties of Cyanate Esters

For over 100 years chemists had puzzled over the existence of true cyanate esters (25, 26) before they were finally synthesized (27, 28) and characterized

Table 12.2. Overall Reaction Yield in Terms of CNBr for Activation of Sepharose 4B (average values)

Initial Amount of Material[a]	Reaction Step	Remarks[b]	Amount of Material After Reaction Step	Overall Reaction Yield (in %)
23,500 μmol CNBr	Incorporation of cyanate ester into resin	Incorporation ≃3,000 μmol Remainder lost by hydrolysis	3,000 μmol	12.8
3,000 μmol *N* on resin	Hydrolysis of cyanate esters during activation	Carbamates: ≃60% Cyanate esters + imidocarbonates: ≃40%	1,200 μmol	5.1
1,200 μmol "active" groups	Coupling of ligand	Usually only about 300 μmol of ligand can be coupled	300 μmol	1.3

[a] Calculations are based on the activation of 1 g dry Sepharose 4B (=25 gr wet gel) using 2.5 g of CNBr (=23,500 μmole).
[b] The assumed reaction yields represent "average values" as calculated from 40 activations performed by the authors.

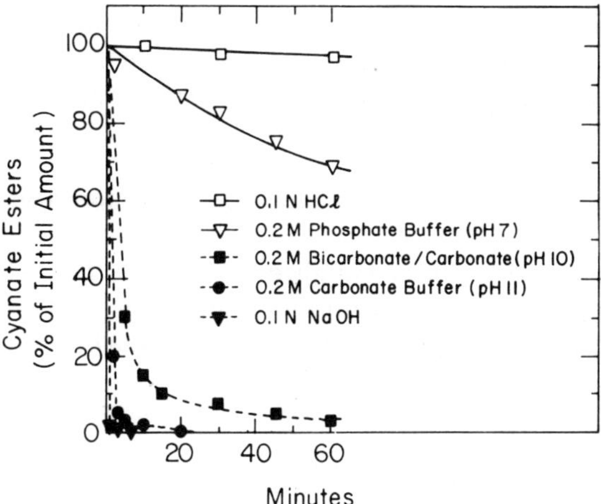

Fig. 12.6. Hydrolysis of cyanate esters on activated Sepharose 4B in solutions of pH 1, 7, 10, 11, 14 at ambient temperatures (7).

(11, 12, 29). Cyanate esters were found to show an exceptionally high reactivity towards nucleophilic attack. Therefore, they are stable only in the absence of nucleophiles and reportedly react with such weak nucleophiles like carboxylic acids (acetic acid, citric acid, etc.), phosphoric acid, and various anions (I^-, SCN^-, NO_2^-, CN^-, etc.) (11, 12, 29). This fact has to be borne in mind, since it limits the choice of possible buffer solutions for activation and coupling.

Using a spectrophotometric procedure for the determination of cyanate esters directly on the resin (7), the stability of cyanate esters on activated Sepharose has been investigated. Cyanate esters were found to be surprisingly stable in 0.1 *N* HCl. When activated Sepharose was allowed to react with 0.1 *N* HCl at ambient temperatures, only about 30% of the cyanate esters decayed in the course of 12 hr. It seems that only concentrated HCl is capable of rapidly hydrolyzing cyanate esters. With increasing pH, the stability of cyanate esters decreased considerably (see Figure 12.6). At pH 10, hydrolysis to carbamates was complete within 1 hr, and in 0.1 *N* NaOH (pH 14) the decay of cyanate esters was practically instantaneous, the approximate half-life time being about 30–60 sec.

Cyanate esters react rapidly with amines (11, 12), forming isourea derivatives:

$$\underset{\text{cyanate ester}}{\text{R—O—C}\equiv\text{N}} + \text{H}_2\text{N—R}' \longrightarrow \underset{\text{isourea derivative}}{\text{R—O—}\overset{\overset{\text{NH}}{\|}}{\text{C}}\text{—NH—R}'} \qquad (9)$$

This reaction is of particular interest. Since coupling is known to proceed via amino groups of the ligand, the isourea derivative can be expected to be the major product of coupling via cyanate esters. No ammonia is released in this reaction.

12.2.2. Brief Survey of the Properties of Imidocarbonates

Imidocarbonates were extensively studied only because they were the products obtained by unsuccessful attempts to synthesize cyanate esters (26, 30, 31). The linear imidocarbonates, as represented by diethylimidocarbonate

$$C_2H_5{-}O{-}\overset{\overset{\displaystyle NH}{\|}}{C}{-}O{-}C_2H_5$$

are relatively stable, basic compounds with a pK_a of about 10. They can be determined quantitatively by titration with acid and are protonated in a neutral aqueous solution. The cyclic imidocarbonates, as represented by 2-imino-1,3-dioxolane

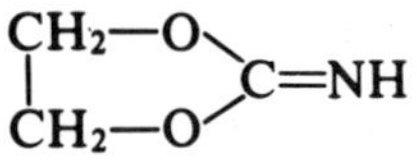

$$\begin{matrix} CH_2{-}O \\ | \\ CH_2{-}O \end{matrix}\! \qquad\qquad\quad \rangle C{=}NH$$

are much more difficult to synthesize. Free 2-imino-1,3-dioxolane was obtained for the first time (32) only in 1964. It seems to be a rather unstable compound, whose properties are not well known. In general, one may expect cyclic imidocarbonates to be less stable and to react more vigorously than linear imidocarbonates.

According to Sandmeyer's synthesis of diethylimidocarbonate (30), the crude product was obtained by reaction in a concentrated solution of alkali. After separation, the wet diethylimidocarbonate was dried by adding solid NaOH. Despite these drastic conditions some diethylimidocarbonate obviously survived hydrolysis to carbamate [see equation (4)]. This shows that linear imidocarbonates are rather stable towards base.

On the other hand, both linear and cyclic imidocarbonates are highly sensitive to hydrolysis by acid (31). Even at pH 4, rapid hydrolysis with the liberation of ammonia and the formation of the corresponding carbonate takes place:

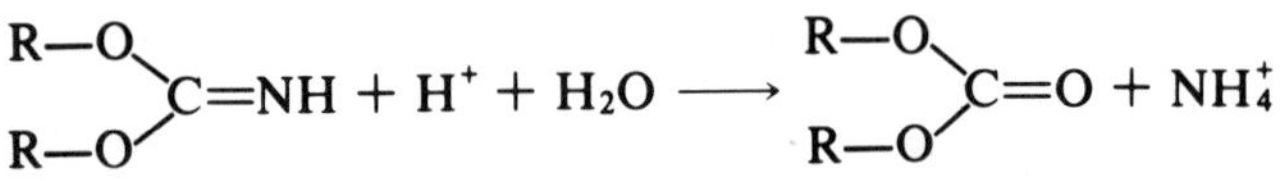

$$\underset{\text{imidocarbonate}}{\begin{matrix} R{-}O \\ R{-}O \end{matrix}\!\rangle C{=}NH} + H^+ + H_2O \longrightarrow \underset{\text{carbonate}}{\begin{matrix} R{-}O \\ R{-}O \end{matrix}\!\rangle C{=}O} + NH_4^+ \qquad (10)$$

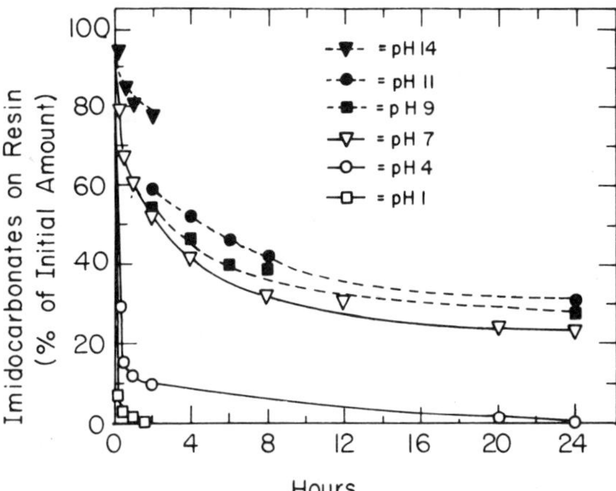

Fig. 12.7. Hydrolysis of imidocarbonates on activated Sepharose 4B in solutions of pH 1, 3, 7, 9, 11, and 14 at ambient temperatures (7). At pH 14, resin solubilization limited the experiment to 2 hr.

Unfortunately, the reaction of imidocarbonates with amines does not seem to lead to a single product. Therefore the mechanism of coupling via the imidocarbonate moiety is still not fully elucidated. This point is discussed in detail below.

Using a procedure based on selective acid hydrolysis, the stability of the imidocarbonate moiety on activated Sepharose and Sephadex was investigated (7). In correspondence with the generally expected behavior it was found that imidocarbonates are extremely sensitive to acid hydrolysis, increasing in stability with increasing pH (Figure 12.7). Surprisingly, imidocarbonates on activated Sepharose are most stable at pH 14. Since the coupling capacity of activated Sepharose decreases rapidly during hydrolysis by base, this result indicates that imidocarbonates cannot be the active species on Sepharose.

12.2.3. Coupling Yields of Imidocarbonates and of Cyanate Esters

A comparison of Figures 12.6 and 12.7 shows that imidocarbonates and cyanate esters behave in a directly opposite fashion during hydrolysis in acid or base. This fortunate circumstance makes it possible to obtain activated resins containing either imidocarbonates or cyanate esters as the sole active species: If a freshly activated resin is hydrolyzed for about 30 min in 0.1 *N* HCl, all imidocarbonates will be removed from the resin and more than 95% of the cyanate esters will remain intact. On the other hand, a few minutes of hydrolysis of an activated resin in 0.1 *N* NaOH will destroy all cyanate esters while a substantial amount of the imidocarbonates remain unchanged.

The use of acid or base hydrolyzed resins containing either imidocarbo-

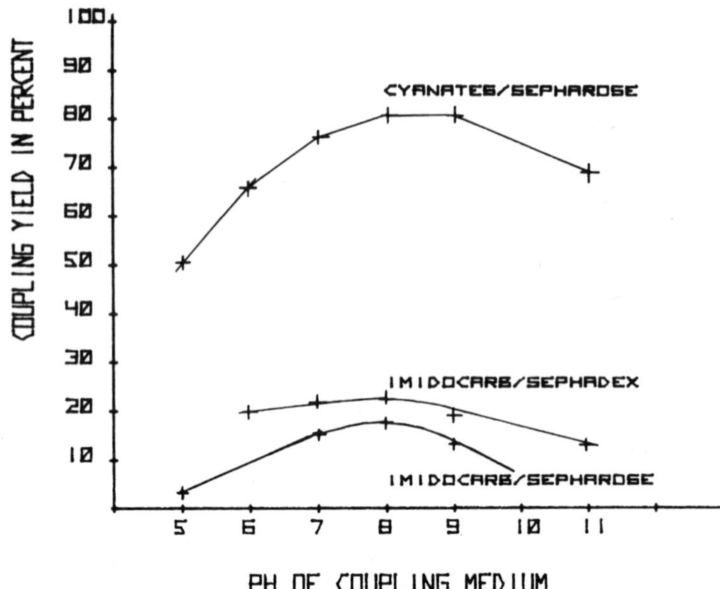

Fig. 12.8. Coupling of methionine to acid-hydrolyzed activated Sepharose 4B (containing only cyanate esters), to base-hydrolyzed activated Sepharose 4B (containing only imidocarbonates), and to activated Sephadex G-50 (fine) as a function of pH (19).

nates or cyanate esters rendered possible several new approaches to the study of coupling. The role played by cyanate esters was investigated, employing acid-hydrolyzed samples of activated Sepharose, to which amino acids (methionine) or small peptides were coupled. It was found that at pH 8.5 the amount of coupled model ligand corresponded to more than 80% of the amount of cyanate esters initially present. By varying the pH of the reaction medium, coupling via cyanate esters was shown to be rather independent of the pH within the range of 7 to 10, as illustrated by Figure 12.8.

When the same kind of experiment was repeated employing base-hydrolyzed samples of activated Sepharose that contained a precisely determined amount of imidocarbonates as sole active species, it was found that at pH 8.5 the amount of coupled model ligand corresponded only to about 15% of the amount of imidocarbonates initially present. Changing the pH of the coupling medium revealed a flat optimum around pH 8, as shown in Figure 12.8.

The average composition of resins activated according to the procedure of March et al. (22), is shown in Table 12.1. Since about 80% of the cyanate esters, but only about 15% of the imidocarbonates, will react with ligand, coupling of ligand to *Sepharose* will be due predominantly to cyanate esters even though imidocarbonates are more abundant on the resin than cyanate esters.

On *Sephadex* the lower relative reactivity of imidocarbonates is easily compensated by their quantitative preponderance. Therefore, coupling can be assumed to proceed mainly via imidocarbonates. Since the exact amount of cyanate esters present on freshly activated Sephadex depends strongly on the experimental conditions during activation, the percentage of ligand coupled via cyanate esters may vary considerably. This fact should be borne in mind when interpreting the results of early investigations into the coupling of ligand to Sephadex.

Based on the assumption that coupling of ligand occurs simultaneously and independently via cyanate esters and via imidocarbonates, one should be able to predict the coupling capacity of activated resins by determining the amounts of cyanate esters and imidocarbonates present on the resin.

$$\boxed{\begin{array}{c}\text{Maximum}\\ \text{coupling capacity}\\ (\mu\text{mol ligand/g dry resin})\end{array}} = 0.80 \cdot \boxed{\begin{array}{c}\text{Cyanate}\\ \text{esters}\\ (\mu\text{mol/g}\\ \text{dry resin})\end{array}} + 0.15 \cdot \boxed{\begin{array}{c}\text{Imidocarbonates}\\ (\mu\text{mol/g}\\ \text{dry resin})\end{array}} \qquad (11)$$

Using equation (11), the coupling capacity of several samples of activated Sepharose was predicted with satisfactory accuracy, as shown by the results summarized in Table 12.3 (7).

The data collected in Table 12.3 show that for Sepharose activated according to the procedure of March et al. (22), about 70% of the coupled ligand is linked to the resin by reaction with cyanate esters. These data can be regarded as direct experimental evidence for the preponderance of cyanate esters as active species on Sepharose. Furthermore, it is evident that contrary to the presently accepted view, imidocarbonates react only sluggishly with ligand.

Only the *upper limit* of the resin's coupling capacity can be predicted by determining the amount of imidocarbonates and cyanate esters present on the resin. A stoichiometrical relation between the amount of active groups and the amount of coupled ligand was until now demonstrated only for small, monovalent model ligands. The exact amount of coupling of macromolecules, which contain many possible points of attachment, cannot be predicted easily because multisite attachments, steric factors, and restrictions of diffusion have to be taken into account. The prediction of a resin's coupling capacity with macromolecules, which is important from a practical point of view, requires further investigation.

12.2.4. Mechanism of Coupling via Cyanate Esters and via Imidocarbonates

In accordance with model reactions (11, 12), the reaction of cyanate esters with amino groups of ligand (*N*-terminal and ϵ-amino group of lysine) can be

Table 12.3. Prediction of Coupling Capacity of Activated Sepharose (7)

	Analytical Data (μmol/g resin)		Coupling Capacity (μmol/g resin)	
Sample	Cyanate Esters	Imidocarbonates	Predicted[a]	Experimentally Determined
Sepharose 4B[b]	225	700	285	263
Sepharose 4B[b] (after hydrolysis in 0.1 *N* HCl)	220	0	176	180
Sepharose 4B[b] (after hydrolysis in 0.1 *N* NaOH)	0	650	98	102
Sepharose 4B[c]	176	550	223	237
Sepharose 4B[d]	303	280	284	262

[a] Equation (11) was used to predict coupling capacity.
[b] Same batch of activated resin.
[c] Different batch of activated resin.
[d] Obtained from Pharmacia (Lot FF 16294).

expected to lead rapidly to the quantitative formation of isourea derivatives [see equation (9)]. Cyanate esters, however, are highly electrophilic and are capable of reacting with several additional amino acid residues. The fact that cyanate esters are involved in coupling of ligand makes it necessary to consider the formation of additional linking bonds. Taking the known reactions of cyanate esters into account, coupling via cysteine, histidine, and tyrosine residues is most probably taking place to a significant extent, and even coupling via glutamic and aspartic acid, serine, threonine, and methionine residues may occur to a small extent under favorable circumstances.

Preliminary experimental evidence supports this view. Since the isourea linkage is hydrolyzed by ammonia with concomitant formation of a guanidine derivative, according to equation (12),

$$\underset{\text{isourea derivative}}{\mathrm{R{-}O{-}\overset{\overset{\displaystyle NH}{\|}}{C}{-}NH{-}R'}} + \mathrm{NH_3} \longrightarrow \mathrm{R{-}OH} + \underset{\text{guanidine derivative}}{\mathrm{H_2N{-}\overset{\overset{\displaystyle NH}{\|}}{C}{-}NH{-}R'}} \qquad (12)$$

the amount of guanidine derivatives formed upon aminolysis of coupled ligand is indicative of the amount of ligand coupled via isourea linkages. When coupled insulin was treated with 0.8 *N* NH_4OH, only about 30% of the stoichiometrically expected amount of homoarginine was formed. Since under identical conditions model reaction with coupled *N*-α-acetyl-lysine produced 80% of the stoichiometrically expected amount of homoarginine, a significant fraction of the coupled insulin must have been linked via residues other than lysine (33, 34).

The mechanism of coupling via the imidocarbonate moiety was investigated in several studies, employing either model compounds or Sephadex. Since all freshly activated resins also contain cyanate esters, these studies employed resins containing both kinds of active groups in varying and undefined proportions. This may explain some of the conflicting experimental data published about the coupling reaction.

During an intensive investigation of the properties of imidocarbonates, Houben et al. (35, 36) described the formation of *N*-substituted imidocarbonates as the sole product of the reaction between diethylimidocarbonate and aromatic amines or amino acid esters:

$$\begin{matrix}\text{Et—O} \\ \text{Et—O}\end{matrix}\!\!\!>\!\text{C=NH} + \text{H}_2\text{N—R} \longrightarrow \begin{matrix}\text{Et—O} \\ \text{Et—O}\end{matrix}\!\!\!>\!\text{C=N—R} + \text{NH}_3 \qquad (13)$$

diethylimidocarbonate *N*-substituted imidocarbonate

In this reaction ammonia is released. Once the *N*-substituted imidocarbonate is formed, hydrolysis could conceivably give rise to *N*-substituted carbamates:

$$\begin{matrix}\text{Et—O} \\ \text{Et—O}\end{matrix}\!\!\!>\!\text{C=N—R} + \text{H}_2\text{O} \longrightarrow \text{Et—O—}\overset{\overset{\text{O}}{\|}}{\text{C}}\text{—NH—R} + \text{EtOH} \qquad (14)$$

N-substituted imidocarbonate *N*-substituted carbamate

When alanine methyl ester was coupled to activated Sepharose, the release of ammonia was reported and no increase of the nitrogen content of the resin could be detected after coupling (1). In accordance with equations (13) and (14), the formation of *N*-substituted imidocarbonates and *N*-substituted carbamates was proposed.

Later the product of model reactions between amino acids and diethylimidocarbonate was reported to be a mixture of *N*-substituted imidocarbonates, *N*-substituted carbamates, and substituted isourea derivatives, as shown in Figure 12.3 (5, 13). The possible formation of isourea derivatives during coupling was further supported by the failure to detect any significant liberation of ammonia during the coupling of glycine to activated Sephadex (37). Using methyl-4,6-*O*-benzylidine-α-D-glucopyranoside, the corresponding isourea derivative was identified by IR spectroscopy as a possible product formed upon coupling of glycine to the activated model compound (4). This reaction is similar to the coupling reaction via cyanate esters:

$$\text{resin}\begin{matrix}-O\\-O\end{matrix}\!\!>C{=}NH + H_2N{-}R \longrightarrow \text{resin}\begin{matrix}-O{-}\overset{\overset{\displaystyle NH}{\|}}{C}{-}NHR\\-OH\end{matrix} \qquad (15)$$

In accordance with equation (15), no ammonia is liberated by the formation of the isourea-linking bond.

In a very detailed study by Bronström et al. (38), activated Sephadex was subjected to coupling conditions in the absence of ligand. Under these conditions the activated resin was found to release nearly the same amount of ammonia as in the presence of ligand, showing that coupling itself does not cause the liberation of ammonia, which had been accredited to the coupling reaction in several previous studies. This result was regarded as very strong evidence for the preponderant formation of the isourea linkage. However, Bronström et al. did not offer an explanation for the release of ammonia from the activated resin under coupling conditions. Since carbamates are hydrolyzed only very slowly at pH 8–9, and since imidocarbonates are supposed to release ammonia only in acidic medium, most of the liberated ammonia was unaccountable.

Further evidence for the formation of the isourea linkage is derived from measurements of the changes of the isoelectric point of proteins due to coupling to soluble polysaccharides, using the technique of isoelectric focusing (39). Since both *N*-substituted imidocarbonates ($pK_a = 4.7$) and *N*-substituted carbamates are unprotonated in the range of pH 6–10, a reduction in the net positive charge of the immobilized protein is expected if indeed *N*-substituted imidocarbonates or carbamates are formed during coupling. That no such reduction occurred supports the formation of an isourea linkage during coupling and confirms the result of an earlier investigation (40), which revealed a basic group of pK_a 10.2 on Sephadex after coupling glycylleucine.

It is now widely accepted that both imidocarbonates and cyanate esters yield isourea linkages when reacting with amino groups of ligand. However, in the case of imidocarbonates the extent and relative importance of the formation of *N*-substituted imidocarbonates and carbamates has not been fully clarified, and in the case of cyanate esters further studies into the extent of coupling via alternative amino acid residues are required in order to fully elucidate the mechanism of coupling.

The isourea linkage differs considerably from linkage via *N*-substituted imidocarbonates and carbamates with respect to charge and stability. As mentioned above, *N*-substituted imidocarbonates and carbamates are unprotonated in the range of pH 6 to 10. A resin with such linkages would be free of the positively charged isourea linkage, hence avoiding interference from "ion exchange interactions." In addition, *N*-substituted carbamates should be more stable to hydrolysis by base than the isourea linkage. Therefore, interesting modifications and new applications of the CNBr procedure might result from a better understanding of the mechanism of coupling.

12.3. THE DESIGN OF ACTIVATION PROCEDURES

Workers in the field have repeatedly stressed that the activation procedure should be tailored to suit the intended application. "Trial and error" and personal preferences have often dictated the choice of a specific procedure. With the more precise understanding of the mechanism of activation that has recently emerged, a more rational approach to the problem of optimization of activation should be possible.

Most of the published activation procedures can be classified into six categories, as shown in Table 12.4, using the following criteria:

1. *Mode of Addition of CNBr.* CNBr may be added to the reaction mixture as dilute aqueous solution, as concentrated solution in organic solvent, or as solid material.
2. *Mode of Controlling pH.* The pH can be kept constant by the continuous addition of base (titration technique) or by the use of a strongly buffered reaction mixture (buffer technique).

Table 12.4. Survey of CNBr Activation Procedures[a]

Mode of pH Control	Mode of Addition of CNBr: Solid	Dilute Aqueous Solution	Concentrated Solution in Organic Solvent
Titration technique	Kagedal (37)[b] Cuatrecasas (45)[d] Wilchek (41)	Axen (1)[c] Porath (42) Axen (5) Stage (24)	Nishikawa (21)[d]
Buffer technique		Porath (23)[d]	March (22)[d] Jensen (54)

[a] For Sepharose, unless stated otherwise.
[b] For Soluble Dextran.
[c] For Sephadex.
[d] Very detailed description.

In general, procedures employing the buffer technique are characterized by short reaction times (1–3 min) compared with procedures employing the titration technique (5–12 min). The buffer technique is reliable and easy to perform, whereas the titration technique requires the availability of an automatic titrator or considerable experimental skill in order to perform the titration manually in a satisfactory way. It seems that the titration technique yields resins with a slightly higher coupling capacity.

The availability of a hood is the only prerequisite required for the performance of an activation, since CNBr is highly poisonous. For all practical purposes, the usual laboratory equipment and ordinary glassware is sufficient. If CNBr activations are performed very frequently, the construction of a special "activation-device" (47) may prove advantageous.

Good results are obtained by adding CNBr as solid material, and several such procedures have been published (37, 41, 45). Since aqueous solutions of CNBr cannot be stored at all, and since solutions of CNBr in organic solvents show a marked tendency to explode, adding solid CNBr could be the method of choice in cases where activations are done only very rarely. Due to the low solubility of CNBr in water, the use of aqueous solutions precludes the achievement of high initial concentrations of CNBr. For the achievement of very high coupling capacities, the use of concentrated solutions of CNBr in organic solvents like dimethylformamide (DMF) seems to be the method of choice.

12.3.1. Titration Technique with Dilute Aqueous Solution of CNBr

The procedure for the use of the titration technique according to Porath et al. (42), as modified by Axen et al. (5) is as follows:

The agarose gel was washed on a G3 glass *sinter*. Excess water was removed under suction for 3 min. To about 10 g of swollen Sepharose 4B (corresponding to about 400 mg dry resin) are added 20 mL of water and 16 mL of CNBr solution (25 mg CNBr/mL), which has to be prepared fresh prior to use. The pH of the reaction mixture was adjusted to 11 and maintained there by the addition of 2 *M* NaOH. (The titration was done by an automatic titrator; however, good results can also be obtained by manual titration using a 10 mL buret—*the authors*). The suspension has to be stirred during activation, and the temperature must be kept at 23–25°C. The time of activation was 6 min. The activated gel was rapidly washed on a G3 glass sinter with 300 mL 0.1 *M* sodium bicarbonate. The washing time was 5–8 min, after which the activated resin was immediately used for the fixation of peptides and proteins.

The reported coupling capacity is:

290 μmol/g dry resin for dipeptide

177 mg chemotrypsin/g dry resin

This procedure, first published in 1967, can be regarded as the prototype of all procedures based on the titration method. At that time Sephadex was still widely used. Due to the lack of experience with Sepharose, the procedure is heavily based on experimental know-how gained from work with Sephadex, and several significant modifications are necessary in order to optimize the procedure for the activation of Sepharose.

Amount of CNBr: Only 0.4 g CNBr/10 g wet gel were employed. This is far below the saturation amount of CNBr (1 g CNBr/10 g wet gel). Therefore, increasing the amount of CNBr can be expected to improve the coupling capacity.

Initial concentration of CNBr: The use of an aqueous solution containing only 25 mg CNBr/mL precluded the attainment of the high initial concentration of CNBr, necessary for optimal activation.

Temperature: Contrary to Sephadex, the activation of Sepharose can be considerably improved if the temperature is kept as low as possible (4–10°C) during the entire process of activation and washing.

Washings: For washing the activated resin, 0.1 *M* sodium bicarbonate was used. The choice of a basic washing medium is based on the desire to prevent the decay of the acid-sensitive imidocarbonates. As seen in Figure 12.6, *the use of a basic washing medium is detrimental in the case of Sepharose*, where the base-sensitive cyanate esters have to be protected from rapid hydrolysis by a neutral or even acidic washing medium. Despite several publications describing the stabilizing effect of acidic washings on activated Sepharose (43, 44, 46), the habit of washing activated Sepharose with basic buffer solutions seems to be ineradicable. Washings with solutions containing acetic acid or citric acid should also be avoided in view of the known reactivity of cyanate esters to carboxylic acids. When the resin is used immediately after activation for coupling, rapid washings with ice-cold water (0°C) seems to be a good compromise for the protection of both imidocarbonates and cyanate esters (see Figures 12.6 and 12.7). Otherwise, washings with 10^{-3} *M* HCl (pH 3) or even 10^{-1} *M* HCl (pH 1) will stabilize the cyanate esters for many hours with the concomitant hydrolysis of all imidocarbonates, preserving about 70% of the initial coupling capacity (See Table 12.3).

12.3.2. Titration Technique with Concentrated Non-Aqueous Solution of CNBr

The use of the titration technique in a nearly optimal fashion can be illustrated by the procedure of Nishikowa et al. (21):

The gel was washed carefully with distilled water (at least 5 times the gel volume) on a sintered glass funnel. After the gel was sucked dry for a few moments to obtain a packed cake, it was transferred to the reaction vessel. Water was added to adjust the volume to 1.2 times the settled bed volume. For working with 10 g of swollen Sepharose 4B, 1 g of CNBr was dissolved in 2 mL of *N*-methyl-2-pyrrolidone (2 mL of purified DMF will give identical results—*the authors*). With stirring, the concentrated solution of CNBr was added to the cooled (10°C) reaction mixture. Some of the CNBr precipitated immediately, forming finely divided microcrystals. Simultaneous with vigorous stirring, the pH was adjusted to 11 and held there with the addition of 2 or 4 *M* NaOH. Stirring has to be sufficiently vigorous to permit quick mixing and to facilitate the rapid resolubilization of CNBr. The time for uptake of base was temperature-dependent, taking 3–6 min at 20–25°C and about 15 min at 10°C. The activated gel was washed on a sintered glass funnel (coarse mesh) with 500 mL of 0.1 *M* bicarbonate buffer and used immediately for coupling.

The reported coupling capacity for ϵ-aminocaproic acid is:

$$20\ \mu\text{mol/mL swollen Sepharose 4B}$$

corresponding to:

$$500\ \mu\text{mol/g dry resin}$$

This procedure seems to be nearly optimal, as indicated by the considerable increase in the resin's coupling capacity as compared with the procedure of Porath et al. Washings with a basic medium should be avoided, as discussed above. The incorporation of cyanate esters into the resin is very rapid, reaching its peak within the first 3 min of activation. During the final stages of the reaction, characterized by a slowdown in base consumption, the concentration of free CNBr is very low and incorporation of additional cyanate esters into the resin does not compensate for the loss of cyanate esters due to hydrolysis. Therefore it is better to stop the activation reaction at the first sign of a slowdown of base consumption.

12.3.3. Buffer Technique with Concentrated Non-Aqueous Solution of CNBr

The use of the buffer technique can be illustrated by the procedure of March et al. (22):

The agarose beads are washed and sucked dry on a glass sinter. 10 g of swollen gel are mixed with 10 mL of water and 20 mL of 2 *M* sodium carbonate. The mixture is cooled to 4°C. With vigorous stirring, 1 mL of a solution of CNBr in acetonitrile (2 g CNBr/mL) is added all at once. (DMF may be substituted for acetonitrile; use 2 mL of a solution containing 1 g CNBr/mL—*the authors*.) After 1–2 min, the slurry is poured onto a coarse sintered-glass funnel and washed with 100–200 mL of 0.1 *M* sodium bicarbonate, water, and the buffer which is to be used in the subsequent coupling reaction. After the last wash, the slurry is filtered under suction to a compact cake and used immediately for coupling.

The reported coupling capacity is:

150 μmol/g dry resin for alanine

125 mg albumin/g resin

This procedure is very easy to perform. Reported coupling yields are lower than for the other procedures, but they can be improved by using ice-cold water as washing medium instead of sodium bicarbonate. The reaction time is very critical. Activation is complete after about 1 min; further prolongation of the reaction considerably reduces the coupling capacity.

12.3.4. The Use of Solid CNBr

The illustrative procedure for the use of solid CNBr according to Cuatrecasas (45), as modified by Wilchek et al. (41), is as follows:

Sepharose 4B was carefully washed on a sintered glass funnel (coarse mesh) with water. Next 10 g of the washed gel was suspended in 30 mL of water at ambient temperature, and 1 g of solid CNBr was added all at once to the suspension. The pH of the solution was brought to pH 11 with 5 *N* NaOH and kept between 10.8 and 11.2 for 8 min by the addition of NaOH. Continuous vigorous stirring assured the dissolution of CNBr during the first 5 min. Washings with cold water under suction were used to terminate the reaction. The activated resin was immediately employed for coupling.

The reported coupling capacity is:

15–20 mg anti-DNP antibody/mL gel

corresponding to:

375–400 mg antibody/g dry resin

The role of temperature during activations employing solid CNBr has not been sufficiently investigated. Cooling slows down the dissolution of the solid CNBr, causing a considerable increase in reaction time. *Contrary to all other procedures*, it seems that when using solid CNBr, cooling does not significantly increase the coupling capacity.

12.4. ANALYTIC PROCEDURES FOR CHARACTERIZING THE ACTIVATED RESIN

12.4.1. Determination of the Total Nitrogen Content

Usually the standard Kjeldahl procedure or Dumas microanalysis have been employed for the determination of the total nitrogen content of activated resins. However, the Kjeldahl procedure is both time-consuming and laborious, and Dumas microanalysis usually requires the services of a microanalytical laboratory. The following procedure does not require any specialized equipment and renders possible the accurate determination of the total nitrogen content of activated resins in a relatively easy fashion (7). This procedure, the standard Kjeldahl procedure, and Dumas microanalysis have been shown to yield identical results for samples of Sepharose and Sephadex (See Table 12.5).

Procedure: A weighted amount of about 15 mg of dry resin is placed in a 50 mL measuring bottle, 0.5 mL of 96% H_2SO_4 are added, and the mixture is heated in an oven to 100–120°C for 15 min. Next, 0.2 mL of 30% H_2O_2 are added slowly and carefully (vigorous reaction). The mixture clears up, and heating is continued at 180°C for another 200 min. After this, the colorless solution is diluted to 50.0 mL with 0.2 *M* sodium acetate (pH 5.5). Aliquots of known volume are used for the determination of NH_4^+ by the ninhydrin reaction (7), the TNBS reaction (48), or any other method. The amount of NH_4^+ found is equivalent to the total nitrogen content of the resin.

12.4.2. Determination of the Imidocarbonate Content

Of all the nitrogen derivatives present on activated resins, only the imidocarbonates will release NH_4^+ upon mild acid hydrolysis in 0.1 *N* HCl. This was first suggested by Axen et al. (2) and was later employed for the development of a procedure for the determination of imidocarbonates directly on the activated resin (7).

Procedure: A sample of activated resin containing about 1–20 μmol of imidocarbonates is placed in a 25 mL measuring bottle, and 5 mL of 0.1 *N* HCl

Table 12.5. Comparison of Results[a] Obtained by Dumas Microanalysis, the Kjeldahl Procedure, and the Proposed Procedure (19)

Content of Sample	Result of Dumas Microanalysis (μmol/100 mg sample)	Result of Kjeldahl Procedure (μmol/100 mg sample)	Proposed Procedure (μmol/100 mg sample)
Pure Sepharose 4B	0.9	1.0	0.8
Activated Sepharose 4B	143	144	144
Commercially activated Sepharose 4B	176	180	178
Activated Sephadex G50 (fine)	154	155	154

[a] Data are the average of three repetitive determinations.

are added. After 30 min of hydrolysis at 40°C (or alternatively, after 60 min of hydrolysis at room temperature) the volume is made up to 25 mL exactly with 0.2 *M* sodium acetate (pH 5.5). After thorough mixing the suspension is allowed to settle or is centrifuged for a short time to remove the insoluble resin. Aliquots of known volume of the clear supernatant are used for the determination of NH_4^+ by the ninhydrin reaction (7), the TNBS reaction (48), or any other method. The amount of NH_4^+ found is equivalent to the amount of imidocarbonates on the resin.

12.4.3. Determination of the Cyanate Ester Content

In 1904, König (49) reported that the reaction between pyridine and cyanogen halides can lead to colored products. This reaction was adopted for the quantitative determination of CN^- (50) by using barbituric acid as the color-forming reagent. The same reaction can be used for the quantitative determination of cyanate esters directly on the activated resin. The reaction pathway is shown in Figure 12.9. With the possible exception of traces of triazines, cyanate esters are the only species present on activated resin giving rise to the characteristic purple color. Since the molar absorption coefficient ϵ of the purple color is about 137,000 L $mol^{-1}cm^{-1}$(7), the reaction is highly sensitive and as little as 5 nmol of cyanate esters can be determined. (Previously [see ref. 10], ϵ was reported as 15,000 L $mol^{-1}cm^{-1}$; this value is erroneous).

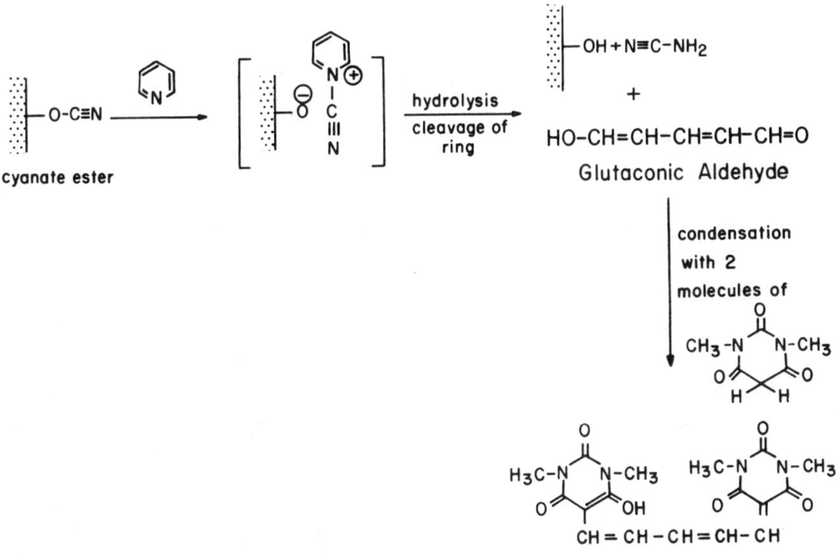

Fig. 12.9. Mechanism of color formation. The vital step in the reaction sequence is the attack of the electrophilic cyanate ester on pyridine, leading to ring cleavage and color formation.

1. as a quick, qualitative test for the presence of activation on samples of Sepharose;
2. for quantitative determinations of cyanate esters directly on the activated resin;
3. as quick test for the presence of excess CNBr in the washings after an activation;
4. pyridine can be used for the destruction of excess CNBr in the washings after activation.

12.4.3a. Preparation of Qualitative Test Reagent

To 14 mL of pyridine are added 6 mL of water and 0.5 g of barbituric acid or *N,N′* dimethylbarbituric acid. A clear, colorless solution is obtained, which darkens slowly. For qualitative tests, a slight coloration of the test reagent is of no importance.

12.4.3b. Qualitative Test for Presence of Free CNBr or Cyanate Esters on Activated Polysaccharides

To 10–20 mg of dry polysaccharide, or to 0.1–1.0 mL of swollen polysaccharide, or to several drops of the aqueous washings of freshly activated material, 1–2 mL of qualitative reagent are added. After slight shaking, the presence of less than 5 nmol of CNBr or cyanate ester can be detected by the formation of a red-purple color, which develops within 30 sec and becomes maximum after 10 min.

12.4.3c. Preparation of Quantitative Test Reagent

First, 500 mg of *N,N′*-dimethylbarbituric acid (recrystallized from water) are suspended in 5 mL of water (only *N,N′*-dimethylbarbituric acid can be used, since barbituric acid does not react quantitatively), then 45 mL of cold distilled pyridine are added. A clear, colorless solution is obtained. Even if stored at −20°C, the mixture slowly darkens, so it seems best to prepare the required amount of reagent mixture fresh prior to use.

12.4.3d. Quantitative Determination of Cyanate Esters

To 5–20 mg of dry activated resin or an equivalent amount of wet resin, 5 mL of reagent are added. With vigorous stirring the mixture is warmed to 40°C for 25 min in a closed test tube. After 25 min the mixture is diluted to any convenient volume with distilled water. For 10 mg of freshly activated resin, dilution to 250–500 mL is usually necessary in order to reduce the optical density of the purple solution to 0.5–1.0. Aliquots of this diluted solution are filtered and used to measure the absorbance at about 588 nm. The amount of cyanate esters in the sample is calculated employing a value of 137,000 L $mol^{-1}cm^{-1}$ as molar absorption coefficient.

12.4.4. Determination of the Amount of Coupled Ligand

12.4.4a. Standard Procedure According to Axen et al. (5)

After coupling, the gel was carefully washed and shrunk with acetone. After removal of the acetone under reduced pressure for 1 hr, the products were dried in vacuo for 48 hr over phosphorus pentoxide. Weighted amounts of polysaccharide conjugates were hydrolyzed in 6 *M* HCl for 24 hr at 110° in closed evacuated vessels. The amount of coupled ligand was determined by amino acid analysis of the hydrolysate. The high carbohydrate content did not disturb the analysis.

Glycyl-leucine served as the model ligand. The determination of coupled ligand was based on leucine, since the recovery of glycine is low owing to the covalent fixation of its amino group to the polymer. The determination of coupled proteins was based on the recovery of aspartic acid, glutamic acid, alanine, and leucine.

For studies of the coupling reaction, a high molar excess of model ligand has to be employed. Coupling of glycyl-leucine was done in the following way: To 20 mg glycyl-leucine in 5 mL sodium bicarbonate solution (0.1 *M*), about 1 g swollen Sepharose 4B were added. Coupling was allowed to proceed at 23°C for 16 hr. After coupling the resins were washed successively with 0.5 *M* sodium bicarbonate, 10^{-3} *M* HCl, 1 *M* sodium chloride, and water.

This generally accepted procedure has the advantage of being suitable for all amino-acid-containing ligands. However, it is both time consuming and laborous. For studies requiring the analysis of a large number of coupled resins, we have developed a procedure that makes possible the determination of coupled ligand directly on the resin, without prior hydrolysis, in a rapid and accurate way. The suggested procedure has the disadvantage of being limited to methionine-containing model ligands: methionine itself, glycyl methionine, and so on.

12.4.4b. Procedure for Determination of Coupling Capacity, Employing Methionine as Model Ligand According to Kohn and Wilchek (7)

Coupling procedure: Coupling medium is prepared by dissolving 150 mg methionine (Met) in 10 mL of 0.2 *M* sodium bicarbonate solution. To 20 mg of dry, activated resin or 0.5 g of wet resin, 1 mL of coupling medium is added. The coupling reaction is allowed to proceed with mild stirring for 4 hr at room temperature. After that, the resin is extensively washed and either stored as wet suspension or dried for further analysis in vacuo over P_2O_5.

Determination of coupled Met: Met can be determined by measuring the amount of H_2O_2 required for its oxidation to the corresponding sulfoxide (51, 52). Only cysteine, histidine, tyrosine, and tryptophen residues interfere. For the determination of Met on Sepharose, the procedure of Albanese (52) has been simplified: To a weighted sample of about 15–50 mg of coupled, dry resin, is added 1 mL of oxidation mixture. (Oxidation mixture: Add 0.2 mL of 30% H_2O_2 to 100 μL 6N H_2SO_4). The samples are allowed to react with stirring at 20°C for 40–60 min. Next, the samples are diluted to about 6 mL with water, and then the iodometric titration of the residual H_2O_2 is performed as described Vogel (53), using 1 mL of KI solution, 0.5 mL of molybdate catalysator solution, and 0.05 *N* standard sodium thiosulfate solution as titrant. The end point must be approached very slowly; it is reached when the starch indicator remains colorless for at least 3 min. The amount of coupled Met is calculated as outlined by Albanese (52). (This procedure requires the use of a microburet with a precision of ± 1 μL. The procedure can be adapted, however, for use with a regular buret).

12.5. SUMMARY

With the numerous applications of CNBr-activated polysaccharide resins for affinity chromatography and immobilized enzymes, it was recognized that considerable deviations from the expected behavior of the designed columns often occur.

Furthermore, the problems of ion exchange interference and hydrophobic interaction due to charged or hydrophobic groups on the resin, and the problem of ligand leakage due to slow hydrolysis of the ligand-resin bond, have complicated the use of CNBr-activated polysaccharide-ligand resins. These complications can be partly avoided or at least minimized by carefully designing the experimental conditions during preparation and usage of the polysaccharide-ligand resins. This, however, requires a thorough understanding of the molecular mechanisms involved in the attachment of ligand to the resin. The aim of this chapter has been to provide an explicit and detailed account of the chemical basis of the use of CNBr for the activation of polysaccharide resins.

REFERENCES

1. R. Axen, J. Porath, and S. Ernback, *Nature* **214,** 1302–1304 (1967).
2. R. Axen and P. Vretblad, *Acta Chem. Scand.* **25,** 2711–2716 (1971).

3. G. J. Bartling, *Biotech. Bioeng.* **14,** 1039–1043 (1972).
4. L. Ahrgren, L. Kagedal, and S. Akerström, *Acta Chem. Scand.* **26,** 285–288 (1972).
5. R. Axen and S. Ernback, *Eur. J. Biochem.* **18,** 351–360 (1971).
6. T. Miron, W. G. Carter, and M. Wilchek, *J. Solid Phase Biochem.* **1,** 225–236 (1976).
7. J. Kohn and M. Wilchek, *Anal. Biochem.,* **115,** 375–382 (1981).
8. M. Wilchek, T. Oka, and Y. J. Topper, *Proc. Nat. Acad. Sci. USA* **72,** 1055–1058 (1975).
9. R. Bywater and L. Kagedal, *Prot. Biol. Fluids* **27,** 789–791 (1979).
10. J. Kohn and M. Wilchek, *Biochem. Biophys. Res. Commun.* **84,** 7–14 (1978).
11. K. A. Jensen, M. Due, A. Holm, and C. Wentrup, *Acta Chem. Scand.* **20,** 2091–2106 (1966).
12. E. Grigat and R. Pütter, *Angew. Chem., Int. Ed.* **6,** 206–218 (1967).
13. R. Axen and P. Vretblad, *Prot. Biol. Fluids* **18,** 383–389 (1970).
14. G. Kay and E. M. Crook, *Nature* **216,** 514–515 (1967).
15. G. Kay and M. D. Lilly, *Biochem. Biophys. Acta* **198,** 276–285 (1970).
16. T. Lang, C. J. Suckling, and H. C. S. Wood, *J. Chem. Soc.* (*Perkin I*) **1977,** 2189–2194 (1977).
17. J. Turkova, in *Journal of Chromatography Library,* Vol. 12, Elsevier, Amsterdam (1978), pp. 159–169.
18. C. R. Lowe, in *Laboratory Techniques in Biochemistry and Molecular Biology,* Vol. 7 (T. S. Work and E. Work, Eds.), North-Holland, Amsterdam (1979), p. 268.
19. J. Kohn and M. Wilchek, unpublished results.
20. R. M. Izatt, J. H. Rytting, L. D. Hansen, and J. J. Christensen, *J. Am Chem. Soc.* **88,** 2641–2653 (1966).
21. A. H. Nishikawa and P. Bailon, *Anal. Biochem.* **64,** 268–275 (1975).
22. S. C. March, I. Parikh, and P. Cuatrecasas, *Anal. Biochem.* **60,** 149–152 (1974).
23. J. Porath, K. Aspberg, H. Drevin, and R. Axen, *J. Chromatog.* **86,** 53–56 (1973).
24. D. E. Stage and M. Mannik, *Biochim. Biophys.* **343,** 382–391 (1974).
25. S. Cloez, *C. R. Hebd. Seances Acad. Sci.* **44,** 482 (1857).
26. J. U. Nef, *Liebigs Ann. Chem.* **287,** 265 (1895).
27. E. Grigat and R. Pütter, *Chem. Ber.* **97,** 3012–3017 (1964).
28. K. A. Jensen, M. Due, and A. Holm, *Acta Chem. Scand.* **19,** 438–442 (1965).
29. D. Martin, *Z. Chem.* **7,** 123–136 (1967).
30. T. Sandmeyer, *Chem. Ber.* **19,** 862–867 (1886).
31. J. Houben and E. Schmidt, *Chem. Ber.* **46,** 2447–2461 (1913).
32. R. W. Addor, *J. Org. Chem.* **29,** 738–742 (1964).
33. Y. J. Topper, T. Oka, B. K. Vonderhaar, and M. Wilchek, *Biochem. Biophys. Res. Commun.* **66,** 793 (1975).
34. M. Wilchek, in E. K. Pye and H. H. Weetall, Eds., *Enzyme Eng.* **3,** 283–289 (1978).
35. J. Houben and R. Zivadinovitsch, *Chem. Ber.* **69,** 2352–2360 (1936).
36. J. Houben and E. Pfannkuch, *J. Prakt. Chem.* **105,** 7 (1922).
37. L. Kagedal and S. Akerström, *Acta Chem. Scand.* **25,** 1855–1859 (1971).

38. K. Bronström, S. Ekman, L. Kagedal, and S. Akerström, *Acta Chem. Scand.* **B28,** 102–108 (1974).
39. B. Svensson, *FEBS Lett.* **29,** 167–169 (1973).
40. R. Axen, P. A. Myrin, and J. C. Janson, *Biopolymers* **9,** 401 (1970).
41. M. Wilchek, V. Bocchini, M. Becker, and D. Givol, *Biochemistry* **10,** 2828–2833 (1971).
42. J. Porath, R. Axen, and S. Ernback, *Nature* **215,** 1491–1492 (1967).
43. I. Johansson, M. Joustra, and H. Lundgren, Swedish Patent 361,046, filed 72-09-07, granted 74-01-24.
44. M. Joustra and R. Axen, *Prot. Biol. Fluids* **23,** 525–529 (1975).
45. P. Cuatrecasas, *J. Biol. Chem.* **245,** 3059–3065 (1970).
46. Pharmacia Fine Chemicals, *Affinity Chromatography—Principles and Methods,* Uppsala (Sweden), June 1979, pp. 13–15.
47. T. Korpela and K. Kurkijärvi, *Anal. Biochem.* **104,** 150–152 (1980).
48. J. R. Whitaker, P. E. Granum, and G. Aasen, *Anal. Biochem.* **108,** 72–75 (1980).
49. W. König, *J. Prakt. Chem.* **69,** 105–110 (1904).
50. E. Asmus and H. Garschagen, *Z. Anal. Chem.* **138,** 414–418 (1953).
51. G. Toennies and T. P. Callan, *J. Biol. Chem.* **129,** 481–490 (1939).
52. A. A. Albanese, J. E. Frankston, and V. Irby, *J. Biol. Chem.* **156,** 293–302 (1944).
53. A. L. Vogel, *A Textbook of Quantitative Inorganic Analysis,* 3rd ed., Wiley, New York (1961), pp. 343–363.
54. H. B. Jensen and T. Miron, *J. Solid Phase Biochem.* **5,** 45–60 (1980).

CHAPTER

13

SOLID PHASE SYNTHESIS AND BIOLOGICAL APPLICATIONS OF POLYDEOXYRIBONUCLEOTIDES

R. BRUCE WALLACE and KEIICHI ITAKURA

Department of Molecular Genetics
City of Hope Research Institute
Duarte, California 91010

Synthetic DNA chemistry is no longer an esoteric discipline without obvious practical applications. In the mid-1970s it became of great practical value because of the sudden development of recombinant DNA techniques. One of the most significant practical applications of synthetic DNA has probably been the synthesis of artificial genes for the production of peptide hormones in *E. coli* (1–4). But this was just the beginning. The availability of synthetic DNA of defined sequences has had a profound effect on the way molecular biologists think about possible research projects, as well as on the conduct of individual projects (5). Indeed, the demand for synthetic DNA is growing rapidly, and it is not unusual to have to wait for the desired polynucleotides to become available. In response to these demands, the most important progress made by DNA chemists has very likely been the development of a simple, rapid, and reliable method for chemical synthesis. Automation of the whole procedure is attractive because it has been very successful in the peptide field (6).

In this chapter, we review (i) the current developments in solid phase synthesis of polydeoxyribonucleotides, which is the key factor for automation, and (ii) the potential uses of solid phase polynucleotide synthesis, with a few examples of its successful applications. To keep the length of the chapter within reasonable limits, we have confined ourselves to recent developments in the fields. If readers are interested in an extensive review of the chemistry and uses of polynucleotides, several review articles have recently appeared (7–9).

13.1. SOLID PHASE SYNTHESIS OF POLYDEOXYRIBONUCLEOTIDES

The concept of solid phase synthesis originated as a new approach to the synthesis of peptides, and most of the developmental work has been directed towards this class of compounds (6). The principle, however, is generally applicable throughout synthetic chemistry, and it has been applied to a variety of chemical reactions (10). It is especially useful for the preparation of oligomers and polymers of known sequences by the stepwise assembly of chemically related monomers.

The general solid phase strategy for the stepwise synthesis of oligomers is: (i) attachment of the first monomer unit to the solid support, (ii) removal of the appropriate protecting group from the first unit bound to the support, and (iii) coupling of the next unit to the first unit on the support. This operation is repeated sequentially until a polymer of the desired length is assembled. Finally, the product is released and all protecting groups are removed. The advantages of solid phase synthesis over conventional solution methods can be listed as follows:

1. The isolation of the product bound to the insoluble resin from soluble reagents and by-products can be simply performed by filtering and washing of the resin. Therefore, the isolation can be done rapidly without the loss of product, which is usually encountered during the purification of synthetic intermediates by liquid phase methods.
2. In order to force the reactions to completion, the soluble reagents can be used in excess and, thus, should furnish high reaction yields.
3. All synthetic operations are carried out in one flask, therefore, the loss of reaction products during the operations should be minimal.
4. The most important feature of solid phase synthesis is the automation of the whole operation. Since the essential operations of solid phase synthesis are mixing, washing, and filtering, it is possible to automate the synthetic process, particularly for polymers such as peptides and nucleotides, which are made of simple monomer units. Indeed, in the case of peptide synthesis, mechanization and automation of the complete operation have been successful (6).

One drawback of solid phase synthesis in the peptide field has been purification of the desired product. Even with a 99% yield of the coupling reaction to form peptide bonds, the final product is always a mixture of various lengths of peptides which are difficult to separate because peptides are made of 20 different amino acids that are polar and nonpolar, basic and acidic, and lipophylic and hydrophylic. On the other hand, polydeoxyribonucleotides are a polymer of only four different deoxynucleotides linked by phosphodiester bonds, therefore, the isolation of the desired product from products that have different phosphate charges is relatively easy.

Another criticism of the solid phase synthesis has been accumulation of side products on the solid support. These side products appear because it is impossible to purify the synthetic intermediates on the support. However, the same limitation is true for the liquid-phase method in the case of polynucleotides. Using present methods, fine purification of fully protected polynucleotides (phosphotriesters) longer than decamers is very inefficient (11). Therefore, fine purification is performed on the phosphodiester compounds after removal of all protecting groups from the phosphotriesters of the desired length (2). Consequently, we believe that solid phase synthesis of polynucleotides has many advantages over the liquid phase method and that it will become the preferred synthetic approach in future studies in molecular biology.

13.1.1. Synthetic Strategies

13.1.1a. Assembly of Polynucleotides

Two strategies for the assembly of polynucleotides on a solid support are illustrated in Figure 13.1. The desired polynucleotides may be assembled by stepwise coupling of mononucleotides, dinucleotides, or trinucleotides using these strategies: (i) attachment of the 3′-hydroxyl group of the terminal nucleoside to the solid support and sequential addition of activated nucleotides to the 5′-hydroxyl group of the nucleoside (3′-5′ elongation), and (ii) attachment of the 5′-hydroxyl group of the terminal nucleoside to the solid support and sequential chain elongation towards the 3′-end (5′-3′ elongation) (Figure 13.1). The advantages and disadvantages of each of these strategies have not been extensively investigated as yet, but the 3′-5′ elongation method is used exclusively in recent syntheses by the methods described in Figure 13.2.

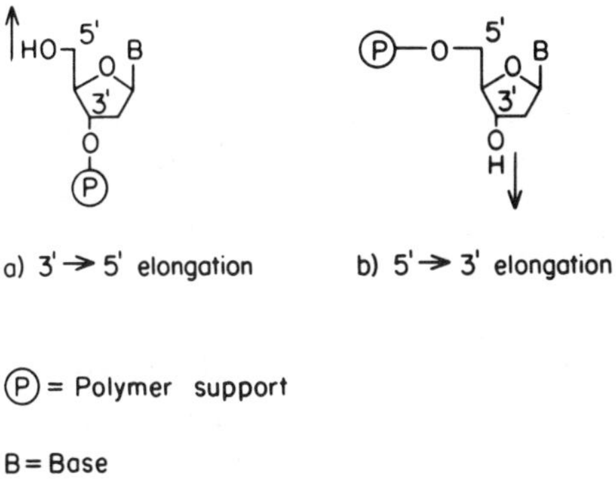

a) 3′ → 5′ elongation   b) 5′ → 3′ elongation

(P) = Polymer support

B = Base

Fig. 13.1. Direction of polynucleotide assembly.

13.1.1b. Coupling Methods

A few years ago it was almost impossible to synthesize oligodeoxyribonucleotides longer than a decamer by any solid phase method (12). This was largely due to the low efficiency of the coupling reaction forming internucleotidic phosphate bonds using the classical phosphodiester approach (13).

$$\mathrm{RO{-}\overset{\overset{\displaystyle O}{\|}}{\underset{\underset{\displaystyle O^-}{|}}{P}}{-}OH} + \mathrm{HOR'} \longrightarrow \mathrm{RO{-}\overset{\overset{\displaystyle O}{\|}}{\underset{\underset{\displaystyle O^-}{|}}{P}}{-}OR'}$$

The recently improved phosphotriester and phosphite-phosphotriester methods (Figure 13.2) have dramatically changed this situation, and various lengths of polynucleotides can now be synthesized rapidly and accurately on a solid support.

Phosphotriester Method. This method is the best established and currently the preferred method for synthesizing polynucleotides. The basic coupling unit is the 3′-phosphodiester component 1, which is activated by various coupling reagents. The activated compound reacts with the 5′-hydroxyl group of another nucleoside to form a phosphotriester bond between the two nucleosides (Figure 13.2a). One cycle of the operation is performed within 90 min using 1% polystyrene crosslinked with divinylbenzene. After the construction of the polynucleotide chains, all protecting groups are removed and natural DNA is generated. Among coupling reagents, 1-(mesitylene-2-sulfonyl)-3-nitro-1,2,4-triazole (MSNT) is the most highly recommended for high coupling yields (14,15). Although this coupling reaction is slower than with the phospite-phosphotriester approach, the method has more flexibility and is better controlled.

Phosphite-Phosphotriester Method. This method has been developed by Letsinger's group (16) and applied to solid phase synthesis by other groups (17–19). The coupling reaction involves formation of the internucleotidic phosphite bonds 5, and oxidation of the phosphites to the phosphotriesters 6. The big advantage of this approach is that the coupling units, phosphite intermediates 4, are much more reactive than those for the phosphotriester approach. The coupling reaction can be completed in 10 min, and one cycle of the monomer addition can be done within 60 min using a silica gel solid support (18, 19). There is no doubt that this method is very promising for the

A) The Phosphotriester Method

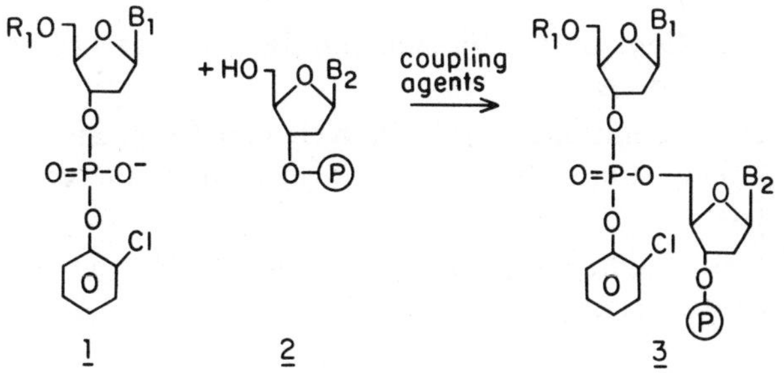

B) The Phosphite-Phosphotriester Method

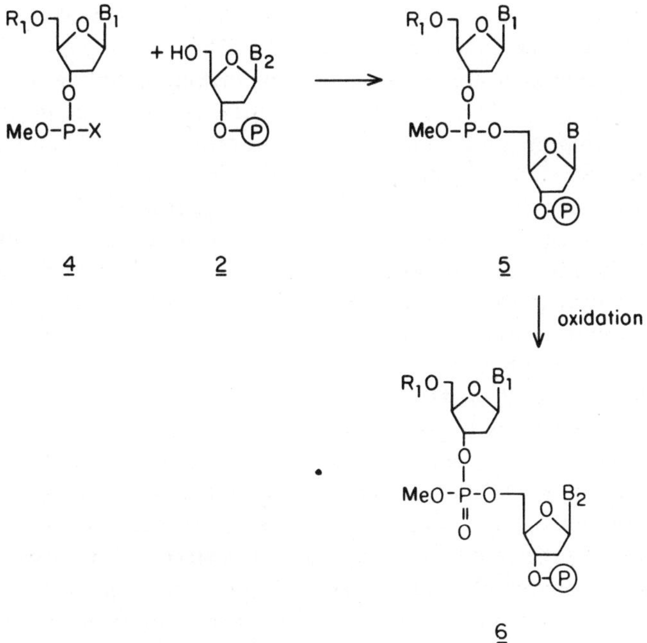

Fig. 13.2. Coupling methods used for solid phase synthesis.

synthesis of polynucleotides on the solid support. Prior to its use as a general method, the following points must be thoroughly examined:

1. Side reactions during the coupling reaction, particularly on bases, should be investigated.
2. Reaction conditions to quantitatively remove the methyl group must be developed in order to obtain high yields of the desired phosphodiesters.
3. A block coupling approach for synthesis of longer polynucleotides must be developed.

13.1.1c. Coupling Units

The choice between monomers and oligomers (di- and trinucleotides) as coupling units is, most likely, dependent upon the length of polynucleotides to be synthesized. With present technology, it is practical to synthesize up to pentadecanucleotides by stepwise monomer addition (18–20) and 40-nucleotides by the block coupling approach (21). The advantages of the monomer coupling approach are the simplicity of making coupling units and their higher reactivity. Also, it is only necessary to synthesize in advance four kinds of coupling units in solution. On the other hand, 16 kinds of dinucleotides and 64 kinds of trinucleotides should be available before starting assembly of the polynucleotides on a solid support by the block coupling approach. Although many papers have recently been published on the simplified method for the synthesis of these blocks (22, 23), it still requires much labor. However, this approach has at least two advantages over the monomer coupling approach. Needless to say, the coupling reaction to form phosphotriester (or phosphite) bonds between nucleosides would never go to completion regardless of the method used. The stepwise method using monomers as coupling units gives a mixture of polynucleotides with one nucleotide difference at the end of the synthesis and encounters the difficulty of purifying the longer polynucleotides. Also, the accumulation of side reactions for each coupling cycle on a solid support may put a limitation on the size to be synthesized by this approach. The strategy involving the coupling of purified oligonucleotide blocks could overcome the problems described above. The desired products are easier to separate from the other shorter polynucleotides lacking an oligomer fragment than from those lacking only a mononucleotide residue. One-third fewer coupling cycles are required for the synthesis of the same length of polynucleotide by the block coupling approach if trimers are used than if monomers are used. Therefore, this strategy should give a better quality of polynucleotides even if the coupling reactions do not go to completion.

Each dimer and trimer can be synthesized easily on a 5 g scale and used for at least 50 coupling reactions after one preparation. It is tentatively concluded that longer polynucleotides should be synthesized by the block coupling approach and short ones by the monomer approach, though we do not yet know the limitations of these approaches.

13.1.2. Solid Support

The unique feature of solid phase synthesis is the solid support itself, and future improvement of the synthesis will probably depend upon finding better supports. A variety of solid supports have been examined, including popcorn polystyrene (24), soluble polystyrene (25), 1% crosslinked polystyrene (26, 27), polydimethylacrylamide (12), macroreticular polystyrene (28), nonporous glass beads (29), and Kel-F-g polystyrene (3). Almost all of these supports were examined using the classical phosphodiester method, and reexamination of these solid supports using the more efficient coupling method described in Figure 13.2 has just begun.

The general requirements for a useful support can be readily listed as follows:

1. The support must provide enough points of reactive sites where the polynucleotide chain can be grown to give a useful yield of polynucleotides per unit, and it must minimize the interactions between growing chains.
2. It must allow rapid, unhindered contact between the growing chain and the reagents.
3. It must be chemically and physically stable to the conditions of the synthesis and be readily separable from the liquid phase at every stage of the synthesis.
4. For automation of the synthetic operation, the support should have a low affinity for protic solvents such as MeOH and water, which are inhibitors of coupling reactions. If it requires a very long time to remove these solvents from the support, coevaporation with an organic solvent such as pyridine is essential.

13.1.2a. Polystyrene Supports

This class of support, copolymers of styrene and divinylbenzene, has been most extensively studied in the peptide field (6). Actually, the first successful supports for solid phase peptide synthesis were copolymers of styrene and divinylbenzene. The degree of crosslinking determines the extent of swelling in the solvent, the effective pore size, and the physical stability of the

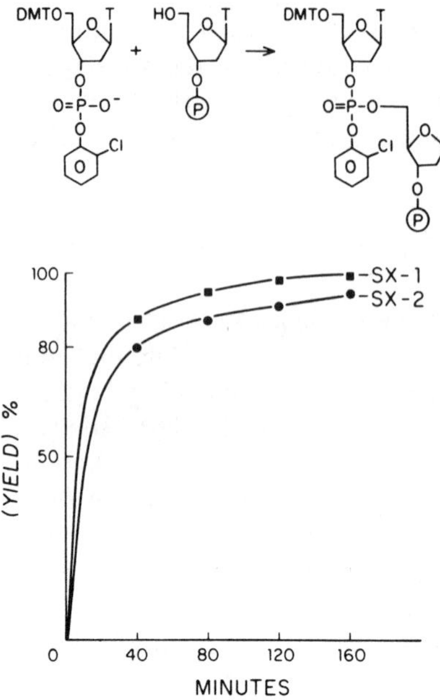

Fig. 13.3. Reaction kinetics for forming a phosphotriester bond on polystyrene copolymers with 1% (SX-1) and 2% (SX-2) crosslinked divinylbenzene.

beads. A highly swollen support, low crosslinked polystyrene with 1% or 2% divinylbenzene, has very recently been reexamined for the synthesis of polynucleotides (20, 31) using the phosphotriester approach (Figure 13.2a).

The reaction kinetic (Figure 13.3) to form a phosphotriester bond on 1% and 2% crosslinked beads has shown that the coupling reaction goes faster on the former than the latter. Furthermore, the coupling reactions on the 1% crosslinked beads generally give higher yields than those on the 2% beads. At the present time, 1% crosslinked resin is considered to be the most satisfactory among crosslinked polystyrene supports. The same conclusion was reached in the peptide field (6). This resin is easily handled and readily separated from reagents and solvents by filtration, and it is readily functionalized for the attachment of the first nucleosides (Figure 13.4). The high lipophillicy of this type of resin is very suitable for coupling reactions in anhydrous conditions. In order to remove water in the matrix, washing the resin with an appropriate anhydrous organic solvent for a short time is all that is required. Therefore, this resin can be readily used for the automation of the synthetic operation.

There may be some minor disadvantages to using a highly swollen solid support; additional solvents and reagents may be required for mixing with the resin than with nonswelling solid supports. An alternative to the use of a highly swollen styrene resin is to use a rigid structural polystyrene that barely swells in solvents. Highly crosslinked copolymers have a rigid structure and

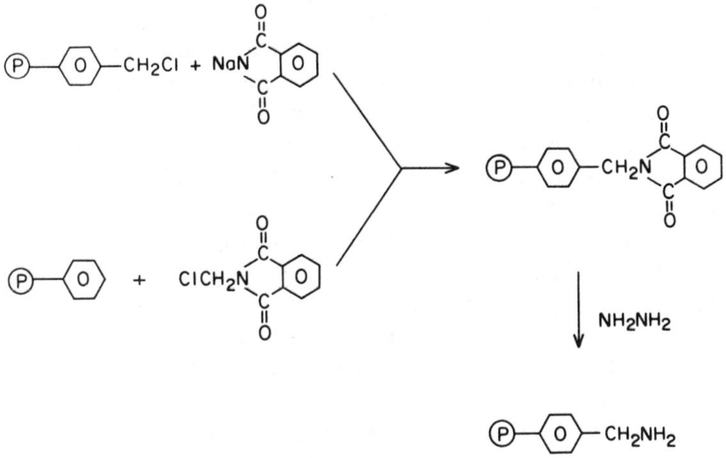

Fig. 13.4. Derivatization of polystyrene resin.

do not swell much in solvents, therefore, it is possible to confine the reaction's outer space. Functionalization of the resin, however, is difficult because of the small surface area; also, this system does not have enough capacity to give reasonable yields of the final products. The macroporous resins are rigid, highly crosslinked copolymers having a porous structure with a relatively great effective surface (28). Commercially available SM-2 macroporous resin was tested, and the results were rather disappointing with slow reactions (32).

Kel-F-g styrene made by grafting polystyrene onto the inert fluorocarbon core with γ irradiation (33) may be a promising alternative to the highly swollen polystyrene resins. The polystyrene in the graft copolymer is bound to a Kel-F core, and a crosslinked reagent such as divinylbenzene is not required. Shavarova and her coworkers first applied a polymer containing 29% (w/w) polystyrene to the polynucleotide field using the classical phosphodiester approach and reported high coupling yields (30). Our group examined similar Kel-F-g styrene that contained only 7.5% polystyrene, using the phosphotriester approach (31), and we found that the coupling yields on the beads were not satisfactory. The Kel-F-g styrene containing higher polystyrene (29%) should be investigated thoroughly, using the recent coupling methods, before any conclusions are drawn on this resin.

13.1.2b. Silica Supports

As nonswelling solid supports, several nonpolystyrene supports have also been suggested. Among the most promising are porous glass beads. In principle, a nonswelling support has some advantages over a swelling support; a higher rate of diffusion (namely, faster reaction rates) is one of the most important. The ideal support in this class should have a large surface area to increase the available reaction site and a large enough pore diameter not to hinder the growing of polynucleotide chains. These requirements are not mutually exclusive. If the support has a greater surface area it has less pore diameter, and vice versa. Therefore, it is essential to find supports that have a sufficient surface area and a reasonable pore diameter.

Matteucci and Caruthers have modified a separation grade silica gel (300 nm pore size) on which oligodeoxyribonucleotides (up to 12 mer) have been successfully synthesized with a high efficiency by the phosphite-phosphotriester approach (18). Using the same silica gel support, Ogilvie and Nemer (17) reported oligoribonucleotide synthesis (hexamer).

The advantages of silica suports are not limited to an efficient reaction. Other potential advantages are: (i) The silanol groups on the surface can be functionalized in various ways, which can then serve as points of attachment for the first nucleoside (Figure 13.5). (ii) Any unreacted silanol groups can be capped very easily by trimethylsilyl chloride, therefore, anhydrous conditions on the surface can be readily attained by flushing with appropriate solvents (18)—this is an essential requirement for the automation of the process. (iii) These nonswelling supports do not require as much solvents for mixing as do swollen supports; accordingly, nonswelling supports would require less of the expensive nucleotide coupling units (this could be overcome in the swollen supports by loading on a higher amount of the first nucleoside).

13.1.2c. Other Supports

Polyacrylamides. Crosslinked polydimethylacrylamide resin was introduced as a well-swollen polymer suitable for use in conjunction with polar reaction media to synthesize peptides (34). Gait et al. have successfully synthesized oligonucleotides of defined sequences using the phosphotriester approach on this resin (35). A similar support, polyacrylmorpholidate resin, has also been used for the synthesis (36, 37), and the longest known polynucleotide with a defined sequence (31 mer) was synthesized on this resin (21). These resins swell very well both in water and in a polar organic solvent such as pyridine, and they are mechanically stable. The coupling reactions on these resins proceed rapidly and without problems. The arguments in favor of polar support materials for reaction involving polar intermediates have been presented previously with special regard to peptide synthesis (34). However, polar supports such as polyamide supports have an inherent disadvantage for the synthesis of polynucleotides. In contrast to the peptide synthesis, an absolute anhydrous condition is an essential requirement for the coupling reaction in polynucleotide synthesis. The high affinity of the polyamide resins to protic solvents (water and alcohols) requires an additional step to remove the small amount of protic solvents in the matrix, either coevaporation of the resins with pyridine (36) or extensive washing of the resins with anhydrous organic solvents (38). In addition, inefficiency of the deblocking of the 4′,4′-dimethoxytrityl (DMT) group from the growing oligonucleotide chains by zinc bromide (discussed below) may hinder the usage of the polyamide resins as an ideal solid support.

Cellulose. Crea and Horn (39) found cellulose to be a useful solid support for the phosphotriester method. Cellulose swells in polar solvents such as pyridine and water, and is mechanically stable and easily manipulated. This resin is not entirely satisfactory, however, because the coupling yields are not high and complete capping of the hydroxyl functions of the resin is difficult. Furthermore, it has an inherent affinity to polar solvents and the coevaporation step with pyridine is essential before coupling reactions can ensue.

Miscellaneous Supports. As described above, various solid supports have been investigated for the synthesis of polynucleotides with the classical phosphodiester method. The evaluation of these supports is not possible because coupling yields are poor with this method, even in solution. With the new coupling approaches described in Figure 13.2 in our hands, only a few supports have been extensively examined thus far and, therefore, important developments in solid phase synthesis will probably hinge upon finding better solid supports.

13.1.2d. Reversible Linkage

With a few exceptions, the polynucleotide chain is elongated from the 3′-end to the 5′-end, and the nucleoside at the 3′-end of the desired sequence to be synthesized is usually attached to the solid support. The most popular way (Figure 13.6), at present, is to derivatize the resin to the primary amino resin (Figures 13.4 and 13.5). Either the 3′-hydoxyl group of the nucleoside 7 or the amino group of the resin 9 is reacted with succinic anhydride to give the carboxyl derivatives 8 or 10, which can be easily converted to the starting support 11 (Figure 6). Norris et al. used a slightly different approach, a secon-

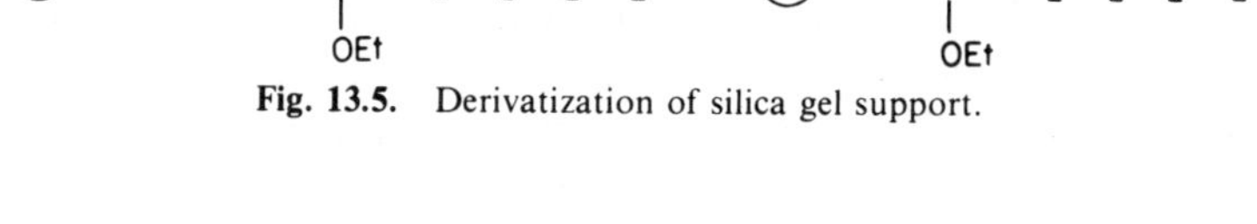

$$\text{(Si)}{-}OH + EtO{-}\underset{OEt}{\overset{OEt}{Si}}{-}CH_2CH_2CH_2NH_2 \longrightarrow \text{(Si)}{-}O{-}\underset{OEt}{\overset{OEt}{Si}}{-}CH_2CH_2CH_2NH_2$$

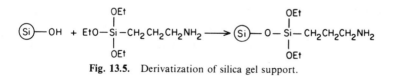

Fig. 13.5. Derivatization of silica gel support.

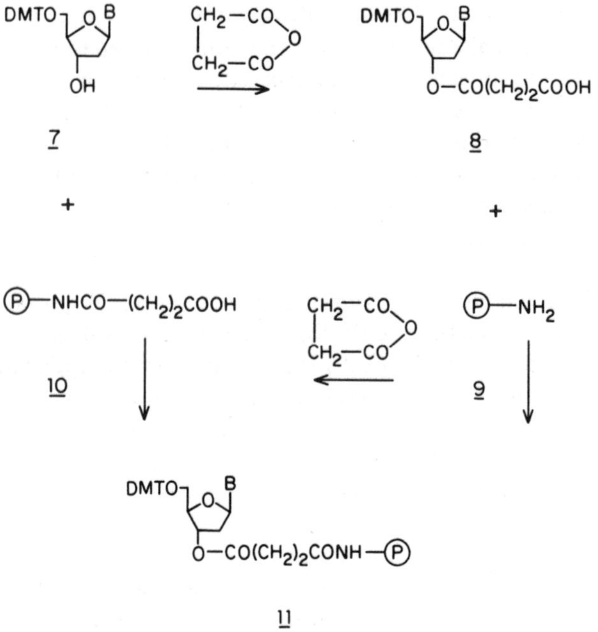

Fig. 13.6. Attachment of the first nucleoside to the resin.

dary amine in place of the primary amine and phthalic anhydride in place of succinic anhydride (37). Anchoring of the first nucleoside by a ribonucleotide was reported for the cellulose support (40). The cleavage of polynucleotides from all these supports can be done simply by treatment with concentrated ammonium hydroxide.

13.1.3. Protection and Deprotection

Almost all functional groups of the nucleotide are protected to avoid side reactions and to increase the solubility of the nucleotides during the synthetic operation. All protecting groups developed for the synthesis of polynucleotides in solution are also used for synthesis on solid supports.

13.1.3a. 5′-Hydroxyl Group

The 4′,4′-Dimethoxytrityl (DMT) group is exclusively used for the protection of the 5′-hydroxyl group under solid phase conditions. This group is removed under acidic conditions, 2% benzenesulfonic acid (BSA) (41) and 10% trichloroacetic acid (42). None of these conditions are satisfactory because a glycosidic bond between *N*-benzoyladenine and deoxyribose is not

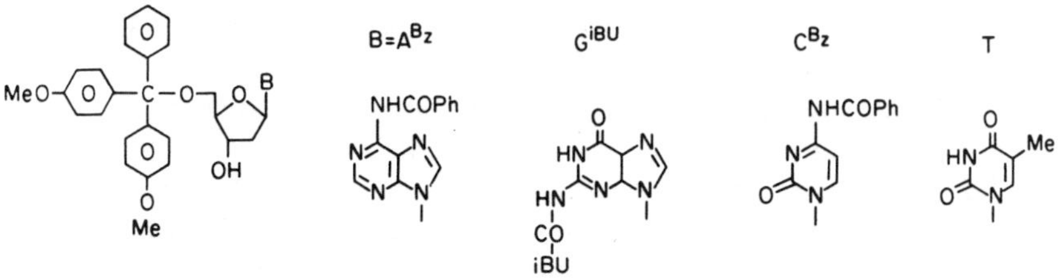

Fig. 13.7. Basic protected nucleosides.

completely stable under those protic acidic conditions. Depurination (42) and anomerization (43) were observed as side reactions. Very recently, two groups (44, 45) independently introduced zinc bromide ($ZnBr_2$) for the removal of the DMT group. This Lewis acid is superior to the protic acids in that no depurination was detected during the deprotection reactions. The complete removal of the DMT group from the growing chain is an indispensable requirement for the synthesis and, furthermore, for the purification of the desired products; this is discussed in detail in Section 13.1.5.

13.1.3b. Amino Group

The following protecting groups have been used for each base that has an amino group on a heterocyclic ring: benzoyl for adenine and cytosine, and isobutyryl for guanine (13; Figure 13.7). For removal of these protecting groups a fairly drastic condition is necessary, concentrated ammonium hydroxide at 50°C for several hours. Actually, these protecting groups were developed for synthesis by the phosphodiester method (13), which requires strong basic conditions to remove another protecting group at the 3′-hydroxyl function, an acetyl group extending the polynucleotide chain. Therefore, the stable protecting groups were required on the amino functions described above. In contrast, the extension of polynucleotide chains using the recent method (Figure 13.2) on a solid support required only acidic conditions. Less stable protecting groups, therefore, could be used for the masking of the amino functions and a drastic deblocking condition could be avoided.

13.1.3c. Phosphate Group

One of the most important advances in the phosphotriester approach, the protection and deprotection of internucleotidic phosphate bonds, owes a great deal to Reese and coworkers (46). They have introduced aryl derivatives, phenyl, and *o* and *p*-chlorophenyl groups, the only satisfactory protecting groups to give high coupling yields. A great number of protecting groups other than aryl derivatives have been examined, but none gave the desired coupling yields (47) except a phenylethyl derivative developed by Pfleiderer's group (48).

For the deprotection of these phenyl derivatives, oximate ion (49) is quite superior to other nucleophilic reagents such as hydroxyl (50) and fluoride ions (51) for the suppression of the undesired phosphotriester bond cleavage (Figure 13.8). Moreover, oximate ion regenerates the original guanine base from the modified guanine residue produced during coupling reactions (52). This is discussed further in the next section.

In the phosphite-phosphotriester approach, the methyl group is used for the successful synthesis of polynucleotides (17–19) and removed by treatment with triethylammonium thiophenoxide (18). One potential side reaction of this deblocking condition could be a simultaneous nucleophilic attack on the deoxyribose 5′-carbon atom by nucleophilic reagents. Extensive studies on the side reactions must be performed before the methyl group can become a general protecting group for the phosphate function.

13.1.4. Side Reactions

It is likely that undesirable side reactions will occur during solid phase synthesis, and they may accumulate on the products, to some extent, because purification of the synthetic intermediates is impossible. The coupling reagents used in the phosphotriester approach are, essentially, sulfonating reagents. Obvious side reactions of coupling reagents are sulfonation of the 5′-hydroxyl group of deoxyribose (15) and the 6-position of the guanine base (52). Reese and Ubasawa isolated the nitrotriazolide derivative 12 by reac-

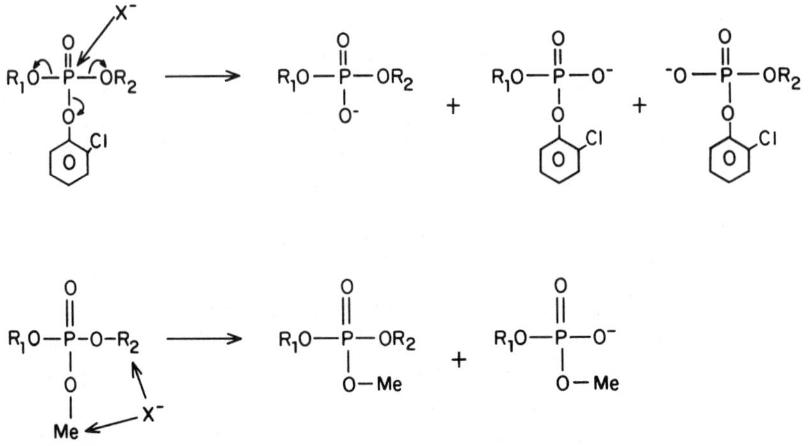

Fig. 13.8. Potential side reactions in deprotection of phosphate protecting groups.

tion of *N*-benzoyl-3′,5′-di-*O*-acetyl-2′-deoxyguanosine with the coupling reagent 1-(mesitylene-2-sulfonyl)3-nitro-1,2,4-triazolode (MSNT) in the presence of diphenyl phosphate (52; Figure 13.9). Sulfonation of the 5′-hydroxyl group terminates chain elongation, giving lower overall yields of the products. Fortunately, modification of the guanine residue is apparently reversed during the unblocking of polynucleotides with oximate ion (49). The ready conversion of the base-modified compounds 12 into amino derivatives by ammonolysis reveals a dangerous aspect of using ammonia to remove aryl protecting groups from the protected polynucleotides. A similar side reaction, base modification of guanine residue, could take place in the synthesis by the phosphite-phosphotriester method; phosphorylation of the base residues followed by substitution with tetrazole (if X = tetrazole in Figure 13.2b). This possibility has not been thoroughly examined.

13.1.5. Purification of Polynucleotides

The final product resulting from the solid phase synthesis is most likely a mixture of polynucleotides, the desired product with a series of shorter ones. Since the synthetic intermediates are not purified and side reactions can be expected during deblocking of the fully protected products, the mixture could be more complicated. Furthermore, it is now possible to synthesize much longer polynucleotides (21) with the solid phase method than with the liquid phase method. A rapid and efficient separation method is of great importance in obtaining satisfactory purity of the final product after removal of all protecting groups.

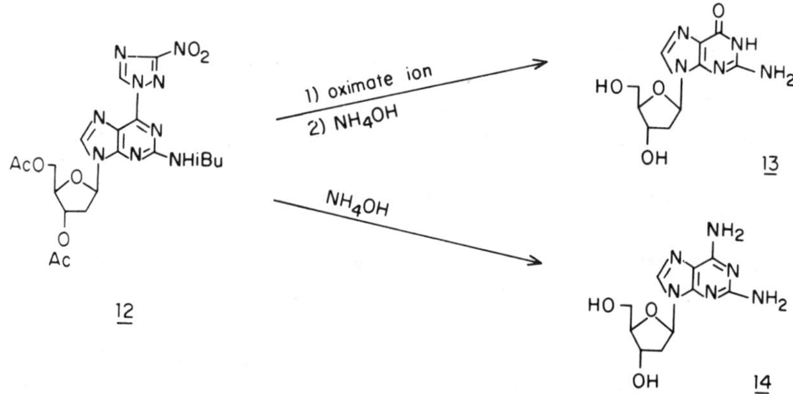

Fig. 13.9. Regeneration of deoxyguanosine by treatment with oximate ion.

13.1.5a. High Performance Liquid Chromatography (HPLC)

Anion Exchange Column. Ion exchange chromatography has been the most widely used technique for the purification of charged compounds such as polynucleotides. Diethylaminoethyl (DEAE) cellulose can be used for resolution of polynucleotides of up to 20 residues long in the presence of 7*M* urea (53). However, this method is very time consuming (2 days). For almost all studies of molecular biology, one O.D. unit of a polynucleotide (approximately 40 μg) is quite sufficient. This amount of material can be purified easily by chromatography on an analytical HPLC anion exchanger column, such as Permaphase AAX (DuPont) (54, 41) or Partisil 10 SAX (Whatman) (38). The desired product is always eluted later than any shorter sequence and can be readily identified. It takes approximately 30 min to collect the desired peak (approximately 1 O.D. unit), which is desalted and used in most studies without further purification. If an extremely pure polynucleotide is required (for example, a primer for DNA or RNA sequencing), it can be purified further by HPLC on a reversed-phase μBondapak C_{18} column. Purification based on two different principles gives an extremely pure sample (38, 41). An icosadecamer is probably the upper limit of chain length for purification using the ion exchangers.

Reversed-Phase Column. Theoretically, only the final product (the desired one) synthesized on the solid support has the 4′,4′-dimethoxytrityl (DMT) group. The other shorter products should not have this group if the deblocking reaction of the DMT group goes to 100% and capping of any unreacted 5′-hydroxyl groups is complete at each cycle of the operation. After the final coupling all protecting groups are removed, except the acid-labile DMT group, and the resulting mixture is analyzed on a reversed-phase, μBondapak C_{18} column (19, 21). The first eluted peak contains polynucleotides corresponding to the failure sequence, and the second peak is the desired product. If necessary after the removal of the DMT group, the desired product is further purified by repeating the reversed-phase chromatography or by gel electrophoresis on polyacrylamide as described below.

13.1.5b. Gel Electrophoresis on Polyacrylamide

As shown by Maxam and Gilbert (55), resolution of DNA fragments with only a one-charge difference is possible on polyacrylamide gel electrophoresis in the presence of 7 *M* urea. Markham et al. (56) analyzed the polynucleotide mixture at various stages in the synthesis of an octadecanucleotide by the solid phase method by analytical gel electrophoresis. The desired products always move more slowly than any shorter sequences and can be easily identified. Using preparative gel electrophoresis, 100 μg of the desired product can be readily resolved. The area of the polyacrylamide gel containing the product is cut out and the product electrophorically eluted from the gel into a dialysis membrane and desalted (57). This method has probably the best resolution among the presently available purification methods for polynucleotides and is applicable to almost any size of synthetic polynucleotide.

13.1.6. Automation

Solid phase polynucleotide synthesis has certain unique features that simplify the manipulation involved in reagent transfer, mixing, filtering, and masking. All synthetic operations can be performed in one reaction vessel, therefore, eliminating the repeated transfer of reagents from one container to another usually encountered in liquid phase synthesis. Furthermore, since the polynucleotides are polymers of four different repeating units, each cycle of the synthesis can be carried out in essentially the same way for each internucleotidic phosphate bond. It is only necessary to change the coupling unit at each cycle according to the sequence of the polynucleotide. A simple hand-operated apparatus was devised to carry out these operations (58). Very recently, Matteucci and Caruthers introduced a semiautomatic device using HPLC-type equipment, and they succeeded in the synthesis of a decamer and an undodecamer using the phosphite-phosphotriester method (18). Completely automated apparatus have become commercially available from Vega Biochemicals and from Biologicals. The actual value of these automatic instruments has not yet been thoroughly examined, but we believe that it will be practical in the very near future to synthesize up to an icosanucleotide overnight using a synthesis machine. It is possible that by the time this review is published, the machine-aided synthesis of polynucleotides will have materialized.

13.2. APPLICATION OF SYNTHETIC DNA TO MODERN BIOLOGY

Recent advances in the chemical synthesis of oligodeoxyribonucleotides have made readily available large quantities of synthetic DNAs of defined sequence. This phenomenon has dramatically increased the number of applications to which oligonucleotides are being used. Originally, due mainly to the pioneering work of Khorana and coworkers, one of the main applications of synthetic DNA was in the construction of genes coding for specific products (13). Later, with the advent of molecular cloning technology, gene synthesis took on a new dimension, for it was then possible to clone the synthetic DNA and obtain expression of these genes in *E. coli* (1–4). Molecular cloning also stimulated other applications of oligonucleotides, such as the synthesis of "linkers" (59, 60) and "adapters" (61) to create or modify restriction endonuclease cleavage sites to facilitate cloning.

Synthetic DNA has also found applications in the structural analysis of DNA itself. The ability to synthesize and purify large quantities of short DNA fragments made it possible to produce and crystallize short DNA duplexes for X-ray crystallographic analysis (62, 63). These studies have led to new insights into the dynamic structure of DNA.

More recently, oligonucleotides have found use in almost every phase of molecular cloning, including construction, identification, and characterization of particular clones, as well as in manipulation of the isolated cloned DNA to modify its structure, DNA sequence, or mode of gene expression. All of these uses of oligonucleotides rely on the specificity with which the synthetic DNA will form Watson–Crick base pairs with various naturally occurring complementary DNAs or RNAs. This is true whether the oligonucleotide is used directly as a hybridization probe or whether it is used as a site-specific primer on a DNA or RNA template.

A discussion of the applications of synthetic DNA in modern biology follows, including several examples of their use.

13.2.1. Gene Synthesis

For biological studies, polynucleotides with defined sequences longer than those accessible by the chemical methods are often required. Khorana and coworkers developed an enzymatic method for joining chemically synthesized oligonucleotides (10–15 bases long) using the inherent nature of oligonucleotide chains to form ordered duplexes by virtue of base pairing (13). These oligonucleotides are phosphorylated at the 5′-hydroxyl group by [γ–^{32}P]ATP with polynucleotide kinase and joined together with DNA ligase through the formation of the phosphodiester bonds between the 5′-phosphate and 3′-hydroxyl groups. Synthesis of various double-stranded DNA has been performed using this approach (64–66). The longest gene synthesized to date is the total synthesis of a human leukocyte interferon gene (66); no fewer than 514 base pairs–more than 1000 nucleotides–have been assembled from 66 oligonucleotide building blocks.

Recent progress in the solid phase synthesis of polydeoxyribonucleotides permits the development of another approach for the construction of double-stranded DNA of desired sequences (68). As described below, two polynucleotides (27 and 25 mer) are synthesized on the solid phase. The sequence of these two fragments are designed to form a 10 base pair duplex through their 3′-termini. When these fragments are hybridized to each other at their 3′-termini ends and used as substrates for a DNA polymerase I reaction with the four deoxynucleoside triphosphates, a 42-base-pair double-stranded DNA is produced, as described below.

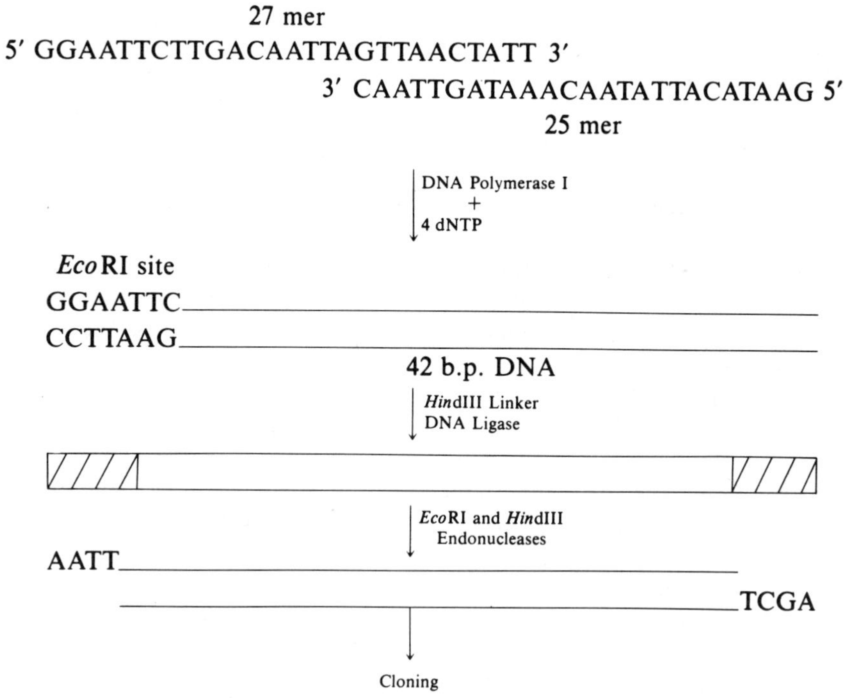

27 mer

5′ GGAATTCTTGACAATTAGTTAACTATT 3′

3′ CAATTGATAAACAATATTACATAAG 5′

25 mer

DNA Polymerase I + 4 dNTP

*Eco*RI site

GGAATTC

CCTTAAG

42 b.p. DNA

*Hin*dIII Linker DNA Ligase

*Eco*RI and *Hin*dIII Endonucleases

AATT

TCGA

Cloning

*Hin*dIII linkers are added to both ends of the resulting duplex DNA by DNA ligase, followed by digestion with restriction endonucleases *Eco*RI and *Hin*dIII and cloned into plasmid vector pBR327. This cloned synthetic sequence represents a consensus *E. coli* promoter and has been shown to be functional by various biological criteria (68). For the chemical synthesis of a 42-base-pair duplex DNA, 84 nucleotides are required by the classical approach. However, the requirement is reduced to only 27 plus 25 nucleotides, or 52 nucleotides, by the new approach. With further improvement of the chemical synthesis of polydeoxyribonucleotides by the solid phase method, this method should become a proven general approach for the synthesis of double-stranded DNA fragments that encode regulatory DNA sequences or moderate-sized proteins.

13.2.2. Oligonucleotides as Primers

Oligonucleotides of defined sequence can be used as specific primers on a template DNA or RNA template containing complementary sequences

(69–72). This priming specificity has been exploited for DNA or RNA sequencing, radiolabeled probe synthesis, and in the production of specifically primed cDNA for cloning. In order to utilize the priming ability of oligonucleotides, one needs to have knowledge of the nucleotide sequence of the priming site. (Note that we are excluding from discussion the use of randomly generated oligonculeotides as primers (73,74)). This information is obtained (or deduced) in several ways, depending on the specific application to which the oligonucleotide is being put.

13.2.2a. Oligonucleotides as Primers for DNA or RNA Sequencing

When a portion of a nucleotide sequence of a template DNA or RNA has been previously determined, oligodeoxyribonucleotides can be synthesized as specific primers for further sequencing (75, 76). Smith et al. (75) sequenced a substantial amount of the yeast iso-1-cytochrome c gene, cloned in the plasmid vector pBR322, using synthetic DNA sequences as short as nonamers as primers for the chain termination method of sequencing developed by Sanger and coworkers (77). Advantage was taken of the specificity with which the oligonucleotide directs the initiation of DNA synthesis and of the ability of DNA polymerase I to strand-displace during chain elongation.

Messing and coworkers (78) have recently described the synthesis and use of a pentadecanucleotide complementary to the genome of the single-stranded DNA bacteriophage vector M13mp7. This sequence is complementary to the viral DNA at a site immediately 3′ to the cluster of five restriction endonuclease cleavage sites useful for cloning in this bacteriophage vector. The pentadecanucleotide serves as a universal primer for the chain termination method of DNA sequencing, initiating DNA synthesis within the vector DNA and proceeding into the cloned DNA, permitting rapid DNA sequence analysis of the cloned DNA. Similarly, Wallace et al. (79) described the synthesis and use of a set of seven universal primers suitable for sequencing DNA cloned at any one or a combination of two of five unique restriction sites in pBR322.

Figure 13.10 shows an example of the sequencing of DNA by the chain termination method using an octanucleotide primer. The specificity of priming by an octanucleotide, and the speed with which such a short sequence can be synthesized by the solid phase method, makes it practical in many circumstances to design and synthesize these short sequences for a specific DNA sequencing task.

13.2.2b. Oligonucleotide-Primed Sequences as Specific Hybridization Probes

The specificity with which oligonucleotides will hybridize to a complementary sequence can be utilized in two ways to create specific hybridization

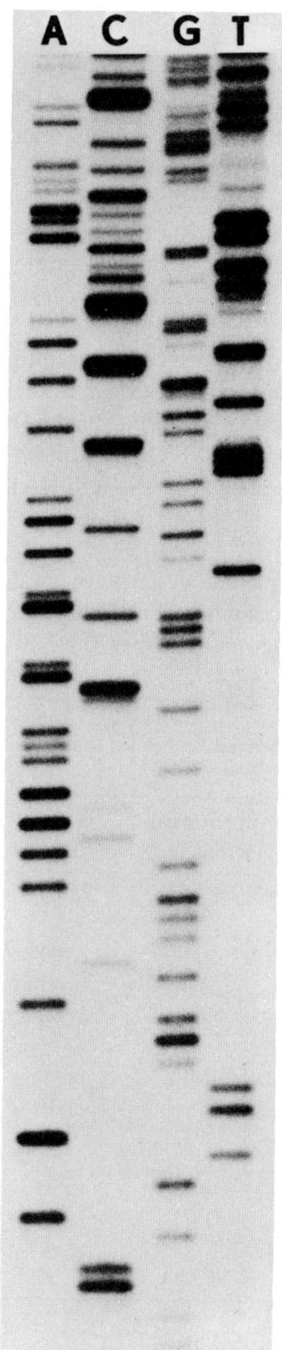

Fig. 13.10. Oligonucleotide-primed sequencing of DNA. Octanucleotide TGGATGGA was synthesized by the solid phase method and used as a primer for sequencing a 1,000 b.p. DNA fragment containing a complementary sequence by the chain termination method as described elsewhere (79, 99).

probes. First, the oligonucleotide can be used as a primer on a DNA or RNA template to produce a radiolabeled primer extension product with DNA polymerase I or AMV reverse transcriptase; the primer extension product can then be used as a specific probe. Second, the oligonucleotide can be radiolabeled and used directly as a hybridization probe (see next section). Most recently these two types of probes have found greatest application in the identification of specific cloned DNAs that code for proteins of known amino acid sequence. Table 13.1 summarizes several recent examples of the successful use of such probes.

The amino acid sequence of a region of a protein of interest can be used to deduce a set of oligonucleotide sequences complementary to a portion of the mRNA coding for that region of the protein. At least two approaches have been used for the design of these oligonucleotide primers which are utilized for the synthesis of a radioactively labeled DNA copy of the mRNA (cDNA) using AMV reverse transcriptase.

Oligonucleotide Primers of Unique Sequence. For primer design, Agarwal and coworkers (80–83) use tetra-peptide sequences composed of tryptophan or methionine or both, each having only a single codon and any of the nine

Table 13.1 Use of Oligonucleotide and Oligonucleotide-derived Probes for the Isolation of Specific Cloned DNAs

Probe	Cloned DNA Coding for:	Approximate Abundance[a]	Reference
1. Oligonucleotide primed cDNA	Rat insulins	1/24	82
	Human fibroblast interferon	1/600	89
	Human leukocyte interferon	1/400	90
	Human HLA	1/3000	106
	Bacteriorhodopsin[b]	—	88
2. Oligonucleotide of unique sequence	Yeast iso-1-cytochrome c	1/3500	94
	Yeast iso-1-cytochrome c	—	95
3. Oligonucleotide of mixed sequence	Human β2-microglobulin	1/535	98
	Mouse H-2K[b]	1/30,000	99

[a] Refers to the fraction of clones which were identified as the desired one rather than the fraction of the total mRNA or genomic complexity accounted for by the desired sequence.

[b] Cloning of this cDNA was reported in Reference 88 as a note added in proof without details.

amino acids having only two codons. The sequence of the oligonucleotide primer to be synthesized is determined by one of two methods. Either the sequence synthesized is complementary to the statistically favored codon for one or more of the amino acids (84), e.g., GAG rather than GAA for glutamic acid (85), or the sequence synthesized contains a dG residue when the third codon position could be C or U (allowing the formation of either a dG-C or a dG-U base pair), or it is a dT when the third codon position could be A or G (allowing the formation of either a dT-A or a dT-G base pair). This latter approach is based on the observation that dG-U and dT-G base pairs are less stable but permissible oppositions (83, 86, 87). Either approach results in the design of a single oligonucleotide of unique sequence. For the identification of bacteriohodopsin mRNA, Chang et al. (88; see Table 13.1) extended this approach to use two oligonucleotides in order to avoid the formation of a dT-G base pair at the 3′-end of the primer RNA duplex.

Oligonucleotide Primers of Mixed Sequences. Goeddel et al. (89) used a mixture of oligonucleotides containing all possible sequences complementary to a region of an mRNA coding for a known amino acid sequence as primer for cDNA synthesis. By choosing a stretch of amino acid sequence for human fibroblast interferon (FIF), they designed six sets of dodecanucleotide primers, each a mixture of four sequences. The pooled oligonucleotide mixture was complementary to all possible codon combinations for the FIF peptide sequence Met-Ser-Tyr-Asn. The radiolabeled cDNA prepared with the primer pool was used to identify a cloned cDNA coding for FIF (see Table 13.1). Similarly, amino acid sequence of human leukocyte interferon (LeIF) was used to design primers for the identification of clone LeIF cDNA (90; see Table 13.1).

13.2.2c. Cloning of Oligonucleotide Primed Sequences

By using oligonucleotides as specific primers, it is possible to produce a specific DNA fragment for cloning. This approach has not received as much attention as might have been expected. The reasons are not entirely clear. The successful application of oligonucleotide probes (below) and oligonucleotide-primed cDNA probes (above) have meant that it is not necessary to clone oligonucleotide-primed cDNA sequences. Instead, one can clone entire collections of cDNA sequences—a "shotgun library"—and screen this library with specific probes.

Winter et al. (91) have used oligonucleotide primers to prepare and clone a full-length double-stranded DNA copy of human influenza virus RNA. The strategy employed was to synthesize a dodecanucleotide primer complementary to the known sequence of the 3′-end of all of the eight RNA molecules composing the viral genome. This primer was used to produce a single-stranded DNA copy of the viral RNA. A second primer (a tridecanucleotide) was synthesized based on the known sequence of the 5′-end of all of viral RNAs. This primer was complementary to the single-stranded DNA and was used to produce a double-stranded DNA suitable for cloning. The strategy allowed successful isolation of a full-length copy of one of the influenza viral genes coding for two nonstructural proteins. Similarly, Goeddel et al. (89) have described the use of specific oligonucleotides as primers on a DNA fragment derived from a human FIF cDNA clone in order to remove the coding sequence for the leader peptide and create a new gene coding for mature human FIF. This technique can be described as "editing" and will likely be a major application of oligonucleotides in the biotechnology field.

13.2.3. Oligonucleotides as Hybridization Probes

Under appropriate conditions, oligonucleotides form stable duplexes when hybridized to DNA or RNA molecules containing complementary sequences. Oligonucleotides can be synthesized, radiolabeled, and used as hybridization probes to characterize or identify specific DNA or RNA sequences. Just as in the case of the oligonucleotide primer approach (above), oligonucleotide design depends on knowing or being able to deduce the nucleotide sequence of the complementary molecules. This information is often obtained from protein sequence data.

13.2.3a. Oligonucleotide Probes of Unique Sequence

In a very unusual case, Stewart and Sherman (92) were able to deduce the nucleotide sequence of a portion of the yeast cytochrome c gene based on the analysis of the amino acid sequence of several frame-shift mutants. This information allowed the design of oligonucleotide probes for the characterization of the yeast cytochrome c mRNA (93), as well as the identification of a cloned DNA containing the cytochrome c gene (94, 95; see Table 13.1). Unfortunately, amino-acid-sequence information for a particular protein rarely allows determination of a unique sequence for the gene coding for the protein.

13.2.3b. Oligonucleotide Probes of Mixed Sequence

In a study of the effect of single-base-pair mismatches on the hybridization behavior of oligonucleotides to ϕX174 DNA, Wallace et al. (96) proposed that synthetic oligonucleotides might be useful as specific probes for cloned

DNA. They observed that the duplexes with a single-base-pair mismatch, which were formed when 11, 14, or 17 base long oligonucleotides were hybridized to OX174 DNA, were significantly less stable (dissociated at lower temperature) than their perfectly matched counterparts. This difference in thermal stability made it possible, by the appropriate choice of hybridization temperature, to virtually eliminate the formation of mismatched duplexes without affecting the formation of perfectly matched ones.

Wallace and coworkers (96) have taken advantage of the hybridization properties of oligonucleotides in developing a methodology for the identification and isolation of specific cloned DNA sequence. The general approach is to chemically synthesize a mixture of oligonucleotides that represent all possible codon combinations for a small portion of the amino acid sequence of a given protein. Within this mixture must be one sequence complementary to the DNA coding for that part of the protein. This complementary oligonucleotide will form a perfectly base-paired duplex with the DNA from the coding region for the protein, whereas the other oligonucleotides in the mixture will form mismatched duplexes. Under stringent hybridization conditions only the perfectly matched duplex will form, allowing the use of the mixture of oligonucleotides as a specific hybridization probe.

Mixed sequence oligonucleotide probes have been used successfully for the isolation of a cloned cDNA encoding human β2-microglobulin (98), as well as a cloned cDNA encoding the mouse H-2K^b transplantation antigen (99; see Table 13.1 for comparison). An example of the use of a synthetic hexadecanucleotide mixture as a probe for the identification of the mouse H-2K^b cDNA clone is shown in Figure 13.11.

There appears to be a significant advantage in using oligonucleotides as probes directly rather than using them as primers for radiolabeled cDNA probe synthesis. First, greater specificity can be obtained using the hybridization approach than using the priming approach. Under appropriate conditions, a mismatched base pair does not allow the formation of oligonucleotide:polynucleotide duplexes (96, 97), whereas base-pair mismatches can be tolerated in, and are often required for, the priming method (82, 84). Second, with the priming approach, the amount of probe obtained is dependent on the amount of mRNA available for template. In isolating cloned sequences for very-low-abundance mRNAs, use of the primer approach would entail the isolation of large amounts of mRNA to produce sufficient cDNA probe for screening. In addition, it may be increasingly more difficult to produce a specific cDNA probe as the abundance of a mRNA decreases. In an examination of Table 13.1, it can be seen that oligonucleotide probes have permitted the detection of cloned DNAs present at one-tenth the abundance yet achieved with the priming approach.

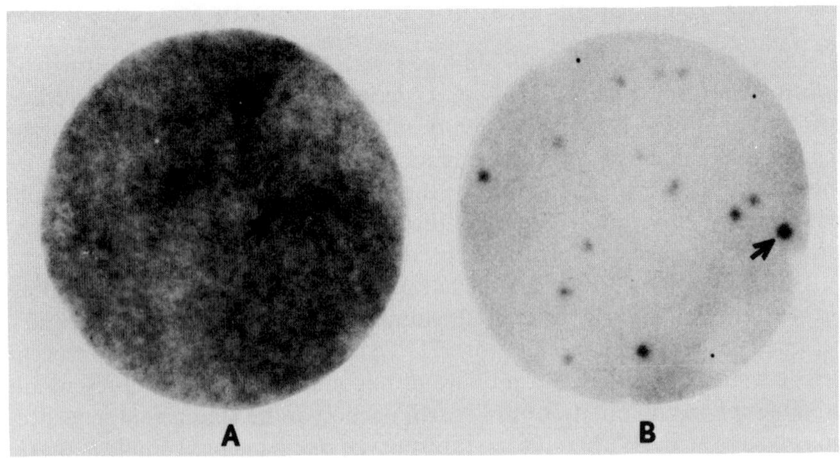

Fig. 13.11. Oligonucleotide of mixed sequence as hybridization probes. A mixture of eight hexadecanucleotide sequences (CC$\genfrac{}{}{0pt}{}{C}{T}TC\genfrac{}{}{0pt}{}{C}{T}TG\genfrac{}{}{0pt}{}{C}{T}$TCCATCC) was synthesized by the solid phase method. This probe was designed to screen for cDNA clones encoding the amino acid sequence Trp-Met-Glu-Gln-Glu-Gly of the mouse transplantation antigen H-2K^b. (*A*) 30,000 independent cDNA clones were screened with the ^{32}P-labeled hexadecanucleotide mixture. One clone (*arrow*) was found to hybridize with the probe. (*B*) The colony found to hybridize in *A* was rescreened with the probe and found to rehybridize. All of the colonies that hybridize here are identical. One clone (*arrow*) has been completely characterized and is described elsewhere (99).

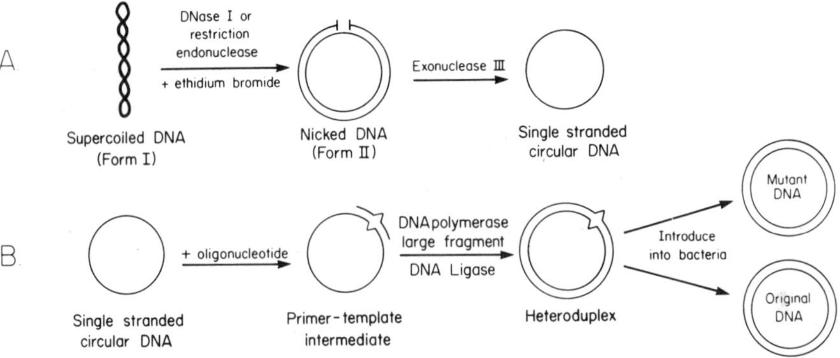

Fig. 13.12. Oligonucleotide-directed mutagenesis. (*A*) A procedure for preparing single-stranded circular DNA from plasmid DNA is described. (*B*) Single-stranded circular DNA either from single stranded DNA phage or from plasmid DNA (see *A*) is used as the template for oligonucleotide-directed mutagenesis as described in the text. The oligonucleotide primer used for the mutagenesis contains fewer nucleotides (deletion), more nucleotides (insertions), or the same number of nucleotides (point mutations) as the complementary region on the template.

13.2.4. Oligonucleotides as Site-Specific Mutagens

Synthetic oligonucleotides provide the most precise and versatile way to introduce specific mutations into DNA. In this technique, an oligonucleotide that contains the desired mutation is synthesized (by point mutation, deletion, or insertion). The oligonucleotide is then used as a primer to direct the synthesis of the complementary strand on a single-stranded circular DNA template (see Figure 13.12b). Specific transition (100–103) and transversion (104, 105) point mutations have been engineered using single-stranded circular DNA isolated from ϕX174 or fd bacteriophage. Wallace and coworkers (106, 107) have developed a technique for introducing mutations (deletions and point mutations) in plasmid DNA. In this technique, single-stranded circular DNA is prepared from covalently closed circular double-stranded plasmid DNA by nicking the DNA with either a restriction endonuclease or DNAse I in the presence of ethidium bromide (108), followed by treatment of the nicked DNA with exonuclease III (see Figure 13.12A). The single-stranded circular DNA then serves as a template for oligonucleotide-directed mutagenesis (Figure 13.12B).

The greatest obstacle to the general application of oligonucleotide-directed mutagenesis is in identifying the desired mutant DNA from a background of unaltered DNA. This is particularly true for cloned DNAs, which do not have an assayable phenotype in *E. coli.* Wasylyk et al. (105) used the fact that mutagenesis of the TATA box of the cloned covalbumin gene (TCTATA–TCTAGA) formed a new *Xba*I recognition site. Screening for the desired mutant clones involved inspection of individual clones for the presence of the *Xba*I site. Most mutageneses will not create or destroy restriction sites.

Wallace and coworkers (106, 107) have described a general procedure not only for introducing specific mutation (point mutation or deletion) in plasmid DNAs using oligonucleotides, but also for identifying the desired mutant DNAs that might be produced at a low frequency. In the latter case, the hybridization specificity of the oligonucleotide is used to identify the desired mutant. Since mismatched oligonucleotide:polynucleotide duplexes will not form under conditions where perfectly matched duplexes will (96, 97), specific mutations can be identified. Figure 13.13 shows an example of the production and identification of a specific A-T to T-A transversion mutation.

Zarucki-Shultz et al. (109) have extended this technique to the introduction of multiple mutations. Since it is often desirable to produce point mutations at more than one site in a DNA molecule, oligonucleotides of mixed sequence were used to direct multiple changes in a cloned ovalbumin gene. Oligonucleotide hybridization was used to identify all mutant DNAs, and DNA sequencing was performed to sort out the various mutants.

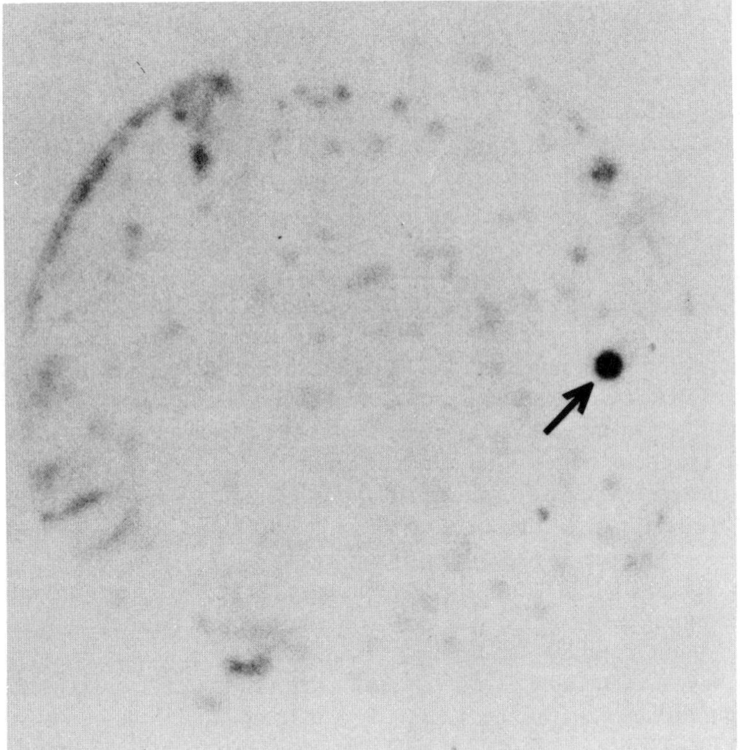

Fig. 13.13. The nonadecanucleotide CTCCTGTGGAGAAGTCTGC was synthesized by the solid phase method and used to direct an A-T to T-A transversion mutation in the human β-globin gene cloned in the plasmid pBR322 (111). After in vitro mutagenesis reactions the DNA was used to transform *E. coli*. The nonadecanucleotide was ^{32}P-labeled and used as a hybridization probe to screen the resultant colonies. Under the conditions of hybridization used, mismatched duplexes do not form, and mutant colonies are identified. In this particular screening, one colony out of the 125 screened hybridized to the probe. The DNA from this colony has been fully characterized and is described elsewhere (107).

It seems clear that oligonucleotide-directed mutagenesis will represent a major application of synthetic DNA in the future. This technique provides a way of changing the amino acid sequence specified by particular cloned genes, thus making it possible to produce various protein analogs. Introduction of large deletions and insertions will make possible the modification of particular cloned DNAs to affect their gene expression. Mutagenesis of cloned DNAs will also play a major role in understanding the function of various DNA sequences in controlling gene expression, post-transcriptional processing, and even the function of the gene product itself.

13.3. SUMMARY

Over the past five years, synthetic oligonucleotides have become indispensable reagents for a multitude of applications in molecular biology. There are sure to be applications of synthetic DNA that we have not yet thought of. Other applications have not yet proved themselves. For example, the ability to discriminate between perfectly matched and mismatched oligonucleotide: polynucleotide duplexes promises to provide a reliable assay to screen human genetic diseases that are due to specific point mutations. It is certain that the need for synthetic sequences will increase tremendously over the next few years. Fortunately, this need can be met by introducing the techniques of DNA synthesis into the biochemical laboratory. That this will be possible is due mainly to the advent of solid phase synthesis and the commercial availability of all of the starting materials. Automation of DNA synthesis will also be important in increasing the supply of oligonucleotides of defined sequence.

REFERENCES

1. K. Itakura, T. Hirose, R. Crea, A. D. Riggs, H. L. Heyneker, F. Bolivar, and H. W. Boyer, *Science* **198,** 1056 (1979).
2. R. Crea, A. Kraszewski, T. Hirose, and K. Itakura, *Proc. Nat. Acad. Sci. USA,* **76,** 106 (1978).
3. D. V. Goeddel, D. G. Kleid, R. Bolivar, H. L. Heyneker, D. G. Yansura, R. Crea, T. Hirose, A. Kraszewski, K. Itakura, and A. D. Riggs, *Proc. Nat. Acad. Sci. USA,* **76,** 106 (1979).
4. D. V. Goeddel, H. L. Heyneker, T. Hozumi, R. Arentzen, K. Itakura, D. G. Yansura, M. J. Ross, G. Miozzari, R. Crea, and P. H. Seeburg, *Nature* **281,** 544 (1979).
5. K. Itakura and A. D. Riggs, *Science* **209,** 1401 (1980).
6. B. W. Erickson and R. B. Merrifield, in *The Protein,* Vol. 2 (H. Neurath, R. L. Hill, and C.-L. Boeder, Eds.), Academic Press, New York (1976), p. 255.
7. H. Kossel and H. Seliger, *Prog. Chem. Org. Nat. Prod.* **32,** 297 (1975).
8. R. Wu, C. P. Bahl, and S. A. Narang, *Prog. Nucleic Acids Res. Mol. Biol.* **21,** 101 (1978).
9. H. Koster, Ed., *Nucleic Acids Res. Symp. Series,* No. 7, 1980.
10. P. Hodge and D. C. Sherrington, Eds., *Polymer-Supported Reactions in Organic Synthesis,* Wiley, New York (1980).
11. Our unpublished results.
12. M. J. Gait and R. C. Sheppard, *Nucleic Acids Res.* **4,** 1135 (1976).
13. H. G. Khorana, *Science* **203,** 614 (1979).
14. S. S. Jones, B. Rayner, C. B. Reese, A. Ubasawa, and M. Ubasawa, *Tetrahedron* **36,** 3075 (1980).
15. M. J. Gait and S. G. Popov, *Tetrahedron Lett.* **21,** 2265 (1980).

16. R. L. Letsinger, J. L. Finnan, G. A. Heavner, and W. B. Lunsford, *J. Amer. Chem. Soc.* **97,** 3278 (1975).
17. K. K. Ogilvie and M. J. Nemer, *Tetrahedron Lett.* **21,** 4159 (1980).
18. M. D. Matteucci and M. H. Caruthers, *Tetrahedron Lett.* **21,** 719 (1980).
19. F. Chow, T. Kempe, and G. Palm, *Nucleic Acids Res.* **9,** 2807 (1981).
20. K. Miyoshi and K. Itakura, *Nucleic Acids Res. Symp. Series,* No. 7 (1980), p. 281.
21. P. Dembek, K. Miyoshi, and K. Itakura, *J. Amer. Chem. Soc.* **106,** 706 (1980); and our recent unpublished results.
22. T. Hirose, R. Crea, and K. Itakura, *Tetrahedron Lett.,* 2449 (1978).
23. G. R. Gough, K. J. Collier, H. L. Weith, and P. T. Gilham, *Nucleic Acids Res.* **7,** 1955 (1979).
24. H. Koster and F. Cramer, *Justus Liebigs Ann. Chem.* **766,** 6 (1972).
25. H. Hayatsu and H. G. Khorana, *J. Amer. Chem. Soc.* **89,** 3880 (1966).
26. L. R. Melby and D. R. Strobach, *J. Amer. Chem. Soc.* **89,** 450 (1967).
27. G. M. Blackburn, M. J. Brown, and M. R. Harris, *J. Amer. Chem. Soc.* **89,** 2438 (1967).
28. H. Koster and F. Cramer, *Justus Liebigs Ann. Chem.* **1974,** 946 (1974).
29. H. Koster, *Tetrahedron Lett.,* 1527 (1972).
30. V. K. Potapov, V. P. Veiko, O. N. Koroleva, and Z. A. Shavarova, *Nucleic Acids Res.* **6,** 648 (1979).
31. K. Miyoshi, R. Arentzen, T. Huang, and K. Itakura, *Nucleic Acids Res.* **8,** 5507 (1980).
32. Our unpublished results.
33. H. A. J. Battaerd and G. W. Tregear, in *Graft Copolymers,* Wiley-Interscience, New York (1967).
34. R. C. Sheppard, in *Peptides* (H. Nesvadba, Ed.), North-Holland, Amsterdam (1971), p. 111.
35. M. J. Gait, M. Singh, and R. C. Sheppard, *Nucleic Acids Res.* **8,** 1081 (1980).
36. K. Miyoshi, T. Huang, and K. Itakura, *Nucleic Acids Res.* **8,** 5491 (1980).
37. K. E. Norris, F. Norris, and K. Brunfeld, *Nculeic Acids Res. Symp. Series,* No. 7 (1980), p. 233.
38. M. L. Duckworth, M. J. Gait, P. Goelet, G. H. Hong, M. Singh, and R. C. Titmas, *Nucleic Acids Res.* **9,** 1691 (1981).
39. R. Crea and T. Horn, *Nucleic Acids Res.* **8,** 2331 (1980).
40. T. Horn, M. P. Vasser, M. E. Struble, and R. Crea, *Nucleic Acids Res. Symp. Series,* No. 7 (1980), p. 225.
41. K. Miyoshi, T. Miyake, T. Hozumi, and K. Itakura, *Nucleic Acids Res.* **8,** 5473 (1980).
42. M. J. Gait, S. G. Popov, M. Singh, and R. C. Titmas, *Nucleic Acids Res. Symp. Series,* No. 7 (1980), p. 243.
43. Our unpublished results.
44. M. D. Matteucci and M. H. Caruthers, *Tetrahedron Lett.* **21,** 3243 (1980).
45. V. Kohli, H. Blocker, and H. Koster, *Tetrahedron Lett.* **21,** 2683 (1980).
46. C. B. Reese, *Tetrahedron* **34,** 3143 (1978).

47. W. T. Markiewicz, E. Biala, R. W. Adamiak, K. Grzeskomiak, R. Kierzek, A. Kraszewski, J. Stawinski, and M. Wiewiorowski, *Nucleic Acids Res. Symp. Series,* No. 7 (1980), p. 115.
48. W. Pfleiderer, E. Uhlmann, R. Charubala, D. Flockerzi, G. Silber, and R. S. Varma, *Nucleic Acids Res. Symp. Series,* No. 7, (1980), p. 61.
49. C. B. Reese, R. C. Titmas, L. Yau, *Tetrahedron Lett.,* 2727 (1978).
50. C. B. Reese and C. Satthill, *Chem. Commun.,* 767 (1968).
51. K. Itakura, N. Katagiri, C. P. Bahl, R. H. Wightman, and S. A. Narang, *J. Amer. Chem. Soc.* **97,** 7237 (1975).
52. C. B. Reese and A. Ubasawa, *Nucleic Acids Res. Symp. Series,* No. 7 (1980), p. 5.
53. R. V. Tomlinson and G. M. Tanner, *Biochemistry* **2,** 697 (1963).
54. J. H. van Boom and J. de Rooj, *J. Chromatography* **131,** 169 (1977).
55. A. Maxam and W. Gilbert, *Proc. Nat. Acad. Sci. U. S.* **74,** 560 (1977).
56. A. F. Markham, M. D. Edge, T. C. Atkinson, A. R. Greene, G. R. Heathcliffe, C. R. Newton, and D. Scanlon, *Nucleic Acids Res.* **8,** 5193 (1980).
57. J. Rossi, W. Ross, T. Egan, D. Lipman, and A. Landy, *J. Mol. Biol.* **128,** 21 (1979).
58. R. C. Pless and R. L. Letsinger, *Nucleic Acids Res.* **2,** 743 (1975).
59. R. H. Scheller, R. E. Dickerson, H. W. Boyer, A. D. Riggs, and K. Itakura, *Science* **196,** 177 (1977).
60. C. P. Bahl, K. J. Marians, R. Wu, J. Stawinsky, and S. A. Narang, *Gene* **1,** 81 (1976).
61. C. P. Bahl, R. Wu, R. Brousseau, A. K. Sood, H. M. Hsiung, and S. A. Narang, *Biochem. Biophys. Res. Commun.* **81,** 695 (1978).
62. A. H. J. Wang, G. J. Quigley, F. J. Kolpa, J. L. Crawford, J. H. van Boom, G. van der Marel, and A. Rich, *Nature* **282,** 680 (1979).
63. H. Drew, T. Takano, S. Tanaka, K. Itakura, and R. E. Dickerson, *Nature* **286,** 567 (1980).
64. K. L. Agarwal, H. Buchi, M. H. Caruthers, N. Gupta, H. G. Khorana, K. Kleppe, A. Kumar, E. Ohtsuka, U. L. Rajibhandary, J. H. van de Sande, V. Sgaramella, H. Weber, and T. Yamada, *Nature* **227,** 27 (1970).
65. H. Koster, H. Blocker, R. Frank, S. Geussenhainer, W. Kaiser, *Hoppe-Seyler's Z. Physiol. Chem.* **356,** 1585 (1975).
66. A. V. Chestukhin, G. M. Dolganov, M. F. Shemyakin, E. M. Khodkova, G. S. Monastyrskaya, and E. D. Sverdlov, *Nucleic Acids Res.* **8,** 6163 (1980).
67. M. D. Edge, A. R. Greene, G. R. Heathcliffe, P. A. Meacok, W. Schuch, D. B. Scanlon, T. C. Atkinson, C. R. Newton, and A. F. Markham, *Nature* **292,** 756 (1981).
68. X. Soberon, J. J. Rossi, G. P. Larson, and K. Itakura, *Nucleic Acids Res. Symp. Series,* No. 9, in press (1981).
69. R. Wu, *Nature New Biol.* **236,** 198 (1972).
70. R. Padmanabhan, R. Padmanabhan, and R. Wu, *Biochem. Biophys. Res. Commun.* **48,** 1295 (1972).
71. R. Wu, C. D. Tu, and R. Padmanabhan, *Biochem. Biophys. Res. Commun.,* **55,** 1092 (1973).

72. R. Padmanabhan, *Biochemistry* **16,** 1996 (1977).
73. J. Taylor, R. Illmensee, and J. Summers, *Biochim. Biophys. Acta* **442,** 324 (1976).
74. A. Berns and R. Jaenisch, *Proc. Nat. Acad. Sci. USA,* **73,** 2448 (1976).
75. M. Smith, D. W. Leung, S. Gillam, C. R. Astell, D. L. Montgomery, and B. D. Hall, *Cell* **16,** 753 (1979).
76. D. Zimmern and P. Kaesberg, *Proc. Nat. Acad. Sci. USA,* **75,** 4257 (1978).
77. F. Sanger, S. Nicklen, and A. R. Coulson, *Proc. Nat. Acad. Sci. USA,* **74,** 5463 (1977).
78. J. Messing, R. Crea, and P. H. Seeburg, *Nucleic Acids Res.* **9,** 309 (1981).
79. R. B. Wallace, M. J. Johnson, S. V. Suggs, K. Miyoshi, R. Bhatt, and K. Itakura, *Gene,* in press (1981).
80. B. E. Noyes, M. Mevarech, R. Stein, and K. L. Agarwal, *Proc. Nat. Acad. Sci. USA,* **76,** 1770 (1979).
81. M. Mevarech, B. E. Noyes, and K. L. Agarwal, *J. Biol. Chem.* **254,** 7472 (1972).
82. S. J. Chan, B. E. Noyes, K. L. Agarwal, and D. F. Steiner, *Proc. Nat. Acad. Sci. USA.* **79,** 5036 (1979).
83. K. L. Agarwal, J. Brunstedt, and B. E. Noyes, *J. Biol. Chem.* **256,** 1023 (1981).
84. M. Houghton, A. G. Stewart, S. M. Doel, J. S. Emtage, M. A. W. Eaton, J. C. Smith, T. P. Patel, H. M. Lewis, A. G. Porter, J. R. Birch, T. Cartwright, and N. H. Carey, *Nucleic Acids Res.* **8,** 1913 (1980).
85. R. Grantham, C. Gautier, M. Gouy, M. Jacobzone, and R. Mercier, *Nucleic Acids Res.* **9,** 43 (1981).
86. O. C. Uhlenbeck, F. H. Martin, and P. Doty, *J. Mol. Biol.* **57,** 217 (1971).
87. S. Gillam, K. Waterman, and M. Smith, *Nucleic Acids Res.* **2,** 625 (1975).
88. S. H. Chang, A. Majundar, R. Dunn, O. Makabe, U. L. Rajbhandary, H. G. Khorana, E. Ohtsuka, T. Tanaka, V. O. Taniyama, and M. Ikehara, *Proc. Nat. Acad. Sci. USA,* **78,** 3398 (1981).
89. D. V. Goeddel, H. M. Shepard, E. Velverton, D. Leung, R. Crea, A. Sloma, and S. Pestka, *Nucleic Acids Res.* **8,** 4057 (1980).
90. D. V. Goeddel, E. Velverton, A. Ullrich, H. L. Heyneker, G. Miozzari, W. Holmes, P. H. Seeburg, T. Dull, L. May, N. Stebbing, R. Crea, S. Maeda, R. McCandliss, A. Sloma, J. M. Tabor, M. Gross, P. C. Familletti, and S. Pestka, *Nature* **287,** 411 (1980).
91. G. Winter, S. Fields, M. J. Gait, and G. G. Brownlee, *Nucleic Acids Res.* **9,** 237 (1981).
92. J. W. Stewart and F. Sherman, in *Molecular and Environmental Aspects of Mutagenesis* (L. Pradash, F. Sherman, M. W. Miller, C. W. Lawrence, and H. W. Taber, Eds.), Thomas, Springfield (1974), pp. 102–107.
93. J. W. Szostak, J. I. Stiles, C. P. Bahl, and R. Wu, *Nature* **265,** 61 (1977).
94. D. L. Montgomery, B. D. Hall, S. Gillam, and M. Smith, *Cell* **14,** 673 (1978).
95. J. W. Szostak, J. I. Stiles, B.-K. Tye, P. Chiu, F. Sherman, and R. Wu, *Methods in Enzymology* **68,** 419 (1979).
96. R. B. Wallace, J. Shaffer, R. F. Murphy, J. Bonner, T. Hirose, and K. Itakura, *Nucleic Acids Res.* **6,** 3543 (1979).

97. R. B. Wallace, M. J. Johnson, T. Hirose, T. Miyake, E. H. Kawashima, and K. Itakura, *Nucleic Acids Res.* **9,** 879 (1981).
98. S. V. Suggs, R. B. Wallace, T. Hirose, E. H. Kawashima, and K. Itakura, *Proc. Nat. Acad. Sci. USA,* in press (1981).
99. A. A. Reyes, M. J. Johnson, M. Schold, H. Ito, Y. Ike, C. Morin, K. Itakura, and R. B. Wallace, *Immunogenetics,* in press (1981).
100. A. Razin, T. Hirose, K. Itakura, and A. D. Riggs, *Proc. Nat. Acad. Sci. USA,* **75,** 4268 (1978).
101. C. A. Hutchinson, S. Phillips, M. H. Edgell, S. Gillam, P. Jahnke, and M. Smith, *J. Bio. Chem.* **253,** 6551 (1978).
102. S. Gillam and M. Smith, *Gene* **8,** 81 (1979).
103. S. Gillam and M. Smith, *Gene* **8,** 99 (1979).
104. S. Gillam, P. Jahnke, C. Astell, S. Phillips, C. A. Hutchinson, and M. Smith, *Nucleic Acids Res.* **6,** 2973 (1979).
105. B. Wasylyk, R. Derbyshire, A. Guy, D. Molko, A. Roget, R. Teoule, and P. Chambon, *Proc. Nat. Acad. Sci. USA,* **77,** 7024 (1980).
106. R. B. Wallace, P. F. Johnson, S. Tanaka, M. Schold, K. Itakura, and J. Abelson, *Science* **209,** 1396 (1980).
107. R. B. Wallace, M. Schold, M. J. Johnson, P. Dembek, and K. Itakura, *Nucleic Acids Res.* **9,** 3647 (1981).
108. D. Shortle and D. Nathans, *Proc. Nat. Acad. Sci. USA,* **75,** 2170 (1978).
109. T. Zarucki-Shultz, S. Y. Tsai, K. Itakura, X. Soberon, R. B. Wallace, M. J. Tsai, S. L. C. Woo, and B. W. O'Malley, submitted for publication *J. Biol. Chem.* (1981).
110. A. K. Sood, D. Pereira, and S. M. Weissman, *Proc. Nat. Acad. Sci. USA* **78,** 616 (1981).
111. R. M. Lawn, A. Efstratiadis, C. O'Connell, and T. Maniatis, *Cell* **21,** 647, (1980).

CHAPTER

14

IMMOBILIZED PROTEIN MODIFICATION REAGENTS

CATHERINE LEWIS and WILLIAM H. SCOUTEN

Chemistry Department
Bucknell University
Lewisburg, PA 17837

The chemical modification of amino acids is one of the most useful approaches to the study of protein molecules. A knowledge of the primary sequence and of the higher-order folding dictated by that sequence does not help one to predict the specificity of a protein from its structure. But by chemically altering specific amino acid side-chains within the protein, one can gain much information about the roles of those residues in the biological properties of a protein. Chemicals for protein modification are used to determine the number of residues of a given type or to alter the enzymatic or conformational properties of a protein by modifying residues at the active site or at some key structural position. The catalytic mechanisms and binding sites of many enzymes, as well as the three-dimensional positional effects of some specific amino acids, have been elucidated in this way (1). Of the many reagents in use, the most successful are those that can be applied without side effect to the protein. Alteration of the protein in a nonspecific way, loss of conformation, or any degree of denaturation must be avoided if the derivatization is to be assessed accurately.

Since the 1950s many reagents have been developed and applied to the modification of amino acids with reactive side-chains. These reagents traditionally are used free in solution. Separation of the reagent from the protein is often laborious and time-consuming and may result in further derivatization. The immobilization of reagents on insoluble supports facilitates the separation process, since only simple filtration is required. More important is the potential of immobilized reagents as topographic probes. The number of residues modified by a specific reagent will depend on how accessible each residue is to the solvent. If the residue is buried in the interior of the protein it is unlikely to be modified unless the protein is unfolded or denatured. Denaturation followed by modification is commonly used. Although it is possible to categorize residues as either buried or on the surface, the distance of buried residues from the surface cannot be measured in this way. If the reagent is immobilized, however, the length of the spacer arm attaching the reagent to the solid support can be used to measure this distance. If there are hydrophobic groups on the arm, the immobilized group is more likely to reach the protein's interior, which is usually hydrophobic and shielded from the aqueous medium.

14.1. DERIVATIZATION OF METHIONINE, TRYPTOPHAN, AND CYSTEINE BY COVALENT LINKAGE TO AN INSOLUBLE SUPPORT

Immobilized reagents were often developed in the process of exploring affinity chromatography techniques for purifying biological molecules. If a protein is retained on a column through a specific covalent linkage and then released in an altered form with only certain residues derivatized, then chemical modification using an immobilized reagent has been achieved. Shechter et al. (2) selectively modified methionine on derivatized acrylamide with such a technique. Treatment of a mixture of free amino acids with chloroacetamidoethyl polyacrylamide (P~$NHCOCH_2Cl$) in 0.1 *M* acetic acid containing 0.1% NaI for 12 days at 37°C resulted in selective and complete removal of methionine from the amino acid mixture. A series of peptides treated under the same conditions resulted in selective removal of only those peptides that contained methionine.

The release of methionyl peptides from the matrix was achieved in several ways. Free methionine or peptides in which the amino group of methionine was involved in the binding, for example P~gly-met, were released as homoserine under incubation conditions without further treatment, but polymer-bound methionyl peptides in which the carboxylic group participated in a peptide bond, for example P~met-val, were not released. To release these other peptides from the matrix, two different methods were employed. Incubation of the polymer for 2 hr at 100°C resulted in conversion of the bound methionine to homoserine and cleavage and liberation of homoserine and free amino acids. Alternatively, the polymer was treated with 2-mercaptoethanol in 0.1 *M* NH_4HCO_3, and methionyl peptides covalently bound to the polymeric support were released quantitatively as the intact peptides, with no cleavage of the peptides and no formation of homoserine. These methods are outlined in Figure 14.1.

Prolonged incubation of several enzymes and proteins with the polymer under the same conditions indicated that the presence of methionine on the surface of the protein was necessary for binding of the protein to the polymer. There was no reaction between the polymer and lysozyme in which the two methionine residues are known to be buried. Proteins in which the me-

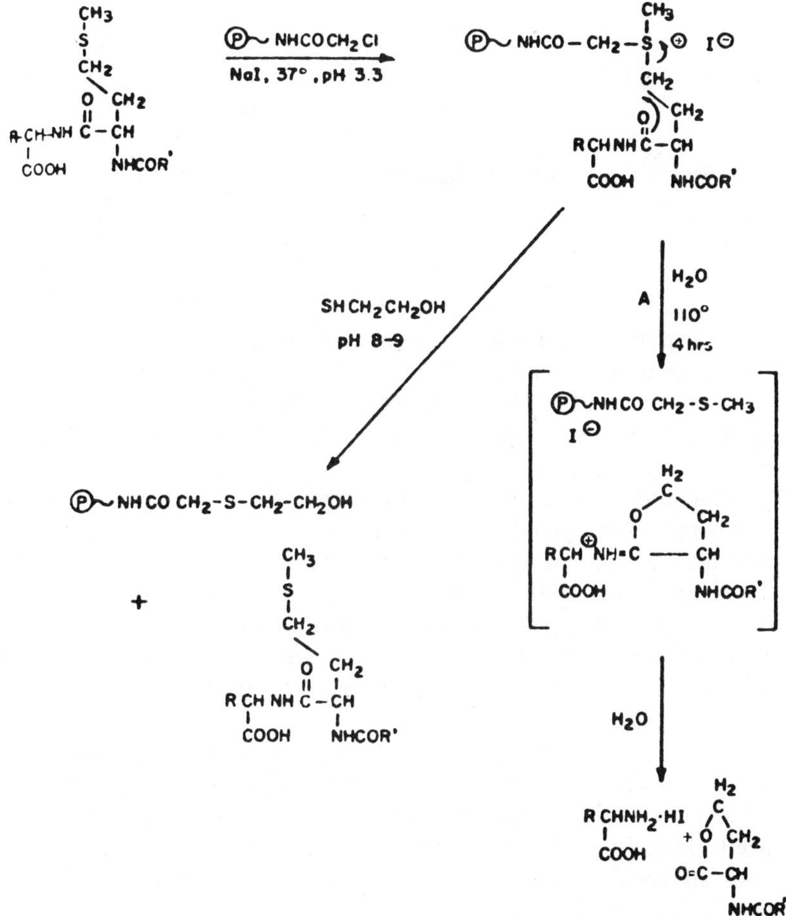

Fig. 14.1. Modification of methionine using chloracetamidoethyl polyacrylamide. Reprinted with permission from Shechter et al., *Biochemistry* **16,** 1424 (1977). Copyright 1977, American Chemical Society (2).

thionine does not affect biological activity showed 40 to 100% of their native activity while bound to the matrix. Binding of the enzyme to the polymer reduced ribonuclease activity to 12%, indicating that the methionines are necessary for full activity.

Although the conditions required in this experiment for release of the modified protein were harsh, that is, boiling of the protein, the specific covalent linkage of methionine to the polymer is a useful technique. Not only does it offer an alternative method for isolating methionyl-containing peptides, but it results in an effective blocking of methionine residues while

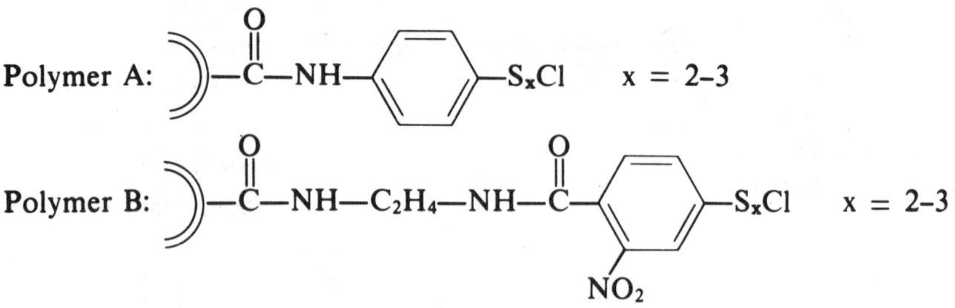

Polymer A: $\rangle\!\!\rangle$—$\overset{O}{\overset{\|}{C}}$—NH—$C_6H_4$—$S_xCl$ x = 2–3

Polymer B: $\rangle\!\!\rangle$—$\overset{O}{\overset{\|}{C}}$—NH—$C_2H_4$—NH—$\overset{O}{\overset{\|}{C}}$—$C_6H_3(NO_2)$—$S_xCl$ x = 2–3

bound to the polymer. But a loss in activity as great as 60% in enzymes that are known not to require methionine for catalysis indicates that some structural changes occurred during the binding and that this method has induced undesired conformational changes as a side effect of the binding and blocking of methionyl residues.

A similar study was carried out using the selective binding of tryptophan to arylsulfenyl polyacrylamides (3). When mixtures of amino acids, excluding cysteine, were reacted with polymers A or B in 2% acetic acid for 30 min at room temperature, all the amino acids except tryptophan were recovered in 100 ± 3% yield. The polymer was then washed with 10% acetic acid and suspended in a stripping buffer, 0.05 *M* dithiothreitol–0.1 *M* Tris, pH 8.5, for 30 min, TLC of the supernatant showed the presence of 2-thiotryptophan and no tryptophan.

Reaction of polymer B with the peptide trp-ala in 10% acetic acid in the presence of all other amino acids except tryptophan and cysteine showed quantitative removal of the peptide. The polymer was then washed with excess acetic acid, suspended in the stripping buffer, and hydrolyzed in acid. The only amino acid found was alanine since tryptophan is known to be completely destroyed by acid hydrolysis.

Human serum albumin was digested with trypsin, lyophilized, and dissolved in 98–100% formic acid. After reaction with polymer A for 1 hr, the polymer was filtered off and washed in excess formic acid and 8 *M* urea. The polymer was then suspended in stripping buffer, acid hydrolyzed, and analyzed for its amino acid content by electrophoresis. The amino acids found were arginine 1.05%, valine 0.9%, and alanine 3%. The theoretical yield from the one tryptic peptide of human serum albumin that contains tryptophan is arginine 1%, valine 1%, alanine 3%, and tryptophan 1%. The modification of tryptophan following immobilization on the derivatized polyacrylamide is shown in Figure 14.2.

The results of this study demonstrate that it is possible to achieve selective binding of tryptophan to an insoluble matrix. Since tryptophan generally occurs in relatively low abundance in proteins, the method is useful as a one-step isolation procedure for peptides that contain tryptophan. Furthermore, the

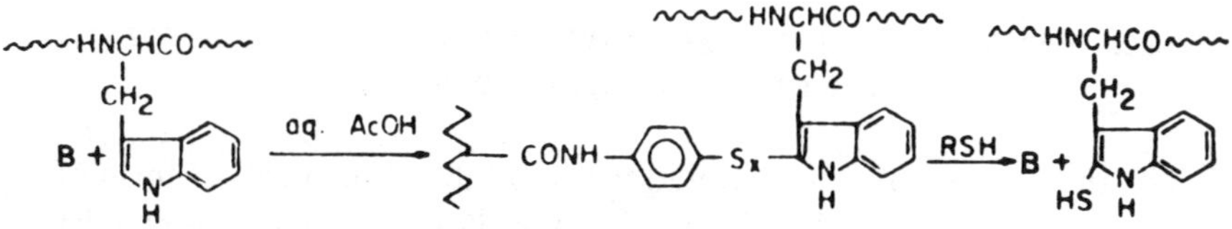

Fig. 14.2. Modification of tryptophan by selective binding to arylsulfanyl polyacrylamides. Reprinted with permission from ref. 3.

polymer was shown to selectively modify tryptophan. However, nonspecific adsorption of peptides occurred, and washing with formic acid and 8 *M* urea was necessary to remove them. Although no reaction of intact proteins with the matrix was reported, it is evident that nonspecific adsorption would occur with these also and that washing with urea would denature them.

Singh et al. (4) have reported the preparation of an immobilized maleimido group, shown in structure **1**, which was used to couple thiol-containing compounds to agarose via cleavable phenyl ester linkages. The derivatized gel had the capacity to bind 2.3 μmol of glutathione/mL gel, and the coupling was demonstrated to be 90% covalent. For proteins, however, nonspecific adsorption and reaction of imidazolyl and amino groups with the gel were observed. Blocking of the thiol groups in the protein resulted in a fivefold to twentyfold decrease in binding for bovine serum albumin (BSA), hemoglobin, and glyceraldehyde 3-phosphate dehydrogenase (GAPDH), indicating a residual binding not due to coupling of sulfhydryls to the gel. Cleavage of the modified maleimido proteins from the agarose by treatment with 1 *M* hydroxylamine resulted in 100% release of BSA but less than 50% release of hemoglobin and GAPDH. It was necessary to wash the gel with 2 *M* guanidine hydrochloride to remove the noncovalently bound hemoglobin and GAPDH that remained.

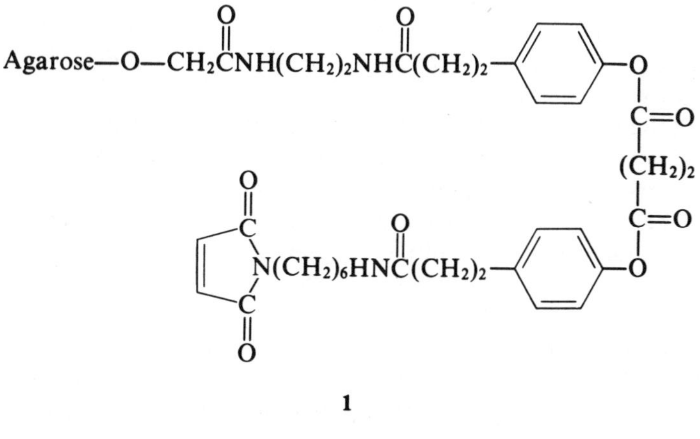

Agarose—O—$CH_2\overset{O}{\overset{\|}{C}}NH(CH_2)_2NH\overset{O}{\overset{\|}{C}}(CH_2)_2$—$C_6H_4$—O—C(=O)—$(CH_2)_2$—C(=O)—O—$C_6H_4$—$(CH_2)_2\overset{O}{\overset{\|}{C}}NH(CH_2)_6N$(maleimide)

1

14.2. ENHANCED INSULIN ACTIVITY FOLLOWING IMMOBILIZATION ON CNBr-ACTIVATED SEPHAROSE

Oka and Topper (5) demonstrated that a soluble "superactive" form of insulin could be prepared by immobilization. Porcine insulin was coupled to Sepharose that had been activated with cyanogen bromide at pH 9.0. Insulin-like material was extracted from the insulin–Sepharose by suspending the immobilized hormone in a sterile medium containing BSA at 37°C for 24 hr at pH 7.5. The suspension was centrifuged and the supernatant filtered. The insulinlike material prepared in this way was five times more effective in stimulating the accumulation of α-aminoisobutyric acid in midpregnancy mouse mammary gland tissue. It also enhanced the normal DNA synthesis in these tissues and the glucose-6-phosphate dehydrogenase and phosphogluconate dehydrogenase activities in mammary epithelial cells that are normally stimulated by insulin. The extracted material was shown to be insulinlike by the elimination of its activity by antibody against insulin and by blockage of its effect by native insulin on insulin-insensitive mammary tissue from virgin mice.

The effect of superinsulin on other tissues has been investigated. Diaphragm tissue from obese mice is known to be unresponsive to insulin, but its response to superinsulin is almost as great as diaphragm tissue from lean mice, which does respond to native insulin (6). These effects were measured by the oxidation of (1-^{14}C)-glucose to $^{14}CO_2$. In contrast, tissue from lean mice showed the same activity with insulin and superinsulin.

The form of superinsulin released from Sepharose is believed to be a substituted guanidine (7). The series of steps leading to the derivatized insulin is shown in Figure 14.3. Steps *A* and *B* are the CNBr-activation of Sepharose and coupling of free amino groups in the protein to the Sepharose via CNBr respectively. The product (structure 2 in the figure) is a Sepharose-bound isourea. Alkyl amines have been shown (8) to retain a charge when coupled to Sepharose by this method. Step *C* is the release of superinsulin from the Sepharose by BSA to form the N_1-N_2-disubstituted guanidine (R = insulin, R′ = BSA).

Although the initial preparation of superinsulin (5) suggested that BSA was required as a substituent, more recent experiments have demonstrated that superinsulin can be prepared with ammonium bicarbonate instead of BSA. The product released from the Sepharose would thus be a monosubstituted guanidine (9). The residue or residues whose guanidination leads to the enhanced biological activity has been shown not to be lysine, but as yet it remains unidentified.

The unusual properties of the insulin released from Sepharose and the success of the method have led to an extension of the technique to other hormones. Superactive forms of both placental lactogen and placental pro-

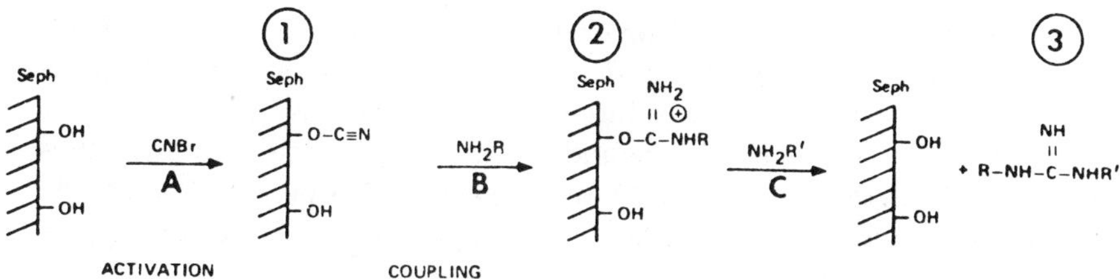

Fig. 14.3. Preparation of "super insulin" by reaction of insulin with CNBr-activated Sepharose. Reprinted with permission from ref. 7.

lactin have been prepared from lactogen-Sepharose and prolactin-Sepharose, using the method of Oka and Topper (10). These superhormones stimulate the synthesis of α-lactalbumin in mammary gland and also the RNA synthesis that is required for α-lactalbumin's synthesis.

14.3. IMMOBILIZED REDUCING AGENTS

Gorecki and Patchornik (11) have reported the use of immobilized dihydrolipoic acid as a powerful reducing agent for disulfide compounds. A common problem in the reduction of disulfide bonds is the ease with which they are reoxidized. Reagents such as mercaptoethanol and thioglycolic acid must be present in large excess, since the equilibrium between disulfides and thiols is close to unity. Dithiothreitol, which forms a sterically favorable dithiolane ring upon oxidation with a low redox potential, is thus superior in that it can maintain the thiol in a reduced state. Removal of the dithiothreitol after reduction often results in reoxidation of the disulfide.

In an effort to make use of the stabilized dithiolane ring and avoid the problems of reoxidation, Gorecki and Patchornik immobilized dihydrolipoic acid on various insoluble supports. Since the redox potential of dihydrolipoic acid (−0.325 V) is lower than that of cysteine (−0.21 V) it could act as a powerful reducing agent for disulfides, as shown in structure **2**.

$$
\begin{array}{lcl}
\text{P}\!-\!\overset{\overset{\displaystyle O}{\|}}{C}\!-\!(CH_2)_4\!-\!\underset{\diagdown S\!-\!\!-\!\!-\!\!-\!S \diagup}{CH\!-\!CH_2\!-\!CH_2} & \nwarrow \; \nearrow & RSH + R'SH \\
 & \big) \big(& \\
\text{P}\!-\!\overset{\overset{\displaystyle O}{\|}}{C}\!-\!(CH_2)_4\!-\!\underset{SH}{\underset{|}{C}}H\!-\!CH_2\!-\!\underset{SH}{\underset{|}{C}}H_2 & \swarrow \; \searrow & RSSR'
\end{array}
$$

2

At pH 7.5–8.5 in 0.05–1.0 *M* buffers, the reduction of disulfides in cysteine, cystamine, oxidized mercaptoethanol, and oxidized glutathione was completed within 10 min. The reaction occurred with simple mixing of the disulfide with polymer suspended in buffer. After reduction, filtration of the polymer left the free thiol in solution. Of the various polymers used, polyacrylamide and Sephadex bound the most lipoic acid and were thus more potent as reducing agents. With the exception of polyacrylamide, the reduced compounds were adsorbed onto the polymers if dilute buffers (less than 0.05 *M*) were used.

This study also reported the successful reduction of a protein without adsorption. Papain, a thiol enzyme that must be activated by thiol reagents before use, showed full activity within a few seconds after incubation with the reducing polymer and its substrate, *N*-α-benzoyl-L-arginine ethyl ester.

The success of immobilized lipoic acid as a biochemical reducing agent is exemplified by the fact that the method has now been patented by Gorecki and Patchornik (12). Using cellulose, Sephadex, derivatives of polyacrylamide, and aminoethyl polystyrene to immobilize the lipoic acid, these investigators have quantitatively reduced many disulfide-containing molecules, including the hormone oxytocin (13). The polymers containing the lipoic acid are reduced with sodium borohydride and are stable within a pH range of 3.0 to 11. They can be used repeatedly without loss of lipoic acid and without degradation of the polymer and can be kept from oxidizing for prolonged periods if stored under nitrogen.

In a similar investigation, Firestone and Scouten (14) used *N*-propyldihydrolipoamide immobilized on porous glass beads as a reducing agent. Oxidized glutathione, at pH 9.5 and under N_2, was reduced within 5 min in 0.1 *M* potassium phosphate buffer containing a tenfold molar excess of reduced lipoamide. Lower ionic strength of the buffer resulted in 60% adsorption of the glutathione to the glass matrix. *N*-propyldihydrolipoamide glass was shown to be entirely effective in activating papain and mercuripapain by reducing and demercurizing them respectively. But it was observed that the enzymes were partially adsorbed to the beads. Since the active site of papain is known to be a hydrophobic trough, it was reasonable to expect a noncovalent attraction between it and the hydrophobic lipoamide moiety. An attempt was made to prevent the adsorption by carrying out the reduction in various concentrations of Triton X 100. Only 50% of the total activity of the papain could be removed by the addition of 1.0% Triton X 100. Further increases in the concentration of Triton X 100 reduced the activity of the enzyme. Similar results were found for chymopapain and bromelain.

14.4. POROUS GLASS BEADS AS AN INSOLUBLE SUPPORT

In spite of the difficulties associated with using glass as a solid support—its friability and the more common problem of protein adsorption—porous glass beads have several advantages. (See also Chapter 11.) Among the most significant are its rigidity and stability in organic solvents. Matrices such as acrylamide shrink and swell, thereby increasing the difficulty of removing materials trapped in the pores. Ionic strength, pressure, flow rates, and bacterial growth are also of less consequence with glass beads.

In addition to their use in the immobilization of *N*-propyldihydrolipoamide discussed above, porous glass beads have been used as an insoluble matrix for acetylimidazole in the acetylation of tyrosines, shown in structure

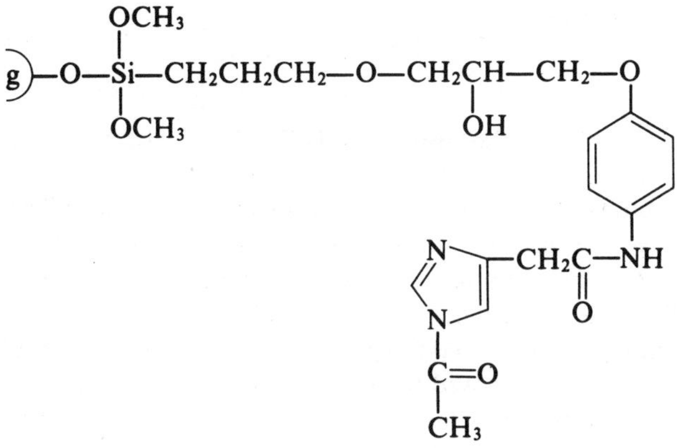

3

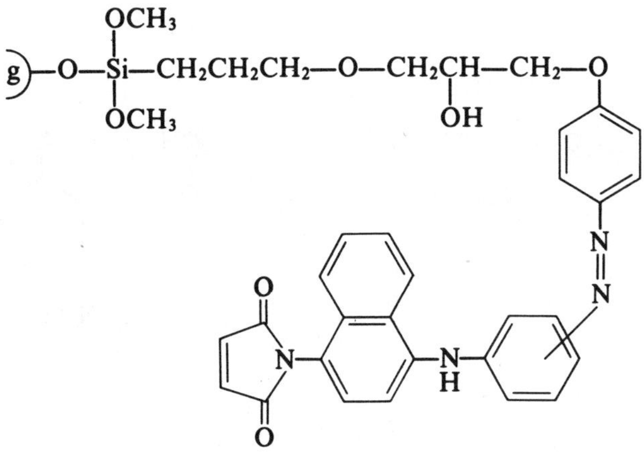

4

3, and for *N*-(1-anilino-naphthyl-4)-maleimide (ANM), a thiol-specific fluorescent label, shown in structure **4** (15). Although protein was not adsorbed to the glass, attempts to acetylate free tyrosine and the tyrosines in insulin were only partly successful, since removal of the acetyl group with hydroxylamine was not achieved following the reaction. Reactions of immobilized ANM with sodium dithionite resulted in the production of two fluorescent derivatives similar to those obtained from the reaction of free ANM with sodium dithionite. Further use was made of the immobilized ANM in showing that the reactive sulfhydryl group of free lipoamide dehydrogenase was buried in the lipoamide dehydrogenase component of the pyruvate dehydrogenase complex (16).

14.5. IMMOBILIZED OXIDIZING AGENTS

Methylene blue, a well-known photocatalyst, was immobilized on glass beads to give the product shown in structure **5** (17). After appropriate washing, the methylene blue beads were reacted with a solution of methionine in 50% acetic acid in the presence of light and air for 6 hr at 37°C. TLC following the reaction showed the presence of both methionine sulfoxide and methionine. Samples exposed to the light in the absence of beads or in the presence of beads but the absence of light showed no methionine sulfoxide. Reduction of the oxidized methionine was achieved with 2-mercaptoethanol in 10% sodium carbonate.

The selective modification of the methionyl residues of lysozyme was carried out using the same beads in 84% acetic acid at 4°C for 26 hr (18). The activity of the enzyme was assayed at various times during the photooxida-

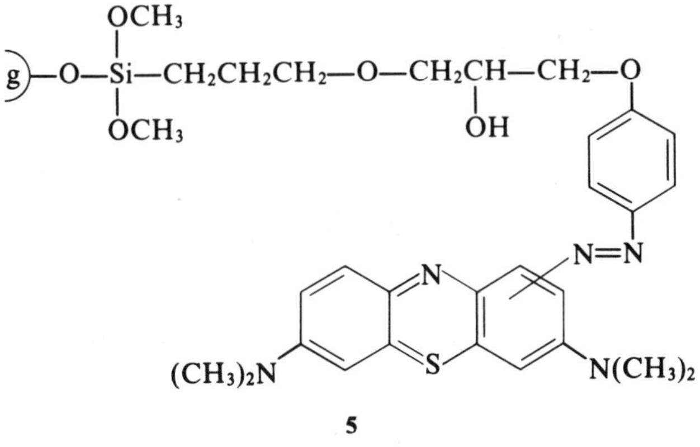

OCH_3

g—O—Si—$CH_2CH_2CH_2$—O—CH_2CH—CH_2—O

OCH_3 OH

N=N

N

$(CH_3)_2N$ S $N(CH_3)_2$

5

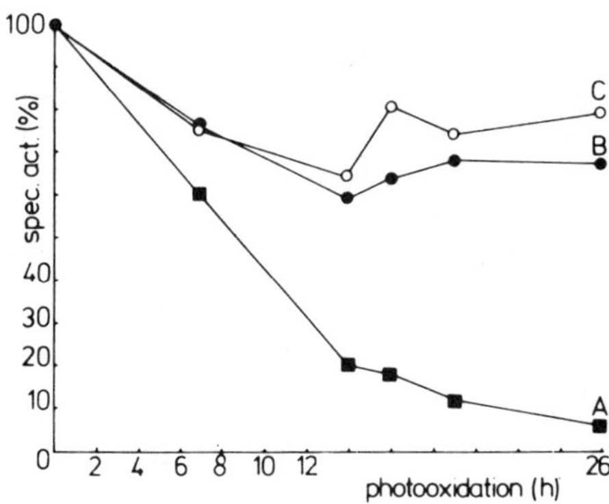

Fig. 14.4. Decrease in percentage specific activity of lysozyme as a function of hours of photooxidation. Reprinted with permission from ref. 18.

tion. The specific activity of the enzyme decreased by 94% when oxidized in the presence of light and the methylene blue beads, as shown by line A in Figure 14.4. The activity of the lysozyme solution containing the beads but shielded from light decreased by 21% (line C, Figure 14.4) and the activity of the solution containing no beads but exposed to light decreased by 33% (line B, Figure 14.4). Analysis of protein content showed no adsorption of protein onto the beads, indicating that the loss of activity was not due to reduced protein concentration.

The results of the photooxidation of methionine using immobilized methylene blue are similar to those using the free photocatalyst, the only significant difference being the increased amount of time required to carry out the reaction with the immobilized reagent. Because methionine plays only a structural role in lysozyme, the activity could not be abolished entirely. Jori et al. (19) were able to decrease the activity to 5%, whereas experiments with the immobilized reagent resulted in a decrease to 6%.

The advantages of using immobilized dyes for photocatalyzed oxidations go beyond the ease with which they can be removed from the reaction mixture. They can be used in solvents in which the unbound dye is insoluble and thus unable to sensitize singlet oxygen formation efficiently. They are more stable when bleached than free dyes and can be reused with little or no loss in effect. Schaap et al. (20) have studied in some detail the efficiency of immobilized Rose Bengal in catalyzing some well-known organic photoxidation reactions. Rose Bengal, like methylene blue a singlet oxygen-sensitizing dye, was coupled to chloromethylated styrene-divinyl benzene copolymer beads. The immobilized dye was used to catalyze the 1,2 cycloaddition of singlet oxygen to alkenes to form 1,2 dioxetanes in CH_2Cl_2. A suspension of free Rose Bengal in CH_2Cl_2 was inefficient in photosensitizing the generation of singlet oxygen. The quantum yield for the formation of singlet oxygen with polymer-bound Rose Bengal in CH_2Cl_2 was calculated to be 0.43.

Rose Bengal immobilized on Sepharose has also been shown to be effective in the photooxidation of amino acids (21). In this study, however, no selective modification was achieved, nor was it demonstrated that protein adsorption did not take place. At pH 8 at 10°C in 20 m*M* Tris-HCl buffer, five free amino acids (cysteine, methionine, histidine, tryptophan, tyrosine) were shown to be oxidized as measured by oxygen uptake. Comparison with results from the free dye in solution showed the same relative order of oxygen uptake when using the immobilized dye but an overall reduction in the amount of oxygen incorporated. Treatment of pig heart citrate synthetase with the immobilized dye showed the same rate constants for photooxidation as those obtained with the free dye.

14.6. IDENTIFICATION OF MEMBRANE PROTEINS BY IMMOBILIZATION

The study of biological membranes has been facilitated by the application of immobilized reagents. Figure 14.5 depicts the procedure used by Eshdat (22)

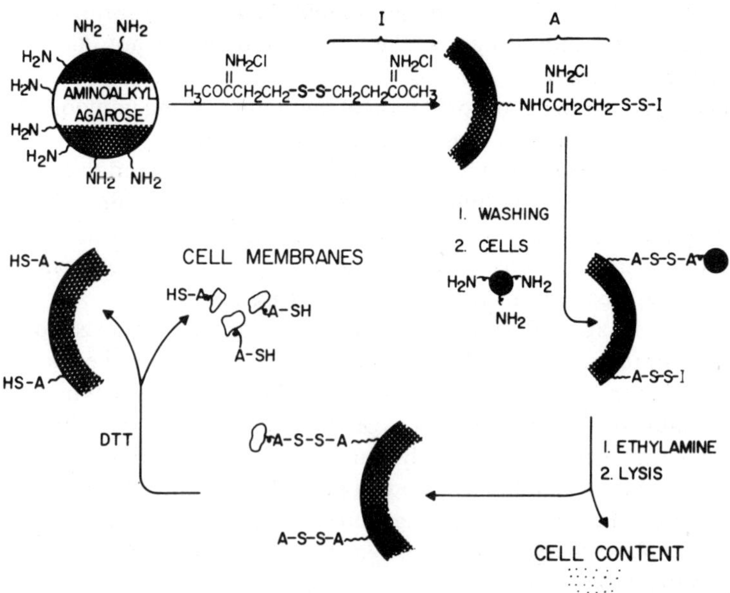

Fig. 14.5. Cell membrane isolation by covalent fractionation on a solid support. Aminoalkyl agarose beads were activated by reaction with imidoester groups (I) of the crosslinking reagent 3,3′-dithiobispropionimidate to form stable amidine groups (A), which link dithioimidate groups to the beads. Reprinted with permission from ref. 22.

to isolate and purify human erythrocyte membranes. Aminoalkyl agarose beads carrying cleavable imidoester groups were reacted with human red blood cells. Following inactivation of excess imidoester groups with ethylamine and washing of the cell-coated beads, the immobilized cells were lysed so that only the membranes, or cell "ghosts," remained attached to the beads. Reduction of the disulfide bonds linking the membranes to the beads released the purified membranes. Identification of the specific outer-surface-membrane proteins—which were bound to the agarose beads through disulfide bonds—was achieved by solubilizing noncovalently attached proteins in sodium dodecyl sulfate (SDS) followed by reduction and gel electrophoresis of the proteins that were covalently attached to the solid support.

Kamio and Nikaido (23) have also used an insoluble support to identify the outer-membrane proteins of *Salmonella typhimurium*. Rather than isolating the membranes, they coupled CNBr-activated dextran directly to intact cells. When solubilized in SDS, the covalently bound dextran–protein complexes were of such high molecular weight that they were prohibited from entering an SDS acrylamide gel. Comparison of the resulting electrophoretic pattern with the pattern of membranes not coupled to dextran showed the absence of particular protein bands corresponding to the proteins that had been coupled to the dextran. Of the four major outer-membrane proteins of *S. typhimurium*, three were coupled to the dextran when intact cells were used. When the outer membranes were removed from the cell, forming vesicles, all four proteins reacted with the dextran, indicating that the limited reaction with intact cells was due to the position of the protein in the membrane. This positional effect was further shown by the reactivity of the inner membrane proteins of *S. typhimurium* with the dextran. When intact cells were coupled to the activated dextran, no proteins reacted; when vesicles were used, all the inner membrane proteins reacted. The lack of reaction of the inner-membrane proteins in intact cells demonstrated the lack of penetration of the CNBr-activated dextran. Since the outer membrane allows diffusion of hydrophilic molecules smaller than 600–700 daltons, immobilization of CNBr on dextran prohibited the reagent from passing through the membrane. In this way the activated dextran was used successfully as a topologic probe.

REFERENCES

1. G. E. Means and R. E. Feeney, *Chemical Modification of Proteins,* Holden Day, San Francisco, 1971.
2. Y. Shechter, M. Rubinstein, and A. Patchornik, *Biochemistry* **16,** 1424 (1977).
3. M. Rubinstein, Y. Shechter, and A. Patchornik, *Biochem. Biophys. Res. Commun.* **70,** 1257 (1976).

4. P. Singh, S. D. Lewis, and J. A. Shafer, *Arch. Biochem. Biophys.* **203,** 774 (1980).
5. T. Oka and Y. Topper, *Proc. Nat. Acad. Sci.* **71,** 1630 (1974).
6. T. Oka and Y. Topper, *Science* **188,** 1317 (1975).
7. M. Wilchek, T. Oka, and Y. Topper, *Proc. Nat. Acad. Sci.* **72,** 1055 (1975).
8. R. Jost, T. Miron, B. K. Vonderhaar, and M. Wilchek, *Biochim. Biophys. Acta* **362,** 75 (1974).
9. Y. Topper, T. Oka, B. K. Vonderhaar, and M. Wilchek, *Biochem. Biophys. Res. Commun.* **66,** 793 (1975).
10. B. K. Vonderhaar and Y. Topper, *Biochem. Biophys. Res. Commun.* **60,** 1323 (1974).
11. M. Gorecki and A. Patchornik, *Biochim. Biophys. Acta* **303,** 36 (1973).
12. M. Gorecki and A. Patchornik, U. S. Patent 3,914,205 (1975).
13. R. Sperling and M. Gorecki, *Biochemistry* **13,** 2347 (1974).
14. W. H. Scouten and G. L. Firestone, *Biochim. Biophys. Acta* **453,** 277 (1976).
15. C. D. Lewis and W. H. Scouten, unpublished results.
16. W. H. Scouten, *Enz. Eng. Symp.* **4,** 368 (1978).
17. C. Lewis and W. Scouten, *J. Chem. Educ.* **53,** 395 (1976).
18. C. Lewis and W. Scouten, *Biochim. Biophys. Acta* **444,** 326 (1976).
19. G. Jori, G. Galiazzo, A. Marzotto, and E. Scoffone, *J. Biol. Chem.* **243,** 4272 (1968).
20. A. P. Schaap, A. L. Thayer, E. C. Blossey, and D. C. Neckers, *J. Amer. Chem. Soc.* **97,** 3741 (1975).
21. N. M. Kaye and P. D. Weitzman, *FEBS Lett.* **62,** 334 (1976).
22. Y. Eshdat and A. Prujansky-Jakobovits, *FEBS Lett.* **101,** 43 (1979).
23. Y. Kamio and H. Nikaido, *Biochim. Biophys. Acta* **464,** 589 (1977).

CHAPTER

15

IMMOBILIZED CELLS

STAFFAN BIRNBAUM, PER-OLOF LARSSON, and KLAUS MOSBACH

Pure and Applied Biochemistry
Chemical Center, University of Lund
P. O. Box 740
S-220-07, Lund, Sweden

Immobilized cells are a natural phenomenon, observable in several facets of life. Microbial films develop on such diverse surfaces as sand, soil, bark, skin, teeth, and intestinal villi. Adhesion is often a prerequisite for microorganisms to invoke pathogenic effects on their host cells (1). Recent years have witnessed increasing interest in immobilized cell systems. Although this interest has mainly focused on the productive potential of such systems, their study is important for the fundamentalist as well as the practitioner because they often present a good model for cell systems as they are found in nature.

Many of the naturally occurring immobilized cell systems have been in use since the early 19th century. It was probably Schuetzenbach who first employed one, though quite unknowingly (2). In 1823 his "quick vinegar" process utilized large wooden vats filled with beechwood shavings onto which a microbial film had developed. These microorganisms oxidized the ethanol in wine to acetic acid and thereby provided an effective, rapid, and reusable method for the production of vinegar. It was not until the time of Louis Pasteur that microorganisms were known to be responsible for the conversion of wine to vinegar. It may therefore be said that Pasteur was actually the first person who consciously described an immobilized cell system (2). He suggested that a mechanical support could be used to hold up the microbial mat that developed on the surface of the wine and thereby prevent the film from sinking. The microbial film could thus be used repeatedly and a cleaner product obtained. Since these first examples of the use of adsorbed cells, many natural immobilized cell systems have been employed, particularly in the treatment of waste water, where activated sludge, trickling bed, and rotating disk devices are frequently used.

The first example of artificial immobilized cells was reported by Hattori and Furusaka in 1960. They adsorbed *Escherichia coli* cells to Dowex 1 and showed their metabolic activity to be intact (3). In 1966 Mosbach and Mosbach (4) were the first to use the technique of immobilization (lichen cells entrapped in polyacrylamide) for the formation of biochemicals. In 1969 Updike et al. proved that polyacrylamide-entrapped cells were viable (5), and in 1970 Mosbach and Larsson used living immobilized cells for steroid production (6). When supplied with nutrients, the cells propagated within their surrounding support which stabilized the preparation and increased its biotransforming capacity. The first industrially operating process based on immobilized cells was devised by Chibata and coworkers at the Tanabe Seiyaku Company in Japan in 1973 for the production of aspartic acid (7).

These initial efforts were followed by an enormous expansion of the field in the mid-1970s, the major driving force being the expectancy of finding new, cheap, and effective industrial catalysts. Inspiration and guidelines for this development came from a related field, immobilized enzymes, which had a seniority of about five years. Today we have a plethora of immobilized cell systems; most are based on microbial cells, but some recent ones are based on plant and animal cells.

"Immobilized cells" may be defined as simply cells that have been associated with a support material. It is usually understood that this association is made intentionally. Microorganisms spontaneously attached to the surfaces of teeth, for example, are not usually designated as immobilized. Intentional immobilization of cells can be achieved by methods such as adsorption to a support, covalent binding to a support, crosslinking, entrapment in a gel, and entrapment in microcapsules (Figure 15.1). Immobilized microorganisms may vary in appearance. In Figure 15.2 this is exemplified by beads (diameter of about 2 mm), spaghettilike fibers, and a membrane. All three preparations represent the gel entrapment method, which is the one most widely used for microorganisms. One of the beads shown in Figure 15.2 was freeze-dried and studied by scanning electron microscopy (Figure 15.3). The micrograph shows the surface of the bead and illustrates how the yeast cells are entrapped by the polymer.

The dynamic character of immobilized cell research is reflected in the rapidly increasing number of reports in the literature. Only a few reports per year appeared in the early 1970s, by the middle of that decade the numbers increased to about 10 per year, and by 1980 they escalated to nearly 200. The remarkable number of reviews, about 30, appearing in 1980 and 1981 suggests that the area is entering a stage of maturity. A selection of reviews covering general aspects of immobilized cells is found in references 2 and 8–25.

The present overview gives a comparatively broad coverage of the field; for example, it includes plant and animal cells (but not organelles). The literature (exclusive of patents) has been surveyed up to mid-1981, and although

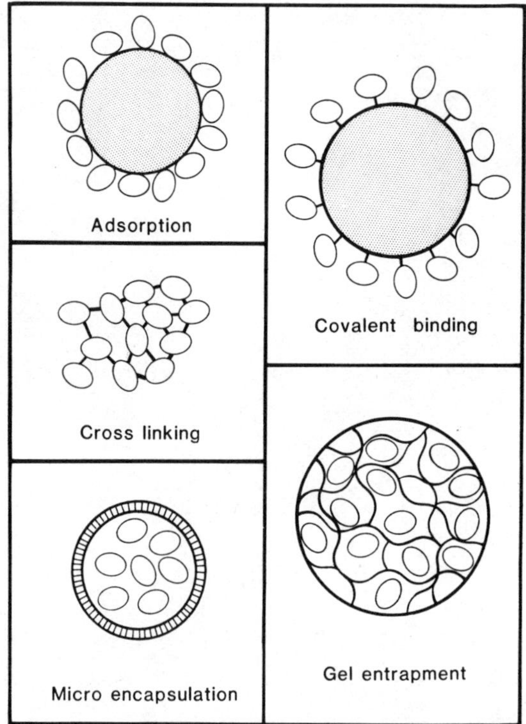

Fig. 15.1. Schematic representation of cell immobilization methods.

it is not complete, it does cover a substantial and representative selection of available publications.

15.1. IMMOBILIZED CELLS VS. IMMOBILIZED ENZYMES

The most prominent feature of immobilized whole-cell systems compared with those of the corresponding enzyme systems is their wide flexibility (Figure 15.4). Cells can be used in a number of metabolic and physiologic states. For example, they may be dead but still catalytically active, alive but resting, or alive and growing. In Figure 15.4 we have also included the term "permeabilized" as many of the immobilized cell systems referred to in this chapter have been treated with permeabilizing agents.

The use of immobilized cells instead of immobilized enzymes offers various advantages. Many of the enzymes used in industry are, for example, of microbial origin and the use of whole microorganisms instead of isolated en-

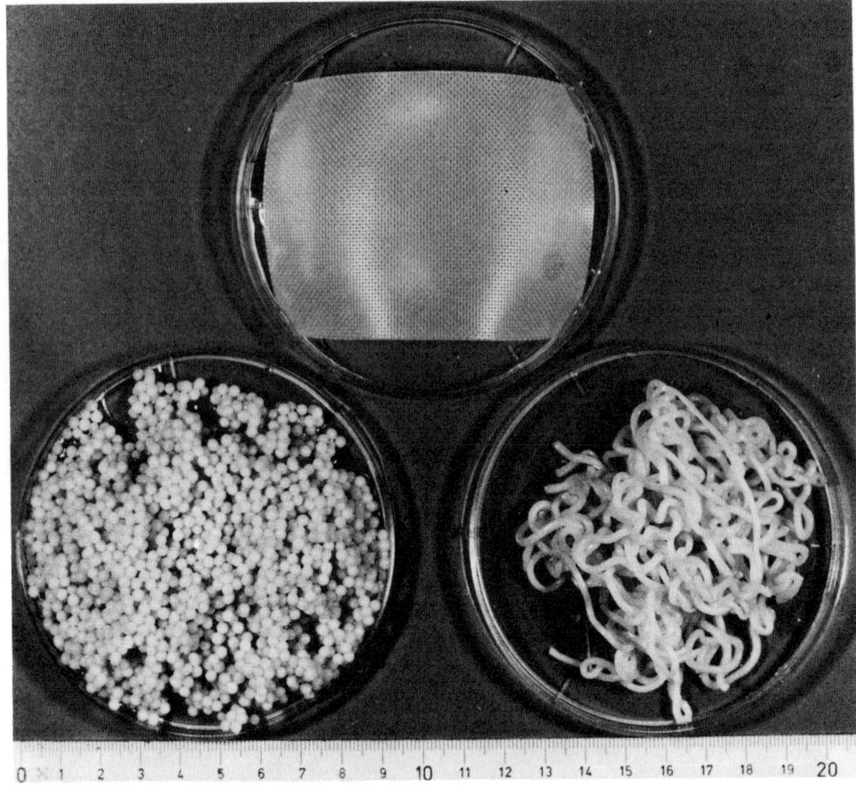

Fig. 15.2. Immobilized cell preparations. Yeast cells immobilized in calcium alginate in beads, in fibers, and in a membrane. The membrane is reinforced with a nylon net.

zymes thus obviates costly and tedious purification procedures. Furthermore, the stability of individual enzymes is often improved if they are retained in their natural environment, that is, the cell structure (26–28). The most distinct advantage of using immobilized cells instead of immobilized enzymes is gained in connection with multistep reactions. Here the immobilized cell already has the necessary enzymes properly arranged, whereas using immobilized enzymes requires elaborate coupling procedures for optimum results. A special case is coenzyme-dependent reactions. Here immobilized cell systems usually have the decided advantage of being able to regenerate coenzymes intracellularly and thus do not present the problems of coenzyme regeneration and coenzyme retention encountered with immobilized enzyme systems.

If immobilized cells are to carry out long reaction sequences, including

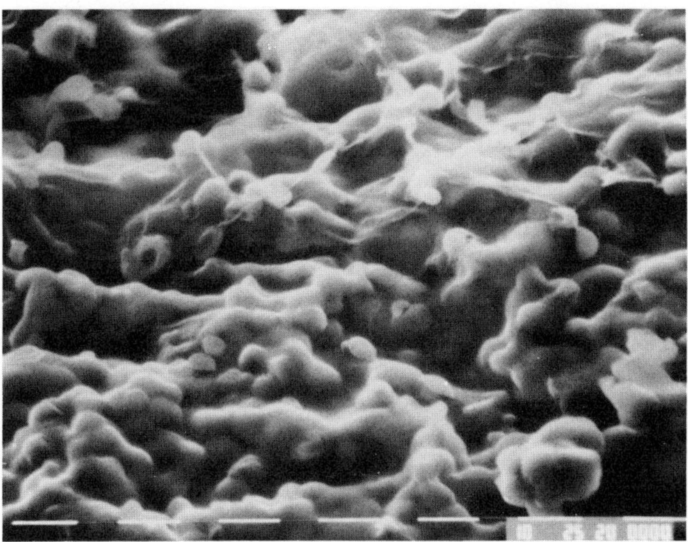

Fig. 15.3. Scanning electron micrograph of the surface of an immobilized yeast preparation. Each white bar along the lower edge represents a length of 10 μm.

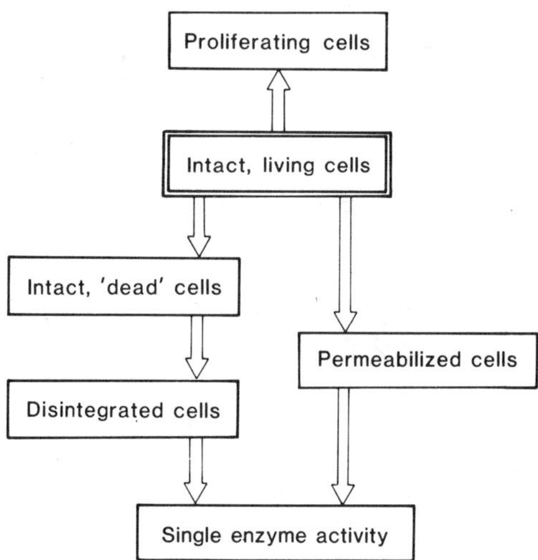

Fig. 15.4. The flexibility obtained when using immobilized cell catalysts.

Table 15.1. Some Examples of Activated Immobilized cells

Activation Principle	Activator	Immobilized Cell	Process	Activation Factor	Reference
Cell proliferation	Nutrients	*Saccharomyces cerevisiae*	Glucose to ethanol	200	31
Cell proliferation/ induction	Nutrients/cortisol	*Arthrobacter simplex*	Cortisol to prednisolone	10	32
Cell proliferation/ induction	Nutrients/ phenylacetic acid	*Escherichia coli*	Penicillin G to 6-APA	19	33
Lysis/permeabilization	Ammonium fumarate	*Escherichia coli*	Fumarate to aspartate	9	7
Lysis/permeabilization	Immobilization (polyacrylamide)	*Gluconobacter melanogenus*	Sorbose to sorbosone	2	34
Permeabilization	Diethyl ether	*Catharanthus roseus*	$NADP^+$ to NADPH	10	35
Permeabilization	Immobilization (polyacrylamide)	*Achromobacter liquidium*	Histidine to urocanic acid	8	36
Permeabilization	Acetone	*Brevibacterium ammoniagenes*	NAD^+ to $NADP^+$	162	37

cofactor regeneration steps, they must be viable. In living immobilized cells, the enzymes responsible for the desired activity can be synthesized *in situ* either by induction in existing cells or by cell proliferation. A preparation of living cells can be given a dose of nutrients to restore declining activity or to boost the activity to many times its original value (29, 30). Table 15.1 lists a selection of "activated" immobilized cell systems. Extensive compilations of such systems can be found in refs. 9 and 11. Besides enzyme induction and cell proliferation, lysis and permeabilization are also considered as methods of activation in the table. An example of activation by nutrients is given in Figure 15.5. A preparation of immobilized *Arthrobacter simplex* was used to convert the steroid cortisol to prednisolone in a number of consecutive batch transformations. The activity increased with each batch and finally reached a level where the time needed for complete conversion was conveniently short.

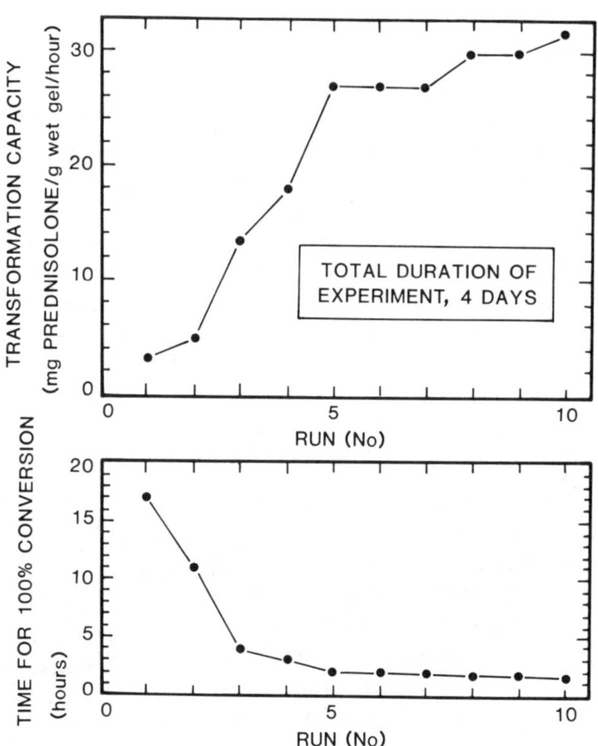

Fig. 15.5. Activation of immobilized cells by nutrients. Immobilized *Arthrobacter simplex* was treated with nutrients and the substrate-inducer cortisol in 10 consecutive batch incubations. The activity was measured as product formation (prednisolone). Reproduced by permission from *Nature* **263,** 796 (1976). Copyright 1976, Macmillan Journals Ltd. (29).

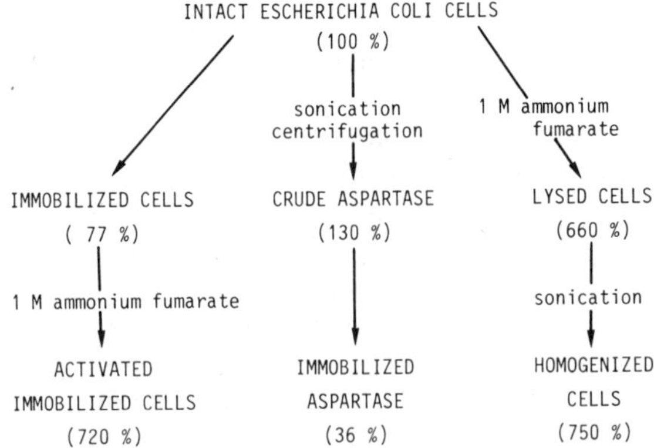

Fig. 15.6. Activation by permeabilization and lysis. The aspartase activity of free intact cells is set to 100%. The cells are immobilized in polyacrylamide (7).

In separate experiments the cause of activation was determined to be due to cell proliferation and induction (29, 32).

An example of activation by lysis, or permeabilization, is shown in Figure 15.6. The scheme illustrates that the full activity of intact cells was not expressed. However, disruption of membranes and other cellular structures by various means rendered the enzymes more accessible, and as a consequence the observed activity increased (7).

Permeabilization of cellular barriers can be developed into a delicate way of attaining access to intracellular enzymes (38). In some instances, the permeabilization procedures may be gentle enough (e.g., a brief exposure to solvents) to preserve many cellular functions. The enzymes are left in their natural surroundings and remain stable, enzyme isolation is unnecessary, and substrate–product diffusion in and out of the cells is less hindered. Interesting possibilities are opened up by the recent observation (39) that NAD^+ or ATP analogs bound to water-soluble polyacrylamide preparations can enter cells treated with organic solvents, take part in an enzyme reaction, and subsequently diffuse out to take part in another enzyme reaction. In a continuously working membrane reactor, polymer-bound NAD^+ was used as a shuttling coenzyme, being reduced by permeabilized cells (formate dehydrogenase) and oxidized by alanine dehydrogenase.

In other instances the use of immobilized pure enzymes is preferable. Whole cells possess a number of enzyme activities that may compete with the one desired. This will lower the yield and also lower the purity of the product. When living cells are used, metabolites are unavoidably excreted into the product stream. This is sometimes unacceptable, for example, in medical applications with extracorporeal shunts containing an immobilized biocatalyst.

At least so far as single enzyme reactions are concerned, it is usually possible to obtain immobilized enzyme preparations with much higher activity per unit volume than it is with the corresponding immobilized cells. The explanation is simply that in immobilized cell preparations, the desired enzyme is diluted with other cell constituents. Consequently, immobilized enzymes are preferred when very high activities per unit volume are needed.

The distinction between immobilized enzymes and immobilized cells is often arbitrary. The cells within an immobilized cell preparation, for example, may lyse and then gradually lose their original morphological characteristics. The desired activity, on the other hand, may remain unaffected. After some time the preparation is virtually indistinguishable from an immobilized crude enzyme preparation. In other instances, the term "cell" is certainly justified, living cells being a clear-cut case.

15.2. THE ADVANTAGES OF IMMOBILIZED CELLS OVER FREE CELLS

The merits as well as the drawbacks of using immobilized cells instead of free cells depend on the intrinsic properties of the cells and the purpose for which they are used. Many general notions of the pros and cons of immobilized cell systems can be held; a number are given in Tables 15.2 and 15.3. The somewhat categorical statements in Table 15.2 are explained in the text below. The drawbacks itemized in Table 15.3 are discussed in the next section.

1–3. Many advantages are a direct consequence of the simple fact that immobilized cell particles are about 1000 times larger than free cells. The size of these particles makes them easy to handle and use in a number of situations (as indicated in the table). The particle size also allows easy recycling in industrial processes as well as in analytical devices. The reusability obviously contributes significantly to the overall economy of a process, the relative importance naturally varying with the procedure employed. The aspartic acid process is an often-cited example. Here the cost of the catalyst was reduced to 10% of the original value on introduction of immobilized cell technology. The overall impact was a reduction of production costs by a substantial 40% (40).

4. The hallmark of immobilized cells (as well as of immobilized enzymes) is their value in continuous processes, again a consequence of the physical size of the particles. Continuous processes utilizing immobilized biocatalysts may rely on advantageously designed reactor configurations, such as packed bed reactors. These reactors have a high catalytic capacity per unit volume and thus reduce the physical size of the production facilities needed

Table 15.2. Advantages of Immobilized Cells over Free Cells[a]

Advantages gained by the physical size of immobilized cell particles (0.3–3 mm).	1. The catalyst is easy to handle, wash, store, and transport.
	2. Dosing the catalyst in batch processes is easy.
	3. The catalyst is easy to retrieve in batch processes.
	4. The catalyst is suitable for continuous processes, notably in advantageous designs such as packed beds.
Advantages gained by the physical confinement of the cells to a support.	5. The product is not contaminated by cells.
	6. The cells are protected against pH shock, microbial contamination, and other traumas.
	7. Cellular activities are stabilized.
	8. Coimmobilization with other catalysts gives favorable proximity effects.
	9. Coimmobilization with magnetic particles gives favorable handling properties.
	10. The viscosity in the bulk phase is lowered.
Advantages gained by the chemical and physical composition of the support.	11. The support material may modify the catalytic properties of the preparation by changing the concentration of reactants and other substances in the cellular environment.
Advantages gained by the reuse of biocatalyst.	12. Less cell cultivation is needed.
	13. There is less spent cell mass to dispose of.

[a] Categorical statements are elaborated in the text.

for a given output compared with conventional homogeneous fermentation processes.

Continuous processes, whether based on free cells or immobilized cells, have the merits of keeping the cells in a defined metabolic state with the possibility of increasing the yield of compounds known to be produced maximally under specific conditions. The constant supply of substrate and the constant removal of products protect the cells from substrate depletion or product accumulation; the latter is particularly significant when the process is sensitive to product inhibition. Immobilization is thus a rational way of making the cells work at their maximum capacity. In living immobilized cells, the possibility of using steady-state conditions may be of particular importance. The nutrient supply might be kept constant at a level that sustains high cellular activities without significant cell proliferation.

5. The confinement of the cells to a support material implies a number of desirable properties, including those established for heterogeneous catalysts in general. The product, for example, is largely free of cells and cell debris, which substantially facilitates work-up procedures. This is an important advantage since work-up costs for many conventional processes in the field of biotechnology can exceed 50% of total production costs.

Not all of the cells, however, are retained by the support. Especially with immobilization by physical adsorption, they may be readily released. This problem is accentuated if living cells are used in the procedure and if nutrients are added that lead to cell proliferation (41, 42). In such situations substantial cell release may occur, the old cells on the support being slowly replaced by new ones. Usually the conditions can be adjusted to ensure that at least 90 to 99% of the catalytic capacity is still associated with the support. The overall process would then be quite satisfactory. The contamination of the product would decrease to 1–10% compared with a process based on free cells, as would the requirement for nutrients. The slow exchange of cells on the support would guarantee a dynamic population of catalytically competent cells. Scanning electron microscopy has been used to illustrate how the cell population changes with time. One study (43), continued for many months, showed that intraparticular cell proliferation gave rise to microcolonies, cells originally localized to the surface layer were gradually detached, and cells in the interior were lysed and replaced by new ones.

When living cells are used in immobilization procedures and cell leakage must be kept to a minimum, certain precautions should be observed:

a. The cells should be immobilized by entrapment.
b. The density of cells in the preparation should not be too high.
c. The preparation should preferentially be reinforced by crosslinking agents.
d. The supply of nutrients should be kept as low as possible.

6. Physical confinement also plays a vital role in protecting the cell from detrimental environmental perturbations. In particular, the entrapping methods provide the cells with a protection against physical, chemical, and microbial hazards. Animal and plant cells are fragile structures and clearly benefit from the physical protection given by the supporting matrix, for example in stirred tank reactors. Accidental local and temporary extremes of temperature, pH, and chemical reagents have less of a damaging effect on immobilized cells because of the general buffering capacity of the solid phase. As for microbial infection, the entrapping matrix can be expected to exclude contaminating microorganisms. Bacteria susceptible to phage infec-

tion decidedly benefit from the confinement in a protecting support. The pore width of the immobilizing matrix should be made small enough to prevent viral entry. Such small pores also offer protection against proteolytic enzymes excreted by contaminating bacteria. The presence of immobilized lytic enzymes has been proposed for "self-sterilizable columns" (44).

7. The support material has a general stabilizing effect on cellular activities that enhances operational as well as storage stability. This phenomenon is rather complex and the reasons for it have been examined in only a few reports (26, 27, 28). Immobilization seems to promote the preservation of the structural integrity of the cells, even if they are killed during the immobilization process, and thereby conserves many of the essential qualities necessary for retaining the desired enzymatic activity. For example, it is conceivable that the spacial positioning of enzyme sequences on membranes is optimal and that any alteration of the cellular organization results in lower activity and less stable preparations.

8. The advantageous properties of immobilized cells may be augmented by inclusion of complementary catalytic elements such as other cells, enzymes, or organic and inorganic catalysts. The immobilization of different catalytic entities in one preparation often gives interesting kinetic advantages due to proximity effects (45). For example, in the reaction sequence:

$$\text{Substrate} \xrightarrow{\text{enzyme}} \text{intermediate} \xrightarrow{\text{cell}} \text{product}$$

the proximity effect rapidly supplies the cell with a favorably high concentration of the intermediate. The principle of coimmobilization is discussed in Section 15.4.7.

9, 10. Coimmobilization of cells and magnetic materials is a way of making the preparations easily retrievable in difficult media, for example in viscous media or in media containing substantial amounts of particulate matter (31). See Section 15.4.7 for further details.

11. The chemical composition of the support may also play a specific role and let the immobilized cells experience a favorable environment due to partitioning effects, thus increasing their effectiveness. For example, production of ethanol by the yeast *Saccharomyces cerevisiae* immobilized in kappa-carrageenan promoted a very high final concentration of alcohol, presumably because the ethanol shunned the very polar gel phase and was therefore enriched in the surrounding aqueous phase (41). The yeast cells could consequently ferment the sugar in a microenvironment containing a comparatively low concentration of the inhibiting ethanol. Another example is offered by steroid conversion using microorganisms immobilized in hydrophobic gels. The hydrophobic gels enriched the otherwise limiting substrate and therefore enhanced the reaction rate (46). A third example refers to immobi-

lized plant cells, which were shown to produce anthraquinones in a very high yield (47), possibly because of a favorable influence of the support in enhancing the secondary metabolism of the immobilized cells.

12, 13. The easy reuse of cells that is possible with immobilized cell systems also has other very important advantages. First, recycling obviates the need of cultivating new cells for every transformation attempted and thereby saves materials, labor, and production facilities. Second, reuse of cells drastically reduces the problems of cell mass disposal and therefore constitutes an environmental bonus.

15.3. LIMITATIONS OF IMMOBILIZED CELLS

Table 15.3 lists a number of situations in which immobilization of cells is unsuitable or the catalytic capacity of the cells cannot be fully realized without special precautions. The table refers in particular to entrapped cells, the most

Table 15.3. Situations in which the Catalytic Potential of Cells is Limited by Immobilization[a]

Circumstances in which Immobilization is Less Favorable	Consequence
1. The cells are unstable to the immobilization procedure used.	The preparation is inactive.
2. The enzyme activity is extracellular.	The activity is lost.
3. The product is intracellular.	The product is inaccessible.
4. The product has a low solubility.	Intraparticular precipitation occurs.
5. The desired activity is closely associated with cellular proliferation.	The catalyst can be used for only a short period of time.
6. The cellular activity is high or the substrate solubility is low (or both).	The potential capacity of the cells cannot be expressed because of limiting mass transfer in the particles.
7. The cellular activity is high and the product is inhibitory.	Same as above.
8. The substrate has a very high molecular weight.	The activity is low because the substrate has difficulty reaching the catalyst.
9. The product has a very high molecular weight.	The activity is low because the product accumulates within the catalyst.

[a] Categorical statements are elaborated in the text.

popular type of preparation. The rather categorical statements in the table are elaborated here.

1. The immobilization procedure may involve chemical agents, solvents, or heat. These factors may be detrimental to the desired activity (48). This is especially critical if the immobilized cells are to participate in complex biosynthetic reactions, which in practice means that the cells must be living. The susceptibility to chemical agents varies according to cell type and physiological state, animal cells and bacterial spores being at the extremes of sensitiveness and imperviousness. Suitable gentle immobilization methods are usually available. If not, activation (or reactivation) can sometimes be brought about by supplying the preparation with nutrients that allow the cells to repair cellular damage or to proliferate and thereby replace the cells destroyed by the immobilization procedure (29).

When only one or a few activities are wanted, rather harsh immobilization techniques may be adequate and in fact desirable—provided, of course, the activities are retained—because such procedures may effectively destroy competing activities, increasing the permeability and thereby considerably improving the overall performance of the catalyst (49). For example, the occurrence of succinic acid as a by-product in the conversion of ammonium fumarate to L-malate can be prevented by immobilization techniques involving treatment with bile extracts, which, in addition, considerably increase the activity of the preparation (50).

2. Immobilization is unsuitable when the desired enzyme activity is extracellular. The enzyme will be lost from the preparation unless a support is chosen with pores narrow enough to retain these molecules. However, the enzyme excretion might be considered a cheap way of producing and distributing the enzyme in a reactor without contaminating the medium with cells (51).

3. If the desired product is an intracellular one, for example, a protein or a fatty material, the product is difficult to harvest unless the cells are reversibly attached by a means such as adsorption. The support can then be reused many times and its main benefit is improving cell cultivation. Alternatively, the immobilized cell is manipulated to release the desired product, a technique of particular interest in genetically engineered strains producing proteins.

4. The use of immobilized entrapped biocatalysts can be problematic when the product has low solubility. If the product concentration exceeds the solubility limit, intraparticular precipitation of the product may result. The problem is accentuated if mass transfer limitations exists (see below).

Another equally undesirable situation arises when the poorly soluble product is a gas, such as carbon dioxide. In weak gels with insufficient tensile strength the gas bubbles that form may disrupt the gel particles (52, 53).

5. When the product is more or less closely associated with cellular proliferation, another problematic situation arises. The immobilized preparation can be used for only a limited time, after which it is crowded with cells. As discussed above, reversible immobilization may be a useful procedure in this case. Another method is intermittent surface sterilization (51, 54). The chosen remedy might include treatment with suitable antibiotics to break the connection between cell growth and product formation.

6. A limitation of the immobilization technique, whether the catalyst is an enzyme or a whole cell, is a reduced reaction rate due to a rate limiting mass transfer of substrates and products in the particulate preparation. This limitation has been discussed thoroughly in several review articles (55–58). The substrate has to make its way to the catalytic site, whether on the surface of the carrier or in the pores of the support. In the latter case we must deal with external as well as internal diffusion. The first diffusion step occurs from the (well-mixed) bulk liquid to the surface of the support through the so-called Nernst diffusion layer. In well-mixed immobilized whole cell reactors this diffusion layer will be of little importance as its thickness could be assumed to be less than 10 percent of the particle thickness so long as the diffusion layers of the individual cells do not overlap (58). When they do (at higher cell densities) the thickness of the diffusion layer outside the particle increases. The substrate subsequently undergoes a partitioning at the surface of the support and, when cells are entrapped, must diffuse through the matrix to the catalytic site. With intact whole cells, the substrate must also pass the cell wall or cell membranes to reach the catalytic site. After the substrate has been converted to product, the diffusion must occur in the reverse order to deliver the product to the bulk liquid.

The transport of substrate and product will be slowed down even further if there is an unfavorable interaction of the substrate or product with the support material, which may occur if, among other causes, the size of the substrate or the product is a substantial fraction of the pore diameter or if the diffusing molecules interact with the support via ionic forces.

If the immobilized cells are very active, the substrate concentration at the catalytic site (i.e., within the cells) will be lower than optimum. The transport of substrate by diffusion through the unstirred layer and through the support matrix is simply not fast enough to keep up with cells working at their maximum rate. Figure 15.7 illustrates the situation. The concentration of substrate is high and constant in the bulk phase and very low in the middle of the particle. The level K_m (Michaelis–Menten constant) indicates the sub-

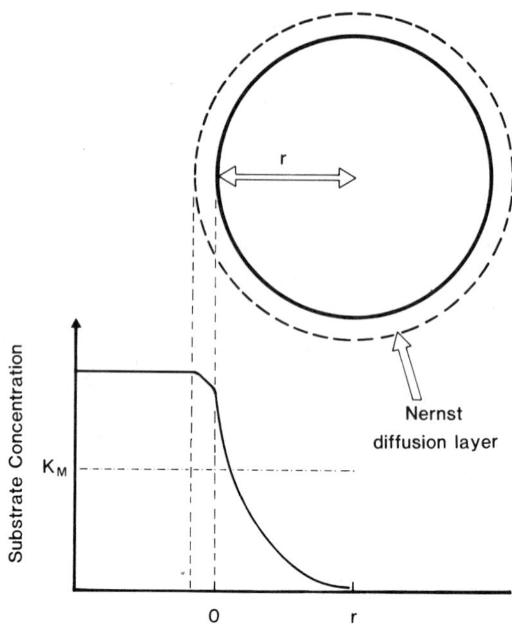

Fig. 15.7. Substrate concentration profile in an immobilized catalyst preparation during catalysis.

strate concentration where the free cell has its half-maximum activity. Assuming that the immobilized cells intrinsically have the same catalytic properties, only the peripheral parts of the particle in the figure have a substrate concentration sufficient to saturate the enzyme systems at work. The cells in the rest of the particle thus operate at suboptimum conditions because of substrate depletion, and the effectiveness of the system is low. To improve the situation, the mass transport rate should be increased. This can be done in a number of ways:

a. Increase the substrate concentration in the bulk solution.
b. Decrease the particle size.
c. Increase the diffusion rate (e.g., by using particles with larger pore sizes).
d. Decrease the cell loading.
e. Use surface bound cells.

Decreasing the cell loading does not primarily serve to increase the mass transport rate, but rather to reduce the *need* for a rapid mass transport. Increasing the substrate concentration seems to be a simple solution but is by no means always easy. A very important example is oxygen, which has a low solubility in fermentation media and therefore very easily becomes a limiting factor.

7. Poor mass transfer lowers the effectiveness of immobilized cells not only because it limits substrate concentration, but also because it contributes to the accumulation of products in the immediate vicinity of the cells. If the product then inhibits the performance of these cells, the loss in effectiveness may be very pronounced. Means of avoiding product accumulation are principally the same as for preventing substrate depletion.

8. Immobilization, at least within particles, is unsuitable when the substrate has a very high molecular weight and, obviously, when the substrate is insoluble, The pores of the catalyst preparation may seriously hamper diffusion or may not allow passage of the substrate molecules to the actual site of catalysis.

9. An analogous situation arises when the product has a high molecular weight and thus does not diffuse readily through the support. The accumulation of product can lead to seriously reduced activity.

15.4. IMMOBILIZATION PROCEDURES

The association of cells with a support can be accomplished by numerous conceptually separate preparative procedures, as shown in Figure 15.1. These methods include (i) adsorption to a support, (ii) covalent attachment to a support, (iii) ionic and/or covalent crosslinking, (iv) lattice entrapment, (v) microcapsule entrapment, and (vi) several miscellaneous procedures. The preparation should fulfil certain criteria, particularly for catalysis, such as high cell loading, high activity, and long operational stability. If the preparation is to be used in industry, additional qualities are desirable. For example, the fabrication of the immobilized cell particle should be simple, cheap, and preferably result in the formation of spherical beads with good mechanical, chemical, biological, and toxicological properties.

Usually the particular immobilization method that is optimal for a specific system cannot be predetermined. Several authors have reported screening studies to evaluate empirically which immobilization procedure is best for the specific system studied and for the properties desired (7, 59–66).

15.4.1. Adsorption to a Support

An array of publications describing various adsorption supports and techniques is summarized in Table 15.4. An excellent in-depth overview has been given by Daniels (106).

Table 15.4. Examples of Immobilized Cells Adsorbed to a Solid Support

Support	Cell	Use	Reference[a]
Dowex 1	*Escherichia coli*	—	3 (67)
DEAE-Sephadex	Mammalian cells	Cell culture, virus production, protein production	68 (69–72, 424)
ECTEOLA-Cellulose	*Aspergillus* sp.	Sucrose inversion	73 (74–76)
DEAE-Cellulose	*Nocardia erythropolis*	Steroid transformation	77 (78, 79)
Amberlite	*Saccharomyces cerevisiae*	Ethanol production	80
Gelatin	Mammalian cells	Cell culture	81 (82, 83)
Glass	*Saccharomyces carlsbergensis*	Ethanol production	84 (85–87)
Ceramic	*Acetobacter aceti*	Acetic acid production	88 (89)
Brick	*Saccharomyces cerevisiae*	Ethanol production	90
Anthracite	*Pseudomonas* sp.	Phenol degradation	91
Sand	Mixed culture	Wastewater treatment	92
Wood	*Saccharomyces cerevisiae*	Ethanol production	93
Pectate	*Saccharomyces cerevisiae*	—	94 (95, 96)
Metallic surfaces	*Pseudomonas fluorescens*	—	97
Chitosan	Mammalian cells	Cell culture	81
Cellophane	Brewer's yeast	Beer production	98
Polystyrene	*Pseudomonas* sp.	—	99 (100)
Polyester	Methanogenic bacteria	Methane production	101
Polyphenylene oxide	*Kluyveromyces lactis*	Lactose hydrolysis	102 (103)
Polyvinyl chloride	*Pseudomonas aeruginosa*	Denitrification and heavy metal removal	104 (105)
Polypropylene	*Pseudomonas aeruginosa*	Denitrification and heavy metal removal	104 (105)

[a] References to other systems using the same support are in parentheses.

The adsorption of cells onto a fixed support depends on a number of variables (99, 106–111). The nature of the cell surface—qualities such as its composition, charge, and age—has pertinent effects on the adhesive properties of the cell. The composition, charge, and shape of the support material influence the attachment of cells to the support and the conditions during adsorption as well as during operation determine the extent of cell retention.

For example, during adsorption such variables as cell concentration, carrier concentration, salt concentration, pH, temperature, and time are important. Since the adsorption of cells to the support depends on noncovalent interactions (such as ionic, hydrogen, van der Waal's, and hydrophobic bonding), the binding of cells to the carrier is relatively weak and cell desorption may readily occur, especially during cell growth. Thus, strict control of the process parameters is imperative. These variables include ionic strength, pH, temperature, and the shear stress generated in the system. The latter characteristic will depend on either the flow velocity in column reactors or the stirring speed in tank reactors. Another drawback of the adsorption procedure, besides the risk for cell release, is the low loading capacity of the carrier since usually only the outer surface of the support is available for catalyst binding.

Nevertheless, the merits of the adsorption method are considerable. Immobilization is simple (merely mixing of the cell suspension with the support), gentle (usually performed under physiological conditions) and cheap (the carrier can often be regenerated). In addition, the system does not experience any internal diffusional restrictions, as is often the case with entrapped cells.

The adsorption method is particularly suitable for extrasensitive cell systems, such as cultured animal cells. Giard et al. (71) have given an illuminating example of the application of mammalian cells adsorbed to a support by cultivating human diploid fibroblasts on microcarriers for subsequent production of interferon. Table 15.5 gives an outline of the steps involved.

Table 15.5. Protocol for Attachment and Culture of Human Fibroblasts on Microcarriers for Subsequent Production of Interferon (112).

Step	Time	Procedure (all at 37°C)
1	0 hr	Inoculate 6×10^8 cells into 1600 mL of initial growth medium[a] containing 15 g sterile microcarriers. Stir at 60 rpm for 1 min every 60 min.
2	6 hr	Dilute culture to 3500 mL with initial growth medium and stir continuously at 60 rpm.
3	3 days	Remove 1000 mL of medium and replace with 2500 mL replenishment medium.[a] Continue stirring at 60 rpm.
4	7 days	Remove 2500 mL of medium and replace with 2500 mL replenishment medium containing 2.5% (v/v) fetal calf serum. Continue stirring at 60 rpm.
5	10 days	Remove medium and wash culture with 1000 mL PBS. Repeat and then wash once with 1000 mL DME.[b] The microcarriers are now ready for the interferon production phase.[a]

[a] As described in ref. 112.
[b] DME = Dulbecco's modification of Eagle's medium.

The adsorption of cells to surfaces has been applied not only to production systems but model systems as well. For example, reversible cell binding to biospecific affinity adsorbents has been used in cell affinity chromatography (113–130). This method has been developed for studying distinct cell surface components. Specific antigen–antibody (116), hormone–receptor (118), and lectin–glycoprotein (114) interactions have been investigated. Thanks to the specificity and reversibility of these complexes a select type of cell can be separated from a mixed cell culture.

15.4.2. Covalent Attachment to a Support

Instead of adsorbing cells to a surface, one can covalently bind them to the support surface. The cells are thereby firmly bound to the carrier, and cell release, which is often a problem for adsorbed cells, is minimized. A comprehensive summary of the covalent coupling procedures on record is given in Table 15.6.

As with adsorbed cells, cells covalently coupled to the outer surface of a support do not experience the internal diffusional limitations that occur with entrapped cells. The intrinsic catalytic activity of cells covalently bound to a support is often lower than that of free cells because of the toxicity of the coupling reagents or procedure. Ideally, the coupling should occur under mild aqueous conditions and the reagents should be directed towards surface structures not involved in the enzyme reactions in question. One method of realizing this goal and thereby minimizing the deleterious effects of the coupling reagents is to activate the carrier before the cells are added. In this manner Jack and Zajic were able to bind *Micrococcus luteus* cells to carbodiimide-activated carboxymethyl cellulose (132). A schematic diagram of the preactivation method is given in Figure 15.8. The immobilized cells retained 75% of their L-histidine ammonia-lyase activity, and their cell membrane showed no appreciable damage although their viability was lost.

Though covalent coupling to a support is the preferred method when immobilizing enzymes, the method is of minor importance when immobilizing whole cells because of (i) the expense of the coupling reagents, (ii) the fact that not every cell type can be covalently coupled under sufficiently mild conditions, and (iii) the poor cell-loading on the particle.

15.4.3. Ionic and/or Covalent Crosslinking

Immobilized cell preparations have also been formed by ionic and/or covalent crosslinking, as shown in Table 15.7. Flocculated cell aggregates are obtained by incubating the cells with polyelectrolytes such as chitosan and other polyamines (143, 156). This method is simple, gentle, and cheap. The

Table 15.6. Examples of Immobilized Cells Covalently Attached to a Solid Support

Support	Coupling Reagent	Cell	Use	References[a]
Glycidylmethacrylate	Glutaraldehyde	*Aspergillus niger*	Gluconic acid production	131
CM-Cellulose	Carbodiimide	*Micrococcus luteus*	Urocanic acid production	132
γ-Aminopropylsilica	Glutaraldehyde	*Saccharomyces carlsbergensis*	Ethanol production	84
Amine functionalized methacrylate	Glutaraldehyde	Yeast cells	Killer toxin production	133 (134, 135)
Polyphenyleneoxide	Glutaraldehyde	*Solanum aviculare*	Glycoalkaloid production	103 (102)
Metal hydroxides	—	*Acetobacter* sp.	Acetic acid production	136 (137–139)
Cellulose	Cyanuric chloride	*Saccharomyces cerevisiae*	Ethanol production	76

[a] References to other systems using the same support are in parentheses.

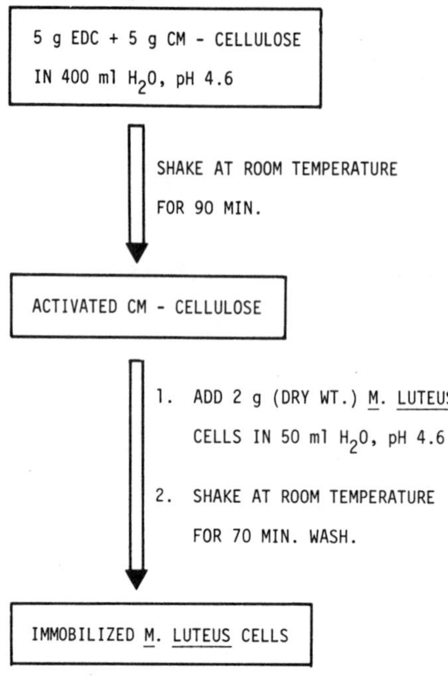

Fig. 15.8. Covalent attachment of cells to carboxymethyl cellulose (132). The amount of 1-ethyl-3-(3′-dimethylamino-propyl)-carbodiimide (EDC) represents a large excess.

factors influencing flocculation depend on several characteristics of the cells, the crosslinking agent, and the environmental conditions.

Alternatively, the cells can be covalently crosslinked with such bifunctional reagents as glutaraldehyde. Several of the commercially available immobilized whole-cell glucose isomerase preparations like Maxazyme (Gist Brocades), Takasweet (Miles Kali Chemie), or Sweetzyme (Novo Industri) are prepared by glutaraldehyde crosslinking (145). Often a spacer or carrier like gelatin (64, 65, 152), albumin (64, 65, 146–151), or polyethyleneimine (153) is included. Yet these preparations begin to resemble entrapped cells, and like entrapped cells, crosslinked cells often have diffusional limitations.

Many bacteria and fungi exhibit mycelial growth, which may be considered a natural crosslinking, forming a kind of immobilized cell preparation with several of the advantageous characteristics described above. Recently, it has been reported that certain crown ethers can induce several non-mycelium-producing bacteria to form filamentous structures (154, 155).

15.4.4. Lattice Entrapment

The most popular method of immobilizing cells is to include them in polymeric matrices. This reduces the problem of cell leakage, as often occurs with ad-

Table 15.7. Examples of Immobilized Cells Ionically and/or Covalently Crosslinked

Crosslinker	Cell	Use	Reference[a]
Anionic and cationic polyelectrolyte	*Arthrobacter* sp.	Glucose isomerization	140 (141, 142)
Chitosan	*Streptomyces* sp.	Glucose isomerization	143
Glutaraldehyde	*Escherichia coli*	Aspartic acid production	7 (144, 145)
Glutaraldehyde and albumin	*Escherichia coli*	Lactose hydrolysis	146 (64, 65, 147–151)
Glutaraldehyde and gelatin	*Actinoplanes missouriensis*	Glucose isomerization	152 (64, 69)
Glutaraldehyde and polyethyleneimine	*Kluveromyces fragilis*	Lactose hydrolysis	153

[a] References to other systems using the same support are in parentheses.

sorbed cells. Entrapment is generally a gentle and simple method; it allows high cell densities and is therefore often preferred when immobilizing cells. The activity of entrapped cell preparations may appear to be lower than comparable surface-bound preparations because of the diffusional limitations (see Section 15.3), but the productivity will be greater because higher cell loading can be achieved. As mentioned above, the envelopment of the cell with a protective gel barrier minimizes the accessibility of the cell to adverse conditions. Further, this method is advantageous in comparison with the other immobilization methods when cell growth occurs since the majority of the daughter cells will remain entrapped within the gel matrix. Table 15.8 lists several entrapment methods, which are discussed in greater detail below.

15.4.4a. Polymerization (Polyacrylamide)

The first and still most widely used procedure to entrap cells is to include them in polyacrylamide gels and related hydrogels. The first example of an entrapped cell preparation, as mentioned earlier, was reported by Mosbach and Mosbach in 1966 (4). The cells are entrapped by mixing them with a monomer and a crosslinker before polymerization. A catalyst system is then added and polymerization occurs. The gel formed by such block polymerization is subsequently granulated or cut into particles of appropriate size (48, 188).

Table 15.8. Examples of Immobilized Cells Entrapped in Polymeric Matrices

Support	Cell	Use	Reference[a]
Polyacrylic acid and its derivatives	*Curvularia lunata*	Steroid transformation	6 (4, 5, 7, 29, 34, 36, 37, 43, 46, 48, 51, 54, 61–63, 157–236)
Agar (agarose)	*Saccharomyces pastorianus*	Sucrose inversion	237 (35, 59, 63, 197, 238–250)
Kappa-carrageenan	*Streptomyces phaeochromogenes*	Glucose isomerization	60 (41, 50, 59, 204, 251–262)
Collagen	*Brevibacterium flavum*	Glutamic acid production	263 (49, 264–274)
Gelatin	*Saccharomyces cerevisiae*	Sucrose inversion	275 (198, 276–280)
Alginate	*Saccharomyces cerevisiae*	Ethanol production	281 (30–32, 45, 47, 59, 61–63, 81, 188, 193, 196, 207, 261, 282–316)
Chitosan	*Escherichia coli*	L-tryptophan production	317
Cellulose	*Actinoplanes missouriensis*	Glucose isomerization	318 (289, 291, 319–325)
Polystyrene	*Candida tropicalis*	Phenol degradation	61
Eudragit	*Escherichia coli*	6-APA production[b]	326
Epoxy resin	*Escherichia coli*	6-APA production[b]	327
Polyurethane	*Arthrobacter simplex*	Steroid transformation	328 (46, 219, 225–227, 229, 329, 330, 383)
Silica hydrogel	*Saccharomyces cerevisiae*	—	85
Polyvinyl alcohol	*Brevibacterium ammoniagenes*	Coenzyme *A* production	200 (194, 331)
Carboxymethyl cellulose	*Candida tropicalis*	Phenol degradation	289

[a] References to other systems using the same support are in parentheses.
[b] 6-APA = 6-aminopenicillanic acid.

In order to minimize the toxic effects of the reagents and the polymerization procedure upon the cells, there are three rules to obey (48):

1. Keep the time of exposure to the chemicals to a minimum.
2. Dissipate the heat generated during polymerization as effectively as possible.
3. Keep the system buffered.

The time of exposure may be shortened by flushing the polymerization mixture with nitrogen gas to remove inhibitory oxygen and/or by using a relatively high catalyst concentration. For efficient cooling the polymerization can be carried out in a sandwich type of polymerization chamber, immersed in a water bath (48). A flow diagram of the procedure for entrapment of cells in polyacrylamide is given in Figure 15.9 (29, 48, 182). Alternatively, beads can be formed by suspension polymerization using an organic phase and an emulsifying agent (332). In the latter case, one can easily hold the temperature at 0–5°C. By avoiding the toxic solvents commonly used in suspension polymerization (332) and choosing dibutylphtalate, Klein and Schara obtained *Candida tropicalis* cells entrapped in spherical beads of polyacrylamide that retained 90 to 100% of the phenol-degrading activity of free cells (209).

Polymerization of acrylamide, acrylic acid, or their derivatives can also be brought about by illumination or irradiation (46, 212–229, 333). A notable example of this technique is the work of Fukui and colleagues (46, 212, 213, 216–220, 225–229, 333). They have used a number of hydrophilic or hydrophobic polymers with terminal acrylic functional groups (46, 212, 216–219, 225–227).

15.4.4b. Thermal Gelation (Agar; Kappa-Carrageenan and Collagen)

Aqueous solutions of several naturally occuring polymers form gels when the temperature is lowered or when the preparation is dried; they are therefore suitable for cell entrapment. Subsequent to gelation these carriers must often be treated with stabilizing agents to increase the mechanical as well as chemical strength of the gel. Agar (agarose), kappa-carrageenan, collagen, and its hydrolytic derivative gelatin are the polymers most often used for this purpose.

Agar (agarose) is extracted from certain species of marine algae. It is well known for its use as support media for cell culture. Agar's popularity is due to its biocompatability (it is included in the "Generally Regarded As Safe" list under the Food and Drug Act, USA), its stability (very few microorganisms can metabolize agar or degrade it), its hysteresis (it gels at a low temperature but melts at a high temperature), and its potency (gelation is

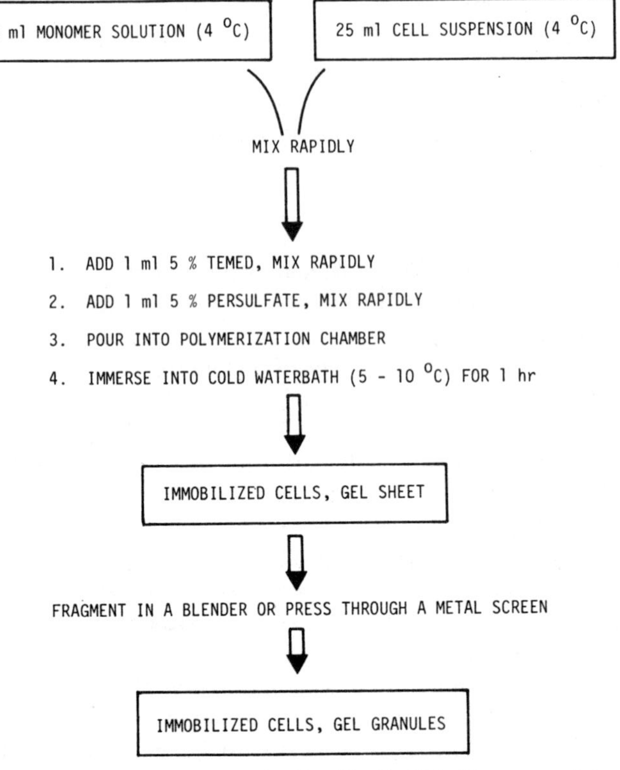

Fig. 15.9. Entrapment of cells in polyacrylamide (29, 48, 182). Monomer solution = 25 mL 0.1 *M* Tris-HCl, pH 7.5, containing 7.13 g acrylamide and 0.38 g *N,N'*-methylene-bis-acrylamide. Cell suspension = 25 mL 0.1 *M* Tris-HCl, pH 7.5 containing 5 g cells (wet weight; centrifuged at 10,000*g*). A polymerization chamber may be constructed from two glass plates (TLC plates; 20 · 20 cm) kept 2 mm apart with the aid of a piece of latex tubing that also serves as a sealant. TEMED = N,N,N',N'-Tetra-methylethylenediamine.

perceptible at concentrations as low as 0.1%) (334). Cells have been included in agar and agarose and have been used for numerous processes (Table 15.8). The preparative procedure for cell entrapment in agar (or agarose) is simple, as the following example shows:

Agar (2 g) is dissolved in 80 mL physiological saline at 100°C and is subsequently cooled to 50°C. Then, 20 mL of cell suspension (50°C) are added and the resulting mixture is allowed the cool below the setting temperature of agar (around 40°C). The cell gel block is then granulated to fragments of appropriate size and washed with saline. Alternatively, beads may be formed by dripping the mixture into a cold organic phase (237). Lower temperatures can be employed, if necessary, by using agarose with a lower setting temperature (35).

Kappa-carrageenan, a water-soluble gum isolated from various species of red seaweed, is an anionic polyelectrolyte owing to its half-ester sulfate groups. Like agar, it sets on cooling and has recently found increased popularity for the entrapment of cells. Chibata and coworkers replaced polyacrylamide with carrageenan in a number of immobilized cell systems (50, 251–254). Though kappa-carrageenan solidifies upon cooling, the continuous presence of cations (such as K^+, Ca^{+2} or NH_4^+) is required unless the gel is treated with hardening agents like glutaraldehyde and hexamethylenediamine (60). This treatment was often observed to increase the operational stability of the preparation (60, 251–253, 256). To entrap cells in kappa-carrageenan, the following procedure is recommended (27):

Kappa-carrageenan (4 g) is dissolved in 80 mL physiological saline at 100°C and is cooled to 50°C. Washed cells are suspended in 20 mL physiological saline (50°C) and mixed with the polymer solution. The mixture is cooled to 10°C and the gel block is fragmented into particles of appropriate size (2 × 2 × 2 mm). The preparation is then washed with a gel-strengthening agent such as 0.3 *M* KCl to produce mechanically firm particles.

Beads may also be formed by dripping the warm cell–polymer mixture directly into 0.3 *M* KCl (20°C) using a needle and syringe (which should also be warm). If necessary, the gel particles may be stabilized by treatment with glutaraldehyde and hexamethyldiamine (252).

To reduce the temperature at which the cell–polymer mixture jells, one must reduce the concentration of kappa carrageenan in the cell–gel mixture. If locust bean gum is added in compensation, gel particles with mechanical properties similar to the original kappa-carrageenan preparation are obtained (60).

Collagen is a natural linear protein polymer that serves as a support and connective matrix in mammalian cell systems and has been extensively used as a support in animal cell cultures. Collagenous carriers have been used for immobilizing a number of cell systems (Table 15.8). Vieth and Venkatasubramanian described the entrapment of *Streptomyces venezuelae* cells in collagen (266). The cells were mixed with a collagen dispersion, the pH raised to 11, and the mixture cast on a Mylar sheet. The cell–collagen preparation was subsequently dried and tanned with glutaraldehyde. The membrane formed may be used as is or fragmented into chips. Modifications of the procedure include lower pH (8.5 and 7.0) and addition of glutaraldehyde before casting (263, 266, 274). Formaldehyde and dialdehyde starches have also been used to improve the mechanical properties of collagen–cell preparations (264, 268).

Gelatin, a hydrolytic derivative of collagen, has been used quite extensively (Table 15.8). The following steps are generally followed to prepare gelatin-entrapped cell derivatives:

Gelatin (e.g., 10%) is dissolved in an appropriate buffer at 95°C and is cooled to 40°C before mixing with a cell suspension (40°C). The mixture is cast on a flat surface to form a membrane upon cooling and/or drying. Alternatively, the mixture is suspended in a water-immiscible solvent to form droplets that solidify upon cooling. A third method is to simply cool the mixture and then disintegrate the gel block into particles of suitable size. Gelatin preparations are usually treated with glutaraldehyde or formaldehyde (e.g., 1% solution for 1 hr) to improve the mechanical and chemical stability. Alternatively the crosslinker could be added to the gelatin–cell suspension.

15.4.4c. Ionotropic Gelation (Alginate and Chitosan)

Several naturally occurring polysaccharides form gels (owing to their polyionic nature) when contacted with certain counter ions. Alginic acid, a linear copolymer of L-guluronic and D-mannuronic acid, is extracted from seaweed.

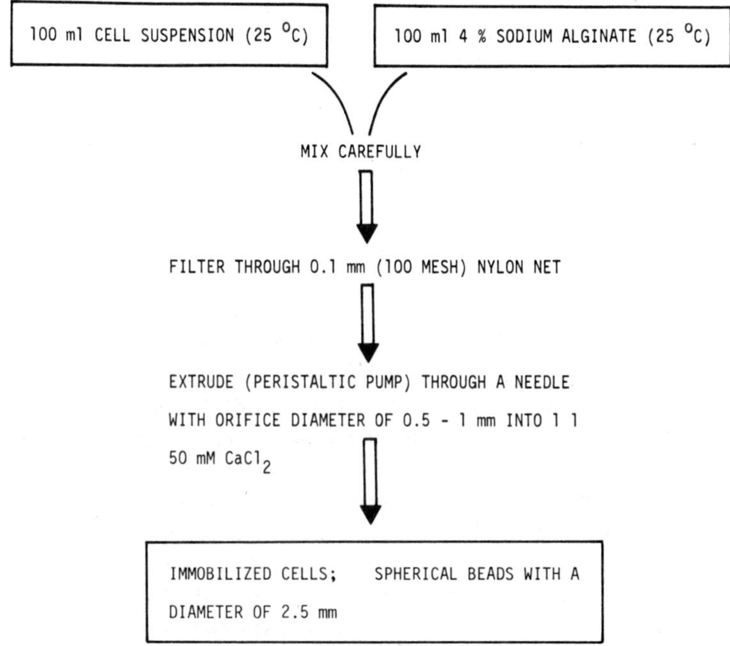

Fig. 15.10. Entrapment of cells in calcium alginate beads. The filtering serves to remove lumps. A useful set-up is a Büchner funnel with a snugly fitted nylon net. A syringe and needle are more convenient than a pump for small-scale (<50 mL) immobilizations.

It has found widespread use for entrapping cells (Table 15.8). The preparation of alginate cell-containing beads is gentle and simple. The cells are mixed with a sodium alginate solution and the mixture is dripped into a solution containing a counter ion, such as Ca^{+2} or Al^{+3}, to form ionically crosslinked beads (61, 281). A schematic diagram of the procedure is given in Figure 15.10. The phase transition from a sol to a gel is a reversible process, and to preserve the gel structure a counter ion should be present. However, difficulties arise when certain complexing agents are present in the medium. For example, phosphate, citrate, EDTA, as well as a number of other substances bind to calcium and thereby remove it from the gel. This causes the carrier to solubilize and hence release the entrapped cells. Methods of circumventing this difficulty by crosslinking the preparation with polyamines have recently been described (42, 284, 306). Three stabilization procedures are schematically shown in Figure 15.11 and the physical consequences of one of these methods are depicted in Figure 15.12. Immobilized cell preparations, stabilized under three different conditions, were cut into thin disks and were subsequently treated with 0.1 *M* phosphate buffer. The peripheral layer of the preparations was obviously stabilized and remained intact, whereas the core was dissolved. The thickness and the strength of the stabilized layer could easily be varied from one that was very thin to a completely homogeneously stabilized preparation (42). Beads treated with agents that dissolve the unstabilized interior but leave the surface intact become, in

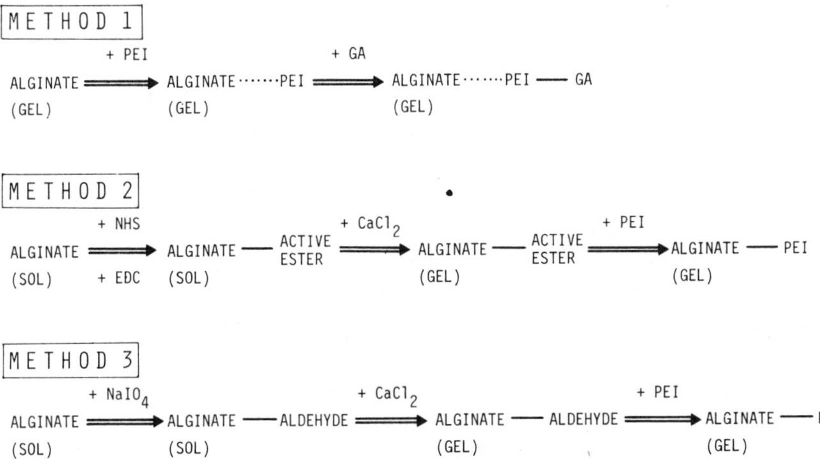

Fig. 15.11. Covalent stabilization of alginate gels. PEI = polyethyleneimine; GA = glutaraldehyde; NHS = *N*-hydroxysuccinimide; EDC = 1-ethyl-3-(3′-dimethylaminopropyl)-carbodiimide. Ionic linkage is indicated by dotted line. Reproduced with permission from ref. 42.

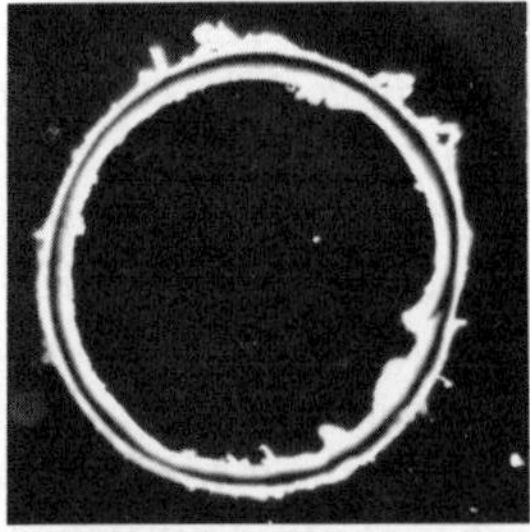
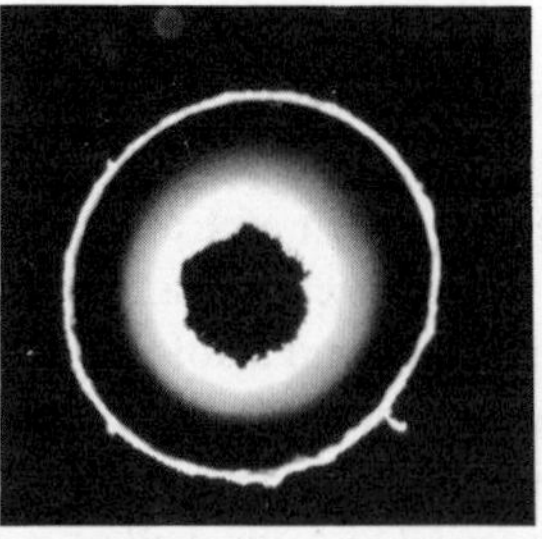
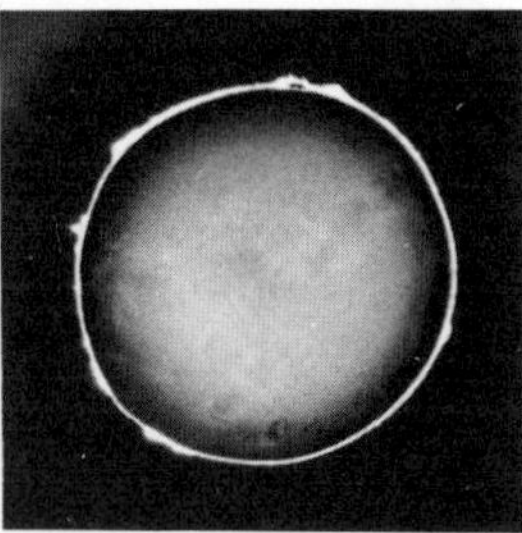

Fig. 15.12. Resistance of stabilized alginate gels in phosphate buffer. An alginate entrapped yeast preparation was treated with 0.5% (w/v) polyethyleneimine for (left to right) 1, 6, and 24 hr and then crosslinked with glutaraldehyde (1%; v/v) for 1 min. The stabilized preparations were subsequently cut into thin disks and treated with 0.1 *M* phosphate buffer to dissolve the nonstabilized regions. The photographs were obtained with a dark-field microscope. Reproduced with permission from ref. 42.

effect, capsules (42, 284). Other methods for stabilizing the gel, both chemically and mechanically, are also available (11, 60, 285).

Chitosan is a glucosamine polymer obtained by deacetylation of chitin, which is extracted from crab shells and similar sources. A solution of chitosan will set when in contact with anions such as phosphate, owing to the cationic nature of the polymer. Furthermore, chitosan will precipitate above pH 7.0 Vorlop et al. mixed *Escherichia coli* cells with chitosan and dripped the mixture into a solidifying bath of sodium tripolyphosphate and then transfered the beads thus formed to a phosphate buffer (317). The gentleness of the method was evidenced by the survival of the cells.

15.4.4d. Precipitation Gels (Cellulose)

The most abundant polysaccharide in nature is cellulose. It is therefore quite understandable that a number of investigators have used cellulose and its derivatives for the entrapment of cells (Table 15.8). Cellulose (or a derivative) is dissolved in an organic phase such as dimethylsulfoxide and *N*-ethylpyridinium chloride, the cells are mixed with the cellulose solution, and the mixture precipitated by dripping it into an aqueous phase (324). The beads are often treated with glutaraldehyde to give additional stability to the gel and to reduce cell leakage. The enzyme systems employed in this case have usually been single-step reactions such as glucose isomerase, invertase, or lactase. The exposure of the cells to nonphysiological conditions during bead formation leaves the cells in a nonviable state, unable to participate in more complex reactions.

Polystyrene and Eudragit are some other synthetic polymers which, like cellulose, undergo precipitation to form insoluble carriers (61, 326).

15.4.4e. Polycondensation Gels (Epoxy and Polyurethane)

Cells can also be entrapped by polycondensation with epoxy- or isocyanate-substituted polymers (Table 15.8). A disadvantage with these procedures is that the materials used are often toxic to the cells, but good activity and desirable support properties (high rigidity, chemical inertness, and porosity) have often been obtained.

Klein and coworkers mixed a water-soluble polyaminoamide with an epoxy polymer and cells and allowed the mixture to condense (327). To obtain a preparation with a desirable pore size, alginate was included in the preparation. After condensation, the alginate was removed by dissolving it with phosphate, thus leaving a porous structure.

The polyurethane method employs polymers bearing terminal isocyanate groups. Fukui and coworkers prepared such water-miscible urethane prepolymers of varying hydrophobicity and simply added them to an aqueous solution containing the cells to be entrapped (46, 219, 227, 229, 328). In this manner they could control the hydrophobic character of the support and prepare a gel optimal for the conditions of the system. An example is the case of steroid transformation, where the substrate is more soluble in an organic phase than in an aqueous one. Klein and coworkers modified this procedure somewhat to obtain polyurethane foams that were completely elastic and highly porous (329).

15.4.5. Microcapsule Entrapment

Cells can also be entrapped by encapsulation. Chang entrapped human erythrocytes within crosslinked protein microcapsules formed by interfacial polymerization (335). Ado et al. immobilized microbial cells within cellulose acetate–butyrate membranes reinforced with chitosan (336, 337). It is interesting that the mechanical strength of the preparation was claimed to be comparable with that of ion exchange resins. Mohan and Li developed liquid-surfactant membrane-encapsulated cell preparations of approximately 20–40 μm in diameter in which each droplet contained about 500 to 600 *Micrococcus denitrificans* cells (338). Encapsulation in polyurea has also been attempted but with limited success, mainly because of the toxic effects of the membrane precursors diamines and isocyanates (7).

Cells have also been encapsulated in completely aqueous systems by alginate crosslinked with polyamines (42, 284).

15.4.6. Miscellaneous Immobilization Procedures

Other methods have been described for the immobilization of whole cells (Table 15.9). For example, cells can be confined simply by using commer-

Table 15.9. Other Cell Immobilization Methods

Support	Cell	Use	Reference[a]
Hollow fiber	*Pseudomonas fluorescens*	Urocanic acid production	339 (314, 340)
Membrane confinement	*Streptococcus faecium*	Arginine analysis	341 (342–344)
Polyester sacs	*Streptomyces* sp.	Glucose isomerization	345
Stainless-steel mesh spheres	*Streptomyces griseus*	Streptomycin production	346 (28)
Aqueous two-phase partitioning	*Saccharomyces cerevisiae*	Ethanol production	347 (348)

[a] References to other systems using the same support are in parentheses.

cially available filters or membranes. Hollow fiber reactors have been used for denitrification of water as well as for the production of urocanic acid (314, 339). Membranes are also used in analytical devices such as cell electrodes to confine the cells to the vicinity of the electrode surface (349–351).

An interesting new technique related to cell immobilization is based on the partitioning of cells to one of two liquid-aqueous phases. The technique is thus an extension of the methodology developed by Albertsson and coworkers for separating enzymes, organelles, and cells (352). The use of two liquid-aqueous phases obviously constitutes a very mild process compared with systems based on one organic liquid phase, (e.g., hexane) and one aqueous phase. The latter system usually leads to death of the cells (207).

Kühn (347) fermented glucose to ethanol by yeast in an aqueous system made up of polyethylene glycol and aqueous dextran. When the system was agitated, the two phases dispersed into microdroplets, allowing a very effective mass transfer to occur. When the agitation ceased, two layers were formed within 10 min. The lower dextran-rich phase contained essentially all the yeast cells. The upper layer, constituting about 90% of the total volume, was pumped to an alcohol stripping unit and then returned. A similar aqueous two-phase system containing yeast cells, cellulase, and glucosidase for the conversion of cellulose to ethanol was recently described by Hahn-Hägerdal et al. (348).

15.4.7. Some Special Techniques for Improving Immobilized Cell Preparations

One of the troublesome aspects of immobilized cell technology, especially of entrapped cells, is limiting mass transfer rates. One way of tackling this problem is to prepare small immobilized cell particles. This can be done by suspending the cells together with the water-soluble gel precursors in an organic phase and stirring the mixture vigorously enough for it to form small droplets. At this stage the setting/polymerization is induced. Suitable, fairly nontoxic solvents are dibutylphthalate (209), tributylcitrate, tributylphosphate, and others (359). Another way of producing small immobilized cell particles is illustrated in Figure 15.13 (353). The cells suspended in alginate, for example, or in carrageenan or chitosan, are pumped through the inner of two concentrically arranged nozzles. A flow of compressed air in the outer nozzle breaks up the cell-suspension jet stream into minute droplets (0.1–1 mm). These then fall into a solidifying bath to form small particles. Similar results are also obtained by vibrating a standard extrusion nozzle through which the cell suspension is pumped (354). Alternatively, the cell suspension

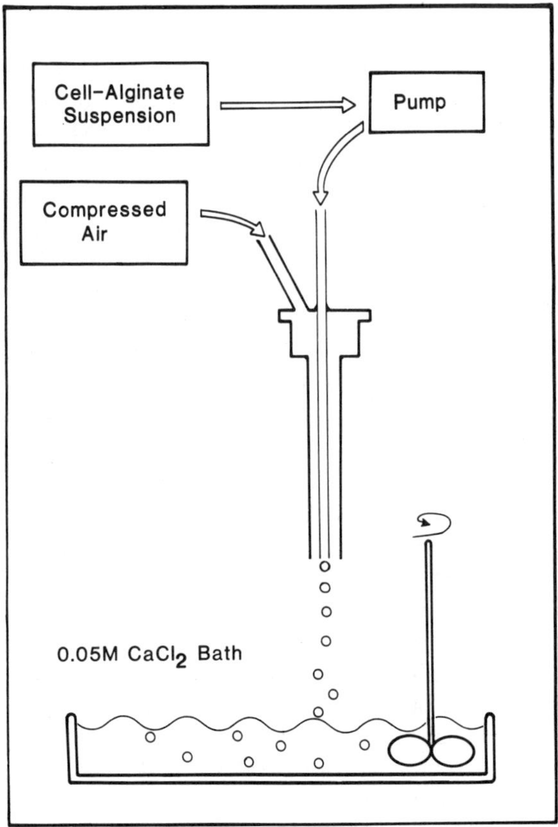

Fig. 15.13. Device for making small spherical immobilized cell particles.

may be forced through a very fine nozzle under high pressure. If alginate or chitosan beads are partly dried they will shrink with a simultaneous substantial increase in mechanical strength (317). At our institute these methods are used to form beads with diameters in the range of 0.1–1 mm.

Entrapment in alginate is an attractive procedure because it is gentle, simple, and readily forms spherical particles. In several applications alginate has been utilized on a temporary basis to aid in the fabrication of spherical partices from other support materials. Klein and Eng (327) mixed alginate with epoxy resin, crosslinker, and cells, dripped the mixture into a calcium chloride solution, and obtained spherical calcium alginate beads. The epoxy resin cured subsequently within this framework to give very hard spherical particles. The alginate was then dissolved by treating the particles with phosphate buffer, resulting in hard porous beads. Other combinations with alginate are also reported, for example, gelatin and agarose (81).

The special technique of coimmobilization of different types of cells has been used in a number of instances. An important example is the coimmobilization of coenzyme-regenerating cells and cells that require exogenous coenzymes for proper functioning. Yeast has been used as an ATP regenerator in CoA synthesis with coimmobilized *Brevibacterium ammoniagenes* (356). The same pair of organisms have also been used for the synthesis of $NADP^+$ (233, 336).

To improve the supply of oxygen, which is often a limiting factor for actively metabolizing immobilized cells, several different techniques have been suggested. These include addition of hydrogen peroxide to the medium in combination with immobilized cells containing coentrapped manganese dioxide (354), catalase (357), or active carbon (358), any of which will break down the peroxide to the desired oxygen. Recent methods involve coimmobilization with algae cells that will photolytically generate oxygen.

Wikström et al. used the method for the oxidative deamination of amino acids to keto acids (359), and Adlercreutz et al. used it for the oxidation of glycerol to dihydroxyacetone (360).

Other applications of cell–cell and cell–enzyme coimmobilizations are the production of ketogulonic acid by *Gluconobacter melanogenus* plus *Pseudomonas syringae,* and the fermentation of ethanol from cellobiose by *Saccharomyces cerevisiae* plus glucosidase (45, 169).

The presence of suspended matter in the medium can also create problems, as it becomes difficult to separate the biocatalyst from solid debris. In order to alleviate this difficulty one can prepare magnetic biocatalysts that are easily retrievable as long as the contaminating material is not magnetic. Larsson and Mosbach have described a "magnetic" immobilized yeast preparation (31). The inclusion of magnetite in the alginate particles did not in terfere with the metabolism of the yeast cells nor did it affect their viability.

The magnetic immobilized yeast was consequently an effective producer of ethanol from glucose. The systems discussed earlier based on one aqueous phase will also easily cope with suspended matter.

If the mechanical characteristics of an immobilized preparation should be inadequate, they can be reinforced by inclusion of a stabilizing component such as nylon or steel nets or some other solid support (32, 303, 361). Atkinson et al. described the use of stainless steel mesh spheres of many sizes, shapes, and densities for entrapping a wide variety of microorganisms (23, 346).

15.5. CHARACTERIZATION

Once an immobilized cell system has been prepared, it is important to obtain a proper understanding of its character whether it is to be used as a potential industrial biocatalyst, as an analytical aid, or simply as an aid for facilitating biological studies. Proper characterization of an immobilized biocatalyst should cover:

1. Chemical properties (functional groups, chemical stability, etc.).
2. Mechanical properties (particle size and shape, mechanical stability).
3. Catalytic properties (kinetic constants, intrinsic activity, stability and biological status).

A comprehensive and recommendable treatise on the characterization of immobilized biocatalysts has recently appeared (362).

15.5.1. Characterization of Chemical Properties

The chemical properties are usually evident from the starting materials and from the process of preparation, but the inclusion of a large amount of cells in a polymer may considerably change the chemical properties. A preparation should therefore be characterized (368) with respect to its polymer content, its functional group content, and its stability and swellability in commonly encountered media. The porosity should also be estimated. Toxicity (leakage of polymers) is also an important factor in some applications.

15.5.2. Characterization of Mechanical Properties

Knowledge of the mechanical properties of a preparation is important for properly predicting its performance in a suggested reactor configuration. The important characteristics are the rigidity, the elasticity, and the resist-

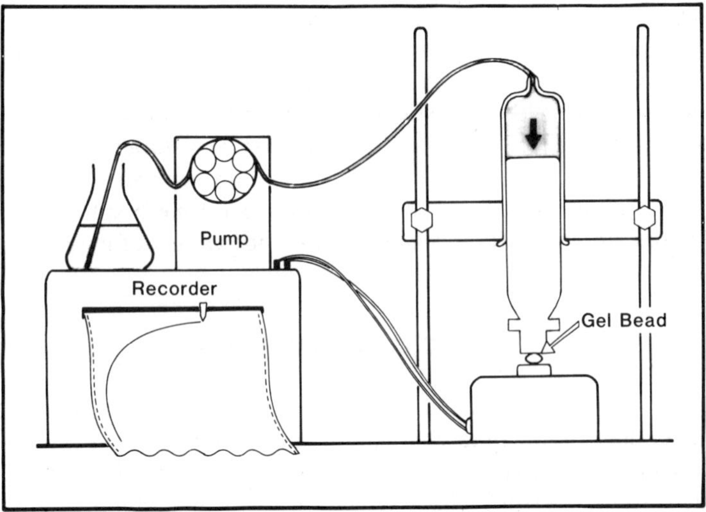

Fig. 15.14. Measuring the mechanical stability. The immobilized cell particle is placed on a force transducer (0–25 *N*). A hydraulically driven plunger (10 mL syringe) compresses the particle at a rate determined by the peristaltic pump, for example, 1.5 mm/min.

ance to abrasion. These factors depend on the size, the shape, the chemical composition, and the cell content of the particles (11, 369).

In the column type of reactor there is a pressure drop over the length of the bed. This pressure drop depends on the velocity and the viscosity of the fluid as well as on the particle size of the preparation and the dimensions of the reactor. If the particles are easily compressible they may cause plugging of the bed with a concomitant sharp reduction of the range of permissible flow rates. The compressability of a preparation to be used in packed bed reactors should therefore be assessed.

Figure 15.14 depicts a device for such measurements. It consists of two parts, a force-transducing plate and a plunger driven hydraulically by a laboratory peristaltic pump (11, 363). The device allows convenient determination of the rigidity (the force needed to compress the bead a defined distance), the critical breaking point (the force required to rupture the particle), as well as the elasticity (the extent to which the particle will resume its original shape and size). Figure 15.15 shows a comparison of the rigidity of three immobilized preparations. Two of the preparations (*B* and *C*) have been taken through a drying and shrinking cycle, which considerably increases the rigidity of the particles. One of the dried preparations (*B*) has been rehydrated to give particles of intermediate rigidity. The procedure should obviously be used primarily for screening purposes and should be supple-

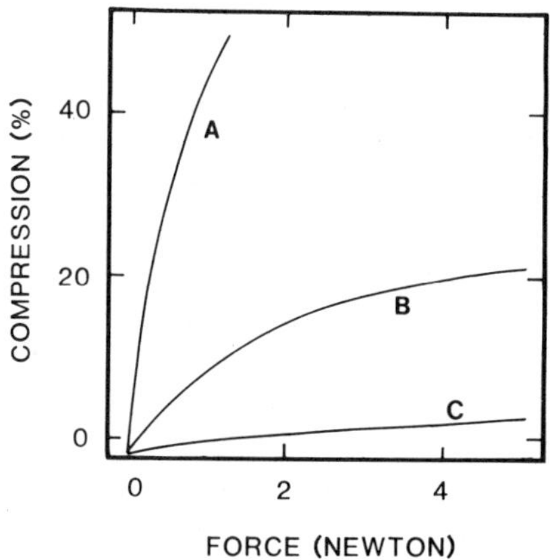

Fig. 15.15. Stability of related immobilized cell preparations. The curves were constructed from data obtained with the device shown in Figure 15.14. A = immobilized yeast. B = dried and rehydrated immobilized yeast. C = dried immobilized yeast.

mented by tests carried out under conditions similar to operating conditions, for example, by actually studying the behavior of a packed bed at different flow rates (282, 288). The stability of gels in a flowing system should be determined from long-time tests, as the gel may very well get compacted but at such a slow rate that the problem may remain unnoticed in short-time tests.

For preparations intended to be used in stirred tank reactors, another parameter needs to be investigated, the resistance to abrasion. In this respect, several variables are important, some related to the immobilized preparation, some to the reactor, and some to the medium. A test procedure has been suggested in which the preparation is stirred under defined conditions and the release of cells, of activity, or of fines, is measured by suitable means (11).

15.5.3. Characterization of the Catalytic Functions

When cells are immobilized, their kinetic properties may be considerably altered (see also section 15.3). The immobilization process may damage the cells and lead to a lower amount of active enzymes. On the other hand, the immobilization may permeabilize diffusion barriers, make the enzyme more available, and in practice make the preparation many times more active; moreover, living immobilized cells may multiply within the support, with a higher activity as a result.

Another characteristic contributing to changed catalytic properties is the partitioning effect of the support material. The effect may influence the concentration of substrates, products, effectors, hydrogen ions, and other substances. If, for example, the support enriches the microenvironment with substrate, the (apparent) K_M-value will decrease. If protons are enriched (negatively charged support) the (apparent) pH-optimum will increase (55).

The most important influence, however, is usually the effect of diffusion or mass transfer. Immobilized cells working under diffusional limitations will experience a microenvironment in which the concentration of substrate and product are different from those in the bulk solution. Diffusional limitations increase the effective K_M value and lead to broader pH-optimum and broader temperature optimum (55–58).

It is clear that these factors make it very difficult to predict the activity of preparations if only the amount of microorganisms added in the immobilization process is taken into account. Contrary to the situation with immobilized enzymes, though, the number of cells that actually get immobilized are usually close to 100% (entrapment of cells).

It thus seems profitable to characterize immobilized cell preparations kinetically to improve the prediction of behavior in later applications. Such characterization should include pH profile, temperature profile, K_M (where applicable), V_{max}, immobilization yield, and other properties influencing effectiveness.

The catalytic stability is another aspect that should be characterized, the operational as well as storage stability. Finally, it is useful to describe the biological and microbiological status of the preparation.

15.5.3a. Maximum Rate, Immobilization Yield, and Effectiveness

The *maximum initial reaction rate*, V'_{max} (units/g = μmol/min·g preparation), constitutes a good starting point when comparing different preparations with respect to their catalytic potential. V'_{max} should be determined under specified conditions (pH, temperature, buffer, and salts). Precautions must be taken in the experimental determination of the maximum initial reaction rate:

1. To eliminate the influence of partitioning effects, the system should be well buffered and a suitable ionic strength chosen.
2. To eliminate the influence of limiting external diffusion, the substrate concentration should be high and the stirring rapid (high pumping rates for packed beds). Whether the stirring is sufficient can be determined by varying the speed of the stirrer and observing whether the reaction rate changes.
3. To eliminate the effects of limiting internal diffusion, the substrate concentration should be high, at least 100 K_M, and the system should be well buffered to avoid internal pH shifts due to the reaction.

The V'_{max} value may be difficult to determine experimentally when the substrate solubility is low or when substrate inhibition occurs. Calculation of

V'_{max} by extrapolation is usually not possible, as the standard kinetic methods for soluble enzymes (e.g., Lineweaver–Burk plots) are not applicable to diffusion-limited systems.

The *immobilization yield* may be determined when V'_{max} is known.

$$\text{immobilization yield} = \frac{V'_{max} \cdot 100}{V_{max}} \%$$

V_{max} refers to the maximum rate of the same amount of free cells. As discussed above, the yield is usually less than 100% because of damage caused by the immobilization. But it may also be greater than 100% because of the changed properties of the cell (see Section 15.1 and ref. 7).

Determining the *effectiveness* of an operating immobilized-cell system is of great interest. In the absence of diffusional limitations and partitioning effects, the effectiveness is 1.0. In real applications it is lower. Knowledge of effectiveness is a very good point of departure when trying to improve the system and make it operate at its highest potential.

The effectiveness of an immobilized cell system is described by two effectiveness factors, the stationary and the operative. Both factors are functions of biocatalyst properties as well as external conditions, such as temperature or substrate concentration, which should consequently be specified.

The stationary effectiveness factor is defined as

$$\eta = \frac{\text{observed initial rate}}{\text{intrinsic initial rate}}$$

The intrinsic initial rate is understood to be the initial reaction rate in the absence of diffusional limitations and partitioning effects. The stationary effectiveness factor thus describes the performance of an immobilized cell system in the initial stage of reaction, that is, before any appreciable amount of products has been formed.

The operative effectiveness factor, on the other hand, describes the performance in terms of the time needed to convert a specified percentage of substrate, for instance 90%, and is therefore directly related to practical applications:

$$\eta_0 = \frac{T_0}{T_1}$$

In the case where the desired conversion is 90%, then T_1 is the time needed for the immobilized catalyst to convert 90% of the substrate and T_0 is the time needed for the immobilized catalyst to convert the same percentage of the substrate in the absence of diffusional limitations and partitioning effects. It is thus possible to determine the effectiveness factors directly.

This is simple when very mild immobilization methods are used, such as entrapment in alginate or agar. The intrinsic properties of the cells are then fully preserved and the intrinsic initial rate and T_0 can be determined in assays with free cells.

In other instances the situation is more complicated. Many immobilization methods may profoundly change the properties of the cells and even the intrinsic properties of the enzymes. As a consequence, the values of the intrinsic initial rate and T_0 are not the same as for free cells. For the determination of the intrinsic initial rate and T_0, conditions must thus be found that ensure the absence of diffusional limitations: First, the influence of external mass transfer and partitioning effects should be eliminated, as described for the V'_{max} determination (with the exception of high substrate concentration). Second, the internal diffusion limitations must be eliminated. This becomes more difficult when the effectiveness factors are to be determined for low substrate concentrations. One method is to reduce the size of the particle preparation carefully until the initial activity no longer increases and the time necessary for specified conversion no longer decreases. At this stage there are no diffusional limitations. The method is particularly simple in cases of reversible immobilization (adsorption, ionic crosslinking, and entrapment in gels or microcapsules that dissolve with suitable treatment).

The stationary and operational effectiveness factors may also be numeri-

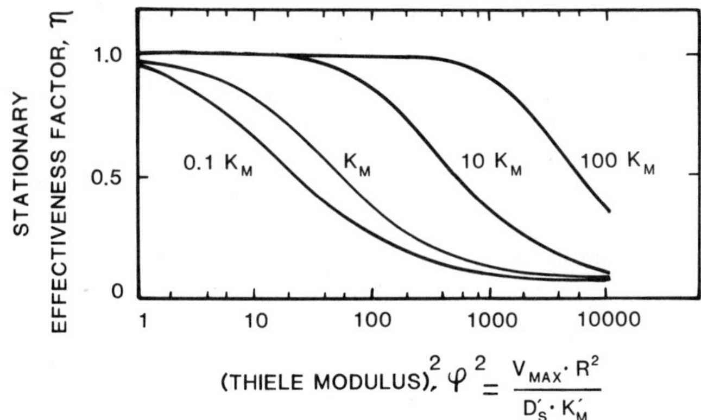

Fig. 15.16. Stationary effectiveness as a function of the square of the Thiele modulus. The curves refer to Michaelis–Menten kinetics in spherical particles in well-mixed reactors (no external diffusion limitation). R = radius, K'_M = intrinsic K_M, D'_S = effective diffusion coefficient. Reproduced with permission from ref. 58.

cally calculated (58). Often the results are presented graphically, as in Figure 15.16 in which the stationary effectiveness factor is given as a function of the Thiele modulus for a Michaelis–Menten type of reaction in a spherical particle. The square of the Thiele modulus is used in the figure (proportional to the enzyme content). As shown, the substrate concentration (expressed in K_M units) strongly influences the stationary effectiveness. The weakness of the numerically derived effectiveness factors is that the Thiele modulus contains many experimentally determined factors. The total error might thus be considerable (58).

15.5.3b. Catalytic Stability

Many reports describe increased catalytic stability for immobilized cells under operative or storage conditions (11). Immobilization has been reported to protect against extreme pH, high temperature, and other environmental perturbations.

As discussed in Section 15.2, this stabilization may be explained by a number of mechanisms. There is, however, a risk of overestimating the stabilizing effect of immobilization. Immobilized biocatalysts may not express their full intrinsic activity due to limiting diffusion of reactants. If only the apparent activity is determined, it may leave a false impression of very high stability, at least for some time.

Figure 15.17 illustrates the situation. After 1 week, the intrinsic activity has decreased to 50%, that is, the half-life is 1 week whereas the apparent or the observed activity is still 90% of its original value. Half-life calculations

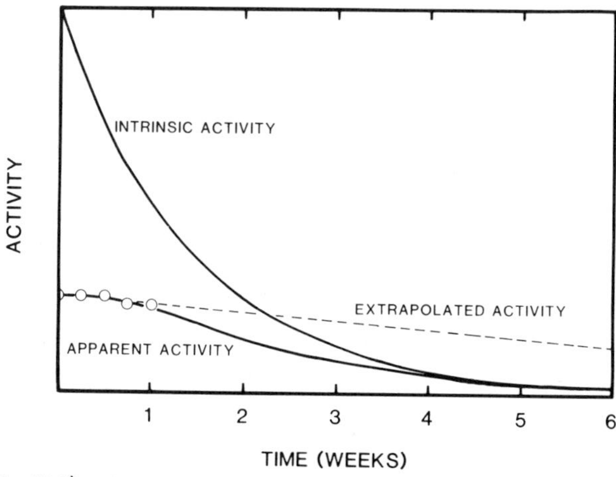

Fig. 15.17. Overestimation of the half-life of immobilized catalysts.

based on extrapolation of the apparent activity after 1 week would be 6 weeks (Figure 15.17), a grossly exaggerated value.

The example thus emphasizes the necessity of following the activity decay at least to the 50% level before ascribing a half-life time to the catalyst. Alternatively, means should be found to actually measure the decay of the intrinsic activity (see Section 15.5.3). Moreover, the term "half-life" should be reserved for exponential decay processes. Even the intrinsic activity of immobilized cells will seldom follow such a decay curve, and especially not if the cells are living. Living cell preparations may initially *increase* their activity, because of cell proliferation (29).

15.5.3c. Biological Characterization

Immobilization of whole cells may leave the cells in a number of physiological states, as illustrated in Figure 15.4. The question whether the cells are living is of special significance when long reaction sequences are to be executed, as they require a more or less intact cellular metabolism. A prerequisite for activation and reactivation of an immobilized cell preparation by intraparticular cell proliferation also requires the presence of living cells. What, then, are the methods for determining viability of immobilized cells (11, 364)?

1. Gentle dissolution of the support followed by plate spreading and colony counting.
2. Activity determination
3. Biomass determination
4. Microscopic inspection

   (2–4: before and after incubation in a nutrient medium.)
5. Oxygen uptake/oxygen evolution.
6. Influence of antibiotics.
7. Penetration of dyes.

The dissolution of the support and determination of the viability of released cells by plate counting techniques is a very attractive method because of its accuracy in determining the percentage of viable cells. The method, however, is fully applicable only when the cells are immobilized in a reversible way, such as adsorbed cells or cells immobilized in agar, alginate, carrageenan, gelatin, or related non-covalently crosslinked gels (32).

The microscopic observation of increased cell number upon addition of nutrients gives a direct qualitative verification of the presence of living cells. Qualitative verification, although in a less direct fashion, can also be obtained from activity and biomass determination—biomass entrapped in gels, for example, can be determined by dry weight, total protein (366), DNA, or, as recently proposed, by heme determination (chemiluminescence; 184). Continued oxygen uptake has been used as an indicator of living cells (177). A drastically changed activity or stability upon the addition of antibiotics may also serve as a marker for living cells (182). Finally, viability, or at least membrane integrity, can be tested with dyes (e.g., trypan blue), a technique commonly used in animal cell culture (81).

15.6. APPLICATIONS WITH IMMOBILIZED MICROORGANISMS

The potential applications of immobilized microorganisms are numerous and in many fields (Figure 15.18). Production of various chemicals is by far the most important area, but if sewage treatment in biological beds is included in the comparison, then water purification rivals chemical production. Conventional biological beds for water purification are not covered here, only immobilized cell systems specialized in the destruction of specific compounds.

15.6.1. Production of Biochemicals

A vast number of compounds may be prepared with immobilized cells, but only a few of the many processes described have so far actually reached the stage of commercial operation. The reason may be in part that processes involving immobilized biocatalysts usually have to compete with already established and optimized processes, such as those used in present-day fermentation technology. A number of processes are being studied by various companies who, for obvious reasons, do not want publicity. The industrially established processes are listed in Table 15.10; the table covers both immobilized enzymes and immobilized cells, the distinction being somewhat

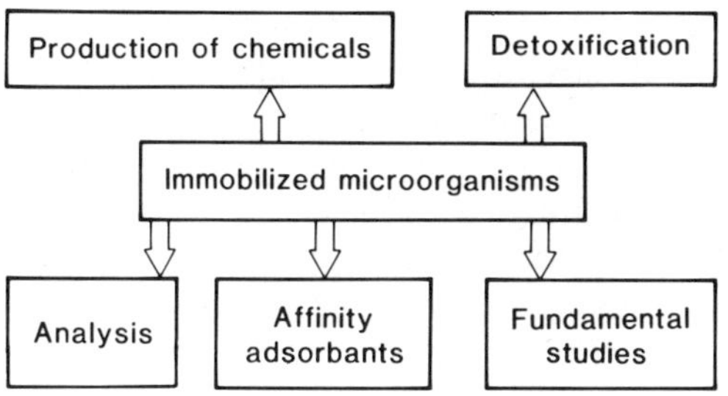

Fig. 15.18. Applications of immobilized microorganisms.

Table 15.10. Immobilized Enzymes and Cells Used in Industry

Product	Immobilized Enzyme (Cell)	Year of Introduction
L-Amino acids	Aminoacylase	1969
Fructose	Glucose isomerase (*Bacillus coagulans*)	1973
6-Aminopenicillanic acid	Penicillin amidase	1973
Aspartic acid	Fumarase (*Escherichia coli*)	1973
Malic acid	Aspartase (*Brevibacterium ammoniagenes*)	1974
Glucose/galactose	β-Galactosidase	1977

arbitrary. The examples of immobilized enzymes or cells in the table refer to preparations in which the cells most certainly are not living and possibly contain only one sort of enzyme activity, the desired one. The industrially established processes and those of potential industrial interest are discussed in more detail in the following subsections.

15.6.1a. Pharmaceuticals

Pharmaceuticals are, in general, high-value compounds produced in moderate quantities. They were therefore identified early as an ideal target for immobilized cell technology. The moderate scale was assumed to require only a limited investment and a relatively unproblematic start-up phase, and the expected increase in yield of valuable products seemed to ensure a profitable operation.

Mosbach and Larsson started to develop immobilized cells for the production of pharmaceutically interesting steroids (6). In one study, 11-β-hydroxylation by the fungus *Curvularia lunata* entrapped in polyacrylamide produced cortisol from Reichstein compound S (Figure 15.19; ref. 6). Later studies included dehydrogenation of cortisol by the bacterium *Arthrobacter simplex* entrapped in polyacrylamide or calcium alginate with formation of prednisolone in excellent yield. It was calculated (29) that even a very small column reactor (1–2 kg of biocatalyst) would suffice to supply the total Swedish demand for dehydrogenated steroids, assuming continuous operation and idealized operating conditions. This very promising although not fully realistic view was based on the very positive consequence of using living immobilized cells. Immobilized living cell preparations retained their activity for several months in the presence of nutrients, or even raised their activity level, as shown in Figure 15.5 (refs. 29, 182).

Treatment with nutrients has also been used to convert immobilized

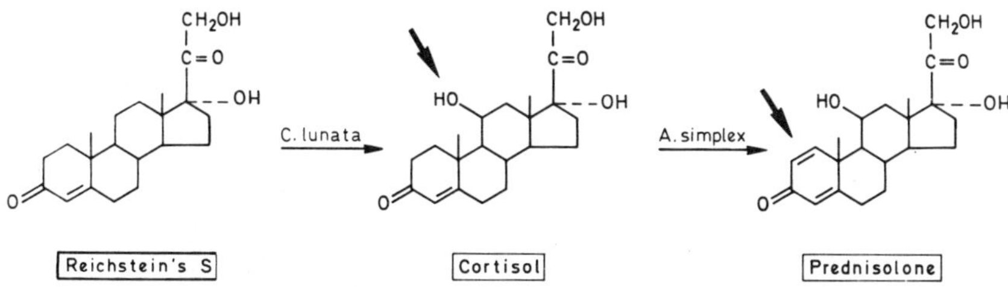

Fig. 15.19. Steroid transformations with immobilized cells. The arrows indicate the result of the 3-keto-steroid-1-dehydrogenase activity and the 11-β-hydroxylase activity. Reproduced with permission from ref. 188.

spores into vegetative cells. The approach may be of particular significance when drastic immobilization conditions are desired; spores may then stand a better chance of survival than vegetative cells. After completion of the immobilization, the addition of nutrients leads to germination and subsequent generation of the desired activity. The concept of such an *in situ* germination of immobilized spores was developed by Ohlson et al. with special reference to steroid hydroxylations (196). Intraparticular spore germination has also been used in the context of "ABE" (Acetone-Butanol-Ethanol) fermentations (301).

Side-chain degradation of sterols is an important reaction because it converts abundantly available sterols such as β-sitosterol and cholesterol to substances useful in the pharmaceutical industry, such as androstadienedione (ADD). So far the efficacy of immobilized cell systems has been rather modest in these systems; one reason may be the very low solubility of sterols in aqueous solutions. The solubility can be increased by suitable derivitization of the sterol. Atrat et al. (79) improved the conversion rates in this way. The poor solubility of steroids in general has been dealt with by Fukui and coworkers in a number of studies using hydrophobic gels to increase the steroid concentration in the vicinity of the cells (218, 227). Other steroid systems are shown in Table 15.11.

Another area of interest in terms of pharmaceutical production is the biosynthesis and the transformation of antibiotics. Suzuki and coworkers have reported the production of bacitracin (189, 206), penicillin G (62), and ampicillin (201) using *Bacillus* sp., *Penicillium chrysogenum*, and *Kluyvera citrophila* cells, respectively, immobilized in polyacrylamide. The conversion of penicillin to 6-aminopenicillanic acid (6-APA), a precursor of semisynthetic penicillins, has been extensively investigated (33, 165, 201, 325, 326, 327, 329). A difficulty is that the immobilized cells have often retained an unwanted competing lactamase activity, which may argue for the use of an immobilized purified enzyme—actually, the usual practice.

Table 15.11. Examples of Steroid Transformation with Immobilized Microorganisms

Type of Transformation	Substrate/Product	Microorganism	Support	Reference[a]
Δ^1-Dehydrogenation	Cortisol/prednisolone	*Arthrobacter simplex*	Alginate	32 (6, 29, 43, 46, 48, 181–183, 188, 216–218, 265, 274, 367)
Δ^1-Hydrogenation	ADD/AD[b]	*Mycobacterium phlei*	DEAE-cellulose	370 (371)
3-Hydroxy/7-hydroxysteroid-dehydrogenation	Dehydrocholic acid/12-ketochenodeoxycholic acid	*Brevibacterium fuscum*	Carrageenan	262
17-Hydroxysteroid-dehydrogenation	Testosterone/AD[b]	*Nocardia rhodocrous*	Polyurethane	46
20-Hydroxysteroid-dehydrogenation	Cortisol/20-hydroxy-derivative	*Mycobacterium globiforme*	Cellulose	371 (372)
9-α-Hydroxylation	AD/9-α-hydroxy-AD[b]	*Corynebacterium* sp.	Polyacrylic acid derivative	373
11-α-Hydroxylation	Progesterone/hydroxyprogesterone	*Rhizopus nigricans*	Agar	63
11-β-Hydroxylation	Reichstein S/cortisol	*Curvularia lunata*	Polyacrylamide	48 (6, 181, 188, 196, 374)
Sterol oxidation	Cholesterol/cholestenone	*Nocardia erythropolis*	DEAE-cellulose	77 (219, 227)
Sterol side-chain degradation	Cholesterol derivative/ADD-derivative[b]	*Mycobacterium phlei*	Polyacrylamide	79 (232, 78)

[a] References to systems with same type of transformation are in parentheses.
[b] AD = Androst-4-ene-3,17-dione; ADD = androsta-1,4-diene-3,17-dione.

15.6.1b. Organic Acids, Including Amino Acids, and Peptides

Table 15.12 lists the organic products achieved with immobilized microorganisms. Chibata and coworkers were the first to design an industrial process based on immobilized microorganisms; they achieved the production of an amino acid, L-aspartic acid:

$$\text{ammonium fumarate} \xrightarrow[E.\ coli]{\text{immobilized}} \text{L-aspartic acid}$$

The system reduced the operational costs by a substantial 40%, thanks to the reuse of biocatalyst (40). Several modifications have improved the productivity of the biocatalyst; they include the substitution of carrageenan for polyacrylamide and stabilization with hexamethylene diamine and glutaraldehyde (8, 251–253, 255).

The first to report the actual production of an amino acid were Slowinski and Charm, who employed polyacrylamide-entrapped *Corynebacterium glutamicum* cells to produce glutamic acid, a compound used as a flavor enhancer. This preparation possessed much more activity than the analogous free cell system (158). The production of glutamic acid meant synthesis from nutrients alone, whereas the aspartic acid process is based on a single enzyme activity. Prolonged *de novo* synthesis from nutrients obviously requires living cells.

The breakdown of amino acids by immobilized cells has also been studied. Franks reported the breakdown of arginine to ornithine and putrescine using *Streptococcus faecalis* cells (157), whereas Chibata's group examined the production of citrulline using *Pseudomonas putida* cells (160).

A related class of compounds is the α-keto acids, which have been proposed as dietary substitutes for amino acids for patients with renal insufficiency. Brodelius et al. used the immobilized yeast *Trigonopsis variabilis* to produce about 20 different α-keto acids by oxidative deamination of amino acids (299, 354). *T. variabilis* contains a D-amino acid oxidase and thus produces L-amino acids as well as the α-keto acid from a racemic amino acid mixture. Other microorganisms, such as the *Providencia* sp., have the reverse specificity and produce keto acids from L-amino acids (358).

A number of immobilized cell systems have also involved the production of organic acids not directly related to amino acids. The most prominent example is L-malic acid, which is used as a substitute for citric acid. The acid can be advantageously produced by immobilized *Brevibacterium flavum*, as amply demonstrated by Chibata and coworkers:

$$\text{fumaric acid} \xrightarrow[\textit{Brevibacterium flavum}]{\text{immobilized}} \text{L-malic acid}$$

Table 15.12. Examples of the Production of Organic Acids, Including Amino Acids, and Peptides, by Means of Immobilized Microorganisms

Product	Reaction	Microorganism	Support	Reference[a]
L-Aspartic acid	Aspartase	*Escherichia coli*	Carrageenan	253 (8, 159, 161, 167, 186, 195, 265, 251, 252, 330, 255)
L-Alanine	Aspartate decarboxylase	*Pseudomonas dacunhae*	Carrageenan	365
L-Glutamic acid	Multistep	*Corynebacterium glutamicum*	Polyacrylamide	158 (263)
L-Tryptophan	Tryptophanase	*Escherichia coli*	Chitosan	317 (190)
L-Ornithine	Two-step	*Streptococcus faecalis*	Polyacrylamide	157
D-Phenylglycine	Hydantoinase	*Bacillus* sp.	Polyacrylamide	208
L-Isoleucine	Multistep	*Serratia marcenscens*	Carrageenan	257
L-Methionine	Amino acylase	*Aspergillus ochraceus*	Crosslinked mycelium	147
L-Tyrosine	β-Tyrosinase	*Erwinia herbicola*	Collagen	268

About 20 α-keto acids (e.g. α-keto-γ-methiolbutyric acid)	D-Amino oxidase	*Trigonopsis variabilis*	Alginate	299 (354, 358, 359)
L-Malic acid	Fumarase	*Brevibacterium flavum*	Carrageenan	50 (166, 168, 175, 195, 252, 254)
Citric acid	Multistep	*Aspergillus niger*	Collagen	270 (272)
Acetic acid	Two-step	*Acetobacter aceti*	Ceramic	88 (139, 137, 136)
Lactic acid	Multistep	Mixed culture	Gelatin	88 (139, 137, 136)
L-Citrulline	Arginine deiminase	*Pseudomonas putida*	Polyacrylamide	160 (157)
Formic acid	Multistep	*Clostridium butyricum*	Polyacrylamide	174
Urocanic acid	Histidine ammonialyase	*Achromobacter liquidium*	Polyacrylamide	36 (132)
2-Ketogluconic acid	Multistep	*Serratia marcescens*	Collagen	265
Glutathione	Glutathione synthetase	*Saccharomyces cerevisiae*	Carrageenan	256 (179, 204, 210, 233)

[a] References to other systems with the same product are in parentheses.

By treating the cells with bile extract, the desired fumarase activity was increased and the generation of an unwanted side-product, succinic acid, was decreased (50).

Many other organic acids have been produced with immobilized microorganisms, as shown in Table 15.12.

15.6.1c. Coenzymes and Related Compounds

A number of coenzyme and related products that have been synthesized by immobilized cells are listed in Table 15.13. As mentioned previously, one of the attributes of immobilized whole-cell catalysts is their retention of the coenzyme regenerating capacity. This advantage eliminates the need to supply the coenzyme exogenously. Thus, for example, the generation of ATP by immobilized cells has been investigated in connection with the production of glutathione (233), $NADP^+$ (233, 336), coenzyme A (356), and CDP-choline (337, 356, 213). As seen in Table 15.13 the generation or regeneration of ATP has been investigated with immobilized yeast cells, immobilized chloroplasts (as well as chromatophores), and immobilized mitochondria.

The first to report on the formation of coenzymes by immobilized cells was probably Shimizu et al. (164). They entrapped *Brevibacterium ammoniagenes* cells in polyacrylamide and produced coenzyme A from pantothenic acid, cysteine, and ATP—a multistep conversion. The synthesis of one mole coenzyme A requires four moles ATP, three of which can be regenerated. Samejina et al. therefore coimmobilized *Saccharomyces cerevisiae* with *Brevibacterium ammoniagenes* and produced coenzyme A without adding ATP to the reaction medium (356).

CDP-choline (cytidine diphosphate choline), a precursor to lecithin, is an important therapeutic agent in the treatment of brain injuries. The transformation of CMP (cytidine monophosphate) to CDP-choline is also a multistep sequence and requires the regeneration of ATP. CDP-choline was produced by immobilized yeast cells without adding ATP to the reaction medium because ATP was continuously regenerated by the glycolytic system of the immobilized yeast (213, 228, 356, 337).

Another coenzyme that has attracted considerable attention is $NADP^+$. Thus, NAD^+ was converted, in good yield, to the much more expensive $NADP^+$ using the NAD-kinase activity of polyacrylamide entrapped *Brevibacterium ammoniagenes* or *Achromobacter aceris* cells (37, 180, 222). As the production of $NADP^+$ also requires ATP, the coenzyme regenerating system of immobilized yeast cells described above was employed (336, 233). The results showed that both coimmobilized and separately immobilized bacteria and yeast cells were successful in eliminating the external requirement for ATP.

Table 15.13. Examples of the Production of Coenzymes and Related Compounds by Immobilized Microorganisms and Organelles

Product	Reaction	Cell or Organelle	Support	Reference[a]
ATP	Phosphorylation	*Saccharomyces cerevisiae*	Microencapsulation	356 (337, 336, 220, 233)
ATP	Phosphorylation	*Rhodopseudomonas capsulata* (chromatophores)	Albumin and glutaraldehyde	64 (285, 150)
ATP	Phosphorylation	Chloroplast	PVA[b]	331 (377)
ATP	Phosphorylation	Mitochondria	Agar	378
$NADP^+$	NAD-Kinase	*Brevibacterium ammoniagenes*	Polyacrylamide	37 (180, 222, 336, 233)
NADH	Hydrogenase	*Alcaligenes eutrophus*	Alginate	290
CDP-choline[c]	Multistep	*Saccharomyces cerevisiae*	Microencapsulation	356 (337, 212, 213, 228)
Coenzyme A	Multistep	*Brevibacterium ammoniagenes*	Polyacrylamide	164 (194, 200, 356)
FAD[d]	Adenyl transferase	*Arthrobacter oxydans*	PVA[b]	200
PLP[e]	Pyridoxine oxidase	*Pseudomonas fluorescens*	PVA[b]	200
2-Ketogulonic acid	Two-step	*Gluconobacter melanogenes*	Polyacrylamide	169
Panthothenic acid	Panthothenic acid synthetase	*Escherichia coli*	Agar	241

[a] References to systems with the same product are in parentheses.
[b] PVA = Polyvinyl alcohol.
[c] CDP = Cytidine diphosphate.
[d] FAD = Flavin adenine dinucleotide.
[e] PLP = Pyridoxal phosphate.

Klibanov and Puglisi recently reported the use of *Alcaligenes eutrophus* cells entrapped in calcium alginate for the hydrogenase catalyzed reduction of NAD^+ to NADH (290). This seems to be a cheap and effective way to obtain reduced adenine nucleotides, and it may be used in the future in combination with other dehydrogenases to produce other valuable reduced compounds.

15.6.1d. Sugars

Together with the oil crisis of 1973, which led to the search for alternative energy sources, there developed a "sugar crisis," as evidenced by the rapid rise in raw sugar prices in the early 1970's. This promoted the search for other sources of sweeteners than cane sugar and beet sugar. The large glucose reserves available, such as corn starch in the USA, promoted the commercialization of the glucose isomerase process. In this process glucose is partly (~42%) converted into fructose, organoleptically a much sweeter sugar.

The glucose isomerase process based on immobilized biocatalysts is now producing about 2×10^6 tons dry weight of high fructose syrup per annum (15), and the volume is expected to double before 1985. The process handles large quantities of low-priced substrates and products, a situation in marked contrast to what was originally proposed to be ideal for immobilized catalysts, namely small volume and high price. The secret of the phenomenal success of the glucose isomerase process is, therefore, not only the traditional advantages attributed to immobilized biocatalysts; it is rather that the process gave the market a new product that was needed and that could not be obtained by other means (379).

The glucose isomerase process is a one-enzyme reaction, isolated enzyme as well as whole cells have been used. The cells, however, are certainly not living in the commercial preparations. The subject has been extensively reviewed (380).

The first report of an immobilized whole cell system for the isomerizaton of glucose was given by Vieth et al. using *Streptomyces phaeochromogenes* cells immobilized in collagen membranes (264). The immobilized complex had an activity of 131 units per gram and was used in a continuous column process for 40 days at 70°C. Saini and Vieth later examined the kinetic parameters of collagen-entrapped *S. venezuelae* cells (49) and found that even at low substrate concentrations the effectiveness factor was 0.95. Hence, the mass transfer limitations for glucose-to-fructose conversion by the immobilized preparation were minimal. Vieth and Venkatasubramanian have thoroughly described the preparation and properties of these systems (266, 267).

Kumakura et al. investigated glucose isomerase using *S. phaeochromogenes* cells entrapped in various acrylic acid derivatives, polymerized by γ-irradiation (214, 215, 221). Linko and coworkers have studied glucose isomerase activity of cellulose-entrapped *Actinoplanes missouriensis* cells (318, 319, 321, 324), and Chibata and coworkers have reported the use of carrageenan as a suitable support for *S. phaeochromogenes* cells (60, 252). A number of crosslinked or flocculated whole cell preparations have also been used (144, 140, 141, 156, 381). In particular, Joosten and coworkers investigated the kinetic parameters of flocculated preparations (140, 381, 142). Gelatin has also been used as a carrier for cells with glucose isomerase activity (277, 152).

As we have seen, the glucose isomerase process based on immobilized biocatalysts is very important economically. It is therefore pertinent to have a closer look at one of the commercially available preparations and its recommended use. The biggest producer is immobilized glucose isomerase for marketing is Novo Industri, Denmark. Their Sweetzyme is an extrudate containing *Bacillus coagulans* cells (382) crosslinked by glutaraldehyde. The crosslinked immobilized cell preparation is supplied as dry hard granules, 0.3–1 mm in diameter and very stable under storage at 4°C (1% loss per month). Figure 15.20 shows a selection of data supplied by the manufacturer. The pH profile has an optimum at pH 8 and the temperature optimum is very high, over 80°C. However, the stability at such a temperature is very low (half-life of a few hours) compared with that at 60°C, which is the temperature chosen for industrial applications. The activity measured by initial conversion rates is about 200 μmol/min per g catalyst. The choice of 60°C in large-scale applications is a compromise among various concerns such as productivity, stability of catalyst, formation of side products, and risk of infection.

In production plants the biocatalyst is commonly used for three half-lives, giving an operational life of about 3000 hr, or four months. In order to allow an even production rate the biocatalyst is distributed among several reactors, about 4 to 10, and the operating cycles of the reactors are usually staggered. For example, a huge plant producing 500 tons per day might have 10 reactors (1.5×5 m). The flow through each reactor decreases during its operational lifetime down to 12.5% of the initial rate to compensate for the diminishing activity, thus ensuring the desired conversion level of 42% fructose. The staggered arrangement of the reactors, however, keeps the flow variations of the whole plant as low as $\pm$ 10% of the average flow.

Figure 15.21 is a schematical outline of a glucose isomerizing plant with four in-line reactors. A number of auxiliary functions are involved in the process.

Many microbial cells contain β-galactosidase activity that can be used for the breakdown of lactose to glucose and galactose, a process of importance

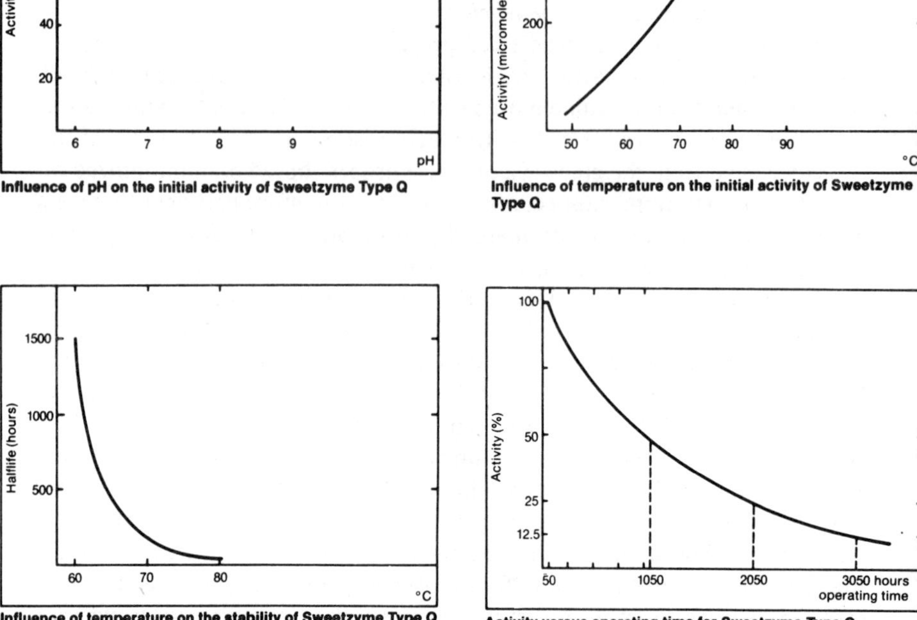

Fig. 15.20. Some properties of the commercial immobilized cell preparation Sweetzyme. Reproduced with permission from ref. 382.

in the dairy industry. Toda was one of the first investigators to examine the β-galactosidase activity of immobilized microorganisms (238). In 1975 he reported entrapping *E. coli* cells in spherical agar pellets and studying the kinetics of a fixed-bed reactor of the immobilized preparation. Linko and coworkers later described a β-galactosidase system using *Kluyveomyces fragilis* or *K. lactis* cells entrapped in cellulose derivatives (321, 323, 324). Jirku et al. applied hydroxyalkylmethacrylate for the fixation of permeabilized *Zymosaccharomyces lactis* cells (134) and polyphenyleneoxide to immobilize permeabilized *Kluyveomyces lactis* cells (102). De Rosa et al. have immobilized *Caldariella acidophila* in various derivatives and obtained thermally stable lactase preparations (203, 322, 151, 383).

Another sugar transformation is the conversion of sucrose to glucose and fructose. Johnson and Ciegler found in a very early work that ECTEOLA-cellulose was the most suitable ion-exchange resin for adsorbing spores of *Aspergillus oryzae*-containing invertase activity (73). Later, Toda and

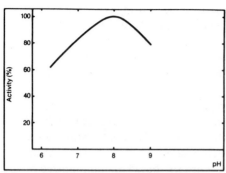

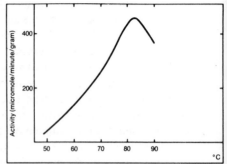

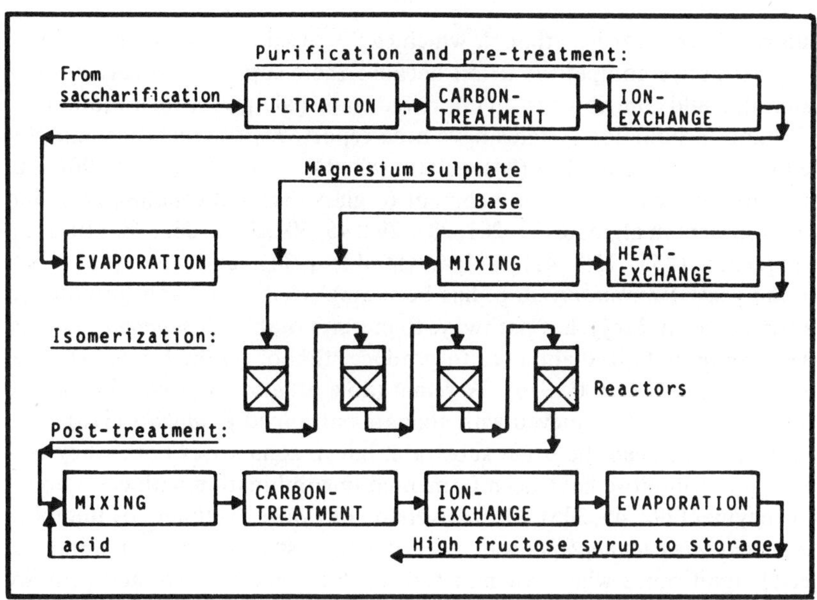

Fig. 15.21. Layout of a glucose isomerization plant. Reproduced with permission from ref. 382.

Shoda described the kinetics of sucrose inversion using agar-entrapped *Saccharomyces pastorianus* cells (237). Linko and coworkers investigated both cellulose and alginate as carriers for entrapping *S. cerevisiae* cells for their invertase activity (321, 324, 291, 294). Using calcium alginate as carrier, they estimated a remarkable half-life of 5.7 years (291, 294). Gelatin-entrapped *S. cerevisiae* cells have also been used for the inversion of sucrose (275, 280). D'Souza and Nadkarni reported an original idea for the conversion of sucrose to fructose and gluconic acid by means of a multistep system including yeast cells (invertase), glucose oxidase, and catalase. The glucose oxidase was coupled to yeast cells with Concanavalin A and the aggregate entrapped in polyacrylamide polymerized by γ irradiation (224).

A number of other systems involving sugars have been reported. Examples are the production of phosphorylated glucose, in both the 1 and 6 position (163), the conversion of sorbitol to L-sorbose (255) as well as L-sorbose to L-sorbosone, important intermediates in the biosynthesis of vitamin C (34), and the oxidation of a number of saccharides (176).

15.6.1e. Fuels and Chemical Feedstocks

The rise in petroleum costs has prompted the search for alternative energy sources, particularly renewable ones. Brazil and other sugar-producing countries have regarded ethanol, which can be used in today's automobiles, as an attractive competitor to oil because their supply of sugar, the raw material, could contribute to an improved trade balance for their economies.

Ethanol production systems have been reported with yeast cells immobilized by adsorption to glass (84), gelatin (82, 83), wood (93), brick (90), and amberlite (80); by covalent attachment to glass (84) and cellulose (76); and by entrapment in alginate (31, 281, 287, 292, 45, 302, 305, 306, 308, 310, 315), carrageenan (255, 259, 41), polyacrylamide (211), and agar (66). Usually the yeast *Saccharomyces p.* has been employed for ethanol production via the Embden–Meyerhof pathway. Bacteria, notably *Zymomonas mobilis,* have been immobilized and used to produce ethanol via the Entner-Doudoroff pathway (384, 100, 316). A remarkable ethanol productivity of 116 g/L/hr was recently achieved with alginate-entrapped *Z. mobilis* (316).

Glucose has been the usual substrate, but in some cases both whey permeate and cellubiose have been fermented in combination with coimmobilized enzymes (292, 45, 305). The conversion of xylose to ethanol by immobilized *Pachysolen tannophilus* was recently demonstrated (311), an ability of special significance when raw materials such as wood, straw, and peat are considered. Minier and Goma reported the combination of immobilized yeast cell fermentation with solvent extraction (90). Consequently, complete fermentation of 407 g glucose/L was achieved. As mentioned in Section 15.4.6, two-aqueous phase partitioning was employed for simultaneous catalyst confinement and product extraction (347, 348).

Hydrogen is another energy alternative that has attracted attention in recent years. A number of bacteria and algae produce hydrogen under anaerobic conditions; consequently immobilized cell systems have been investigated for the production of hydrogen.

The bacterium *Clostridium butyricum* has been extensively studied. Suzuki and coworkers immobilized the bacterium in polyacrylamide or agar (with or without acetylcellulose filters) for continuous hydrogen evolution from glucose as well as wastewater (170, 385, 386). In one study, hydrogen formation from entrapped *C. butyricum* was conveniently studied with an enzyme-transistor probe (387). These hydrogen production systems have been employed in combination with fuel cells to obtain a continuous current of up to 0.8 A and a cell voltage (five cells) of 2.2 V over a 10-day period (174, 243, 244, 385, 386). By using immobilized chloroplasts in combination with *C. butyricum* in the presence of an electron carrier, the photolysis of water was achieved and employed together with the hydrogen oxygen fuel cell to obtain a photocurrent (243, 247, 250). In a subsequent publication, Suzuki and colleagues reported replacing the immobilized chloroplasts, which have a very short half-life, with immobilized *Chlorella vulgaris* cells and thereby continuously producing hydrogen for six days, using $NADP^+$ as an electron carrier (388).

Photosynthetic bacteria of the family *Rhodospirillaceae* have been immobilized and used to photometabolize a number of organic substrates (such as malate, wastewater, glucose, and acid-hydrolyzed cellulose) to produce H_2 under anaerobic conditions (239, 246, 249). An economic evaluation estimated the cost of producing 1000 ft^3 of H_2, using immobilized *Rhodospirillium rubrum* cells, as about $10 (246). Present H_2 cost is $3 to $6 per 1000 ft^3. Thus, if H_2 production employing immobilized whole cells is to be competitive in the future, the system needs to be further optimized or the cost of H_2 produced by conventional methods must increase or both. Weetall and Krampitz used immobilized *Anacystis nidulans* cells in combination with agar entrapped *R. rubrum* cells for the biophotolysis of water (242). Since *R. rubrum* is incapable of oxidizing H_2O, *A. nidulans* cells were employed for producing molecular oxygen and reducing exogenous $NADP^+$. *R. rubrum*, in turn, reoxidized NADPH to produce molecular hydrogen.

Other microorganisms have also been considered. One is the blue-green algae *Anabaena* sp., immobilized with glass beads as well as agar and used to produce hydrogen (245, 389). A photocurrent of 15–20 mA was continuously obtained from a fuel cell containing these immobilized cells (245). Another blue-green algae, *Mastigocladus laminosus,* was immobilized in calcium alginate onto a SnO_2 electrode and subsequently produced a photocurrent (295). Chloroplasts were immobilized together with hydrogenase and produced H_2 in the presence of ferrodoxin (303).

Methane production has also been a topic of interest since the oil crisis. A number of reports have appeared describing the potential application of immobilized cells to these systems (197, 304, 101). Scherer et al. entrapped whole cells of *Methanosarcina barkeri* cells in calcium alginate for the conversion of methanol to methane (304). A half-life of four days was observed in a buffer medium and the conversion rate was 9.5 mmol/hr/g dry cells, equivalent to about 16 gas-volume/reactor-volume/day.

The feedstock for the chemical industry is, as a rule, produced from petroleum. With rising oil prices alternative production means, including the use of immobilized cells, may become economical. Häggström and Molin utilized the well-known "ABE" system of *Clostridium acetobutylicum* for the continuous production of acetone, butanol, and ethanol using immobilized cells (301). Krowel et al. have looked at a very similar system for the production of butanol and isopropanol (293). The production of 2,3-butanediol by similar means has been investigated (258).

15.6.1f. Miscellaneous Products

A number of compounds produced by immobilized cells do not fall into the aforenamed groups. For example, Chibata and coworkers have investigated the conversion of glycerol to dihydroxyacetone using polyacrylamide entrapped *Acetobacter xylinum* cells (192). Holst et al. employed *Gluconobacter oxydans* cells (having high catalase activity) entrapped in calcium alginate for the same conversion (357). Since the conversion step requires oxygen, an improved oxygenation method was tried. When hydrogen peroxide was added to the substrate solution, a substantial enhancement of the productivity rate was observed. Dihydroxyacetone production by polyacrylamide-entrapped *G. oxydans* cells has also been reported (390).

An important concern in agriculture is nitrogen fixation. In the past, the extensive use of ammonia-based nitrogenous fertilizers compensated for the nitrogen depletion of cultivated land that occurred mostly as a result of intensive cereal-grain production. The Haber process is employed to produce ammonia synthetically from nitrogen and hydrogen, but since it is directly dependent on fossil fuel, a search for alternative means of obtaining ammonia is of interest. One proposed method is to utilize immobilized cells having high nitrogenase activity. Thus, *Azotobacter vinelandii* cells were adsorbed to an anion exchanger and retained a nitrogenase activity of 4.2 μmol/min/g gel (74). In a subsequent report the same cell strain was covalently coupled to cellulose and used continuously for nearly one month without substantial decrease in nitrogenase activity (76). Other reports have also investigated nitrogen fixation, either with *Klebsiella pneumoniae* cells immobilized with collagen (391) or with *Azotobacter chroococcum* cells immobilized in agar (248).

As discussed earlier, the production of high molecular weight products, such as proteins, by immobilized cells, in particular by entrapped cells, represents a formidable challenge due to the diffusional limitations inherent in the immobilization technique. In spite of these difficulties, two reports of enzyme production by entrapped microbial cells have appeared (54, 51). α-Amylase was continuously produced by *Bacillus subtilis* cells entrapped in 5% polyacrylamide in a batch system (54). The stability and productivity of the immobilized cells were greater than the analogous free cell system. In order to reduce cell leakage, the gel particles were intermittently washed with 70% ethanol to sterilize the peripheral region of the gel. In a subsequent study on protease production by *Streptomyces fradiae* cells entrapped in polyacrylamide, the surface sterilization step was found to have a lengthening effect on the stability of the preparation (51). Thus, an immobilized cell preparation with high productivity (12,000 U/mL/hr) and long stability ($t_{1/2}$ = 30 days) was obtained.

Other products that have been obtained from immobilized cells include the conversion of methanol to formaldehyde by polyacrylamide-entrapped yeast cells (199) and the production of L-menthol from DL-menthylsuccinate, an important compound in the cosmetic industry (229). In the latter case *Rhodotorula nimuter var. texensis* cells entrapped in hydrophobic gels carried out the conversion in a water-saturated organic phase.

15.6.2. Analysis Systems

Immobilized enzymes are extensively used in analysis, often integrated with a transducer to form devices such as enzyme electrodes (392, 393, 396) and enzyme thermistors (394, 392). The application of immobilized cells in analysis is of more recent date, nevertheless several reviews have appeared (395, 351, 397) and several applications have been made (see Table 15.14). Analytical systems based on immobilized cells differ in several respects from those of immobilized enzymes:

1. The specificity is usually lower with cells, especially if the cells are living and thus contain a multitude of enzymes. In general, this is a severe limitation, as specific compound determination cannot be made in a complex mixture because of the risk of interference. But the low specificity is sometimes beneficial in the sense that a single immobilized cell preparation can be used for the determination of many compounds provided that they do not appear in the same sample. Alternatively, the low specificity can be used to obtain a cumulative measure of the metabolizable substances in a sample. The measure of BOD (biological oxygen demand) is a prominent example.
2. The time needed to obtain a steady-state response is usually longer than for enzymes, owing to various diffusion barriers present in the cell. But with the use of initial slopes in the response curve, the time required for analysis can be as short as one minute (235)!
3. Intact cells contain coenzymes and coenzyme regenerating systems. Addition of coenzymes—an expensive necessity in many enzyme-based systems—is therefore obviated (411, 412).
4. The sensitivity of immobilized cell systems is usually lower than for immobilized enzyme systems owing to the higher catalyst density possible with the latter systems. In some instances, however, immobilized cells should allow very sensitive analysis. Examples of this are substances that promote or inhibit growth or metabolism, such as vitamins (414), toxins (animal cells) and mutagens (410).
5. The lifetime of an analytical system based on immobilized living cells is sometimes considerably extended by the inclusion of nutrients, which restore the desired activity (399, 400).

The first reports of immobilized cell systems in analysis used conventional immobilization supports in order to retain the cells in the vicinity of the transducer (i.e., oxygen electrodes, pH electrodes, ammonia-gas electrodes, thermistors, and potentiometric platinum electrodes) that was used to measure the cellular response. More recently, special filters and membranes have also been used for this purpose (see also Table 15.14). The BOD measure-

Table 15.14. Examples of Analytical Systems Using Immobilized Cells

Substance	Cell	Support	Transducer	Sensitivity	Reference
Acetic acid	*Trichosporon brassicae*	Acetyl cellulose filter	Oxygen electrode	5 mg/mL	398
Amino acids					
L-Arginine	*Streptococcus faecium*	Dialysis membrane	Ammonia gas electrode	50 μM	341
L-Aspartate	*Bacterium cadaveris*	Dialysis membrane	Ammonia gas electrode	50 μM	399
L-Cysteine	*Proteus morganii*	Dialysis membrane	H_2S electrode	0.3 mM	400
Glutamic acid	*Escherichia coli*	Cellophane membrane	CO_2 electrode	8 mg/mL	350
L-Glutamine	*Sarcina flava*	Dialysis membrane	Ammonia gas electrode	20 μM	344
L-Glutamine	Porcine kidney (animal cells)	Dialysis membrane	Ammonia gas electrode	50 μM	343
L-Histidine	*Pseudomonas* sp.	Cellulose acetate–nitrate filter	Ammonia gas electrode	100 μM	401
Phenylalanine	*Leuconostoc mesenteroides*	Acetyl cellulose filter	Oxygen electrode	0.1 μg/mL	402
L-Serine	*Clostridium acidiurici*	Dialysis membrane	Ammonia gas electrode	50 μM	342
Ammonia	Nitrifying bacteria	Acetyl cellulose filter	Oxygen electrode	0.05 mg/L	403
Antidiuretic hormone	Toad bladder (animal cells)	—	Sodium electrode	10 ng/mL	404
Arsenate	*Saccharomyces cerevisiae*	Polyacrylamide	Thermistor	2 mM	171
Assimilable sugars	*Brevibacterium lactofermentum*	Cellophane membrane	Oxygen electrode	0.2 mM	405
BOD[a]	Soil bacteria	Polyacrylamide	Oxygen electrode	100 ppm	172
BOD	*Clostridium butyricum*	Polyacrylamide	Platinum electrode	10 ppm	173

BOD	*Trichosporon cutaneum*	Acetyl cellulose filter	Oxygen electrode	10 ppm	406
BOD	*Hansenula anomala*	Dialysis membrane	Oxygen electrode	10 μM	349
Cephalosporins	*Citrobacter freudii*	Collagen	pH electrode	62.5 μg/mL	407
2,4-Dinitrophenol	*Saccharomyces cerevisiae*	Polyacrylamide	Thermistor	1 mM	171
Ethanol	*Trichosporon brassicae*	Acetyl cellulose filter	Oxygen electrode	2 mg/L	408
Formic acid	*Clostridium butyricum*	Acetyl cellulose filter	Platinum electrode	10 mg/L	202
Glucose	*Saccharomyces cerevisiae*	Polyacrylamide	Thermistor	1 mM	171
Glucose	*Pseudomonas fluorescens*	Collagen	Oxygen electrode	2 mg/L	271
Methane	*Methylomonas flagellata*	Acetyl cellulose filter	Oxygen electrode	13.1 μM	409
Mutagens	*Bacillus subtilis*	Acetyl cellulose filter	Oxygen electrode	1.6 μg/mL	410
Nicotinic acid	*Lactobacillus arabinosus*	Agar	pH electrode	50 ng/mL	240
Nitrate	*Azotobacter vinelandii*	Dialysis membrane	Ammonia gas electrode	10 μM	411
Nitrilotriacetic acid	*Pseudomonas* sp.	Dialysis membrane	Ammonia gas electrode	100 μM	412
Nystatin	*Saccharomyces cerevisiae*	Collagen	Oxygen electrode	0.5 U/mL	413
Steroid	*Nocardia opaca*	Polyacrylamide	Glassy carbon electrode	2.5 μm	234, 235
Thiamine	*Saccharomyces cerevisiae*	Alginate	Oxygen electrode	0.07 μg/mL	414

[a] BOD = Biological oxygen demand.

ments listed in the table are very fast (a couple of hours) compared with standard procedures (usually five days).

15.6.3. Detoxification

A major and increasing concern of modern society is to prevent waste materials and toxic substances from polluting the environment. Several large-scale systems for the treatment of municipal wastewater involve adsorbed cells. In the following discussion we concern ourselves with the removal of specific compounds found in industrial and municipal water systems.

High concentrations of nitrate and nitrite in drinking water are often a serious problem in areas with intensive agriculture. Considerable efforts have been made to devise methods for removal of the compounds from the water. Mohan and Li and their colleagues were among the first groups to report an immobilized cell system for nitrate and nitrite reduction (338). They employed *Micrococcus denitrificans* cells encapsulated in liquid-surfactant membranes for this purpose. Holló et al. reported the use of *Pseudomonas aeroginosa* cells adsorbed to polyvinyl chloride or polypropylene for denitrification (104, 105); simultaneously they observed the removal of heavy metals such as plutonium, lead, and zinc. Nilsson et al. employed *Pseudomonas denitrificans* cells entrapped in calcium alginate for denitrification (30); they showed that the system could be run continuously and could be reactivated with nutrients. In a subsequent report, the performance of this system was compared with a system in which the bacteria were membrane-confined in a hollow-fiber device (314). The large-scale denitrification of drinking water, employing sand filters with adsorbed microorganisms, has recently been proposed (415).

The oxidative degradation of phenol had been studied extensively by Klein and coworkers using *Candida tropicalis* cells immobilized in various supports (61, 187, 209, 289, 236). In one of these studies the kinetics of the system was investigated in depth (236). Sommerville et al. examined the oxidation of benzene using polyacrylamide entrapped *Pseudomonas putida* cells (177). They found that the activity of the preparation could be increased by treating the preparation with benzene and succinate or with iron salts. The increase in activity was concluded to be a result of increase in cell number.

The utilization of immobilized cells for the removal of more specific compounds has also been reported. Kinoshita et al. described a system employing polyacrylamide-entrapped *Achromobacter guttatus* cells for hydrolysis of ϵ-aminocaproic acid, a waste product of the nylon industry (162). A few years later the degradation of β-lactam antibiotics by polyacrylamide-entrapped *Escherichia coli* cells was reported (178). Klibanov and Huber have recently examined the use of *Alcaligenes eutrophus* cells entrapped in either calcium alginate or kappa-carrageenan for the detritiation of contaminated water from nuclear power plants (261). The use of flocculated *Stemphylium loti* cells was investigated for cyanide degradation (416). Flocculation was found to be preferable to entrapment in polyacrylamide, calcium alginate, or kappa-carrageenan. Finally, Ghose and Kannan have investigated the use of *Sarcina ureae* cells entrapped in cellulose triacetate fibers for the hydrolysis of urea (320).

15.6.4. Affinity Adsorbent

Affinity chromatography is based on the biospecific interaction between the support-bound ligand and the chromatographed molecules. Ideally, the bound ligand should be well characterized and attached to the support in a defined way. Although generally commendable, this approach has its limitations. It can be difficult to synthesize and attach a specific ligand, or a polyvalent ligand may be required to provide enough binding strength. In such cases the immobilization of crude systems such as membranes, cell walls, and even whole cells is a pragmatic way of obtaining useful adsorbents.

This has been verified by Genaud et al. (417) and Bétail et al. (418), who immobilized yeast cells, *Candida lipolylica*, and erythrocytes and used them as adsorbents for the purification of lectins from *Ricinus communis* seeds. Ramstorp et al. immobilized *Streptococcus mutans* cells in polyacrylamide and used them to purify a substance from carrots (probably a dextran) that had the ability to agglutinate free *Streptococcus mutans* cells (419). It was suggested that the isolated material in question (and many similar dietary substances) may be of importance for the ecology of the oral microflora. Mattiasson et al. immobilized *Staphylococcus aureus* and several strains of β-hemolytic streptococci and showed that the cells had an affinity for immunoglobulins and other serum proteins (355).

15.6.5. Fundamental Studies

The association of microorganisms with a solid support should be a useful combination in the study of various fundamental properties of microorganisms as well as of cells from higher animals. There are, however, only a few reports where this has been the main objective. We feel that the possibilities of the approach have been overlooked or have not received the attention they deserve.

The idea of using immobilized cells as a research tool was conceived in the work of Updike et al. (5), who used immobilized microorganisms in flowing stream configuration for biochemical studies.

Koshcheenko et al. used immobilized cells to elucidate metabolic reaction pathways (371). In their studies, *Mycobacterium globiforme* cells were adsorbed to cellulose and packed into a column, suitable combinations of steroid substrates and electron acceptors were applied, and the products in the effluent were determined. The probable reaction pathway for the conversion of cortisol to its 20-hydroxy derivative was deduced in a facile manner and was shown to involve three consecutively acting enzymes, Δ^1-dehydrogenase, 20β-hydroxysteroid dehydrogenase, and Δ^1-hydrogenase.

Sekiguchi et al. entrapped *Penicillum urticae* in polyacrylamide and found that the immobilization facilitated the isolation of a new intermediate in the biosynthesis of the antibiotic patulin, namely isopatulin (421). The identification of the isopatulin intermediate was important for clarifying reaction mechanisms as well as the biosynthetic route.

Combinations of immobilized cells and transducers, such as oxygen electrodes and thermistors, are useful devices not only for analyical purposes (Section 15.6.2) but might also constitute convenient tools in metabolic studies. Examples were offered by Mattiasson et al. (171) and Danielsson et al. (422), who used a "microbe thermistor" (yeast) to rapidly demonstrate the metabolic response to substrates, inhibitors, and uncouplers of oxidative phosphorylation. By coupling immobilized *Brevibacterium fermentum* with an oxygen electrode, means were found to rapidly screen a number of sugars with respect to their fermentability (405).

Immobilization should be a very convenient technique when specific surface structures of microorganisms are identified. Such structures play a crucial role when, for example, pathogenic microorganisms adhere to host cells. The identification of these elements may provide clues to how their pathogenicity can be diminished. The technique for achieving this understanding may be the same as that described in the preceding section (15.6.4), in which the immobilized cells are packed into a column and used as an adsorbent.

Finally, immobilized microorganisms could probably be used for the convenient study of synchronous growth, phage infection, lysis, and cell fusion, to name a few potential areas of application.

15.7. APPLICATIONS OF IMMOBILIZED PLANT AND ANIMAL CELLS

Recently, the immobilization technique has been extended to include both plant and animal cells. Plant and animal cells produce a number of valuable compounds, primarily medicinal products, which normally cannot be obtained from microbial cells. Culturing of plant and animal cells is a recent development and has not been applied or established for industrial purposes, with the exception of virus production. It is reasonable to assume that the inherent advantages of immobilization may be even better realized with plant and animal cells that is today possible with immobilized microbial systems.

Figure 15.22 gives a survey of the various routes to obtain immobilized plant cells, and Table 15.15 gives a comprehensive list of immobilized plant cells. A review in this area has also appeared (10). In 1979 Brodelius et al. investigated three separate plant cell systems immobilized in calcium alginate (283). It was shown that the *de novo* synthesis of anthraquinones by *Morinda citrifolia* cells was nearly 10 times that of a comparable free cell system (47). The authors also demonstrated that entrapped *Catharanthus roseus* cells were able to produce and excrete ajmalicine alkaloids. The product is usually stored within the vacuoles of the cell; it was therefore speculated that the chloroform used to extract it could have altered the membrane permeability of the cells. An "enhanced production" effect by chloroform was observed by Jones and Veliky for the 5β-hydroxylation of digitoxigenin to periplogenin by calcium-alginate entrapped *Daucus carota* cells (309). The production of ajmalicine as well as NADPH recycling activity of simultaneously permeabilized and immobilized *C. roseus* cells has been investigated by Felix et al. (35). Brodelius and Nilsson screened a number of supports for immobilization of *C. roseus* and found calcium alginate to be the best (59). The synthesis of ajmalicine from distant precursors as well as the *de novo* synthesis of ajmalicine was 76% and 40% higher, respectively, for calcium-alginate-entrapped cells than for free cells when incubated under growth-limiting conditions. Transformation of glycosides (e.g., digitoxin to digoxin) has been carried out by entrapped *Digitalis lanata* cells (283, 423). Other immobilized plant cell systems investigated include the conversion of gitoxigenin to 5β-hydroxygitoxigenin by *Daucus carota* cells (313) and the production of solasodine by immobilized *Solanum aviculare* cells (103).

The advent of the microcarrier technique is one of the most celebrated developments in both animal cell cultivation and cell immobilization. In 1967

Table 15.15. Examples of Immobilized Plant Cells

Cell	Product	Support	Reference
Morinda citrifolia	Anthraquinone	Calcium alginate	283
Catharanthus roseus	Ajmalicine	Calcium alginate	283
Digitalis lanata	Digoxin	Calcium alginate	283
Digitalis lanata	β-Methyldigoxin	Calcium alginate	423
Daucus carota	Periplogenin	Calcium alginate	309
Daucus carota	5-β-Hydroxygitoxigenin	Calcium alginate	313
Solanum aviculare	Solasodine	Polyphenyleneoxide	103
Vicia faba (protoplasts)	—	Calcium alginate	300

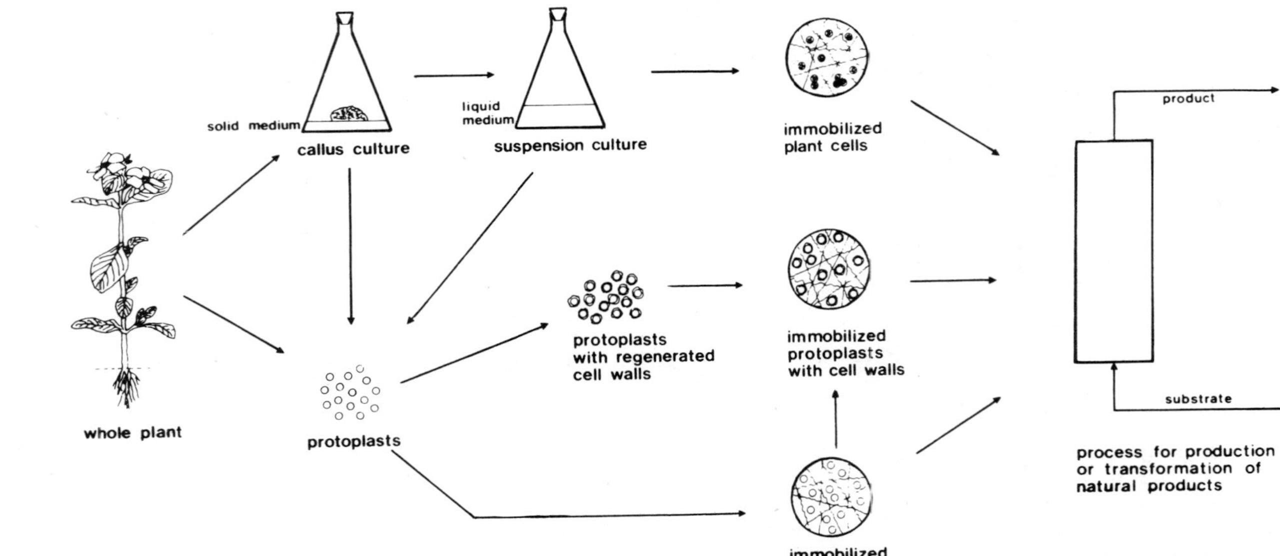

Fig. 15.22. Various routes to immobilized plant cells. Reproduced with permission from ref. 47.

van Wezel demonstrated that DEAE-Sephadex A-50 beads could be used for large-scale culture of anchorage-dependent animal cells, including diploid human fibroblasts (68). Two direct advantages were consequences of this development. First, the working volume of these systems was drastically decreased as the surface-area-to-volume ratio was increased by roughly 50 times (424). Second, anchorage-dependent cells could now be used in suspension culture systems. Thus, handling procedures are simplified (the risk of contamination is thereby decreased), cell yield per volume is increased, process control is improved, and labor requirements are reduced. Furthermore, reactors designed for suspension cultures can be utilized. Animal cells cultured on microcarriers have been applied for the production of viruses and proteins (424). In addition, these systems can be used for model studies on animal cells in culture.

At present there are three types of commercially available microcarriers; those based on dextran, polyacrylamide, or polystryene. Recently, a new generation of microcarriers has been developed based on collagen, a more natural surface for cell attachment (81). Nilsson and Mosbach prepared microcarriers of gelatin (a derivative of collagen) and observed the attachment and growth of two primary cell cultures, S 157N and DMH W1073. In addition, a gentle method for harvesting the cells intact was devised by dissolving the carrier with either collagenase or dispase. Other carriers based on the attachment of anchorage-dependent cells to collagen surfaces have been reported (424, 425).

Only a few reports of other methods of immobilizing animal cells have appeared (81, 284, 426, 355). Nilsson and Mosbach entrapped a number of cell types in alginate or agarose or both combined (81). Those preliminary studies showed that the cell membrane of hepatocytes remained intact on immobilization, that entrapped islets of Langerhans secreted insulin, and that immobilized adipocytes were capable of metabilizing glucose to fatty acids and could be stimulated by the addition of glucose. No proliferation of cultured cells was, however, observed after immobilization.

Lim and Sun published an in-depth report of islets of Langerhans encapsulated in alginate stabilized with polyamines (284). These preparations remained morphologically and functionally intact for four months in vitro. Furthermore, implantation of these microencapsulated islets into diabetic rats corrected the diabetic state for 2 to 3 weeks.

15.8. CONCLUSIONS AND OUTLOOK

As we have seen, applications of immobilized cell technology in industry are still rather limited. Apart from the fact that the use of immobilized cells does not always offer any immediate advantages, it should be borne in mind that it takes time for the necessary technical knowledge to spread and to lead to processes sophisticated enough to compete with existing optimized processes.

We feel certain that the large number of studies currently in progress will result in an increasing number of practical applications in the near future. This will particularily be the case the more it is realized that the immobilizing matrix is not just a support but also a medium capable of modifying the catalytic microenvironment, thus giving it unique properties.

The choice of a specific matrix composition can have favorable partitioning effects, as was shown for ethanol fermentation using a carrageenan support and steroid transformation using a hydrophobic support. The carrier may also be used as a size discriminator to allow access only for molecules smaller than a predetermined size and excluding larger molecules and contaminating viruses and microorganisms.

Coimmobilization is an approach that could be developed to high sophistication. The proximal arrangement of different cells, of cells and enzymes, and of cells and organic or inorganic catalysts may form very favorable catalytic composites. An example is coimmobilization of cells with active carbon for the *in situ* generation of oxygen from hydrogen peroxide.

The traditional supports may be modified to suit demanding applications. One example is the coimmobilization of cells with magnetic particles. "Magnetic" immobilized microorganisms should be useful in situations where contaminating solids of a particulate nature are encountered. The technique based on two-phase liquid partitioning systems is also well-suited to deal with such situations.

To protect a valuable product, affinity ligands could be included in the support matrix to adsorb the product as soon as it is excreted from the cells. Conversely, the affinity ligands could be used to adsorb and neutralize damaging hydrolytic enzymes.

An area of considerable potential is the use of immobilized plant and animal cells. Entrapment of the fragile cells from plant suspension cultures (many of which are known to produce or transform interesting pharmaceuticals) appears highly attractive, as the method offers simultaneous protection against shearing forces. The same holds for animal cells, which are of interest because of their capacity to form or transform hormones, interferons, and antibodies.

Finally, we point to the obvious merging of enzyme technology (including immobilized cell technology) and genetic engineering. Preliminary results obtained in our laboratory in conjunction with other groups have shown that *E. coli* cells transformed by recombinant DNA techniques to produce interferon or human growth hormone can be advantageously put to use in the immobilized form. The same promising results are obtained with immobilized *Bacillus subtilis* geared to excrete proinsulin.

ACKNOWLEDGEMENTS

We gratefully acknowledge the advice given by Professor Volker Kasche and the financial support of the Swedish Board for Technical Development and the Biotechnology Research Foundation (Sweden).

REFERENCES

1. K. Elliot, M. O'Connor, and J. Whelan, Eds., *Adhesion and Microorganism Pathogenicity,* Pitman Medical, London (1981).
2. B. J. Abbott, in *Annual Reports of Fermentation Processes,* Vol. 1 (D. Perlman and G. T. Tsao, Eds.), Academic Press, New York (1977), p. 205.
3. T. Hattori and C. Furusaka, *J. Biochem.* **48,** 831 (1960).
4. K. Mosbach and R. Mosbach, *Acta Chem. Scand.* **20,** 2807 (1966).
5. S. J. Updike, D. R. Harris, and E. Shrago, *Nature* **224,** 1122 (1969).
6. K. Mosbach and P.-O. Larsson, *Biochem. Bioeng.* **12,** 19 (1970).
7. I. Chibata, T. Tosa, and T. Sato, *Appl. Microbiol.* **27,** 878 (1974).
8. I. Chibata, *Immobilized Enzymes—Research and Developments,* Wiley, New York (1978).
9. S. Ohlson, *Immobilized Whole-Cell Catalysts and HPLAC,* Ph.D. Thesis, University of Lund, Sweden, LUTKDH/(TKBK-1005)/1-81/(1980).
10. P. Brodelius and K. Mosbach, *Advances in Applied Microbiology,* Vol 28 (A. I. Laskin, Ed.), Academic Press, New York (1982), p. 1.
11. J. Klein and F. Wagner, in *Characterization of Immobilized Biocatalysts,* Dechema Band 84 (K. Buchholz, Ed.), Verlag Chemie, Weinheim and New York (1979), pp. 265–335.
12. P. Dunnill, *Phil. Trans. R. Soc. Lond.,* **B 290,** 409 (1980).
13. I. Chibata and T. Tosa, *TIBS* **5,** 88 (1980).
14. I. Chibata and T. Tosa, in *Advances in Applied Microbiology,* Vol. 22 (D. Perlman, Ed.), Academic Press, New York (1977), p. 1.
15. C. Bucke and A. Wiseman, *Chem. Ind.* 4 April 1981, p. 234.
16. B. J. Abbott, in *Advances in Applied Microbiology,* Vol. 20 (D. Perlman, Ed.), Academic Press, New York (1976), p. 203.
17. T. R. Jack and J. E. Zajic, in *Advances in Biochemical Engineering,* Vol. 5, Springer Verlag, Berlin (1977), p. 125.
18. B. J. Abbott, in *Annual Reports on Fermentation Processes,* Vol. 2 (D. Perlman Ed.), Academic Press, New York (1978), p. 91.
19. P. Linko, in *Enzyme Engineering in Food Processing,* Vol. 2 (P. Linko and J. Larinkari, Eds.), Applied Science, London (1980), p. 27.

20. K. Venkatasubramanian and W. R. Vieth, in *Process in Industrial Microbiology,* Vol. 15 (M. J. Bull, Ed.), Elsevier, New York (1979), p. 61.
21. G. Durand and J. M. Navarro, *Proc. Biochem.,* Sept. 1978, p. 14.
22. M. D. Lilly, in *Biotechnology,* Dechema Band 82, Verlag Chemie, Weinheim (1978), p. 165.
23. B. Atkinson, G. M. Black, and A. Pinches, *Proc. Biochem.,* May 1980, p. 24.
24. R. A. Messing, *Appl. Biochem. Biotechnol.* **6,** 167 (1981).
25. P. S. J. Cheetham, in *Topics in Enzyme and Fermentation Technology,* Vol. 4 (A. Wiseman, Ed.), Ellis Horwood, Chichester (1980), p. 189.
26. T. Tosa, T. Sato, Y. Nishida, and I. Chibata, *Biochim. Biophys. Acta* **483,** 193 (1977).
27. I. Chibata and T. Tosa, in *Annual Review of Biophysics and Bioengineering,* Vol. 10 (L. J. Mullins et al., Eds.), Annual Reviews, Palo Alto (1981), p. 197.
28. A. M. Klibanov, *Anal. Biochem.* **93,** 1 (1979).
29. P.-O. Larsson, S. Ohlson, and K. Mosbach, *Nature* **263,** 796 (1976).
30. I. Nilsson, S. Ohlson, L. Häggström, N. Molin, and K. Mosbach, *Eur. J. Appl. Microbiol. Biotechnol.* **10,** 261 (1980).
31. P.-O. Larsson and K. Mosbach, *Biotechnol. Lett.* **1,** 501 (1979).
32. S. Ohlson, P.-O. Larsson, and K. Mosbach, *Eur. J. Appl. Microbiol. Biotechnol.* **7,** 103 (1979).
33. P.-O. Larsson, K. Mosbach, and S. Ohlson, *U.S. Patent* 4,246,346 (1981).
34. C. K. A. Martin and D. Perlman, *Biotechnol. Bioeng.* **18,** 217 (1976).
35. H. R. Felix, P. Brodelius, and K. Mosbach, *Anal. Biochem.* **116,** 462 (1981).
36. K. Yamamoto, T. Sato, T. Tosa, and I. Chibata, *Biotechnol. Bioeng.* **16,** 1601 (1974).
37. K. Murata, J. Kato, and I. Chibata, *Biotechnol. Bioeng.* **21,** 887 (1979).
38. H. R. Felix, *Anal. Biochem.* **120,** 211 (1982).
39. S. Chand and K. Mosbach, Manuscript in preparation.
40. I. Chibata, in *Enzyme Engineering in Food Processing,* Vol. 2 (P. Linko and J. Larinkari, Eds.), Applied Science, London (1980), p. 1.
41. M. Wada, J. Kato, and I. Chibata, *Eur. J. Appl. Microbiol. Biotechnol.* **11,** 67 (1981).
42. S. Birnbaum, R. Pendleton, P.-O. Larsson, and K. Mosbach, *Biotechnol. Lett.* **3,** 393 (1981).
43. K. A. Koshcheenko, G. V. Sukhodolskaya, V. S. Tyurin, and G. K. Skryabin, *Eur. J. Appl. Microbiol. Biotechnol.* **12,** 161 (1981).
44. B. Mattiasson, *Biotechnol. Bioeng.* **19,** 777 (1977).
45. B. Hägerdal and K. Mosbach, in *Enzyme Engineering in Food Processing,* Vol. 2 (P. Linko and J. Larinkari, Eds.), Applied Science, London (1980), p. 129.
46. S. Fukui, S. A. Ahmed, T. Omata, and A. Tanaka, *Eur. J. Appl. Microbiol. Biotechnol.* **10,** 289 (1980).
47. P. Brodelius, B. Deus, K. Mosbach, and M. H. Zenk, in *Enzyme Engineering,* Vol. 5 (H. H. Weetall and G. P. Royer, Eds.), Plenum Press, New York (1980), p. 373.
48. P.-O. Larsson and K. Mosbach, in *Methods in Enzymology,* Vol. 44 (K. Mosbach, Ed.), Academic Press, New York (1976), p. 183.

49. R. Saini and W. R. Vieth, *J. Appl. Chem. Biotechnol.* **25,** 115 (1975).
50. I. Takata, K. Yamamoto, T. Tosa, and I. Chibata, *Enzyme Microb. Technol.* **2,** 30 (1980).
51. T. Kokubu, I. Karube, and S. Suzuki, *Biotechnol. Bioeng.* **23,** 29 (1981).
52. P. G. Krouwel and N. W. F. Kossen, *Biotechnol. Bioeng.* **22,** 681 (1980).
53. P. G. Krouwel and N. W. F. Kossen, *Biotechnol. Bioeng.,* **23,** 651 (1981).
54. T. Kokubu, I. Karube, and S. Suzuki, *Eur. J. Appl. Microbiol. Biotechnol.* **5,** 233 (1978).
55. L. Goldstein, in *Methods in Enzymology,* Vol. 44 (K. Mosbach, Ed.), Academic Press, New York (1976), p. 397.
56. J. M. Engasser and C. Horvath, in *Applied Biochemistry and Bioengineering,* Vol. 1 (L. B. Wingard, Jr., E. Katchalski-Katzir, and L. Goldstein, Eds.), Academic Press, New York, 1976, p. 127.
57. K. J. Laidler and P. S. Bunting, in *Methods in Enzymology,* Vol. 64 (D. L. Purich, Ed.), Academic Press, New York (1980), p. 227.
58. V. Kasche and K. Buchholz, in *Characterization of Immobilized Biocatalysts,* Dechema Band 84 (K. Buchholz, Ed.), Verlag Chemie, Weinheim and New York (1979), pp. 208–243.
59. P. Brodelius and K. Nilsson, *FEBS Lett.* **122,** 312 (1980).
60. I. Takata, T. Tosa, and I. Chibata, *J. Solid Phase Biochem.* **2,** 225 (1977).
61. U. Hackel, J. Klein, R. Megnet, and F. Wagner, *Eur. J. Appl. Microbiol.* **1,** 291 (1975).
62. Y. Morikawa, I. Karube, and S. Suzuki, *Biotechnol. Bioeng.* **21,** 261 (1979).
63. I. S. Maddox, P. Dunnill, and M. D. Lilly, *Biotechnol. Bioeng.* **23,** 345 (1981).
64. V. Larreta Garde, B. Thomasset, A. Tanaka, G. Gellf, and D. Thomas, *Eur. J. Appl. Microbiol. Biotechnol.* **11,** 133 (1981).
65. M. F. Cocquempot, B. Thomasset, J. N. Barbotin, G. Gellf, and D. Thomas, *Eur. J. Appl. Microbiol. Biotechnol.* **11,** 193 (1981).
66. I. B. Holcberg and P. Margalith, *Eur. J. Appl. Microbiol. Biotechnol.* **13,** 133 (1981).
67. T. Hattori and C. Furusaka, *J. Biochem.* **50,** 312 (1961).
68. A. L. van Wezel, *Nature* **216,** 64 (1967).
69. D. W. Levine, D. I. C. Wang, and W. G. Thilly, *Biotechnol. Bioeng.* **21,** 821 (1979).
70. Y. Beaudry, J. P. Quillon, J. Frappa, R. Deloince, and R. Fontanges, *Biotechnol. Bioeng.* **21,** 2351 (1979).
71. D. J. Giard, D. H. Loeb, W. G. Thilly, D. I. C. Wang, and D. W. Levine, *Biotechnol. Bioeng.* **21,** 433 (1979).
72. C. L. Crespi and W. G. Thilly, *Biotechnol. Bioeng.* **23,** 983 (1981).
73. D. E. Johnson and A. Ciegler, *Arch. Biochem. Biophys.* **130,** 384 (1969).
74. E. Seyhan and D. J. Kirwan, *Biotechnol. Bioeng.* **21,** 271 (1979).
75. K. DeNicola and D. J. Kirwan, *Biotechnol. Bioeng.* **22,** 1283 (1980).
76. J. L. Gainer, D. J. Kirwan, J. A. Foster, and E. Seyhan, *Biotechnol. Bioeng., Symp.* **10,** 35 (1980).
77. P. Atrat, E. Hüller, C. Hörhold, *Z. Allg. Mikrobiol.* **20,** 79 (1980).

78. P. Atrat, E. Hüller, C. Hörhold, M. J. Buchar, A. Y. Arinbasarova, and K. A. Koschtschejenko, *Z. Allg. Mikrobiol.* **20,** 159 (1980).
79. P. Atrat, C. Hörhold, M. J. Buchar, and K. A. Koschtschejenko, *Z. Allg. Mikrobiol.* **20,** 239 (1980).
80. A. J. Daugulis, N. M. Brown, W. R. Cluett, and D. B. Dunlop, *Biotechnol. Lett.* **3,** 651 (1981).
81. K. Nilsson and K. Mosbach, *FEBS Lett.* **118,** 145 (1980).
82. O. C. Sitton and J. L. Gaddy, *Biotechnol. Bioeng.* **22,** 1735 (1980).
83. O. C. Sitton, G. C. Magruder, N. L. Book, and J. L. Gaddy, *Biotechnol. Bioeng., Symp.* **10,** 213 (1980).
84. J. M. Navarro and G. Durand, *Eur. J. Appl. Microbiol.* **4,** 243 (1977).
85. P. G. Rouxhet, J. L. van Haecht, J. Didelez, P. Gerard, and M. Briquet, *Enzyme Microb. Technol.* **3,** 49 (1981).
86. J. P. Whiteside and R. E. Spier, *Biotechnol. Bioeng.* **23,** 551 (1981).
87. B. Arkles and W. S. Brinigar, *J. Biol. Chem.* **250,** 8856 (1975).
88. C. Ghommidh, J. M. Navarro, and G. Durand, *Biotechnol. Lett.* **3,** 93 (1981).
89. A. Marcipar, N. Cochet, L. Brackenridge, and J. M. Lebeault, *Biotechnol. Lett.* **1,** 65 (1979).
90. M. Minier and G. Goma, *Biotechnol. Lett.* **3,** 405 (1981).
91. C. D. Scott and C. W. Hancher, *Biotechnol. Bioeng.* **18,** 1393 (1976).
92. Anonymous, *Chem. Eng.* **83,** 87 (1976).
93. M. Moo-Young, J. Lamptey, and C. W. Robinson, *Biotechnol. Lett.* **2,** 541 (1980).
94. M. Vijayalakshmi, A. Marcipar, E. Segard, and G. B. Broun, *Ann. New York Acad. Sci.* **326,** 249 (1979).
95. M. A. Vijayalakshmi, A. Marcipar, and N. Cochet, *J. Polymer Sci., Polymer Symp.* **68,** 57 (1980).
96. M. A. Vijayalakshmi, D. Picque, and G. B. Broun, in *Enzyme Engineering,* Vol. 5 (H. H. Weetall and G. P. Royer, Eds.), Plenum Press, New York (1980), p. 451.
97. J. E. Duddridge, C. A. Kent, J. F. Laws, P. C. Miller, and T. R. Bott, in *Abstracts of Communications,* Second European Congress of Biotechnology, Society of Chemical Industry, London, 1981, p. 227.
98. A. P. Kolpakchi, V. S. Isaeva, A. Yu. Zhvirblyanskaya, E. N. Kazantsev, E. N. Serova, and N. N. Rattel, *Prikl. Biokhim. Mikrobiol.* **12,** 866 (1976); through *Chem. Abstr.* **86,** 53873 (1977).
99. M. Fletcher, *Can. J. Microbiol.* **23,** 1 (1977).
100. E. J. Arcuri, R. M. Worden, and S. E. Shumate II, *Biotechnol. Lett.* **2,** 499 (1980).
101. L. van der Berg and K. J. Kennedy, *Biotechnol. Lett.* **3,** 165 (1981).
102. V. Jirku, J. Turková, B. Veruovic, and V. Kubánek, *Biotechnol. Lett.* **2,** 451 (1980).
103. V. Jirku, T. Macek, T. Vanek, V. Krumphanzl, and V. Kubánek, *Biotechnol. Lett.* **3,** 447 (1981).
104. J. Holló, J. Tóth, R. P. Tengerdy, and J. E. Johnson, in *Amer. Chem. Soc., Symp. Ser. No. 106* (K. Venkatasubramanian, Ed.) (1979), p. 73.

105. R. P. Tengerdy, J. E. Johnson, J. Holló, and J. Töth, *Appl. Biochem. Biotechnol.* **6,** 3 (1981).
106. S. L. Daniels, in *Developments in Industrial Microbiology,* Vol. 13, (E. D. Murray, Ed.), SIM, Washington (1972), p. 211.
107. F. B. Kolot, *Proc. Biochem.,* Oct.–Nov. 1980, p. 2.
108. F. B. Kolot, *Proc. Biochem.,* Aug.–Sept. 1981, p. 2.
109. F. B. Kolot, in *Developments in Industrial Microbiology,* Vol. 21, SIM, Washington (1980), p. 295.
110. W. D. Murray and L. van den Berg, *J. Appl. Bacteriol.* **51,** 257 (1981).
111. F. B. Kolot, *Proc. Biochem.,* Oct.–Nov. 1981, p. 30.
112. *Separation News,* No. 2, 1981, Pharmacia Fine Chemicals.
113. P. Truffa-Bachi and L. Wofsy, *Proc. Nat. Acad. Sci. USA* **66,** 685 (1970).
114. G. M. Edelman, U. Rutishauser, and C. F. Millette, *Proc. Nat. Acad. Sci. USA* **68,** 2153 (1971).
115. M. Horisberger, *Biotechnol. Bioeng.* **18,** 1647 (1976).
116. H. Wigzell and B. Andersson, *J. Exp. Med.* **129,** 23 (1969).
117. J. M. Davie and W. E. Paul, *Cell. Immunol.* **1,** 404 (1970).
118. D. D. Soderman, J. Germershausen, and H. M. Katzen, *Proc. Nat. Acad. Sci. USA* **70,** 792 (1973).
119. B. R. Venter, J. C. Venter, and N. O. Kaplan, *Proc. Nat. Acad. Sci. USA* **73,** 2013 (1976).
120. U. Hellström, M.-L. Dillner, S. Hammarström, and P. Perlmann, *J. Exp. Med.* **144,** 1381 (1976).
121. U. S. Rutishauser and G. M. Edelman, in *Methods of Cell Separation,* Vol. 1 (N. Catsimpoolas, Ed.), Plenum Press, New York (1977), p. 193.
122. U. Rutishauser, C. F. Millette, and G. M. Edelman, *Proc. Nat. Acad. Sci. USA* **69,** 1596 (1972).
123. U. Rutishauser and G. M. Edelman, *Proc. Nat. Acad. Sci. USA* **69,** 3774 (1972).
124. D. J. Dvorak, E. Gipps, and C. Kidson, *Nature* **271,** 564 (1978).
125. U. Rutishauser, P. D'Eustachio, and G. M. Edelman, *Proc. Nat. Acad. Sci. USA* **70,** 3894 (1973).
126. Z. Eshhar, T. Waks, and M. Bustin, in *Methods in Enzymology,* Vol. 34 (W. B. Jakoby and M. Wilchek, Eds.), Academic Press, New York (1974), p. 750.
127. J. B. Robbins and R. Schneerson, in *Methods in Enzymology,* Vol. 34 (W. B. Jakoby and M. Wilchek, Eds.), Academic Press, New York (1974), p. 703.
128. G. M. Edelman and U. Rutishauser, in *Methods in Enzymology,* Vol. 34 (W. B. Jakoby and M. Wilchek, Eds.), Academic Press, New York (1974), p. 195.
129. V. Ghetie, G. Mota, and J. Sjöquist, *J. Immunol. Methods* **21,** 133 (1978).
130. J. J. Killion and G. M. Kollmorgen, *Nature* **259** 674 (1976).
131. R. P. Nelson, *U. S. Patent* 3,957,580 (1976).
132. T. R. Jack and J. E. Zajic, *Biotechnol. Bioeng.* **19,** 631 (1977).
133. V. Jirku, J. Turková, and V. Krumphanzl, *Biotechnol. Lett.* **2,** 509 (1980).
134. V. Jirku, J. Turková, A. Kuchynková, and V. Krumphanzl, *Eur. J. Appl. Microbiol. Biotechnol.* **6,** 217 (1979).
135. V. E. Gulaya, J. Turková, V. Jirku, A. Frydrychová, J. Čoupek, and S. N. Ananchenko, *Eur. J. Appl. Microbiol. Biotechnol.* **8,** 43 (1979).

136. J. F. Kennedy, J. D. Humphreys, S. A. Barker, and R. N. Greenshields, *Enzyme Microb. Technol.* **2,** 209 (1980).
137. J. F. Kennedy, in *Amer. Chem. Soc., Symp. Ser., No. 106* (K. Venkatasubramanian, Ed.) (1979), p. 119.
138. J. F. Kennedy, S. A. Barker, and J. D. Humphreys, *Nature* **261,** 242 (1976).
139. J. F. Kennedy, in *Enzyme Engineering,* Vol. 4 (G. B. Broun, G. Manecke, and L. B. Wingard Jr., Eds.), Plenum Press, New York (1978), p. 323.
140. J. G. Boersma, K. Vellenga, H. G. J. de Wilt, and G. E. H. Joosten, *Biotechnol. Bioeng.* **21,** 1711 (1979).
141. S. J. Bungard, R. Reagan, P. J. Rodgers, and K. R. Wyncoll, in *Amer. Chem. Soc., Symp. Ser. No. 106* (K. Venkatasubramanian, Ed.) (1979), p. 139.
142. M. A. van Keulen, K. Vellenga, and G. E. H. Joosten, *Biotechnol. Bioeng.* **23,** 1437 (1981).
143. N. Tsumura and T. Kasumi, *Abstracts of Papers,* 5th International Fermentation Symposium, Berlin (1976), p. 291.
144. O. J. Lantero, Jr., in *Enzyme Engineering,* Vol. 4 (G. B. Broun, G. Manecke, and L. B. Wingard Jr., Eds.), Plenum Press, New York (1978), p. 349.
145. J. Holló, E. László, and Á. Hoschke, *Starch* **33,** 361 (1981).
146. D. Petre, C. Noel, and D. Thomas, *Biotechnol. Bioeng.* **20,** 127 (1978).
147. K.-I. Hirano, I. Karube, and S. Suzuki, *Biotechnol. Bioeng.* **19,** 311 (1977).
148. J. N. Barbotin and B. Thomasset, *Biochimie* **62,** 359 (1980).
149. E. Drioli, G. Iorio, R. Molinari, M. de Rosa, A. Gambacorta, and E. Esposito, *Biotechnol. Bioeng.* **23,** 221 (1981).
150. V. Larreta Garde, G. Gellf, and D. Thomas, *Eur. J. Biochem.* **116,** 337 (1981).
151. M. de Rosa, A. Gambacorta, L. Lama, B. Nicolaus, and V. Buonocore, *Biotechnol. Lett.* **3,** 183 (1981).
152. J. V. Hupkes and R. van Tilburg, *Starch* **28,** 356 (1976).
153. S. Gestrelius, *U. S. Patent* 4,288,552 (1981).
154. W.-W. Tso, *Biotechnol. Lett.* **2,** 519 (1980).
155. W.-W. Tso and W.-P. Fung, *Biotechnol. Lett.* **3,** 421 (1981).
156. C. K. Lee and M. E. Long, *U. S. Patent* 3,821,086 (1974).
157. N. E. Franks, *Biochim. Biophys. Acta* **252,** 246 (1971).
158. W. Slowinski and S. E. Charm, *Biotechnol. Bioeng.* **15,** 973 (1973).
159. T. Tosa, T. Sato, T. Mori, and I. Chibata, *Appl. Microbiol.* **27,** 886 (1974).
160. K. Yamamoto, T. Sato, T. Tosa, and I. Chibata, *Biotechnol. Bioeng.* **16,** 1589 (1974).
161. T. Sato, T. Mori, T. Tosa, I. Chibata, M. Furui, K. Yamashita, and A. Sumi, *Biotechnol. Bioeng.* **17,** 1797 (1975).
162. S. Kinoshita, M. Muranaka, and H. Okada, *J. Ferment. Technol.* **53,** 223 (1975).
163. S.-ur-R. Saif, Y. Tani, and K. Ogata, *J. Ferment. Technol.* **53,** 380 (1975).
164. S. Shimizu, H. Morioka, Y. Tani, and K. Ogata, *J. Ferment. Technol.* **53,** 77 (1975).
165. T. Sato, T. Tosa, and I. Chibata, *Eur. J. Appl. Microbiol.* **2,** 153 (1976).
166. I. Chibata, T. Tosa, and K. Yamamoto, in *Enzyme Engineering,* Vol. 3 (E. K. Pye and H. H. Weetall, Eds.), Plenum Press, New York (1978), p. 463.

167. I. Chibata, T. Tosa, and T. Sato, in *Methods in Enzymology,* Vol. 44 (K. Mosbach, Ed.), Academic Press, New York (1976), p. 739.
168. K. Yamamoto, T. Tosa, K. Yamashita, and I. Chibata, *Eur. J. Appl. Microbiol.* **3,** 169 (1976).
169. C. K. A. Martin and D. Perlman, *Eur. J. Appl. Microbiol.* **3,** 91 (1976).
170. I. Karube, T. Matsunaga, S. Tsuru, and S. Suzuki, *Biochim. Biophys. Acta* **444,** 338 (1976).
171. B. Mattiasson, P.-O. Larsson, and K. Mosbach, *Nature* **268,** 519 (1977).
172. I. Karube, S. Mitsuda, T. Matsunaga, and S. Suzuki, *J. Ferment. Technol.* **55,** 243 (1977).
173. I. Karube, T. Matsunaga, and S. Suzuki, *J. Solid Phase Biochem.* **2,** 97 (1977).
174. I. Karube, T. Matsunaga, S. Tsuru, and S. Suzuki, *Biotechnol. Bioeng.* **19,** 1727 (1977).
175. K. Yamamoto, T. Tosa, K. Yamashita, and I. Chibata, *Biotechnol. Bioeng.* **19,** 1101 (1977).
176. G. W. Schnarr, W. A. Szarek, and J. K. N. Jones, *Appl. Environ. Microbiol.* **33,** 732 (1977).
177. H. J. Somerville, J. R. Mason, and R. N. Ruffell, *Eur. J. Appl. Microbiol.* **4,** 75 (1977).
178. W. Pache, *Eur. J. Appl. Microbiol. Biotechnol.* **5,** 171 (1978).
179. K. Murata, K. Tani, J. Kato, and I. Chibata, *Eur. J. Appl. Microbiol. Biotechnol.* **6,** 23 (1978).
180. T. Uchida, T. Watanabe, J. Kato, and I. Chibata, *Biotechnol. Bioeng.* **20,** 255 (1978).
181. P.-O. Larsson, S. Ohlson, and K. Mosbach, in *Enzyme Engineering,* Vol. 4 (G. B. Broun, G. Manecke, and L. B. Wingard, Jr., Eds.), Plenum Press, New York (1978), p. 317.
182. S. Ohlson, P.-O. Larsson, and K. Mosbach, *Biotechnol. Bioeng.* **20,** 1267 (1978).
183. H. S. Yang and J. F. Studebaker, *Biotechnol. Bioeng.* **20,** 17 (1978).
184. J. R. Mason, H. J. Somerville, and S. J. Pirt, *J. Appl. Chem. Biotechnol.* **28,** 770 (1978).
185. J. R. Mason, S. J. Pirt, and H. J. Somerville, in *Enzyme Engineering,* Vol. 4 (G. B. Broun, G. Manecke, and L. B. Wingard Jr., Eds.), Plenum Press, New York (1978), p. 343.
186. H. Škodová, J. Chaloupka, and J. Skoda, *Biotechnol. Bioeng.* **23,** 2151 (1981).
187. J. Klein, U. Hackel, P. Schara, P. Washausen, F. Wagner, and C. K. A. Martin, in *Enzyme Engineering,* Vol. 4 (G. B. Broun, G. Manecke, and L. B. Wingard Jr., Eds.), Plenum Press, New York (1978), p. 339.
188. P.-O. Larsson, S. Ohlson, and K. Mosbach, in *Applied Biochemistry and Bioengineering,* Vol. 2 (L. B. Wingard Jr., E. Katchalski-Katzir, and L. Goldstein, Eds.), Academic Press, New York (1979), p. 291.
189. Y. Morikawa, K. Ochiai, I. Karube, and S. Suzuki, *Antimicrob. Agents Chemother.* **15,** 126 (1979).
190. P. Decottignies-Le Maréchal, R. Calderón-Seguin, J. P. Vandecasteele, and R. Azerad, *Eur. J. Appl. Microbiol. Biotechnol.* **7,** 33 (1979).

191. K. Murata, T. Uchida, K. Tani. J. Kato, and I. Chibata, *Eur. J. Appl. Microbiol. Biotechnol.* **7,** 45 (1979).
192. K. Nabe, N. Izuo, S. Yamada, and I. Chibata, *Appl. Environ. Microbiol.* **38,** 1056 (1979).
193. S. Suzuki and I. Karube, in *Amer. Chem. Soc., Symp. Ser. No. 106* (K. Venkatasubramanian, Ed.) (1979), p. 59.
194. S. Shimizu, Y. Tani, and H. Yamada, in *Amer. Chem. Soc., Symp. Ser.* No. 106 (K. Venkatasubramanian, Ed.) (1979), p. 87.
195. I. Chibata, in *Amer. Chem. Soc., Symp. Ser.* No. 106 (K. Venkatasubramanian, Ed.) (1979), p. 187.
196. S. Ohlson, S. Flygare, P.-O. Larsson, and K. Mosbach, *Eur. J. Appl. Microbiol. Biotechnol.* **10,** 1 (1980).
197. I. Karube, S. Kuriyama, T. Matsunaga, and S. Suzuki, *Biotechnol. Bioeng.* **22,** 847 (1980).
198. W. Tischer, W. Tiemeyer, and H. Simon, *Biochimie* **62,** 331 (1980).
199. R. Couderc and J. Baratti, *Biotechnol. Bioeng.* **22,** 1155 (1980).
200. H. Yamada, S. Shimizu, Y. Tani, and T. Hino, in *Enzyme Engineering,* Vol. 5 (H. H. Weetall and G. P. Royer, Eds.), Plenum Press, New York (1980), p. 405.
201. Y. Morikawa, I. Karube, and S. Suzuki, *Eur. J. Appl. Microbiol. Biotechnol.* **10,** 23 (1980).
202. T. Matsunaga, I. Karube, and S. Suzuki, *Eur. J. Appl. Microbiol. Biotechnol.* **10,** 235 (1980).
203. M. de Rosa, A. Gambacorta, B. Nicolaus, V. Buonocore, and E. Poerio, *Biotechnol. Lett.* **2,** 29 (1980).
204. K. Murata, K. Tani, J. Kato, and I. Chibata, *Biochimie* **62,** 347 (1980).
205. T. Matsunaga, I. Karube, and S. Suzuki, *Biotechnol. Bioeng.* **22,** 2607 (1980).
206. Y. Morikawa, I. Karube, and S. Suzuki, *Biotechnol. Bioeng.* **22,** 1015 (1980).
207. J. M. C. Duarte and M. D. Lilly, in *Enzyme Engineering,* Vol. 5 (H. H. Weetall and G. P. Royer, Eds.), Plenum Press, New York (1980), p. 363.
208. H. Yamada, S. Shimizu, H. Shimada, Y. Tani, S. Takahashi, and T. Ohashi, *Biochimie* **62,** 395 (1980).
209. J. Klein and P. Schara, *J. Solid Phase Biochem.* **5,** 61 (1980).
210. K. Murata, K. Tani, J. Kato, and I. Chibata, *Eur. J. Appl. Microbiol. Biotechnol.* **11,** 72 (1981).
211. M. H. Siess and C. Divies, *Eur. J. Appl. Microbiol. Biotechnol.* **12,** 10 (1981).
212. S. Fukui, A. Tanaka, and G. Gellf, in *Enzyme Engineering,* Vol. 4 (G. B. Broun, G. Manecke, and L. B. Wingard, Jr., Eds.), Plenum Press, New York (1978), p. 299.
213. A. Kimura, Y. Tatsutomi, N. Mizushima, A. Tanaka, R. Matsuno, and H. Fukuda, *Eur. J. Appl. Microbiol. Biotechnol.* **5,** 13 (1978).
214. M. Kumakura, M. Yoshida, and I. Kaetsu, *J. Solid-Phase Biochem.* **3,** 175 (1978).
215. M. Kumakura, M. Yoshida, and I. Kaetsu, *Eur. J. Appl. Microbiol. Biotechnol.* **6,** 13 (1978).
216. K. Sonomoto, A. Tanaka, T. Omata, T. Yamane, and S. Fukui, *Eur. J. Appl. Microbiol. Biotechnol.* **6,** 325 (1979).

217. T. Omata, A. Tanaka, T. Yamane, and S. Fukui, *Eur. J. Appl. Microbiol. Biotechnol.* **6,** 207 (1979).
218. T. Yamané, H. Nakatani, E. Sada, T. Omata, A. Tanaka, and S. Fukui, *Biotechnol. Bioeng.* **21,** 2133 (1979).
219. T. Omata, T. Iida, A. Tanaka, and S. Fukui, *Eur. J. Appl. Microbiol. Biotechnol.* **8,** 143 (1979).
220. M. Asada, K. Morimoto, K. Nakanishi, R. Matsuno, A. Tanaka, A. Kimura, and T. Kamikubo, *Agric. Biol. Chem.* **43,** 1773 (1979).
221. M. Kumakura, M. Yoshida, and I. Kaetsu, *Biotechnol. Bioeng.* **21,** 679 (1979).
222. T. Hayashi, Y. Tanaka, and K. Kawashima, *Biotechnol. Bioeng.* **21,** 1019 (1979).
223. S. F. D'Souza and G. B. Nadkarni, *Biotechnol. Bioeng.* **22,** 2191 (1980).
224. S. F. D'Souza and G. B. Nadkarni, *Biotechnol. Bioeng.* **22,** 2179 (1980).
225. S. Fukui, T. Omata, T. Yamane, and A. Tanaka, in *Enzyme Engineering,* Vol. 5 (H. H. Weetall and G. P. Royer, Eds.), Plenum Press, New York, 1980, p. 347.
226. S. Fukui, K. Sonomoto, N. Itoh, and A. Tanaka, *Biochimie* **62,** 381 (1980).
227. T. Omata, A. Tanaka, and S. Fukui, *J. Ferment. Technol.* **58,** 339 (1980).
228. A. Kimura, Y. Tatsutomi, R. Matsuno, A. Tanaka, and H. Fukuda, *Eur. J. Appl. Microbiol. Biotechnol.* **11,** 78 (1981).
229. T. Omata, N. Iwamoto, T. Kimura, A. Tanaka, and S. Fukui, *Eur. J. Appl. Microbiol. Biotechnol.* **11,** 199 (1981).
230. Y. Ikariyama, M. Aizawa, and S. Suzuki, *J. Solid Phase Biochem.* **4,** 69 (1979).
231. I. Karube, K. Aizawa, S. Ikeda, and S. Suzuki, *Biotechnol. Bioeng.* **21,** 253 (1979).
232. P. Atrat, E. Hüller, and C. Hörhold, *Eur. J. Appl. Microbiol. Biotechnol.* **12,** 157 (1981).
233. K. Murata, K. Tani, J. Kato, and I. Chibata, *Enzyme Microb. Technol.* **3,** 233 (1981).
234. U. Wollenberger, F. Scheller, and P. Atrat, *Anal. Lett.* **13,** 1201 (1980).
235. U. Wollenberger, F. Scheller, and P. Atrat, *Anal. Lett.* **13,** 825 (1980).
236. J. Klein and P. Schara, *Appl. Biochem. Biotechnol.* **6,** 91 (1981).
237. K. Toda and M. Shoda, *Biotechnol. Bioeng.* **17,** 481 (1975).
238. K. Toda, *Biotechnol. Bioeng.* **17,** 1729 (1975).
239. M. A. Bennett and H. H. Weetall, *J. Solid Phase Biochem.* **1,** 137 (1976).
240. T. Matsunaga, I. Karube, and S. Suzuki, *Anal. Chim. Acta* **99,** 233 (1978).
241. Y. Kawabata and A. L. Demain, in *Amer. Chem. Soc., Symp. Ser. No. 106* (K. Venkatasubramanian, Ed.), (1979), p. 133.
242. H. H. Weetall and L. O. Krampitz, *J. Solid Phase Biochem.* **5,** 115 (1980).
243. S. Suzuki, I. Karube, T. Matsunaga, and H. Kayano, in *Enzyme Engineering,* Vol. 5 (H. H. Weetall and G. P. Royer, Eds.), Plenum Press, New York (1980), p. 143.
244. S. Suzuki, I. Karube, T. Matsunaga, S. Kuriyama, N. Suzuki, T. Shirogami, and T. Takamura, *Biochimie* **62,** 353 (1980).
245. H. Kayano, I. Karube, T. Matsunaga, S. Suzuki, and O. Nakayama, *Eur. J. Appl. Microbiol. Biotechnol.* **12,** 1 (1981).
246. H. H. Weetall, B. P. Sharma, and C. C. Detar, *Biotechnol. Bioeng.* **23,** 605 (1981).

247. I. Karube, T. Matsunaga, T. Otsuka, H. Kayano, and S. Suzuki, *Biochim. Biophys. Acta* **637,** 490 (1981).
248. I. Karube, T. Matsunaga, Y. Otomine, and S. Suzuki, *Enzyme Microb. Technol.* **3,** 309 (1981).
249. M. Vincenzini, W. Balloni, D. Mannelli, and G. Florenzano, *Experientia* **37,** 710 (1981).
250. H. Kayano, T. Matsunaga, I. Karube, and S. Suzuki, *Biotechnol. Bioeng.* **23,** 2283 (1981).
251. Y. Nishida, T. Sato, T. Tosa, and I. Chibata, *Enzyme Microb. Technol.* **1,** 95 (1979).
252. T. Tosa, T. Sato, T. Mori, K. Yamamoto, I. Takata, Y. Nishida, and I. Chibata, *Biotechnol. Bioeng.* **21,** 1697 (1979).
253. T. Sato, Y. Nishida, T. Tosa, and I. Chibata, *Biochim. Biophys. Acta* **570,** 179 (1979).
254. I. Takata, K. Yamamoto, T. Tosa, and I. Chibata, *Eur. J. Appl. Microbiol. Biotechnol.* **7,** 161 (1979).
255. M. Wada, J. Kato, and I. Chibata, *Eur. J. Appl. Microbiol. Biotechnol.* **8,** 241 (1979).
256. K. Murata, K. Tani, J. Kato, and I. Chibata, *Eur. J. Appl. Microbiol. Biotechnol.* **10,** 11 (1980).
257. M. Wada, T. Uchida, J. Kato, and I. Chibata, *Biotechnol. Bioeng.* **22,** 1175 (1980).
258. J. W. Chua, A. Erarslan, S. Kinoshita, and H. Taguchi, *J. Ferment. Technol.* **58,** 123 (1980).
259. M. Wada, J. Kato, and I. Chibata, *Eur. J. Appl. Microbiol. Biotechnol.* **10,** 275 (1980).
260. M. Wada, J. Kato, and I. Chibata, *J. Ferment. Technol.* **58,** 327 (1980).
261. A. M. Klibanov and J. Huber, *Biotechnol. Bioeng.* **23,** 1537 (1981).
262. H. Sawada, S. Kinoshita, T. Yoshida, and H. Taguchi, *J. Ferment. Technol.* **59,** 111 (1981).
263. A. Constantinides, D. Bhatia, and W. R. Vieth, *Biotechnol. Bioeng.* **23,** 899 (1981).
264. W. R. Vieth, S. S. Wang, and R. Saini, *Biotechnol. Bioeng.* **15,** 565 (1973).
265. K. Venkatasubramanian, A. Constantinides, and W. R. Vieth, in *Enzyme Engineering,* Vol. 3 (E. K. Pye and H. H. Weetall, Eds.), Plenum Press, New York (1978), p. 29.
266. W. R. Vieth and K. Venkatasubramanian, in *Methods in Enzymology,* Vol. 44 (K. Mosbach, Ed.), Academic Press, New York (1976), p. 243.
267. W. R. Vieth and K. Venkatasubramanian, in *Methods in Enzymology,* Vol. 44 (K. Mosbach, Ed.), Academic Press, New York (1976), p. 768.
268. H. Yamada, K. Yamada, H. Kumagai, T. Hino, and S. Okamura, in *Enzyme Engineering,* Vol. 3 (E. K. Pye and H. H. Weetall, Eds.), Plenum Press, New York (1978), p. 57.
269. S. Suzuki and I. Karube, in *Enzyme Engineering,* Vol. 4 (G. B. Broun, G. Manecke, and L. B. Wingard Jr., Eds.), Plenum Press, New York (1978), p. 329.
270. W. R. Vieth and K. Venkatasubramanian, in *Enzyme Engineering,* Vol. 4 (G. B. Broun, G. Manecke, and L. B. Wingard Jr., Eds.), Plenum Press, New York (1978), p. 307.

271. I. Karube, S. Mitsuda, and S. Suzuki, *Eur. J. Appl. Microbiol. Biotechnol.* **7,** 343 (1979).
272. W. R. Vieth and K. Venkatasubramanian, in *Amer. Chem. Soc., Symp. Ser. No. 106* (K. Venkatasubramanian, Ed.) (1979), p. 1.
273. S. Suzuki and I. Karube, in *Amer. Chem. Soc., Symp. Ser. No. 106* (K. Venkatasubramanian, Ed.) (1979), p. 221.
274. A. Constantinides, *Biotechnol. Bioeng.* **22,** 119 (1980).
275. L. Gianfreda, P. Parascandola, and V. Scardi, *Eur. J. Appl. Microbiol. Biotechnol.* **11,** 6 (1980).
276. J. Tramper, H. C. van der Plas, A. van der Kaaden, F. Müller, and W. J. Middelhoven, *Biotechnol. Lett.* **1,** 397 (1979).
277. J. A. Roels and R. van Tilberg, in *Amer. Chem. Soc., Symp. Ser. No. 106* (K. Venkatasubramanian, Ed.) (1979), p. 147.
278. S. Bachman, L. Gebicka, and Z. Gasyna, *Starch* **33,** 63 (1981).
279. M. Marek, O. Valentová, K. Demnerová, J. Jizba, M. Blumauerová, and J. Kás, *Biotechnol. Lett.* **3,** 327 (1981).
280. P. Parascandola and V. Scardi, *Biotechnol. Lett.* **3,** 369 (1981).
281. M. Kierstan and C. Bucke, *Biotechnol. Bioeng.* **19,** 387 (1977).
282. P. S. J. Cheetham, *Enzyme Microb. Technol.* **1,** 183 (1979).
283. P. Brodelius, B. Deus, K. Mosbach, and M. H. Zenk, *FEBS Lett.* **103,** 93 (1979).
284. F. Lim and A. M. Sun, *Science* **210,** 908 (1980).
285. F. Paul and P. M. Vignais, *Enzyme Microb. Technol.* **2,** 281 (1980).
286. H. Dallyn, W. C. Falloon, and P. G. Bean, *Lab. Pract.,* **26,** 773 (1977).
287. F. H. White and A. D. Porino, *J. Inst. Brew.* **84,** 228 (1978).
288. P. S. J. Cheetham, K. W. Blunt, and C. Bucke, *Biotechnol. Bioeng.* **21,** 2155 (1979).
289. J. Klein, U. Hackel, and F. Wagner, in *Amer. Chem. Soc., Symp. Ser. No. 106* (K. Venkatasubramanian, Ed.) (1979), p. 101.
290. A. M. Klibanov and A. V. Puglisi, *Biotechnol. Lett.* **2,** 445 (1980).
291. Y.-Y. Linko, L. Weckström, and P. Linko, in *Enzyme Engineering in Food Processing,* Vol. 2 (P. Linko and J. Larinkari, Eds.), Applied Science, London (1980), p. 81.
292. B. Hägerdal, *Acta Chem. Scand. B* **34,** 611 (1980).
293. P. G. Krouwel, W. F. M. van der Laan, and N. W. F. Kossen, *Biotechnol. Lett.* **2,** 253 (1980).
294. Y.-Y. Linko, L. Weckström, and P. Linko, in *Enzyme Engineering,* Vol. 5 (H. H. Weetall and G. P. Royer, Eds.), Plenum Press, New York (1980), p. 355.
295. H. Ochiai, H. Shibata, Y. Sawa, and T. Katoh, *Proc. Nat. Acad. Sci. USA* **77,** 2442 (1980).
296. N. Makiguchi, M. Arita, and Y. Asai, *J. Ferment. Technol.* **58,** 17 (1980).
297. N. Makiguchi, M. Arita, and Y. Asai, *J. Ferment. Technol.* **58,** 167 (1980).
298. N. Makiguchi, M. Arita, and Y. Asai, *J. Ferment. Technol.* **56,** 333 (1980).
299. P. Brodelius, B. Hägerdal, and K. Mosbach, in *Enzyme Engineering,* Vol. 5 (H. H. Weetall and G. P. Royer, Eds.), Plenum, New York (1980), p. 383.
300. P. Scheurich, H. Schnabl, U. Zimmermann, and J. Klein, *Biochim. Biophys. Acta* **598,** 645 (1980).
301. L. Häggström and N. Molin, *Biotechnol. Lett.* **5,** 241 (1980).

302. Y.-Y. Linko and P. Linko, *Biotechnol. Lett.* **3,** 21 (1981).
303. P. E. Gisby and D. O. Hall, *Nature* **287,** 251 (1980).
304. P. Scherer, M. Kluge, J. Klein, and H. Sahn, *Biotechnol. Bioeng.* **23,** 1057 (1981).
305. Y.-Y. Linko, H. Jalanka, and P. Linko, *Biotechnol. Lett.* **3,** 263 (1981).
306. I. A. Veliky and R. E. Williams, *Biotechnol. Lett.* **3,** 275 (1981).
307. V. L. Garde, B. Thomasset, and J.-N. Barbotin, *Enzyme Microb. Technol.* **3,** 216 (1981).
308. D. Williams and D. M. Munnecke, *Biotechnol. Bioeng.* **23,** 1813 (1981).
309. A. Jones and I. A. Veliky, *Eur. J. Appl. Microbiol. Biotechnol.* **13,** 84 (1981).
310. T. Shiotani and T. Yamane, *Eur. J. Appl. Microbiol. Biotechnol.* **13,** 96 (1981).
311. R. Maleszka, I. A. Veliky, and H. Schneider, *Biotechnol. Lett.* **3,** 415 (1981).
312. C. F. Mandenius, B. Danielsson, and B. Mattiasson, *Biotechnol. Lett.* **3,** 629 (1981).
313. I. A. Veliky and A. Jones, *Biotechnol. Lett.* **3,** 551 (1981).
314. B. Mattiasson, M. Ramstorp, I. Nilsson, and B. Hahn-Hägerdal, *Biotechnol. Lett.* **3,** 561 (1981).
315. G. H. Cho, C. Y. Choi, Y. D. Choi, and M. H. Han, *Biotechnol. Lett.* **3,** 667 (1981).
316. A. Margaritis, P. K. Bajpai, and J. B. Wallace, *Biotechnol. Lett.* **3,** 613 (1981).
317. K.-D. Vorlop and J. Klein, *Biotechnol. Lett.* **3,** 9 (1981).
318. Y.-Y. Linko, R. Viskari, L. Pohjola, and P. Linko, *J. Solid Phase Biochem.* **2,** 203 (1978).
319. Y.-Y. Linko, R. Viskari, L. Pohjola, and P. Linko, in *Enzyme Engineering,* Vol. 4 (G. B. Broun, G. Manecke, and L. B. Wingard, Jr., Eds), Plenum Press, New York (1978), p. 345.
320. T. K. Ghose and V. Kannan, *Enzyme Microb. Technol.* **1,** 47 (1979).
321. Y.-Y. Linko, K. Poutanen, L. Weckström and P. Linko, *Enzyme Microb. Technol.* **1,** 26 (1979).
322. M. De Rosa, A. Gambacorta, E. Esposito, E. Drioli, and S. Gaeta, *Biochemie* **62,** 517 (1980).
323. L. Weckström, Y.-Y. Linko and P. Linko, in *Enzyme Engineering in Food Processing,* Vol. 2 (P. Linko and J. Larinkari, Eds.), Applied Science, London (1980), p. 148.
324. P. Linko, K. Poutanen, L. Weckström and Y.-Y. Linko, *Biochimie* **62,** 387 (1980).
325. S. Giovenco and A. Maimone, in *Abstracts of Communications,* 2nd European Congress of Biotechnology, Society of Chemical Industry, London, 1981, p. 131.
326. J. Klein and F. Wagner, in *Enzyme Engineering,* Vol. 5 (H. H. Weetall and G. P. Royer, Eds.), Plenum Press, New York (1980), p. 335.
327. J. Klein and H. Eng, *Biotechnol. Lett.* **1,** 171 (1979).
328. A. Tanaka, I.-N. Jin, S. Kawamoto, and S. Fukui, *Eur. J. Appl. Microbiol. Biotechnol.* **7,** 351 (1979).
329. J. Klein and M. Kluge, *Biotechnol. Lett.* **3,** 65 (1981),
330. M. C. Fusee, W. E. Swann, and G. J. Calton, *Appl. Environ. Microbiol.* **42,** 672 (1981).

331. H. Ochiai, H. Shibata, T. Matsuo, K. Hashinokuchi, and I. Inamura, *Agric. Biol. Chem.* **42,** 683 (1978).
332. H. Nilsson, R. Mosbach, and K. Mosbach, *Biochim. Biophys. Acta* **268,** 253 (1972).
333. A. Tanaka, S. Yasuhara, S. Fukui, T. Iida, and E. Hasegawa, *J. Ferment. Technol.* **55,** 71 (1977).
334. R. L. Davidson, Ed., *Handbook of Water-Soluble Gums and Resins,* McGraw-Hill, New York (1980).
335. T. M. S. Chang, *Artificial Cells,* Charles C. Thomas, Springfield (1972).
336. Y. Ado, K. Kimura, and H. Samejima, in *Enzyme Engineering,* Vol. 5 (H. H. Weetall and G. P. Royer, Eds.), Plenum Press, New York (1980), p. 295.
337. Y. Ado, Y. Suzuki, T. Tadokoro, K. Kimura, and H. Samejima, *J. Solid Phase Biochem.* **4,** 43 (1979).
338. R. R. Mohan and N. N. Li, *Biotechnol. Bioeng.* **17,** 1137 (1975).
339. J. K. Kan and M. L. Shuler, *Biotechnol. Bioeng.* **20,** 217 (1978).
340. I. A. Webster, M. L. Shuler, and P. R. Rony, *Biotechnol. Bioeng.* **21,** 1725 (1979).
341. G. A. Rechnitz, R. K. Kobos, S. J. Riechel, and C. R. Gebauer, *Anal. Chim. Acta* **94,** 357 (1977).
342. C. L. Di Paolantonio, M. A. Arnold, and G. A. Rechnitz, *Anal. Chim. Acta* **128,** 121 (1981).
343. G. A. Rechnitz, M. A. Arnold, and M. E. Meyerhoff, *Nature* **278,** 466 (1979).
344. G. A. Rechnitz, T. L. Riechel, R. K. Kobos, and M. E. Meyerhoff, *Science* **199,** 440 (1978).
345. T. K. Ghose and S. Chand, *J. Ferment. Technol.* **56,** 315 (1978).
346. B. Atkinson, G. M. Black, P. J. S. Lewis, and A. Pinches, *Biotechnol. Bioeng.* **21,** 193 (1979).
347. I. Kühn, *Biotechnol. Bioeng.* **22,** 2393 (1980).
348. B. Hahn-Hägerdahl, B. Mattiasson, and P. Å. Albertsson, *Biotechnol. Lett.* **3,** 53 (1981).
349. J. Kulys and K. Kadziauskiene, *Biotechnol. Bioeng.* **22,** 221 (1980).
350. M. Hikuma, T. Yasuda, I. Karube, and S. Suzuki, *Ann. N. Y. Acad. Sci.* **369.** 307 (1981).
351. G. A. Rechnitz, *Science* **214,** 287 (1981).
352. P.-Å. Albertsson, *Partition of Cell Particles and Macromolecules,* 2nd., Almqvist & Wiksell, Stockholm (1971).
353. R. E. Sparks, R. M. Salemme, P. M. Meier, M. H. Litt, and O. Lindan, *Trans. Amer. Soc. Artif. Int. Organs* **15,** 353 (1969).
354. P. Brodelius, K. Nilsson and K. Mosbach, *Appl. Biochem. Biotechnol.* **6,** 293 (1981).
355. B. Mattiasson, M. Ramstorp, K. Widebäck, and G. Kronvall, *J. Appl. Biochem.* **2,** 321 (1980).
356. H. Samejima, K. Kimura, Y. Ado, Y. Suzuki, and T. Tadokoro, in *Enzyme Engineering,* Vol. 4 (G. B. Broun, G. Manecke, and L. B. Wingard Jr., Eds.), Plenum Press, New York (1978), p. 237.
357. O. Holst, S. O. Enfors, and B. Mattiasson, *Eur. J. Appl. Microbiol. Biotechnol.* **14,** 64 (1982).

358. E. Szwajcer, P. Brodelius, and K. Mosbach, *Enzyme Microb. Technol.* **4,** 409 (1982).
359. P. Wikström, E. Szwajcer, P. Brodelius and K. Mosbach, *Biotechnol. Lett.* **4,** 153 (1982).
360. P. Adlercreutz, O. Holst, and B. Mattiasson, *Enzyme Microb. Technol.* **4,** 395 (1982).
361. N. N. Zueva, V. N. Shcherbakova, V. Y. Yakovleva, Y. S. Nikitin, I. V. Avsyuk, C. T. T. Mai, and I. V. Berezin, *Appl. Biochem. Microbiol.* **16,** 918 (1981).
362. K. Buchholz, Ed., *Characterization of Immobilized Biocatalysts,* Dechema Band 84, Verlag Chemie, Weinheim and New York (1979).
363. J. Klein, P. Washausen, M. Kluge, and H. Eng, in *Enzyme Engineering,* Vol. 5 (H. H. Weetall and G. P. Royer, Eds.), Plenum Press, New York (1980), p. 359.
364. J. Klein and F. Wagner, in *Biotechnology,* Dechema Band 82, Verlag Chemie, Weinheim (1978), p. 1.
365. K. Yamamoto, T. Tosa, and I. Chibata, *Biotechnol. Bioeng.* **22,** 2045 (1980).
366. A. Freeman, T. Blank, and Y. Aharonowitz, *Eur. J. Appl. Microbiol. Biotechnol.* **14,** 64 (1982).
367. N. E. Voishvillo, A. V. Kamernitskii, A. Ya. Khaikova, I. G. Leont'ev, V. N. Paukov, and L. A. Nakhapetyan, *Bulletin of the Academy of Sciences of the USSR* **25,** 1303 (1976).
368. G. Manecke, E. Ehrenthal, and J. Schlünsen, in *Characterization of Immobilized Biocatalysts,* Dechema Band 84, (K. Buchholz, Ed.), Verlag Chemie, Weinheim and New York (1979), p. 49.
369. K. Buchholz in *Characterization of Immobilized Biocatalysts,* Dechema Band 84, (K. Buchholz, Ed.), Verlag Chemie, Weinheim and New York (1979), p. 111.
370. P. Atrat and H. Groh, *Z. Allg. Mikrobiol.* **21,** 3 (1981).
371. K. A. Koshcheenko, A. Yu. Arinbasarova, and G. K. Skryabin, *Appl. Biochem. Microbiol.* **15,** 491 (1980).
372. K. A. Koshcheenko, S. A. Gulevskaya, G. V. Sukhodol'skaya, K. A. Lusta, B. A. Fikhte, and G. K. Skryabin, *Dokl. Biol. Sci.* **233,** 122 (1978).
373. A. Tanaka, K. Sonomoto, M. Hoq, N. Usui, K. Nomura, and S. Fukui, in *6th Enzyme Engineering Conference, Program and Abstracts,* Japanese Society of Enzyme Engineering (1981), p. 62.
374. M. I. Bukhar, N. V. Vdovina, L. A. Krasnova, and K. A. Koshcheenko, *Appl. Biochem. Microbiol.* **16,** 366 (1981).
375. A. L. Compere and W. L. Griffith, in *Developments in Industrial Microbiology,* Vol. 17 (E. D. Murray, Ed.), SIM, Washington (1972), p. 247.
376. S.-L. Stenros, Y.-Y. Linko, and P. Linko, in *6th Enzyme Engineering Conference, Program and Abstracts,* Japanese Society of Enzyme Engineering (1981), p. 79.
377. M. F. Cocquempot, D. Thomas, M. L. Champigny, and A. Moyse, *Eur. J. Appl. Microbiol. Biotechnol.* **8,** 37 (1979).
378. H. Matsuoka, S. Suzuki, and M. Aizawa, *Biotechnol. Bioeng.* **23,** 1103 (1981).
379. R. D. Sweigart, in *Applied Biochemistry and Bioengineering,* Vol. 2 (L. B.

Wingard, Jr., E. Katchalski-Katzir, and L. Goldstein, Eds.), Academic Press, New York (1979), p. 209.

380. R. L. Antrim, W. Colilla, and B. J. Schnyder, in *Applied Biochemistry and Bioengineering,* Vol. 2 (L. B. Wingard, Jr., E. Katchalski-Katzir, and L. Goldstein, Eds.), Academic Press, New York (1979), p. 97.

381. A. Kikkert, K. Vellenga, H. G. J. de Wilt, and G. E. H. Joosten, *Biotechnol. Bioeng.* **23,** 1087 (1981).

382. "Sweetzyme®," *Technical Bulletin,* Novo Industri, 1981.

383. E. Drioli, G. Iorio, R. Santoro, M. de Rosa, A. Gambacorta, and B. Nicolaus, *J. Mol. Catal.* **14,** 247 (1982).

384. W. Grote, K. J. Lee, and P. L. Rogers, *Biotechnol. Lett.* **2,** 481 (1980).

385. I. Karube, S. Suzuki, T. Matsunaga, and S. Kuriyama, *Ann. N. Y. Acad. Sci.* **369,** 91 (1981).

386. S. Suzuki, I. Karube, and T. Matsunaga, *Biotechnol. Bioeng. Symp. No. 8* (1978), p. 501.

387. F. Winquist, B. Danielsson, I. Lundström, and K. Mosbach, in *Fermentation, 2nd Rotenburger Symposium 1980* (R. M. Lafferty, Ed.), Springer Verlag, Vienna (1981).

388. H. Kayano, T. Matsunaga, I. Karube, and S. Suzuki, *Biochim. Biophys, Acta* **638,** 80 (1981).

389. G. R. Lambert, A. Daday, and G. D. Smith, *FEBS Lett.* **101,** 125 (1979).

390. A. H. Sonaer and M. A. Caglar, in *Abstracts of Communication,* Second European Congress of Biotechnology, Society of Chemical Industry, London (1981), p. 157.

391. K. Venkatasubramanian and Y. Toda, *Biotechnol. Bioeng., Symp. No. 10,* (1980), p. 237.

392. J. Everse, C. L. Ginsburgh, and N. O. Kaplan, in *Methods of Biochemical Analysis,* Vol. 25 (D. Glick, Ed.), John Wiley, New York (1979), p. 135.

393. G. G. Guilbault, in *Methods in Enzymology,* Vol. 44 (K. Mosbach, Ed.), Academic Press, New York (1976), p. 579.

394. K. Mosbach and B. Danielsson, *Anal. Chem.* **53,** 83A (1981).

395. B. Mattiasson, in *Amer. Chem. Soc., Symp. Ser. No. 106* (K. Venkatasubramanian, Ed.) (1979), p. 203.

396. H. Nilsson, *Analytical Applications of Immobilized Enzymes with Special Emphasis on Enzyme Electrodes,* Ph.D. Thesis, University of Lund, Sweden (1977).

397. S. Suzuki and I. Karube, in *Applied Biochemistry and Bioengineering,* Vol. 3, (L. B. Wingard Jr., E. Katchalski-Katzir, and L. Goldstein, Eds.), Academic Press, New York (1981), p. 145.

398. M. Hikuma, T. Kubo, T. Yasuda, I. Karube, and S. Suzuki, *Anal. Chim. Acta* **109,** 33 (1979).

399. R. K. Kobos and G. A. Rechnitz, *Anal. Lett.* **10,** 751 (1977).

400. M. A. Jensen and G. A. Rechnitz, *Anal. Chim. Acta* **101,** 125 (1978).

401. R. R. Walters, B. E. Moriarty, and R. P. Buck, *Anal. Chem.* **52,** 1680 (1980).

402. T. Matsunaga, I. Karube, N. Teraoka, and S. Suzuki, *Anal. Chim. Acta* **127,** 245 (1981).

403. M. Hikuma, T. Kubo, T. Yasuda, I. Karube, and S. Suzuki, *Anal. Chem.* **52,** 1020 (1980).
404. S. Updike and I. Treichel, *Anal. Chem.* **51,** 1643 (1979).
405. M. Hikuma, H. Obana, T. Yasuda, I. Karube, and S. Suzuki, *Enzyme Microb. Technol.* **2,** 234 (1980).
406. M. Hikuma, H. Suzuki, T. Yasuda, I. Karube, and S. Suzuki, *Eur. J. Appl. Microbiol. Biotechnol.* **8,** 289 (1979).
407. K. Matsumoto, H. Seijo, T. Watanabe, I. Karube, I. Satoh, and S. Suzuki, *Anal. Chim. Acta,* **105,** 429 (1979).
408. M. Hikuma, T. Kubo, T. Yasuda, I. Karube, and S. Suzuki, *Biotechnol. Bioeng.* **21,** 1845 (1979).
409. T. Okada, I. Karube, and S. Suzuki, *Eur. J. Appl. Microbiol. Biotechnol.* **12,** 102 (1981).
410. I. Karube, T. Matsunaga, T. Nakahara, S. Suzuki, and T. Kada, *Anal. Chem.* **53,** 1024 (1981).
411. R. K. Kobos, D. J. Rice, and D. S. Flournoy, *Anal. Chem.* **51,** 1122 (1979).
412. R. K. Kobos and H. Y. Pyon, *Biotechnol. Bioeng.* **23,** 627 (1981).
413. I. Karube, T. Matsunaga, and S. Suzuki, *Anal. Chim. Acta* **109,** 39 (1979).
414. B. Mattiasson, P.-O. Larsson, L. Lindahl, and P. Sahlin, *Enzyme Microb. Technol.* **4,** 153 (1982).
415. P. Dunnill, *Chem. Ind.* 4 April 1981, p. 204.
416. N. Nazly and C. J. Knowles, *Biotechnol. Lett.* **3,** 363 (1981).
417. L. Genaud, J. Guillot, M. Damez, and M. Coulet, *J. Immunol. Methods* **22,** 339 (1978).
418. G. Bétail, M. Coulet, L. Genaud, J. Guillot, and M. Scandariato, *C. R. Soc. Biol.* **169,** 561 (1975).
419. M. Ramstorp, P. Carlsson, D. Bratthall, and B. Mattiasson, *Caries Res.* **16,** 423 (1982).
420. B. Mattiasson, M. Ramstorp, in *Enzyme Engineering,* Vol. 5 (H. H. Weetall and G. P. Royer, Eds.), Plenum Press, New York (1980), p. 401.
421. J. Sekiguchi, G. M. Gaucher, and Y. Yamada, *Tetrahedron Lett.* **1,** 41 (1979).
422. B. Danielsson, B. Mattiasson, and K. Mosbach, in *Applied Biochemistry and Bioengineering,* Vol. 3 (L. B. Wingard Jr., E. Katchalski-Katzir, and L. Goldstein, Eds.), Academic Press, New York (1981), p. 97.
423. A. W. Alfermann, I. Schuller, and E. Reinhard, *Planta Med.* **40,** 218 (1980).
424. *Microcarrier Cell Culture, Principles and Methods,* Pharmacia Fine Chemicals, 1981.
425. L. Civerchia-Perez, B. Faris, G. LaPointe, J. Beldekas, H. Leibowitz, and C. Franzblau, *Proc. Nat. Acad. Sci. USA* **77,** 2064 (1980).
426. G. Pilwat, P. Washausen, J. Klein, and U. Zimmermann, *Z. Naturforsch.* **35C,** 352 (1980).

INDEX

Vol. 58. **Analysis of Pesticide Residues.** Edited by H. Anson Moye

Vol. 59. **Affinity Chromatography.** By William H. Scouten

Vol. 60. **Quality Control in Analytical Chemistry.** By G. Kateman and F. W. Pijpers

Vol. 61. **Direct Characterization of Fineparticles.** By Brian H. Kaye

Vol. 62. **Flow Injection Analysis.** By J. Ruzicka and E. H. Hansen

Vol. 63. **Applied Electron Spectroscopy for Chemical Analysis.** Edited by Hassan Windawi and Floyd Ho

Vol. 64. **Analytical Aspects of Environmental Chemistry.** Edited by David F. S. Natusch and Philip K. Hopke

Vol. 65. **The Interpretation of Analytical Chemical Data by the Use of Cluster Analysis.** By D. Luc Massart and Leonard Kaufman

Vol. 66. **Solid Phase Biochemistry: Analytical and Synthetic Aspects.** Edited by William H. Scouten